TEACHER'S PLANNING GUIDE
ACTIVITIES • ALTERNATIVES • ASSESSMENT

With Blackline Masters

SEE companion **Teacher's Answer Edition** for answers to all questions and exercises in the student edition.

HRW

ALGEBRA ONE INTERACTIONS

COURSE 2

Integrating

Mathematics
Technology
Explorations
Applications
Assessment

HOLT, RINEHART AND WINSTON
Harcourt Brace & Company
Austin • New York • Orlando • Atlanta • San Francisco • Boston • Dallas • Toronto • London

Associate Director: Gary Standafer

Managing Editor: Ron Huenerfauth

Project Editors: Charles McClelland, Charles Sims, Andrew Roberts, Joel Riemer

Copyeditor: Laurie Baker

Editorial Permissions: Ann Farrar

Design: Diane Motz, *Art Director*
Marc Cooper, *Design*
Julie Ray, *Design Staff*

Image Services: Elaine Tate, *Art Buyer Supervisor*
Michele Rumpf, *Image Service Staff*

Studio Photography Sam Dudgeon

Production: Donna Lewis, *Senior Production Coordinator*
Amber Martin, *Production Assistant*
George Prevelige, *Manufacturing Manager*

Printed in the United States of America

1 2 3 4 5 6 7 125 03 02 01 00 99 98 97

ISBN: 0-03-095357-X

What Is the Philosophy and Purpose of *HRW Algebra One Interactions*?

Implement Curriculum Reform

The reform movement in school mathematics has caused a shift not only in the way mathematics is taught, but also in the particular mathematics courses offered. Many of the traditional mathematics courses such as general mathematics and pre-algebra are being eliminated from the high school curriculum. Algebra one, a course at one time reserved for average and above-average students, is becoming a required course for all high school students. However, publishers' algebra textbooks have catered only to the college bound student. Unable to cope with the abstract presentation of mathematical topics in these textbooks, the remainder of the student population was relegated to a computational skills-based program called general mathematics. Another popular remedial course, pre-algebra, veiled the computational skills practice with the introduction of algebraic symbols and a smattering of geometry topics. This structure in the mathematics curriculum provided students with the necessary number of credits for graduation and, in a time long past, provided students with the minimum mathematics skills needed for non-professional careers.

Prepare Students for the Future

General math and pre-algebra no longer provide a student with the math skills needed to enter into today's technological society. Today, an understanding of algebraic concepts is a minimal requirement for entry into most careers and is a gateway to all forms of higher mathematical study.

The goal is clear and noble and certainly in the best interest of the student. However, the algebra teacher is now faced with a broad base of students of divergent learning levels and learning styles.

Support the Teacher

How can the teacher manage a classroom setting that has suddenly undergone such change? It is obvious that teaching practices and teaching materials must undergo change to accommodate the new algebra classroom. *HRW Algebra One Interactions* is a program designed specifically to accommodate the divergent needs of the students found in this new setting.

Make Algebra Accessible to All

HRW Algebra One Interactions makes algebra accessible to everyone. The lessons in this program are written in such a way that students are presented first with numbers and data, then with tables and graphs, and finally with abstract algebraic concepts. The instruction typically begins with the presentation of real-life data or with the use of manipulatives. In other words, instruction on important mathematical concepts begins with the concrete. As the instruction progresses, the students move to the tabular, then to the graphical, and then to the symbolic. All levels of student abilities are accounted for in this approach. This is the essence of the instruction found in the student lessons. Labels and separate textbook features that describe students in terms of different ability labels will not be found in the student book. The provisions for different learning levels are

embedded in the instructional design of the student lessons. For example, students are often asked to look for patterns by organizing observational data into tables; then they make conjectures from the data. As students generalize or synthesize their observations as conjectures, they progress from the concrete to the abstract. Students who do not possess acute memorization skills will be able to grasp algebraic thinking that was once reserved only for the college-bound student. Students historically labeled as advanced or accelerated will find the instruction more relevant and will gain a deeper understanding of the concepts. They will be more likely to show an interest in mathematically oriented subjects such as chemistry, physics, logic, economics, statistics, and computer science.

Extend Instructional Time

HRW Algebra One Interactions is a two-book algebra one program designed to be taken over a two-year period of time. Twice as much instructional time allows the teacher to take a hands-on, exploratory, and activity-oriented approach to instruction. Manipulatives and calculators will be used extensively, and students will be actively involved in using mathematics in real-world applications. Much group work and many projects will be incorporated into instruction. The program includes all the topics of an algebra one program. It contains a greater coverage of geometry topics and integrates the mathematics of transformations, statistics, probability, and maximum and minimum. *HRW Algebra One Interactions* also includes all the pre-algebra skills necessary for success in algebra.

Maximize Curriculum Options

Many districts will use this program as a paced or extended algebra one program for students who begin their algebra one study in grade nine and finish at the end of grade ten. Having completed this algebra one, these students will have the prerequisite skills to move into geometry or possibly algebra two. They may or may not take an additional math course before graduation.

Other districts will use *HRW Algebra One Interactions* as an extended algebra one program for students who begin their algebra one study in grade eight. These students will complete the program at the end of grade nine. This allows students to take as many as three other mathematics courses before graduation, much as the traditional curriculum is organized.

And finally, students who are ready to begin a two-year algebra one program in the seventh grade will find *HRW Algebra One Interactions* the best way to complete an algebra one curriculum by the end of grade eight.

Algebra Explorations: Functions and Graphs CD-ROM

Algebra Explorations: Functions and Graphs is an easy to use CD-ROM program that allows students to explore functions in a dynamic learning environment.

Algebra Explorations features:

- interactive presentations using data tables, a grapher, an equation editor, a notebook, and a calculator
- applications asking students to make predictions, write observations, manipulate variables, and note conclusions
- a Teacher Management System with tracking information, assessment helps, and methods for creating student reports and portfolio directories
- available for Macintosh® or Windows®

Table of Contents

Exploration 1: Graphing Functions This exploration introduces students to the basic concepts of the Cartesian coordinate system.

Exploration 2: Linear Functions—Slope Students are introduced to slope by using the theme of Egyptian pyramid building.

Exploration 3: Linear Functions—Intercepts In the final activity, students use their knowledge of slope and intercepts to describe the flight paths of objects in outer space.

Exploration 4: Exponential Functions The exploration begins with an examination of earthquakes and the Richter scale. Students visit the lab of Dr. Nenchal to discover patterns.

Exploration 5: Reciprocal Functions Students play a harp to discover the reciprocal relationship between the length of a string and pitch.

Exploration 6: Quadratic Functions The relationship between weight of an automobile and braking distance provides the introduction to quadratic functions.

Exploration 7: Absolute-Value Functions An alien creature in a special undersea diving pod introduces the concept of an absolute-value function.

Exploration 8: Stretches Students stretch an absolute-value function to help a rescue helicopter find a lost hiker.

Exploration 9: Reflections Students use the reflection of a linear function to operate a special spying device.

Exploration 10: Translations Students use their knowledge of translations of functions to operate a special mathematical pinball machine.

Exploration 11: Combining Transformations Students use their knowledge of transformations of functions to create special creatures that are morphed from known animals based on the transformation performed.

Exploration 12: On Your Own Students are given the opportunity to use the grapher and data table to explore topics of their own interest.

Use the CD-ROM with lessons as follows.

Exploration	*Course 1* Lessons	*Course 2* Lessons
1	2.6, 8.1	2.2
2	8.2–8.5	2.1, 2.2
3	8.2–8.5	2.1, 2.2
4	10.1	6.5, 6.6
5	10.3	10.2
6	10.2	8.1
7	10.4, 10.5	5.1, 5.2
8	10.6	5.2, 5.3
9	10.6	5.2, 5.4
10	10.6	5.2, 5.5
11	10.6	5.2, 5.6
12	10.1–10.6	5.2–5.6

Key to Support Material

Planning Resources

Guides for lesson planning.

Teaching Transparencies Directory—contains reduced full-color reproductions of every transparency. Each reproduction includes its instructional page from the transparency package.

Solution Key—contains complete solutions for every exercise in the Pupil's Edition.

Block-Scheduling Handbook—provides a description for individual work, group activities, and ongoing assessment to assist you in various block-scheduling situations.

Technology Handbook—provides a guide for using hand-held graphics calculators and computer software.

Resources for Every Lesson

Available for each lesson to meet individual student needs.

Practice Workbook—provides additional practice for all the concepts taught in each lesson. Also available as a blackline master at the end of each lesson in the *Teacher's Planning Guide*.

Reteaching Masters—presents lesson concepts with alternative strategies to ensure that all students have understood the lesson content.

Lesson Activity Masters—are located at the end of each lesson in the *Teacher's Planning Guide* and contain activities designed to extend lesson content.

Problem Solving/Critical Thinking Masters—contains stimulating problems or simulations that extend and enrich the lesson material.

Technology Masters—contains computer and calculator activities that offer additional practice or alternative technology to that provided in the lesson.

Spanish Resources—contains a Spanish translation of lesson objectives, definitions and properties, and lesson review material with accompanying practice.

Basic Skills Practice Masters—are located at the end of each lesson in the *Teacher's Planning Guide* and contain additional practice in basic pre-algebra and algebra skills.

Teaching Transparencies—are full color transparencies each accompanied by a instructional page for the teacher. Visuals include tables, graphs, charts, four-color art, technical drawings, spreadsheets and statistical graphs.

Cooperative-Learning Activities—contains detailed cooperative-learning guidance for explorations in each lesson in the pupil book.

Resources for Every Chapter

Available for each chapter to meet individual student needs.

Tech Prep Mathematics—consists of tech prep projects and applications related to the chapter content.

Lab Activities and Long-Term Projects—are hands-on activities that help give concrete meaning to abstract mathematical concepts and long-term projects that relate to the chapter material to engage the student for more than one day.

Writing Activities for Your Portfolio—contains creative-writing activities appropriate for a student's portfolio. These activities are related to the math content and applications found in the individual chapters.

Algebra Explorations CD-ROM—provides a computer-based interactive environment for exploring functions and graphs.

Assessment Resources

Provides a wide variety of assessment opportunities.

Chapter Assessment and Diagnostic Test—contains one two-page multiple-choice test per chapter and one two-page free-response test per chapter as well as a pre-book diagnostic test for algebra.

Mid-Chapter Assessment—consists of both multiple-choice and free-response questions.

Assessing Prior Knowledge/Quiz—is a combination quiz and warm-up for each lesson.

Alternative Assessment—consists of various opportunities for authentic and performance-based assessment, as well as assessment consisting of game-type activities.

Assessment Software—is a set of program disks with item disks for each course and a User's Guide. Available for Windows® and Macintosh®, this program will generate test items from a base of over 1250 static items.

Assessment Software: Item Listing—includes the test questions and answers from *Course 1* and *Course 2* that are found in the assessment software. There are over 1250 test items per course consisting of multiple-choice, quantitative-comparison, and short-answer questions.

Using Your *Teacher's Planning Guide*

Welcome to your *Teacher's Planning Guide*, which we believe to be the most advanced and most useful teaching aid ever developed. Compare this with the more traditional wrap-around teacher's editions, and you will quickly see its advantages.

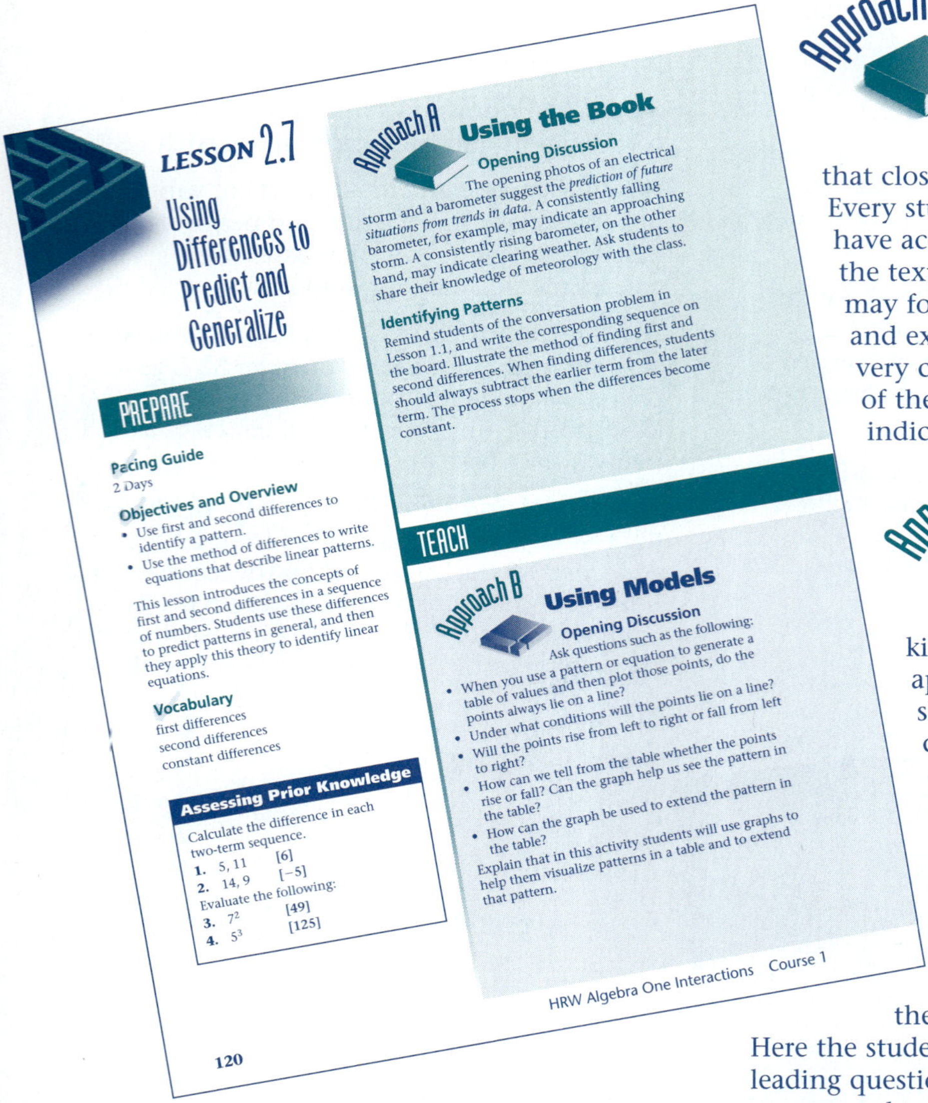
LESSON 2.7
Using Differences to Predict and Generalize

PREPARE

Pacing Guide
2 Days

Objectives and Overview
- Use first and second differences to identify a pattern.
- Use the method of differences to write equations that describe linear patterns.

This lesson introduces the concepts of first and second differences in a sequence of numbers. Students use these differences to predict patterns in general, and then they apply this theory to identify linear equations.

Vocabulary
first differences
second differences
constant differences

Assessing Prior Knowledge
Calculate the difference in each two-term sequence.
1. 5, 11 [6]
2. 14, 9 [−5]
Evaluate the following:
3. 7^2 [49]
4. 5^3 [125]

Approach A **Using the Book**

Opening Discussion
The opening photos of an electrical storm and a barometer suggest the *prediction of future situations from trends in data.* A consistently falling barometer, for example, may indicate an approaching storm. A consistently rising barometer, on the other hand, may indicate clearing weather. Ask students to share their knowledge of meteorology with the class.

Identifying Patterns
Remind students of the conversation problem in Lesson 1.1, and write the corresponding sequence on the board. Illustrate the method of finding first and second differences. When finding differences, students should always subtract the earlier term from the later term. The process stops when the differences become constant.

TEACH

Approach B **Using Models**

Opening Discussion
Ask questions such as the following:
- When you use a pattern or equation to generate a table of values and then plot those points, do the points always lie on a line?
- Under what conditions will the points lie on a line?
- Will the points rise from left to right or fall from left to right?
- How can we tell from the table whether the points rise or fall? Can the graph help us see the pattern in the table?
- How can the graph be used to extend the pattern in the table?

Explain that in this activity students will use graphs to help them visualize patterns in a table and to extend that pattern.

120 HRW Algebra One Interactions Course 1

Two Pathways—You Choose

The most obvious, and innnovative, aspect of your planning guide is shown by the horizontal bar that divides the lesson pages into two parts. For every lesson in *HRW Algebra One Interactions*, you are provided with two different teaching approaches.

Approach A: *Using the Book*

This approach is focused on a presentational style that closely follows the Pupil's Edition. Every student is assumed to have, or to have access to, his or her own copy of the textbook. Under this approach you may follow the models, explorations, and examples found in the textbook very closely. Appropriate places for use of the *Teaching Transparencies* are indicated.

Approach B: *Using Models*

Some teachers prefer to use more kinesthetic or hands-on approaches—which are not just for slow learners. Approach B, while conveying the same content and covering the same objectives as the Pupil's Edition, gives you many ideas for model-based approaches.

The explorations in Approach B are more student-driven than those in Approach A, where the steps are spelled out in detail. Here the students are more on their own. Given leading questions and hints, they are encouraged to participate to a greater degree in the design of the explorations.

Detail

The level of detail possible in the more spacious format of this *Planning Guide* is much greater than in traditional wrap-around formats. Here you will benefit from more extensive presentations of teaching tips, activities, examples, and ongoing assessment answers.

Three-Part Structure

Prepare

The first section in every lesson will save you time in setting goals for you and your class. Each contains the following useful features:

- Pacing Guide
- Objectives and Overview
- Vocabulary
- Assessing Prior Knowledge

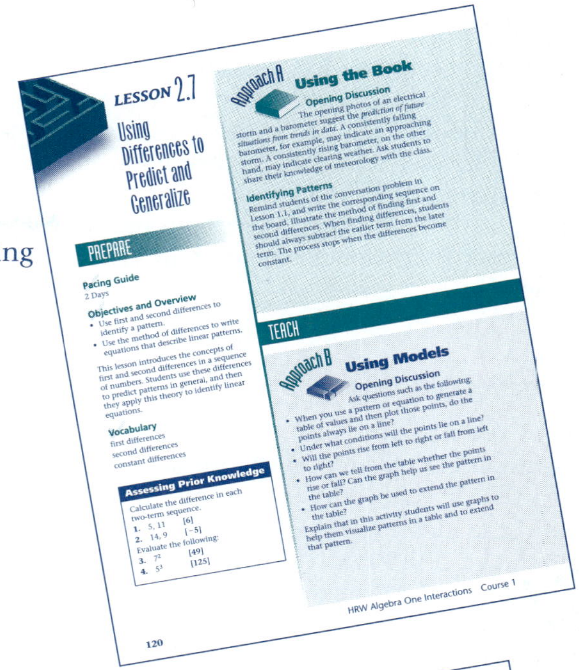
LESSON 2.7
Using Differences to Predict and Generalize

PREPARE

Pacing Guide
2 Days

Objectives and Overview
- Use first and second differences to identify a pattern.
- Use the method of differences to write equations that describe linear patterns.

This lesson introduces the concepts of first and second differences in a sequence of numbers. Students use these differences to predict patterns in general, and then they apply this theory to identify linear equations.

Vocabulary
first differences
second differences
constant differences

Assessing Prior Knowledge
Calculate the difference in each two-term sequence.
1. 5, 11 [6]
2. 14, 9 [−5]
Evaluate the following:
3. 7^2 [49]
4. 5^3 [125]

Approach A Using the Book
Opening Discussion
The opening photos of an electrical storm and a barometer suggest the *prediction of future situations from trends in data*. A consistently falling barometer, for example, may indicate an approaching storm. A consistently rising barometer, on the other hand, may indicate clearing weather. Ask students to share their knowledge of meteorology with the class.

Identifying Patterns
Remind students of the conversation problem in Lesson 1.1, and write the corresponding sequence on the board. Illustrate the method of finding first and second differences. When finding differences, students should always subtract the earlier term from the later term. The process stops when the differences become constant.

TEACH

Approach B Using Models
Opening Discussion
Ask questions such as the following:
- When you use a pattern or equation to generate a table of values and then plot those points, do the points always lie on a line?
- Under what conditions will the points lie on a line?
- Will the points rise from left to right or fall from left to right?
- How can we tell from the table whether the points rise or fall? Can the graph help us see the pattern in the table?
- How can the graph be used to extend the pattern in the table?

Explain that in this activity students will use graphs to help them visualize patterns in a table and to extend that pattern.

120 HRW Algebra One Interactions Course 1

Teach

The main body of the instructional material is identified by the horizontal bar dividing the pages into the two different approaches. Choose Approach A, Using the Book, if you want to stay close to the text for your instruction. Or choose Approach B, Using Models, if you prefer a more kinesthetic, hands-on approach.

Approach A Using the Book, continued
Additional Example 2
The table below gives the height, for the first 4 seconds, of a cannonball dropped from 500 feet above the ground. Find the first and second differences of the height values. Then extend the table to 10 seconds. About how many seconds after it is dropped does the cannonball strike the ground?

Time	0	1	2	3	4
Height	500	484	436	356	244

[First differences: −16, −48, −80, −112; Second Differences: constant −32. Height values to 10 seconds: 500, 484, 436, 356, 244, 100, −76, −284, −524, −796, −1100. The cannonball strikes the ground between 5 and 6 seconds after it is dropped. Height values below 0 are meaningless.]

Ongoing Assessment, page 114
The rocket reaches its maximum height halfway between the time it is launched and the time it hits the ground.

Ongoing Assessment, page 115
Try This
7.5 seconds

Teaching Tip
The Path of the Rocket A common misconception that students have when viewing a graph such as the one in Example 2 is that the shape of the graph is the same as that of the rocket's path. Point out that the horizontal axis is time, not the horizontal distance traveled by the rocket. One of many possibilities is that the rocket was shot straight up and fell straight back down. The horizontal distance that it traveled cannot be determined from the information given.

Inclusion Strategy
Technology Have students work in pairs at a computer. Try to pair students so that at least one student has experience working on the computer. The student with less experience should be at the keyboard, while the other student coaches.

Have the students enter the values on page 115 in columns A and B of a spreadsheet. After entering the formulas for cells C2 and D2, use the fill down feature to copy the formulas into cells C3 through C6 and D3 through D6.

Next have students experiment by changing the numbers in column B. Challenge them to try to create (1) a sequence whose second differences are constant and positive, (2) a sequence whose first differences are constant and positive, (3) a sequence whose first differences are constant and negative, (4) a sequence whose first differences are zero, and (5) a sequence whose second differences are not constant.

Ongoing Assessment, page 115
Compute the value in C3 by subtracting the value in B3 from the value in B4.
Compute the value in D2 by subtracting the value in C2 from the value in C3.

Identifying Linear Patterns
An example of an ocean liner traveling at a constant speed is given. The constant differences in the table show the distance traveled by the ship each hour. This information can be used to find the linear equation for the distance traveled.

Critical Thinking, page 115
Because the distance at time 0 is 0 knots and because first differences show that each additional hour adds 15 knots, the values in the table satisfy the formula $d = 15t$. Remind students that distance = rate × time.

continued on top of next page

TEACH

Approach B Using Models, continued
Point out that with this method, the horizontal bars on the graph must be of equal length. In this case, each is 1 unit long.

Using a Table
Using just the table for the conversation problem, demonstrate the method for finding the differences in a table of values, as shown on page 112 of the textbook. Write the first differences in a row below the given values. Then, in a third row, find the second differences.

Explain to students that they can extend the table by using the first and second differences as follows:
- Extend the second differences. (They are all 1 in this example.)
- Work backward from the second differences to extend the first differences.
- Work backward from the first differences to extend the sequence.

Encourage student participation in working with sequences.

Cooperative Learning
Have students work in pairs on the problems in Examples 1 and 2. Write these problems on the board or make paper copies for students so that they do not see the solutions.

For each exercise, one student should be responsible for drawing the graph and extending the pattern from the graph, while the other student is responsible for extending the sequence by using only the table. The partners should compare their work on Example 1 and then switch roles to complete Example 2.

For Example 1, demonstrate how to make a sequence into a table of values. Make the number of the term the x-value (starting with 1) and make the term itself the y-value.

For Example 2, delete the column for $t = 14$ from the table.

Teaching Tip
Graphing It may take students quite awhile to determine the appropriate scales for the graphs that they need to make in the Cooperative Learning situation described at left. To save time, you may wish to make coordinate grids for each problem and give one copy to each pair of students. Then students will be able to easily plot the points and focus their attention on the patterns.

Exploration
Present the exploration from page 116 of the textbook by writing the appropriate data on the board or the overhead: Walt and his grandson rent a rototiller for $20, plus $7 per hour.

Have students create a table for the cost for renting the rototiller. Then have them write an equation for the cost. [$c = 20 + 7h$]

continued on bottom of next page

122 HRW Algebra One Interactions Course 1 Lesson 2.7 123

Assess, Reteach, and Enrich

The Assess section of your planning guide includes the following features:

- Selected Answers—a list of exercises for which student answers are provided
- Error Analysis
- Alternative Assessment
- Portfolio Activity

The Reteach section is for students who need further instruction.

The Enrich section contains a number of ideas that add interest and relevancy to your lesson—ranging from historical information to applications and interdisciplinary connections.

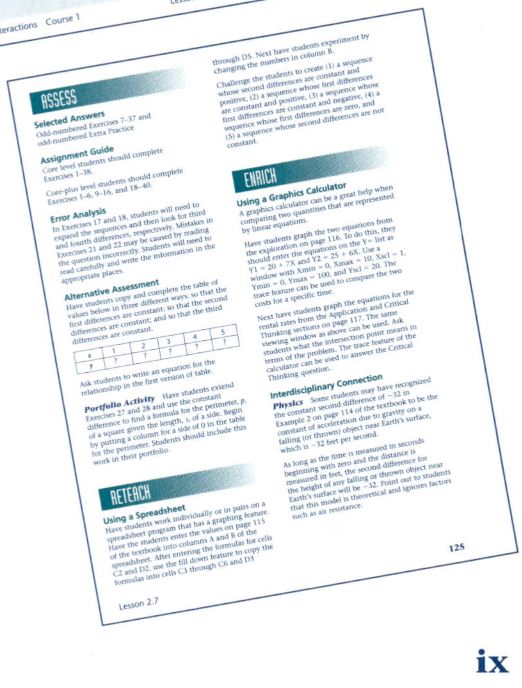
ASSESS

Selected Answers
Odd-numbered Exercises 7–37 and odd-numbered Extra Practice

Assignment Guide
Core level students should complete Exercises 1–38.
Core-plus level students should complete Exercises 1–6, 9–16, and 18–40.

Error Analysis
In Exercises 17 and 18, students will need to expand the sequences and then look for third and fourth differences, respectively. Mistakes in Exercises 21 and 22 may be caused by reading the question incorrectly. Students will need to read carefully and write the information in the appropriate places.

Alternative Assessment
Have students copy and complete the table of values below in three different ways: so that the first differences are constant; so that the second differences are constant; and so that the third differences are constant.

x	1	2	3	4	5
y	?	?	?	?	?

Ask students to write an equation for the relationship in the first version of table.

Portfolio Activity Have students extend Exercises 27 and 28 and use the constant difference to find a formula for the perimeter, p, of a square given the length, s, of a side. Begin by putting a column for a side of 0 in the table for the perimeter. Students should include this work in their portfolio.

RETEACH

Using a Spreadsheet
Have students work individually or in pairs on a spreadsheet program that has a graphing feature. Have the students enter the values on page 115 of the textbook into columns A and B of the spreadsheet. After entering the formulas for cells C2 and D2, use the fill down feature to copy the formulas into cells C3 through C6 and D3 through D5. Next have students experiment by changing the numbers in column B.

Challenge the students to create (1) a sequence whose second differences are constant and positive, (2) a sequence whose first differences are constant and positive, (3) a sequence whose first differences are constant and negative, (4) a sequence whose first differences are zero, and (5) a sequence whose second differences are not constant.

ENRICH

Using a Graphics Calculator
A graphics calculator can be a great help when comparing two quantities that are represented by linear equations.

Have students graph the two equations from the exploration on page 116. To do this, they should enter the equations on the Y= list as Y1 = 20 + 7X and Y2 = 25 + 6X. Use a window with Xmin = 0, Xmax = 10, Xscl = 1, Ymin = 0, Ymax = 100, and Yscl = 20. The trace feature can be used to compare the two costs for a specific time.

Next have students graph the equations for the rental rates from the Application and Critical Thinking sections on page 117. The same viewing window as above can be used. Ask students what the intersection point means in terms of the problem. The trace feature of the calculator can be used to answer the Critical Thinking question.

Interdisciplinary Connection
Physics Some students may have recognized the constant second difference of −32 in Example 2 on page 114 of the textbook to be the constant of acceleration due to gravity on a falling (or thrown) object near Earth's surface, which is −32 feet per second.

As long as the time is measured in seconds beginning with zero and the distance is measured in feet, the second difference for the height of any falling or thrown object near Earth's surface will be −32. Point out to students that this model is theoretical and ignores factors such as air resistance.

Lesson 2.7 125

Key Blackline Masters

For your convenience, we have included three blackline masters, with answers, for each lesson in this book. They are yours to photocopy and use as you see fit. These blackline masters, which are taken from the ancillary package for the *HRW Algebra One Interactions* program, are the ones that we believe you will find most useful.

Practice Masters

As a teacher, you know the importance practice to establish mastery of the ideas that students learn in class and to ensure long-term retention of these ideas. For this reason, we have included one **Practice Master** for each lesson. The Practice Masters provide additional practice of the skills and concepts taught in the lessons. There are 10 to 35 practice items on each page.

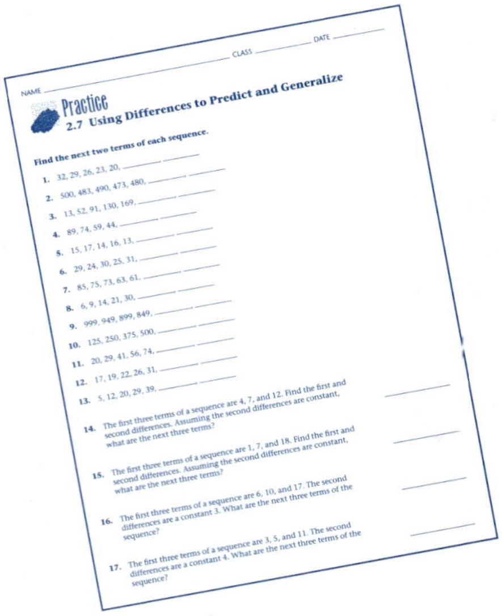
NAME ______ CLASS ______ DATE ______

Practice
2.7 Using Differences to Predict and Generalize

Find the next two terms of each sequence.

1. 32, 29, 26, 23, 20, ______, ______
2. 500, 483, 490, 473, 480, ______, ______
3. 13, 52, 91, 130, 169, ______, ______
4. 89, 74, 59, 44, ______, ______
5. 15, 17, 14, 16, 13, ______, ______
6. 29, 24, 30, 25, 31, ______, ______
7. 85, 75, 73, 63, 61, ______, ______
8. 6, 9, 14, 21, 30, ______, ______
9. 999, 949, 899, 849, ______, ______
10. 125, 250, 375, 500, ______, ______
11. 20, 29, 41, 56, 74, ______, ______
12. 17, 19, 22, 26, 31, ______, ______
13. 5, 12, 20, 29, 39, ______, ______
14. The first three terms of a sequence are 4, 7, and 12. Find the first and second differences. Assuming the second differences are constant, what are the next three terms? ______
15. The first three terms of a sequence are 1, 7, and 18. Find the first and second differences. Assuming the second differences are constant, what are the next three terms? ______
16. The first three terms of a sequence are 6, 10, and 17. The second differences are a constant 3. What are the next three terms of the sequence? ______
17. The first three terms of a sequence are 3, 5, and 11. The second differences are a constant 4. What are the next three terms of the sequence? ______

Lesson Activity Masters

For full comprehension and appreciation of the concepts they learn, students need the kind of hands-on involvement that will make these concepts part of their lives. The **Lesson Activity Masters** included in this planning guide connect mathematics to other disciplines, provide family involvement, and address "hot topics" in mathematics education.

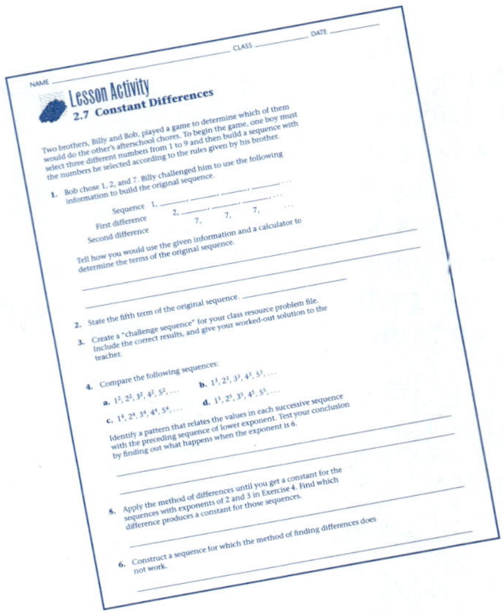
NAME ______ CLASS ______ DATE ______

Lesson Activity
2.7 Constant Differences

Two brothers, Billy and Bob, played a game to determine which of them would do the other's afterschool chores. To begin the game, one boy must select three different numbers from 1 to 9 and then build a sequence with the numbers he selected according to the rules given by his brother.

1. Bob chose 1, 2, and 7. Billy challenged him to use the following information to build the original sequence.

   Sequence 1, ______, ______, ______, ______, ...
   First difference 2, ______, ______, ______, ...
   Second difference 7, 7, 7, ...

   Tell how you would use the given information and a calculator to determine the terms of the original sequence.

2. State the fifth term of the original sequence. ______
3. Create a "challenge sequence" for your class resource problem file. Include the correct results, and give your worked-out solution to the teacher.
4. Compare the following sequences:
   a. $1^2, 2^2, 3^2, 4^2, 5^2, \ldots$
   b. $1^3, 2^3, 3^3, 4^3, 5^3, \ldots$
   c. $1^4, 2^4, 3^4, 4^4, 5^4, \ldots$
   d. $1^5, 2^5, 3^5, 4^5, 5^5, \ldots$

   Identify a pattern that relates the values in each successive sequence with the preceding sequence of lower exponent. Test your conclusion by finding out what happens when the exponent is 6.
5. Apply the method of differences until you get a constant for the sequences with exponents of 2 and 3 in Exercise 4. Find which difference produces a constant for those sequences.
6. Construct a sequence for which the method of finding differences does not work.

Basic Skills Practice Masters

In addition to practice that is related to specific mathematical topics, there are many basic mathematical skills that students must master in order to acquire ease and fluency in the handling of mathematical ideas. The **Basic Skills Practice Masters** (one per lesson) provides both instruction and practice in basic (pre-algebra) skills. There are three kinds of masters:

- Part 1 masters provide instruction, using an example-and-solution format, and practice exercises.
- Part 2 masters provide practice of the content taught in Part 1, using a standardized-test format.
- Cumulative Review masters provide test-format practice of all skills taught and practiced in preceding masters.

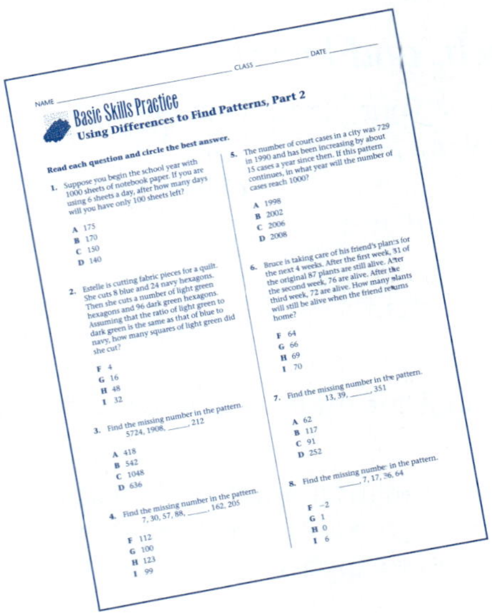
NAME ______ CLASS ______ DATE ______

Basic Skills Practice
Using Differences to Find Patterns, Part 2

Read each question and circle the best answer.

1. Suppose you begin the school year with 1000 sheets of notebook paper. If you are using 6 sheets a day, after how many days will you have only 100 sheets left?
   A 175
   B 170
   C 150
   D 140
2. Estelle is cutting fabric pieces for a quilt. She cuts 8 blue and 24 navy hexagons. Then she cuts a number of light green hexagons and 96 dark green hexagons. Assuming that the ratio of light green to dark green is the same as that of blue to navy, how many squares of light green did she cut?
   F 4
   G 16
   H 48
   I 32
3. Find the missing number in the pattern.
   5724, 1908, ______, 212
   A 418
   B 542
   C 1048
   D 636
4. Find the missing number in the pattern.
   7, 30, 57, 88, ______, 162, 205
   F 112
   G 100
   H 123
   I 99
5. The number of court cases in a city was 729 in 1990 and has been increasing by about 15 cases a year since then. If this pattern continues, in what year will the number of cases reach 1000?
   A 1998
   B 2002
   C 2006
   D 2008
6. Bruce is taking care of his friend's plants for the next 4 weeks. After the first week, 81 of the original 87 plants are still alive. After the second week, 76 are alive. After the third week, 72 are alive. How many plants will still be alive when the friend returns home?
   F 64
   G 66
   H 69
   I 70
7. Find the missing number in the pattern.
   13, 39, ______, 351
   A 62
   B 117
   C 91
   D 252
8. Find the missing number in the pattern.
   ______, 7, 17, 36, 64
   F −2
   G 1
   H 0
   I 6

Active Learning in Your Classroom

How Your Planning Guide Can Help

HRW Algebra One Interactions' instructional philosophy is focused on student involvement in the learning process. In this planning guide, you will find helpful suggestions for implementing this philosophy in your classroom, as well as suggestions for activities over and above those given in the Pupil's Edition.

Discovery . . .

Student discovery of mathematical facts, so often ignored in the past, is potentially one of the most empowering aspects of one's mathematics education. In addition to the activities offered in the Pupil's Edtion, this planning guide will show you additional ways to lead your students along paths to discoveries—and to increased retention and confidence.

Hands-on Involvement: Models . . .

Models are useful, not just to the more concrete learner, but also to the student for whom abstract ideas come easily. The integration of concrete experience with abstract ideas makes for a deeper and more satisfying mathematical experience.

Technology . . .

HRW Algebra One Interactions makes extensive use of calculator and computer technology to increase students' mathematical scope and enjoyment. In your planning guide, you will find suggestions for additional ways to take advantage of the creative possibilities of the latest technologies. Your students will love it!

Cooperative Learning . . .

In cooperative learning, everyone benefits. The dynamics of a learning group allow each student to be both an individual and a team player and to experience the synergy of the group, that all-important something which is more than the sum of group's parts.

. . . And Fun!

Mathematics can and, we think should, be enjoyed in many different ways. The experiences in *HRW Algebra One Interactions* include not only the pleasures of intellectual discovery, but also those of social interaction, kinetic experiences, games, and the shear joy of mathematical empowerment.

Technology

Interactive Learning

Students learn best if they are active participants in the learning process. Technology allows today's mathematics classroom to become a laboratory where students explore and experiment with mathematical concepts rather than just memorize isolated concepts. Students make generalizations and reach conclusions about mathematical concepts and relationships and apply them to real-life situations. *HRW Algebra One Interactions* encourages instruction that utilizes technology with numerous explorations and examples. Where appropriate, students explore, work examples, solve exercises, make and test conjectures, and confirm mathematical ideas for themselves.

Changes in the Classroom

Teachers are finding that technology encourages cooperative learning. Cooperative-learning groups allow students to compare results, brainstorm, and reach conclusions based on group results. As in real life, where scientists or financial analysts often consult each other, students in the technology-oriented classroom learn to communicate and consult with each other. Students learn that such consultations are not "cheating," but are rather a method of sharing information that will be used to solve a problem.

Another change in the classroom will be the role of the teacher from a dispenser of facts and formulas to a guide along the exploration trail. Teachers will become monitors who keep students headed in the right directions. As students gain confidence in their ability, they may ask more questions and even higher-level questions. Teachers no longer always need to have the "right" answer—teachers can suggest further exploration or research. In this way, teachers model real-life situations in which experts with the right answer are not always available or in which there is no right answer.

Real-World Applications

HRW Algebra One Interactions integrates technology throughout the text. This use of technology allows students to explore real-world applications that would not be practical to explore without the use of such technology. Students are able to study realistic rather than contrived data. For instance, when calculating by hand, most students study only simple interest. The use of technology allows students to explore compound interest, which is the way that interest is usually calculated in the financial world today.

Real data about a student's local town or neighborhood can be used to find the average price or rental cost of a home. Students can study data that are more closely related to their own lives, such as the cost of a 4-year college degree based on current costs and projections for the time students will actually be in college. The same can be done for projecting the cost of new cars when they graduate from high school.

X_1	Y_1	Y_2
1	2	0.5
2	4	2
3	6	4.5
3.5	2	6.125
		2

$Y_1 = 2X$

Mathematics is no longer just a study of symbols and equations. Mathematics becomes a useful tool to help students in their own lives.

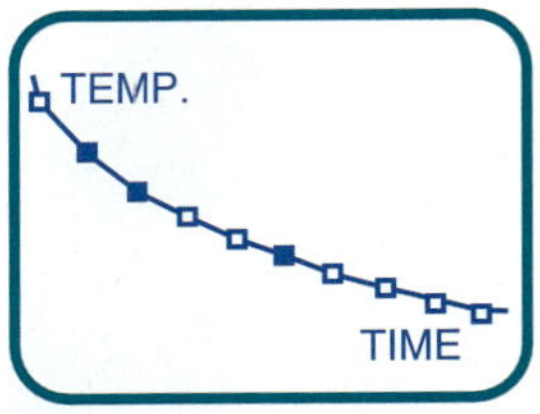

Exploratory Lessons

Exploratory lessons occur throughout *HRW Algebra One Interactions*. Many of these exploratory lessons utilize the power of technology as a tool of exploration. For example, in Lesson 8.3, students explore the concept of a constant. A graphics calculator is used to help students make a conjecture about how the value of b affects the graph of $y = 2x + b$.

The use of technology in simulations allows students to concentrate on the simulation being studied rather than on the generation of random numbers.

Expository Lessons

The power of technology is often used to help students analyze functions that would be difficult or impossible to study without the use of technology. In Lesson 10.2, students use a spreadsheet to analyze the height of a rocket each second of its flight.

In Lesson 9.1, students use a graphics calculator to solve a system of linear equations in both a tabular and graphic format. In Lesson 9.5, a graphics calculator is used to graph the boundary lines of and to find solutions to a system of linear inequalities.

Graphics Calculators

Graphics calculators perform many specialized and complex mathematical actions, including graphing. These calculators use certain keys to access full-screen menus that show additional operations. The scope of capabilities is wide, including traditional arithmetic calculations, matrix operations, statistics mode, function graphing, parametric graphing, polar graphing, sequence graphing, and more. Graphics calculators are capable of running user-defined programs. When using the graphing capabilities of the calculators, the use of a range-setting window adds greatly to the usefulness of the technology. The trace and zoom functions allow students to find the coordinates of points on the curve and to enlarge sections of a graph.

Spreadsheets

A computer spreadsheet program is another valuable technological tool used in the mathematics classroom. By utilizing a spreadsheet, students can enter, store, and sort the type of real-world data that would have been impractical in yesterday's classroom. Spreadsheets facilitate complex mathematics operations and data analysis. Students learn important mathematical concepts by entering formulas to generate simple or complex lists of data that are used to solve problems.

Algebra Explorations CD-ROM

You can draw your students into the power of algebra through compelling graphics and sound with an effective instructional tool, ***Algebra Explorations***. The program, designed to be used with *HRW Algebra One Interactions*, combines mathematical content with visualization tools to create a dynamic learning environment. This innovative piece of technology is a multimedia software program available on CD-ROM that incorporates animations, simulations, and full-motion video sequences of instruction to motivate and reinforce concepts. ***Algebra Explorations*** offers an integration of lesson exposition with a set of math visualization tools including a calculator, a data table, a grapher, a notebook, and an equation editor.

The *Algebra Explorations* CD-ROM program is designed for use by individuals or small groups. Although not a mandatory component of *Algebra One Interactions*, the program will enrich its content and can be used for motivation, reteaching, remediation, enrichment, and practice. During the lesson, the user will be able to interact with the instruction by performing tasks with math visualization tools. For example, the program may ask a student to supply missing data in a data table, to write a conjecture in the notebook, or to graph data by using the grapher tool. Student responses can be cut from the math function tools and pasted into the notebook. The interactive tasks will be tailored specifically to the goals of the lesson presentation.

Explorations and Cooperative Learning

Facilitator of Learning

An increasing number of educators are realizing that more learning takes place when students construct knowledge for themselves. This approach necessitates changes in both the form of instructional materials and the vehicles used to deliver instruction.

Instead of presenting students with rules, theorems, principles, and worked-out examples, this approach calls for students to be given questions to investigate, problems to explore, and conjectures to verify or disprove. Using this approach, a teacher is no longer the source of all information. Instead, the teacher acts as a facilitator, presenting the questions to explore and pointing to areas that need further discussion and clarification.

An activity-oriented approach gives teachers several advantages. Perhaps the most important of these is the ability to provide for learning styles other than verbal and visual. Students who need hands-on, tactile experiences—the kinesthetic learners—are no longer penalized by materials and instructional approaches that emphasize listening and recording information.

Students who need to work in collaboration with others to understand algebra concepts—the interpersonal learners—now have the opportunities that occur during small-group instruction. In fact, all students benefit from the increased written and verbal communication that occurs when active learning is implemented in the classroom.

So what specific form should constructivist materials take in algebra? Although many variations are possible, materials suitable for active learning usually involve identification of patterns, making and testing conjectures, and exploring alternative approaches to problems. Students should be encouraged to try any approach that occurs to them as they work on a problem. Concrete models, graphic representations, and tabular methods such as those facilitated by spreadsheets all have an appropriate place in today's algebra classroom.

In many classroom situations, a key decision will be how and when to use active learning. Some topics and situations will continue to require direct instruction; others will open opportunities for both students and teachers to experience the formation of richer knowledge that results from students' active and personal involvement in their own learning.

New Tools

Throughout *HRW Algebra Interactions*, students are provided many opportunities for hands-on learning through the use of a wide variety of manipulatives. Probability simulations allow for the use of cards, number cubes, and marbles. Geometric concepts are enhanced by utilizing three-dimensional models. The use of algebra tiles enhances understanding of computation with integers in *Course 1* as well as simplifying and factoring polynomials in *Course 2.*

Rapid and continuing changes in technology provide teachers and students in the algebra classroom with more choices for strategies and procedures than in past years. Graphics calculators, spreadsheets, and computer programs are excellent tools for use in student explorations. These tools allow students to experiment with and explore many concepts that would otherwise be too tedious and time consuming.

Communicating in *HRW Algebra One Interactions*

The exploratory approach in *HRW Algebra One Interactions* provides many opportunities for students to communicate. Whether in a traditional classroom setting or in cooperative-learning groups, students communicate orally as they complete explorations, answer Critical Thinking questions, and explain results of simulations and projects. Students communicate in writing as they complete written assignments, keep a journal, and build a portfolio of their work. Students communicate graphically as they make drawings to illustrate a concept or as they display data in a graph.

Active Learning in *HRW Algebra One Interactions*

In the past, many algebra textbooks included little, if any, material appropriate for active-learning situations. Thus, teachers who realized a need for this type of approach had to search out or create special materials.

In marked contrast, *HRW Algebra One Interactions* has been structured to include the frequent use of activity-style lessons and of activities. In lessons beginning with the word *Exploring*, the entire instructional focus is centered around exploration and discovery. Also, many of the "non-exploration" lessons contain explorations that involve students in active-learning experiences. Most are appropriate for small-group work; many allow for use of alternative technologies; and all give students a chance to discover new ideas on their own.

As students begin a set of exercises at the end of a lesson, they may notice that they do not start with a set of simple-minded practice or drill problems. Instead, each exercise set starts with Communicate questions. Here, students can talk or write about key ideas as a check on what they have learned from the lesson. Students may be asked to compare and contrast two different concepts, explain how they would solve a particular problem, or discuss why a particular answer is not reasonable for a given problem.

Students for whom active learning is a new approach may at first be confused by the way in which material is presented. They may have come to expect that they are simply to repeat whatever the teacher or the textbook has explained. But in *HRW Algebra One Interactions,* the textbook doesn't "give it all away." Students must explore and think a bit to get at the central concepts. After a few experiences, students will come to enjoy doing algebra in their own way, rather than being tied to someone else's thinking processes.

Real-World Applications

Teachers who implement activity-style learning can expect to see gradual improvement in long-term recall of information, clearer understanding of connections among concepts, more creative approaches to problem solving, and more sophisticated ways of using and communicating mathematical ideas.

In addition, teachers will be preparing students for future real-world problem-solving situations. Few, if any, of the problems and decisions that face people in their personal and professional lives come in tidy, textbook packaging. "Real" problems are vaguely defined and messy, with missing or unnecessary information, and require difficult compromises to be made. Students who have learned to tackle algebraic problems by actively exploring approaches, collaborating with others in group or team approaches, and comparing results to look for alternative strategies will be equipped to transfer these methodologies to personal and professional challenges later in their lives.

Activity-style instruction that makes frequent use of explorations and investigations gives students more than just a knowledge of algebraic concepts and skills. Careful and consistent implementation of these techniques will prepare all students to face and solve future challenges.

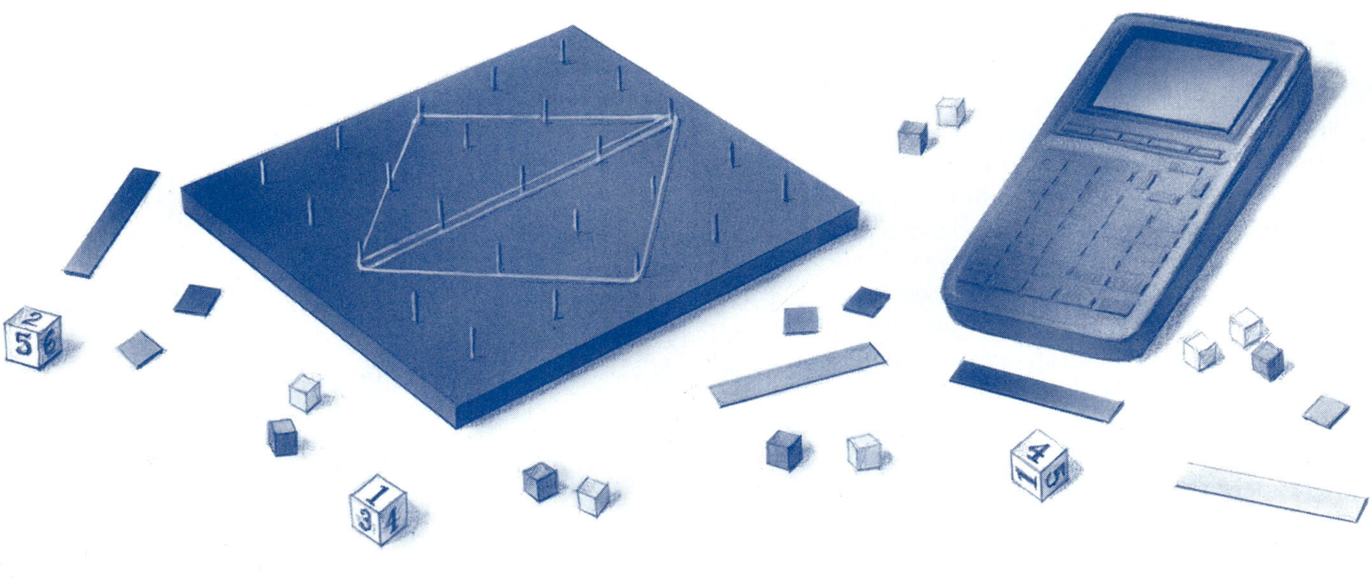

Assessment

Assessment Goals

An essential aspect of any learning environment, such as an algebra classroom, is the process of assessing or evaluating what students have learned. Informally this has been done by using paper-and-pencil tests given by the teacher on a regular basis to measure students' performance against the material being studied. Formal evaluations with standardized tests are generally conducted over a period of years to establish performance records for both individuals and groups of students within a school or school district. Both types of tests are very good at measuring the ability of a student to use a particular mathematical skill or to recall a specific fact. They fall short, however, in evaluating other key goals of learning mathematics, such as being able to solve problems, to reason critically, to understand concepts, and to communicate mathematically, both verbally and in writing. Other techniques, usually referred to as alternative assessment, are needed to evaluate students' performance on these *process* goals of instruction.

The goals of an alternative assessment program are to provide a means of evaluating students' progress in non-skill areas of learning mathematics. Thus, the design and structure of alternative assessment techniques must be quite different from the skill-oriented, paper-and-pencil tests of the past.

Types of Alternative Assessment

In the world outside of school, a person's work is evaluated by what that person can do, that is, by the results the person achieves and not by taking a test. For example, a musician may demonstrate skill by making music, a pilot by flying an airplane, a writer by writing a book, and a surgeon by performing an operation. Students learn to think mathematically and to solve problems on a continuous basis over a long period of time as they study mathematics at many grade levels. Students, too, can demonstrate what they have learned and understand by collecting a representative sample of their best work in a portfolio. A portfolio should illustrate achievements in problem solving, critical thinking, writing about mathematics, mathematical applications or connections, and any other activity that demonstrates an understanding of both concepts and skills.

Specific examples of the kinds of work that students can include in their portfolio are solutions to nonroutine problems, graphs, tables or charts, computer printouts, group reports or reports of individual research, simulations, and examples of artwork or models. Each entry should be dated and should be chosen to show the students' growth in mathematical competence and maturity.

A portfolio is just one way for students to demonstrate their performance on a mathematical task. Performance assessment can also be achieved in other ways, such as by asking students questions and evaluating their answers, by observing their work in cooperative learning groups, by having students give verbal presentations, by working on extended projects and investigations, and by keeping journals.

Peer assessment and self-evaluation are also valuable methods of assessing students' performance. Students should be able to critique their classmates' work and their own work against standards set by the teacher. In order to evaluate their own work, students need to know the teacher's goals of instruction and the criteria (scoring rubrics) that have been established for evaluating performance against the goals. Students can help to design their own self-assessment forms that they then fill out on a regular basis and give to the teacher. They can also help to construct test items that are incorporated into tests given to their classmates. This work is ideally done in small groups of four students. The teacher can then choose items from each group to construct the test for the entire class. Another alternative testing technique is to have students work on take-home tests that pose more open-ended and non routine questions and problems. Students can devote more time to such tests and, in doing so, demonstrate their understanding of concepts and skills and their ability to do mathematics independently.

Scoring

The use of alternative assessment techniques implies the need to have a set of standards against which students' work is judged. Numerical grades are no longer sufficient because growth in understanding and problem solving cannot be measured by a single number or letter grade. Instead, scoring rubrics or criteria can be devised that allow the teacher more flexibility to recognize and comment on all aspects of a student's work, pointing out both strengths and weaknesses that need to be corrected.

Portfolio Holistic Scoring Guide

An individual portfolio is likely to be characterized by some, but not all, of the descriptors for a particular level. Therefore, the overall score should be the level at which the appropriate descriptors for a portfolio are clustered.

		NOVICE	APPRENTICE	PROFICIENT	DISTINGUISHED
PROBLEM SOLVING	Understanding/ Strategies	• Indicates a basic understanding of problems and uses strategies	• Indicates an understanding of problems and selects appropriate strategies	• Indicates a broad understanding of problems with alternate strategies	• Indicates a comprehensive understanding of problems with efficient, sophisticated strategies
	Execution/ Extensions	• Implements strategies with minor mathematical errors in the solution without observations or extentions	• Accurately implements strategies with solutions, with limited observations or extentions	• Accurately and efficiently implements and analyzes strategies with correct solutions with extensions	• Accurately and efficiently implements and evaluates sophisticated strategies with correct solutions and includes analysis, justifications and extensions
REASONING		• Uses mathematical reasoning	• Uses appropriate mathematical reasoning	• Uses perceptive mathematical reasoning	• Uses perceptive, creative, and complex mathematical reasoning
MATHEMATICAL COMMUNICATION	Language	• Uses appropriate mathematical language some of the time	• Uses appropriate mathematical reasoning	• Uses precise and appropriate mathematical language most of the time	• Uses sophisicated, precise, and appropriate mathematical language
	Representations	• Uses few mathematical representations	• Uses a variety of mathematical representations accurately and appropriately	• Uses a wide variety of mathematical representations accurately and appropriately; uses multiple representations with some entries	• Uses a wide variety of mathematical representations accurately and appropriately; uses multiple representations within entries and states their connections
UNDERSTANDING/ CONNECTING CORE CONCEPTS		• Indicates a basic understanding of core concepts	• Indicates an understanding of core concepts with limited connections	• Indicates a broad understanding of some core concepts with connections	• Indicates a comprehensive understanding of core concepts with connections throughout
TYPES AND TOOLS		• Includes few types; uses few tools	• Includes a variety of types; uses tools appropriately	• Includes a wide variety of types; uses a wide variety of tools appropriately	• Includes all types; uses a wide variety of tools appropriately and insightfully

PERFORMANCE DESCRIPTORS	WORKSPACE/ANNOTATIONS
PROBLEM SOLVING • Understands the features of a problem (understands the question, restates the problem in own words) • Explores (draws a diagram, constructs a model and/or chart, records data, looks for patterns) • Selects an appropriate strategy (guesses and checks, makes an exhaustive list, solves a simpler but simular problem, works backward, estimates a solution) • Solves (implements a strategy with an accurate solution) • Reviews, revises, and extends (verifies, explores, analyzes, evaluates strategies/solutions; formulates a rule)	
REASONING • Observes data, records and recognizes pattern, makes mathematical conjectures (indictive reasoning) • Validates mathematical conjectures through logical arguments or counter-examples; constructs valid arguments (deductive reasoning)	
MATHEMATICAL COMMUNICATION • Provides quality explanations and expresses concepts, ideas, and reflections clearly • Uses appropriate mathematical notation and terminology • Provides various mathematical representations (models, graphs, charts, diagrams, words, pictures, numerals, symbols, equations)	

UNDERSTANDING/CONNECTING CORE CONCEPTS

- Demonstrates an understanding of core concepts
- Recognizes, makes, or applies the connections among the mathematical core concepts to other disciplines, and to the real world

Place an X on each continuum to indicate the degree of understanding demonstrated for each core concept.

DEGREE OF UNDERSTANDING OF CORE CONCEPTS

Basic

NUMBER								
MATHEMATICAL PROCEDURES								
SPACE & DIMENSIONALITY								
MEASUREMENT								
CHANGE								
MATHEMATICAL STRUCTURE								
DATA: STATISTICS AND PROBABILITY								

PORTFOLIO CONTENTS

- Table of Contents
- Letter to Reviewer
- 5–7 Best Entries

BREADTH OF ENTRIES

TYPES

0 INVESTIGATIONS/DISCOVERY
0 APPLICATIONS
0 NON-ROUTINE PROBLEMS
0 PROJECTS
0 INTERDISCIPLINARY
0 WRITING

TOOLS

0 CALCULATORS
0 COMPUTER AND OTHER TECHNOLOGY
0 MODELS MANIPULATIVES
0 MEASUREMENT INSTRUMENTS
0 OTHERS
0 GROUP ENTRY

A scoring rubric can be created for each major instructional goal, such as being able to solve problems or communicate mathematically. A rubric generally consists of four or five short descriptive paragraphs that can be used to evaluate a piece of work. For example, if a five-point paragraph scale is used, a rating of 5 may denote that the student has completed all aspects of the assignment and has a comprehensive understanding of problems. The content of paragraph 5 specifies the details of what constitutes the rating of highly satisfactory. On the other hand, a rating of 1 designates an essentially unsatisfactory performance, and paragraph 1 would detail what is unsatisfactory. The other three paragraphs provide an opportunity for the teacher to recognize significant accomplishments by the student as well as aspects of the work that need improvement. Thus, scoring rubrics are a far more realistic and educationally substantive way to evaluate a student's performance than a single grade, which is usually determined by an answer being right or wrong.

The guide pictured, the Kentucky Mathematics Portfolio, was developed by the Kentucky Department of Education for use by school districts throughout that state. This guide is illustrated to show an example of an excellent and effective holistic scoring guide currently in use by math teachers who are practicing performance assessment in their classrooms. The scorer uses the Workspace/Annotations section of the guide to gather evidence about a student's mathematical ability. The top of the guide is then used to assign a single performance rating, based on an overall view of the full contents of the student's portfolio. The lower right-hand corner of the Holistic Scoring Guide lists the Types and Tools for the Breadth of Entries that are appropriate for a student to place in their portfolio.

HRW Algebra One Interactions provides numerous opportunities in each chapter of the student edition that correspond to the Breadth of Entries. They are an integral part of the student textbook. Investigations/discovery are found in all of the explorations found in the student lessons of the program. Interdisciplinary topics from astronomy to zoology and applications from arts to transportation are designated throughout the lessons. Projects are found at the end of every chapter. Communicate exercises offer excellent writing selections for the portfolio and non-routine problems have been integrated throughout the exercise sets. *HRW Algebra One Interactions* makes extensive use of tools such as graphics calculators and algebra tiles as well as CD-ROM technology that students can incorporate into activities for their portfolio.

Assessment and *HRW Algebra One Interactions*

Throughout the textbook, students are asked to explain their work; describe what they are doing; compare and contrast different approaches; analyze a problem; make sketches, graphs, tables, and other models; hypothesize, conjecture, and look for counterexamples; and make and prove generalizations.

All of these activities, including the more traditional responses to routine problems, provide the teacher with a wealth of assessment opportunities to see how well students are progressing in their understanding and knowledge of algebra. The assessment task can be aligned with the major process goals of instruction and scoring rubrics established for each goal. For example, a teacher may decide to organize his or her assessment tasks in the following general areas of doing mathematics:

- Problem solving
- Communication
- Reasoning
- Connections

Within each of these areas, specific goals can be written and shared with students. In this way, the assessment process becomes an integral part, not only of evaluating students' progress, but also of the instructional process itself. The results of assessment can and should be used to modify the instructional approach to enhance learning for all students.

In addition to the many opportunities for performance assessment found in *HRW Algebra One Interactions*, a variety of assessment types are integrated into the chapter-end material. The Chapter Review, the Chapter Assessment, and the Cumulative Assessment include both traditional and alternative assessment. All Chapter Tests include both multiple choice and open-ended type questions. The Cumulative Reviews are formatted in the style of college preparatory exams. In additional to multiple-choice and free-response questions, each Cumulative Assessment contains quantitative-comparison questions that emphasize the concepts of equalities, inequalities, and estimation. Other types of college entrance exam questions found in the Cumulative Review are student-produced response questions with gridded solutions. These Cumulative Assessments expose students to the new types of assessment that they will encounter when they take the latest form of college entrance examinations.

Pacing Guide

Lesson	Traditional (days)	Block (days)
Chapter 1		
Lesson 1.0	2	1
Lesson 1.1	2	1
Lesson 1.2	2	1
Lesson 1.3	2	1
Mid-Chapter Assess.		
Lesson 1.4	2	1
Lesson 1.5	2	1
Lesson 1.6	2	1
Chapter Assess.	1	0.5
Chapter 2		
Lesson 2.1	3	1.5
Lesson 2.2	2	1
Lesson 2.3	2	1
Lesson 2.4	2	1
Mid-Chapter Assess.		
Lesson 2.5	2	1
Lesson 2.6	2	1
Lesson 2.7	2	1
Lesson 2.8	2	1
Chapter Assess.	1	1
Chapter 3		
Lesson 3.1	2	1
Lesson 3.2	2	1
Lesson 3.3	2	1
Mid-Chapter Assess.		
Lesson 3.4	2	1
Lesson 3.5	2	1
Chapter Assess.	1	1
Chapter 4		
Lesson 4.1	2	1
Lesson 4.2	3	1.5
Lesson 4.3	3	1.5
Mid-Chapter Assess.		
Lesson 4.4	3	1.5
Lesson 4.5	3	1.5
Lesson 4.6	3	1.5
Lesson 4.7	3	1.5
Chapter Assess.	1	0.5
Chapter 5		
Lesson 5.1	2	1
Lesson 5.2	2	1
Lesson 5.3	2	1
Mid-Chapter Assess.		
Lesson 5.4	2	1
Lesson 5.5	2	1
Lesson 5.6	2	1
Chapter Assess.	1	1

Lesson	Traditional (days)	Block (days)
Chapter 6		
Lesson 6.1	2	1
Lesson 6.2	2	1
Lesson 6.3	2	1
Mid-Chapter Assess.		
Lesson 6.4	2	1
Lesson 6.5	2	1
Lesson 6.6	2	1
Chapter Assess.	1	0.5
Chapter 7		
Lesson 7.1	3	1.5
Lesson 7.2	3	1.5
Lesson 7.3	3	1.5
Lesson 7.4	2	1
Mid-Chapter Assess.		
Lesson 7.5	2	1
Lesson 7.6	2	1
Lesson 7.7	3	1.5
Chapter Assess.	1	0.5
Chapter 8		
Lesson 8.1	2	1
Lesson 8.2	2	1
Lesson 8.3	2	1
Mid-Chapter Assess.		
Lesson 8.4	2	1
Lesson 8.5	2	1
Lesson 8.6	2	1
Chapter Assess.	1	0.5
Chapter 9		
Lesson 9.1	3	1.5
Lesson 9.2	3	1.5
Lesson 9.3	3	1.5
Lesson 9.4	3	1.5
Mid-Chapter Assess.		
Lesson 9.5	3	1.5
Lesson 9.6	2	1
Lesson 9.7	2	1
Lesson 9.8	2	1
Chapter Assess.	1	0.5
Chapter 10		
Lesson 10.1	2	1
Lesson 10.2	2	1
Lesson 10.3	2	1
Lesson 10.4	2	1
Mid-Chapter Assess.		
Lesson 10.5	2	1
Lesson 10.6	2	1
Lesson 10.7	2	1
Chapter Assess.	1	0.5

Table Of Contents

Chapter 1 Functions, Equations, and Inequalities

Chapter 2 Linear Functions and Systems

Chapter 3 Matrices

Chapter 4 Probability and Statistics

Chapter 5 Transformations

Chapter 6 Exponents

Chapter 7 Polynomials and Factoring

Chapter 8 Quadratic Functions

Chapter 9 Radicals and Coordinate Geometry

Chapter 10 Rational Functions

LESSON 1.0 *Exploring* Functions

PREPARE

Pacing Guide
2 Days

Objectives and Overview
- Understand the concept of a function.
- Use tables, equations, and graphs to represent functions.
- Identify the independent variable, dependent variable, domain, and range of a function.

This lesson introduces functions in an informal way. Functions are explored through tables, equations, and graphs.

Vocabulary

constant	function
variable	domain
dependent variable	range
independent variable	reasonable domain

Assessing Prior Knowledge

Simplify each expression.

1. $40(2) - 34$ [46]

2. $27(18) - 593$ [−107]

Give the coordinates of each point.

3. A [(2, 2)]

4. B [(−2, 0)]

5. C [(1, −2)]

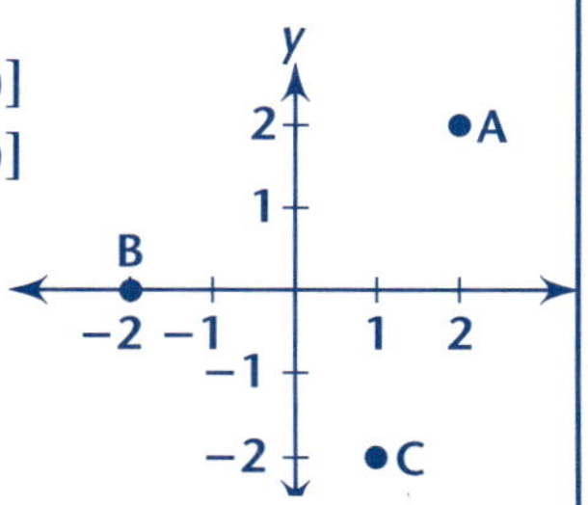

Approach A: Using the Book

Opening Discussion
Refer to the World Cuisine Festival example. Remind students that profit is the money made after subtracting expenses. This example examines the relationship between the number of tickets sold and the profit. Functions are a way to express a real-world relationship such as this, in which one quantity affects another.

Brainstorming
Have students think of other real-world examples in which one quantity affects another. Examples include the following: the amount of sunlight affects the temperature of the pavement, the number of students in the class affects the time it takes to pass out a test, and the amount of gasoline in a car's tank affects how far it can be driven without buying more gas. Point out that these examples can all be represented by functions.

continued on top of next page

TEACH

Approach B: Using Models

Opening Discussion
Write the following problem on the board: A group of students is sponsoring a World Cuisine Festival. The cost for food and other expenses is \$474. Tickets will be sold for \$8.00 each. How many tickets do the students need to sell to make a profit of \$1000?

Remind students that profit is the money made after subtracting expenses. This example can be better understood by examining the relationship between the number of tickets sold and the profit. Functions are a way to express a real-world relationship such as this, in which one quantity affects another.

Brainstorming
Have students think of other real-world examples in which one quantity affects another. Examples include the following: the amount of sunlight affects the temperature of the pavement, the number of students in the class affects the time it

continued on bottom of next page

Approach A

Using the Book, continued

Setting Up the Explorations

Teacher-Guided Explorations If you wish, you can guide the students through the textbook explorations. You can use overhead transparencies or the chalkboard to display the table and graph.

At each stage of the explorations, invite student input and encourage discussion. Students should keep a record of the exploration responses in their notebook. For a greater level of participation, you can ask student volunteers to fill in the table for Exploration 1 at the chalkboard.

Cooperative-Learning Groups Divide the class into groups of four. Each group should choose one member to read the exploration steps from the textbook, one to draw the graphs, one to keep notes of the group's results, and one to write the final report with input from all group members. These responsibilities can be reassigned for each exploration. All group members should be involved in a discussion of the activity as the exploration proceeds. Each student should sign the completed report.

Exploration 1

The information from the World Cuisine Festival example is organized in a table. The table is then used to introduce the concepts of constant, variable, dependent variable, and independent variable.

Students should work carefully to follow the pattern in the process column. This column will be useful for writing the equation in Exploration 2. When choosing letters to represent the variable, students should pick letters that will help them remember what the variable stands for.

TEACH

Approach B

Using Models, continued

takes to pass out a test, and the amount of gasoline in a car's tank affects how far it can be driven without buying more gas. Point out that these examples can all be represented by functions.

Using a Spreadsheet

Creating a Model This model requires a computer with spreadsheet software. If possible, use an overhead device to project the computer screen. Begin with a blank spreadsheet and work with the class to create a spreadsheet that models the number of tickets and the profit.

	A	B	C
1	Expenses	Number of tickets	Profit
2	474	50	−74
3	474	100	326
4	474	150	726
5	474	200	1126
6	474	250	1526

Ask students to give the formulas for the various cells. Use the fill-down feature when appropriate.

Using the Model Use this model to introduce the concepts of constant, variable, dependent variable, and independent variable. Work through the activity by asking questions that students can answer by looking at the spreadsheet. For example, to introduce the term *constant,* ask which quantity stays the same. Students should understand the expenses are constant, while the profit and the number of tickets are variable. Show students how to represent the variable quantities with letters. The profit depends on the number of tickets sold; hence, it is the dependent variable, while the number of tickets sold is the independent variable. (Compare Exploration 1 on page 5.)

Definition of *Function*

A function is defined as a set of ordered pairs. Remind students of the definition of and common notation for an ordered pair. Note that

Be sure to invite participation of each student in the World Cuisine Festival example.

Ongoing Assessment, page 5

2. department expenses, $474, and the price of a ticket, $8

4. The amount of profit depends on the number of tickets sold. As the number of tickets changes, the expenses stay the same, but the profit changes. Thus, the profit is dependent on the number of tickets sold.

6. The independent variable is the one that represents the number of tickets sold.

Definition of *Function*

A function is defined as a set of ordered pairs. Remind students of the definition of and common notation for an ordered pair. Note that the independent variable is placed first in the ordered pair.

Teaching Tip

The definition of function used here is an informal one. A mathematically precise definition of function is a set of ordered pairs

continued on top of next page

the independent variable is placed first in the ordered pair. Remind students that ordered pairs can be graphed on the coordinate plane.

Teaching Tip

The definition of function used here is an informal one. A mathematically precise definition of function is a set of ordered pairs such that each first coordinate is paired with exactly one second coordinate. For example, a function could not contain both (3, 2) and (3, 6). This definition ensures that each value of the independent variable has one and only one corresponding value of the dependent variable.

Exploration 1

In this exploration, students work to write a formula for the function describing the profit of the World Cuisine Festival. You should lead students through this exploration, using the spreadsheet as a tool to answer questions. (For ideas on guiding the students, see Exploration 2 on page 6 of the Pupil's Edition of the textbook.)

Begin by asking students to find the profit for various numbers of tickets, and then ask them to find the number of tickets for a given profit. Have students describe how to find the profit when you know the number of tickets sold. They can then use this description to write an equation for the profit in terms of the number of tickets sold.

Exploration 2

In this exploration, students see how the World Cuisine Festival example is represented by a graph. The graph is used to introduce the terms *domain, range,* and *reasonable domain.* (For ideas on guiding the students, see Exploration 3 on page 7 of the Pupil's Edition of the textbook.)

Demonstrate how to use the spreadsheet program to generate a graph of the relationship between the number of tickets and the profit.

continued on bottom of next page

Approach A — Using the Book, continued

such that each first coordinate is paired with exactly one second coordinate. For example, a function could not contain both (3, 2) and (3, 6). This definition ensures that each value of the independent variable has one and only one corresponding value of the dependent variable.

Exploration 2

This exploration leads students to write an equation to represent the function from the World Cuisine Festival example. Notice that the equation is written with the variables x and y. These letters are commonly used to represent the independent and dependent variables of a function, respectively.

Ongoing Assessment, page 6

3. Multiply 8 by the number of tickets sold, and subtract the expenses.

Exploration 3

The graph of the function representing the profit in the World Cuisine Festival example is examined. Students learn to interpret the graph and use the graph to learn the definitions of domain, range, and reasonable domain. Students should understand that the reasonable domain is the part of the domain of the function that makes sense in the context of the real-world situation. In this situation, the number of tickets must be a positive whole number.

Use **Transparency 1** for discussion.

Ongoing Assessment, page 7

Part 1, Step 4. Find 800 on the vertical axis. From that point move horizontally to the graph of the function. The corresponding x-value is the number of tickets (122 tickets).

Part 2, Step 4. The reasonable domain is the set of positive integers.

TEACH

Approach B — Using Models, continued

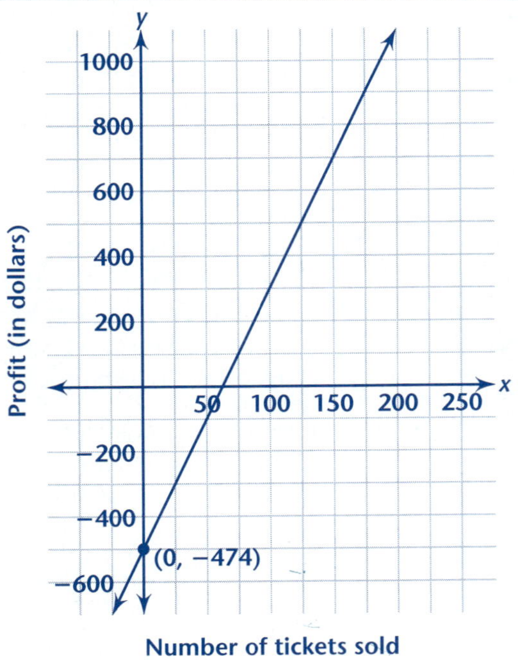

Ask students which axis represents the dependent and which represents the independent variables. Have them use the graph to find the profit for a given number of tickets and the number of tickets that need to be sold for a given profit. Introduce the terms *domain* and *range,* and have students use the graph to identify them for this function. Then introduce the term *reasonable domain.* Students should understand that the reasonable domain is the part of the domain of the function that makes sense in the real-world context. In this case, the reasonable domain is the positive whole numbers. Ask students why negative numbers and fractions are omitted.

ASSESS

Selected Answers

Odd-numbered Exercises 1–27

Assignment Guide

Core level students should complete Exercises 1–27.

Core-plus level students should complete Exercises 1–27.

Error Analysis

Students may forget to change 8% to decimal form in Exercises 2–7. You may need to review the guess-and-check strategy for Exercise 11. Students who have difficulty writing equations should be encouraged to describe the process in words and then to translate the words into an equation.

Alternative Assessment

Finding Real-World Applications
Write the equation $y = 0.04x + 100$ on the chalkboard. Have students make up a real-world situation that could be represented by this function. They should describe the independent and dependent variables and then make a table for the function. What is the reasonable domain for the situation? For extra credit, students can make a graph to represent the function.

Portfolio Activity Ask students to include their work for Exercises 12–18 in their portfolio. Have them make a graph for the situation as well.

RETEACH

Comparing Tables, Equations, and Graphs

Ask student volunteers to report their answers for Exercises 2–11. Add a process column to the table. Describe the process in words. How is the process related to the equation? Work with students to create a graph of the weekly sales versus the total income. How can the graph be used to answer Exercise 11? Brainstorm with the class to write a list of the advantages and disadvantages of each of the three forms of the function.

Have students work in small groups to make a table, write an equation, and draw a graph of the function that represents the total income earned by a person who receives a salary of $100 per week plus a commission of 15% of their sales.

ENRICH

Interdisciplinary Connection

Journalism Graphs of functions are often used in journalism. Have students look through magazines and newspapers for graphs similar to those in the exercises. They should identify the independent variable, dependent variable, and the reasonable domain for the situation. Then have them describe the function in words.

The Mathematical Definition of *Function*

The definition of function used in this lesson is an informal one. A mathematically precise definition of function is a set of ordered pairs such that each first coordinate is paired with exactly one second coordinate. For example, a function could not contain both (3, 2) and (3, 6). This definition ensures that each value of the independent variable has one and only one corresponding value of the dependent variable. Note that a function may contain the ordered pairs (2, 3) and (6, 3).

Give students several sets of ordered pairs, and have them determine if the set is a function. Then have them graph each set of ordered pairs and describe the difference between the graph of a function and the graph of a set of points that is not a function. Students should realize that the graph of a function cannot have two or more points that are aligned vertically. This property is the basis of a visual test, called the *vertical line test,* used to determine if a graph is the graph of a function. The vertical line test states that if a vertical line passes through two or more points on a graph, then the graph is not the graph of a function. Otherwise, the graph is the graph of a function. Have students practice drawing graphs that represent functions and ones that do not.

LESSON 1.1

Addition and Subtraction Equations

PREPARE

Pacing Guide
2 Days

Objectives and Overview
- Use the Addition and Subtraction Properties of Equality to solve equations.

This lesson should be review for most students. Real-world problems are modeled by one-step equations. One-step equations are solved by using the Subtraction Property of Equality or the Addition Property of Equality.

Vocabulary
Subtraction Property of Equality
Addition Property of Equality
neutral pair

Assessing Prior Knowledge

Simplify each expression.		
1.	$4 - 6$	[-2]
2.	$-3 + 7$	[4]
3.	$-6 - 5$	[-11]
Evaluate for $x = 3$.		
4.	$x + 7$	[10]
5.	$x - 12$	[-9]

Approach A: Using the Book

Opening Discussion
Many people use equations to solve problems. Discuss the clay novelties example. Write the equation to represent this situation, and discuss how the variable x is used to represent the unknown amount of increase.

Brainstorming
Once you write an equation, what can you do to isolate the variable? Have students list the techniques they remember from previous courses. Examples include adding the same number to both sides, subtracting the same number from both sides, multiplying both sides by the same number, or dividing both sides by the same nonzero number.

Subtraction Property of Equality
Lead students through the solution to the clay novelties problem. If students are not familiar with algebra tiles, explain that squares with a plus sign

TEACH

Approach B: Using Models

Opening Discussion
Many people use equations to solve problems in their work. Discuss how a craftsperson might use an addition equation to represent the increase in price of an item sold for $8 if the original cost of the item was $3. [**$x + 3 = 8$, where x represents the increase in price**]

Brainstorming
Once you write an equation, what can you do to isolate the variable? Have students list the techniques they remember from previous courses. Examples include adding the same number to both sides, subtracting the same number from both sides, multiplying both sides by the same number, or dividing both sides by the same nonzero number.

Making Algebra Tiles
If algebra tiles are not available, you can create them from two sheets of different-colored overhead transparency film. Cut some rectangles and 1-inch

represent positive units and squares with a minus sign represent negative units. Stress that the objective is to isolate the variable on one side so that its value can be seen. Subtracting equal quantities keeps the equation in balance. After completing the problem, read the Subtraction Property of Equality and have students copy it into their notebook or journal.

Using Graphing Technology to Check

If possible, demonstrate how to use a graphics calculator to check the solution to an equation. Write the corresponding functions from each side of the equation, and enter them in the Y= list of the calculator. Use the trace or intersect features to locate the intersection point. The solution to the equation is the x-coordinate of the point of intersection.

Example 1

In this example, one-step equations are solved by using the Subtraction Property of Equality. Point out the justifications for each step. Also show students how to check their solution by using substitution. You can use algebra tiles on the overhead or draw algebra-tile models on the chalkboard to help students with little or no experience with these types of equations. Note that the second equation requires students to subtract a greater number from a lesser one. You may want to review subtraction of integers.

Additional Example 1

Solve each equation.

A. $x + 5 = 11$ [6]

B. $x + 16 = 7$ [−9]

Ongoing Assessment, page 11

Try This

a. −15

b. 8

Addition Property of Equality

Write the equation $x - 3 = 2$. Ask students what needs to be done to solve this equation. **[Three must be added to each side of the equation.]**

continued on top of next page

squares from one sheet. Mark them with plus signs. Cut the other sheet into rectangles and 1-inch squares and mark them with minus signs. The rectangles represent x and $-x$; the squares represent positive and negative units.

Using Algebra Tiles to Represent and Solve One-Step Equations

Creating a Model On an overhead projector, show students how to represent the equation $x + 3 = 8$ with algebra tiles.

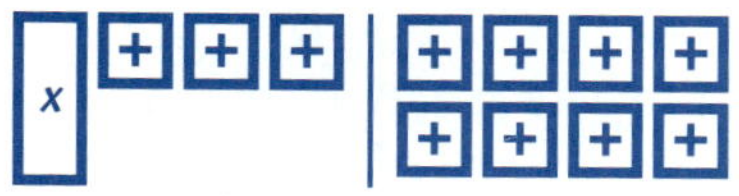

Point out that the square tiles with a + symbol represent a positive unit and the square tiles with a − sign represent a negative unit. The vertical segment represents the equal sign.

A One-Step Addition Equation
Demonstrate that subtraction is modeled by removing tiles. In order to keep the equation in balance, the same number of tiles must be removed from each side of the equation. The objective is to isolate the variable so that its value can be determined. Remove three positive tiles from each side of the equation.

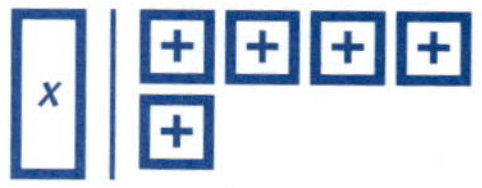

The remaining tiles are used to write the solution, $x = 5$. The increase in price is $5 per item.

A One-Step Subtraction Equation
Represent the subtraction equation $x - 5 = 1$ with the tiles. Then solve the equation by using the tiles.

How do we get the x-tile alone this time? **[Add 5 positive tiles to each side of the equation.]**

continued on bottom of next page

Approach A

Using the Book, continued

This can be done because of the Addition Property of Equality. Read or have a student read the property on page 12, and have students copy it into their notebook or journal. Have students read through the modeling and solving of the equation with algebra tiles, or demonstrate the method by drawing algebra tiles on the chalkboard or by using algebra tiles on the overhead. Remind students that a **neutral pair** has a value of zero, and hence neutral pairs can be removed from either side of an equation.

Ongoing Assessment, page 12

Substitute 5 for x in the original equation, or graph $y = x - 3$ and $y = 2$ on a graphics calculator and find the intersection.

Example 2

In this example, a one-step subtraction equation is solved by using the Addition Property of Equality.

Discuss how each step of the process is justified. Show students how to check their solutions by using substitution. You can use tiles to model this process for students with little or no experience in solving equations.

Additional Example 2

Solve the equation $x - 4 = -3$. [1]

Example 3

Here a real-world situation is modeled by two variables and an addition expression.

The value of one of the variables is given, and the equation is solved to find the value of the other variable. Encourage students to translate a situation into a simple English sentence before writing the equation. They should also clearly state what the variables represent.

Additional Example 3

Sonia gets an employee discount of $10 off the retail price of each item she purchases that costs $50 or more.

TEACH

Approach B

Using Models, continued

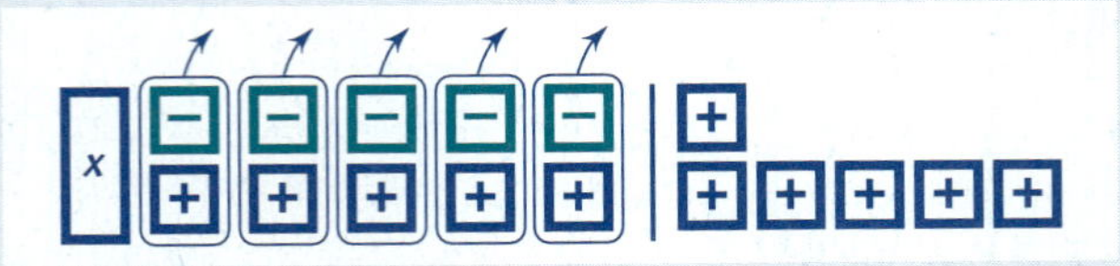

Demonstrate how the 5 positive 1-tiles and 5 negative 1-tiles on the left-hand side can be grouped together in **neutral pairs.** These pairs each have a value of zero, so they can be removed from the equation without affecting the balance. Remove the 5 neutral pairs and read the solution, $x = 6$.

Cooperative Learning

Have students work with a partner to solve the following equations with algebra tiles. Each pair will need one x-tile and 10 each of the positive 1-tiles and negative 1-tiles. The pairs should take turns manipulating the tiles and recording the solutions.

A. $x + 3 = 9$ [6] **B.** $x + 7 = 9$ [2]
C. $x - 3 = 6$ [9] **D.** $x - 5 = -3$ [2]

Addition and Subtraction Properties of Equality

Explain that the properties students used to either add or subtract the same number of tiles to each side of the tile models are called the Addition Property of Equality and the Subtraction Property of Equality.

Write these properties on the chalkboard or overhead, and have students write them in their journal. (Compare pages 11 and 12 of the textbook.) Use the equation $x + 12 = 5$ to demonstrate how to solve one-step equations by using these properties without tiles. Justify each step as you go along and include a check by substitution. (Compare page 11 of the textbook.) Have students solve the equation $x - 6 = -3$ at their desk for practice.

A. Write an equation to describe the relationship between the retail price and the discounted price. **[$r - 10 = d$, where r represents the retail price and d represents the discounted price]**

B. What is the equation if the discounted price of the item is $123? Solve the equation to determine the retail price. **[$r - 10 = 123$; $r = 133$; the retail price is $133.]**

Ongoing Assessment, page 13
Try This

a. $s = r - 3$

b. $51 = r - 3$; the retail price is $54.

Closing the Lesson

Have students work with a partner. Each student should make up one subtraction equation and one addition equation. They should then exchange papers and solve the problems. Have them exchange papers again and check the solutions by using substitution or a graphics calculator.

Students can work together to create and solve problems.

Using Graphing Technology to Check

If possible, demonstrate how to use a graphics calculator to check the solution to the equation that students just solved. Write the corresponding functions from each side of the equal sign: $y = x - 6$ and $y = -3$. Enter them in the Y= list of the calculator. Use the trace or intersect features to locate the intersection point. The solution to the equation is the x-coordinate of the point of intersection.

Application

Cooperative Learning Have students work with their cooperative-learning partners to solve the applied problem below. They should write out what each variable represents and show all their work, including a check of the solution.

Sonia gets an employee discount of $10 off the retail price of each item she purchases that costs $50 or more.

A. Write an equation to describe the relationship between the retail price and the discounted price. **[$r - 10 = d$]**

B. What is the equation if the discounted price of the item is $123? Solve the equation to determine the retail price. **[$r - 10 = 123$; $r = 133$; the retail price is $133.]**

Closing the Lesson

Work as a class to write a procedure for solving a one-step addition equation and another for solving a one-step subtraction equation. Have students copy these procedures into their notebook or journal.

ASSESS

Selected Answers

Odd-numbered Exercises 5–55 and odd-numbered Extra Practice

Assignment Guide

Core level students should complete Exercises 1–57.

Core-plus level students should complete Exercises 1–10 and 15–59.

Error Analysis

Some students may make errors when adding and subtracting positive and negative integers. You may need to review integer addition and subtraction with these students. Students who have trouble writing an equation to represent a real-world situation should be encouraged to write the problem as a simple English sentence and then to translate that sentence into mathematical terms.

Alternative Assessment

Finding Applications Have students make up an applied problem that can be solved with a one-step subtraction or addition equation. They should write an equation to represent the situation and solve it, showing all the steps clearly and including a check. They should then interpret their answer in the context of the problem.

Portfolio Activity Have students pick one subtraction equation and one addition equation from the homework problems. They should write a brief description of the method for solving each equation with the aid of algebra tiles and without. This work can be included in their portfolio.

RETEACH

Using a Balance

One-step addition equations that do not involve negative numbers can be modeled with a balance. To do this activity, you will need a scientific balance, a small covered box that fits on one side of the balance, and about 10 small weights of the same size. Adjust the balance so that it is zeroed with the empty box on one side and nothing on the other side. Have one student put a number of weights in the box without the others seeing it and two additional weights on the same side of the balance, outside the box. Have another student place weights on the opposite side until the two sides are balanced. How can we find the number of weights in the box without looking? Explain that the box represents the variable and that each weight represents one unit. Write the equation represented by the balance as "$x + 2 =$ the number of weights on the other side." Demonstrate how you can use the Subtraction Property of Equality to take away two weights from each side of the balance. Point out that both sides remain balanced. The number of weights remaining on the side opposite the box should be equal to the number of weights inside the box. Open the box to show that this is the case. Give more students a chance to model some additional equations while the other students write the corresponding equations on the chalkboard.

ENRICH

Interdisciplinary Connection

Business There are many examples in this lesson that demonstrate how one-step addition and subtraction equations can be used to represent situations involving the retail price and sales price of an item in a store. Another example could be the price of an item with and without sales tax included. Ask students to make up an example for one of these situations. They should explain the problem, write the equation, solve the equation, and interpret their solution in the context of the problem.

Writing a Computer Program

Have students write a program for their graphics calculator that solves one-step equations of the form $x + a = b$ when a and b are entered. They can then share their program with the rest of the class and explain how it works.

Practice

1.1 Addition and Subtraction Equations

Solve and check each equation.

1. $x + 8 = 20$ ______

2. $y - 5 = 40$ ______

3. $z + 7 = -15$ ______

4. $y - 8 = 25$ ______

5. $x + 2 = -19$ ______

6. $-7 = w + 23$ ______

7. $x - 12 = 42$ ______

8. $x - 36 = 0$ ______

9. $x - 7 = -28$ ______

10. $5 = z - 32$ ______

11. $y + 2 = -29$ ______

12. $-1.9 = 3 + y$ ______

13. $16 = x - 0.1$ ______

14. $3.4 = 0.8 + z$ ______

15. $x - 3.4 = 8$ ______

16. $x + 4 = -4.8$ ______

17. $7.3 = y + 5$ ______

18. $z - 1.2 = 4.9$ ______

19. $-24 = y - 7$ ______

20. $z - 4 = -1.4$ ______

21. $x - 7 = 7$ ______

22. $z + 8 = -50$ ______

23. $4 = y + 13$ ______

24. $0.7 = z + 0.5$ ______

25. $x - 0.7 = 10$ ______

26. $19 = z - 1.1$ ______

27. $24 = y + 10.5$ ______

28. $x - 90 = 9$ ______

29. $x - 3.2 = -0.2$ ______

30. $x + 8.6 = 14$ ______

31. $w + 7 = -10$ ______

32. $2.4 = y + 9$ ______

33. $-36 = y + 19$ ______

34. $w - 9 = -15$ ______

35. $w - 7.5 = 75$ ______

36. $z + 6 = -50$ ______

37. $y + 5.5 = -1.1$ ______

38. $-30 = w + 3.9$ ______

39. $16 = y + 16$ ______

40. $w - 10 = -3.6$ ______

41. $w - 8 = 85$ ______

42. $z + 7 = 40$ ______

43. $y + 3.7 = -0.1$ ______

44. $-10 = w + 8.6$ ______

45. $-25 = y + 15$ ______

46. $w - 6 = -11$ ______

Lesson Activity

1.1 Modeling Addition and Subtraction Equations

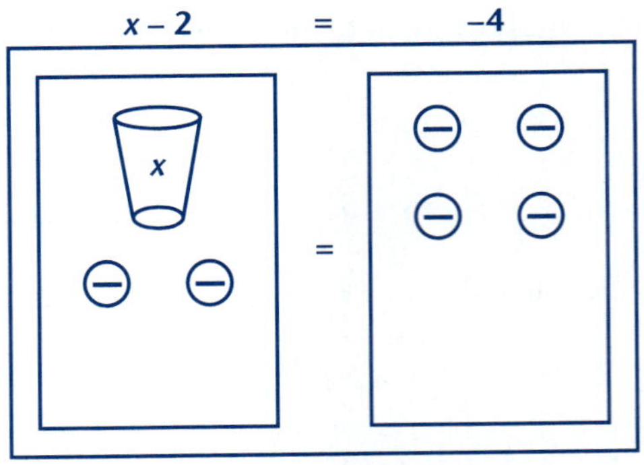

You can use cups, counters, and an equation mat to model the solutions to addition and subtraction equations. Use a large piece of paper to make an equation mat. Use a cup to represent the variable. Choose one color of counters to represent positive integers and a different color to represent negative integers. In the examples below, the circles representing the positive counters are marked with a plus sign (+) and those representing the negative counters are marked with a minus sign (−). You may want to use stickers to mark your counters with a plus or a minus sign and to mark your cup with the variable.

The model of the equation $x - 2 = -4$ is shown above at right. Place one x-cup and 2 negative counters on the left side of the mat. Place 4 negative counters on the right side of the mat.

Isolate the x-cup by adding 2 positive counters to each side of the equation. $x - 2 + 2 = -4 + 2$

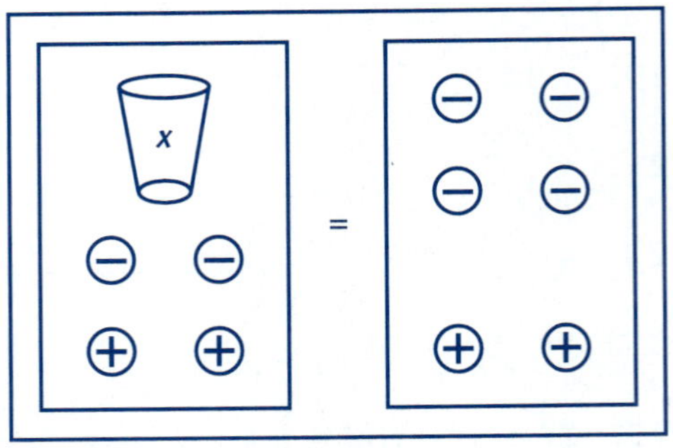

Remove neutral pairs of counters. There are 2 negative counters remaining. $x = -2$

Use cups and counters to model the solution to each equation.

1. $x - 3 = 3$ ______________________

2. $x + 2 = 4$ ______________________

3. $x - 5 = -1$ ______________________

4. $x + 4 = -2$ ______________________

5. $3 + x = -5$ ______________________

6. $x + 6 = 5$ ______________________

7. $-3 = x - 7$ ______________________

8. $5 = x + 8$ ______________________

NAME ________________ CLASS ________ DATE ________

Basic Skills Practice
Writing Equations and Inequalities, Part 1

Translating words into mathematical symbols can help solve problems. Let a variable such as x or n stand for the unknown number.

An **equation** is a mathematical sentence which states that two expressions are equal.

$$n + 6 = 14$$

Number sentences do not always contain an equal sign. An **inequality** is a mathematical sentence that uses $<$, $>$, $\leq$, or $\geq$ to show the relationship between two expressions.

$$15 + n \geq 26$$

Example: Write an equation or inequality for each word sentence.

1. A number is nine more than twenty. $n = 9 + 20$
2. A number times four minus six is ten. $4n - 6 = 10$
3. A number divided by six is less than two. $n \div 6 < 2$ or $\frac{n}{6} < 2$
4. What number added to 478 gives 1019? $n + 478 = 1019$

Write an equation or inequality for each sentence. Use *n* for the variable.

1. A number plus one is greater than three. ________

2. Two less than a number is seven. ________

3. Ten times a number is fifty. ________

4. A number is less than twelve. ________

5. A number and thirty totals sixty-four. ________

6. A number increased by one is one-half of twenty-four. ________

7. Seventeen is 42 minus one-half of a number. ________

8. The product of 12 and a number, divided by 6, is greater than 58. ________

9. Eighteen times a number is 36. ________

10. Four more than a number is less than 38. ________

11. Twenty divided by the difference of 15 and a number is 3. ________

12. Eight times the sum of a number and 3 is 35. ________

13. Thirteen and twice a number is less than or equal to 61. ________

14. Four more than 5 times a number is less than or equal to 39. ________

LESSON 1.2

Multiplication and Division Equations

PREPARE

Pacing Guide
2 Days

Objectives and Overview
- Use the Multiplication and Division Properties of Equality to solve equations.

Real-world problems are modeled. One-step equations are solved by using the Multiplication Property of Equality, the Division Property of Equality, or the Reciprocal Property. Proportions are introduced as well.

Vocabulary
Multiplication Property of Equality
Division Property of Equality
Reciprocal Property
ratio
equivalent ratios
proportion

Assessing Prior Knowledge

Write the reciprocal of each number.

1. $\frac{1}{3}$ [3] **2.** $-\frac{3}{4}$ $\left[-\frac{4}{3}\right]$

3. $\frac{5}{3}$ $\left[\frac{3}{5}\right]$

Evaluate for $x = 5$.

4. $\frac{1}{10}x$ $\left[\frac{1}{2}\right]$ **5.** $\frac{x}{2}$ [2.5]

6. $0.8x$ [4]

Approach A: Using the Book

Opening Discussion
Many people use equations to solve problems. Discuss the pizza example. What operation is needed in this situation? **[multiplication]** Write an equation to represent the total profit that the band boosters earn.

Brainstorming
Once you write an equation, what can you do to isolate the variable? Have students list the techniques that they learned in Lesson 1.1 or that they remember from previous courses. Examples include adding the same number to both sides, subtracting the same number from both sides, multiplying both sides by the same number, or dividing both sides by the same nonzero number.

Division Property of Equality
Lead students through the solution to the pizza problem. If they are not familiar with algebra tiles,

TEACH

Approach B: Using Models

Opening Discussion
Many people use equations to solve problems in their work. Consider a craftsperson who wants to write an equation that represents a profit of \$12 earned from selling x items at \$3 each. What operation is needed in this situation? **[multiplication]** Write the equation to represent the profit that the craftsperson earns. **[$3x = 12$]** Discuss some other applications of equations with students.

Brainstorming
Once you write an equation, what can you do to isolate the variable? Have students list the techniques that they learned in Lesson 1.1 or that they remember from previous courses. Examples include adding the same number to both sides, subtracting the same number from both sides, multiplying both sides by the same number, or dividing both sides by the same nonzero number.

explain that blue squares (or squares with a plus sign) represent positive units and that yellow squares (or squares with a minus sign) represent negative units.

The object is to isolate the variable on one side so that its value can be seen. To do this, the x-tiles are separated, and each one is matched up with an equal number of positive (or negative) tiles on the other side. You can then read the value of each individual x-tile. After completing the problem, read the Division Property of Equality, and have students copy it into their notebook or journal.

Example 1

The fact that the sum of the angles of a triangle is 180° is used to write a one-step equation involving multiplication. The Division Property of Equality is used to solve the equation.

Students need to add like terms to simplify the equation before dividing. Point out that to answer the question, the value for x must be substituted into the expressions for the measures of angles B and C. Have students add the three measures together to verify that the sum is 180°.

Additional Example 1

The perimeter of the triangle below is 65 cm. Find the length of each side.

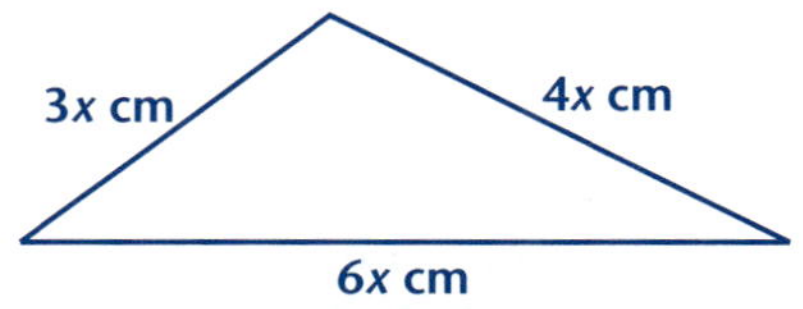

[15 cm, 20 cm, and 30 cm]

Multiplication Property of Equality

Write the equation $\frac{y}{6} = 5$ on the chalkboard or overhead. Ask students what must be done to solve this equation. **[Multiply each side of the equation by 6.]**

This can be done because of the Multiplication Property of Equality. Read or have a student read

continued on top of next page

Using Algebra Tiles to Represent and Solve One-Step Multiplication Equations

Creating a Model Using an overhead projector and algebra tiles, show students how to represent the equation $3x = 12$ with tiles.

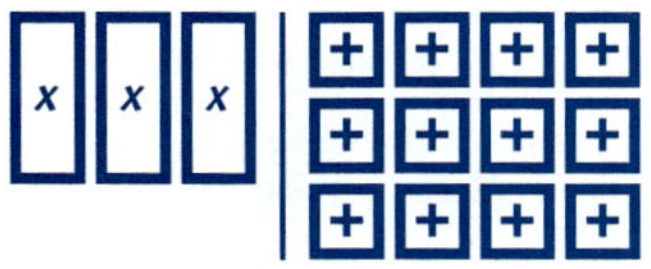

The square tiles with a plus sign represent a positive unit and the square tiles with a minus sign represent a negative unit. The vertical segment represents the equal sign. (For ideas on how to make algebra tiles, see Lesson 1.1, Using Models.)

Using the Model Demonstrate how division is modeled by separating the tiles into groups with one x-tile and an equal number of square tiles. The same number of square tiles must be grouped with each x-tile. The objective is to isolate the variable so that its value can be determined.

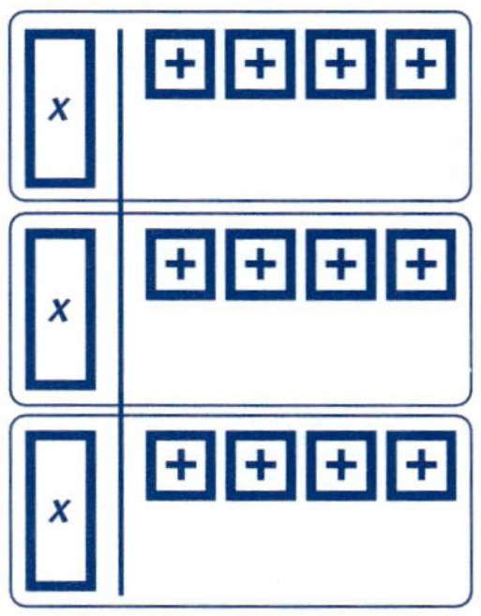

The remaining tiles are used to write the solution, $x = 4$. The craftsperson makes a profit of $12 by selling 4 items.

Cooperative Learning

Have students work with a partner to solve the following equations by using algebra tiles. Each pair will need 3 x-tiles, 9 positive unit tiles, and 12 negative unit tiles. The pairs should take turns manipulating the tiles and recording the solutions.

A. $3x = 9$ [3] **B.** $2x = -12$ [−6]

continued on bottom of next page

Approach A

Using the Book, continued

the property, and have students copy it into their notebook or journal.

Reciprocal Property

How would you solve an equation like $\frac{2}{3}x = 8$? You can divide both sides of the equation by $\frac{2}{3}$ or multiply both sides of the equation by the reciprocal of $\frac{2}{3}$, which is $\frac{3}{2}$. The Reciprocal Property ensures that when a coefficient is multiplied by its reciprocal, the result is 1. Read or have a student read the Reciprocal Property, and have students copy it into their notebook or journal.

Example 2

In this example, a one-step equation is solved by using the Multiplication Property of Equality. A second equation is solved by using the Reciprocal Property and the Multiplication Property of Equality.

Have students contrast these two problems to identify why the different methods were used. Point out that each step of the process is justified. Show students how to check their solutions by using substitution.

Additional Example 2

Solve.

A. $\frac{w}{5} = 7$ [35]

B. $\frac{4}{5}x = 8$ [10]

Example 3

In this example, a real-world situation is modeled by two variables and a multiplication expression. The value of one of the variables is given, and the equation is solved to find the value of the other variable.

Encourage students to translate the situation into a simple English sentence before writing the equation, as was done in this example. Have them clearly state what the variables represent.

TEACH

Approach B

Using Models, continued

Division Property of Equality

Explain that the property students used to divide the square tiles into small groups is called the Division Property of Equality. Write the definition of this property on the chalkboard or overhead and have students write it in their journal. (Compare page 17.) Use the equation $-3x = 15$ to demonstrate how to solve one-step equations without tiles by using the Division Property of Equality. Justify each step as you go along and include a check by substitution. (Compare page 17.) Have students solve the equation $-9x = -3$ at their desk for practice. $\left[\frac{1}{3}\right]$

The Multiplication Property of Equality

How do you solve a division equation such as $\frac{x}{4} = -2$? Multiply each side of the equation by 4 to isolate the variable, and simplify. Demonstrate this for students at the chalkboard or overhead, justifying each step. Explain that the property which allows us to multiply both sides of an equation by the same number is called the Multiplication Property of Equality. Write the definition of this property on the chalkboard and have students copy it into their notebook or journal. (Compare page 17.)

Another Way to Solve a Multiplication Equation

Lead students through this activity, which develops the method for solving one-step multiplication equations with fractional coefficients by using the Reciprocal Property. Write the equation $\frac{2}{3}x = 8$ on the chalkboard. How do we solve a multiplication equation? **[Divide both sides by the coefficient of the variable.]** In this case, divide both sides by $\frac{2}{3}$.

$$\frac{\frac{2}{3}}{\frac{2}{3}}x = \frac{8}{\frac{2}{3}}$$

How do you divide a fraction by a fraction? **[Multiply by the reciprocal of the fraction you are dividing by. $\left(\frac{2}{3} \bullet \frac{3}{2}\right)x = 8 \bullet \frac{3}{2}$;**

Additional Example 3

Bruce is shopping for a sweater. A sign in the store reads "25% off all sweaters." Bruce figures that the sale price of a sweater is 75% of the original price.

A. Write an equation to describe the sale price in terms of the regular price. [$s = 0.75r$]

B. What is the regular price of a sweater if the sale price is $29.25? [$39.00]

Critical Thinking, page 19

If the regular price is 100%, then 30% off the regular price means that you pay 100% − 30%, or 70%, of the regular price.

Ongoing Assessment, page 19
Try This

a. $e = 16h$

b. 31.25 hours

Solving Proportions

Read or have a student read this section to remind students about the concept of a **ratio** and to introduce the concept of a **proportion** as an equation containing two or more equivalent ratios.

Have students give another example of the numbers of boys and girls that could be in the school choir.

Note that all of the proportions in this lesson have the variable in the numerator of one ratio.

Example 4

This example demonstrates a method for solving a proportion by using the Multiplication Property of Equality.

Students may have previously solved proportions by using means and extremes, a method which is developed in the Critical Thinking question.

Use **Transparency 2** for discussion.

continued on top of next page

$x = \frac{24}{2}$, **or 12]** How can we achieve the same result without the complex fraction? When the coefficient is a fraction, instead of dividing both sides of the equation by the coefficient, multiply each side of the equation by the reciprocal of the coefficient. Have students practice this method by solving $\frac{3}{4}x = -9$. **[−12]**

Proportions

Remind students about the concept of a **ratio.** Explain that equivalent ratios are ratios that have the same value. Give some examples, such as 2 to 3, $\frac{2}{3}$, $\frac{6}{9}$, and $\frac{10}{15}$. If the ratio of boys to girls in the school choir is 4 to 5, have students give some examples of the numbers of boys and girls that could be in the choir. **[20 boys and 25 girls, or 12 boys and 15 girls]** Introduce the concept of a **proportion** as an equation that contains two or more equivalent ratios. Note that all of the proportions in this lesson have the variable in the numerator of one of the ratios.

Solving a Proportion

If the ratio of boys to girls is 4 to 5 as above, how many boys are in the choir if there are 65 girls? Let x represent the number of boys in the choir. Then the ratio of boys to girls is $\frac{x}{65}$. This gives the proportion $\frac{4}{5} = \frac{x}{65}$. Demonstrate how to solve this proportion by multiplying both sides of the equation by 65. **[There are 52 boys.]**

Applications

Cooperative Learning Have students work with their cooperative-learning partners to solve the application problems below. They should write out what each variable represents and show all their work, including a check of the solution. You can have three of the groups work at the chalkboard so that all students can check their work.

A. Find the angle measures in triangle *ABC*.

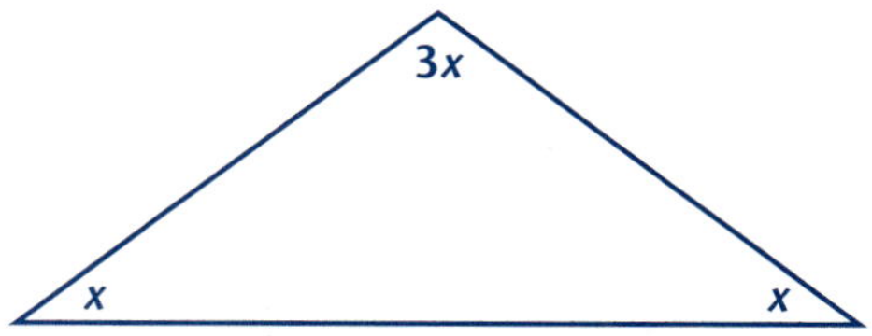

continued on bottom of next page

Approach A: **Using the Book**, continued

Additional Example 4

The ratio of employees with college degrees to employees without college degrees in a small company is 7 to 3. If there are 63 employees without college degrees in the company, how may employees have college degrees? **[147]**

Ongoing Assessment, page 20
Try This
245

Critical Thinking, page 20
If $\frac{a}{b} = \frac{c}{d}$ is a true proportion, then according to the Multiplication Property of Equality, we can multiply each side by b to obtain $a = \frac{bc}{d}$. Then use the property again, and multiply both sides of this equation by d to obtain $ad = bc$.

Closing the Lesson

Have students work with a partner. Each student should make up one multiplication equation, one division equation, and one proportion problem. They should then exchange papers and solve the problems. Have them exchange papers again and check the solutions by using substitution.

TEACH

Approach B: **Using Models**, continued

Remember that the sum of the angles of a triangle is 180°. **[36°, 36°, 108°]**

B. Bruce is shopping for a sweater. A sign in the store reads "25% off all sweaters." Bruce figures that the sale price of a sweater is 75% of the original price. Write a multiplication equation to describe the sale price in terms of the regular price. **[$s = 0.75r$]** What is the regular price of a sweater if the sale price is $29.25? **[The regular price is $39.00.]**

C. The ratio of employees with college degrees to employees without college degrees in a small company is 7 to 3. If there are 63 employees without college degrees in the company, how may employees have college degrees? **[147]**

Inclusion Strategies

Before beginning the application problems above, give each pair a sheet of transparency paper, and assign them to do one of the problems on the paper. Explain that you will be choosing one of the pairs to demonstrate their solution on the overhead.

Closing the Lesson

Work as a class to list all of the different types of one-step equations that they have seen in this chapter. Have them write a one-sentence description of the method they would use to solve each type of equation.

ASSESS

Selected Answers

Odd-numbered Exercises 7–81 and odd-numbered Extra Practice

Assignment Guide

Core level students should complete Exercises 1–82.

Core-plus level students should complete Exercises 1–35 and 54–73.

Error Analysis

Watch for students who make errors when multiplying and dividing positive and negative integers. You may need to review integer multiplication and division with these students. Students who have trouble writing an equation to represent a real-world situation should be encouraged to write the problem as a simple English sentence and then to translate that sentence into mathematical terms.

Alternative Assessment

Applications Have students make up an application problem that can be solved by using a one-step multiplication or division equation. They should write an equation to represent the situation and solve it, showing all the steps clearly and including a check. They should then interpret their answer in the context of the problem.

Portfolio Activity Have students pick one multiplication equation, one division equation, and one proportion from the homework problems. They should write a brief description of the method for solving each equation. This work can be included in their portfolio.

RETEACH

Cooperative Learning Have students work with a partner. They should read through the lesson, alternating paragraphs. The goal is to identify four different kinds of problems in the lesson. **[multiplication equations with integer coefficients, multiplication equations with fractional coefficients, division equations, and proportions]** For each type of problem that they identify, the students should write the identifying characteristics and the method for finding the solution. Their work can be organized in a table that they can use for reference later when solving equations. This activity can be extended by including the two types of equations studied in Lesson 1.1.

ENRICH

Interdisciplinary Connection

Journalism Journalists often make use of percents and ratios in their writing. Have students find an example of this in the local newspaper. Then have them use these examples to make up several problems that can be solved by using the techniques learned in this lesson.

Writing a Computer Program

Have students write a program for their graphics calculator that solves one-step equations of the form $ax = b$ when a and b are entered. They can then share their program and explain how it works to the rest of the class.

NAME ______________________ CLASS ____________ DATE ____________

Practice

1.2 Multiplication and Division Equations

Solve and check each equation.

1. $8x = 64$ ____________

2. $-5x = 2.5$ ____________

3. $4y = 64$ ____________

4. $-9x = -45$ ____________

5. $1.6 = -4w$ ____________

6. $-100 = -20z$ ____________

7. $\frac{t}{5} = 8$ ____________

8. $\frac{p}{-2} = 15$ ____________

9. $\frac{z}{-6} = -12$ ____________

10. $\frac{2}{9}w = -6$ ____________

11. $\frac{3}{5}x = 12$ ____________

12. $\frac{r}{3} = -1.2$ ____________

13. $\frac{7}{10}y = 21$ ____________

14. $\frac{5}{8}x = -0.8$ ____________

15. $\frac{w}{-5} = -3.5$ ____________

16. $-1.5x = 9$ ____________

17. $-5.5x = 1.1$ ____________

18. $-30y = 900$ ____________

19. $-0.88x = 11$ ____________

20. $-5.2w = -15.6$ ____________

21. $-9 = 4.5y$ ____________

22. $\frac{5}{9}y = -10$ ____________

23. $-7x = \frac{14}{15}$ ____________

24. $\frac{3}{4} = -0.8t$ ____________

Solve each proportion.

25. $\frac{x}{9} = \frac{2}{3}$ ____________

26. $\frac{-x}{7} = \frac{4}{7}$ ____________

27. $\frac{x}{3.5} = -\frac{2}{5}$ ____________

28. $\frac{x}{20} = \frac{4}{5}$ ____________

29. $\frac{-x}{6} = \frac{3}{4}$ ____________

30. $\frac{y}{2.4} = \frac{-3}{4}$ ____________

31. $\frac{x}{16} = \frac{-7}{4}$ ____________

32. $\frac{x}{9} = \frac{4}{12}$ ____________

33. $\frac{x}{2} = \frac{5}{6}$ ____________

34. $\frac{-x}{6} = \frac{5}{3}$ ____________

35. $\frac{y}{6} = \frac{3}{8}$ ____________

36. $\frac{x}{-25} = \frac{3}{5}$ ____________

37. $\frac{-x}{-30} = \frac{2}{5}$ ____________

38. $\frac{t}{-2.4} = \frac{3}{2}$ ____________

39. $\frac{-b}{2.4} = \frac{-5}{6}$ ____________

40. $\frac{y}{-3} = \frac{12}{-1.5}$ ____________

41. $\frac{-c}{14.4} = \frac{-2}{3}$ ____________

42. $\frac{k}{-1.4} = \frac{-2.5}{7}$ ____________

NAME ______________________ CLASS __________ DATE __________

Lesson Activity

1.2 Modeling Multiplication Equations

You can use cups, counters, and an equation mat to model the solutions to multiplication and division equations.

The model of the equation $2x = -4$ is shown at right. Place 2 cups on the left side of the mat. Place 4 negative counters on the right side of the mat.

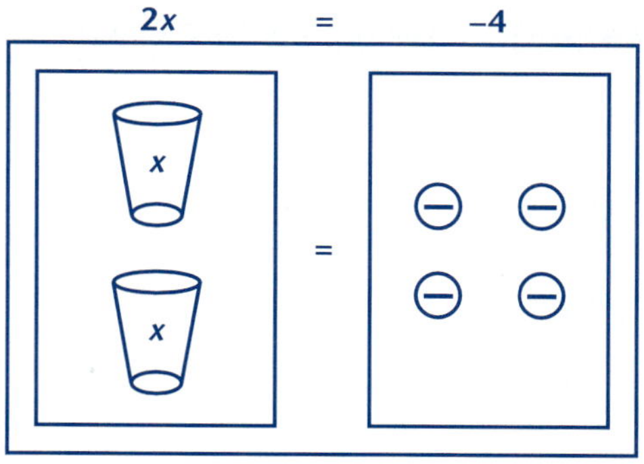

Place an equal number of counters in each cup. The number of counters in each cup is the solution. There are 2 negative counters in each cup.

$$\frac{2x}{2} = \frac{-4}{2}$$

$$x = -2$$

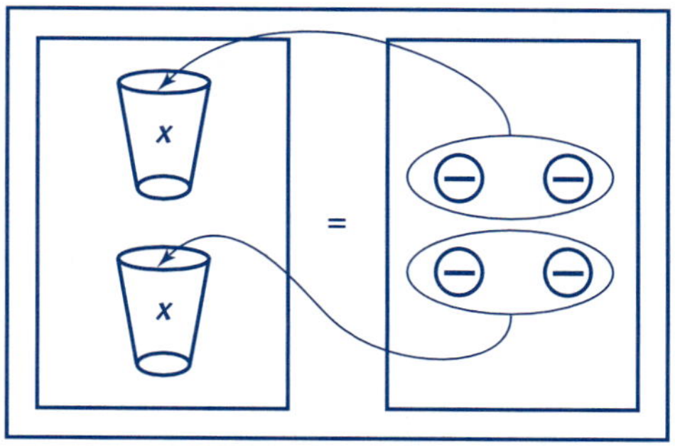

Use cups and counters to model the solution to each equation.

1. $3x = 9$ ______________

2. $2x = 8$ ______________

3. $2x = -10$ ______________

4. $3x = -12$ ______________

5. $3x = 24$ ______________

6. $15 = 3x$ ______________

7. $4x = 20$ ______________

8. $-16 = 2x$ ______________

9. $-18 = 2x$ ______________

10. $-21 = 3x$ ______________

11. $4x = -8$ ______________

12. $4 = 2x$ ______________

13. Describe how you can model the multiplication equation $-3x = 9$ with cups and counters. ______________

____

Basic Skills Practice

Writing Equations and Inequalities, Part 2

Read each question and circle the best answer.

1. Sixty-eight divided by a number is seventeen. Which equation matches this sentence?

A $68n = 17$

B $\frac{68}{n} = 17$

C $68 + 17 = n$

D $n - 68 = 17$

2. Which equation states that three times a number minus seven is twenty-three?

F $3n + 7 = 23$

G $7(3n) = 23$

H $3n - 7 = 23$

I $3 + n - 7 = 23$

3. Which sentence is equivalent to $2n + 5 = 15$?

A Five added to twice a number is fifteen.

B Twice a number is fifteen.

C A number plus five is fifteen.

D Fifteen is a number plus five.

4. A number plus two is greater than or equal to three. Which inequality matches this sentence?

F $n + 2 > 3$

G $n \geq 3$

H $n + 2 \leq 3$

I $n + 2 \geq 3$

5. Angie's grandfather is three times as old as her brother. If Angie's grandfather is 66, which equation could you use to find Angie's brother's age?

A $3n = 66$

B $3 + n = 66$

C $3 \times 66 = n$

D $66 - n = 3$

6. Gene swims 3600 meters in 30 minutes. Which equation could be used to find his average speed per minute?

F $30 \times 3600 = n$

G $3600 - 30 = n$

H $\frac{3600}{30} = n$

I $\frac{30}{3600} = n$

7. During 1996, the company showed a profit of at least $1500 each month. Which inequality reflects this statement?

A $n > \$1500$

B $n \leq \$1500$

C $n \geq \$1500$

D $n < \$1500$

8. The longest Viking ship that has been found is a little less than 95 feet long. Which inequality shows the length of the Viking ships found so far?

F $n + 95$

G $n > 95$

H $n < 95$

I $n \leq 95$

9. On a math test, you could score 100 points and 20 bonus points. Your score on the test was 87, which included 12 bonus points. Which inequality shows the score you would receive with the bonus points.

A $n + 20 = 100$

B $n + 87 = 100$

C $n + 20 = 87$

D $n + 12 = 87$

LESSON 1.3

Two-Step Equations

PREPARE

Pacing Guide
2 Days

Objectives and Overview
- Use properties of equality to solve two-step equations.

Problems involving money, perimeter, and surface area are modeled by two-step equations and solved.

Vocabulary
two-step equation

Assessing Prior Knowledge

Solve each equation.

1.	$-4x = 12$	[−3]
2.	$\frac{5}{3}x = 10$	[6]
3.	$x + 5 = 16$	[11]
4.	$x - 2.8 = 3.2$	[6]
5.	$\frac{x}{2} = -8$	[−16]

Approach A Using the Book

Opening Discussion
Use the film example to introduce the idea of a **two-step equation.** These equations can be recognized by the presence of two operations. What operations are needed to determine the cost of film? **[You need to multiply the number of rolls ordered by 4 and then add the shipping charge.]** The equation representing this situation involves addition and multiplication and thus requires two steps to solve.

Brainstorming
Have the class brainstorm to find other applications that involve both addition and multiplication.

Example 1
A two-step equation is written and solved to determine the number of rolls of film ordered when given the total cost of the order.

continued on top of next page

TEACH

Approach B Using Models

Opening Discussion
A mail-order film company charges \$3 for each roll of 24-exposure, 35-mm color film. There is also a \$2.00 shipping charge on each order. Use this example to introduce the idea of a **two-step equation.** These equations can be recognized by the presence of two operations. What operations are needed to determine the cost of film? **[Multiply the number of rolls ordered by 3, and add the \$2 shipping charge.]** The equation representing this situation involves addition and multiplication and thus requires two steps to solve.

Using a Verbal Model
To help students organize the information, it is helpful to use a verbal model before writing an equation. The film situation can be written in words as follows:

The total cost is \$3 times the number of rolls plus \$2.

continued on bottom of next page

Approach A — Using the Book, continued

Point out the use of the verbal model to help students translate the situation into mathematical terms. Also note the choice of variable letters to help students remember what each variable stands for.

Additional Example 1

William wants to set up a fish tank in his room. The tank itself costs $45 and each fish costs $3.

A. Write an equation to describe the cost in terms of the number of fish in the tank. [$c = 3n + 45$]

B. How many fish did William get for his tank if the bill was $72? [9]

Ongoing Assessment, page 24
Try This

a. 3

b. 2

On Their Own

Have students work at their desk to solve the following equations:

A. $3x - 5 = 10$ [5] **B.** $4x + 8 = -4$ [−3]

Exploration

In this exploration, a two-step equation is written to represent the perimeter of a square that has been decreased or increased by a constant.

You can lead the class through this exploration or have students complete it in small groups. Review the concept of perimeter with students before beginning this exploration. In Step 4, when using a graphics calculator to check the solution, enter the functions as Y = 43 and Y = 4X − 5. An appropriate viewing window is Xmin = 0, Xmax = 15, Ymin = 0, Ymax = 50.

On Their Own

Draw a large square with sides of 10 units on the chalkboard or overhead. Have students answer the following questions:

TEACH

Approach B — Using Models, continued

Which parts of the model need to be variables? Let c be the total cost and n be the number of rolls. The verbal model can then be translated into the equation $c = 3n + 2$. If the total bill was $14, how many rolls were ordered? The equation becomes $3n + 2 = 14$. The equation needs to be solved for n.

Using Algebra Tiles to Represent and Solve Two-Step Equations

Creating a Model Using an overhead projector and algebra tiles, show students how to represent the equation above with the tiles.

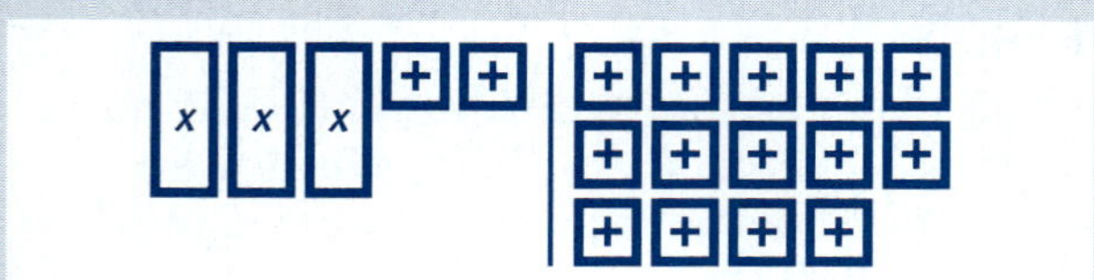

Using the Model The objective is to isolate the variable so that its value can be determined. First remove any square tiles from the variable side of the model and remove the same number and kind of tiles from the other side to maintain the balance.

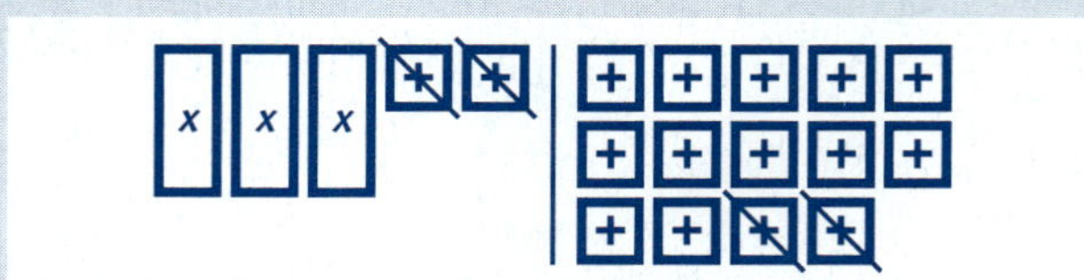

Next divide the square tiles into three equal groups, one group for each x-tile.

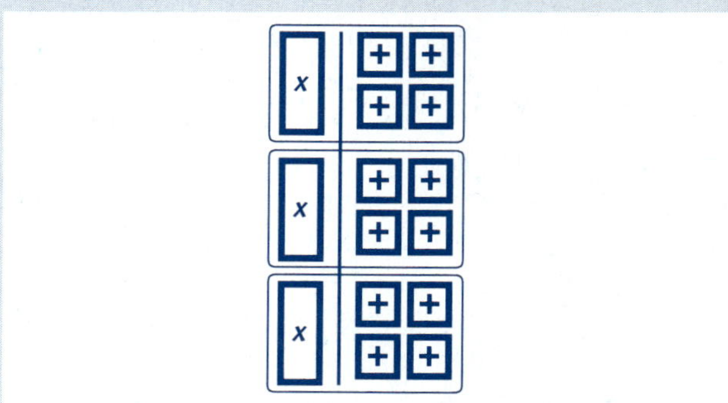

The solution can now be read from one of the groups, $n = 4$; 4 rolls of film were ordered.

A. What is the perimeter of the square? **[$p = 40$]**

Invite student discussion of perimeter problems.

B. If the perimeter of the square is decreased by 8 units, what are the lengths of the sides of the smaller square? **[8]**

Ongoing Assessment, page 24

5. 12.5

Example 2

In this example, a two-step equation is written to represent the surface area of a rectangular prism in terms of its height. The surface area is then substituted, and the equation is solved to find the height of the prism.

You may need to review the terms *rectangular prism, surface area, base,* and *lateral face* with students. Have students check that a figure with these dimensions has the correct surface area.

Use **Transparency 3** for discussion or review.

continued on top of next page

Practice With Tiles

Have students work with a partner to solve the equations below with algebra tiles. Each pair will need 4 x-tiles, 15 positive tiles, and 12 negative tiles. The partners should take turns manipulating the tiles and recording the solutions.

A. $3x - 5 = 10$ **[5]** **B.** $4x + 8 = -4$ **[−3]**

Solving Equations Algebraically

Have one group describe the procedure they used to solve problem A above. As they describe each manipulation of the tiles, write the corresponding algebraic manipulation of the equation. After the solution is given, go back and write down the justification for each step. Students should be able to solve these equations without the tiles, but encourage them to visualize the tiles to get started. Give students a few additional problems to try on their own without the tiles.

Cooperative Learning

Have students work in groups of four on the following two activities, which you can write on the overhead or chalkboard. One student should read the activity and direct the group, one student should write the equation, one student should solve the equation, and the fourth student should check and record the solution. Review the concepts of perimeter, lateral area, and surface area of a rectangular prism.

The Perimeter of a Square

Give each group a large cut-out square and a ruler. Explain that this square was cut out of a larger square. The perimeter of the smaller square is 5 cm less then the perimeter of the larger square. (Compare page 24.)

Write an equation for the perimeter of the larger square in terms of the length of its sides. **[$p = 4s$]**

Write an equation for the perimeter of the smaller square. **[$p = 4s - 5$]**

continued on bottom of next page

Approach A

Using the Book, continued

Additional Example 2

The base of this rectangular prism has a length of 5 feet and a width of 4 feet.

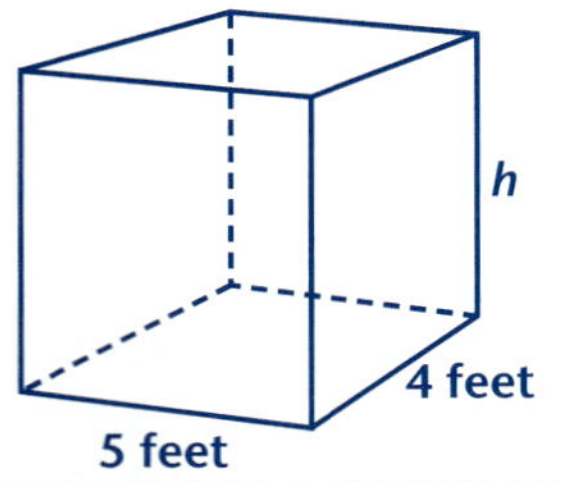

A. Write an equation for the surface area of the prism in terms of its height. **[$SA = 18h + 40$]**

B. What is the height of the prism if the surface area is 202 square feet? **[9 feet]**

Teaching Tip

Technology An appropriate viewing window for the graph is Xmin = 0, Xmax = 15, Ymin = 0, Ymax = 300. Use the trace feature or the intersect feature (found on the calculate menu) to find the intersection. To set up the table, let Xmin = 6 and ΔX = 0.5.

Closing the Lesson

Write several two-step equations on the chalkboard. Have volunteers describe what they would do to solve the problems. Include problems that involve negative numbers, decimals, and fractions.

TEACH

Approach B

Using Models, continued

Measure the perimeter of the small square. Use this to solve the equation and find the lengths of the sides of the large square from which the small one was cut. **[Answers will vary. Check student's work.]**

The Height of a Rectangular Prism

Give each group a drawing of a rectangular prism that has a square base with sides of 5 feet. (Compare page 25.) If possible, give each group a physical model of a rectangular prism to help them visualize the figure.

Write an equation for the surface area of the prism in terms of the height. **[$SA = 20h + 50$]**

What is the height of the prism if the surface area is 200 square feet? **[7.5 feet]**

Using Technology to Find a Solution

At the conclusion of the group activities, show students how they can find solutions to equations by using a graphics calculator. For example, to find the height of the prism, graph $y = 20x + 50$ and $y = 200$, and look for the x-coordinate of the intersection. An appropriate viewing window for the graph is Xmin = 0, Xmax = 15, Ymin = 0, Ymax = 300. Use the trace feature or the intersect feature (found on the calculate menu) to find the point of intersection. This solution can be confirmed by creating a table. To set up the table, let Xmin = 6 and ΔX = 0.5. The solution is the x-value that corresponds to a y-value of 200 for both functions. (Compare page 25.)

Closing the Lesson

Write several two-step equations on the chalkboard. Have volunteers describe what they would do to solve the problems. Include problems that involve negative numbers, decimals, and fractions.

ASSESS

Selected Answers

Odd-numbered Exercises 5–43 and odd-numbered Extra Practice

Assignment Guide

Core level students should complete Exercises 1–43.

Core-plus level students should complete Exercises 1–8 and 22–45.

Error Analysis

Watch for students who try to solve two-step equations by using the order of operations. Explain that when solving equations, the order is reversed. First you add or subtract, and then you multiply or divide. Students having trouble with Exercise 32 may need to review the Distributive Property.

Alternative Assessment

Letter Writing Write a letter to a friend who missed this lesson. Explain the major content of the lesson. Include several worked-out examples.

Portfolio Activity Have students work with a partner to critique each other's answers to Exercises 1–4. They should then edit and rewrite their own responses to be included in their portfolio.

RETEACH

Using an Algorithm

Work as a class to write a series of steps that can be used to solve a two-step equation. The steps should work for any of the problems in Exercises 5–32. Have each student write the steps in their notebook or journal, and then give them several new equations to solve. For each one, they should follow their algorithm carefully and label each step.

ENRICH

Interdisciplinary Connection

Physics The distance traveled by an object moving away from a fixed location at a constant speed is described by the equation $d = vt + d_0$, where v is the constant speed or velocity of the object in miles per hour, t is the time in hours, and d_0 is the initial distance from the fixed location. Suppose that a college student driving from Boston to Los Angeles stops at a rest area 80 miles from Boston. She wants to travel at an average speed of 55 mph until she stops at a hotel for the night. Write an expression that represents the student's distance from Boston in terms of the number of hours she has driven since stopping at the rest area. **[$d = 55t + 80$]** If the hotel is 300 miles from Boston, how many more hours will she have to travel before reaching the hotel? **[4 hours]**

Using a Spreadsheet to Study Investments

After completing Exercises 35 and 36, students can investigate this situation in more detail by using a spreadsheet. Have them make a column for the amount in each account, the interest earned by each account in one year, and the total interest earned in one year. After determining the formulas for the first row, they should use the fill down feature to complete the spreadsheet. Have them pose and then answer several questions involving this information.

NAME ______________________ CLASS __________ DATE __________

Practice

1.3 Exploring Two-Step Equations

Solve and check each equation.

1. $3x + 7 = 13$ __________

2. $12y - 5 = 31$ __________

3. $-5z + 7 = -13$ __________

4. $-7y + 28 = 7$ __________

5. $6x + 48 = 60$ __________

6. $-7 - 12w = 29$ __________

7. $-5x - 12 = 3$ __________

8. $8x - 36 = -20$ __________

9. $7x - 7 = 49$ __________

10. $5 - 8z = 6.6$ __________

11. $-4y + 9 = -15$ __________

12. $-42 = 3 + 5y$ __________

13. $26 = 9x - 1$ __________

14. $-25.6 = 8 + 8z$ __________

15. $1.2x - 4 = 2$ __________

16. $6(4 - 3x) = -48$ __________

17. $3 = -3y + 5$ __________

18. $0.8z - 10 = -2$ __________

19. $-3.5 = 3y - 7.1$ __________

20. $5z - 4.2 = 35.8$ __________

21. $15 - 2.3x = -19.5$ __________

22. $6z + 8.8 = 23.8$ __________

23. $-0.8 = -7y + 1.3$ __________

24. $-70 = 1 - 8z$ __________

25. $\frac{1}{3}x + 3 = 5$ __________

26. $\frac{3}{4}y - 6 = 0$ __________

27. $\frac{y}{4} + 6 = 12$ __________

28. $\frac{z}{8.4} - 9 = -4$ __________

29. $1.5 - 28x = -32.1$ __________

30. $6.6z + 5 = 64.4$ __________

31. $-12.5x + 8 = -67$ __________

32. $9.5y + 8.5 = 56$ __________

33. $-0.75 - 2x = -1.25$ __________

34. $9x - 9.5 = -7.25$ __________

35. $\frac{z}{3.4} + 6.5 = 15.5$ __________

36. $17 = \frac{5}{6}y - 8$ __________

37. $1.8 - 8.5x = -74.7$ __________

38. $7w + 10.5 = 66.5$ __________

39. $1.9 - 2.8x = -23.3$ __________

40. $-6.6z + 12 = -21$ __________

NAME ______________________ CLASS ____________ DATE ____________

Lesson Activity

1.3 Modeling Two-Step Equations

You can use cups, counters, and an equation mat to model the solutions of two-step equations.

The model of the equation $2x + 1 = -5$ is shown at right. Place 2 cups and 1 positive counter on the left side of the mat. Place 5 negative counters on the right side of the mat.

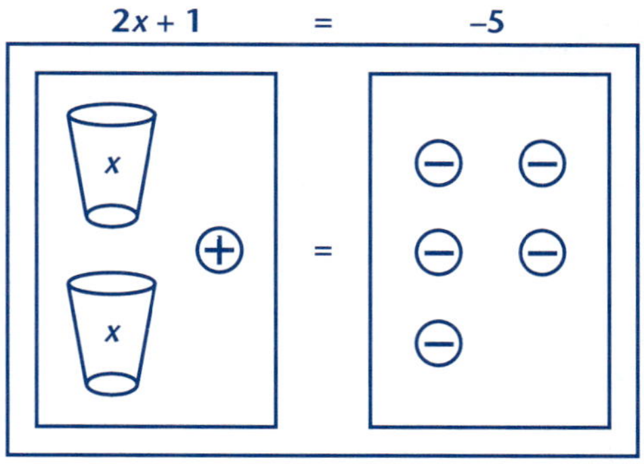

Isolate the 2 x-cups by adding 1 negative counter to each side. (Remember that adding a negative integer is the same as subtracting a positive integer.)

$$2x + 1 - 1 = -5 - 1$$

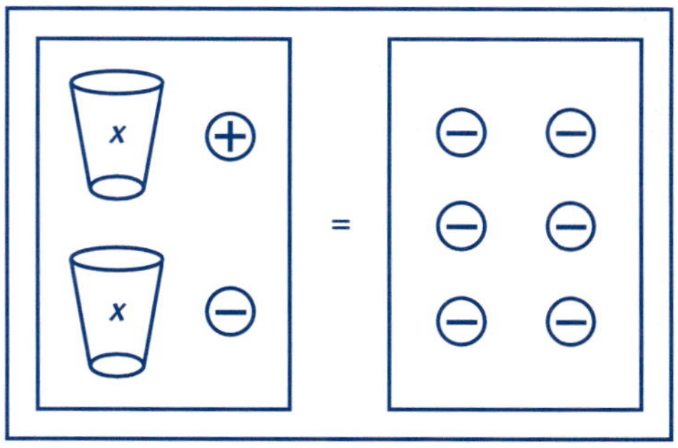

Remove neutral pairs.

$$2x = -6$$

Place an equal number of counters in each cup. The number of counters in each cup is the solution.

$$\frac{2x}{2} = \frac{-6}{2}$$

There are 3 negative counters in each cup.

$$x = -3$$

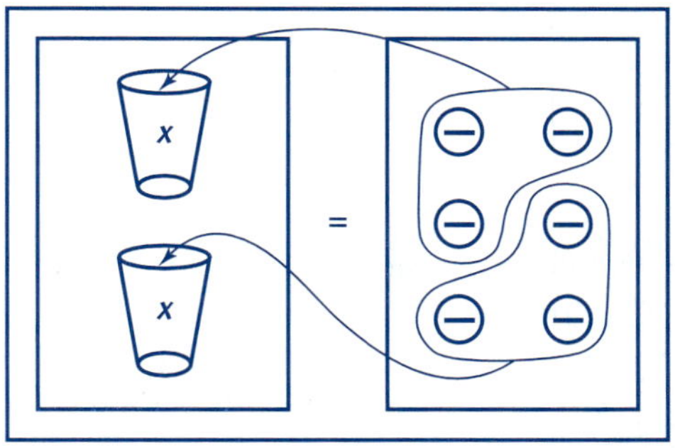

Use cups and counters to model the solution to each equation.

1. $2x + 4 = 2$ ____________ **2.** $2x - 3 = 3$ ____________

3. $3x - 1 = 5$ ____________ **4.** $3x + 2 = 8$ ____________

5. $4 + 2x = 12$ ____________ **6.** $3 + 3x = -6$ ____________

7. $4x + 1 = 5$ ____________ **8.** $2x - 6 = 0$ ____________

9. $5 + 3x = -4$ ____________ **10.** $-3 + 4x = 5$ ____________

NAME ______________________ CLASS ____________ DATE ____________

Basic Skills Practice

Generating Formulas, Part 1

A **formula** is an algebraic sentence. Mathematical formulas are used in science, business, and economics. Letters are chosen to represent various quantities.

To use a formula in solving a problem:

- Choose a formula that fits the problem.
- Replace the variables with given values.
- Compute or solve.

Example 1: In physics, the formula $d = rt$ gives the distance, d, that an object travels in terms of its rate, r, and the time, t, that it travels. What distance would a car travel in 3 hours at a speed of 50 kilometers per hour?

Choose → $d = rt$
Replace → $d = 50 \times 3$
Compute → $d = 150$ kilometers

Problem solving also generates formulas, equations, or number sentences.

Example 2: If girl scout cookies cost $6 for a large box and $3 for a small box, find the total amount received from each girl scout.

l = number of large boxes
s = number of small boxes
t = total amount

$$(\$6 \times l) + (\$3 \times s) = t$$

Scout	No. large boxes	+	No. small boxes	=	Total
Julie	$6 × 5	+	$3 × 4	=	$42
Angie	$6 × 3	+	$3 × 6	=	$36
Melissa	$6 × 1	+	$3 × 2	=	$12

Inequalities may also be used in problem solving.

Example 3: Mr. Rhodes told the student council that in order to have a dance, at least 175 people must attend. The fire laws, however, allow no more than 350 people in the room where the dance will be held. Write an inequality to describe this situation.

$$175 \leq x \leq 350$$

Solve each problem.

1. The speed limit on Loop 1 in Austin is a maximum of 65 mph and a minimum of 45 mph. Write a number sentence to express this situation. ____________

2. The formula for the area of a rectangle can be found by multiplying the length by the width, $A = lw$. What is the area of a rectangle whose length is 30 meters and whose width is 25 meters? ____________

3. Melissa said that Jean is one-fourth as old as Mr. Adams. If Jean is 9 years old, how old is Mr. Adams? ____________

LESSON 1.4

Multistep Equations

PREPARE

Pacing Guide
2 Days

Objectives and Overview
- Use properties of equality and the Distributive Property to solve multistep equations.

Multistep equations are equations with variables on both sides of the equal sign. These equations are solved by four different methods: algebra tiles, algebraic properties, graphs, and graphics calculator tables. The Distributive Property is used to find the measure of the base of a trapezoid. Multistep equations with no solution and multistep equations with an infinite number of solutions are also examined.

Vocabulary
multistep equation
Distributive Property

Assessing Prior Knowledge

Solve each equation.

1.	$-4x + 2 = 14$	$[-3]$
2.	$\frac{5}{3}x - 7 = 8$	$[9]$
3.	$3x + 4 = 6$	$[\frac{2}{3}]$
4.	$0.2x + 2.1 = 3.5$	$[7]$

Approach A: Using the Book

Opening Discussion
Use the carpet-cleaning example to introduce the type of application that requires a **multistep equation.** These equations can be recognized by the presence of a variable expression on each side of the equal sign. What will the variable represent in this case? **[the number of square feet of carpet]** Since this situation compares two carpet prices, the variable will appear on each side of the equation.

Modeling With Algebra Tiles
Before solving the carpet-cleaning problem, read through or demonstrate the method for solving a multistep equation with algebra tiles. Remind students that the goal is to isolate one *x*-tile.

As you discuss the tile manipulations, write the corresponding equations and properties on the chalkboard or overhead. The first step is to remove

continued on top of next page

TEACH

Approach B: Using Models

Opening Discussion
Allen's Carpet Cleaners charges $80.00 plus $0.10 per square foot to clean carpet. Sharon's Steam Cleaners charges $30.00 plus $0.18 per square foot. How do you decide which company to hire? Use this example to introduce applications that require a **multistep equation,** an equation with a variable expression on each side of the equal sign. What will the variable represent in this case? **[the number of square feet of carpet]** Two carpet-cleaning prices are compared, so the variable will appear on each side of the equation.

Using Algebra Tiles to Represent and Solve Multistep Equations
Creating a Model Using an overhead projector and algebra tiles, model $3x - 5 = x - 1$ for students.

continued on bottom of next page

Approach A — Using the Book, continued

an equal number of x-tiles from each side of the equation until the remaining x-tiles are on one side. Now the equation is a two-step equation that can be solved by using the techniques from the previous lesson.

Ongoing Assessment, page 29

$3(-3) + 4 = (-3) - 2$

$-5 = -5$ True

Example 1

In this example, the carpet-cleaning application is solved by using the properties of equality and then is checked with a graphics calculator.

Ask students to summarize what is being done at each step of the solution. Point out some advantages and disadvantages of the algebraic, graphing, and table methods. For example, the graphing and table methods may take longer because you have to consider the window values for the graph and the starting value and increment for the table.

These methods are ideal for checking, however, because you will already have an answer to help you set these parameters. The graphing method is the most helpful in terms of interpreting the problem. By examining the graphs, it is clear that Sharon's Steam Cleaners is less expensive for areas up to 625 square feet and that Allen's Carpet Cleaners is less expensive for areas over 625 square feet. This information can also be seen from the table. The equation, however gives only the square footage for which the two companies charges are equal.

Teaching Tip

Technology The graph in Example 1 can be created on a graphics calculator by entering the equations $Y_1 = 0.10X + 80$ and $Y_2 = 0.18X + 30$ on the Y= list. Use the window values Xmin = 0, Xmax = 1000, Ymin = 0, Ymax =

TEACH

Approach B — Using Models, continued

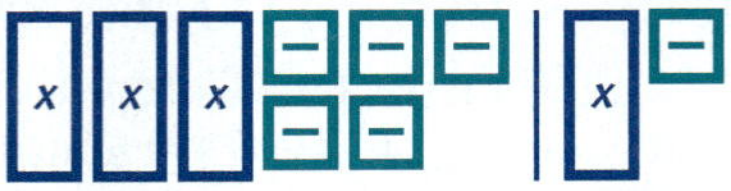

The tiles with a plus sign represent positive units and the tiles with a negative sign represent negative units. The vertical segment represents the equal sign. (For ideas on how to make algebra tiles, see Lesson 1.1, Using Models.)

Using the Model Isolate the variable so that its value can be determined. First remove an equal number of x-tiles from each side until the remaining x-tiles are on only one side.

Next add 5 positive unit tiles to each side and then remove the 5 neutral pairs on the left side and 1 neutral pair on the right side.

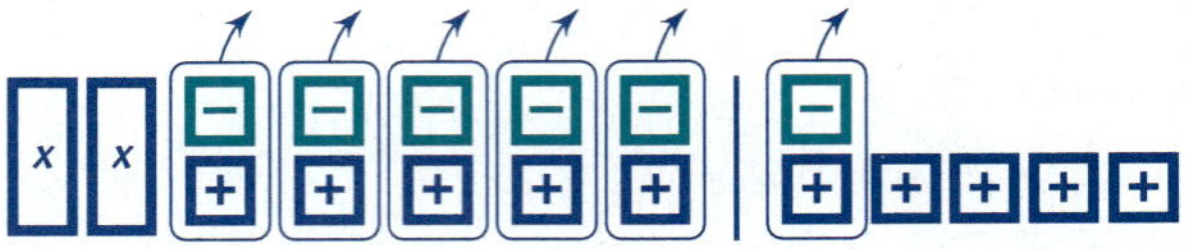

Dividing the square tiles into equal groups, with one group for each x-tile, gives $x = 2$.

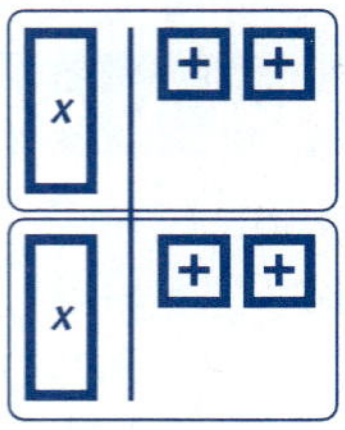

Practice With Tiles

Have students work with a partner to solve the equations below by using algebra tiles.

A. $4x + 3 = 2x - 5$ [−4]

B. $x - 3 = 2x + 7$ [−10]

200. Find the intersection by using the trace or intersect features of the calculator. To create the table, use Xmin = 622 and ΔX = 1.

Additional Example 1

Two local video stores offer a frequent-buyers program. At Amelia's Video, you pay a yearly fee of $40 and $8.00 per video. At Action Video, you pay a yearly fee of $18 and $10.00 per video.

A. Write two equations to show that the total charge for each store depends on the number of videos purchased each year. **[$y = 8x + 40$; $y = 10x + 18$]**

B. Write and solve an equation that can be used to find the number of videos for which the charges are equal. **[$8x + 40 = 10x + 18$; $x = 11$; the charges are equal when 11 videos are purchased each year.]**

C. Use a graphics calculator to create a table and to construct a graph to check your solution.

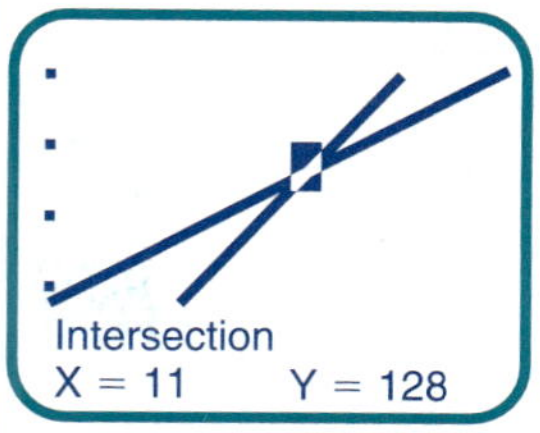

X	Y1	Y2
8	104	98
9	112	108
10	120	118
11	128	128
12	136	138
13	144	148
14	152	158

Ongoing Assessment, page 30
Try This

50 bags

The Distributive Property

Write the Distributive Property on the chalkboard or overhead. Explain that in this lesson, the formula on the left will be used. Give some concrete examples of the formula, beginning with ones in which *a*, *b*, and *c* are replaced by numbers and then including ones in which *b* or *c* is a variable such as *x*. Ask students for some other examples. Note that the

continued on top of next page

Solving Equations Algebraically

Have one group describe the procedure that they used to solve problem A. As they describe each manipulation of the tiles, write the corresponding algebraic manipulation on the chalkboard or overhead. After the solution is given, write the justification for each step. Note that the Subtraction Property of Equality is used to subtract *x*-terms as well as constants from each side of an equation.

Students should be able to solve these equations without the tiles, but encourage them to visualize the tiles to get started.

The Carpet-Cleaning Solution

Refer back to the carpet-cleaning problem at the beginning of the lesson. For how many square feet will the charges be equal? Write equations for students to show that the total charge for each company depends on the number of square feet of carpet cleaned. Write an equation that can be solved to find the number of square feet for which the charges are equal. Have students solve this equation and check their answer by using graphics calculators. **[625 square feet]** (Compare page 30.)

Point out some advantages and disadvantages of the algebraic, graphing, and table methods. For example, the graphing and table methods may take longer, but these methods are ideal for checking, because you already have an answer to help you set the parameters. Graphing is the most helpful in terms of interpreting the problem. From the graphs, it is clear that Sharon's Steam Cleaners is less expensive for areas up to 625 square feet and that Allen's Carpet Cleaners is less expensive for areas over 625 square feet. This information can also be seen from the table. The equation, however, gives only the square footage for which the two companies' charges are equal.

Teaching Tip

Technology The graph in the carpet-cleaning problem can be created on a graphics calculator. Enter $Y_1 = 0.10X + 80$ and $Y_2 = 0.18X + 30$ on the Y= list. Use the window values Xmin = 0, Xmax = 1000, Ymin = 0, Ymax = 200. Find the intersection by using the trace or intersect feature. To create the table, use Xmin = 622 and ΔX = 1.

continued on bottom of next page

Approach A

Using the Book, continued

Distributive Property has other uses that will be explored in later lessons.

Example 2

In this example, values are substituted into the formula for the area of a trapezoid. The resulting equation is then solved to find the length of one of the bases.

You may want to remind students of the definition a trapezoid before beginning this example.

Additional Example 2

An equilateral triangle has side lengths of $3x - 4$. Find x if the perimeter of the triangle is 60. **[8]**

Unusual Solutions

When solving an equation, do you always get a unique answer? Consider a situation such as the carpet-cleaning example. What if one company's price is always less than the other company?

Have students think about the following questions, which they should be able to answer after completing Example 3: What would the graph of the costs look like? What kind of solution would the equation have?

Example 3

Two linear functions that have no common solution are examined.

Students see that when they try to find a common solution, they get a false statement. Have students graph these functions by hand to see that they form parallel lines and thus have no point of intersection.

Additional Example 3

Pizza Quick charges \$2.00 for delivery and \$7 per pizza. Penelope's Pizza charges \$7.00 per pizza plus a \$3.00 delivery charge. Write and solve an equation to show that the charges will never be the same. **[$7x + 2 = 7x + 3$; this equation leads to $2 = 3$, so it has no solution. Thus, the charges will never be the same.]**

TEACH

Approach B

Using Models, continued

An Area Model of the Distributive Property

Another property that is used to simplify equations is the Distributive Property. This property allows you to eliminate parentheses. To help students get a feeling for this property, lead them through the following demonstration:

Draw this divided rectangle on the chalkboard or overhead.

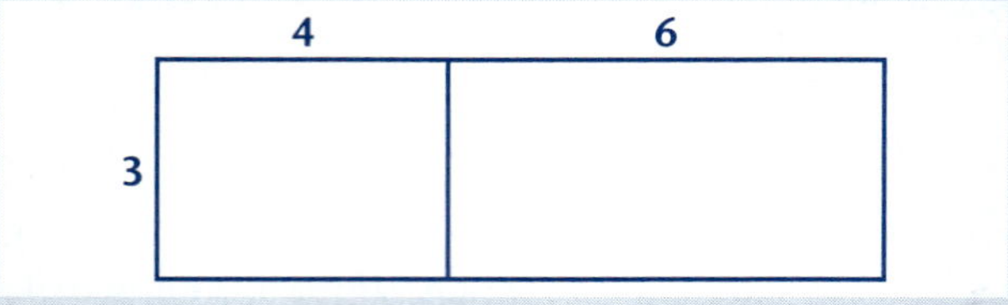

Remind students that the area of a rectangle is found by multiplying its length by its width. Then ask if there are two ways to calculate the area of this rectangle. One way is to look at it as one large rectangle. In this way the area is $3(4 + 6) = 3 \cdot 10 = 30$ square units. Another way to look at the area is as the sum of two smaller areas. In this case the area is $3 \cdot 4 + 3 \cdot 6 = 12 + 18 = 30$ square units. Thus, $3(4 + 6) = 3 \cdot 4 + 3 \cdot 6$. Draw a similar divided rectangle, and have students write two expressions for its area. What is the pattern? Write the formula for the Distributive Property on the chalkboard. (Compare page 30.)

The Trapezoid Example

Give students the formula for the area of a trapezoid, $A = \frac{1}{2}h(b_1 + b_2)$. Lead them through a demonstration of the following problem: Find the measure of b_2 if $h = 6$, $b_1 = 7$, and $A = 51$. **[10]** (Compare Example 2 on page 31.)

Unusual Solutions

Remind students that when they graph the functions from each side of an equation, the solution to the equation is the x-coordinate of the point of intersection. So far the graphs they have seen in this book are lines. A pair of lines can intersect at one point, which is the unique solution to the corresponding equations. What

Example 4

A multistep equation that has an infinite number of solutions is examined.

Point out that after applying the Distributive Property, like terms are combined on either side of the equation before the Subtraction Property of Equality is used. When the equation is solved, all variables are eliminated, and you get a true statement. Have students pick numbers to substitute into the original equation in order to verify that every real number is a solution.

Additional Example 4

Solve the equation $5(x + 2) - 3(x - 1) = 11 + 2(1 + x)$. **[The solution is all real numbers.]**

Critical Thinking, page 32

The graphs are identical.

Closing the Lesson

Write several multistep equations on the chalkboard. Have volunteers describe what they would do to solve the equations. Include equations whose solution involves the use of the Distributive Property.

else can happen with two lines? **[They can be parallel.]** What does that imply about the solution to the corresponding equation? **[There is none.]** The lines can also be the same (concurrent). How many solutions are there in this case? **[There are an infinite number of solutions; each real number is the *x*-value of some point on the two lines.]**

Cooperative Learning

In this activity students discover how to identify whether a multistep equation has no solution or infinitely many. Have students work in groups of four on the following activity, which you can write on the overhead or chalkboard. One student should read the activity and direct the group, one student should graph the functions, another student should write and solve the corresponding equation, and the fourth student should record the group's responses. (Compare Examples 3 and 4 on pages 31 and 32.)

Part I—no solution

Graph the lines $y = 2x + 3$ and $y = 2x - 5$. How are they related? **[They are parallel.]** How many solutions does the equation $2x + 3 = 2x - 5$ have? **[none]** Solve this equation by using the properties of equality. What happens? **[You get a false statement, $3 = -5$.]** How can you use this information to identify an equation with no solution? **[If you try to solve the equation and get a false statement, the equation has no solution.]**

Part II—an infinite number of solutions

Graph the lines $y = 4x + 6$ and $y = 2(2x + 3)$. How are they related? **[They are the same line.]** How many solutions does the equation $4x + 6 = 2(2x + 3)$ have? **[an infinite number]** Solve this equation by using the properties of equality. What happens? **[You get a true statement with no variables, $6 = 6$.]** How can you use this information to identify an equation with an infinite number of solutions? **[If you try to solve the equation and you get a true statement with no variables, the equation has an infinite number of solutions.]**

ASSESS

Selected Answers

Odd-numbered Exercises 7–35 and 39–51 and odd-numbered Extra Practice

Assignment Guide

Core level students should complete Exercises 1–51.

Core-plus level students should complete Exercises 1–9, 18–25, and 28–51.

Error Analysis

Students with little or no experience with the Distributive Property will probably have trouble with several of these exercises. You may want to give these students additional practice at simplifying expressions by using the Distributive Property before assigning those exercises. Students also need to know how to combine like terms.

Alternative Assessment

Solutions of Multistep Equations Have students write a paper in which they discuss multistep equations that have one solution, that have no solution, and that have all real numbers as their solution. They should include examples of each type of multistep equation, discuss the differences between the corresponding graphs of each type of equation, and give the results when the equation is solved algebraically.

Portfolio Activity Have students choose two of the homework equations to include in their portfolio. They should demonstrate how to solve each equation in at least two ways.

RETEACH

Cooperative Learning

Have students work in groups of three. Each student should start with one piece of paper and write a multistep equation on it. The students in each group should pass their paper to the right, and each student should use algebra to solve the problem. Then they should pass the papers again and use graphing technology to check the solution. Finally, they should pass the papers a third time and check the answer with a graphics calculator table.

ENRICH

Interdisciplinary Connection

Economics The break-even point is the point at which the cost to manufacture a product equals the revenue (money brought in) earned by the product. The break-even point can be found by writing a cost equation and a revenue equation and then combining them to form a single equation, which is then solved. Consider the following example:

A small printing company is thinking of adding T-shirt printing to its services. The cost of a used T-shirt-printing machine is $400. The T-shirts and ink cost $3 per shirt. The company plans to charge the customer $12 per shirt. The cost equation is $y = 3x + 400$. The revenue equation is $y = 12x$. To find the break-even point, solve the equation $3x + 400 = 12x$. This means that the company has to sell 45 shirts before it starts to make a profit on shirt sales.

A Unique Solution for Two-Step Equations

Have students investigate the number of possible solutions of two-step equations, which were explored in Lesson 1.3. Can they have no solution? an infinite number of solutions? Suggest that they graph each side of an equation. They should realize that one of the graphs is a horizontal line and that the other graph is not a horizontal line. Thus, the two must intersect at exactly one point. Because of this, a two-step equation always has one and only one solution.

NAME ______________________ CLASS __________ DATE __________

Practice

1.4 Multistep Equations

Solve and check each equation.

1. $5x + 19 = 6x + 12$ ____________

2. $8y - 10 = 6y - 2$ ____________

3. $2a - 9 = -2a + 15$ ____________

4. $9m - 13 = 8m + 9$ ____________

5. $x - 7 = 3x + 7$ ____________

6. $12y - 20 = 12 - 4y$ ____________

7. $14 - 6v = 4v + 6$ ____________

8. $19 - 3y = 4y - 2$ ____________

9. $6 + 3y = 4y - 8$ ____________

10. $-13 - 3z = 12 + 2z$ ____________

11. $7w - 19 = 5w - 5$ ____________

12. $28 + 2a = 5a + 7$ ____________

13. $5x - 40 = 8 - x$ ____________

14. $m - 18 = 3m + 4$ ____________

15. $2(x + 1) = 4x - 4$ ____________

16. $5(2x + 5) = 4x - 7$ ____________

17. $2r - 12 = 2(6 - 5r)$ ____________

18. $9y - 2 = 5(2y + 4)$ ____________

19. $z - 4(z + 2) = -6 - 2z$ ____________

20. $6t - 2(5 + 4t) = 5 + 2t - 7$ ____________

21. $12n + 24 = 2(n - 8)$ ____________

22. $3y + 4 = 3(6 - y)$ ____________

23. $2(3x - 1) = 3(x + 2)$ ____________

24. $10y - 8 + 2y = 7y + 17$ ____________

25. $10d - 22 + 6d = 12d + 10$ ____________

26. $23x + 34 = -14 - 6x + 5x$ ____________

27. $24 - 2s = 20(2s - 3)$ ____________

28. $12 - 4(2w - 11) = -3(2w + 4)$ ____________

29. $7 + 5(3q - 4) = 7(q - 1)$ ____________

30. $3(y + 2) + y = 15 - (y + 4)$ ____________

31. $0.2w - 4 = 0.8 - w$ ____________

32. $1.2z = 2.1z - 9$ ____________

33. $12 + 4.2w = 3w$ ____________

34. $2.5(j + 2) = 1.5(1 + j)$ ____________

35. $5.5 - 2.9m = -2.1 - 3m$ ____________

36. $x - 0.03 = 4.11 - 0.2x$ ____________

37. $9 - 2(8q - 1) = 4(q - 4)$ ____________

38. $6(y + 2) + 2y = 10 - 3(8y + 2)$ ____________

39. $0.6w - 6 = 0.9 - 9w$ ____________

40. $19.8z = 17.8z - 18$ ____________

41. $14 - 3.6w = -2w$ ____________

42. $6.5(j - 5) = 2(0.75 - j)$ ____________

43. $6.5(4 - 0.8m) = 18.2 + 7.8m$ ____________

44. $1.75x - 3.25 = 2.25 - 0.75x$ ____________

45. $0.8w - 15 = 0.4 - 0.2w$ ____________

46. $1.75z = 7.75z - 12$ ____________

NAME ______________________ CLASS ____________ DATE ____________

Lesson Activity

1.4 Modeling Multistep Equations

You can use cups, counters, and an equation mat to model the solutions to multistep equations.

The model of the equation $2x - 1 = x + 4$ is shown at right. Place 2 cups and 1 negative counter on the left side of the mat. Place 1 cup and 4 positive counters on the right side of the mat.

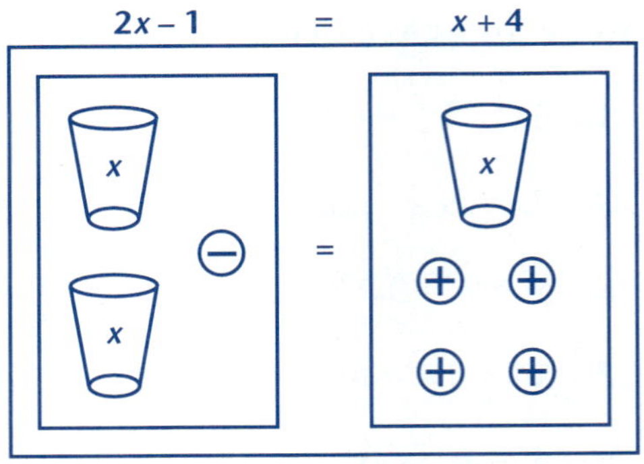

Remove the same number of cups from each side.

$$2x - x - 1 = x + 4 - x$$

$$x - 1 = 4$$

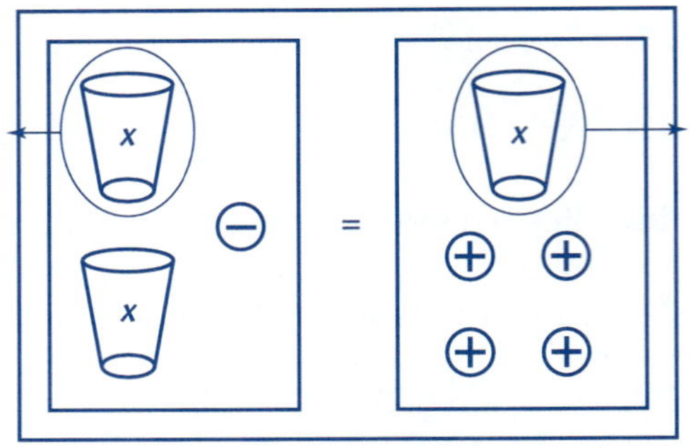

Add 1 positive counter to each side. $x - 1 + 1 = 4 + 1$

Remove neutral pairs. $x = 5$

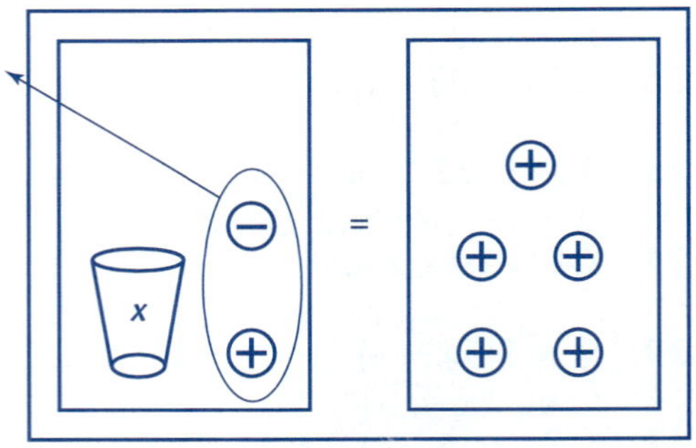

Use cups and counters to model the solution to each equation.

1. $2x + 4 = x - 2$ ______________
2. $x - 3 = 2x - 3$ ______________
3. $3x - 1 = 2x - 5$ ______________
4. $2x - 2 = x - 8$ ______________
5. $-4 + 2x = x + 3$ ______________
6. $5 + 3x = 2x + 6$ ______________
7. $x + 1 = 2x + 5$ ______________
8. $2x + 6 = x - 2$ ______________
9. $4 + 3x = -3 + 2x$ ______________
10. $-3 + x = 2x - 5$ ______________

NAME ______________________ CLASS ____________ DATE ____________

Basic Skills Practice

Generating Formulas, Part 2

Read each question and circle the best answer.

1. A cookie recipe calls for $3\frac{1}{2}$ cups of flour. Angie has only a $\frac{1}{2}$-cup measure. How many times must she fill the measuring cup to obtain the required amount of flour?

A 4

B $5\frac{1}{2}$

C 6

D 7

2. Gene spends \$3.50 each day for lunch. Which equation describes the total cost of lunch, *t,* for any number of days, *x*?

F $\frac{\$3.50}{x} = t$

G $\$3.50(x) = t$

H $\$3.50 + x = t$

I $x + t = \$3.50$

3. The cost of a long distance call is \$1.75 for the first 3 minutes and \$0.10 for each additional minute. Chuck called home and talked for 7 minutes. Which equation would be used to find the cost, *c,* of the call?

A $\$1.75 \times 0.10 = c$

B $\$1.75 + 4 = c$

C $\$1.75 + (4 \times 0.10) = c$

D $\$1.75 + 0.10 = c$

4. The Jacksons are planning a trip. Each day they will drive at least 300 miles but not more than 375 miles. Which number sentence best expresses this situation if *n* represents miles per day?

F $300 < n < 375$

G $300 > n > 375$

H $300 \leq n \leq 375$

I $300 \geq n \geq 375$

5. The possible outcomes of tossing three coins are shown by the following diagrams. What is the total number of possible outcomes of this experiment?

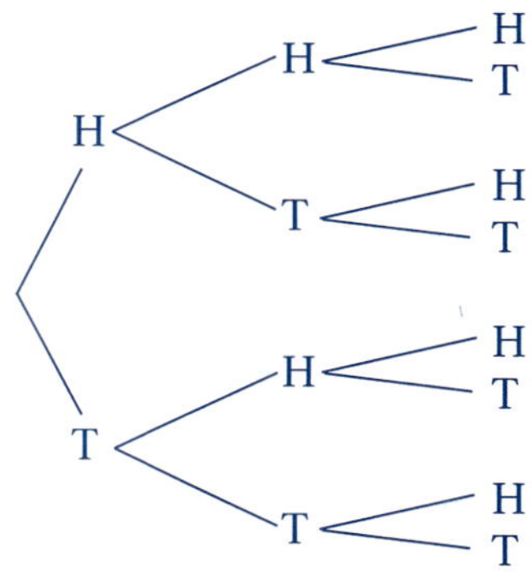

A 2

B 4

C 8

D 12

The pressure at a certain ocean depth can be represented by the formula $P = 15 + \frac{1}{2}D$, where *P* stands for the pressure in pounds per square inch and *D* is the depth in feet. Use this formula to answer each question.

6. What is the ocean depth, in feet, if the pressure is 35 pounds per square inch? ____________

7. If the ocean depth is 60 feet, what is the pressure, in pounds per square inch? ____________

8. What is the ocean depth, in feet, if the pressure is 50 pounds per square inch? ____________

LESSON 1.5

Solving Inequalities

PREPARE

Pacing Guide

2 Days

Objectives and Overview

- Use properties of inequality to solve inequalities.
- Graph solutions of inequalities on a number line.

Inequalities are introduced. The properties of inequalities are given and used to solve inequalities. Solutions are graphed on a number line.

Vocabulary

Addition Property of Inequality
Subtraction Property of Inequality
Multiplication Property of Inequality
Division Property of Inequality

Assessing Prior Knowledge

Write an inequality to represent each situation.

1. There are fewer than 12 cookies left. [$c < 12$]

2. At least 9 people are going on the field trip. [$p \geq 9$]

3. No more than 8 people may ride in the elevator at one time. [$n \leq 8$]

4. No more than 5 members of the same class are allowed to run for student council. [$m \leq 5$]

5. A minimum of 3 leaders are needed for the expedition. [$l \geq 3$]

Approach A

Using the Book

Opening Discussion

Some problems have more than one answer. In these cases, you can often find the solutions by finding a minimum, a maximum, or a range of answers. These types of problems can be represented by inequalities.

Brainstorming

Have students think of phrases that are used to describe inequalities. Examples include *less than, greater than, no less than, no more than, more than, at least, at most, a minimum of,* and *a maximum of.*

Using an Inequality to Represent a Real-World Situation

Have students take turns reading aloud the drama club example. What key words indicate an inequality? [***at least***] Note the use of the verbal model to summarize the information before the equation is written.

TEACH

Approach B

Using Models

Opening Discussion

Some problems have more than one answer. In these cases, you can often find the solutions by finding a minimum, a maximum, or a range of answers. These types of problems can be represented by inequalities.

Brainstorming

Have the class think of phrases that are used to describe inequalities. Examples include *less than, greater than, more than, no more than, at least, at most,* and *a minimum of.*

Representing an Inequality on a Number Line

Have the class make up some inequalities in words by using the phrases they found, such as the number of people who have pets in this class is *greater than* 3. Write these inequalities on the board. Under each, demonstrate how the inequality can be represented on a number line. Remind students that the numbers on a number line always increase from left to right. Also, an

Properties of Inequality

Have students continue to take turns reading the properties of inequality. How are these different from the properties of equality? The first two are not different, but the last two are different in the case of multiplication or division by a negative number.

Ongoing Assessment, page 36

Multiplying or dividing the expressions on each side of an inequality by a negative number reverses the direction of the inequality.

Journal Entry

Have students write a brief description of the properties of inequality in their journal. Rather than listing each property, encourage students to summarize the properties by comparing them with the properties of equality.

Example 1

The drama club example is solved by using the properties of inequality. Note the interpretation of the solution in the context of the problem.

Additional Example 1

A parking garage charges \$2.00 for valet parking and \$1.50 for each hour parked. John has \$13 in his pocket. Find the number of hours that John can park by solving the inequality $13 \geq 1.5x + 2$. **[$x \leq 7.\overline{3}$; no more than 7 hours]**

continued on top of next page

open circle on the number line means that the number is not included in the solution. A shaded circle means that the number is included. Finally, go back to each example and write the inequality with a variable and an inequality symbol.

Cooperative Learning

This activity will help students understand the properties of inequality and how they compare with the properties of equality. Have students work in groups of four. One student should lead the activity, one student should write the inequality and compute the new expressions, another student should keep track of the results, and the fourth student should write down the properties at the conclusion with input from all other group members.

Start with a true inequality involving only numbers, such as $5 < 6$. Experiment by adding, subtracting, multiplying, and dividing each side of the inequality by the same positive or negative number. Does the inequality still hold? If not, what happens?

Repeat this experiment with other inequalities. Write down what you believe to be the Addition, Subtraction, Multiplication, and Division Properties of Inequality.

Properties of Inequality

For each of the four properties of inequality, have one or more groups read their version. Students should realize that the Addition and Subtraction Properties are essentially the same as those for equality, while the Multiplication and Division Properties differ only when the number multiplying or dividing the inequality is negative, in which case the direction of the inequality sign is reversed. (Compare page 36.)

Journal Entry

Have students write a brief description of the properties of inequality in their journal. Rather than list each property, encourage students to summarize the properties by comparing them with the properties of equality.

continued on bottom of next page

Approach A — Using the Book, continued

Ongoing Assessment, page 37

Try This

Let t = number of T-shirts.

$2t + 28{,}000 \leq 40{,}000$

$2t + 28{,}000 - 28{,}000 \leq 40{,}000 - 28{,}000$

$2t \leq 12{,}000$

$t \leq 6{,}000$

The maximum number of T-shirts that can be given away is 6000.

Representing an Inequality on a Number Line

Remind students that numbers always increase from left to right on a number line. Students with little or no experience with number lines may need some additional practice.

Example 2

Here an inequality is solved and graphed on a number line.

Note that the direction of the inequality symbol is reversed when dividing by −2. Show students how they can pick a number in the shaded region and substitute it into the original inequality to test whether the direction of the shading is correct.

Additional Example 2

Solve $\frac{x}{-6} + 3 \leq 4$, and graph the solution on a number line.

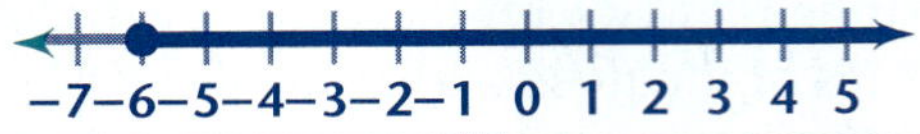

[$x \geq -6$]

Closing the Lesson

Have students work with a partner to complete Exercises 1–3. Ask several pairs to share their responses with the class so that other students can judge their own work.

TEACH

Approach B — Using Models, continued

Solving Inequalities

When you solve an inequality, you do not get a single answer, but a whole set of answers. The answer to an inequality can be expressed by using the inequality symbols or by graphing it on a number line, as was done previously. To solve a complex inequality, use the properties of inequality to simplify it to the point where you have a single variable on one side of the inequality sign and a number on the other side.

Demonstrate how to solve the inequality $-2x + 3 < 15$. Write down the properties used for each step, stressing that the direction of the inequality is reversed when dividing by −2. Graph the solution. Have students test whether the direction of shading is correct by picking a number in the shaded region and verifying that it satisfies the original inequality. (Compare page 38.)

A Real-World Example

A parking garage charges $2.00 for valet parking and $1.50 for each hour parked. John has $13 in his pocket. Find the number of hours that John can park. Lead students through the solution of this problem. Help them assign the variable and write the inequality $13 \geq 1.5x + 2$. Ask students to solve the inequality at their desk. **[$x \leq 7.\overline{3}$]** Then ask students to consider the solution in context. **[John can park for no more than 7 hours.]**

Closing the Lesson

Write several inequalities on the board. Ask students to describe the procedure that they would use to solve them. Would the inequality need to be reversed? Would the graph contain an open or shaded circle? Which direction would be shaded on the graph?

ASSESS

Selected Answers

Odd-numbered Exercises 9–65 and odd-numbered Extra Practice

Assignment Guide

Core level students should complete Exercises 1–64.

Core-plus level students should complete Exercises 1–7, 10–25, and 31–70.

Error Analysis

Watch for students who forget to change the direction of the inequality when necessary. Also, some students may change the direction when subtracting a number from both sides of the inequality. Stress that the sign changes only when dividing or multiplying by a negative number. When graphing, students often confuse the need for an open or shaded circle. It might help to make the following connection: A shaded endpoint is made in two steps—drawing a circle, and then shading it. Similarly, the symbols requiring shaded endpoints ($\leq$, $\geq$) are made in two steps—drawing a sideways V, and then underlining it.

Alternative Assessment

Applying Inequalities Ask students to make up a problem about the sale of an item that involves a fixed cost and a per item cost. They should pick a desired profit, write an inequality, and solve it to find the number of items that must be sold to make at least the desired profit.

Portfolio Activity Have students include their work for Exercises 44–46 in their portfolio. Initially, they should complete Exercises 45 and 46 by using algebra to solve the related inequalities. Then have them graph the two equations from Exercise 44 in the coordinate plane and write a paragraph about how the graph relates to the solution of the problem.

RETEACH

Inequalities on a Graphics Calculator

Consider the inequality $2x + 1 \leq 3$. Form the corresponding equations, $y = 2x + 1$ and $y = 3$, and graph them on a graphics calculator. When is $2x + 1 = 3$ true? **[at the intersection point, when $x = 1$]** Have students compare the relative values of the expressions $2x + 1$ and 3 to the left and right of the intersection. Which part of the graph satisfies the original inequality? Explain. **[the part of the graph to the left of the intersection, because the graph of $y = 2x + 1$ is below the graph of $y = 3$ for this part]** What are the x-values for the part of the graph that represents the solution to the inequality? **[$x \leq 1$]** Have students solve this inequality by using algebraic properties to verify the graphic solution. Give students several additional inequalities to solve. Have them check their solutions with a graphics calculator.

ENRICH

Interdisciplinary Connection

Geometry The triangle inequality is an inequality that relates the side lengths of every triangle. Have students research and write a paper about this famous inequality. They should include some examples.

Compound Inequalities

The inequalities seen thus far have graphs that are rays. Some inequalities have graphs that are segments. These are called compound inequalities. Show students how to graph the inequality $2 \leq x \leq 5$ on a number line. The easiest way to read this inequality is "x is between 2 and 5 inclusive." To solve a more complicated compound inequality, such as $25 < 2x + 15 \leq 55$, you can use the properties of inequalities on all three parts of the inequality. With this example, you first subtract 15 from each of the three parts to get $10 < 2x \leq 40$ and then divide by 2 to get $5 < x \leq 20$. Have students graph this solution on a number line. Note that the left endpoint is open, while the right endpoint is shaded.

NAME ______________________ CLASS ____________ DATE ____________

Practice

1.5 Solving Inequalities

Solve each inequality. Graph your solution on a number line.

1. $x + 6 > 8$

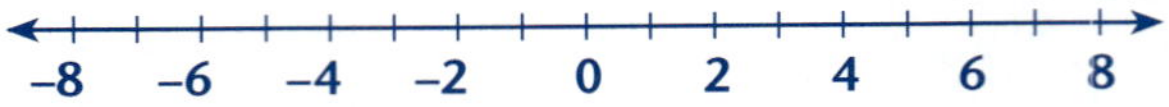

____

2. $x - 2 \leq -7$

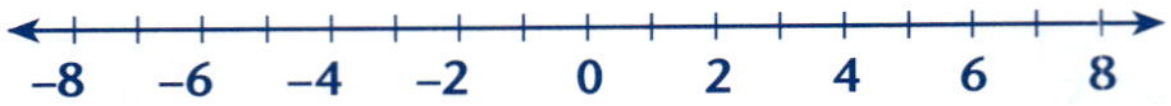

____

3. $5t < 35$

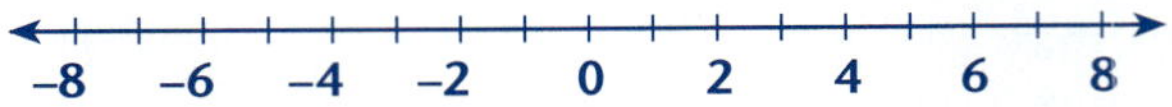

____

4. $\frac{m}{3} \leq 2$

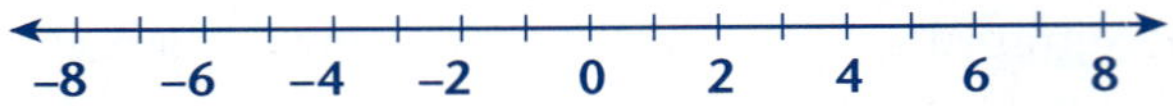

____

5. $8x - 3 > -19$

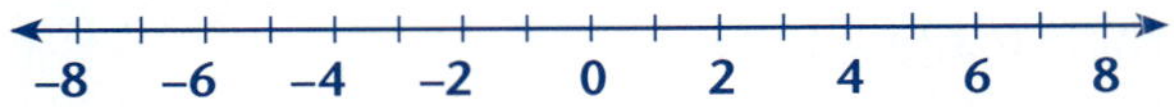

____

6. $4.5 - 2x \geq 6.5$

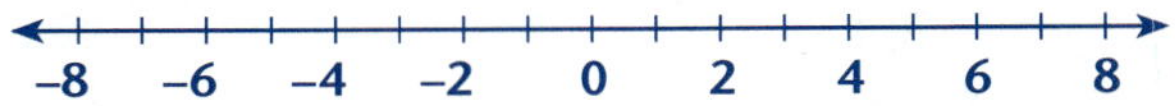

____

7. $\frac{x}{-4} + 3 \leq 2$

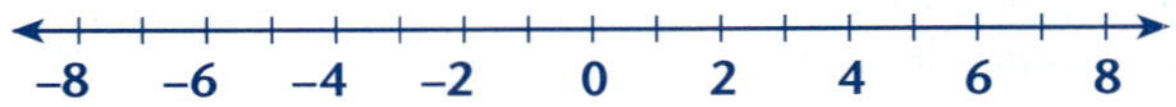

____

8. $\frac{b}{9} < \frac{2}{3}$

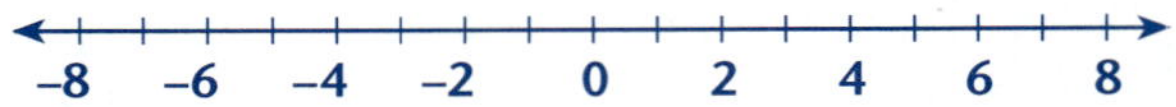

____

9. $\frac{k}{6} - \frac{1}{3} \geq \frac{1}{2}$

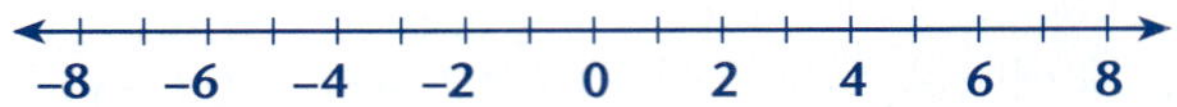

____

10. $9x - 6 \leq 12$

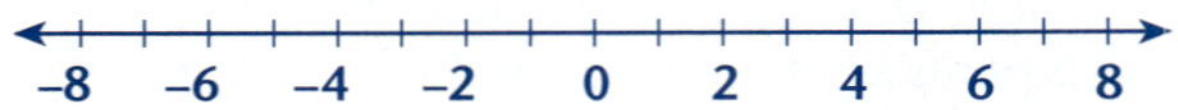

____

11. $1.2x + 5.6 < 2.8$

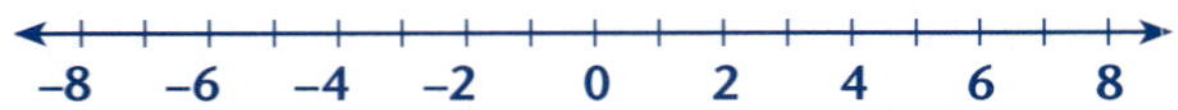

____

12. $14 \leq 12 + 8x$

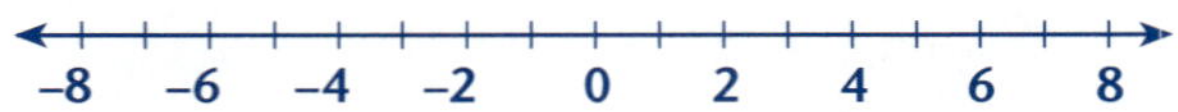

____

NAME ______________________ CLASS ____________ DATE ____________

Lesson Activity

1.5 Modeling Inequalities

You can use cups, counters, and an inequality mat to model the solutions to inequalities.

The model of the inequality $-2x > 6$ is shown at right. Place 2 negative cups on the left side of the mat. Place 6 positive counters on the right side of the mat.

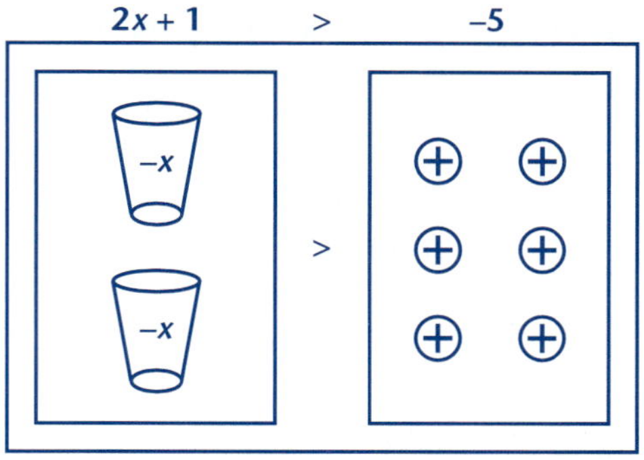

You cannot solve for $-x$, so add 2 positive cups to each side. $\quad -2x + 2x > 6 + 2x$

Remove neutral pairs. $\quad 0 > 6 + 2x$

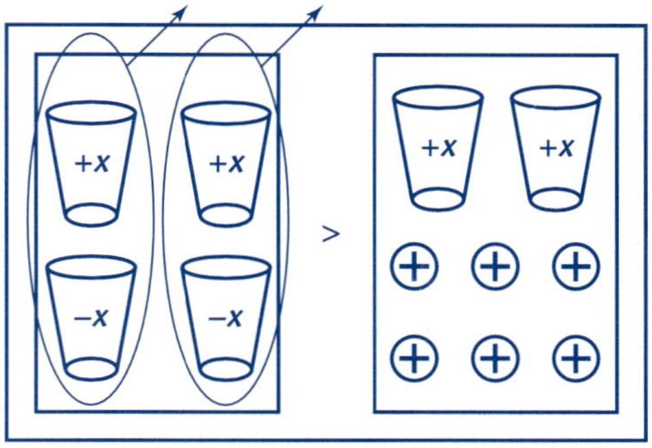

Add 6 negative counters to each side. $\quad 0 - 6 > 6 + 2x - 6$

Remove neutral pairs. $\quad -6 > 2x$

Place an equal number of counters in each cup. $\quad \frac{-6}{2} > \frac{2x}{2}$

There are 3 negative counters in each cup. $\quad -3 > x$ *or* $x < -3$

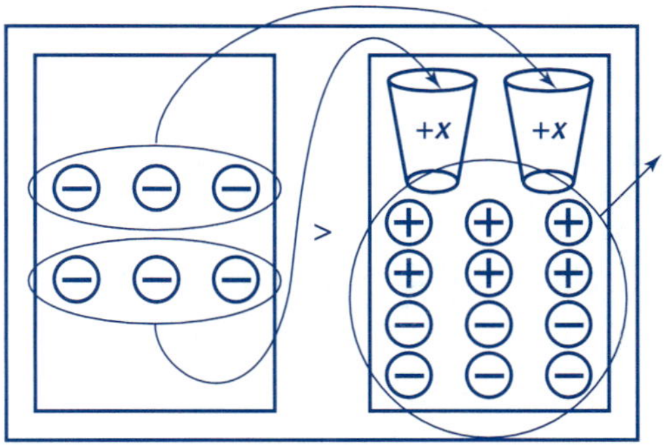

Use cups and counters to model the solution to each inequality.

1. $-3x < 6$ ____________ **2.** $-2x > 4$ ____________

3. $9 \geq -3x$ ____________ **4.** $-x \leq -5$ ____________

5. $-3x > -3$ ____________ **6.** $-8 \leq -2x$ ____________

7. $-10 \leq -2x$ ____________ **8.** $-4 > -4x$ ____________

NAME ______________________ CLASS __________ DATE __________

Basic Skills Practice
Cumulative Review

Read each question and circle the best answer.

1. Arthur has 6 times as many stamps as Gene. If Arthur has 108 stamps, which equation would you use to find out how many stamps Gene has?

A $6 + n = 128$
B $6n = 108$
C $6 \times 128 = n$
D $6 + 108 = n$

2. Which sentence best states that the product of seven and four is greater than nine?

F $7 \times 4 > 9$
G $7 + 4 > 9$
H $\frac{4}{9} > 7$
I $7 \times 4 < 9$

3. Which set is a true solution for $-3 < n < 5$?

A $\{-3, -2, -1, 0, 1, 2, 3, 4, 5\}$
B $\{-3, -2, -1, 0, 1, 2, 3\}$
C $\{-2, -1, 0, 1, 2, 3, 4\}$
D $\{-2, -1, 0, 1, 2, 3, 4, 5\}$

4. Which statement is false?

F $7 + 3 < 11$
G $-26 \geq 0$
H $-9 < 0$
I $8 < 9 \times 2$

5. The maximum grade on a test is 65. The minimum grade is 40. If n represents a grade, which sentence best expresses this situation?

A $65 \leq n \leq 40$
B $40 \leq n \geq 65$
C $65 \geq n \geq 40$
D $40 \geq n \geq 65$

6. The corner store receives a new delivery of 512 candy bars. Now there are 2115 candy bars in stock. Which equation would you use to determine how many candy bars the store had before the delivery?

F $2115 + 512 = n$
G $\frac{2115}{512} = n$
H $n + 512 = 2115$
I $512n = 2115$

7. Complete to form an equivalent ratio.

$$\frac{3}{13} = \frac{9}{x}$$

A 112
B 39
C 27
D 12

Solve each problem.

8. Pat uses 4 ounces of yeast for every 2 loaves of bread that he bakes. How many pounds of yeast does he need for 40 loaves of bread? __________

9. What number is 4% of 15? __________

LESSON 1.6 Absolute-Value Equations and Inequalities

PREPARE

Pacing Guide
2 Days

Objectives and Overview
- Use the definition of absolute value to solve equations and inequalities involving absolute value.

Both the geometric and algebraic definitions of absolute value are examined and used to understand problems concerning error. Absolute-value equations and inequalities are solved by using the algebraic definition of absolute value and are checked with a graphics calculator.

Vocabulary
absolute value

Assessing Prior Knowledge

Solve each equation or inequality.

1. $x - 2.71 = 5.05$ [7.76]

2. $-(2x - 7) = 4$ [1.5]

3. $3x - 4 \geq 14$ [$x \geq 6$]

4. $2 - x > 5$ [$x < -3$]

5. $-(x + 3) \geq 4$ [$x \leq -7$]

Approach A: Using the Book

Opening Discussion
Introduce the idea of allowable error by using the Houdini example. Note that in this application, it is not important whether the guess is too high or too low; only the quantity of the error is important.

Absolute value is used in applications in which the quantity of the value, not its sign, is desired. The notation 130 ± 3, read "130 plus or minus 3," is a shorthand notation for the two separate expressions $130 + 3$ and $130 - 3$ and is often used to describe the range of a value, such as possible error. An absolute-value inequality will be examined in detail in Example 2.

Brainstorming
Have the class think of other applications in which error or tolerance level may be calculated. Examples include the following: in manufacturing, the

continued on top of next page

TEACH

Approach B: Using Models

Opening Discussion
Introduce the idea of allowable error by using the following example. A fortune-teller claims that she can guess your age within 4 years. If she does not guess your age within the acceptable range, you do not have to pay to have your fortune told. Note that in this application, it is not important whether the guess is too high or too low; only the quantity of the error is important. Absolute value is used in applications in which the quantity of the value, not its sign, is desired.

Brainstorming
Have the class think of other applications in which error or tolerance level may be calculated. Examples include the following: in manufacturing, the measurement of a part needs to be within a specified range; in political polling, the percentages are given within a certain number of percentage points; and in scientific measurements, the instruments are accurate within a certain range.

continued on bottom of next page

Approach A — **Using the Book**, continued

measurement of a part needs to be within a specified range; in political polling, the percentages are given within a certain number of percentage points; and in scientific measurements, the instruments are accurate within a certain range.

A Geometric Interpretation of Absolute Value

Read or have a student read the geometric definition of absolute value. Ask students to give some additional examples and to explain their examples by using the number line. For example, $|-3| = 3$ because -3 is 3 units from 0 on the number line.

The Algebraic Definition of *Absolute Value*

Here the definition of absolute value is given. Encourage students to read the second line as "If x is a negative integer, then $|x|$ is the opposite of x."

Use **Transparency 4** for discussion or review of absolute value.

Teaching Tip

Definition of Absolute Value The definition of absolute value (page 42) is often difficult for students to understand. The most important thing for students to realize is that the absolute value of a number is always positive or zero. The absolute value of a positive number is the same number. The absolute value of a negative number is the same as the number with the minus sign covered up or "mentally erased."

In order to help students understand the definition, have them consider the following:

The opposite of 3 is -3.

The opposite of -3 is $-(-3)$, or 3.

Substitute 3 and -3 into the definition:

TEACH

Approach B — **Using Models**, continued

A Geometric Interpretation of Absolute Value

The absolute value of a number is the distance from 0 to the number on a number line. Illustrate $|4| = 4$ and $|-4| = 4$ by plotting the points 4 and -4 on the number line and verifying that in each case the distance from the point to 0 is 4. Ask students to give some additional examples of absolute values and to explain their examples by using the number line. For example, $|-3| = 3$ because -3 is 3 units from 0 on the number line. Mention that absolute value is entered into a calculator by using the ABS key. To evaluate $|5|$, enter ABS(5).

A Number-Line Model for Absolute-Value Equations and Inequalities

Creating the Model The expression $|a - b|$ gives the distance between a and b on a number line. Have students verify this fact by picking values for a and b, plotting them on the number line, and counting the distance between them. Then have them substitute the values for a and b into the expression and simplify. Note that the subtraction must be done before the absolute value is taken. Have students switch the order of the values for a and b and re-evaluate the expression to verify that the answer remains the same.

Using the Model Work with the class to develop the inequality which represents the possible correct guesses that the fortuneteller can make for a person whose age is 41, $|x - 41| \leq 4$. Write a sentence to express this result in terms of the number-line model. The distance between x and 41 on a number line is less than or equal to 4 units. Plot 41, mark off 4 units to the right and 4 units to the left of 41. Then shade the solution.

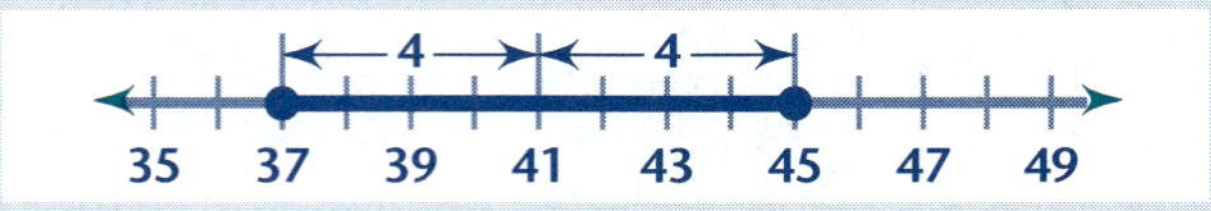

For any number 3 or -3,
3 is a positive number; therefore, the absolute value of 3 is 3, or $|3| = 3$; and
-3 is a negative number; therefore, the absolute value of -3 is the opposite of -3, or $|3| = -(-3)$.

Then point out to students that the definition is used in the same way for all numbers.

On Their Own

Have students find the absolute value of the numbers below. If graphing calculators are available, students may find the answers by using the ABS command. After the students have finished their work and checked their answers, discuss how the results agree with the definition.

A. 7	[7]	**B.** -7	[7]
C. 0	[0]	**D.** 15	[15]
E. -3	[3]	**F.** -20	[20]

Critical Thinking, page 42

Both cases in the definition lead to a non-negative number. Also, the absolute value is a distance; hence, it is not negative.

Example 1

In this example, an absolute-value inequality is solved by using the algebraic definition.

Case 2 of this example can be especially difficult for students. Stress that the equation is a direct application of the definition of absolute value when the quantity inside the absolute-value sign is negative. Since $|x| = -x$, replacing the x with $2x - 5$ gives $|2x - 5| = -(2x - 5)$.

Additional Example 1

Solve $|4x + 6| = 12$. **[1.5 or −4.5]**

Absolute Value and Error

Return to the Houdini example. Why does the inequality $|x - 130| \leq 3$ represent the allowable error for Houdini? The expression

continued on top of next page

Explain that x can have the value of any of the points in the shaded area. Have students pick a point in the shaded area and verify that it is a solution. Use the number-line model to write the solution, $37 \leq x \leq 45$.

Next show students how to solve the equation $|2x + 3| = 5$. First rewrite the expression with subtraction $|2x - (-3)| = 5$. The distance between $2x$ and -3 is 5. Start at -3, and count 5 spaces in either direction to locate the value of $2x$.

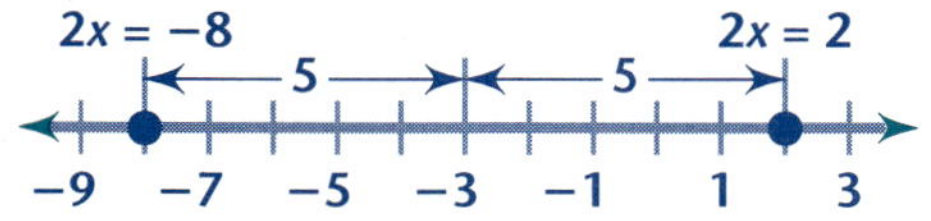

This gives $2x = -8$ or $2x = 2$. Solve these algebraically to get $x = -4$ or $x = 1$. Check these solutions by substitution.

Cooperative Learning

This activity will help students understand absolute-value equations and inequalities. Have students work in groups of four. The groups should solve the problems below, which you may write on the chalkboard or overhead.

A. $|x - 2| \leq 3$ $[-1 \leq x \leq 5]$

B. $|x + 2| = 1$ $[x = -3 \text{ or } x = -1]$

C. $|3x - 1| = 7$ $[x = -2 \text{ or } x = \frac{8}{3}]$

D. $|5x - 12| \geq 8$ $[x \leq \frac{4}{5} \text{ or } x \geq 4]$

One student should write the equation or inequality in terms of distance, one student should graph the equation or inequality on the number line, one student should write the solution, and the fourth student should check the solution by using substitution. Group members should exchange roles for each problem.

The Algebraic Definition of *Absolute Value*

Write the algebraic definition of absolute value on the chalkboard or overhead.

If x is a positive number, then $|x| = x$.

If x is a negative number, then $|x| = -x$.

This definition can be difficult for students to understand. Encourage them to read the second

continued on bottom of next page

Approach A Using the Book, continued

$x - 130$ is the difference between the persons weight and the guess. Whether the guess is above or below the weight is not significant; only the difference is significant. Thus, the absolute-value sign and the inequality symbol for less than or equal to are used. In general, the error of a measurement (or a guess) from the actual value is given by $|\text{measurement} - \text{true value}|$.

Example 2

The inequality for the Houdini example is solved. The method is similar to that used to solve the absolute-value equation in Example 1.

Additional Example 2

In a cereal factory, the cereal in each box needs to weigh within 0.15 ounces of the weight printed on the box, which is 14.6 ounces. Write and solve an absolute-value inequality to express the possible weights of the cereal.

[$|x - 14.6| \le 0.15$; $14.45 \le x \le 14.75$; the cereal must weigh between 14.45 ounces and 14.75 ounces.]

Example 3

A verbal description about distance is translated into an absolute-value inequality and solved.

Encourage students to draw a number line that represents this situation. The method for solving the inequality is the same as that used in Example 2. Note, however, that this solution cannot be written as a single expression.

Additional Example 3

What values are greater than 1.5 units from -1?
[$x < -2.5$ or $x > 0.5$]

Ongoing Assessment, page 44
Try This
$-10 \le x \le 2$

TEACH

Approach B Using Models, continued

line as "If x is a negative number, then $|x|$ is the opposite of x." This will help them see that, in this case, $-x$ is a positive number (the opposite of a negative number). Have students substitute numbers into the definition to see the results.

Using the Algebraic Definition

Demonstrate for students how to solve inequality A from the Cooperative Learning section above by using the algebraic definition.
If $x - 2$ is a positive number, then $|x - 2| = x - 2$.
If $x - 2$ is a negative number, then $|x - 2| = -(x - 2)$.

Since you do not know if $x - 2$ is positive or negative, you must solve the inequality for both cases. In the first case, you solve the inequality $x - 2 \le 3$ to get $x \le 5$; in the second case, you solve the inequality $-(x - 2) \le 3$ to get $x \ge -1$. Combining these two inequalities gives the same solution as that found by using the number line.

Solving or Checking Inequalities With a Graphics Calculator

Solving or checking an inequality on a graphics calculator is similar to solving or checking an equality. Two functions are created from the two sides of the inequality and are entered in the Y= list. The two functions are equal at the x-coordinates of the intersection. At other points, the expression whose graph is higher has a greater value. Show students how to use the graphics calculator to check the solution to inequality D from the Cooperative Learning section. Use the window Xmin = -2, Xmax = 6, Ymin = -2, Ymax = 10. (Compare page 44.)

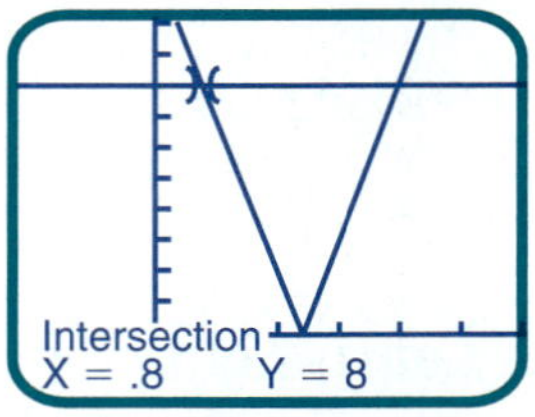

ASSESS

Selected Answers

Odd-numbered Exercises 5–75 and odd-numbered Extra Practice

Assignment Guide

Core level students should complete Exercises 1–75.

Core-plus level students should complete Exercises 1–4 and 14–76.

Error Analysis

Some students may have difficulty writing the two equations or inequalities from an absolute-value equation or inequality. Encourage these students to study the examples and follow the steps used in the examples when completing their own work. Watch for students who try to combine inequalities such as $x < 2$ and $x > 5$ into a single expression. Explain that a pair of inequalities can be put together only when the x-value is between the two numbers.

Use **Transparency 5** to discuss Exercises 57–59.

Alternative Assessment

Relating Solutions to the Definition Have students write a paper that describes the method for solving an absolute-value equation. How does this method relate to the algebraic definition of absolute value? Students' papers should include at least one worked-out example.

Portfolio Activity Have students pick from among the groups of exercises on manufacturing, design, and packaging. They should edit and include this work in their portfolio. If it is not already included, students should check their work with a graphics calculator and include a sketch of the graph.

RETEACH

Estimating a Centimeter

Have students work in groups to complete this activity. Each group should ask 20 people to draw a segment that they believe to be 1 centimeter long. They should then measure each segment to the nearest tenth of a centimeter and record the values in a table. In the next column of the table, they should find the error of each guess by subtracting it from 1 (the actual value). In the third column, they should find the absolute error by taking the absolute value of each error in the second column. They can then see the range of errors. If you were going to set up a booth at a fair and pay anyone who gets within a certain measurement of 1 centimeter, what value would you pick for an acceptable level of error? Write an absolute-value inequality to represent this situation, and solve it to find the acceptable range of lengths.

ENRICH

Interdisciplinary Connection

Government Margins of error are often given for political polls. Considering the margin of error can be an important factor in predicting an election. Here is an example. Two candidates are running for a local office. A newspaper conducted a poll that has an error of $\pm 5\%$. The results of the poll are 43% for Brown and 52% for Givins; the remainder of the voters are undecided. Can you conclude that Givins will win the election? Have students use what they have learned to find the possible percentages for each candidate: 38% to 48% for Brown, and 47% to 57% for Givins. Thus, it is possible for Brown to get 48% while Givins gets only 47%. In this case, Brown would win the election.

From Number Lines to Error Inequalities

Give students number-line graphs that consist of a single interval. Have them make up a problem involving error that corresponds to the graph. Ask them to write the absolute-value inequality represented by the graph.

Practice
1.6 Absolute-Value Equations and Inequalities

Solve each equation or inequality. Check each solution.

1. $|x + 3| = 5$ ____________

2. $|x + 3| = 7$ ____________

3. $|x - 6| = 1$ ____________

4. $|4x - 3| = 5$ ____________

5. $|6x + 2| = 14$ ____________

6. $|6x + 5| = 17$ ____________

7. $|x - 1.2| = 7.6$ ____________

8. $|2x - 1.7| = 0.5$ ____________

9. $|x - 2| \geq 6$ ____________

10. $|x + 8| < 5$ ____________

11. $|2x + 4| > 7$ ____________

12. $|4x - 7| \geq 5$ ____________

13. $|0.5x + 1| > 16$ ____________

14. $|x - 4.5| < 6.8$ ____________

Write an equation or inequality for each situation. Solve the equation or inequality. Graph the solution on a number line.

15. The distance between a number and 5 is equal to 2.

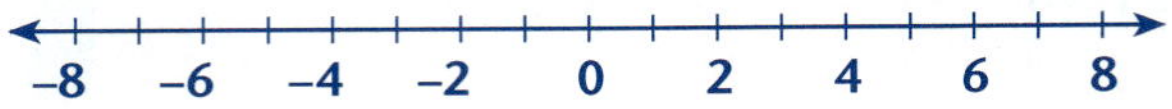

____

16. The distance between a number and −3 is more than 4.

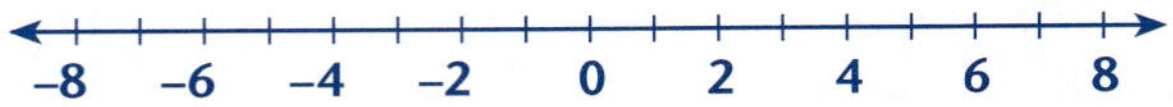

____

17. The distance between a number and 7 is equal to 1.5.

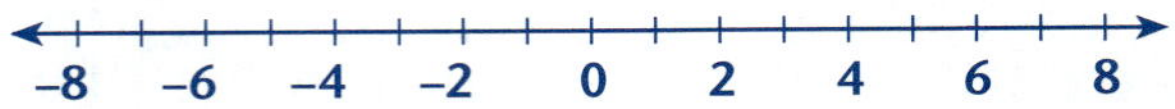

____

18. The distance between a number and 1 is less than 6.

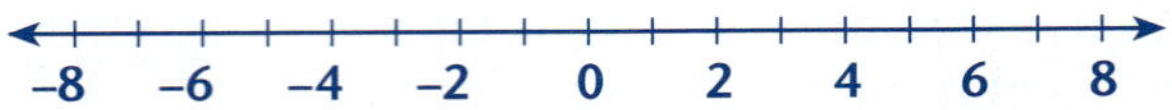

____

19. The distance between a number and −2 is more than 2.5.

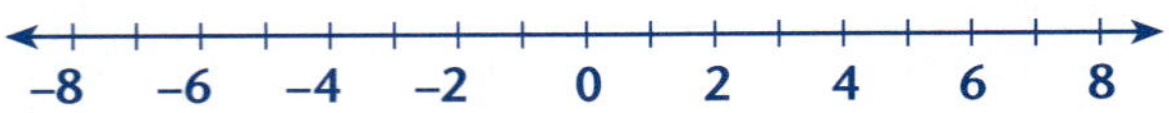

____

20. The distance between a number and −3 is at least 2.

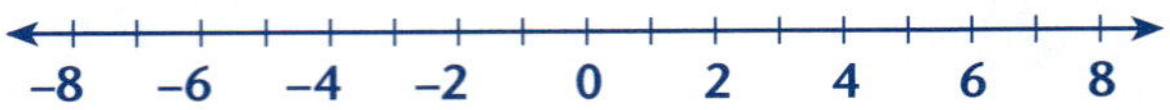

____

NAME ______________________ CLASS ______________ DATE ______________

Lesson Activity

1.6 Modeling Absolute-Value Inequalities

You can use cups, counters, and an inequality mat to model the solution to absolute-value inequalities.

There are two inequalities to model for the absolute-value inequality $|x| > 6$.

For the first inequality, simply remove the absolute-value symbol.

$$x > 6$$

For the second inequality, remove the absolute-value symbol, reverse the inequality sign and replace each counter with its opposite.

$$x < -6$$

The solution to the absolute-value inequality is $x < -6$ or $x > 6$.

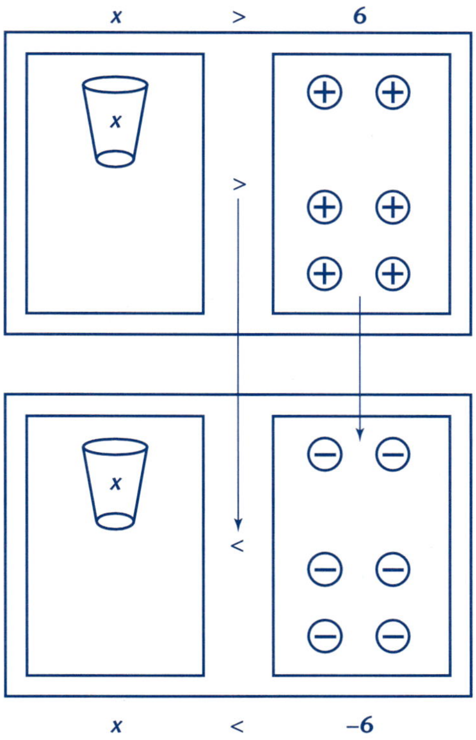

Use cups and counters to model the solution to each absolute-value inequality.

1. $|x| < 4$ ______________

2. $|x| \geq 1$ ______________

3. $|x| < 2$ ______________

4. $|x| \leq 5$ ______________

5. $|x| < 1$ ______________

6. $|x| \geq 2$ ______________

7. $|x| > 5$ ______________

8. $|x| \leq 6$ ______________

9. $|x| > 4$ ______________

10. $|x| \geq 7$ ______________

11. Describe how you would use a model to solve the absolute-value inequality $|x - 2| > 4$. ______________

12. Describe how you would use a model to solve the absolute-value inequality $|3x| \leq 6$. ______________

NAME ______________________ CLASS ____________ DATE ____________

Basic Skills Practice

Solving Equations by Addition or Subtraction, Part 1

You can solve an equation by using an inverse operation to get the variable by itself. Add or subtract the same quantity to both sides of the equation to isolate the variable.

Example 1:

$$n + 3 = 7$$
$$n + 3 - 3 = 7 - 3$$
$$n = 4$$

Check:

$$n + 3 \stackrel{?}{=} 7$$
$$4 + 3 \stackrel{?}{=} 7$$
$$7 = 7$$

Example 2:

$$n - 5 = 12$$
$$n - 5 + 5 = 12 + 5$$
$$n = 17$$

Check:

$$n - 5 \stackrel{?}{=} 12$$
$$17 - 5 \stackrel{?}{=} 12$$
$$12 = 12$$

When solving word problems:
- Define a variable.
- Write an equation.
- Solve and check.

Example 3: Sixty-one points are needed to win a game. Julie has 47 points. How many more points does she need to win?

Define → Let n represent the points needed to win.

Write → $n + 47 = 61$

Solve → $n + 47 - 47 = 61 - 47$

$n = 14$

Check → $n + 47 = 61$

$14 + 47 \stackrel{?}{=} 61$

$61 = 61$

Julie needs 14 more points.

Solve each equation for *n* and check the solution.

1. $n + 8 = 17$ ____________

2. $n + 6 = 15$ ____________

3. $n + 16 = 73$ ____________

4. $n + 27 = 109$ ____________

5. $n + 391 = 487$ ____________

6. $n + 195 = 1178$ ____________

7. $n - 8 = 36$ ____________

8. $n - 6 = 24$ ____________

9. $n - 13 = 43$ ____________

10. $n - 37 = 35$ ____________

11. $n - 190 = 234$ ____________

12. $n - 406 = 292$ ____________

13. $n + 484 = 581$ ____________

14. $n - 80 = 74$ ____________

15. $n - 50 = 222$ ____________

16. $n + 64 = 71$ ____________

17. $n + 37 = 2134$ ____________

18. $n - 152 = 634$ ____________

LESSON 2.1

Exploring Slope

PREPARE

Pacing Guide
3 Days

Objectives and Overview
- Determine the slope of a line through two points by using a graph.
- Determine the slope of a line through two points by using the slope formula.

This lesson should be a review for most students. The geometric formula for slope is given and used to describe the rate of speed of an object, to graph a line, and to study horizontal and vertical lines. The difference formula for slope is developed and then used to find the slope of the line through two given points.

Vocabulary
slope

Assessing Prior Knowledge

Simplify each expression.

1. $\frac{-4}{2}$ [-2]

2. $\frac{9}{-27}$ $\left[-\frac{1}{3}\right]$

3. $-6 - 10$ [-16]

4. $5 - 12$ [-7]

5. $3 - (-2)$ [5]

6. $\frac{2 - 1}{5 - 6}$ [-1]

Approach A: Using the Book

Opening Discussion
Refer to the picture of the hang glider to recall the concept of slope. Students will probably be familiar with the common usage of the word *slope* to mean an incline or decline (a ski slope for instance). Explain that the term *slope* in algebra is a number that indicates the steepness of a line. The common usage of the word can help students remember the meaning of the mathematical term.

Brainstorming
Have students think of other examples of things that have slope in the real world. Examples include a rooftop, a playground slide, and a wheelchair ramp. Point out that each example rises or falls.

Definition of *Slope*
Read or have a student volunteer read the definition of slope from page 56. Draw students' attention to the triangle illustrations over each of the four hills.

continued on top of next page

TEACH

Approach B: Using Models

Opening Discussion
Why is a downhill ski trail called a ski slope? What does it mean to say that you are climbing a steep slope? Students will probably be familiar with the common usage of the word *slope* to mean an incline or decline. Explain that the term *slope* in algebra is a number that indicates the steepness of a line. The common usage of the word can help students remember the meaning of the mathematical term.

Brainstorming
Have students think of other examples of things that have slope in the real world. Examples include a rooftop, a playground slide, and a wheelchair ramp. Point out that these examples include objects that rise and objects that fall.

continued on bottom of next page

Approach A

Using the Book, continued

Point out the rise and run for each hill, and calculate each slope. Make sure that students understand which directions are considered positive and which are negative.

Ongoing Assessment, page 57

A rise and run that are both positive or both negative result in a positive slope. A negative rise and a positive run or a positive rise and a negative run result in a negative slope.

Setting Up the Exploration

Teacher-Guided Explorations If you wish, you can guide the students through the textbook explorations. You can use overhead transparencies or the chalkboard to display the graphs.

At each stage of the exploration, invite student input and encourage discussion. Students should keep a record of the exploration responses in their notebook. For a greater level of participation, you can ask student volunteers to draw the graphs for Explorations 2 and 4 on the chalkboard.

Cooperative-Learning Groups Divide the class into groups of four. Each group should choose one member to read the exploration steps from the textbook, one to draw the graphs, one to do the calculations of slope, and one to keep notes and summarize the group results. These responsibilities can be reassigned for each exploration.

All group members should be involved in a discussion of the activity as the exploration proceeds. Each student should sign the completed report.

Inviting Participation

Have students write their own answers to the Ongoing Assessment and Critical Thinking questions as the explorations proceed. Explain that you will chose several individuals to read their responses at the completion of the lesson. This will encourage students by giving them time to prepare their responses.

TEACH

Approach B

Using Models, continued

The Geometric or Graphic Interpretation of Slope

Discuss the hills with triangle overlays from pages 56 and 57 to introduce the geometric definition of slope as the ratio of rise to run. Draw students' attention to the triangle illustrations over each of the four hills. Point out the rise and run for each hill, and calculate each slope.

Make sure that students understand that in mathematical graphs directions to the right and up are considered positive and directions to the left and down are considered negative.

Cooperative Learning

Before starting the explorations, divide the class into groups of four. Students should take turns drawing the graphs and calculating the slope, and each should contribute suggestions. Each student should be assigned to write a brief statement or summary of the findings for one of the explorations.

Alternate Exploration 1

This exploration enables students to see the connection between the slope of a line and the rate of speed. Give each group a copy of the graph below, which represents the distance, d, that an ant has crawled in t minutes.

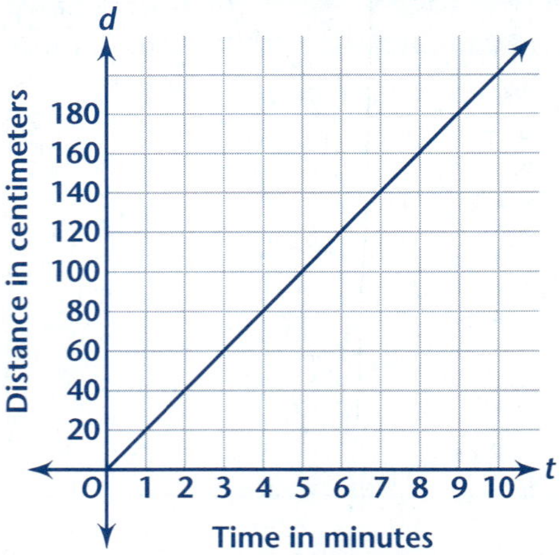

Have students work in groups and use the graph to answer the following questions, which you may write on the overhead or chalkboard. (For

Exploration 1

A time and distance graph is examined to see the relationship between the slope of the line and the rate of speed.

In Step 2, calculate the speed by using the distances at 4, 10, and 16 minutes to show that the speed is constant. When calculating the slope from the rise and run, use different pairs of points in order to demonstrate that the slope is the same regardless of the two points that are chosen. Have students check their equation by substituting the coordinates of some point.

Ongoing Assessment, page 57

3. 30; the slope describes the speed of the hiker in yards per minute.

Exploration 2

In Part I, a line with positive slope is graphed by using a point on the line and the slope. In Step 3, point out that another point can be found by using a fraction equivalent to the slope, $\frac{2}{3}$, to find a different rise and run.

In Part II, a line with negative slope is graphed by using a point on the line and the slope. Stress that $-\frac{2}{3}$, $\frac{-2}{3}$, and $\frac{2}{-3}$ are equivalent ways of writing a negative fraction. The last two equivalent fractions lead to different combinations of rise and run that can be used to graph the same line.

Ongoing Assessment, top of page 58

Part I, Step 5. Graph the point. Then use the slope to find the rise and run. From the point that you graphed, move up a distance equal to the rise, move right a distance equal to the run, and plot a second point. Draw a line through the two points.

Part II, Step 5. Graph the point. Then write the slope with either a negative numerator or negative denominator. Use the slope to find a rise and run. From the point that you graphed, move up or down a distance equal to the rise, move right or left a distance equal to the run, and plot a second point. Draw a line through the two points.

continued on top of next page

ideas on guiding students, see Exploration 1 on page 57 of the Pupil's Edition of the textbook.)

- Describe the movement of the ant. How does the speed of the ant compare with the slope of the line on the graph? **[The ant moved 20 centimeters in 1 minute, for a rate of speed of 20 centimeters per minute. This is the same as the slope of the graph.]**
- Write an equation to describe the relationship between the distance that the ant traveled and the time since it began to move. **[$d = 20t$]**

Student Reports

Choose one group to report their findings for this exploration. The student who wrote the summary for that group should read or paraphrase the group's findings for both steps and invite discussion from other groups.

Using an Erasable Graph

To help students focus their attention on the concepts of the next three explorations, it would be helpful to provide each group with a large rectangular coordinate system that has a permanent scale and axes. The group can then easily graph and regraph lines without having to redraw the axes and scales each time.

An erasable graph can easily be made by covering a large piece of cardboard with contact paper. Look for contact paper that has a grid or light-colored checkerboard design so that you do not have to draw the grid. Draw the axes and mark the scales in increments of 1 with a permanent marker. The points and lines can then be drawn with erasable markers and erased with a damp cloth or paper towel. These grids can be used again and again for teacher demonstration and for student reports.

Exploration 2

In this exploration, students discover a procedure for using the slope to graph a line through a given point. Have students work in their groups and use an erasable graph to answer the following questions, which you may write on the overhead or chalkboard. (For ideas on guiding students, see Exploration 2 on page 58 of the Pupil's Edition of the textbook.)

continued on bottom of next page

Approach A Using the Book, continued

On Their Own

Graph the line through the point (2, 1) and with slope of $\frac{1}{2}$.

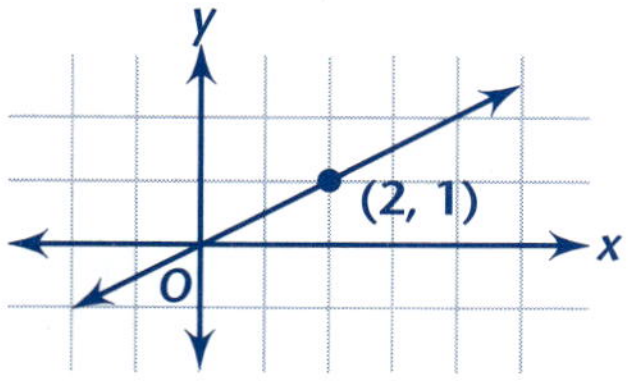

Exploration 3

In Part I, students learn that the rise, and hence the slope of a horizontal line, is always zero. In Part II, graphs of vertical line segments are examined to see that the slope of a vertical line is undefined.

Ongoing Assessment, bottom of page 52 and top of page 59

Part I, Step 4. The slope of any horizontal line is zero.

Part II, Step 5. The slope of any vertical line is undefined.

Critical Thinking, page 59

If the x-coordinates are equal, the points are on a vertical line. If the y-coordinates are equal, the points are on a horizontal line. If neither the x- or y-coordinates are the same, the line is neither vertical nor horizontal.

Exploration 4

The ratio of rise to run between two points is seen to be equivalent to the ratio of the difference in the y-coordinates to the difference in the x-coordinates. This leads to the definition of slope as a difference ratio. Students should realize that when calculating the differences, they must subtract coordinates in the same order.

TEACH

Approach B Using Models, continued

- For a line with positive slope, you are given a point on the line and the slope of the line. How can you use this information to graph the line? **[Graph the point. Use the slope to find the rise and run. From the point that you graphed, move up a distance equal to the rise, move right a distance equal to the run, and plot a second point. Draw a line through the two points.]**
- For a line with negative slope, you are given a point on the line and the slope of the line. How can you use this information to graph the line? **[Graph the point. Write the slope with either a negative numerator or negative denominator. Use the slope to find a rise and run. From the point that you graphed, move up or down a distance equal to the rise, move right or left a distance equal to the run, and plot a second point. Draw a line through the two points.]**

On Their Own

Ask students to graph a line through the point (−2, 4) and with a slope of $-\frac{3}{4}$.

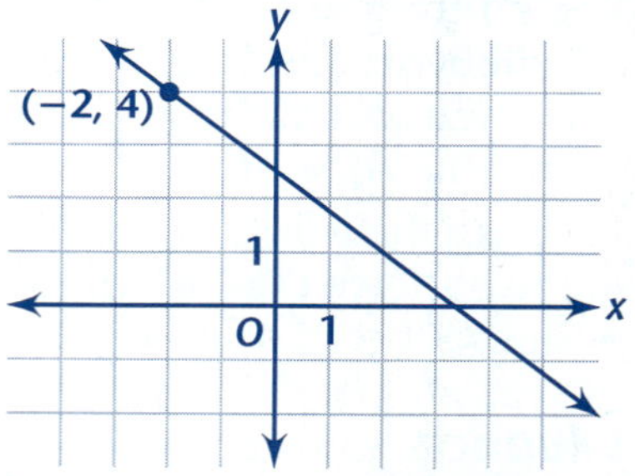

Exploration 3

The slopes of horizontal and vertical line segments are calculated by using rise over run in order to help students see that these types of lines have special slopes. Have students work in their groups and use an erasable graph to answer the following questions, which you may write on the overhead or chalkboard. (For ideas on guiding students, see Exploration 3 on page 58 of the Pupil's Edition of the textbook.)

Ongoing Assessment, bottom of page 59

5. The ratio of the differences is equal to the slope of the line.

Extension

These problems demonstrate the method for finding the slope of a line from the coordinates of two points without graphing.

Application

Direct variation is applied to a time and distance problem. The results are graphed.

Ongoing Assessment, page 60

Try This

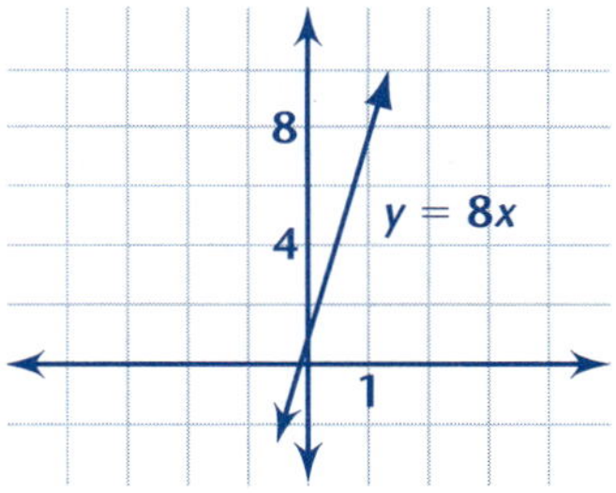

Critical Thinking, page 60

The first and third ratios give the slope of the line. The second and forth ratios are incorrect. When calculating the slope of a line, the coordinates must be subtracted in the same order.

Closing the Lesson

Write the points $A(3, -5)$ and $B(-2, 7)$ on the board. Ask students to graph the points and to find the slope of the line through the points. $\left[-\frac{12}{5}\right]$

- Experiment with horizontal lines to see what you can conclude about the slope of any horizontal line. **[The slope of any horizontal line is zero.]**
- Experiment with vertical lines. What can you conclude about the slope of any vertical line? **[The slope of any vertical line is undefined.]**
- Write the coordinates of two points on a horizontal line. How can you determine whether a set of coordinates forms a horizontal line or a vertical line without drawing the graph? **[If the *x*-coordinates are equal, the points are on a vertical line. If the *y*-coordinates are equal, the points are on a horizontal line.]**

Exploration 4

In this exploration students learn that the slope of a line through two points can be calculated by the ratio $\frac{\text{difference in } y\text{-coordinates}}{\text{difference in } x\text{-coordinates}}$.

Have students work in their groups and use the erasable graphs to examine two points on a line and to investigate the relationship between the difference in *x*-coordinates and the run and between the difference in *y*-coordinates and the rise. They can start by calculating these quantities for the line that goes through the points (1, 2) and (3, 5). (For ideas on guiding students, see Exploration 4 on page 59 of the Pupil's Edition of the textbook.)

Journal Entry

Have students write a procedure for finding the slope of a line when given the coordinates of two points. The rule should include finding the differences in the *x*- and *y*-coordinates.

Closing the Lesson

Write the points $A(9, 2)$ and $B(5, -1)$ on the chalkboard. Have half of the class calculate the slope by using differences with *A* as the first point, while the other half uses *B* as the first point. Compare answers and verify that each method gives $\frac{3}{4}$. What happens if *A* is used first to find the difference in *y*-coordinates and *B* is used first to find the difference in *x*-coordinates? **[You get an incorrect slope of $-\frac{3}{4}$.]**

ASSESS

Selected Answers

Odd-numbered Exercises 7–61 and odd-numbered Extra Practice

Assignment Guide

Core level students should complete Exercises 1–61.

Core-plus level students should complete Exercises 1–11, 16–21, and 31–65.

Error Analysis

Students who have difficulty with Exercises 7–27 may need to review subtraction of integers. For Exercises 31–42, encourage all students to check their graphs to see that a line with a negative slope falls from left to right, while a line with a positive slope rises from left to right.

Alternative Assessment

Are the Points on Horizontal Lines? Ask students to answer the problem below. A good response should include examples and graphs.

Demonstrate at least three different ways to determine whether two given points are on a horizontal line. **[Graph them; see whether the *y*-coordinates are equal; use the definition of slope as a difference ratio to see whether the slope is zero.]**

Portfolio Activity Ask students to include their work for either Exercises 37–42 or Exercises 43–48 in their portfolio.

RETEACH

Using a Geoboard to Model Slope

Geoboards whose pegs are arranged in square arrays can be used to model the rectangular coordinate system. Use two rubber bands for the axes to form the four quadrants. The pegs represent the integer-valued points. Use different-colored rubber bands to connect two points, modeling the line through the two points. Show students how the rise and run can be found by counting the horizontal and vertical spaces as you move from one end of the rubber band to the other.

Cooperative Learning You can have students work in pairs. They should make a rubber-band line, and then one student should find the slope by reading the rise and run from the geoboard, while the other student calculates the slope by using the coordinates of the ends of the rubber band and the difference ratio. Next, have one student name a point and slope, while the other student graphs the line on the geoboard.

ENRICH

Interdisciplinary Connection

Meteorology Have students find out more about slope winds and how hang gliders use them. Ask them to include their findings in a report and present it to the class.

Exploring the Slope of a Secant Line

A secant line is a line that joins any two points of a curve. Have students draw the graph of the curve $y = x^2$ from $x = -4$ to $x = 4$ and then draw the secant line joining the points $(-4, 16)$ and $(-3, 9)$. What is the slope of this line? Repeat this for the points $(-3, 9)$ and $(-2, 4)$, then $(-2, 4)$ and $(-1, 1)$, then $(-1, 1)$ and $(0, 0)$, and so on, finishing with the points $(3, 9)$ and $(4, 16)$. What happens to the slopes of these secant lines? **[They start at −7 and increase to 7.]** How can the rising and falling of the curve be connected to the sign of the slope of the secant line? **[The curve falls when the secant line has a negative slope and rises when the secant line has a positive slope.]**

Have students experiment by drawing other curves and calculating the slopes of secant lines drawn through points with consecutive *x*-coordinates.

NAME ______________________ CLASS ____________ DATE ____________

Practice
2.1 Exploring Slope

Determine the slope of each line.

1.

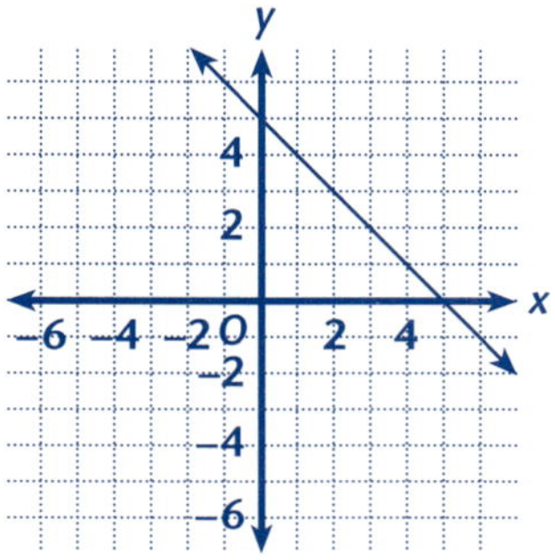

2.

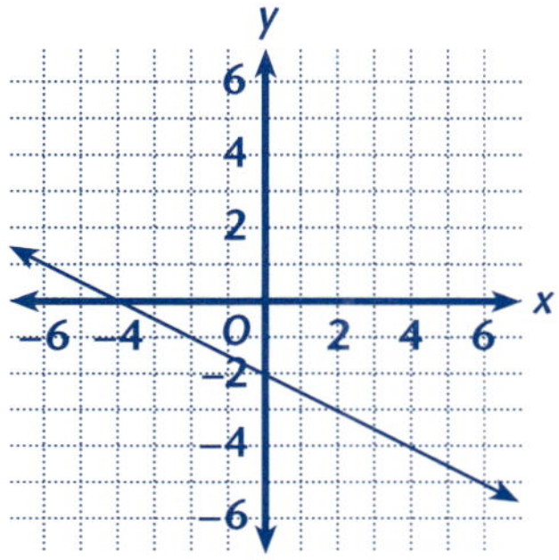

3.

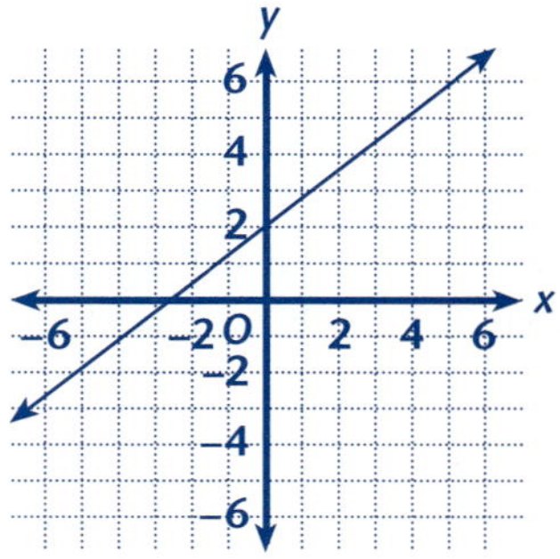

Determine the slope of the line through the two given points.

4. (−5, 2) and (3, 1) ____________

5. (0, −4) and (6, −3) ____________

6. (4, −4) and (−4, 4) ____________

7. (9, 2) and (−2, 1) ____________

8. (5, −6) and (−7, −4) ____________

9. (8, 10) and (−11, −9) ____________

Graph each line described below.

10. slope −1 through the point (2, 3)

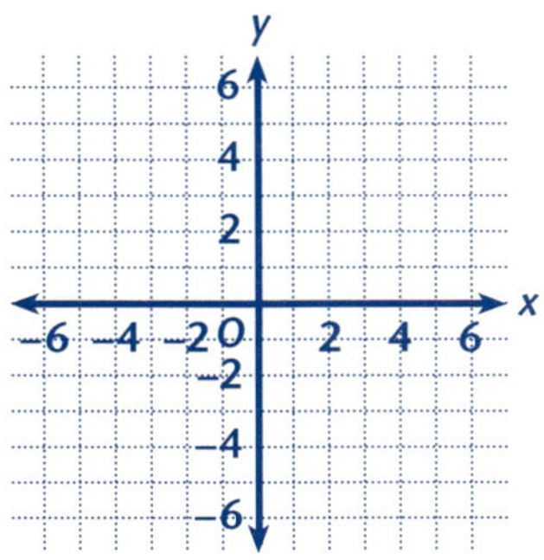

11. slope $\frac{1}{4}$ through the point (−2, 2)

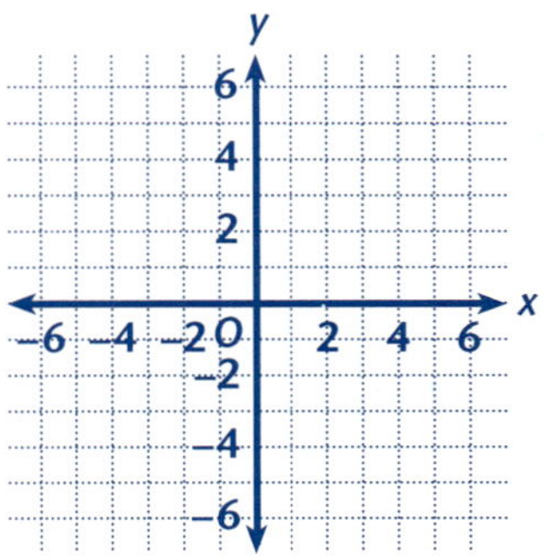

12. through the point (−3, −5) with slope 0

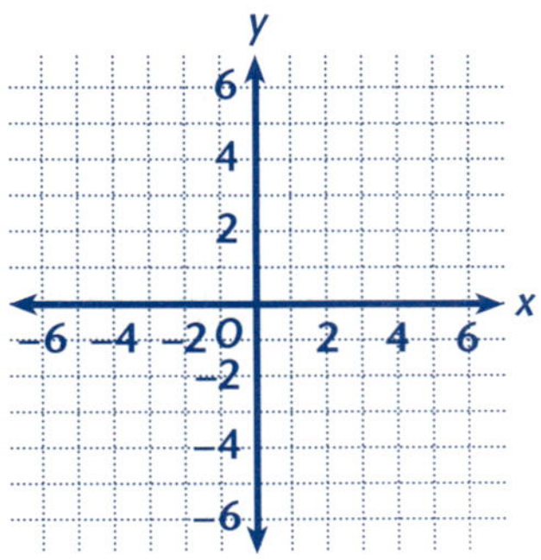

13. slope 0.5 through the point (−1, −4)

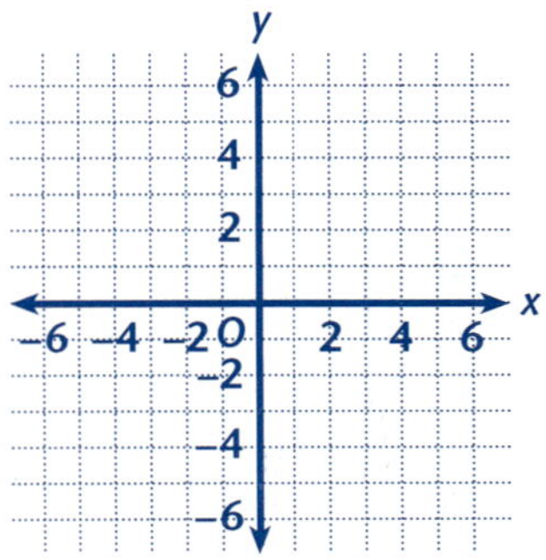

NAME ______________________ CLASS __________ DATE __________

Lesson Activity

2.1 Exploring Slope on a Geoboard

You can use a geoboard as a coordinate plane. Use the bottom of the geoboard as the x-axis and the left-hand side of the geoboard as the y-axis. You can then use rubber bands to show the graphs of segments with a given slope. The segment graphed at the right has a rise of 1 and a run of 2, so it has a slope of $\frac{1}{2}$.

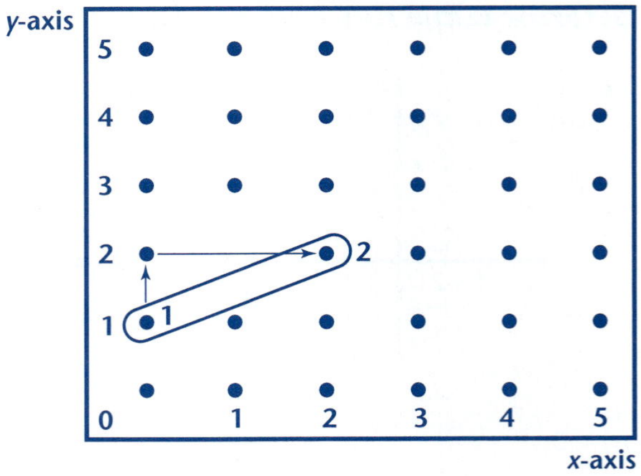

1. Use a geoboard to show at least two other segments with a slope of $\frac{1}{2}$.

Use a geoboard to show at least three segments with each slope given. Describe how you modeled the slope of the segments.

2. $-\frac{1}{2}$ ______________________

3. 1 ______________________

4. $\frac{1}{3}$ ______________________

5. $-\frac{1}{3}$ ______________________

6. 2 ______________________

7. $\frac{1}{4}$ ______________________

8. $-\frac{1}{4}$ ______________________

9. $\frac{2}{3}$ ______________________

10. $-\frac{2}{3}$ ______________________

11. $\frac{3}{4}$ ______________________

12. $-\frac{3}{4}$ ______________________

13. -3 ______________________

NAME ______________________ CLASS __________ DATE __________

Basic Skills Practice

Solving Equations by Addition or Subtraction, Part 2

Read each question and circle the best answer.

1. Coach Hamilton's team ran 20 laps fewer than Coach Faris's team. Coach Faris's team ran 45 laps. How many laps did Coach Hamilton's team run?

A 25
B 30
C 40
D 65

2. Sara won 9 more games of table tennis than her friend, Chuck. Chuck won 15 games. How many games did Sara win?

F 6
G 9
H 15
I 24

3. What is the solution to the equation?

$$n + 213 = 1974$$

A 213
B 1761
C 1974
D 2187

4. Solve the following equation for n.

$$n - 64 = 76$$

F 140
G 76
H 64
I 12

5. It takes Tina 16 minutes to jog 2 miles. Her friend Pamela can jog the same distance in 4 minutes less time. Which equation describes this situation?

A $n + 16 = 2$
B $4 + 2 = n$
C $16 - 4 = n$
D $16 - n = 4$

6. How long does it take Pamela to jog 2 miles?

F 20
G 18
H 12
I 6

Solve each problem.

7. The odometer in the school van read 794 on Monday. On Friday, it read 1031. How many miles was the van driven that week? __________

8. Sami bought a mountain bike for $329. She used money in a savings account and a paycheck for $138. How much had Sami saved for the mountain bike? __________

9. Acme Construction had $9498.53 in its checking account. After the weekly deposit, the account balance was $17,478.29. How much was the deposit? __________

LESSON 2.2 Linear Functions and Graphs

PREPARE

Pacing Guide
2 Days

Objectives and Overview
- Graph the equation of a line.

This lesson should be a review for most students. The slope-intercept form of a line is introduced. A method for writing an equation of a line when two points are known is given. A line from a real-world problem is represented by an equation in standard form and then graphed. The line of best fit, or regression line, and the correlation coefficient are introduced.

Vocabulary

slope-intercept form	line of best fit
y-intercept	regression line
standard form	correlation coefficient
x-intercept	

Assessing Prior Knowledge

Find the slope of the line that contains the given points.

1. (2, −3) , (5, 1) $\left[\frac{4}{3}\right]$

2. (3, −4), (1, −4) [0]

3. (12, 125), (8, 101) [6]

4. (4, 6), (−4, 2) $\left[\frac{1}{2}\right]$

Approach A: Using the Book

Opening Discussion

In the last lesson, a linear function was used to model movement at a constant rate of speed. What was the relationship between the constant rate of speed and the graph? **[The speed was the slope of the graph.]** Linear functions can be used to model other situations that involve a constant rate of change. Direct students' attention to the go-cart example. What is the constant rate of change in this example? **[$2 per lap]**

Brainstorming

Have students think of other real-world examples that involve a constant rate of change. Examples include the amount of water in a pool that is being filled by a hose, a train moving at a constant speed, and the cost of a bag of oranges that are $0.35 each. All of these examples can be modeled by linear functions.

TEACH

Approach B: Using Models

Opening Discussion

In the last lesson, a linear function was used to model movement at a constant rate of speed. What was the relationship between the constant rate of speed and the graph? **[The speed was the slope of the graph.]**

Linear functions can be used to model other situations that involve a constant rate of change. Consider this example: Suppose that a spool which holds electric cable weighs 25 pounds. The cable itself weighs 3 pounds per yard and is being unrolled. What is the constant rate of change in this example? **[3 pounds per yard]**

Brainstorming

Have students think of other real-world examples that involve a constant rate of change. Examples include the amount of water in a pool that is being filled by a hose, a train moving at a constant speed, and the cost of a bag of oranges that are $0.35 each. All of these examples can be modeled by linear functions.

Ongoing Assessment, page 64

The rate is equal to the slope.

The coordinates of the y-intercept are (0, 4).

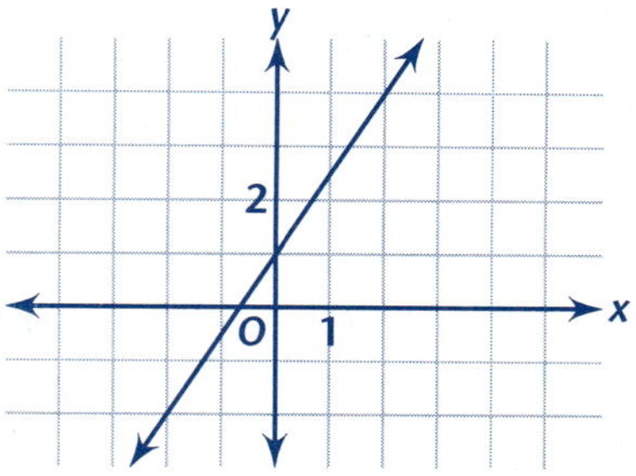

The Slope-Intercept Formula

Discuss the slope-intercept formula by using the go-cart example. You can either read and discuss the example from the book or work through the example at the chalkboard without reference to the book. Students should be able to produce the table, equation, and graph from the description of the rental charges.

Example 1

The slope and y-intercept are read from a linear equation and used to graph the equation. The equation must first be converted to slope-intercept form. Students should understand that the x-coordinate of a point on the y-axis is zero.

Additional Example 1

Graph the line $y = \frac{3}{2}x + 1$.

Ongoing Assessment, page 65

Try This

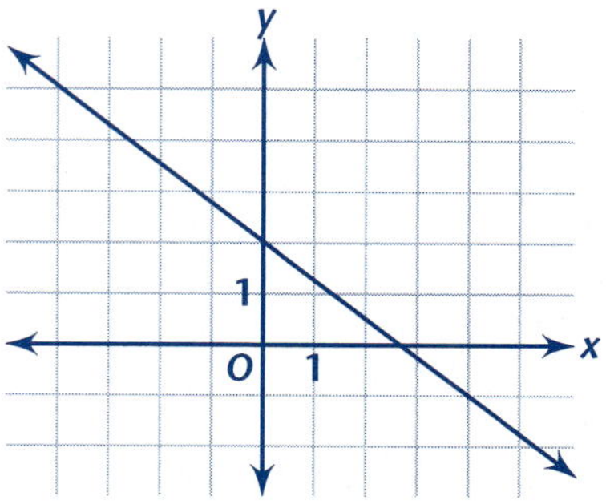

continued on top of next page

The Slope-Intercept Formula

Consider the electric-cable example. Work with students to make a table of values for the number of yards of cable on the spool and the weight of the spool (including the cable). Demonstrate how the equation $y = 3x + 25$ represents the situation. Graph the equation. Stress the connection between the rate of change and the slope of the graph. Define the ***y*-intercept,** and show how the initial weight of the spool is the y-intercept. Use this example to write the definition of **slope-intercept form.**

Cooperative Learning

This activity is especially well-suited to classes for which this material is largely review. Divide the class into four groups. Assign each group one of the topics listed below. Each group is responsible for researching and giving a short presentation on their topic. Extensive use should be made of graphs and other visuals made on large poster board. Each member of the group should present some part of the topic while the other members support the presentation. The group should encourage questions from the other students. Some of the topics require the group to give the rest of the students an exercise to do at their desk that the group then reviews. The visuals used for these presentations can be displayed around the room. (Compare pages 65–69.)

1. Graphing a Line in Slope-Intercept Form

Demonstrate the method for graphing a line when given its equation in slope-intercept form. Examples should be given for a line with positive slope and a line with negative slope. Write a problem on the board for the other students to solve. After sufficient time has passed, review the solution to the problem. If possible, show students how to check their solutions by making a table of values on a graphics calculator.

2. Writing an Equation From Two Points

Demonstrate the method for writing an equation of a line in standard form when given the coordinates of two points on the line. You

continued on bottom of next page

Approach A Using the Book, continued

Critical Thinking, page 65

As the slope increases, the graph becomes steeper in the positive direction. As the slope decreases, the graph flattens out to a horizontal line and then gets steeper in the negative direction. If the y-intercept increases, the graph moves up on the y-axis. If the y-intercept decreases, the graph moves down on the y-axis.

Writing an Equation From Two Points

Ask a volunteer to read the procedure for writing an equation from two points. Work together to write the procedure as a list of steps. This list can be used in Example 2.

Example 2

The coordinates of two points are identified from a real-world situation. These points are used to write a linear function that models the situation.

Why can this situation be modeled by a linear function? [**The electrician charges a constant rate per hour plus a base fee. This information fits the slope-intercept form of a linear equation.**] When determining what the x- and y-coordinates represent, it may help to remind students that the x-coordinate is the independent variable, or the input, while the y-coordinate is the dependent variable, or the output. In this case, Mario wants to input the number of hours he worked to determine how much to charge.

Additional Example 2

A cable TV company charges a basic monthly service amount plus a fee for each extra channel ordered. With 2 extra channels, Jose paid $21, and with 6 extra channels, Jose paid $33. Write an equation to describe the charge in terms of the number of extra channels ordered.
[$y = 3x + 15$]

TEACH

Approach B Using Models, continued

may want to write the procedure as a series of steps and follow the steps as you go through an example. Write a problem on the board for the other students to solve. After sufficient time has passed, review the solution to the problem.

3. Standard Form

Define the standard form of a linear equation and give examples. Show how an equation in standard form can be graphed by determining its intercepts. Write an equation in standard form on the board for the other students to graph. After sufficient time has passed, review the graph.

4. Line of Best Fit

Find a set of data that is nearly linear. Make a table and a scatter plot of the data. Define the line of best fit, or regression line. Demonstrate how to find the regression line with a graphics calculator. Write the equation of the regression line, and graph it on the same axes as the data.

Technology

To find the regression line and correlation coefficient on a graphics calculator, first enter

Have students use technology to model the line of best fit.

Standard Form

Explain that standard form is another way to write a linear equation. Pay special attention to the conditions on *A, B,* and *C*. Show students how the equation from Example 2 can be written in standard form. **[$12x - y = -25$]** Ask students for other examples of linear equations in standard form.

Example 3

A real-world example is given, and a linear equation is written for it in standard form. The equation is then graphed by finding the intercepts. Ask what the intercepts mean in the context of the problem.

Additional Example 3

Refer to Example 3. For each wreath Kayla makes, she needs 8 yards of ribbon, and for each basket she needs 3 yards of ribbon. She has 480 yards of ribbon and would like to use all of it.

A. Write an equation to model the number of yards of ribbon used. **[$3x - 8y = 480$]**

B. Graph the equation.

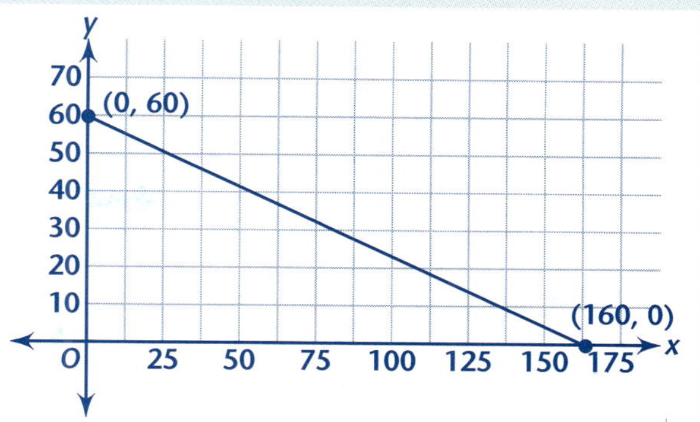

Inclusion Strategy

Categorizing Some students find real-world problems easier to solve if they categorize them.

Problems similar to Example 1 are often called constant-rate-of-change problems. These problems can be identified by a constant rate of change and an initial value, and they are modeled by a linear equation in slope-intercept form.

continued on top of next page

the data into the statistics lists. Then choose LinReg from the statistical calculate menu.

Teaching Tip

Here are a few ideas to help students focus on another group's presentation.

1. Write out a set of evaluation criteria, and have each student evaluate each presentation.
2. Give each student a one-question quiz after each presentation.
3. Require that students take notes on each presentation and hand them in at the end.

Real-World Examples

After the first three groups have given their presentations, it would be a good idea to give students practice on some real-world examples. You can lead the class through these two problems or have students work the problems in their groups or individually.

Write an equation to model each situation.

A. A veterinarian charges a fee for each visit plus a per-hour fee. A farmer paid the veterinarian \$130 for a 2-hour visit and \$280 for a 5-hour visit. **[$y = 50x + 30$]**

B. From 50 pounds of clay, a potter wishes to make plates that require 1.5 pounds of clay each and bowls that require 2.25 pounds each. **[$1.5x + 2.25y = 50$]**

The Correlation Coefficient

After the fourth group presentation, discuss the correlation coefficient. Represented by the letter r and found on the last line of the graphics calculator LinReg screen, this number is a measure of how well the line of best fit estimates the data. If r is close to 1 or -1, the fit is good. If r is closer to zero, the data is either not related, or its relationship can be better described by a different type of function. Have students look at the correlation coefficient for the data found by the fourth group and discuss what it says about the data. Point out that if the correlation coefficient is negative, the slope of the graph is also negative.

continued on bottom of next page

Approach A

Using the Book, continued

The type of problem in Example 3 is commonly called a linear-combination problem. These problems combine two or more types of things, such as wreaths and baskets, and are modeled by a linear equation in standard form.

A Line of Best Fit

Read through and discuss this problem thoroughly. While a line of best fit does not go through the points exactly, the line of best fit, or regression line, can be used to estimate the data. If possible, have students enter the data into graphics calculators to produce the scatter plot and regression line. The years should be entered as years after 1900, so 1920 should be entered as 20.

Closing the Lesson

Write the coordinates of two points on the chalkboard. Have students determine the equation for the line through the two points by using the slope formula and the slope-intercept form. Then have them enter the points in a graphics calculator with the x-coordinates in one list and the y-coordinates in the other. The regression line should be the same as the one that students found by hand. The correlation should be close to 1 or -1. Ask students to explain these findings. **[The line of best fit is a perfect model because there is exactly one line between any two points.]**

TEACH

Approach B

Using Models, continued

Closing the Lesson

Write the coordinates for two points on the chalkboard. Have students determine the equation for the line through the two points by using the slope formula and the slope-intercept form. Then have them enter the points in a graphics calculator with the x-coordinates in one list and the y-coordinates in the other. The regression line should be the same as the one students found by hand. The correlation should be close to 1 or -1. Ask students to explain these findings. **[The line of best fit is a perfect model because there is only one line between any two points.]**

ASSESS

Selected Answers

Odd-numbered Exercises 7–49 and odd-numbered Extra Practice

Assignment Guide

Core level students should complete Exercises 1–50.

Core-plus level students should complete Exercises 1–5, 7–16, 19–23, and 25–52.

Error Analysis

Students who have trouble with Exercises 18–23 probably need to review how to solve an equation with one variable. It might help students to draw a graph for Exercises 28 and 29. You may also need to review the equations for horizontal and vertical lines. In Exercise 43, students should realize that the possible solutions must be integer values.

Alternative Assessment

Further Research on the Olympics
Have students gather Olympic data from an almanac or other source. They should make a poster showing the data in table form and in a graph. The poster should include the equation and graph of the line of best fit. Students can find the correlation coefficient for the data and discuss its meaning as well.

Portfolio Activity Have students look through their homework from this chapter to find an example of one application that used the slope-intercept form of a linear equation and one application that used the standard form of a linear equation. Students should write the equation and draw the graph for each application and write a summary report to include in their portfolio. In the report they should discuss why that particular form of an equation was used.

RETEACH

Cooperative Learning Assign each student a partner. First have pairs of students alternate reading paragraphs from the lesson. Together they should make a review outline of the important material. Then have them examine and rework their responses to Exercises 35–40. Together they should write the equations for these problems and graph each equation. Both students should sign the completed work.

ENRICH

Interdisciplinary Connection

Business Linear programming is a method used by businesses to help them determine the number of items to manufacture given various constraints and goals. Example 3 on page 67 of the textbook illustrates how one goal is represented in a linear programming application. Namely, Kayla's goal is to make $600. Many other considerations may affect the decision of how many of each type of item to manufacture, such as the availability and cost of materials and the cost of labor or storage space. Each of these factors can also be represented on the graph. Ask students to find out more about linear programming and to present their findings to the class. This project can be extended to utilize the material on graphing systems of linear inequalities from Lesson 2.7.

The Correlation Coefficient

Have students do research to find out more about the correlation coefficient. What does it say about a set of data, and how is it used by mathematicians and statisticians? Students should include some sketches of scatter plots with various correlation coefficients. Can they make one whose correlation is close to zero? What does this reveal about the data?

NAME ______________________ CLASS ____________ DATE ____________

Practice

2.2 Linear Functions and Graphs

Use *x*- and *y*-intercepts to draw the graph of each line.

1. $2x + y = 2$

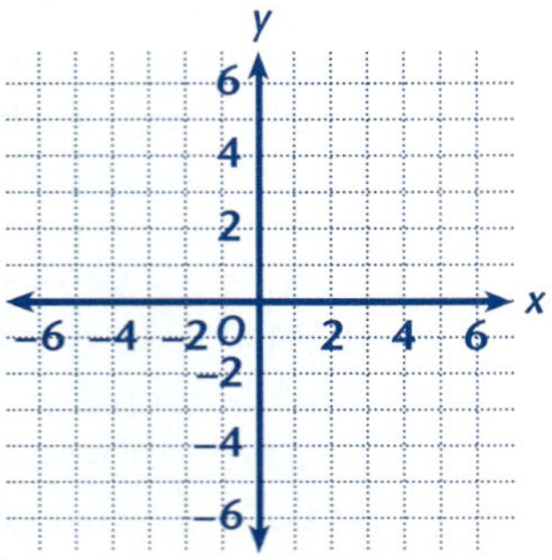

2. $x - 3y = 6$

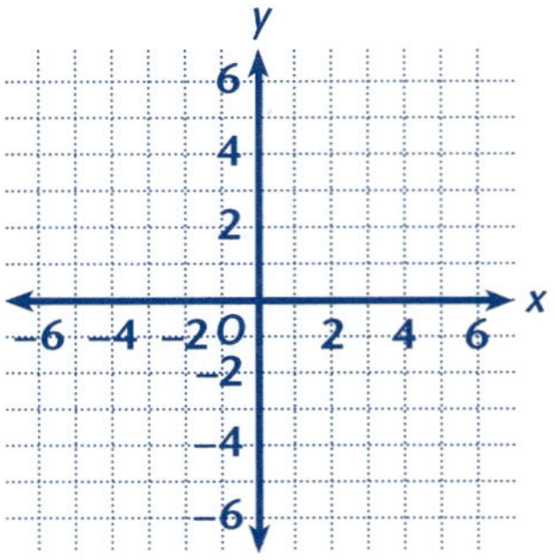

3. $5x - 1.5y = 7.5$

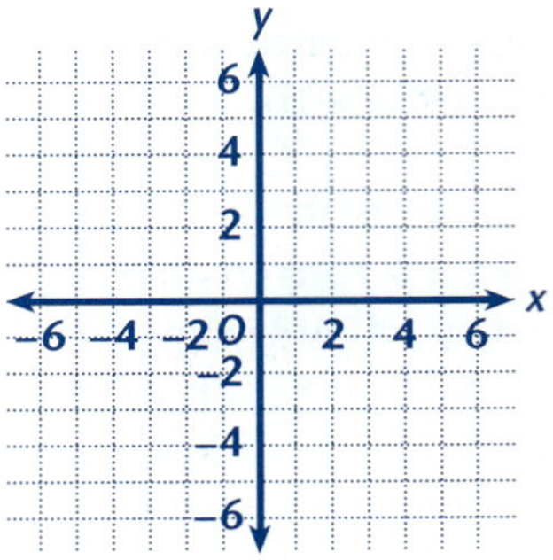

Convert the equation to slope-intercept form. Then state the slope and graph the equation.

4. $2x - y = 1$

____

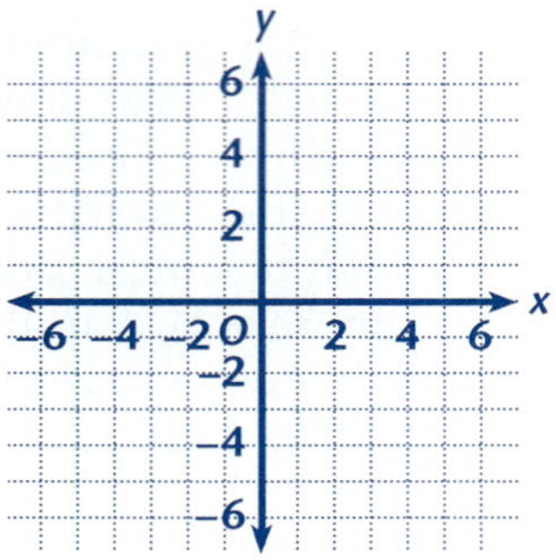

5. $x - 4y = 2$

____

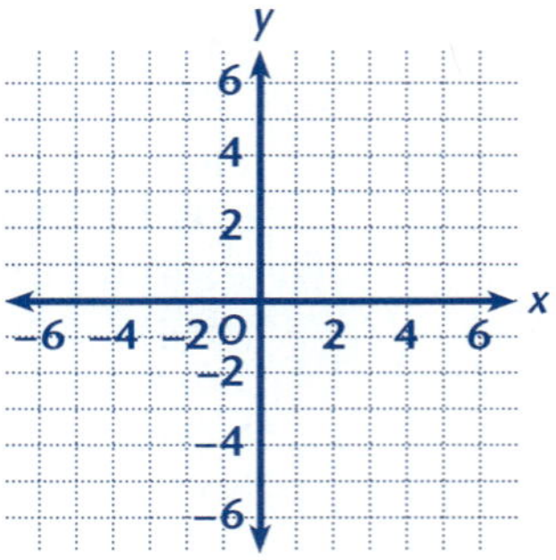

6. $3x - 6y = 6$

____

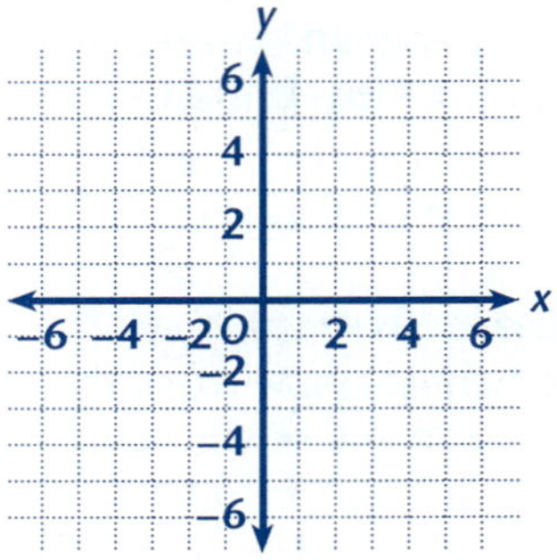

Write the equation of each line described.

7. slope 2 with y-intercept -1 ______________________

8. through $(-2, 1)$ with slope 0 ______________________

9. slope $\frac{1}{2}$ through $(5, 0)$ ______________________

10. through $(3, -5)$ with undefined slope ______________________

11. through $(-4, 2)$ and $(2, 6)$ ______________________

12. with x-intercept $(4, 0)$ and y-intercept $(0, 1)$ ______________________

NAME ______________________ CLASS ____________ DATE ____________

Lesson Activity

2.2 Geoboard Functions

You can use a geoboard as a coordinate plane to graph linear equations. For example, the graph of the equation $y = x + 1$ shown at the right was graphed by using a y-intercept of 1, and a slope of 1.

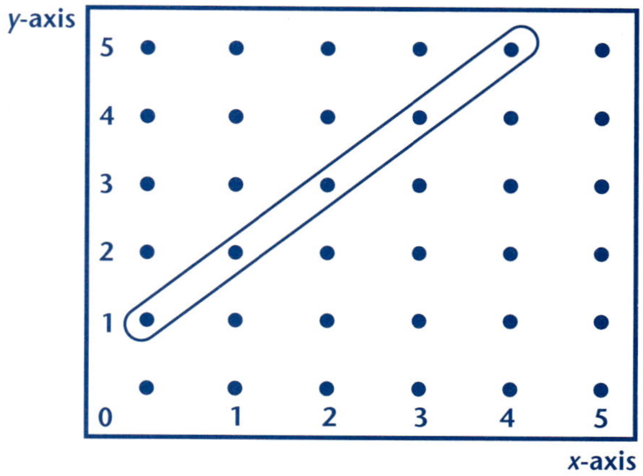

1. Use a geoboard to graph a line with a y-intercept of 2 and a slope of 1. What is the equation of this line?

____

2. Use a geoboard to graph a line with a y-intercept of 3 and a slope of $-\frac{1}{2}$. What is the equation of this line?

____

Give the y-intercept and slope of each line. Then use a geoboard to graph each line.

3. $y = x + 4$ ______________________

4. $y = -x + 5$ ______________________

5. $y = -x + 4$ ______________________

6. $y = \frac{1}{2}x + 1$ ______________________

7. $y = -\frac{1}{2}x + 2$ ______________________

8. $y = \frac{1}{3}x + 1$ ______________________

9. Use a geoboard to graph $y = 2x + 1$ and $y = 2x + 2$ on the same coordinate plane. What is different about the two lines? What is the same? ______________________

NAME ______________________ CLASS __________ DATE __________

Basic Skills Practice

Solving Equations by Multiplication or Division, Part 1

You can solve an equation by using the inverse operation to get the variable by itself. Multiply or divide both sides of the equation by the same quantity.

Example 1: $5n = 15$

$$\frac{5n}{5} = \frac{15}{5}$$

$$n = 3$$

Check: $5n = 15$

$$5 \times 3 \stackrel{?}{=} 15$$

$$15 = 15$$

Example 2: $\frac{n}{5} = 3$

$$\frac{5}{1}\left(\frac{n}{5}\right) = \frac{3}{1}\left(\frac{5}{1}\right)$$

$$n = 15$$

Check: $\frac{n}{5} = 3$

$$\frac{15}{5} \stackrel{?}{=} 15$$

$$3 = 3$$

When solving word problems:
- Define a variable.
- Write an equation.
- Solve and check.

Example 3: James bought 6 concert tickets for \$168. How much did each ticket cost?

Define → Let n represent the cost of one ticket.

Write → $6n = 168$

Solve → $\frac{6n}{6} = \frac{168}{6}$

$$n = 28$$

Check → $6 \times 28 \stackrel{?}{=} 168$

$$168 = 168$$

Each ticket cost \$28.

Solve each equation and check the solution.

1. $7n = 56$ ______________

2. $9n = 153$ ______________

3. $\frac{n}{9} = 15$ ______________

4. $\frac{n}{12} = 7$ ______________

5. $7n = 217$ ______________

6. $57n = 171$ ______________

7. $\frac{n}{32} = 18$ ______________

8. $\frac{n}{15} = 8$ ______________

Answer the following questions in the space provided.

9. During recent road construction, traffic on the freeway was slowed to $\frac{1}{5}$ the normal speed limit. If traffic was moving at about 12 mph, what is the normal speed limit? ______________

10. At the end of the semester, Soon Yi typed 96 words per minute. This was 4 times faster than her speed at the beginning of the course. What was her rate at the beginning of the course? ______________

LESSON 2.3

Graphing Systems of Equations

PREPARE

Pacing Guide
2 Days

Objectives and Overview
- Create tables, and graph systems of equations.
- Find an approximate solution to a system by inspecting a graph.

This lesson introduces students to systems of linear equations. Students learn that the solution to the system is the point of intersection of the graphs of the equations. The solution is estimated by using a graph or table of values and is checked by substitution.

Vocabulary
common solution
system of equations
approximate solution

Assessing Prior Knowledge

Write an equation for each of the following statements:

1. The cost of renting a car is $35 per day plus $0.12 per mile. [$c = 0.12m + 35$]
2. Aaron spent a total of $65 for tapes at $5 each and CDs at $8 each. [$5t + 8c = 65$]
3. During the summer Joe earned $600 and received an allowance of $10 per week. [$a = 10w + 600$]

Approach A: Using the Book

Opening Discussion
Mathematics can be used to help people make decisions. Read, or have a student read the opening problem. Should the Smiths buy the VCR or get cable TV? Since there are two different options, they must be represented by two different equations. This leads to what is called a *system of equations*.

Brainstorming
Have students think of other situations in which they must choose between two options. Examples include renting an apartment or buying a condo with a monthly maintenance fee and renting a car from one company that has unlimited mileage or from one that has a basic rate and a per-mile charge.

Making a Table and Graph
Work through the Smiths' problem by leading the students through the solution while you draw the table and graphs on the overhead or chalkboard.

continued on top of next page

TEACH

Approach B: Using Models

Opening Discussion
Mathematics can be used to help people make decisions. Suppose that you are trying to decide whether to buy a VCR or to order cable TV to watch movies. The VCR costs $185, and if you purchase it, you plan to spend $20 per month on tapes. Cable costs $45 for installation and $34 per month. Since there are two different options, they must be represented by two different equations. This leads to what is called a *system of equations*.

Brainstorming
Have students think of other situations in which they must choose between two options. Examples include renting an apartment or buying a condo with a monthly maintenance fee and renting a car from one company that has unlimited mileage or from one that has a basic rate and a per-mile charge.

continued on bottom of next page

Approach A — Using the Book, continued

You can work through the Smiths' problem by drawing graphs on the overhead.

Point out that the solution can be seen from the table of values or from the graphs; only one is necessary to solve the problem. Students should be able to make the graphs fairly quickly by hand by using the slope and *y*-intercept. Have students substitute 10 into both equations to check that the coordinates of the intersection point satisfy both equations and thus form the common solution to the equations.

Use **Transparency 6** for discussion.

Ongoing Assessment, page 73

Cable

D: m≥0; c≥45

VCR

D: m≥0; R: c≥185

The equations are graphed only in the first quadrant because cost is nonnegative.

Critical Thinking, page 73

A VCR would be the best choice because the cost would be less after 10 months.

TEACH

Approach B — Using Models, continued

Cooperative Learning

Divide the class into groups of four. This activity allows students to discover that the point of intersection of two lines is a common solution to the equations of both lines.

Give each group an erasable graph, a yardstick, and two small cardboard tiles, one with the letter *T* and the other with the letter *R*. (See Lesson 2.1, Using Models, for directions on making an erasable graph.) These will be used for the rabbit and turtle race in Example 2.

One student uses a straightedge to draw two lines on the graph whose intersection has integer-valued coordinates. The coordinates of the intersection are determined, and then two other students each determine the equation of one of the lines. The fourth student substitutes the coordinates of the point of intersection into the two equations in order to check the solution. This procedure should be repeated at least once, with students exchanging roles.

Finally, the group should work together to summarize their findings. Choose several groups to read their summary to the entire class.

Defining Common Solution and System of Equations

Explain that the coordinates of the point of intersection of the two lines form the common solution of the equations. The equations together are called a system of equations. The solution to a system of equations is the ordered pair, (x, y), that satisfies both equations. This ordered pair is also the point of intersection of the graphs of the two equations.

Using Verbal Models to Set Up Systems of Equations

When solving a real-world problem by using a system of equations, it can be difficult to organize the information and represent it with equations. Encourage students to translate the information into a simple English sentence before writing the equations. Also, students should state clearly what their variables

A System of Equations

Define a system of equations and relate the definition to the problem about the Smith family. Students should understand that a solution to the system is a pair of coordinates that satisfies each equation in the system. The solution is the intersection of the graphs of the equations in the system.

Example 1

A common solution to two linear equations is found by examining either a hand-drawn or calculator-generated graph.

The equations are first changed to slope-intercept form so that they can be entered into the calculator or easily graphed by using the slope and y-intercept.

Teaching Tip

Technology The lines from Example 1 can be graphed in the standard viewing window. Caution students to use parentheses when entering fractions such as $\frac{1}{2}$. Students can use the trace or intersect feature of a graphics calculator to find the point of intersection.

Additional Example 1

Use a graphics calculator or graphics software to graph the equations $3x - 5y = 1$ and $x + y = -5$. Find the common solution by examining the graph. **[(−3, −2)]**

Ongoing Assessment, page 74
Try This

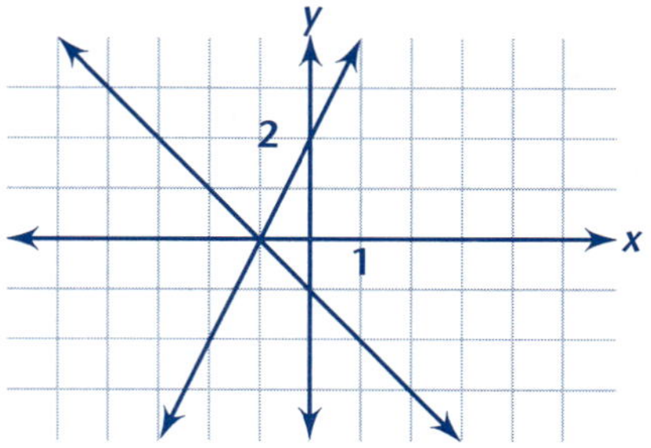

[−1, 0]

continued on top of next page

represent. Solve the problem in the example below by using these techniques. (Compare Example 1 on page 74.)

Aternative Example 1

Betty's Formals rents tuxedos for \$65 and dresses for \$80. One weekend Betty rented 32 outfits for a total of \$2275. How many of each type of outfit did she rent? **[19 tuxedos and 13 dresses]**

First, organize the information.
Let x represent the number of tuxedos rented.
Let y represent the number of dresses rented.

Write equations by using verbal models.

number of tuxedos	+	number of dresses	= 32
x	+	y	= 32
cost for all tuxedos	+	cost for all dresses	= 2275
$65x$	+	$80y$	= 2275

This gives the following system:

$$\begin{cases} x + y = 32 \\ 65x + 80y = 2275 \end{cases}$$

Explain that the brace symbol is used to indicate that the equations form a system.

Solve the system. Change the two equations to slope-intercept form, and graph them by hand. Estimate the point of intersection to find the solution. Make a table of values for the two equations as well in order to show students how they could also have used a table to solve the system.

Check the solution. Substitute the values into the original equations.

If time permits, have students solve the Smiths' problem by using verbal models and the procedure illustrated above. (Compare pages 72 and 73.)

Alternate Example 2

This example uses a physical model to demonstrate common solutions.

Have students place a T marker at 20 inches on the yardstick and an R marker at 0 inches, which represents the starting point. Tell them that they will model a race between a turtle and a rabbit. Have them suppose that the rabbit can run five times faster than the turtle. Explain that the

continued on bottom of next page

Approach A

Using the Book, continued

Example 2

A real-world problem is modeled by a system of linear equations and then solved.

Stress the steps that are used in the solution of this example: organize, write, solve, and check.

Additional Example 2

In his greenhouse, Irving wants to raise two types of plants. He has room for a total of 42 trays of seedlings. One type of seedling costs \$3 per tray, and the other costs \$4 per tray. How many trays of each type of seedling can he buy with \$148? **[20 of the \$3 trays and 22 of the \$4 trays]**

Ongoing Assessment, page 75

Changing the equations to slope-intercept form lets you make the table or graph by finding the y-intercept and then using the rise and run given by the slope to find another point.

Approximate Solution

Point out that while the table of values in Example 2 gives the exact solution to the problem, the graph alone can be used only to estimate the solution. This estimate is called an *approximate solution.*

Sometimes you can verify that these solutions are in fact exact either by making a table of values or by using a check, as was done in Example 2. However, sometimes an estimate is not the exact solution. In this case, the approximate solution can be shown to be reasonable by checking it in the original equation.

TEACH

Approach B

Using Models, continued

respective distances of the turtle and rabbit from the starting point during the race are represented by the following system:

$$\begin{cases} d = t + 20 \\ d = 5t \end{cases}$$

Tell students to use the model to discover the point at which the rabbit will catch up with the turtle. Challenge the groups to be the first to discover the answer. Start the race! **[The rabbit will catch up with the turtle at 25 inches from the starting point.]**

Now have students graph the system on erasable graphs, or have them use graphics calculators to verify their answer.

Approximate Solution

Point out that while the table of values in the formalwear example gives the exact solution to the problem, the graph alone can give only an estimate of the solution. This estimate is called an *approximate solution*. Sometimes you can verify that these solutions are exact either by making a table of values or by using a check, as was done in that example. However, many times an estimate is not an exact solution. In this case, the approximate solution can be shown to be reasonable by checking it in the original equation.

Alternate Example 3

Use the system below to demonstrate the method for finding approximate solutions by using erasable graphs or graphing technology. Then check the reasonableness of the solution by using substitution.

Solve by graphing.

$$\begin{cases} 3x - y = -4 \\ 6y = 2x - 21 \end{cases} \quad \mathbf{[(-2.81, -4.44)]}$$

If using erasable graphs, have students draw dotted lines from the point of intersection to the x-axis and y-axis in order to approximate the solution.

Example 3

An approximate solution to a system of equations is found by graphing.

Both graph paper and a graphics calculator are shown. If using graph paper, students may choose to find the intercepts of the graph and use these to draw the graph rather than rewriting the equations in slope-intercept form. Point out the brace notation used to indicate a system of equations.

Additional Example 3

Solve by graphing.

$\begin{cases} 3x - y = -4 \\ 6y = 2x - 21 \end{cases}$ **[(−2.81, −4.44)]**

Closing the Lesson

Give students some extra practice with systems of equations by having them do Exercises 6 and 7 from page 77 in class. Have two students work at the chalkboard so that all students can check their work.

Teaching Tip

Technology The lines from the previous example can be graphed in the standard viewing window of a graphics calculator. Caution students to use parentheses when entering fractions such as $\frac{1}{2}$. Students can use the trace or intersect feature of a graphics calculator to find the point of intersection. If using the trace feature, you may want to show students how to adjust the window so that the x-coordinates of the points are incremented by tenths. One way to do this is to choose the ZDecimal window.

Closing the Lesson

Give students some extra practice with systems of equations by having them do Exercises 6 and 7 from page 71 in class. Have two students work at the chalkboard so that all students can check their work.

ASSESS

Selected Answers

Odd-numbered Exercises 7–43 and odd-numbered Extra Practice

Assignment Guide

Core level students should complete Exercises 1–41.

Core-plus level students should complete Exercises 1–9 and 16–48.

Error Analysis

Problems can occur when students do not draw graphs carefully. Require students to use graph paper and straightedges when making graphs by hand. Allowing students to check their graphs with a graphics calculator can help them obtain more accurate results. For Exercise 30, students should realize that two scatter plots are needed, one with years and education degrees and one with years and engineering degrees.

Alternative Assessment

A Problem About Speed and Distance
Have students solve the following problem: One train leaves station A and heads for station B at a rate of 50 mph. Another train leaves station B at the same time and heads for station A at a rate of 30 mph. If the two stations are 320 miles apart, how many hours will elapse before the trains pass each other? **[4 hours]**

Portfolio Assessment Exercises 25–28 tie together all of the major concepts of the first three lessons in this chapter: finding slope, writing equations for lines, and using graphs to solve systems of equations. Ask students to write these problems in report form, including a graph, and to add it to their portfolio.

RETEACH

Have students find out about the cable costs in their area. The costs should include an installation fee and a monthly charge. Also have them find the cost of a VCR and the charges for renting videos. Students should then write a system of equations to represent this situation, solve it, and interpret the results. What other factors might influence whether someone gets a VCR or cable TV? Students can present their findings to the class.

ENRICH

Technology

Graphics calculators can be used to find approximate solutions to systems of equations with a high degree of accuracy. After entering the two functions in the Y= list and graphing them, you can repeatedly zoom in on the intersection point until the desired accuracy is reached. Or, after approximating the intersection of the graph, add an additional function Y1 − Y2. This function should equal zero at the point of intersection of the first two functions. Students can use the table features of the calculator, starting with the approximated x-value from the graph and using increasingly small increments, to search for the point of intersection. The x-coordinate of the point of intersection will be between the two x-values for which the sign of Y1−Y2 changes (Y1 − Y2 must be zero between these two x-values).

Interdisciplinary Connection

Sociology Systems of linear equations can be used to examine many situations in which gender differences are apparent. For example, a system of linear equations can be used to predict the year in which the number of female doctors will be equal to the number of male doctors. Have students gather some data that covers at least 10 years and is given separately for men and women. They should make two scatter plots of the data and find the lines of best fit. Then have them approximate the intersection of these lines. Finally, students should write their results in the form of a report and share their report with the class.

NAME ____________ CLASS ________ DATE ________

Practice

2.3 Graphing Systems of Equations

Solve by graphing. Round approximate solutions to the nearest tenth. Check algebraically.

1. $\begin{cases} x + y = 4 \\ 2x - y = 5 \end{cases}$ ____________

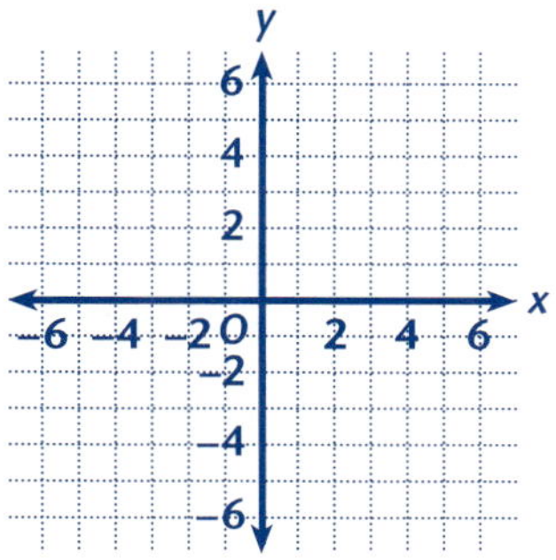

2. $\begin{cases} x + y = 0 \\ 3x - 2y = 10 \end{cases}$ ____________

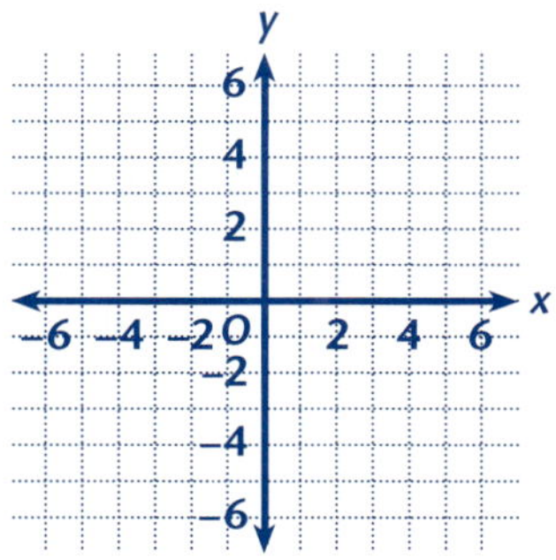

3. $\begin{cases} 2x + y = 7 \\ x + y = 3 \end{cases}$ ____________

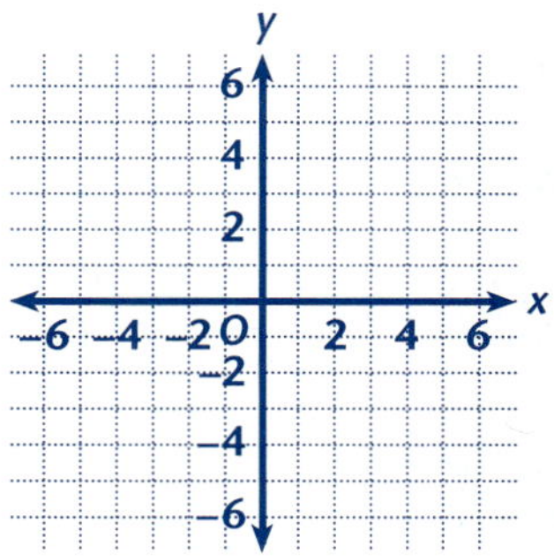

4. $\begin{cases} x + y = 1 \\ 2x - 2y = 6 \end{cases}$ ____________

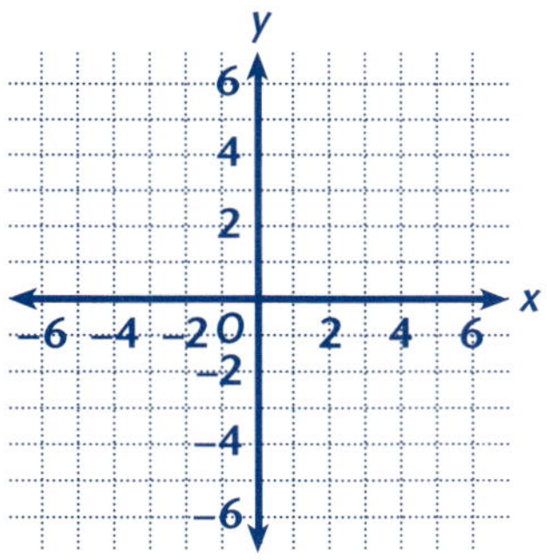

5. $\begin{cases} 3x + 2y = 9 \\ 4x - y = 1 \end{cases}$ ____________

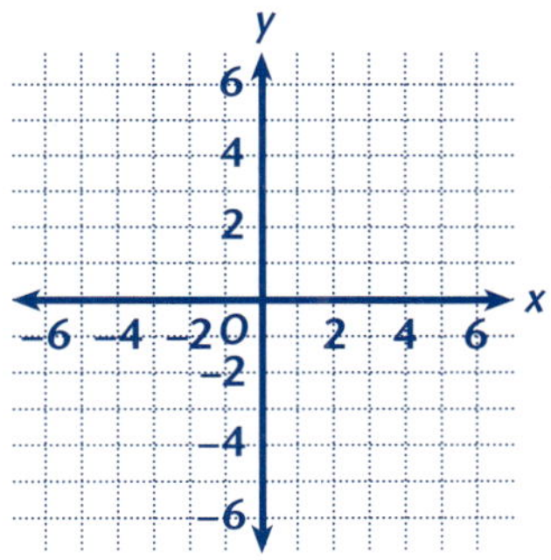

6. $\begin{cases} 3x - 4y = -4 \\ 6x - 2y = 1 \end{cases}$ ____________

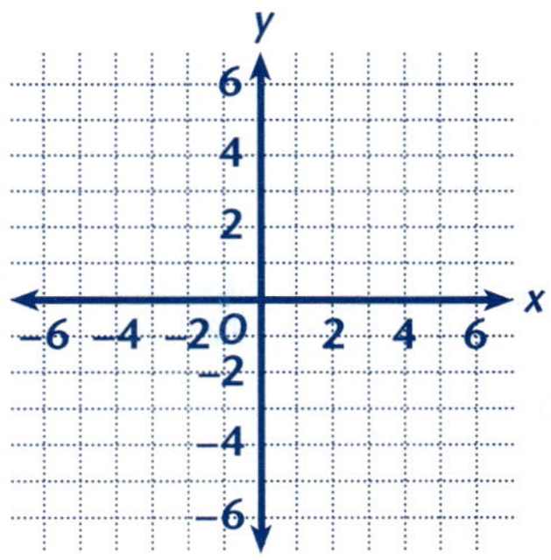

Algebraically determine whether the point (1, 4) is a solution for each pair of equations.

7. $\begin{cases} y = x + 3 \\ y = 2x - 2 \end{cases}$ ____________

8. $\begin{cases} y = 3x + 1 \\ y = -x + 5 \end{cases}$ ____________

9. $\begin{cases} y = 5x - 1 \\ y = -2x + 6 \end{cases}$ ____________

Algebraically determine whether the point (−2, 6) is a solution for each pair of equations.

10. $\begin{cases} y = x + 8 \\ y = 4x - 2 \end{cases}$ ____________

11. $\begin{cases} x + y = 4 \\ x - y = 8 \end{cases}$ ____________

12. $\begin{cases} 4x + y = -2 \\ y = -x + 4 \end{cases}$ ____________

NAME ____________________ CLASS __________ DATE __________

Lesson Activity

2.3 Exploring Solutions to Systems of Linear Equations

Graph the system of equations on the grid provided. Use your graph to answer the questions.

$$\begin{cases} y = -3x + 4 \\ y = x - 4 \end{cases}$$

1. What is the solution of this system of equations?

____

2. What figure is formed by the two given lines and the line $y = -5$?

____

3. What is the area of the figure? ____________________

4. Which of the two given lines forms a large right triangle with the y-axis and the line $y = -5$?

____

5. What is the area of this right triangle? ____________________

6. What figure is formed by the two given lines and the line $y = 1$?

____

7. How is this figure related to the one you identified in Exercise 2?

____

8. Use the grid provided to find two integers whose sum is -8 and whose difference is 6.

____

NAME ______________________ CLASS ____________ DATE ____________

Basic Skills Practice

Solving Equations by Multiplication or Division, Part 2

Read each question and circle the best answer.

1. The school will sponsor up to 16 hockey teams. There are 112 students signed up for hockey. How many teams of 8 players can be formed?

A 8
B 14
C 16
D 18

2. The Hamiltons rent a cabin at the lake for 12 days at $29 a day. They rent a boat for $50 a day. What is the total amount of rent for the cabin?

F $948
G $600
H $348
I $398

3. Solve the equation $14n = 350$.

A 14
B 25
C 350
D 4900

4. Solve $\frac{n}{27} = 754$.

F 20,358
G 279
H 27.9
I 27

5. On a highway, the fastest car is traveling 75 mph. This is 1.5 times as fast as the slowest car. At what speed is the slowest car traveling?

A 135 mph
B 90 mph
C 50 mph
D 15 mph

6. Eight adult tickets to a movie cost $56. Children's tickets cost $3.50 each. What is the cost of each adult ticket?

F $3.50
G $5.75
H $7.00
I $10.50

Solve each problem.

7. Three of the toll booths on the Sam Houston Tollway were closed during rush hour this morning. There were at least 17 cars in line at each of the other 4 tollbooths. How many cars were waiting to pay tolls? ____________

8. If it takes an average of 45 seconds for each car to go through the toll booth, how long will the 17th car wait in line for its turn to go through the toll booth? Give your answer in minutes; remember that there are 60 seconds in a minute. ____________

9. If the three toll booths stayed closed all day and 4268 cars went through the remaining toll booths, how many cars went through each of the 4 toll booths? Assume an equal number of cars went through each open toll booth. ____________

LESSON 2.4 *Exploring* Substitution Methods

PREPARE

Pacing Guide
2 Days

Objectives and Overview
- Graph a system of equations, and estimate a solution from the graph by inspection.
- Use substitution to find an exact solution to a system of equations.

A system of linear equations whose solution is not an integer value is examined to see that the graph cannot be used to find the exact solution. The substitution method is explored and used to solve some real-world applications.

Vocabulary
substitution method

Assessing Prior Knowledge

Solve each equation for x.

1. $8x + 2(3x - 5) = 60$ [5]

2. $3(2x + 4) - 5x = 10$ [−2]

Approach A: Using the Book

Opening Discussion
When you use the graph of a system of equations to find the solution, is your answer exact? Most integer-valued solutions can be found by using a graph and then checking the solution in the original equations. But what if the solution is $\left(\frac{2}{3}, \frac{3}{7}\right)$? Would you be able to get this exact answer by examining a graph? This lesson introduces a method of finding the exact solution to a system of linear equations.

Setting Up the Exploration
Teacher-Guided Explorations If you wish, you can guide students through the textbook explorations. You can draw the graph for Exploration 1 on the overhead or chalkboard. At each stage of the exploration, invite student input and encourage discussion. Students should keep a record of the exploration responses in their notebook. For a greater level of participation, you can ask student

TEACH

Approach B: Using Models

Opening Discussion
When you look at the graph of a system of equations to find the solution, is your answer exact? Most solutions with integer values can be found by using a graph and then checking the solutions in the original equations. But what if the solution is $\left(\frac{2}{3}, \frac{3}{7}\right)$? Would you be able to get this exact answer by examining a graph? This lesson introduces a method of finding the exact solution to a system of linear equations.

Exploration 1
If possible, students should use graphics calculators to graph the functions. Write the problem and do the work on the chalkboard or overhead. Encourage student participation and discussion. (Compare Exploration 1 on page 79.)

Graph the following system of equations:

$$\begin{cases} 10x + 2y = 10 \\ x = 3 \end{cases} \quad [(3, -10)]$$

volunteers to work the problems for Explorations 2 and 3 at the chalkboard while you give directions.

Cooperative-Learning Groups Divide the students into groups of four. Each group should use one piece of paper and take turns working out the exploration steps. All group members should be involved in a discussion of the activity as the exploration proceeds. Each student should sign the completed report.

Exploration 1

Students use a graph to solve a system of equations that includes a vertical line. After discovering that the graphic solution is not exact, students learn that they can use substitution to find the exact solution.

You may want to review the equations and graphs of horizontal and vertical lines before beginning this exploration.

Ongoing Assessment, page 79

2. While we can make a good guess, the *exact* value of the y-coordinate is not obvious.

4. Substitute the known value into the other equation, and solve for the unknown value.

6. substitution; $(3, -2.5)$

On Their Own

Have students solve this system of equations at their seats. You can have one student complete the problem at the board for all students to check their work.

$$\begin{cases} 2x - 3y = 5 \\ y = -2 \end{cases} \qquad \left[\left(-\frac{1}{2}, -2\right)\right]$$

Exploration 2

Students solve one equation for one variable and substitute the variable expression into the other equation to solve a system of equations. They also solve the first equation for the second

continued on top of next page

To do this, students must begin by solving the first equation for y.

$$\begin{cases} y = -5x + 5 \\ x = 3 \end{cases}$$

Have them approximate the solution. Is this exact? How can we use algebra to find the exact solution? Substitute 3 for x in $10x + 2y = 10$, and solve the resulting equation for y.

Exploration 2

Graph the following system of equations:

$$\begin{cases} 2x + y = 3 \\ 5x - 2y = 1 \end{cases} \quad \left[\left(\frac{7}{9}, \frac{13}{9}\right)\right]$$

To do this, students must first solve each equation for y.

$$\begin{cases} y = -2x + 3 \\ y = \frac{5}{2}x - \frac{1}{2} \end{cases}$$

Show students that because the y-coordinates must be equal at the point of intersection, the two right-hand sides of the equations can be set equal to one another. Solve this equation to get the x-value of the intersection, and then substitute it into either original equation to solve for y.

Exploration 3

Write the problem and do the work on the chalkboard or overhead. Encourage student participation and input. (Compare Exploration 3 on page 80.)

How can the method above be improved? Solving both equations for y is time-consuming. Demonstrate how this can be avoided by solving only the first equation for y.

$$\begin{cases} y = -2x + 3 \\ 5x - 2y = 1 \end{cases}$$

Since the y-values are equal at the point of intersection, the right-hand side of the top equation can be substituted for y in the second equation: $5x - 2(-2x + 3) = 1$. Have students solve this equation and then find the corresponding y-value to see that this process gives the same solution for the system.

continued on bottom of next page

Approach A

Using the Book, continued

variable and use substitution to verify that the results are the same.

Students who have had little or no experience in substituting one variable expression into another may need to have this step demonstrated. Encourage all students to use parentheses when substituting a variable expression. Point out that to find the value of the other variable in Steps 1b and 2b, the value of the known variable should be substituted into one of the two original equations.

Ongoing Assessment, top of page 80

4. Solve the equation for one of the variables, substitute that expression for the variable in the other equation, and solve. When one value is known, substitute it into either of the original equations to find the other value.

Exploration 3

A system of equations is solved by substitution. Point out that it is best to solve for a variable whose coefficient is 1. Students can solve for the other variable, but the computations are more complex.

Ongoing Assessment, middle of page 80

4. Solve the equation for one variable, and substitute that expression for the variable into the other equation. Solve this equation to find one value. Substitute the known value into either of the original equations to find the value of the other variable.

On Their Own

Have students solve this system of equations at their desk. You can have one student complete the problem at the board for all students to check their work.

$$\begin{cases} 2x + y = 3 \\ 5x - 2y = 1 \end{cases} \quad \left[\left(\frac{7}{9}, \frac{13}{9}\right)\right]$$

TEACH

Approach B

Using Models, continued

On Their Own

Have students solve this system of equations at their desk. You can have one student complete the problem at the board for all students to check their work.

$$\begin{cases} 2x - 3y = 5 \\ y = 2x \end{cases} \quad \left[\left(-\frac{5}{4}, -\frac{5}{2}\right)\right]$$

Cooperative Learning

Writing an Algorithm Have students work in pairs to write an algorithm, or series of steps, that can be followed to solve a system of equations by substitution. They should test the algorithm by solving Exercises 7 and 11 from page 77 by substitution. (These problems were solved previously by graphing.)

Student Reports

Choose several students to read their algorithm to the class. Ask other students to critique the algorithms. Work together as a class to refine the algorithms.

Alternate Application

The value of investment stocks is often given in fractional form. Lead students through the solution to the problem below. Show students how to organize their work by defining variables, writing the equations, solving the system by substitution, and checking their solutions. (Compare page 81.)

One share of BBN Telephone and 5 shares of Western Auto Parts are worth $37.75. Another investor has 4 shares of BBN Telephone and 12 shares of Western Auto Parts worth $99. What is the value of one share of each stock? **[BBN: $8.25; Western: $5.50]**

Journal Entry

Ask students to write a procedure for using substitution to solve a system of linear equations. What do you look for in the system in order to determine whether substitution is an appropriate method? **[The coefficient of one of the variables in one of the equations should be one.]**

Applications

Cooperative Learning You can have students read and discuss the applications with a partner. Point out that students should check the answers by substituting them into both of the original equations. Then have students make up a similar problem and solve it by using the substitution method. Have students share their problems with the class.

Closing the Lesson

Have students work with their partner to solve the problem below by using the substitution method. Ask one pair to do their work at the board so that all students may assess their own work. (Compare page 81.)

Two shares of Diversity Software and 1 share of Coburn Electronics are valued at $21. Six shares of Diversity Software and 5 shares of Coburn Electronics are worth $87. What is the value of one share of each stock? **[Diversity: $4.50; Coburn: $12]**

ASSESS

Selected Answers

Odd-numbered Exercises 5–39 and odd-numbered Extra Practice

Assignment Guide

Core level students should complete Exercises 1–39.

Core-plus level students should complete Exercises 1–9 and 14–43.

Error Analysis

When using the substitution method, incorrect answers are often caused by algebraic error. Remind students to put parentheses around the expression being substituted, and review the Distributive Property. Students may also have difficulty finding the second value when the first value is a fraction. You may need to have students review addition and subtraction of fractions.

Alternative Assessment

Report on the Substitution Method
Have students write a report that explains the substitution method to a student who was absent when this lesson was presented. The report should include examples and an explanation of why the substitution method may be preferred over the graphing method for some systems of equations.

Portfolio Activity Have students solve Exercise 29 by using both the graphic and substitution methods, and have them include this work in their portfolio.

RETEACH

Have students make up numeration problems similar to Exercises 21 and 22. They can then exchange papers and solve the problems.

ENRICH

Systems of Three Linear Equations

Show students how substitution can be used to solve systems of three linear equations in three variables if one of the values is known. Use Exercises 40 and 41 as examples. Discuss the geometric interpretation of these problems. The three equations can be represented by planes in space, and the solution is the common intersection of these planes. For example, the two walls that form a corner of the room and the floor are three planes with one point of intersection, the bottom corner of the room.

Interdisciplinary Connection

Stocks Have students obtain a copy of the stock market pages from a local newspaper. After learning how to read the pages, students should use them to make up and solve some problems similar to the application. These problems can be shared with and solved by other students in the class.

NAME ______________________ CLASS __________ DATE __________

Practice

2.4 Exploring Substitution Methods

Solve and check each system by using the substitution method.

1. $\begin{cases} y = 2x \\ 2x + y = -12 \end{cases}$ ______________

2. $\begin{cases} y = x + 3 \\ 3x + y = 11 \end{cases}$ ______________

3. $\begin{cases} y = 2x + 1 \\ x + 3y = 31 \end{cases}$ ______________

4. $\begin{cases} x + y = 3 \\ 4x - 2y = 18 \end{cases}$ ______________

5. $\begin{cases} 2x - 3y = -25 \\ 3x + y = 1 \end{cases}$ ______________

6. $\begin{cases} x - 2y = -3 \\ 4x - 3y = 8 \end{cases}$ ______________

7. $\begin{cases} 2x + 5y = -7 \\ 3x - y = -2 \end{cases}$ ______________

8. $\begin{cases} 2x - y = -11 \\ 3x - 6y = 6 \end{cases}$ ______________

9. $\begin{cases} x - y = 4 \\ 2x - 3y = 6 \end{cases}$ ______________

10. $\begin{cases} 3x + y = -3 \\ x - 3y = 11 \end{cases}$ ______________

11. $\begin{cases} x - y = 10 \\ 2x + 3y = 5 \end{cases}$ ______________

12. $\begin{cases} 2x + y = 2 \\ 4x - 2y = -4 \end{cases}$ ______________

13. $\begin{cases} x = y - 4.2 \\ 2x - 3y = -9 \end{cases}$ ______________

14. $\begin{cases} -2x - y = 4 \\ x + y = -3 \end{cases}$ ______________

15. $\begin{cases} 4x - y = -2 \\ -8x + y = 3 \end{cases}$ ______________

16. $\begin{cases} 2x - 2y = 2 \\ 3x + y = -9 \end{cases}$ ______________

First graph each system and estimate the solution. Then use the substitution method to get an exact solution.

17. $\begin{cases} y = 2x \\ 2x + y = 7 \end{cases}$ ______________

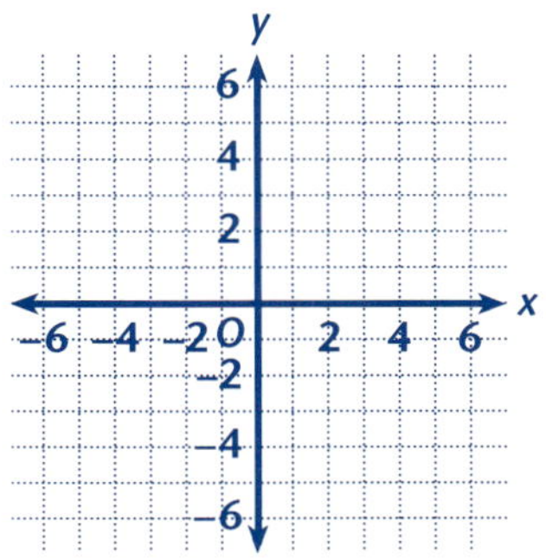

18. $\begin{cases} x + y = 2 \\ x - 2y = 0 \end{cases}$ ______________

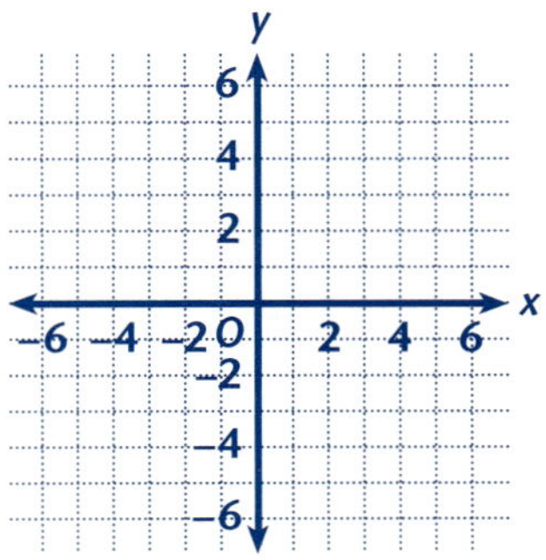

NAME ______________________ CLASS ____________ DATE ____________

Lesson Activity

2.4 Practicing the Substitution Method

Follow the steps below to solve the system of equations. $\begin{cases} 3x + 2y = 80 \\ x - y = 45 \end{cases}$

1. Solve $x - y = 45$ for x. ______________________

2. Substitute your answer for x in $3x + 2y = 80$. ______________________

3. Solve this new equation for y. ______________________

4. Use this value of y to find x in the equation you wrote in Exercise 1. ______________________

5. Prove that your values for x and y are correct by substituting them in the original equations.

____

Solve the system of equations by using three different methods. Then answer the questions. $\begin{cases} 5x - y = -0.5 \\ x + y = 15.5 \end{cases}$

6. Solve by creating a table of values for at least 3 values of x and y.

x	$y = 5x + 0.5$	$y = 15.5 - x$

7. Solve by graphing on the grid provided.

8. Solve by using the substitution method.

____

9. Which method do you prefer to use to solve systems of linear equations and why?

____

10. Which method do you think gives the most exact answer and why?

____

____

NAME ______________________ CLASS ____________ DATE ____________

Basic Skills Practice

Solving Two-Step Equations, Part 1

If an equation contains more than one operation, undo the addition or subtraction first. Then undo the multiplication or division. These steps are performed in the reverse order of the order of operations.

Example 1:

$$2n - 7 = 29$$
$$2n - 7 + 7 = 29 + 7$$
$$\frac{2n}{2} = \frac{36}{2}$$
$$n = 18$$

Check:

$$2n - 7 = 29$$
$$2 \times 18 - 7 \stackrel{?}{=} 29$$
$$36 - 7 \stackrel{?}{=} 29$$
$$29 = 29$$

Example 2:

$$\frac{n}{2} + 4 = 10$$
$$\frac{n}{2} + 4 - 4 = 10 - 4$$
$$2 \times \frac{n}{2} = 6 \times 2$$
$$n = 12$$

Check:

$$\frac{n}{2} + 4 = 10$$
$$\frac{12}{2} + 4 \stackrel{?}{=} 10$$
$$6 + 4 \stackrel{?}{=} 10$$
$$10 = 10$$

Solve and check.

1. $2n + 5 = 9$ ____________

2. $4n + 6 = 18$ ____________

3. $3n - 2 = 10$ ____________

4. $3n - 6 = 24$ ____________

5. $\frac{n}{2} + 5 = 10$ ____________

6. $\frac{n}{6} + 6 = 8$ ____________

7. $\frac{n}{8} + 4 = 6$ ____________

8. $10n - 15 = 45$ ____________

9. $6n + 11 = 19$ ____________

10. $5n + 5 = 40$ ____________

Write an equation and solve.

11. Abraham Lincoln's Gettysburg Address delivered in 1863 refers to 1776 as *"Four score and seven years ago."* Write an equation and solve for the number of years, a score(s), represents.

Equation: ____________ s: ____________

12. Four times a number plus seven is fifteen. Write an equation and solve.

Equation: ____________ n: ____________

13. In January, 12 tune-ups were done at Harold's Garage. The number of snow tire changes was 6 less than 4 times the number of tune-ups. How many snow tire changes were made?

Equation: ____________ x: ____________

LESSON 2.5 The Elimination Method

PREPARE

Pacing Guide
2 Days

Objectives and Overview
- Find an exact solution to a system of equations by using the elimination method.

Students are introduced to the elimination method with an example in which the coefficients of one of the variables are opposites. Later, students learn to multiply one or both equations by constants in order to produce equations in which the coefficients of one of the variables are opposites.

Vocabulary
elimination method

Assessing Prior Knowledge

Simplify each expression.

1.	$3(2x)$	**[$6x$]**
2.	$-2(4x)$	**[$-8x$]**
3.	$-4(-3x)$	**[$12x$]**
4.	$-1(-10x)$	**[$10x$]**

Approach A: Using the Book

Opening Discussion
"In mathematics there are often different ways to solve the same problem. What methods of solving systems of equations have you learned so far?" **[graphing and substitution]** "How do you know which method to use?" **[Choose the more efficient method.]**

Brainstorming
Have students brainstorm some of the advantages and disadvantages of the graphing and substitution methods of solving systems of linear equations. Examples of advantages for the graphing method include the following: it is helpful as a check of an algebraic solution, and if equations are already in slope-intercept form, it can be a fast method; disadvantages are that it does not give an exact solution and that the equation must be rewritten in slope-intercept form. Advantages of the substitution method include the following: it can be very fast, and

TEACH

Approach B: Using Models

Opening Discussion
In mathematics there are often different ways to solve the same problem. What methods of solving systems of equations have we learned so far? **[graphing and substitution]** How do we know which method to use? **[Some methods are more efficient than others for particular types of problems.]**

Brainstorming
Have students brainstorm some of the advantages and disadvantages of the graphing and substitution methods of solving systems of linear equations. Examples of advantages for the graphing method include the following: it is helpful as a check of an algebraic solution, and if equations are already in slope-intercept form, it can be a fast method; disadvantages are that it does not give an exact solution and that the equation must be rewritten in slope-intercept form. Advantages for the substitution method include the following: it can be very fast, and

it gives exact answers; disadvantages are that one of the variables must have a coefficient of 1 or −1 and that it can be cumbersome if fractions, decimals, or very large numbers are involved.

Car Rental

Read the car-rental example, and work with the class to write the system of equations. Ask students why neither graphing nor substitution is an appropriate method for solving this system. **[Graphing: solution is probably not an integer value, so the solution given by this method would be approximate; substitution: none of the variables has a coefficient of 1 or −1.]** In the exploration that follows, students will develop an alternative method to solve this system.

Exploration

This exploration leads students through the elimination method for a system of equations in which the coefficients of one of the variables are opposites.

Then students extend the method to a case in which one of the equations must be multiplied by a constant before a variable can be eliminated.

Teaching Tip

Students may feel uncomfortable adding the terms of the two equations together. Stress that you are adding equal amounts (each side of the second equation) to each side of the first equation.

Ongoing Assessment, page 84

4. Addition of opposites allows one of the variables to be eliminated when the terms on each side of the equal signs are added. The resulting equation can then be solved for the remaining variable.

Using the Elimination Method

Explain to students that the method they learned in the exploration is called the elimination method. Lead the class through the

continued on top of next page

it gives exact answers; disadvantages are that one of the variables must have a coefficient of 1 or −1 and that it can be cumbersome if fractions, decimals, or very large numbers are involved.

Using an Algorithm

Explain to students that an algorithm is a sequence of steps that can be followed to achieve a desired result. Algorithms are very important in computer science. Programmers must be able to write precise instructions that can be followed exactly. In this exercise, students will follow an algorithm for the elimination method. Do not teach this method beforehand. Students should work together to interpret each step and follow the directions to solve the problems by elimination.

Exploration

Students should work in groups of three to complete this exploration. Before beginning, remind students of the Multiplication Property of Equality and the Addition Property of Equality. Demonstrate how the first can be used to multiply each term of an equation by a constant and how the second can be used to add the corresponding sides of two equations together. (Compare the exploration on page 84.)

Copy the summary of the elimination method from page 86 onto the chalkboard or overhead. Each group should go through each step to solve the two systems below. One student should read the step, one student should describe what needs to be done to the system, and one student should complete the step.

System 1: $\begin{cases} x + 2y = 6 \\ 5x - 2y = 30 \end{cases}$ **[(6, 0)]**

System 2: $\begin{cases} 2x - 3y = 12 \\ 4x + y = 10 \end{cases}$ **[(3, −2)]**

Student Reports

Ask one or more groups to report their findings to the class. What is the purpose of Step 2? **[The purpose is to alter the equations so that when you add them, one variable will be eliminated.]** What is the purpose of Step 3? **[to find the value of one of the variables]** When would you want to use this method? **[when the**

continued on bottom of next page

Approach A Using the Book, continued

solution to the car-rental problem. Discuss why x was the variable chosen to be eliminated. Have students check the solution in each of the original equations.

Ongoing Assessment, page 85

The x can be eliminated more easily because the coefficient of x in the second equation is a multiple of the coefficient of x in the first equation.

Example 1

Elimination is used to solve a system of equations.

In this case, both equations must be multiplied by a constant in order to obtain terms whose coefficients are opposites. Explain that the desired coefficients are the least common multiple of the original coefficient and its opposite. Since careless errors are common in these problems, stress to students that they should always check their answer in the original equations.

Additional Example 1

Solve the system from Example 1 by eliminating the y-terms.

Ongoing Assessment, page 85 Try This

$a = 4, b = 3$

Journal Entry

Read and discuss the summary box. Have students rewrite these steps in their own words and include them in their journal.

Example 2

Four different systems of equations are examined, and a method is chosen to solve each one. While different answers are possible, stress that some methods are more efficient than others in particular situations. The skill of

TEACH

Approach B Using Models, continued

coefficients of one variable can easily be converted so that they are opposites]

On Their Own

Have students solve the system below. Give them the hint that they need to multiply one equation by a negative number.

$$\begin{cases} 3x - 5y = 1 \\ 2x - 10y = 6 \end{cases} \quad \left[\left(-1, -\tfrac{4}{5}\right)\right]$$

Multiplying Both Equations by Constants

Use the problem below to demonstrate that both equations may sometimes need to be multiplied by a constant in order to use the elimination method. (Compare Example 1 on page 85.) Then have students rework this problem by eliminating the other variable.

$$\begin{cases} 2x + 3y = 7 \\ 3x + 5y = 11 \end{cases} \quad [(2, 1)]$$

Cooperative Learning

Which Method? Have students work in their group to solve each problem below. One student should use the elimination method, one student should use the substitution method, and one student should use the graphing method. If some students find that the system is too difficult to solve using their method, they can stop after a fair attempt. At the end of a problem, each student should report their findings to the rest of the group, and the group should come to a consensus about which method was the best for that problem. After completing all of the problems, the group should write a summary describing which types of problems each method is most efficient for.

System 1: $\begin{cases} 3x + 4y = 12 \\ 2x - 4y = 3 \end{cases}$

System 2: $\begin{cases} 4x + 3y = 1 \\ y = 5 - x \end{cases}$

System 3: $\begin{cases} 29x + 130y = -332 \\ 19x - 54y = 200 \end{cases}$

System 4: $\begin{cases} 4x + 3y = 5 \\ 2x - 5y = 12 \end{cases}$

choosing the appropriate method is an important one for students to master. It is a good idea to take the time to examine the system and chose the most efficient method any time a system of equations is solved.

Ongoing Assessment, page 86
Try This

The choice of method may vary. Sample answer: I chose the substitution method because the coefficient of y in the first equation is 1.

Additional Example 2

Choose a method for solving the following systems of equations. Explain why you choose each method.

System 1: $\begin{cases} 3x + 4y = 12 \\ 7x - 4y = 3 \end{cases}$

System 2: $\begin{cases} 2x + 7y = 1 \\ 5x = y - 3 \end{cases}$

System 3: $\begin{cases} \frac{2}{3x} - \frac{5}{7}y = 13 \\ \frac{1}{2}x + 8y = \frac{12}{5} \end{cases}$

System 4: $\begin{cases} 4x + 3y = 5 \\ 2x - 5y = 12 \end{cases}$

[There may be more than one correct method for each set of equations.

1. **Use the elimination method because $4y$ and $-4y$ are opposites.**
2. **Use the substitution method. Because y has a coefficient of 1, the equivalent expression can be substituted.**
3. **Use technology. Because the coefficients are fractions, ordinary algebra or manual graphing techniques would be cumbersome.**
4. **Use the elimination method. No term has a coefficient of 1 or -1.]**

Critical Thinking, page 86

AND; the solution must satisfy the first equation AND the second equation.

[There may be more than one correct method for each set of equations.

1. **Use the elimination method because $4y$ and $-4y$ are opposites. $\left(5, -\frac{3}{4}\right)$**
2. **Use the substitution method. Because y has a coefficient of 1, the equivalent expression can be substituted. $(-14, 19)$**
3. **Use technology. Because the coefficients are large, ordinary algebra or manual graphing techniques would be cumbersome. $(2, -3)$**
4. **Use the elimination method. No term has a coefficient of 1 or -1. $\left(-\frac{19}{13}, \frac{122}{13}\right)$]**

Closing the Lesson

Ask each group to read their summary. Did any of the groups disagree on the best method? Invite questions and discussion.

ASSESS

Selected Answers

Odd-numbered Exercises 11–43 and odd-numbered Extra Practice

Assignment Guide

Core level students should complete Exercises 1–44.

Core-plus level students should complete Exercises 1–15, 17, and 19–45.

Error Analysis

Careless errors are common when using the elimination method. Encourage students to do their work neatly and in an organized fashion. Also, students should arrange the equations so that the like terms are aligned. Students may not know how to interpret the results of Exercise 19 because the elimination method produces the equation $0 = 0$. Explain that this equation is true for any value of x, and hence there are an infinite number of solutions to the system. Have students graph these equations to see that they are the same line.

Alternative Assessment

Finding the Best Method Ask students to write a brief description of each method that they have learned for solving systems of equations. They should then look through their homework to find one problem that is most easily solved with each method. Why was each method chosen?

Portfolio Activity Have students review their solutions to Exercises 18–23. They should choose two systems of equations that they solved by different methods to include in their portfolio. In addition, students should explain why they solved each problem by using that method.

RETEACH

Students can use the summary box on page 86 to help them write their own steps for solving a system of equations by elimination. Ask several students to share their steps with the class so that all students can check their work. Then have students use their steps to solve the systems in the even-numbered Communicate exercises from page 87. They should number each step and follow the steps carefully.

ENRICH

Using Matrices and Technology

Another way to solve systems of equations is to use matrices. A matrix is an array of numbers. Matrices can be multiplied, added, and subtracted. Although students will not learn these skills until later, they can use graphing calculators to perform these operations.

To solve a system of linear equations by using matrices, first change the equations into standard form.

$$\begin{cases} 5x + 6y = 14 \\ 3x + 5y = 7 \end{cases}$$

Then write the coefficient matrix,

$$A = \begin{bmatrix} 5 & 6 \\ 3 & 5 \end{bmatrix},$$

and enter it as matrix A in the graphics calculator. Then enter the matrix

$$B = \begin{bmatrix} 14 \\ 7 \end{bmatrix}$$

in the calculator. On the home screen, type $A^{-1}B$ (read "A inverse B"), and press ENTER. The solution is displayed as a matrix with x first and y second: $x = 4$ and $y = -1$.

This method is especially useful when the coefficients are fractions, decimals, or large numbers. Have students solve system D from Example 2 by using this method.

Interdisciplinary Connection

History The person named Gauss mentioned in Exercise 17 is Carl Friedrich Gauss, a famous French mathematician. Have students do research to find out more about Gauss. They should write their findings in the form of a report and present it to the class.

NAME ______________________ CLASS ____________ DATE ____________

Practice

2.5 The Elimination Method

Solve each system of equations by elimination, and check your solutions.

1. $\begin{cases} x + y = 10 \\ x - y = 2 \end{cases}$ ____________

2. $\begin{cases} x - 3y = -13 \\ -x + 4y = 15 \end{cases}$ ____________

3. $\begin{cases} 4x + 2y = 2 \\ 5x + 2y = 4 \end{cases}$ ____________

4. $\begin{cases} 5x - 2y = 3 \\ 5x + y = -9 \end{cases}$ ____________

5. $\begin{cases} 3x + 2y = 8 \\ 2x - 6y = 42 \end{cases}$ ____________

6. $\begin{cases} x + 2y = -1 \\ 4x + 3y = -9 \end{cases}$ ____________

7. $\begin{cases} 3x + 4y = 22 \\ x - 5y = -37 \end{cases}$ ____________

8. $\begin{cases} 2x + y = -1 \\ 7x - 5y = 5 \end{cases}$ ____________

9. $\begin{cases} 2x - 5y = -19 \\ 3x + 2y = 0 \end{cases}$ ____________

10. $\begin{cases} 4x + 3y = 27 \\ 3x + 4y = 29 \end{cases}$ ____________

11. $\begin{cases} 2x - 3y = 6 \\ x + 3y = 12 \end{cases}$ ____________

12. $\begin{cases} x + 5y = -13 \\ 2x - 5y = -19 \end{cases}$ ____________

13. $\begin{cases} x + y = 0 \\ 2x - y = 12 \end{cases}$ ____________

14. $\begin{cases} 2x + 2y = -8 \\ 3x - 3y = 18 \end{cases}$ ____________

15. $\begin{cases} x - 2y = 11 \\ 2x + 2y = -8 \end{cases}$ ____________

16. $\begin{cases} 2x - y = -6 \\ 2x + 3y = 14 \end{cases}$ ____________

Solve each system of equations by using any method.

17. $\begin{cases} 2x - 7y = 5 \\ 5x - 4y = -1 \end{cases}$ ____________

18. $\begin{cases} 5x - 7y = -16 \\ 4x - 2y = -20 \end{cases}$ ____________

19. $\begin{cases} x = 5y \\ 3x - 4y = 11 \end{cases}$ ____________

20. $\begin{cases} 0.3x - 0.5y = 2 \\ 0.3x + y = -4 \end{cases}$ ____________

21. $\begin{cases} \frac{1}{2}x + \frac{1}{3}y = 4 \\ -3y = x - 22 \end{cases}$ ____________

22. $\begin{cases} \frac{2}{3}x + \frac{1}{4}y = 2 \\ 2x + y = 4 \end{cases}$ ____________

23. $\begin{cases} 7x - 2y = 6 \\ 9x - 3y = 6 \end{cases}$ ____________

24. $\begin{cases} 0.7x - 1.6y = 15 \\ 0.3x + 0.4y = 1 \end{cases}$ ____________

NAME ______________________ CLASS ____________ DATE ____________

Lesson Activity

2.5 Applying the Elimination Method

Match the system of linear equations with the method of elimination that could be used to solve the system.

	Method of Elimination		System of Linear Equations
______	**1.** Multiply each side of the first equation by 2 to eliminate y. Solve for x to get $x = 3$.	**A.**	$\begin{cases} 2x + 3y = 6 \\ -4x + 6y = 36 \end{cases}$
______	**2.** Multiply each side of the first equation by 2 to eliminate x. Solve for y to get $y = 4$.	**B.**	$\begin{cases} x + 2y = -2 \\ 4x - 4y = 40 \end{cases}$
______	**3.** Multiply each side of the first equation by -4 to eliminate x. Solve for y to get $y = -4$.	**C.**	$\begin{cases} 3x + 2y = 5 \\ 2x - 4y = 14 \end{cases}$
______	**4.** Multiply each side of the first equation by -3 and each side of the second equation by 2 to eliminate x. Solve for y to get $y = 3$.	**D.**	$\begin{cases} 2x - 5y = -5 \\ 3x + 2y = 21 \end{cases}$

Write equations to represent each problem. Then use the graphing, substitution, or elimination method to solve each problem. Explain why you chose the method you used.

5. Sandy bought shares of stock in two different companies. She bought 5 shares of company A's stock and 10 shares of company B's stock for a total of \$150. If each share of company A's stock was \$3 more than 2 shares of company B's stock, how much did each of the stocks cost?

____

____

____

6. BabySitters, Inc. charges an hourly rate for babysitting services. They also charge a fixed rate for each child being cared for at the same time. The last two bills that they sent to one of their customers were as follows:

Bill 1: 3 children for 5 hours................Total due: \$27.50
Bill 2: 3 children for 9 hours................Total due: \$43.50

How much does BabySitters, Inc. charge per hour and per child?

____

____

____

NAME ______________________ CLASS ____________ DATE ____________

Basic Skills Practice

Solving Two-Step Equations, Part 2

Read each question and circle the best answer.

1. Four times a number decreased by 3 equals 25. Find the number.

A 3
B 7
C 22
D 28

2. There were 220 people, including spectators and players, at a game. If there are 10 times as many spectators as players, which equation would you use to find the number of players?

F $10n = 220$
G $n = \frac{220}{10}$
H $10n + n = 220$
I $220 + 10n = n$

3. Six less than twice a number is 14. Find the number.

A 6
B 10
C 14
D 20

4. Melissa bought three pairs of jeans for $144. One pair cost $9 more than each of the other two. What did she pay for each pair?

F $36, $36, $36
G $45, $48, $51
H $45, $45, $54
I $48, $48, $48

5. Arthur earned $10 more than twice what his sister Susan earned. Together they earned $250. How much did each one earn?

A $125, $125
B $140, $110
C $150, $100
D $170, $80

6. $2n + 6 = 28$. What is the value of n?

F 10
G 11
H 12
I 14

Solve each problem.

7. Angela ordered 263 sugar cones for the Amy's Ice Cream on West Sixth Street. The number of sugar cones was 8 more than 3 times the number of waffle cones. How many waffle cones were ordered? ____________

8. Maggie makes sundaes and smoothies at Amy's. Last week, she made 96 smoothies. This was 4 fewer than 5 times the number of sundaes she made. How many sundaes did Maggie make last week? ____________

9. A chirping cricket can act as a thermometer. Thirty-seven more than one-fourth of the number of chirps in a minute is about equal to the air temperature in degrees Fahrenheit. What is the temperature, in degrees Fahrenheit, if a cricket is chirping at about 40 chirps per minute? ____________

LESSON 2.6

Exploring Non-Unique Solutions

PREPARE

Pacing Guide

2 Days

Objectives and Overview

- Identify consistent and inconsistent systems of equations.
- Compare slopes and intercepts of independent and dependent systems.

The terms *inconsistent system*, *consistent system*, *independent system*, and *dependent system* are taught. Students discover the properties of each type of system.

Vocabulary

inconsistent system	dependent system
consistent system	independent system

Assessing Prior Knowledge

Graph the system of equations.

$$\begin{cases} x - y = 2 \\ x + y = 3 \end{cases}$$

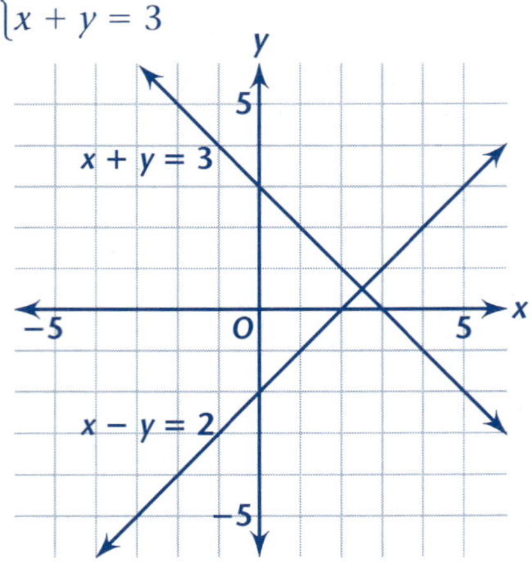

Approach A: Using the Book

Opening Discussion

Two lines in a plane can intersect at one point. In this case, the corresponding system of equations has one solution. What else could occur? **[The lines could be parallel or could be the same (concurrent).]** What are the solutions to the system in these cases?

Inconsistent Systems

Use the marriage-age example to introduce the concept of an inconsistent system of equations. Read or have a student volunteer read through the textbook explanation carefully. Stress that an inconsistent system has no solution, that its graph is a pair of parallel lines, that the equations have the same slope, and that if you try to solve the system algebraically, you get an equation that is false.

Use **Transparency 7** for discussion.

TEACH

Approach B: Using Models

Opening Discussion

Two lines in a plane can intersect at one point. In this case, the corresponding system of equations has one solution. What else can occur? **[The lines could be parallel or the same (concurrent).]** What are the solutions to the system in these cases?

Cooperative Learning

Divide the class into groups of four. Give each group an erasable graph (see Lesson 2.1, Using Models, for directions on making an erasable graph). One student should use a straightedge to draw the lines on the graph. The calculations will be easier if the lines drawn have y-intercepts and intersections with integer-valued coordinates. One student should write the equations for the lines. Another student should solve the system of equations algebraically, and the fourth student should keep track of the group's results and write a summary statement. Each student should be involved at each stage of the process, and the students should change roles for the next exploration.

Critical Thinking, page 91

Answers will vary. Sample answer: Suppose that two airplanes, A and B, are traveling from one city to another city at 500 mph each. For the sake of safety, airplane B will fly 10 miles behind airplane A. Then the distance traveled by airplane A and airplane B are given by the following system:

$$\begin{cases} d = 500t \\ d = 500t - 10 \end{cases}$$

Attempting to solve this system leads to a false statement.

$500t = 500t - 10$

$0 = -10$ False

Therefore, the system is inconsistent.

Consistent Systems

Explain that a system of equations with at least one solution is called a **consistent system.** Students will see in the explorations that these systems can be **dependent** or **independent.**

Inclusion Strategies

English Language Development Have students write definitions for the common usage of the terms *consistent, inconsistent, dependent,* and *independent.* Discuss their meanings and how they relate to the mathematical definitions of the terms. This exercise can help students remember the mathematical meaning of these words.

Exploration 1

This exploration introduces the concept of a dependent system.

Students should understand that a dependent system is a type of consistent system. You may need to review the point-slope formula for the equation of a line prior to beginning this exploration.

Have students work in pairs or small groups on this exploration. Each student should work individually to draw the graphs, write the equations, and solve the system algebraically. The members of the group should then meet to discuss their results and answer the questions.

continued on top of next page

Inconsistent Systems

This exploration allows students to examine the case in which the two lines are parallel. Have students work in their groups and use the erasable graph to investigate the following questions, which you may write on the overhead or chalkboard. (See pages 90 and 91.)

Draw two lines that are parallel. Compare their slopes and y-intercepts. **[The slopes are the same; the y-intercepts are different.]**

Find the equations for both lines. Solve the system of equations. What happens? **[You get a false statement.]** How many solutions does this system of equations have? **[none]**

Inconsistent Systems

Ask several groups to report their findings to the class. Explain that the type of system they investigated is called an **inconsistent system.** Stress that an inconsistent system has no solution, that its graph is a pair of parallel lines, that the equations have the same slope with different y-intercepts, and that if you try to solve the system algebraically, you get an equation that is false.

Exploration 1

This exploration allows students to examine the case in which the two equations in the system are the same line. Have students work in their group and use the erasable graph to investigate the following questions, which you may write on the overhead or chalkboard. (For ideas on guiding the students, see Exploration 1 on page 85 of the Pupil's Edition of the textbook.)

Draw a line on the erasable graph, and create two different equations for the line by choosing two different pairs of points on the line and writing an equation for each pair in point-slope form, $y - y_1 = m(x - x_1)$.

Solve the system of equations algebraically. What happens? **[You get an equation that is always true.]** Do the two lines have any points in common? Explain. **[The lines are the same, so they have every point—that is, an infinite number of points—in common.]** How many

continued on bottom of next page

Approach A — Using the Book, continued

Ongoing Assessment, page 91

1. The lines are the same.

4. The variables are eliminated; yes; yes; yes; all ordered pairs that satisfy one of the equations also satisfy the other and are thus a solution to the system.

Exploration 2

In this exploration, students work to determine the relationship among the slopes and y-intercepts of lines that intersect, are parallel, or are the same.

This should allow students to determine the number of solutions to an equation by finding its slope and y-intercept.

It may help students to draw an example before answering the questions in each step. Note that in Step 1b, the y-intercepts can be different or, if the lines intersect at their y-intercepts, the same.

Use **Transparency 8** for a discussion of the types of systems.

Ongoing Assessment, page 92

Independent systems have lines with different slopes. Both dependent and inconsistent systems have lines with the same slope. Independent systems have lines that intersect at one point. Dependent systems have only one line, and inconsistent systems have parallel lines. Independent and dependent systems have solutions and are thus consistent, while inconsistent systems have no solutions. Independent systems can have the same or different y-intercepts, while dependent systems have the same y-intercepts and inconsistent systems have different y-intercepts.

Cultural Connection

This reading gives students some insight into the way in which systems of equations were represented by early civilizations. Have students determine the type of system represented by the rods.

TEACH

Approach B — Using Models, continued

solutions does the system have? **[an infinite number]**

Consistent Systems

Have one or more groups read their results for Exploration 2. Explain that a system of equations that has at least one solution is called a **consistent system.** The type of consistent system seen in Exploration 2 is called a **dependent system.** Stress that a dependent system has an infinite number of solutions, that the two equations have the same slope, the same y-intercept, and the same graph, and that if you solve the system algebraically, you get an equation that is always true.

Independent Systems

The type of system that has exactly one solution is called an **independent system.** An independent system is also a consistent system. What does the graph of an independent system look like? **[intersecting lines]** What is the relationship between the slopes and y-intercepts of the equations? Make a few sketches to help students answer this question. **[The slopes are different; the y-intercepts may be the same or they may be different.]**

Closing the Lesson

Work together as a class to produce a table on the chalkboard or overhead of the characteristics of independent, dependent, and inconsistent systems. All students can copy the completed table into their notebook or journal for later reference. (See the summary box on page 92.)

ASSESS

Selected Answers

Odd-numbered Exercises 7–43 and odd-numbered Extra Practice

Assignment Guide

Core level students should complete Exercises 1–44.

Core-plus level students should complete Exercises 1–14 and 17–45.

Error Analysis

In Exercises 7–15, students may have trouble interpreting results such as $0 = 0$ or $3 = 0$. Remind them that if the statement is always true, there are an infinite number of solutions to the system of equations and that if the statement is false, there are no solutions. Students should be able to answer Exercises 16–21 without solving the system. Remind them to convert the equations to slope-intercept form before reading the slope and y-intercept.

Alternative Assessment

Forming Systems With Given Properties Give students the equation of a line. Have them use what they learned in this lesson to write a second equation to form each of the following types of systems.

A. an inconsistent system

B. a dependent system

C. an independent system

They should check their systems by solving them by graphing.

Portfolio Assessment Exercises 7–21 tie together the major concepts of the first six lessons in this chapter. Have students solve the three systems that were created above algebraically as well. Ask them to write these problems in report form and to add them to their portfolio.

RETEACH

Concept Maps

Have students make a concept map such as the one shown here for the different types of systems. They can add information about the solutions, y-intercepts, slopes, and intersections for each type of system in each rectangle. Then have them use these maps to categorize some additional systems of equations.

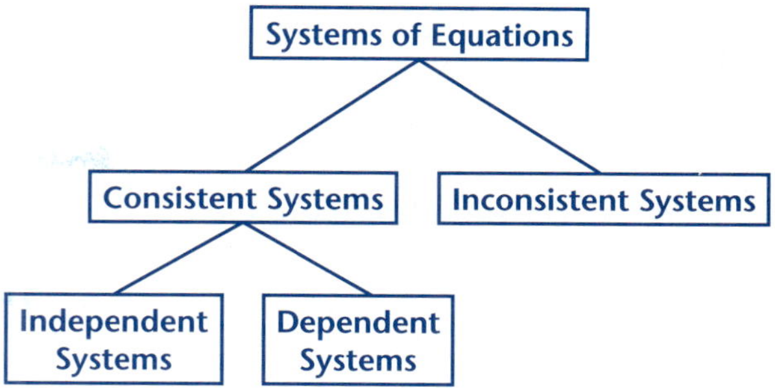

ENRICH

Characterizing Systems in Standard Form

Have students go through the examples and homework to make a list of three examples of each type of system: inconsistent, dependent, and independent. Then, if necessary, rewrite the equations in standard form. Have students study these examples and look for patterns. Is there a way to recognize the type of system from the equations? Have them test their conjecture on other systems. **[In a dependent system, one of the equations can be multiplied by a constant to produce the other equation. In an inconsistent system, one of the equations can be multiplied by a constant to produce coefficients that are the same and constants that are different. All other systems are independent.]**

Interdisciplinary Connection

Science The table below shows the temperatures at which water boils and freezes in three different temperature scales—Fahrenheit (°F), Celsius (°C), and Kelvin (K).

	°F	°C	K
Boiling temp.	212	100	373
Freezing temp.	32	0	273

Have students use the table to find one equation that converts degrees Celsius to degrees Fahrenheit and one equation that converts kelvins to degrees Fahrenheit. Have students describe this system by using the language of this lesson and interpret their findings in the context of the problem. **[The system is inconsistent, and hence has no solution. The lines are parallel. There is no temperature that has the same value in degrees Celsius and in kelvins.]**

NAME ______________________ CLASS __________ DATE __________

Practice

2.6 Non-Unique Solutions

Solve each system algebraically.

1. $\begin{cases} x + y = 5 \\ -2x - 2y = -10 \end{cases}$ ____________

2. $\begin{cases} x + 2y = 6 \\ x + 2y = -4 \end{cases}$ ____________

3. $\begin{cases} 2x - y = 1 \\ x + y = 5 \end{cases}$ ____________

4. $\begin{cases} 3x + y = 5 \\ 3x + y = -2 \end{cases}$ ____________

5. $\begin{cases} 2y = -x - 4 \\ 2x = -4y - 8 \end{cases}$ ____________

6. $\begin{cases} 4x - 2y = -2 \\ y = 2x + 1 \end{cases}$ ____________

7. $\begin{cases} 3x - 2y = 6 \\ -6x + 4y = -12 \end{cases}$ ____________

8. $\begin{cases} x - y = 2 \\ -x + y = -2 \end{cases}$ ____________

9. $\begin{cases} 4x + y = 10 \\ y = -4x + 5 \end{cases}$ ____________

10. $\begin{cases} 2x + y = 3 \\ 4x = 6 - y \end{cases}$ ____________

Determine whether each system is dependent, independent, or inconsistent.

11. $\begin{cases} 2x + y = 8 \\ y = -2x + 8 \end{cases}$ ____________

12. $\begin{cases} 3x - y = 3 \\ y = 3x - 9 \end{cases}$ ____________

13. $\begin{cases} y + 4 = -5x \\ y = 6x - 7 \end{cases}$ ____________

14. $\begin{cases} 3x - 2y = -5 \\ 4y = 6x + 10 \end{cases}$ ____________

15. $\begin{cases} 6x - 2y = -10 \\ y = 3x + 2 \end{cases}$ ____________

16. $\begin{cases} y + 6x = 18 \\ y = -7x + 2 \end{cases}$ ____________

17. $\begin{cases} y + 4x = 3 \\ 2y = 8x + 6 \end{cases}$ ____________

18. $\begin{cases} x - y = 3 \\ 2y = 2x + 6 \end{cases}$ ____________

19. $\begin{cases} y = 3x + 1 \\ 3x = 1 - y \end{cases}$ ____________

20. $\begin{cases} 3x = y + 2 \\ 3y = x + 2 \end{cases}$ ____________

21. $\begin{cases} 4 = x - y \\ 8 + 2y = 2x \end{cases}$ ____________

22. $\begin{cases} x + 5y = 10 \\ y = 5x - 10 \end{cases}$ ____________

23. $\begin{cases} 4x - 2y = 1 \\ 2x - 4y = -1 \end{cases}$ ____________

24. $\begin{cases} 5x - 2y = 4 \\ 5x - 4 = 2y \end{cases}$ ____________

NAME ________________ CLASS ________ DATE ________

Lesson Activity

2.6 Comparing Linear Systems and Solutions

Write an equation to complete each type of linear system. Then graph each system on the grid provided.

1. Independent system
$2x + y = 3$

____

2. Dependent system
$2x + y = 3$

____

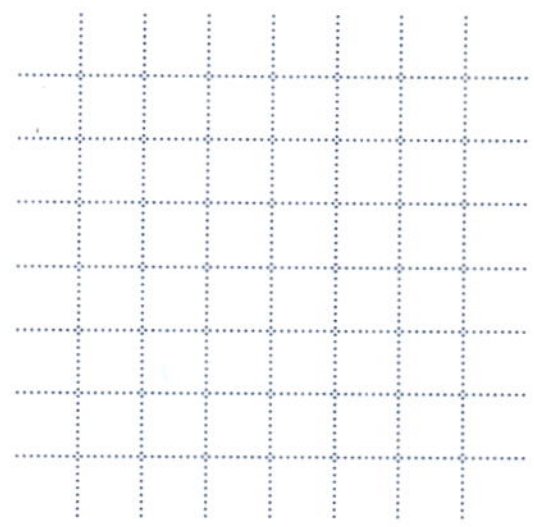

3. Inconsistent system
$2x + y = 3$

____

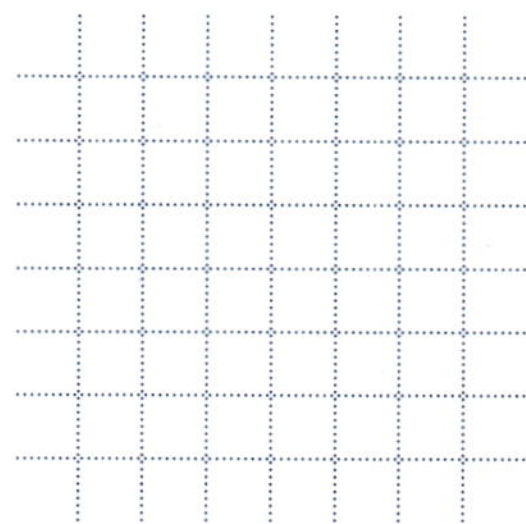

4. Study the systems and graphs you created for Exercises 1–3. Then explain the characteristics of the solution(s) for each system. Include examples where appropriate.

____

____

Tell whether each statement is always true, sometimes true, or never true.

5. The equations in an independent system have different y-intercepts. ________

6. There is only one solution to an inconsistent system. ________

7. If $y = 4$ in both equations of a dependent system, then the value of x is the same in both equations of the system. ________

NAME ______________________ CLASS __________ DATE __________

Basic Skills Practice
Cumulative Review

Read each question and circle the best answer.

1. The odometer on Colby's car read 893 on Monday. On Friday it read 1233. How many miles was the car driven that week?

A 2126
B 893
C 340
D 300

2. June had $565.15 in her checking account. After June made a deposit, the account balance was $783.96. How much money did June deposit?

F $218.81
G $490.06
H $565.15
I $783.96

3. The length of a rectangle is 5 centimeters more than its width. The perimeter of the rectangle is 70 centimeters. Find its width and length.

A 25 cm, 10 cm
B 15 cm, 20 cm
C 10 cm, 30 cm
D 30 cm, 40 cm

4. Miss James traveled 250 kilometers in 5 hours. How far did she travel in 1 hour?

F 75 km
G 60 km
H 50 km
I 40 km

5. Thomas had $15 in the bank. He worked 10 hours. When he put his paycheck in the bank, his balance was $75. How much did he earn per hour when he worked?

A $4
B $5
C $5.25
D $6

6. A customer buys a gallon of maple syrup for $24.50. Sales tax is 6%. What is the total cost?

F $24.56
G $24.60
H $25.97
I $26.00

Solve each problem.

7. A manufacturer found that 2% of the socks produced were unsuitable for marketing and had to be rejected. Of 3500 pairs, how many pairs were rejected? __________

8. Misha plans to install wall-to-wall carpet. A rectangular hall on the first floor is 8.5 yards long and 3.0 yards wide. What is the area covered by the carpet in square yards? __________

9. In a track meet, Quinton scored 9 points. This was $\frac{1}{3}$ of his team's final score. How many points did the team score? __________

LESSON 2.7

Graphing Linear Inequalities

PREPARE

Pacing Guide

2 Days

Objectives and Overview

- Determine the boundary line of the solution of a linear inequality.
- Graph the solution to a linear inequality.
- Graph the solution to a system of linear inequalities.

The concept of linear inequalities is presented, and a method for graphing linear inequalities is introduced. Students then learn that a system of inequalities can be solved by finding the intersection of the graphs of each inequality.

Vocabulary

linear inequality
boundary line
intersection

Assessing Prior Knowledge

Write the inequality for each.

1. Barbara spent at least \$12. [$b \geq 12$]
2. Sam made more than 2 dozen cookies. [$c > 24$]

Approach A

Using the Book

Opening Discussion

Situations in the real-world do not always lead to equations. Sometimes they need to be represented by an inequality. Discuss some of the examples of unequal situations presented in the Why feature. **Linear inequalities** allow us to find solutions to unequal situations involving more than one variable. Point out that linear inequalities, unlike the inequalities seen before, are graphed on a coordinate plane.

Brainstorming

Have students think of other examples of unequal quantities in the real world. Examples include the following: you do not have enough money for train fare; the temperature was higher today than yesterday; and to pass the test, you must have a score of at least 60.

continued on top of next page

TEACH

Approach B

Using Models

Opening Discussion

Situations in the real-world do not always lead to equations. Sometimes they need to be represented by an inequality. Discuss some of the examples of unequal situations, such as a grocery bill that was more than \$25 or a paper that must be at least 10 pages long to earn a B. **Linear inequalities** allow us to find solutions to unequal situations that involve more than one variable. Point out that linear inequalities, unlike the inequalities seen before, are graphed in the coordinate plane.

Brainstorming

Have students think of other examples of unequal quantities in the real-world. Examples include the following: you do not have enough money for train fare; the temperature was higher today than yesterday; and to pass the test, you must have a score of at least 60.

continued on bottom of next page

Approach A

Using the Book, continued

Exploration

This exploration can be done individually, with a partner, or as a whole class. If the class works together, have each student pick a different point to test.

The line $y = x$ is graphed. The x- and y-coordinates for points above and below the line are examined. Students should realize that for points above the line, y is greater, and for points below the line, x is greater. This allows students to see that the line divides the plane into two half-planes with unique characteristics.

Ongoing Assessment, page 96

2. The y-coordinate is always greater than the x-coordinate; $y > x$.
3. The y-coordinate is always less than the x-coordinate; $y < x$.

Boundary Lines

Read or have a student volunteer read the paragraph about boundary lines. Students should understand that the solution to the inequality includes the boundary line if the symbol is $\le$ or $\ge$ and does not include the boundary line if the symbol is $<$ or $>$. Explain that the boundary line is dashed when it is not included in the solution.

Ongoing Assessment, page 96

point R

Example 1

This example demonstrates the method for graphing an inequality.

Point out that the equation of the boundary line is formed by replacing the inequality symbol with an equal sign. Note that the line is graphed by using x- and y-intercepts. This method is faster than converting to slope-intercept form in this case. Stress that the shaded region contains points for which the inequality is true.

TEACH

Approach B

Using Models, continued

Cooperative Learning

This activity allows students to discover that an equation divides the coordinate plane into two regions and that the points on each side of the line satisfy an inequality.

Divide the class into groups of three. One person from each group should make a graph of the equation $y = 2x + 3$, and one person should make a table with the following three columns: (x, y), $2x + 3$, and y. The third student is responsible for writing the group's summary. Students should each pick one part of the graph—above the line, below the line, or on the line—and choose three points in their region. They should evaluate $2x + 3$ and then fill in the table for each of their three points. At the completion of the activity, the group should compare the columns for $2x + 3$ and y and write down their conclusions.

Graphing Inequalities

Ask several of the groups to report their findings to the class. Students should have realized that for the points above the line, y was greater than $2x + 3$; for the points below the line, y was less than $2x + 3$; and for the points on the line, y was equal to $2x + 3$. Show students how these results can be written as inequalities, and demonstrate how to shade the graph to represent the solution of $y \le 2x + 3$ or of $y \ge 2x + 3$.

The Boundary Line

Explain that the line that divides the plane in the graph of a linear inequality is called the *boundary line*. This line is drawn as a solid line when the points on the line satisfy the inequality ($\le$ or $\ge$) and as a dashed line when the points on the line are not part of the solution ($<$ or $>$).

Ask students what type of boundary line the graph of $2x - 6y \le 12$ would have. **[solid]** Have students graph this boundary line (point out that it is easy to graph by using intercepts), and show them how to pick a test point to

Additional Example 1

Graph the following inequality to find the solution:

$2x - 6y \leq 12$

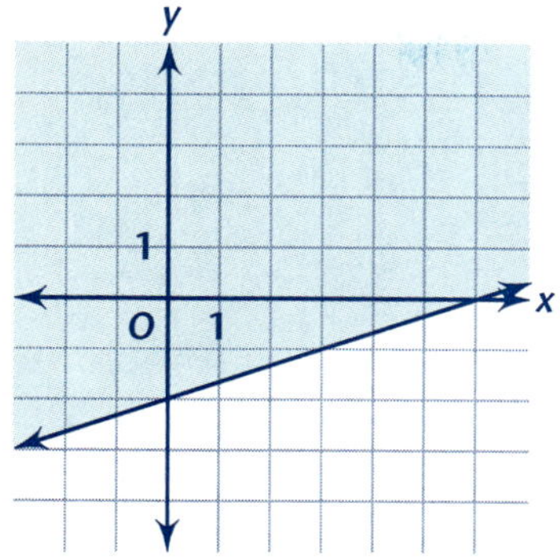

Systems of Linear Inequalities

Read and study the graphs that introduce a system of linear inequalities. Explain that each inequality in the system is graphed individually on the same set of axes. The solution to the system contains the points that are in the solution for both of the inequalities. You can have students verify this fact by choosing points in each of the four regions defined by the lines and testing them in the inequalities.

Use **Transparency 9** for discussion.

Example 2

This example explains the method used to solve a system of linear inequalities by graphing.

The lines could also have been graphed by converting the equations to slope-intercept form. The point (0, 0) was chosen as a test point because it is easy to substitute. In both cases, (0, 0) made the inequality true, so this point is included in the solution. Note that separate graphs are not necessary.

Ongoing Assessment, page 98

(1, 1) and (0, 0) are clearly in the overlapping region; and 1 and 0 are numbers that can be used to quickly evaluate an equation.

continued on top of next page

determine which side of the line to shade. Stress that if the test point satisfies the inequality, the side containing the point is shaded.

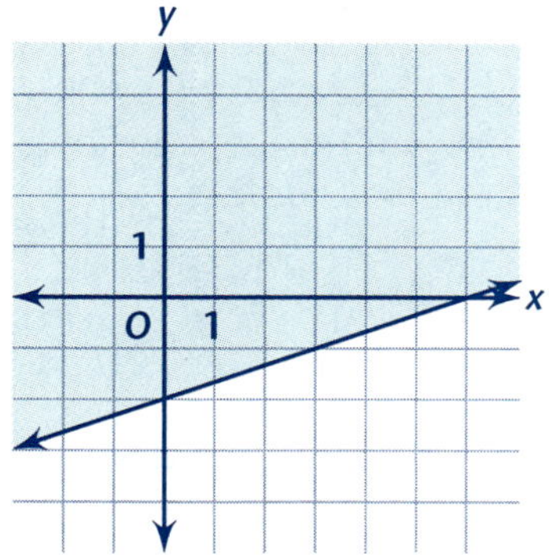

Systems of Linear Inequalities

Like solutions to systems of equations, the solutions to systems of inequalities must satisfy each inequality.

Using Transparency Sheets

This activity allows students to discover that the solution to a system of linear inequalities is the intersection of the solutions to each inequality in the system.

Students should complete this activity in their cooperative-learning groups. Give each group several sheets of overhead transparency paper, each with a coordinate system of the same size and scale.

Two students should each graph one of the inequalities of the system on separate transparency sheets. The third student is responsible for writing a summary of the group's results with input from the other group members.

Superimpose the two graphs so that the coordinate axes are aligned. What region on the graph forms the solution of the inequality? Test some points in this region to check your guess. How is this region related to the solution set of each individual inequality? **[It is the region where the solutions to the two inequalities overlap.]**

continued on bottom of next page

Approach A

Using the Book, continued

Additional Example 2

Solve by graphing.

$\begin{cases} 3x - 5y \le 15 \\ 3x + 6y > 9 \end{cases}$

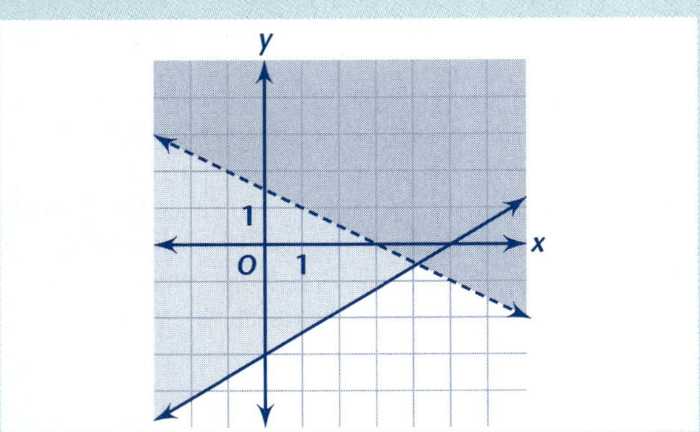

On Their Own

Solve by graphing.

$\begin{cases} y < -2x + 4 \\ y > 3x - 4 \end{cases}$

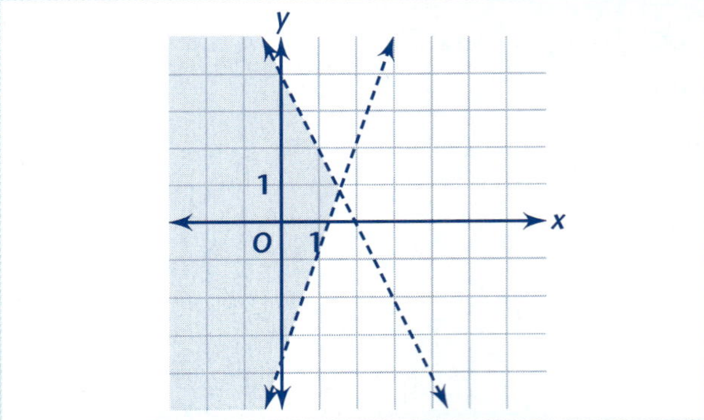

Example 3

A real-world example is solved by using a system of linear inequalities.

In this case both of the variables must be positive, so the graph is restricted to the first quadrant. Have students pick one of the points in the solution set and verify that it meets both conditions.

TEACH

Approach B

Using Models, continued

Use transparency sheets to graph inequalites.

Student Reports

Have one or more groups report their findings to the class. Brainstorm with the class to produce a list of steps for solving a system of linear inequalities. Have each student copy this list in their notebook or journal for later reference. Then have them use the steps to solve the following system by graphing:

$\begin{cases} 3x - 5y \le 15 \\ 3x + 6y > 9 \end{cases}$

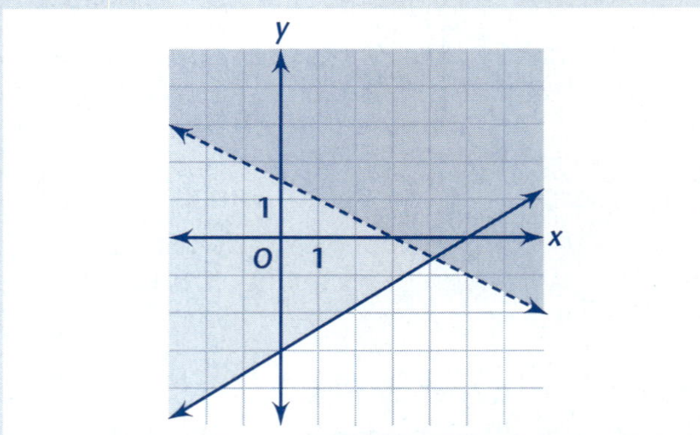

Ongoing Assessment, page 99

Kwame cannot work a negative number of hours.

Additional Example 3

Nick works two jobs during the summer. He baby-sits for \$4.00 per hour and works as a lifeguard for \$7.50 per hour. He can work no more than 25 hours a week, but he wants to make at least \$120 per week. Write a system of inequalities that represents all the combinations of hours and jobs that Nick can work. Graph the solution.

$$\begin{cases} x + y \leq 25 \\ 4x + 7.5y \geq 120 \end{cases}$$

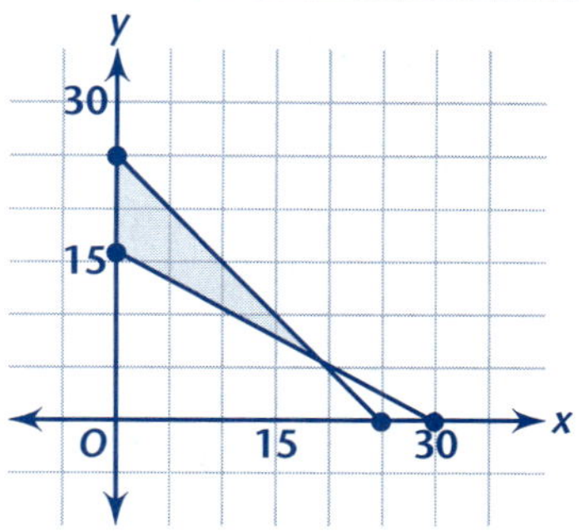

Critical Thinking, page 100

18 hours if he works 0 hours mowing and 18 hours at the library; \$210 by working 0 hours mowing and 30 hours at the library

Closing the Lesson

Give students some extra practice with systems of equations by having them do Exercises 8 and 9 from page 101 in class. Have two students work at the chalkboard so that all students can check their work.

An Application

Have students work in their cooperative-learning groups to solve the problem below. One person should be responsible for writing the inequalities, and the others should each graph one of the inequalities.

Nick works two jobs during the summer. He baby-sits for \$4.00 per hour and works as a lifeguard for \$7.50 per hour. He can work no more than 25 hours per week, but he wants to make at least \$120 per week. Write a system of inequalities that represents all the combinations of hours and jobs that Nick can work. Graph the solution.

$$\left[\begin{cases} x + y \leq 25 \\ 4x + 7.5y \geq 120 \end{cases}\right]$$

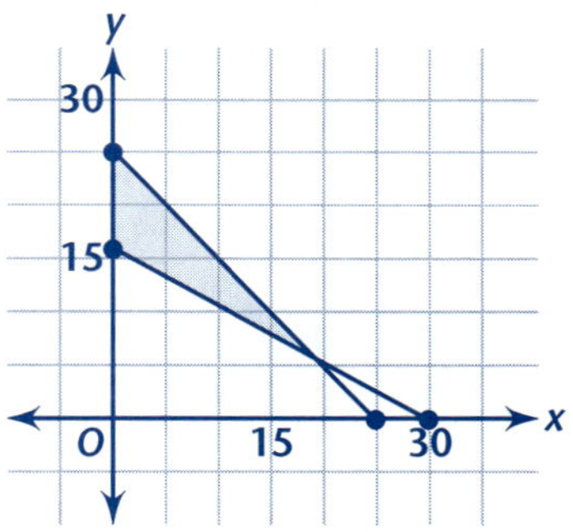

Closing the Lesson

Ask one or more groups to report their findings to the class. Ask for some examples of job combinations that Nick could do to accomplish his goals. What combination allows him to work the least number of hours and earn at least \$120? **[16 hours life-guarding and 0 hours baby-sitting.]** What is the greatest amount of money he can earn and still work 25 hours or less? **[\$187.50, if he works 25 hours as a life-guard and 0 hours baby-sitting]**

ASSESS

Selected Answers

Odd-numbered Exercises 9–35 and odd-numbered Extra Practice

Assignment Guide

Core level students should complete Exercises 1–36.

Core-plus level students should complete Exercises 1–7 and 10–39.

Error Analysis

Students sometimes shade the wrong region. Stress that if the test point satisfies the inequality, shade the region containing the point. You may need to review the equations for horizontal and vertical lines before students complete this assignment.

Alternative Assessment

Paper on Solving a System of Linear Inequalities Have students write a paper that describes a method for solving a system of linear inequalities by graphing. The paper should include different types of examples with worked-out solutions.

Portfolio Activity Have students edit their solution to Exercise 7 and include it in their portfolio. In addition, they should give several examples of plant combinations that are in the solution.

RETEACH

Have students choose two jobs that they would like to do next summer and find out how much each job pays per hour. Then they should set some conditions on the number of hours that they will work per week and on the total amount of money that they wish to earn each week. They should write a system of inequalities to represent the situation and solve it by graphing. Then have them write several combinations of hours for each job that will satisfy the conditions.

ENRICH

Interdisciplinary Connection

Business Linear programming is a method used by businesses to help them determine the number of items to manufacture given various constraints and goals. The constraints and goals are represented by inequalities and form a system of inequalities. Businesses study the solution to help them make decisions. Many considerations, such as the availability and cost of materials and the cost of labor or storage space, may factor into the decision of how many of each type of item to manufacture. Each of these considerations can be represented on the graph. Ask students to find out more about linear programming and to present their findings to the class.

Technology

Some graphics calculators and graphing software can represent systems of inequalities. If this technology is available, ask one or more students to learn how to use it and to demonstrate it for the rest of the class. They should come to the front of the room and encourage student input and discussion. The presentation should include examples and opportunities for the other students to solve an example.

NAME ______________________ CLASS __________ DATE __________

Practice

2.7 Graphing Linear Inequalities

Graph the common solution of each system on your own graph paper. Choose a point from the solution and use it to check both inequalities.

1. $\begin{cases} y > x + 4 \\ y \le 2 \end{cases}$ ____________

2. $\begin{cases} x + y \le 0 \\ y > -1 \end{cases}$ ____________

3. $\begin{cases} y > 3 \\ x \le -2 \end{cases}$ ____________

4. $\begin{cases} y \ge x - 2 \\ y > -x + 2 \end{cases}$ ____________

5. $\begin{cases} y \le x - 3 \\ y < -2x + 1 \end{cases}$ ____________

6. $\begin{cases} y \ge 3x \\ -2y \le 5x \end{cases}$ ____________

7. $\begin{cases} 4x + y > 2 \\ 2x - y \le 4 \end{cases}$ ____________

8. $\begin{cases} x + 2y < 4 \\ 3x - y < 2 \end{cases}$ ____________

9. $\begin{cases} 5x - y > 3 \\ 4x + y > 0 \end{cases}$ ____________

10. $\begin{cases} 3x - 5y \le -10 \\ x + 4y > 4 \end{cases}$ ____________

11. $\begin{cases} 2x + 3y \ge 0 \\ 3x - y \ge -3 \end{cases}$ ____________

12. $\begin{cases} 3x + 2y \le 6 \\ x - y \ge -3 \end{cases}$ ____________

Write the system of inequalities that represents the shaded region. Use the points shown to find equations for the boundary lines.

13.

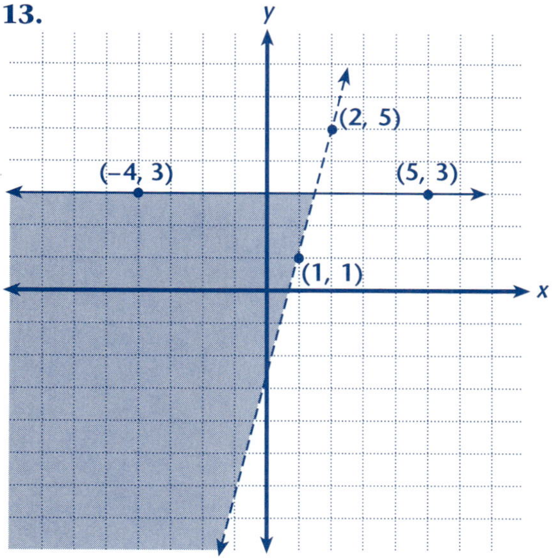

____

14.

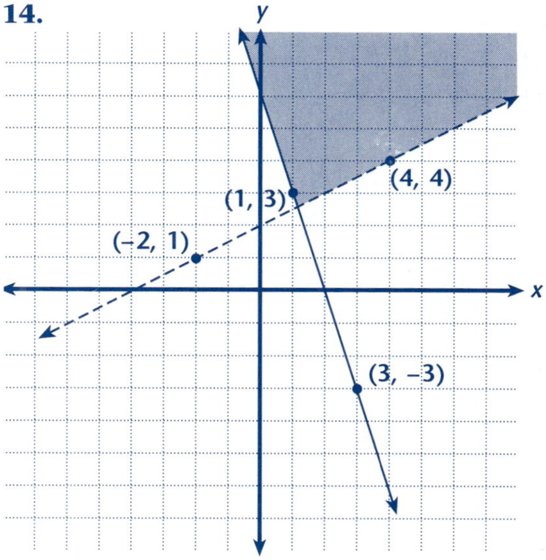

____

NAME ______________________ CLASS __________ DATE __________

Lesson Activity

2.7 Graphing Linear Inequalities: Single Inequalities and Systems

Graph each inequality on the grid provided to find the solution set. Then answer the questions.

1. $2x + y < 6$

2. $4x - 2y \geq 8$

3. Which graph has a boundary represented by a dashed line? Which has a solid line for a boundary? Why?

____

Name two possible solutions for each inequality.

4. $2x + y < 6$ ______________________

5. $4x - 2y > 8$ ______________________

WINTER SALE BEGINS NOVEMBER 1!!!

Blue jeans	Flannel shirts
Reg. \$50	Reg. \$30
Sale \$37.50	Sale \$15

6. Hattie has \$150 to spend and wants to buy her clothes before November 1. Nadine has \$90 to spend and can wait until after November 1. Write a system of equations and graph them on the grid provided to find all of the combinations of blue jeans and shirts that each can purchase with the amount of money that each has to spend.

____

____

NAME ____________________ CLASS __________ DATE __________

Basic Skills Practice

Graphing Points on a Line, Part 1

Example 1: Graph the solution to $x + 7 = 11$.
First solve the problem: $x + 7 = 11$.
Think: $x = 11 - 7 = 4$ (This is an inverse operation.) $x = 4$

Graph:

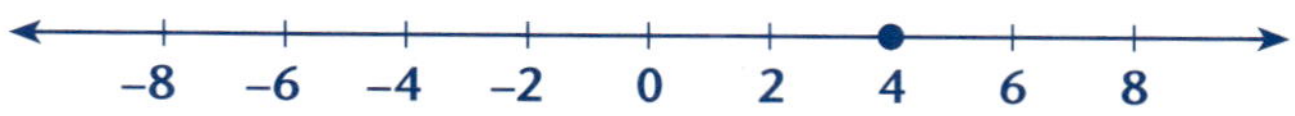

Example 2: Graph the solution of $x > 2$.
Solve: $x > 2$.
Think: x can be any number greater than 2.

Graph:

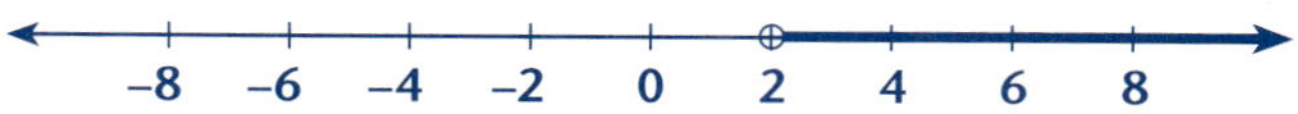

The open circle shows that 2 is NOT a solution.

Example 3: Graph the solution of $x \leq -4$.
Solve: $x \leq -4$.
Think: x can be -4 or any number less than -4.

Graph:

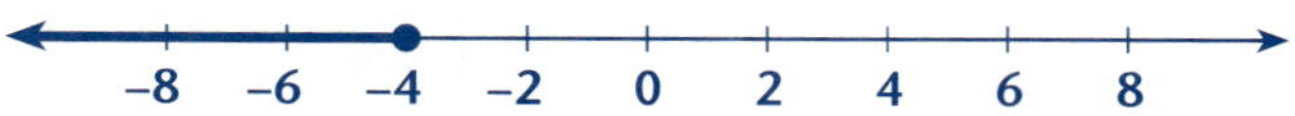

The closed circle shows that -4 is included in the solution.

Graph the following equations:

1. $x + 5 = 8$

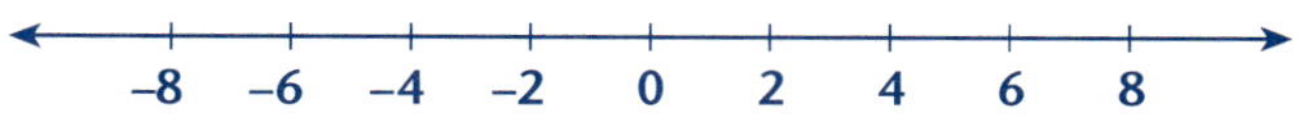

2. $x - 4 = 2$

3. $x < 3$

4. $x > -4$

5. $x \geq -2$

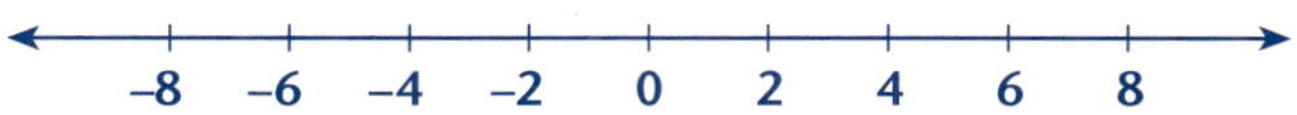

6. $x \leq 5$

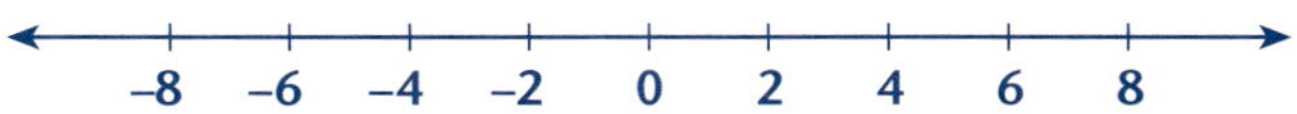

7. $x > -1$

LESSON 2.8

Classic Applications

PREPARE

Pacing Guide

2 Days

Objectives and Overview

- Solve traditional application problems.

This lesson introduces students to five types of classical problems that can be solved using systems of linear equations: mixture, money, numeration, travel, and age.

Assessing Prior Knowledge

Write an equation to describe each problem. You may use either one or two variables.

1. The dimes in Joan's pocket are worth \$2.50. [$0.10d = 2.50$]
2. The dimes and nickels in Sam's pocket are worth \$3.00. [$0.05n + 0.10d = 3$]
3. Aaron has 8 ounces of peanuts and cashew nuts. [$p + c = 8$]
4. The sum of the digits of a two-digit number is 6. [$t + u = 6$]
5. At r miles per hour, Eileen drove 150 miles in 3 hours. [$150 = 3r$]

Approach A: Using the Book

Opening Discussion

Use the car photos and captions to introduce the term *classic.* Something that is a classic is an important part of our culture. Adults are expected to be familiar with classics. For example, most adults in this country could recognize a Ford Model-T or a '65 Mustang. Because of their usefulness, classic problems in mathematics are done by nearly every algebra student, and it is expected that students who have taken algebra will have some familiarity with these problems.

Brainstorming

Have students think of examples of classics from other disciplines. Examples include the following: classic children's books, *Charlotte's Web;* classic Christmas songs, "Jingle Bells"; and classic American foods, hamburger.

TEACH

Approach B: Using Models

Opening Discussion

What is a classic? Think of a classic car, for example. A classic is usually something that was unique or different when it was first introduced. Something that is a classic is an important part of our culture. Adults are expected to be familiar with classics. For example, most adults in this country could recognize a Ford Model-T or a '65 Mustang. Because of their usefulness, classic problems in mathematics are done by nearly every algebra student, and it is expected that students who have taken algebra will have some familiarity with these problems.

Brainstorming

Have students think of examples of classics from other disciplines. Examples include the following: classic children's books, *Charlotte's Web;* classic Christmas songs, "Jingle Bells"; and classic American foods, hamburger.

Inclusion Strategies

Visual Models Encourage students to draw pictures and use tables to represent the situations in these problems. Tables are especially useful with mixture, money, and travel problems for organizing the information and then writing the equations.

Example 1

This example introduces students to mixture problems.

Students can draw beakers and label them as in the photo. This will help them organize the information. The table is very helpful for most students. Point out that the entries in the first two columns in the table are multiplied together to get the entries in the third column. The last two columns are read from top to bottom to get the equations.

Use **Transparency 10** for discussion.

Additional Example 1

How many ounces of a 10% acid solution should be mixed with a 35% acid solution to produce 50 ounces of a 25% solution? **[20 ounces of 10% solution and 30 ounces of 35% solution]**

Example 2

This is an example of a classic coin problem.

Stress that one equation involves the number of coins and that the other involves the value of the coins. Note that the units must be consistent. In this example, all of the money values were written in cents.

Additional Example 2

In a coin bank there are 120 nickels and dimes worth $10.15. How many of each type of coin are in the bank? **[37 nickels and 83 dimes]**

Example 3

This is an example of a classic numeration problem.

continued on top of next page

Cooperative Learning

This activity requires the separation of the class into two different types of groups: expert groups and study groups. First divide the class into five groups, each of which works to become experts at one of the five types of classic application problems. The class is then reorganized into study groups containing at least one expert on each type of problem. Every student in the study group is responsible for teaching the other group members the application problem on which they are experts.

Teaching Tip

To provide motivation and assessment for this activity, explain that after the activity you will give a quiz that involves writing a system of equations for each of the five types of problems. When grading the quiz, figure the class's average score on each application. Award bonus points to the expert group whose application had the highest class average. Then calculate each study group's average quiz score. Award bonus points to the study group that had the highest quiz average.

Five Application Problems

Assign each of the expert groups one of the types of application problems. Give the group a handout that includes one example of their type of problem that is worked out in detail and several examples with answers for students to work as practice. Explain that the goal of this part of the activity is for each member to become an expert at their group's type of application problem. Later they will help other students learn this application.

Instruct students to pay careful attention to the way in which the problem is set up. The variables usually represent similar things in each problem of that type. If a table is used, students should learn how it is set up and how to use it to solve that type of problem.

After a specified amount of time, have students meet with their study groups. Each member should be given time to explain and discuss their application. They should help the other group members with a practice problem and answer any questions.

continued on bottom of next page

Approach A

Using the Book, continued

Stress that the variables should represent the digits of the number. Show students how to write a two-digit number in expanded form and how to write the expanded form of a two-digit number with its digits reversed.

Additional Example 3

The sum of the digits of a two-digit number is 8. Twice the original two-digit number is 17 more than the number with its digits reversed. Find the original two-digit number. **[35]**

Ongoing Assessment, page 106
Try This
75

Example 4

This is an example of a classic travel problem that involves travel with and against the wind.

A similar procedure can be used to solve a problem that involves a boat traveling with and against a current. Stress that the problem utilizes the formula *rate* • *time* = *distance.*

These three quantities form the columns of the table. Point out how the equations were made simpler by dividing by the constant factor.

Additional Example 4

A family canoed upstream against the current for 3 hours to reach a picnic area that was 6 miles from their campground. The return trip, traveling with the current, took 2 hours. Find the rate of the current and the rate of the canoe with no current. **[The rate of the current is 0.5 miles per hour, and the rate of the canoe with no current is 2.5 miles per hour.]**

Critical Thinking, page 107
The rate of the plane would remain the same.

TEACH

Approach B

Using Models, continued

Here is a description of each type of problem, including the common assignments of variables, useful tables, and an example. For other examples, see pages 104–107 in the Pupil's Edition or any other second-year algebra text.

1. Mixture Problems

Two solutions containing different percents of acid are combined to make a given amount of solution with a given acidity. The variables represent the amount of each type of acid that is to be combined. Make a table with the following row headings: First solution, Second solution, and Final solution. The column headings should be as follows: Percent of acid, Amount of solution, and Amount of acid. The two equations involve the total amount of solution and the total amount of acid. When writing the equation, the percents should be changed to decimals.

Example How many ounces of a 10% acid solution should be mixed with a 35% acid solution to produce 50 ounces of a 25% solution? **[20 ounces of 10% solution and 30 ounces of 35% solution]**

2. Money Problems

There is a mixture of two types of coins. You are given the total value of the money and the total number of coins. The variables represent the number of each type of coin. Make a table with rows for each type of coin and columns for the number of each coin, the value of each coin, and the total value in cents. The two equations involve the number of coins and their total value.

Example In a coin bank there are 120 nickels and dimes worth $10.15. How many of each type of coin are in the bank? **[37 nickels and 83 dimes]**

3. Numeration Problems

This is a word problem that involves the digits of a two-digit number and the number formed by reversing or otherwise altering the digits. The variables, say t and u, stand for the tens and units digit of the original number. This number

Example 5

This is an example of a classic age problem. It may help students if they model the equations with words initially and then translate the words into variables.

Additional Example 5

Sarah is 30 years older than her son. Next year, she will be 7 times her son's age. How old is each now? **[Sarah: 34; son: 4]**

Closing the Lesson

Work with the class to write the names of the five types of application problems from this lesson on the chalkboard. Then write the characteristics of each type under each name.

can then be written as $10t + u$. The number with the digits reversed can be written as $10u + t$.

Example The sum of the digits of a two-digit number is 8. Twice the original two-digit number is 17 more than the number with its digits reversed. Find the original two-digit number. [35]

4. Travel Problems

This problem involves a plane, boat, or similar object moving with the wind or current in one direction and then returning against the wind or current in the other direction. The variables represent the constant speed of the wind or current and the speed of the object with no wind or current. Make a table with the following row headings: With wind (or current) and Against wind (or current). The column headings should be as follows: Rate, Time, and Distance. Use the relationship *rate* • *time* = *distance* to write the equations.

Example A family canoed upstream against the current for 3 hours to reach a picnic area that was 6 miles from their campground. The return trip, traveling with the current, took 2 hours. Find the rate of the current and the rate of the canoe with no current. **[The rate of the current is 0.5 miles per hour, and the rate of the canoe with no current is 2.5 miles per hour.]**

5. Age Problems

This problem uses information that relates the current ages of two people and their ages a fixed number of years before or after the current time. The variables represent the current ages. One equation relates the current ages and the other relates the ages in the past or future.

Example Sarah is 30 years older than her son. Next year, she will be 7 times her son's age. How old is each now? **[Sarah: 34; son: 4]**

ASSESS

Selected Answers

Odd-numbered Exercises 7–23 and odd-numbered Extra Practice

Assignment Guide

Core level students should complete Exercises 1–24.

Core-plus level students should complete Exercises 1–26.

Error Analysis

Students sometimes make errors in the initial set-up of these problems. Encourage them to identify the type of problem and to follow the steps illustrated in the examples for that type of problem.

Alternative Assessment

Examples of Classic Problems Have students choose two or three types of classic problems. For each problem, they should explain its unique characteristics, give common examples of the variables, describe any tables, and give the characteristics of each equation. They should then write a worked-out example of each type of problem.

Portfolio Activity Have students look through their homework from this lesson to find an example of each type of classic problem. They should rewrite the solution to these problems clearly in their portfolio and include a sentence about how they recognized the type of problem.

RETEACH

Write each type of classic application on the chalkboard and work with students to list the identifying characteristics of the problem, the common assignment of the variable, a skeleton of any table used, hints for writing the equations, and any other useful information for each type. Students should copy this work into their notebook or journal for later reference. Then give students a new example of each type, and have them work with a partner to solve the examples.

ENRICH

Writing Classic Problems

Have students work with a partner to make up examples of these classic problems. It is usually easier if they start with the answer and then work backward to formulate the question. Offer to include particularly good questions on the next quiz or test.

Interdisciplinary Connection

Aviation Students who have flown cross-country or overseas may have realized that the flight is longer in one direction than it is in the other. This is due to winds. Have students pick a destination that they would like to fly to and find out the air time for flying to that destination and from that destination. They should also find out about the wind patterns along the flight path. Have them write their findings in a report and present it to the class.

NAME ______________________ CLASS ____________ DATE ____________

Practice

2.8 Classic Applications

1. The tens digit of a 2-digit number is twice the units digit. The sum of the digits is 12. Find the original number. ____________

2. Olga has 5 times as many dimes as nickels. If she has $3.30, how many of each coin does she have? ____________

3. Milk that is 4% butterfat is mixed with milk that is 1% butterfat to obtain 18 gallons of milk that is 2% butterfat. How many gallons of each type of milk are needed? ____________

4. A banker invested a portion of $10,000 at 5% interest per year and the remainder at 4% interest per year. If the banker received $470 in interest after 1 year, how much was invested at each rate? ____________

5. The sum of Nora's age and her grandmother's age is 71. Four times Nora's age is 6 less than the grandmother's age. Find their ages. ____________

6. Sailing with the current, a boat takes 3 hours to travel 48 miles. The return trip, against the current, takes 4 hours. Find the average speed of the boat for the entire trip and the speed of the current. ____________

7. A coin bank contains nickels, dimes, and quarters totaling $5.45. If there are twice as many quarters as dimes and 11 more nickels than quarters, how many of each coin are in the bank? ____________

8. Find the 2-digit number whose tens digit is 3 less than the units digit. The original number is 6 more than 4 times the sum of the digits. ____________

9. A daughter is 28 years younger than her father. In 5 years the father will be 3 times as old as his daughter. How old is each now? ____________

10. A chemist has a solution that is 20% peroxide and another solution that is 70% peroxide. She wishes to make 100 L of a solution that is 35% peroxide. How much of each solution should she use?

____

NAME ____________________ CLASS __________ DATE __________

Lesson Activity
2.8 Classic Applications

Write the equations, method of solution, and solution to each of the following:

1. Joe can row a boat a distance of 12.75 km in 1.5 hours when he is rowing with the current. It takes him 1.7 hours to row the same distance against the current. What is the rate of the current? At what rate can Joe row in still water?

Equations ______________ Method ______________

______________ Solution ______________

2. How many quarts of a 25% alcohol solution would you mix with a 30% alcohol solution in order to make 15 quarts of a solution that is 28% alcohol?

Equations ______________ Method ______________

______________ Solution ______________

3. Find the 2-digit number that has both of the following characteristics:
a. The tens digits is 1 less than the units digit.
b. The value of the number is 1 more than 5 times the sum of its digits.

Equations ______________ Method ______________

______________ Solution ______________

4. Mike has 95 coins that have a total value $10.75. The coins consist of quarters and nickels. How many of each type of coin does he have?

Equations ______________ Method ______________

______________ Solution ______________

5. If you add 5 to the sum of Paula's age and her mother's age, the total is 48. Paula's mother is 5 years less than 3 times the age of Paula. How old are Paula and her mother?

Equations ______________ Method ______________

______________ Solution ______________

6. Make up your own classical application problem, and give it to a classmate to solve. (Hint: Know the solution before you create the problem.)

____

____

NAME ______________________ CLASS __________ DATE __________

Basic Skills Practice

Graphing Points on a Line, Part 2

Read each question and circle the best answer.

1. Monday's average temperature was 80° F; the temperature varied by 5° F throughout the day. Which number line represents Monday's temperatures?

A 

B 

C 

D 

2. The Sanchez family is taking a vacation to Colorado. They want to save at least $750 before their trip for expenses. Which number line shows the amount of money that the Sanchez family would like to save?

F 

G 

H 

I 

3. Dr. Hughes advised her patient that his weight should not vary more than 5 lb in either direction from 170 lb. Which number line represents this weight range?

A 

B 

C 

D 

4. The Crunchie Cookie Company asks its truck drivers to drive at least 55 mph but slower than 65 mph. Which number line shows the range of speeds that the company wants its drivers to maintain?

F 

G 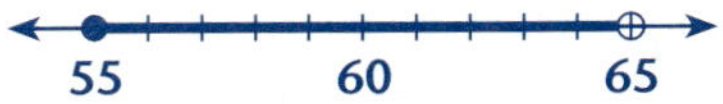

H 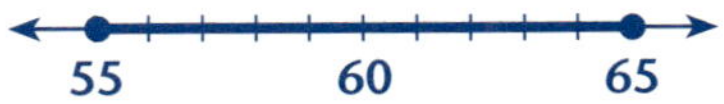

I 

5. Which inequality represents the number line?

A $x \geq 60$

B $x < 60$

C $x \leq 60$

D $x > 60$

6. Jeremy's mother told him not to spend more than $40 for a new pair of jeans. Which number line shows the price range that Jeremy should stay within?

F 

G 

H 

I 

Solve each problem.

7. Shopping for school clothes, Jayna found a jacket that cost $80. The cost of the jacket is $10 more than two pairs of jeans that she also wants to buy. What is the cost of each pair of jeans? __________

8. At a radar checkpoint, 5 cars were traveling at the following speeds: 65 mph, 72 mph, 62 mph, 58 mph, and 70 mph. What was the average speed of the cars, in miles per hour? __________

LESSON 3.1 Using Matrices to Display Data

PREPARE

Pacing Guide

2 Days

Objectives and Overview

- Interpret data from given information, and write the data in table or matrix form.
- Determine the dimensions of a matrix, and locate an entry by its address.
- Determine whether two matrices are equal.

Students learn how to arrange information in matrix form and how to name the matrix. They also learn how to determine the dimensions of a matrix and how to find entries of a matrix by using row and column addresses. They learn the definition of square matrix and of equal matrices.

Vocabulary

matrix
entry
address

Assessing Prior Knowledge

Solve each equation.

1.	$11x = 66$	[6]
2.	$0.3x = 27$	[90]
3.	$4x + 3 = -17$	[−5]
4.	$4(x - 2) = -6$	[0.5]

Approach A Using the Book

Opening Discussion

Graphs, tables, and spreadsheets are valuable tools in problem solving. What are some other tools? What are some ways that data in newspapers or magazines are organized or presented?

Brainstorming

Have students brainstorm different ways to organize data. Have them discuss how they would publish the statistics of the school's sports teams in the school newspaper. What would be a good way to show sports data so that it is easy to read and understand? How could they find the statistics of a specific player easily?

Organizing Data

Use the transportation data on pages 122 and 123 of the Pupil's Edition or other data to show students how to organize data into a table of rows and columns. Emphasize the use of brackets to denote matrices and the use of capital letters to name matrices. Point out

TEACH

Approach B Using Models

Opening Discussion

Use examples of graphs, tables, and spreadsheets to demonstrate that they are common tools for problem solving. Name some other tools for problem solving. How are data in newspapers or magazines organized or presented? What types of data are normally presented in table form in a newspaper?

Brainstorming

Have students brainstorm different ways to organize data. Have them discuss how they would publish the statistics of the school's sports teams in the school newspaper. Ask them to look at a table of sports statistics and to describe how to read the entries in the table. What would be a good way to show sports data so that it is easy to read and understand? How are table entries found by a computer program?

Using a Graphics Calculator

A graphics calculator can motivate students to enjoy using matrices to display data.

that all entries in a matrix are located by their matrix address, which lists the row first and the column second. Emphasize that this order is never changed, so confusion is avoided. Point out that labels are not part of the matrix.

Ongoing Assessment, page 123

c_{22}

Example 1

In this example, a matrix of airline distances is given and questions about the entries of the matrix are answered.

The definition of **square matrix** is also given. For question D, point out to students that in general, entry d_{36} does not equal entry d_{63}, so it is important to be consistent in listing the row first and the column second when giving a matrix address. In this particular example, however, since the distance between two cities is the same both ways, d_{36} does equal d_{63}.

Use **Transparency 11** for discussion or review of this example.

Additional Example 1

Information from road maps is given in table form as shown below. Mileage is given as the shortest distance between cities using the roads on the map. The table is easily changed into the format of a matrix. Answer the questions below about the location of the information in the matrix.

	Atlanta	Boise	Chicago	Denver	El Paso
Atlanta	0	2223	708	1430	1439
Boise	2223	0	1711	835	1204
Chicago	708	1711	0	1021	1464
Denver	1430	835	1021	0	692
El Paso	1439	1204	1464	692	0

$$M = \begin{bmatrix} 0 & 2223 & 708 & 1430 & 1439 \\ 2223 & 0 & 1711 & 835 & 1204 \\ 708 & 1711 & 0 & 1021 & 1464 \\ 1430 & 835 & 1021 & 0 & 692 \\ 1439 & 1204 & 1464 & 692 & 0 \end{bmatrix}$$

A. What do the entries in matrix M represent? **[Each entry represents the distance in miles between the cities whose names**

continued on top of next page

Provide students with graphics calculators that have matrix functions. Ask them to follow these (or similar) steps in order to display a matrix on the calculator screen. (1) Press the matrix key. (2) Choose Edit mode. (3) Choose [*A*] as the name of the matrix. (4) Enter the dimensions of the matrix. (5) Enter the numbers listed in the matrix. (6) Display matrix *A* on the home screen by choosing [*A*] from the matrix menu.

As students enter the values into the graphics calculator, they should notice that the matrix display is automatically filled, usually by rows. That is, row 1 fills first, then row 2, and so on. Therefore, the entries should be made by row, not by column. Students should also notice that if they entered the wrong dimensions, the matrix display will not correspond to the matrix they are trying to enter.

Working With a Graphics Calculator

Have students enter the data from the table into a graphics calculator as a matrix. It represents the medal count, by nation, of the top five countries competing in the 1992 Summer Olympic Games.

	Gold	Silver	Bronze	Total
Unified Team	45	38	29	112
United States	37	34	37	108
Germany	33	21	28	82
China	16	22	16	54
Cuba	14	6	11	31

Teaching Tip

Matrix Addresses You may want to reinforce the idea that the addresses of the entries of a matrix are given as row by column.

As students enter the data into the graphics calculator, they may see the address of the entry at the bottom of the screen. This address is listed as *row, column,* which is the usual addressing method for entries of a matrix. Have students use this prompt to help them enter the values.

After the matrix is entered into the graphics calculator, ask them to scroll across a row of the matrix using the arrow key. They should see the row component of the address at the bottom of the screen change. If they scroll down a column,

continued on bottom of next page

Approach A

Using the Book, continued

appear in the row and column headings.]

B. What are the row and column dimensions of M? **[5 by 5]**

C. How many miles is it from Chicago to El Paso? What is the matrix address of this entry? **[1464 miles; m_{35} or m_{53}]**

D. What does entry m_{42} represent? **[The entry at m_{42} is 835 miles, which is the distance between Denver and Boise.]**

E. What does entry m_{33} represent? **[The entry at m_{33} is 0, which means that there are 0 miles between Chicago and Chicago.]**

Ongoing Assessment, page 124
Try This

4139 kilometers; d_{52}, d_{25}

Critical Thinking, page 124

The data is easy to read; the data is easier to understand; the entries can be found easily by using the row-by-column form.

Teaching Tip

Using Technology You may want to highlight the use of matrices to store large amounts of data in computer programs. Matrices are convenient because the addresses of their entries are easy to access with certain algorithms. When a program stores data in an address, it may use a nested looping structure that "fills" the addresses sequentially by row and column. The entries are just as easy to access by recalling the row and column numbers.

Matrix Equality

Emphasize that two matrices are equal if and only if their dimensions are the same and each corresponding entry is the same for each address in the matrix.

TEACH

Approach B

Using Models, continued

they should see the column component of the address change.

Cooperative Learning

Matrices Help students understand how matrices are used to organize data by dividing the class into groups of four.

Ask one student in the group to find an example of data in a newspaper or magazine. Have the second student organize the data into table or matrix form, including appropriate labels for each row or column. Ask the third student to enter the data into a graphics calculator and to display it. Have the fourth student write a report of the group's activities, including any problems encountered in presenting the data or in entering the data into the calculator.

Each student should discuss the contents of the report, verify the entries, and sign the report.

Matrix Equality

Cooperative Learning Have students work in pairs. Discuss matrix equality on page 125 of the Pupil's Edition with students. Write the following matrix pairs on the chalkboard.

1. $\begin{bmatrix} 7 & ? & ? \\ 5 & ? & 3 \\ ? & 1 & ? \end{bmatrix} \quad \begin{bmatrix} ? & 4 & 6 \\ ? & 9 & ? \\ 8 & ? & 2 \end{bmatrix}$

2. $\begin{bmatrix} ? & 5 & ? & 2 \\ 7 & ? & 3 & ? \end{bmatrix} \quad \begin{bmatrix} 11 & ? & 17 & ? \\ ? & 13 & ? & 19 \end{bmatrix}$

3. $\begin{bmatrix} ? & ? \\ 5 & 10 \end{bmatrix} \quad \begin{bmatrix} 15 & 20 \\ ? & ? \end{bmatrix}$

Tell students to replace the question marks with numbers so that the matrices are equal.

Using Technology Write the matrices from Example 2 on page 125 of the Pupil's Edition on the chalkboard. Have the students determine if the matrices are equal. One student should enter matrix S, while the other student enters matrix T. Then they can compare the displays.

Point out that for two row-by-column dimensions to be the same, the number of rows has to be the same, and the number of columns has be the same. A dimension of 3×5 is not equal to a dimension of 5×3.

Two matrices may have the same number of entries, and each may be an acceptable way to store the same data, but they are not equal unless the dimensions match.

Example 2

This example involves two matrices with mathematical expressions as entries.

The expressions need to be evaluated to determine if the entries in corresponding positions of the two matrices are equal. Point out that the dimensions need to match and each pair of corresponding entries has to be equal for the two matrices to be equal, as they are in this example.

Additional Example 2

Are the following matrices equal? Explain why or why not.

$$S = \begin{bmatrix} 4^2 & (10+4) & 8 \\ \sqrt{25} & -8(5) & \frac{49}{7} \end{bmatrix}$$

$$T = \begin{bmatrix} 16 & 5 \\ 14 & -40 \\ 8 & 7 \end{bmatrix}$$

[No; matrix *S* has dimensions 2 by 3, while matrix *T* has dimensions 3 by 2; the dimensions are not equal, so the matrices are not equal.]

Example 3

In this example, two matrices are given with entries that are variable expressions.

Because the given matrices are equal, each pair of corresponding entries is equal. Therefore, algebra can be used to find the value of the expressions. Point out that the equations formed

continued on top of next page

Matrices With Variable Entries

Ask students to use the store function of a graphics calculator to store 2 for X, 3 for Y, and 4 for Z. Have students work in pairs to enter the following matrix into a graphics calculator:

$$\begin{bmatrix} 3X & X+Y \\ 2XY & Y-Z \end{bmatrix}$$

Have one student enter the matrix with variables into the calculator. (Note: The display will show the value of the expression, not the variables.) Ask the other student to enter an equal matrix with the same values of X, Y, and Z into the calculator. Then have them compare the displays of each matrix and answer the following questions: Do the matrices have the same dimensions? Do the matrices have the same entries? Are the matrices equal?

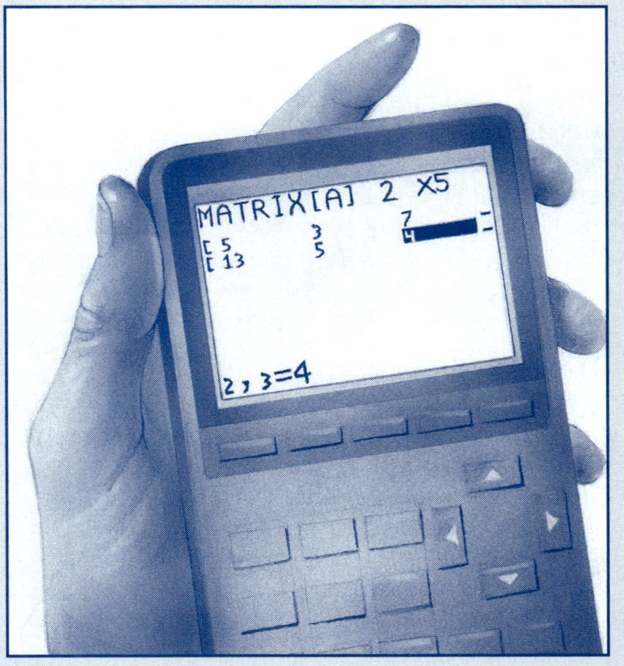

Graphics Calculators can be used to store, display, and manipulate matrices.

Teaching Tip

Matrix Display on a Graphics Calculator

Students working with graphics calculators to display matrices may find that the entire matrix display does not fit on the screen at one time.

continued on bottom of next page

Approach A — Using the Book, continued

from the individual entries make up a system of equations with more than one variable. Substitution, elimination, or graphing can be used to solve the equations.

Additional Example 3

Matrices U and V are equal. Find the values for x, y, and z.

$$U = \begin{bmatrix} 2x - 4 & x - 3 \\ 3xy & -4y \end{bmatrix}$$

$$V = \begin{bmatrix} 12 & -5y \\ -12z & 2y + 6 \end{bmatrix}$$

$[x = 8; y = -1; z = 2]$

Inclusion Strategy

Write a Matrix Arrange the classroom into rows of desks. Give each student a large sheet of paper on which the desk arrangement is drawn. Have each student take turns entering his or her age on a copy of the arrangement on the overhead projector. Ask students to copy each student's age onto their paper to form a class matrix of ages. What are the dimensions of the matrix? What age is listed in row 2, column 1?

Ongoing Assessment, page 125

Since $3z = 2xy$, you find the value for z by substituting the values for x and y into the expression $2xy$. This gives $3z = 2(4)(3)$. Therefore,

$$z = \frac{2(4)(3)}{3} = 8.$$

Closing the Lesson

Journal Writing Summarize the lesson by having students write in their journal the definition of a matrix and how the dimensions of a matrix are found. Ask them to describe how to determine whether two matrices are equal and to give some of the advantages of representing data with a matrix.

TEACH

Approach B — Using Models, continued

Point out that if the matrix is too large for the screen, the arrow keys are used to scroll to the right, left, up, or down to see the rest of the display. This is especially important when matrix entries have long decimal values that cannot be shown on one screen.

Journal Entry

Have students list in their journal the steps necessary to use a graphics calculator to organize data in matrix form. Ask them to include sample matrices and to discuss the dimensions of a matrix and of equal matrices.

ASSESS

Selected Answers

Odd-numbered Exercises 5–27 and odd-numbered Extra Practice

Assignment Guide

Core level students should complete Exercises 1–33.

Core-plus level students should complete Exercises 1–33.

Error Analysis

Students who have trouble with Exercise 9 may have difficulty interpreting the schematic representation of the air routes between the cities (called a *network*). Point out that the arrow in the diagram implies that there is a direct route between two cities *in the direction of the arrow.* Thus, since there is only one arrow from city C to city B, there is only one direct route from city C to city B. Help students understand networks better by having them create another matrix representing the air routes that go through one other city first. For this matrix, AB would be 0 because there is no way to get from A to B by passing through one other city first; AC would be 4.

Alternative Assessment

Organizing Data Ask students to suppose that they have 12 sports scores to organize into matrix form so that they can be displayed for the school newspaper. Ask them to use real data if it is available. Have them list all possible matrices to display the sports scores and give the dimension of each matrix. **[There are 6 possibilities—12 × 1, 1 × 12, 6 × 2, 2 × 6, 4 × 3, and 3 × 4.]**

Portfolio Activity Have students find an example of a table or matrix in a newspaper or magazine and display it for the class. Have them give the dimensions and label the row and column addresses. Ask them to write a paragraph explaining what the data is and why a matrix is a good way to display this particular set of data.

RETEACH

Ask students to tell which of the matrices below is equal to matrix Z.

$$Z = \begin{bmatrix} |8| & \sqrt{16} \\ -3(2) & 10-3 \\ -6 & -8-2 \end{bmatrix}$$

$$A = \begin{bmatrix} 8 & -6 & -6 \\ 4 & 7 & -10 \end{bmatrix} \qquad B = \begin{bmatrix} 8 & 4 \\ -6 & 7 \\ -6 & -10 \end{bmatrix}$$

$$C = \begin{bmatrix} 8 & 4 \\ -5 & 7 \\ -6 & -6 \end{bmatrix}$$

[*B* is equal to *Z*.]

ENRICH

Preview the next lesson by having students write two matrices with the same dimension and predict their sum. Then have them check their predictions by entering the matrices into a graphics calculator. Use matrix A and matrix B as names. Ask them to add the matrices with the calculator (usually by using the command $[A] + [B]$, where $[A]$ and $[B]$ are matrix names and + is normal addition).

Interdisciplinary Connection

Statistics A statistician is responsible for representing data in a form that is easy to use. When data needs to be presented in a concise, easy-to-read format, a matrix is often used.

NAME ______________________ CLASS __________ DATE __________

Practice

3.1 Using Matrices to Display Data

The matrix shows windchill factors for various combinations of wind speed and temperature.

Thermometer reading (degrees Fahrenheit)

Wind speed (mph)

$$\begin{array}{r} \\ \mathbf{5} \\ \mathbf{10} \\ \mathbf{15} \\ \mathbf{20} \\ \mathbf{25} \end{array}\begin{array}{c} \begin{array}{rrrrrrrr} \mathbf{35} & \mathbf{30} & \mathbf{25} & \mathbf{20} & \mathbf{15} & \mathbf{10} & \mathbf{5} & \mathbf{0} \end{array} \\ \begin{bmatrix} 33 & 27 & 21 & 19 & 12 & 7 & 0 & -5 \\ 22 & 16 & 10 & 3 & -3 & -9 & -15 & -22 \\ 16 & 9 & 2 & -5 & -11 & -18 & -25 & -31 \\ 12 & 4 & -3 & -10 & -17 & -24 & -31 & -39 \\ 8 & 1 & -7 & -15 & -22 & -29 & -36 & -44 \end{bmatrix} \end{array} = \mathbf{D}$$

1. What are the row and column dimensions of D? ______________________

2. What is the entry in the d_{32} matrix address? ______________________

3. What is the entry in the d_{46} matrix address? What does this entry represent? ______________________

4. What is the windchill factor if the temperature is 5°F, and the wind is blowing at 10 mph? ______________________

5. If the wind chill factor is −17, find the temperature and wind speed. ______________________

The table shows road mileages between some cities in the United States.

Cities	Distance	Cities	Distance
Boston to Cleveland	639	Chicago to Boston	983
Chicago to Cleveland	192	Chicago to Denver	1050
Dallas to Denver	801	Dallas to Cleveland	1205
Denver to Boston	1991	Cleveland to Denver	1369
Dallas to Chicago	931	Boston to Dallas	1815

6. Form a matrix to display the mileage between pairs of cities. Label each row and each column with the city names in alphabetical order. Name the matrix N.

7. The distance between which two cities is greatest? ______________________

NAME ______________________ CLASS __________ DATE __________

Lesson Activity

3.1 Using Matrices to Display Data

Use the data in matrix L to answer Exercises 1–4.

	Native speakers (millions)	Total no. of speakers worldwide (millions)
Mandarin	827	930
Hindi	327	400
Spanish	326	371
English	319	463

$$\begin{bmatrix} 827 & 930 \\ 327 & 400 \\ 326 & 371 \\ 319 & 463 \end{bmatrix} = L$$

1. What do the entries in the matrix represent? ______________________

2. What are the dimensions of matrix L? ______________________

3. Which entry tells how many native Spanish speakers there are in the world? ______________________

4. What information can you learn from entry l_{32}? ______________________

5. Create a matrix to show the following information:

There are 184 million native Bengali speakers in the world. There are 2 million fewer native Arabic speakers. The total number of Bengali speakers worldwide is 192 million, which is 22 million fewer than the total number of Arabic speakers.

Inches can be represented in terms of other units of measure. Use matrix I to answer Exercises 6–10.

ft	in.
yd	cm
mi	m

$$I = \begin{bmatrix} 0.83 & 1 \\ 0.028 & 2.54 \\ 0.000016 & 0.025 \end{bmatrix}$$

6. Which entry tells how many miles are in an inch? ______________________

7. What information do you learn from entry i_{22}? ______________________

8. How are the table and the matrix related? ______________________

____

9. Why is there a 1 in entry i_{12}? ______________________

10. What are the dimensions of the table and the matrix? ______________________

NAME ______________________________ CLASS ____________ DATE ____________

Basic Skills Practice

Graphing Points on a Coordinate Grid, Part 1

A coordinate grid is made up of the intersection of two lines. The vertical line is called the *y*-axis, and the horizontal line is called the *x*-axis. The lines intersect at the origin. The coordinates of the origin are (0, 0).

Example 1: Write the ordered pair for each point.

Point *A* From the origin, move right 4 units and down 2 units. The ordered pair is (4, −2).

Point *B* From the origin, move left 3 units and up 5 units. The ordered pair is (−3, 5).

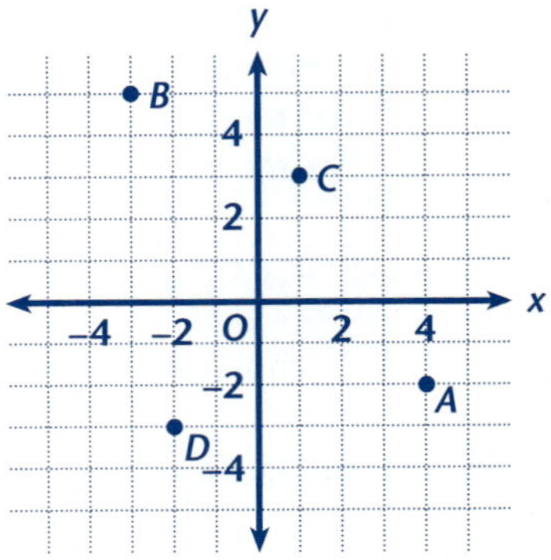

Example 2: Write the point named by each ordered pair.

(1, 3) From the origin, move right 1 unit and up 3 units. Point *C* is located at (1, 3).

(−2, −3) From the origin, move left 2 units and down 3 units. Point *D* is located at (−2, −3).

Use the coordinate grid at right for Exercises 1–8. Write the ordered pair.

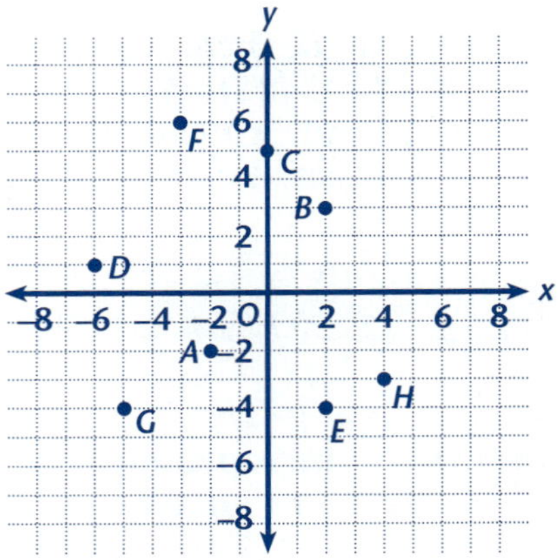

1. point *A* __________ **2.** point *B* __________

3. point *C* __________ **4.** point *D* __________

5. point *E* __________ **6.** point *F* __________

7. point *G* __________ **8.** point *H* __________

Use the coordinate grid at right for Exercises 9–20. Write the point named by the ordered pair.

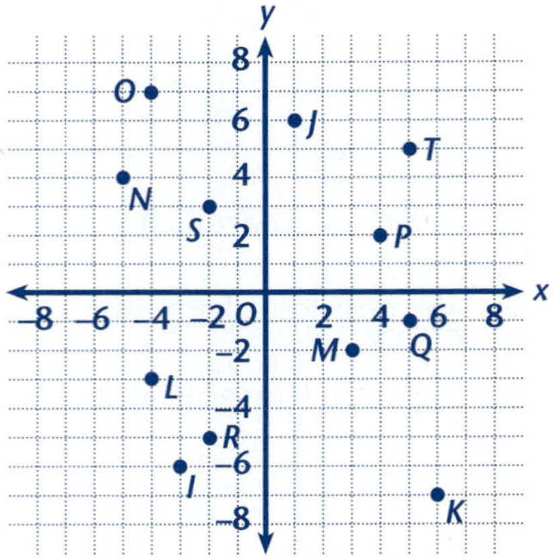

9. (6, −7) __________ **10.** (−3, −6) __________

11. (5, 5) __________ **12.** (−4, 7) __________

13. (−2, 3) __________ **14.** (1, 6) __________

15. (−4, −3) __________ **16.** (3, −2) __________

17. (−2, −5) __________ **18.** (4, 2) __________

19. (5, −1) __________ **20.** (−5, 4) __________

LESSON 3.2

Adding and Subtracting Matrices

PREPARE

Pacing Guide
2 Days

Objectives and Overview
- Add and subtract matrices.
- Determine whether there is an additive identity for matrices.

Student use a real-world application of matrices to learn how to add and subtract matrices. They review the Commutative Property for Addition and the Associative Property for Addition and show that matrix addition is also commutative and associative.

Vocabulary
identity matrix
zero matrix

Assessing Prior Knowledge

Add or subtract.

1.	$5 + (-3.7)$	[1.3]
2.	$-7.2 - 8.8$	[−16]
3.	$\frac{3}{8} + \left(-\frac{3}{4}\right)$	$\left[-\frac{3}{8}\right]$
4.	$0 + (-0.09)$	[−0.09]

Approach A: Using the Book

Opening Discussion
Point out that if data are organized in a matrix and the data change, you will need a mathematical way of making the change.

Ask students to imagine working in a store that sells videos, games, CDs, DATs, laser discs, and tapes. How might they use a matrix to display an inventory of each item? If an inventory is made of the stock at the beginning of every month, how would the matrix change? What mathematical methods would be helpful in representing the amount of stock available after an inventory?

Brainstorming
Ask students for examples of data that need to be updated periodically. For example, a bank keeps track of accounts with a matrix, and a newspaper prints the statistics for a baseball team in the same matrix format every week. As the players get more

continued on top of next page

TEACH

Approach B: Using Models

Opening Discussion
Point out that if a store keeps track of its inventory in a matrix, then it makes sense to be able to add or subtract the matrix as items are continually bought and restocked.

Ask students to imagine working in a store that sells videos, games, CDs, DATs, laser discs, and tapes. How might they use a matrix to display an inventory of each item? If an inventory is made of the stock at the beginning of every month, how would the matrix change? What mathematical procedures would be helpful in representing the amount of stock available after an inventory?

Brainstorming
Ask students for examples of data that are organized in a matrix. When the data change, you will need a way to combine the information in the old and new matrices. For example, a bank keeps track of accounts with a matrix, and a newspaper prints the

continued on bottom of next page

Approach A

Using the Book, continued

hits, score more runs, etc., the statistics are updated, and the new matrix is printed in the same format as the old matrix.

Adding and Subtracting Matrices

Students may understand adding and subtracting matrices better if the matrices represent real-world data, such as the examples on page 129 of the Pupil's Edition. Point out that the dimensions for each matrix must be the same and that the data in each matrix must represent exactly the same information in order for adding or subtracting matrices to work.

Example 1

In this example, two matrices are given, and one matrix is subtracted from the other.

One represents the stock in a shop and the other represents the sales of the stock. Make sure students see that each matrix is organized *in exactly the same way* with the same dimensions. For example, video data is located in the first row, first column *for both matrices*. Subtracting the two matrices would not make sense if the matrices were not organized the same way. Also point out that the matrix representing the difference of the two matrices will also be organized in the same way and have the same dimensions.

Additional Example 1

The number of students enrolled in June for the next school year is given by class and gender in the table below and in matrix form on the next page.

June Enrollment

	Girls	Boys
Freshmen	142	91
Sophomores	155	176
Juniors	123	99
Seniors	89	101

TEACH

Approach B

Using Models, continued

statistics for a baseball team in the same matrix every week. As players get more hits, score more runs, etc., the statistics are updated.

Using a Graphics Calculator

A graphics calculator can help students add or subtract matrices and can help students verify that matrix addition is commutative and associative.

Provide students with graphics calculators with matrix functions. Ask them to follow these steps to add or subtract three matrices with a calculator: (1) Enter the matrices into the calculator. (Note: See Lesson 3.1 of this planning guide for instructions). Choose [*A*] as the name of the first matrix, [*B*] as the name of the second matrix, and [*C*] as the name of the third matrix. (2) Return to the home screen. (3) Choose the appropriate combination of matrices *A*, *B*, and *C* to perform the additions or subtractions. Use normal addition or subtraction keys as operators. For example $A + B - C$ would be entered as $[A] + [B] - [C]$. Point out that they should use parentheses with the matrix names if they are verifying the Associative Property.

Working With a Graphics Calculator

Have students enter the following matrices into a graphics calculator and then perform the operations shown:

$$A = \begin{bmatrix} 8 & 4 \\ -2 & 3 \\ 1 & 2 \end{bmatrix} \quad B = \begin{bmatrix} 1 & -2 \\ 3 & 2 \\ -4 & -1 \end{bmatrix} \quad C = \begin{bmatrix} 5 & -1 \\ 0 & 6 \\ 2 & -3 \end{bmatrix}$$

A. $A + B$ **B.** $B - A$ **C.** $C + A - B$

$$\left[\textbf{A.} \begin{bmatrix} 9 & 2 \\ 1 & 5 \\ -3 & 1 \end{bmatrix} \quad \textbf{B.} \begin{bmatrix} -7 & -6 \\ 5 & -1 \\ -5 & -3 \end{bmatrix} \quad \textbf{C.} \begin{bmatrix} 12 & 5 \\ -5 & 7 \\ 7 & 0 \end{bmatrix} \right]$$

Addition and Subtraction Properties

A calculator can be used to explore and develop conjectures about the algebra of adding and subtracting matrices.

Using the matrices in the exercises above, have students verify that matrix addition is commutative ($A + B = B + A$ and

$$A = \begin{bmatrix} 142 & 91 \\ 155 & 176 \\ 123 & 99 \\ 89 & 101 \end{bmatrix}$$

During the summer, several students moved out of the school district, so the enrollment needed to be adjusted to reflect the *loss* of students, according to the following matrix:

$$B = \begin{bmatrix} 1 & 0 \\ 4 & 1 \\ 0 & 3 \\ 1 & 2 \end{bmatrix}$$

Since the number of students has decreased, how is enrollment for the next year found? **[Subtract each entry in *B* from its corresponding entry in *A*. The result will be a new matrix *C*, which shows the enrollment for the next year.**

$$C = \begin{bmatrix} 141 & 91 \\ 151 & 175 \\ 123 & 96 \\ 88 & 99 \end{bmatrix}$$

	Enrollment	
	Girls	**Boys**
Freshmen	141	91
Sophomores	151	175
Juniors	123	96
Seniors	88	99

Example 2

In this example, students add a matrix to the result in Example 1.

Point out that the matrix used to update the stock still has the same dimensions and the same organization (labels) as the original matrix. Emphasize that because addition is commutative, adding the matrices as $C + D$ gives the same result as adding them as $D + C$.

Additional Example 2

Refer to Additional Example 1. During the summer, several students moved into the school district, so the enrollment needed to be adjusted

continued on top of next page

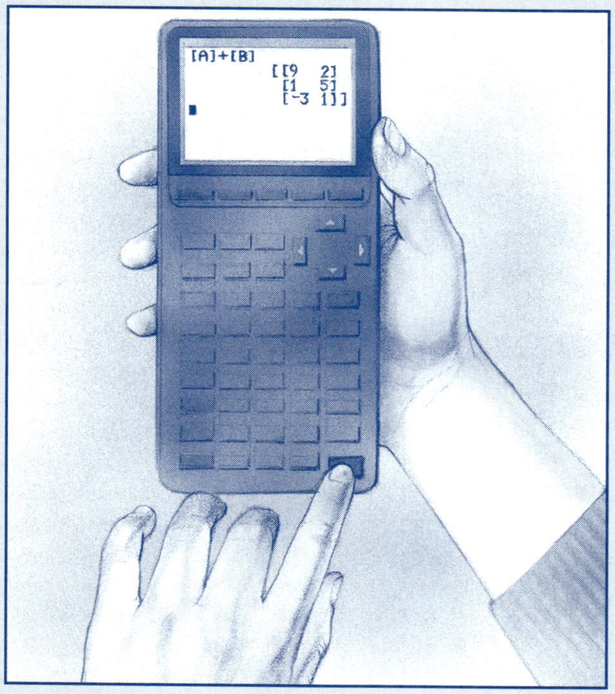

Students can study addition and subtraction of matrices using a graphics calculator.

$B + C = C + B$). Have them also use the three matrices to verify that matrix addition is associative ($(A + B) + C = A + (B + C)$). Point out that a calculator is programmed to follow the correct order of operations, so using parentheses is necessary to verify the properties. Also have students use a calculator to verify that matrix subtraction is not commutative ($A - B \neq B - A$).

Cooperative Learning

Matrices Help students understand the properties of matrices by dividing the class into groups of four. Ask one student in each group to write three matrices with the same dimensions. Have the second student use the three matrices to show that matrix addition is commutative. Ask the third student to use the three matrices to show that matrix addition is associative. Have the fourth student show that matrix subtraction is not commutative. The students should discuss the results. Have students switch roles and repeat the activity with three more matrices.

continued on bottom of next page

Approach A Using the Book, continued

to reflect the *gain* of students, according to the following matrix:

$$D = \begin{bmatrix} 8 & 12 \\ 10 & 0 \\ 1 & 20 \\ 2 & 7 \end{bmatrix}$$

How would you update your enrollment matrix from Additional Example 1 to show the final enrollment for the next year? [**Add each entry in the new matrix, *D*, to the corresponding entry in *C* to get the new enrollment matrix, *E*, for the next year.**]

$$E = \begin{bmatrix} 149 & 103 \\ 161 & 175 \\ 124 & 116 \\ 90 & 106 \end{bmatrix}$$

	Enrollment	
	Girls	**Boys**
Freshmen	149	103
Sophomores	161	175
Juniors	124	116
Seniors	90	106

Ongoing Assessment, page 131

Try This

$$\begin{bmatrix} 4 & 4 \\ 2 & 3 \end{bmatrix}$$

Critical Thinking, page 131

In a large business, as a matter of fact, the inventory may be updated every time a purchase is made. In order to manage a large inventory, computers are used to update the inventory and manage the stock.

TEACH

Approach B Using Models, continued

Identity Matrix for Addition

Have students work in pairs and use the following matrix to explore the identity matrix for addition, which is the zero matrix:

$$\begin{bmatrix} 1.1 & -3.2 & 4 \\ -2 & 1.4 & -\frac{1}{2} \end{bmatrix}$$

Have one student enter the data as matrix *A* in the calculator. Ask the other student to enter the corresponding zero matrix as matrix *B*. Have them find $A + B$ and $B + A$ and answer the following questions: Are the dimensions the same? What is the sum? Is the sum the same as the original matrix?

Using Technology

Have students use spreadsheets to help them understand how corresponding entries are added in matrices with the same dimensions. Ask them to put the following data into a spreadsheet. Then label matrix *A* as the medal count after the first day of the 1996 Summer Olympics and label matrix *B* as the corresponding medal count after the second day.

Medal Count
After Day 1 of the 1996 Summer Olympics

	Gold	Silver	Bronze	Total
Germany	0	2	3	5
China	2	1	0	3
U.S.	0	2	1	3
Belgium	1	0	1	2
France	1	0	1	2

$$A = \begin{bmatrix} 0 & 2 & 3 & 5 \\ 2 & 1 & 0 & 3 \\ 0 & 2 & 1 & 3 \\ 1 & 0 & 1 & 2 \\ 1 & 0 & 1 & 2 \end{bmatrix}$$

Exploration: Addition and Subtraction Properties

In this exploration, students explore addition and subtraction of matrices.

Students discover that addition of matrices is both commutative and associative, while subtraction is neither. Also, matrices must have the same dimensions to be subtracted.

Use **Transparency 12** for discussion.

Ongoing Assessment, page 131

2, 4. Matrices must have the same dimensions in order to be added or subtracted.

Inclusion Strategy

Ask students to write a 3×2 matrix with positive and negative entries on a large sheet of paper. Then ask them to work in groups of three. Label the group's matrices A, B, and C. Ask students to find $A + B - C$, $B + A - C$, and $A + C - B$. Which two answers are the same? **[the first and the second]**

The Identity Matrix for Addition

The identity matrix for addition, or the zero matrix, is a matrix filled with zeros. Point out that every matrix has a corresponding zero matrix. It has the same dimensions as the original matrix, and all of its entries are zero. When added to the matrix, the result does not change the original matrix. Have students compare the zero matrix with the additive identity, 0.

Ongoing Assessment, bottom of page 131

They both have zeroes as entries, but the zero matrix for C is 2×2 and the zero matrix for D is 2×3.

Ongoing Assessment, page 132

Matrix addition is commutative.

$$L + M = \begin{bmatrix} 2 & 3 \\ 6 & -1 \end{bmatrix} + \begin{bmatrix} 4 & 5 \\ -2 & 0 \end{bmatrix}$$

$$= \begin{bmatrix} 2+4 & 3+5 \\ 6+(-2) & -1+0 \end{bmatrix} = \begin{bmatrix} 6 & 8 \\ 4 & -1 \end{bmatrix}$$

continued on top of next page

Medal Count
After Day 2 of the 1996 Summer Olympics

	Gold	Silver	Bronze	Total
Germany	0	4	5	9
China	3	1	1	5
U.S.	2	6	2	10
Belgium	2	0	1	3
France	3	2	3	8

$$B = \begin{bmatrix} 0 & 4 & 5 & 9 \\ 3 & 1 & 1 & 5 \\ 2 & 6 & 2 & 10 \\ 2 & 0 & 1 & 3 \\ 3 & 2 & 3 & 8 \end{bmatrix}$$

Have students use the spreadsheet to find matrix C, the number of medals won on the second day.

The medals won during the second day by these five countries are entries in matrix C; $C = B - A$.

Medal Count for Day 2
of the 1996 Summer Olympics

	Gold	Silver	Bronze	Total
Germany	0	2	2	4
China	1	0	1	2
U.S.	2	4	1	7
Belgium	1	0	0	1
France	2	2	2	6

$$C = \begin{bmatrix} 0 & 2 & 2 & 4 \\ 1 & 0 & 1 & 2 \\ 2 & 4 & 1 & 7 \\ 1 & 0 & 0 & 1 \\ 2 & 2 & 2 & 6 \end{bmatrix}$$

continued on bottom of next page

Approach A: Using the Book, continued

and

$$M + L = \begin{bmatrix} 4 & 5 \\ -2 & 0 \end{bmatrix} + \begin{bmatrix} 2 & 3 \\ 6 & -1 \end{bmatrix}$$
$$= \begin{bmatrix} 4+2 & 5+3 \\ -2+6 & 0+(-1) \end{bmatrix} = \begin{bmatrix} 6 & 8 \\ 4 & -1 \end{bmatrix}$$

So, $L + M = M + L$.

Matrix addition is associative.

$$L + (M + N) = \begin{bmatrix} 2 & 3 \\ 6 & -1 \end{bmatrix} + \left(\begin{bmatrix} 4 & 5 \\ -2 & 0 \end{bmatrix} + \begin{bmatrix} -1 & 4 \\ 7 & -5 \end{bmatrix}\right)$$
$$= \begin{bmatrix} 2 & 3 \\ 6 & -1 \end{bmatrix} + \begin{bmatrix} 4-1 & 5+4 \\ -2+7 & 0-5 \end{bmatrix}$$
$$= \begin{bmatrix} 2 & 3 \\ 6 & -1 \end{bmatrix} + \begin{bmatrix} 3 & 9 \\ 5 & -5 \end{bmatrix}$$
$$= \begin{bmatrix} 5 & 12 \\ 11 & -6 \end{bmatrix}$$

and

$$(L + M) + N = \left(\begin{bmatrix} 2 & 3 \\ 6 & -1 \end{bmatrix} + \begin{bmatrix} 4 & 5 \\ -2 & 0 \end{bmatrix}\right) + \begin{bmatrix} -1 & 4 \\ 7 & -5 \end{bmatrix}$$
$$= \begin{bmatrix} 2+4 & 3+5 \\ 6-2 & -1+0 \end{bmatrix} + \begin{bmatrix} -1 & 4 \\ 7 & -5 \end{bmatrix}$$
$$= \begin{bmatrix} 6 & 8 \\ 4 & -1 \end{bmatrix} + \begin{bmatrix} -1 & 4 \\ 7 & -5 \end{bmatrix}$$
$$= \begin{bmatrix} 5 & 12 \\ 11 & -6 \end{bmatrix}$$

So, $L + (M + N) = (L + M) + N$.

Closing the Lesson

Inviting Student Participation
Summarize the lesson by having three students write matrices with the same dimensions on the chalkboard. Then have a fourth student list a property of the matrices. Have successive students write properties, sums, or differences for the matrices until all operations and properties of the matrices have been reviewed. Repeat the process with three new matrices until all students have participated.

TEACH

Approach B: Using Models, continued

Journal Entry

Have students use examples to discuss matrix addition and subtraction in their journal. Ask them to include a discussion of the Commutative Property and the Associative Property as they apply to matrix addition.

Teaching Tip

Matrix Display on a Graphics Calculator Students using a graphics calculator to display matrices may find it difficult to use the calculator to do matrix addition or subtraction. Students should start with a clear home screen and then choose the first matrix by name from the matrix menu. After choosing the matrix, the screen reverts back to the home screen so that addition, +, or subtraction, −, can be entered as the operation. Then students need to choose the matrix menu again so that the other matrix can by chosen by name.

ASSESS

Selected Answers

Odd-numbered Exercises 9–39 and odd-numbered Extra Practice

Assignment Guide

Core level students should complete Exercises 1–40.

Core-plus level students should complete Exercises 1–41.

Error Analysis

Students who have trouble interpreting how matrices are used to organize data may have trouble answering Exercise 25. Point out that the total number of free throws made during the playoffs is found by adding the entire column of numbers. Students first find the matrix of points scored during the playoff series by subtracting the first matrix from the second (see Exercise 23). Then they have to add the numbers in the free-throw column.

Alternative Assessment

Portfolio Activity Have students choose several problems in the exercise set that they enjoyed. Have them show the matrices and a complete solution, with steps. Ask them to explain why they chose these problems.

RETEACH

Use the following matrices to help students review the concepts of the lesson:

$A = \begin{bmatrix} 4 & -2 \\ \frac{3}{4} & 1 \end{bmatrix}$ $B = \begin{bmatrix} -3 & 5 \\ -\frac{2}{4} & -2 \end{bmatrix}$

$C = \begin{bmatrix} w & x \\ y & z \end{bmatrix}$

1. Find $A + B$. $\begin{bmatrix} \mathbf{1} & \mathbf{3} \\ \mathbf{\frac{1}{4}} & \mathbf{-1} \end{bmatrix}$

2. Find $A - B$. $\begin{bmatrix} \mathbf{7} & \mathbf{-7} \\ \mathbf{\frac{5}{4}} & \mathbf{3} \end{bmatrix}$

3. Does $A - B = B - A$? **[no]**

4. Find C such that $B + C$ is the zero matrix.
$\begin{bmatrix} \mathbf{3} & \mathbf{-5} \\ \mathbf{\frac{2}{4}} & \mathbf{2} \end{bmatrix}$

ENRICH

Using Technology Have students explore the matrix functions of a graphics calculator by entering the following sequence of commands (in boldface) in the Matrix Math menu or an equivalent sequence of commands.

{2,2} —> dim [A]	*"—>" is the store key; creates matrix A with dimensions 2 × 2.*
[A]	*lists matrix A*
Fill(1,[A])	*fills matrix A with 1s*
randM(2,2)	*creates a 2 × 2 matrix and fills it with random numbers*

Interdisciplinary Connection

Computer Programming A computer programmer has to know how to enter data efficiently into a computer. One technique is to use control loops to enter data into an array (a matrix). The dimensions of the array are specified in the program, and the addresses of the matrix entries are given as row-by-column entries. The program fills the addresses of the matrix quickly. If a second matrix is also entered, the computer can be used to add the two matrices quickly.

NAME ______________________________ CLASS ______________ DATE ______________

Practice

3.2 Adding and Subtracting Matrices

Perform each of the matrix operations. If an operation is not possible, explain why.

1. $\begin{bmatrix} 1 & 6 \\ 2 & 9 \end{bmatrix} + \begin{bmatrix} 3 & 0 \\ 4 & 11 \end{bmatrix}$

2. $\begin{bmatrix} 9 & 4 & 7 \\ 0 & 3 & 13 \end{bmatrix} + \begin{bmatrix} 15 & 8 & 2 \\ 17 & 1 & 5 \end{bmatrix}$

3. $\begin{bmatrix} -4 & 10 \\ -13 & -8 \\ 17 & 20 \end{bmatrix} - \begin{bmatrix} -5 & -31 \\ 4 & -17 \\ 16 & -23 \end{bmatrix}$

4. $\begin{bmatrix} -7 & -33 & 38 \\ 44 & 2 & -28 \\ 9 & -9 & 26 \end{bmatrix} - \begin{bmatrix} -5 & 33 & -14 \\ -13 & 40 & 50 \\ 16 & -21 & -6 \end{bmatrix}$

5. $\begin{bmatrix} 2.3 \\ -1.9 \\ 6 \end{bmatrix} + \begin{bmatrix} -5.2 \\ -3 \\ -1.7 \end{bmatrix} - \begin{bmatrix} -4.6 \\ 0.2 \\ 10 \end{bmatrix}$

6. $\begin{bmatrix} 1.2 & 3.4 & 7 \\ -0.9 & 9.2 & -1.3 \end{bmatrix} - \begin{bmatrix} -2 & 1.5 \\ 6 & 7 \end{bmatrix}$

7. $\begin{bmatrix} \frac{1}{3} & \frac{1}{4} \\ -\frac{3}{4} & \frac{1}{2} \end{bmatrix} + \begin{bmatrix} \frac{2}{3} & -1 \\ \frac{1}{2} & \frac{1}{4} \end{bmatrix}$

8. $\begin{bmatrix} 0 & -2 & 5 \\ 1.5 & 6 & -7 \\ -10 & 3 & 5.2 \end{bmatrix} + \begin{bmatrix} -5 & 1 & 3 \\ 5 & 6.3 & 5 \\ 9 & 7 & 0.9 \end{bmatrix}$

9. $[2 \quad 3.1 \quad 4.5] + \begin{bmatrix} 1 \\ 2.4 \\ 7.6 \end{bmatrix}$

10. $\begin{bmatrix} 7.3 & -0.1 \\ -8 & 6.2 \end{bmatrix} - \begin{bmatrix} -5 & -3.7 \\ 4.9 & 1.4 \end{bmatrix} + \begin{bmatrix} 0.8 & 5.8 \\ -2.6 & 13 \end{bmatrix}$

Lesson Activity

3.2 Adding and Subtracting Matrices

Matrices A and B show how much it costs to buy a share of each stock listed and how much it costs to sell a share of each stock one month later.

Buying Stocks: Jan.
Trifax Co.
Compu Count Inc.
Video Antics

$$A = \begin{bmatrix} 15\frac{1}{2} \\ 12\frac{5}{8} \\ 71\frac{3}{4} \end{bmatrix}$$

Selling Stocks: Feb.
Trifax Co.
Compu Count Inc.
Video Antics

$$B = \begin{bmatrix} 12\frac{3}{4} \\ 14 \\ 71\frac{3}{4} \end{bmatrix}$$

1. Create a new matrix that shows the profit or loss for this investor in February.

2. Describe the investment activity represented by the matrix you created.

____

3. Matrix C shows gas, electric, and water bills for the first three months of 1996. Matrix D shows the same bills for the first three months of 1997. $E = C + D$. Use the information in the matrices to rewrite matrices C and D so that the entries indicate the actual amount of each bill.

Matrix C

	Gas	**Electric**	**Water**
Jan. 96	$2a + 3$	$b + 4$	$3c - 2$
Feb. 96	$a - 2$	$2b + 1$	$c + 5$
Mar. 96	$2a - 1$	$3b + 3$	$2c + 3$

Matrix D

	Gas	**Electric**	**Water**
Jan. 97	$5a - 2$	$2b - 3$	$c + 6$
Feb. 97	$2a + 2$	$b + 2$	$2c - 3$
Mar. 97	$2a + 2$	$b - 1$	$c - 2$

Matrix E

	Gas	**Electric**	**Water**
Jan.	106	61	24
Feb.	45	63	17
Mar.	61	82	16

______________________ ______________________

Matrix C Matrix D

NAME ______________________________ CLASS ______________ DATE ______________

Basic Skills Practice

Graphing Points on a Coordinate Grid, Part 2

Read each question and circle the best answer.

Use the graph to answer Exercises 1 and 2.

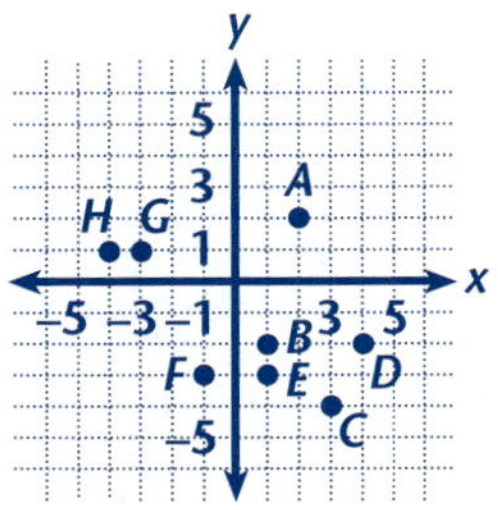

1. For which point is $x > 1$ and $y < -2$?

A point *A*
B point *B*
C point *C*
D point *D*

2. For which point is $x < -3$ and $y \geq 1$?

F point *E*
G point *F*
H point *G*
I point *H*

3. What is the ordered pair for point *A*?

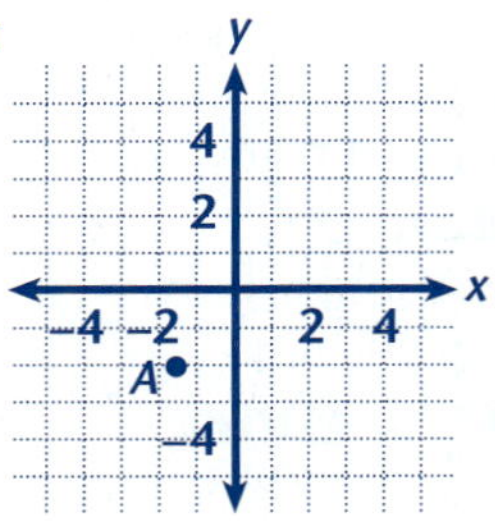

A $(-1.5, -2)$
B $(-2, -2)$
C $(-2, -1.5)$
D $(1.5, 2)$

4. Which is an equation for line *m*?

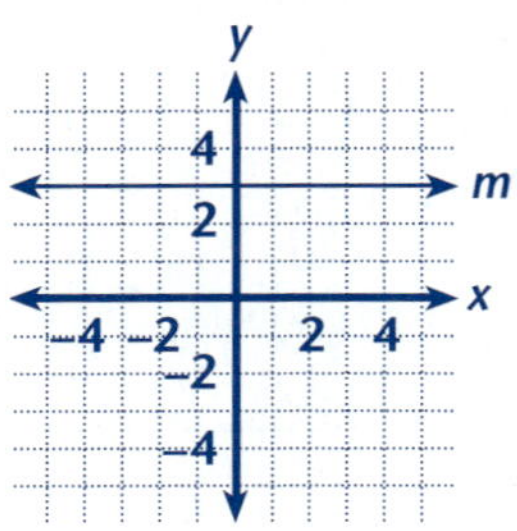

F $x = 3$
G $y = 3$
H $y = -3$
I $x = -3$

Use the graph to answer Exercises 5 and 6.

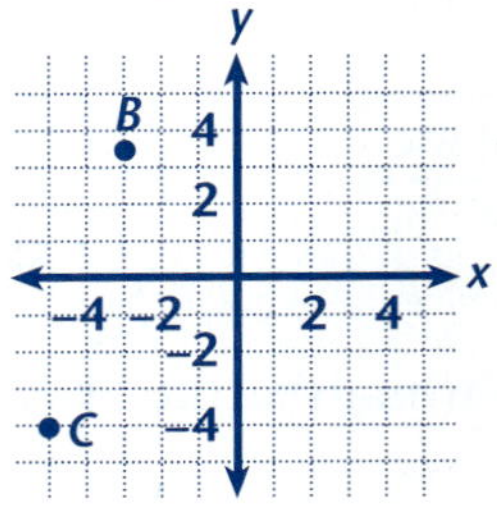

5. What are the coordinates of point *B*?

A $(-3, 4)$
B $\left(3, -3\frac{1}{2}\right)$
C $\left(-3, 3\frac{1}{2}\right)$
D $\left(-3\frac{1}{2}, 3\right)$

6. What are the coordinates of point *C*?

F $(-5, -4)$
G $(-5, 4)$
H $(5, -4)$
I $(-4, -5)$

7. For which point is $x > -1$ and $y \leq -2$?

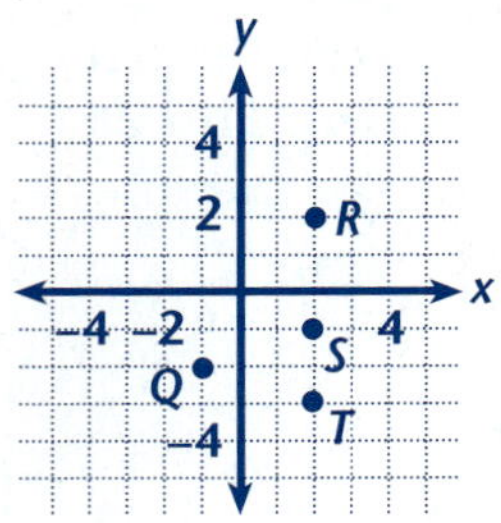

A point *Q*
B point *R*
C point *S*
D point *T*

8. What is the ordered pair for point *S*?

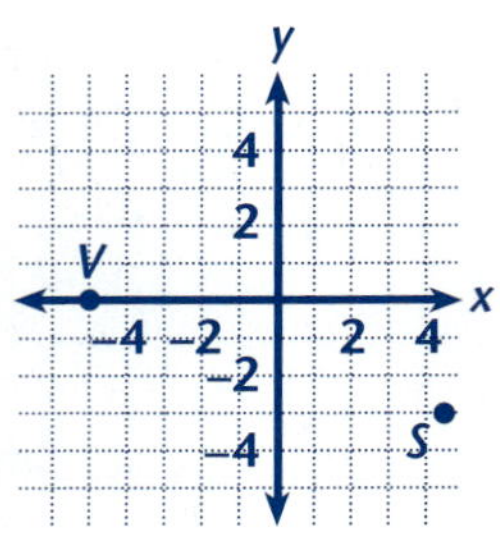

F $(4.5, 3)$
G $(-4.5, 3)$
H $(3, -4.5)$
I $(4.5, -3)$

LESSON 3.3

Exploring Matrix Multiplication

PREPARE

Pacing Guide

2 Days

Objectives and Overview

- Multiply a matrix by a scalar, and multiply matrices.
- Determine the identity matrix for multiplication.

Students explore how to multiply each entry in a matrix by a constant, a process called *scalar multiplication*. They also explore how to multiply two matrices and how to interpret the product matrix. The identity matrix for multiplication is defined and explored.

Vocabulary

scalar multiplication
scalar
product matrix
identity matrix

Assessing Prior Knowledge

Simplify each expression.

1.	$-2(8) + 3.1(7)$	[5.7]
2.	$12(4) + 10(6)$	[108]
3.	$4(1) + 7(0)$	[4]
4.	$4(x) + (-5)(x - 1)$	$[-x + 5]$

Approach A: Using the Book

Opening Discussion

Discuss how repeated addition of the same matrix can be shortened by using multiplication. Would this be the same as multiplying two matrices? **[no]**

Brainstorming

Suppose that Sally entered inventory data from her toy store into the left matrix. She wants to expand the store with 10% more dolls, 20% more trucks, and 50% fewer models in each aisle. This information is stored in the right matrix.

$$\begin{array}{l} \\ \textbf{Aisle 1} \\ \textbf{Aisle 2} \\ \textbf{Aisle 3} \end{array} \begin{array}{c} \begin{array}{ccc} \textbf{Dolls} & \textbf{Trucks} & \textbf{Models} \end{array} \\ \begin{bmatrix} 40 & 30 & 14 \\ 20 & 15 & 30 \\ 30 & 20 & 4 \end{bmatrix} \end{array} \qquad \begin{array}{l} \\ \textbf{Dolls} \\ \textbf{Trucks} \\ \textbf{Models} \end{array} \begin{array}{c} \textbf{Toys} \\ \begin{bmatrix} 1.1 \\ 1.2 \\ 0.5 \end{bmatrix} \end{array}$$

Have students brainstorm how they could use matrices to find the total number of toys in each aisle after the change. **[Multiply each row of the**

continued on top of next page

TEACH

Approach B: Using Models

Opening Discussion

Have students think about how to shorten repeated addition of the same matrix. Would this be the same as multiplying two matrices? **[no]**

Brainstorming

Suppose that the left matrix below represents the inventory at a toy store and that the right matrix represents a plan to increase the inventory 10% for dolls, to increase the inventory 20% for trucks, and to decrease the number of models by 50%.

$$\begin{array}{l} \\ \textbf{Aisle 1} \\ \textbf{Aisle 2} \\ \textbf{Aisle 3} \end{array} \begin{array}{c} \begin{array}{ccc} \textbf{Dolls} & \textbf{Trucks} & \textbf{Models} \end{array} \\ \begin{bmatrix} 40 & 30 & 14 \\ 20 & 15 & 30 \\ 30 & 20 & 4 \end{bmatrix} \end{array} \qquad \begin{array}{l} \textbf{Dolls} \\ \textbf{Trucks} \\ \textbf{Models} \end{array} \begin{bmatrix} 1.1 \\ 1.2 \\ 0.5 \end{bmatrix}$$

Have students brainstorm how they could use matrices to find the total number of toys in each aisle after the change. **[Multiply each row of the**

continued on bottom of next page

Approach A

Using the Book, continued

left matrix by the matrix on the right entry-by-entry, then add. The results are Aisle 1: 87, Aisle 2: 55, Aisle 3: 59.]

Setting Up the Explorations

Teacher-Guided Explorations If you wish, you can guide the students through the textbook explorations by using the overhead projector. Follow each exploration step by step. Make sure students understand how each entry in the solution matrix is determined. Explain how to interpret the product matrices of application problems.

Cooperative-Learning Groups Divide the class into groups of four. Ask one student in each group to read the steps from the textbook. Have another student write the matrices used in the explorations. Have a third student answer the questions about the product matrices. Have a fourth student write a report of the group's activities. Ask each student to sign the report.

Scalar Multiplication

Point out that scalar multiplication is a short way to perform repeated addition of the same matrix and that it is related to normal repeated addition. Emphasize that a scalar is a real number, not a matrix, that is multiplied by each entry in the matrix.

Exploration 1

In this exploration, students multiply two matrices, each with dimensions of 2 by 2.

Emphasize that the number of columns in the left matrix must equal the number of rows in the right matrix in order for them to be multiplied. If the matrices apply to real-world situations, as they do in this exploration, then the left matrix column labels must match the right matrix row labels. The column labels for the product matrix are the same as the column labels for the right

TEACH

Approach B

Using Models, continued

left matrix by the matrix on the right entry-by-entry, then add. The results are Aisle 1: 87, Aisle 2: 55, Aisle 3: 59.]

Multiplying Two Matrices

If graphics calculators are available, you may want to show students how to multiply two matrices by using the calculator's matrix functions. Have students make conjectures about what entries should be in the product matrix. Also enter two matrices that cannot be multiplied because the number of columns in the left matrix does not equal the number of rows of the right matrix. Show the error message given. Also use the calculator to show that the multiplication of two matrices is not commutative.

Cooperative Learning

Cooperative-Learning Groups Before starting the explorations, divide the class into groups of four. Ask one student to read the steps from the textbook. Have another student write the matrices used in the explorations. Have a third student answer the questions about the product matrices. Have a fourth student write a report of the group's activities. Ask each student to sign the completed report.

Exploration 1

Have students investigate the answers to the following questions, which may be listed on the chalkboard or on the overhead. (For ideas on guiding students, see Exploration 1 on page 139 of the Pupil's Edition of the textbook.)

- For the matrices below, explain why the matrix on the left cannot be multiplied by the matrix on the right.

$$\begin{array}{r} \\ \text{Sam} \\ \text{Kim} \end{array} \begin{array}{c} \begin{array}{cc} \text{Cars} & \text{Trucks} \end{array} \\ \begin{bmatrix} 8 & 15 \\ 12 & 10 \end{bmatrix} \end{array} \qquad \begin{array}{r} \\ \text{Sam} \\ \text{Kim} \end{array} \begin{array}{c} \begin{array}{cc} \text{Wheels} & \text{Nails} \end{array} \\ \begin{bmatrix} 4 & 6 \\ 7 & 9 \end{bmatrix} \end{array}$$

[These two matrices cannot be multiplied because the column and row labels do not match up correctly; the right matrix row labels must be the same as the left matrix column labels.]

matrix, and the row labels for the product matrix are the same as the row labels for the left matrix.

Ongoing Assessment, page 139

4. $\begin{bmatrix} 122 & 191 \\ 108 & 174 \end{bmatrix}$

Follow-Up Demonstration (Optional)

If you need closure for this exploration, have students determine how many cars and trucks they plan to make and the number of wheels and nails they will need for construction without using matrices. Use just the information given. Then present the matrix form shown in the exploration. Reemphasize that the data must be arranged correctly in the matrices so that they can be multiplied.

On Their Own

The following matrices show the number of cars, trucks, wheels, and nails needed for next year's service club fair. Find the product.

$$\begin{array}{r} \\ \textbf{Sam} \\ \textbf{Kim} \end{array}\begin{array}{c} \begin{array}{cc}\textbf{Cars} & \textbf{Trucks}\end{array} \\ \begin{bmatrix} 10 & 4 \\ 15 & 16 \end{bmatrix} \end{array} \qquad \begin{array}{r} \\ \textbf{Cars} \\ \textbf{Trucks} \end{array}\begin{array}{c} \begin{array}{cc}\textbf{Wheels} & \textbf{Nails}\end{array} \\ \begin{bmatrix} 4 & 7 \\ 6 & 9 \end{bmatrix} \end{array}$$

$$\left[\begin{array}{r} \\ \textbf{Sam} \\ \textbf{Kim} \end{array}\begin{array}{c} \begin{array}{cc}\textbf{Wheels} & \textbf{Nails}\end{array} \\ \begin{bmatrix} 64 & \mathbf{106} \\ \mathbf{156} & \mathbf{249} \end{bmatrix} \end{array}\right]$$

Exploration 2

In this exploration, students learn a procedure for multiplying two matrices.

Point out that they must first check that the number of columns in the left matrix equals the number of rows in the right matrix. Emphasize that if the left matrix has dimensions $m \times n$ and the right matrix has dimensions $n \times p$, then the product matrix will have dimensions $m \times p$. Each entry, p_{rc}, in the product matrix is found by multiplying row r by column c, as shown in the exploration. For example, p_{21} comes from multiplying row 2 of the left matrix by column 1 of the right matrix.

Use **Transparency 13** for discussion.

continued on top of next page

- Can only square matrices be multiplied? [No; **as long as the number of columns in the left matrix equals the number of rows in the right matrix and the appropriate labels match, the two matrices can be multiplied.**]

Student Reports

One or more members of each group should report their group's findings to the class. Ask them to list their calculations for each entry of the product matrix. The reporter should state the group's conclusions, including the answers to the exploration questions, and the entire group should answer questions from other groups.

Follow-Up Demonstration (Optional)

For this exploration, you may want to demonstrate how to find the number of cars and trucks Kim and Sam plan to make and the number of wheels and nails they will need to construct them without using matrices. Then present the matrix form shown in the exploration as a quicker way to calculate the numbers as long as the matrices are compatible for multiplication.

On Their Own

The following matrices show the number of cars, trucks, wheels, and nails needed for next year's service club fair. Find the product.

$$\begin{array}{r} \\ \textbf{Sam} \\ \textbf{Kim} \end{array}\begin{array}{c} \begin{array}{cc}\textbf{Cars} & \textbf{Trucks}\end{array} \\ \begin{bmatrix} 10 & 4 \\ 15 & 16 \end{bmatrix} \end{array} \qquad \begin{array}{r} \\ \textbf{Cars} \\ \textbf{Trucks} \end{array}\begin{array}{c} \begin{array}{cc}\textbf{Wheels} & \textbf{Nails}\end{array} \\ \begin{bmatrix} 4 & 7 \\ 6 & 9 \end{bmatrix} \end{array}$$

$$\left[\begin{array}{r} \\ \textbf{Sam} \\ \textbf{Kim} \end{array}\begin{array}{c} \begin{array}{cc}\textbf{Wheels} & \textbf{Nails}\end{array} \\ \begin{bmatrix} 64 & \mathbf{106} \\ \mathbf{156} & \mathbf{249} \end{bmatrix} \end{array}\right]$$

Exploration 2

Have students answer the following questions about the procedure used to multiply two matrices. (For ideas on guiding students, see Exploration 2 on page 140 of the Pupil's Edition of the textbook.)

- If an entry of the product matrix has the address p_{24}, does this imply that the left and right matrices have dimensions of 2×4? [No; **however, it does imply that the left matrix has at least 2 columns and that the right matrix has at least 4 rows.**]

continued on bottom of next page

Approach A — Using the Book, continued

Follow-Up Demonstration (Optional)

If you need closure for this exploration, restate the procedure for multiplying two matrices. Use examples and encourage students to give the entries for the product matrix.

Ongoing Assessment, page 140

4. Answers may vary. Sample answer: First check that the number of columns in the left matrix equals the number of rows in the right matrix. If they are equal, then multiply row 1 of the left matrix by column 1 of the right matrix to get an entry in row 1, column 1 of the product matrix. The entry in row 1, column 2 comes from multiplying row 1 of the left matrix by column 2 of the right matrix. Continue this pattern until all entries of the product matrix have been found.

On Their Own

Find each product.

1. $\begin{bmatrix} 2 & -2 \\ 3 & -4 \end{bmatrix} \bullet \begin{bmatrix} -1 & 5 \\ 3 & -2 \end{bmatrix}$ $\left[\begin{matrix} -8 & 14 \\ -15 & 23 \end{matrix}\right]$

2. $\begin{bmatrix} -2 & 1 \\ 4 & -3 \end{bmatrix} \bullet \begin{bmatrix} 4 & 6 \\ -1 & -2 \end{bmatrix}$ $\left[\begin{matrix} -9 & -14 \\ 19 & 30 \end{matrix}\right]$

3. Can the following matrices be multiplied? Why or why not?

$A = \begin{bmatrix} 3 & 2 & 1 \\ -2 & 4 & 4 \end{bmatrix} \qquad B = \begin{bmatrix} 1 & 4 \\ -1 & 3 \end{bmatrix}$

[*AB* is not possible because the number of columns in *A* does not equal the number of rows in *B*. However, *BA* is possible.]

Application

Matrices are commonly used in stock orders, as shown in the Pupil's Edition. Emphasize the need to match the number of columns in the left matrix with the number of rows in the right matrix before the matrices can be multiplied.

TEACH

Approach B — Using Models, continued

- Suppose that the product matrix has one entry. What does this imply about the left and right matrices? **[The left matrix has one row and the right matrix has one column.]**
- Give a general statement about the dimensions of the product matrix of an $m \times n$ matrix and an $n \times p$ matrix. **[The dimensions of the product matrix will be $m \times p$.]**

Follow-Up Demonstration (Optional)

Review the procedure for multiplying two matrices. Use examples, then show the following diagram to help students see that a row of the left matrix is multiplied by a column of the right matrix:

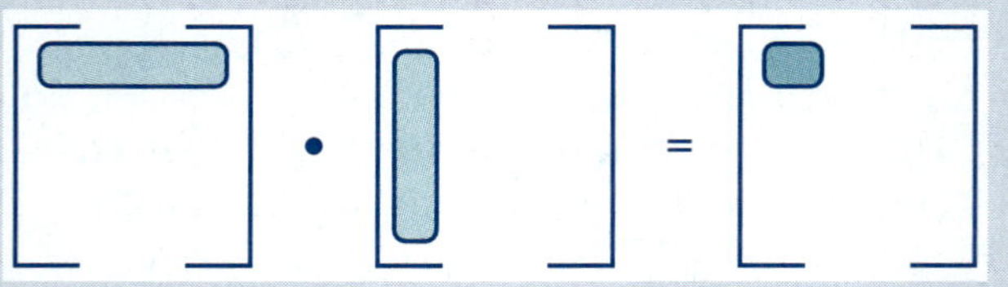

On Their Own

Find each product.

1. $\begin{bmatrix} 2 & -2 \\ 3 & -4 \end{bmatrix} \bullet \begin{bmatrix} -1 & 5 \\ 3 & -2 \end{bmatrix}$ $\left[\begin{matrix} -8 & 14 \\ -15 & 23 \end{matrix}\right]$

2. $\begin{bmatrix} -2 & 1 \\ 4 & -3 \end{bmatrix} \bullet \begin{bmatrix} 4 & 6 \\ -1 & -2 \end{bmatrix}$ $\left[\begin{matrix} -9 & -14 \\ 19 & 30 \end{matrix}\right]$

3. Can the following matrices be multiplied? Why or why not?

$A = \begin{bmatrix} 3 & 2 & 1 \\ -2 & 4 & 4 \end{bmatrix} \qquad B = \begin{bmatrix} 1 & 4 \\ -1 & 3 \end{bmatrix}$

[*AB* is not possible because the number of columns in *A* does not equal the number of columns in *B*. However, *BA* is possible.]

Exploration 3

Have students investigate the following questions about the identity matrix. You may want to display the questions on the chalkboard or overhead. (For ideas on guiding students, see Exploration 3 on page 141 of the Pupil's Edition of the textbook.)

Exploration 3

The multiplicative identity matrix is explored.

Students see that the product of a square matrix and its multiplicative identity is the same square matrix. Students learn that multiplication of matrices is commutative when one of the matrices is the identity matrix. Reemphasize that, in general, the multiplication of matrices is *not* commutative.

Ongoing Assessment, page 141

5. The multiplication identity matrix must be a square matrix because it is used to multiply from the left and from the right. Therefore, the number of rows and columns of the identity matrix must be the same.

Follow-Up Demonstration (Optional)

To bring closure to the exploration, you may want to redo Exploration 3 on page 141 of the Pupil's Edition for the class, using different matrices for L and N. This should reinforce students' conclusions from Steps 4 and 5.

On Their Own

Find each product, given

$L = \begin{bmatrix} 2 & -4 \\ -1 & 3 \end{bmatrix}$, $I = \begin{bmatrix} 1 & 0 \\ 0 & 1 \end{bmatrix}$, and $N = \begin{bmatrix} -1 & 2 \\ -3 & 5 \end{bmatrix}$.

1. LI $\begin{bmatrix} 2 & -4 \\ -1 & 3 \end{bmatrix}$

2. IL $\begin{bmatrix} 2 & -4 \\ -1 & 3 \end{bmatrix}$

3. LN $\begin{bmatrix} 10 & -16 \\ -8 & 13 \end{bmatrix}$

4. Does $LN = NL$? **[no]**

Closing the Lesson

Summarize the lesson by reviewing how to multiply two matrices. Tell students to check that the number of columns in the left matrix equals the number of rows in the right matrix. Then each entry is found by multiplying a row from the left matrix by a column from the right matrix. Ask students to explain why the following multiplication is not possible:

$\begin{bmatrix} -2 & -4 \\ -1 & 5 \end{bmatrix} \bullet [1 \quad 4 \quad -2]$

[The matrix dimensions are not compatible.]

Can the identity matrix for multiplication have any dimensions other than 2×2? **[yes, as long as it is square]**

When is the product of two matrices commutative? **[when the product of the matrices equals the identity matrix]**

Follow-Up Demonstration (Optional)

To bring closure to the exploration, you may want to demonstrate each of the questions answered above with a pair of matrices. This should reinforce students' conclusions.

On Their Own

Find each product, given

$L = \begin{bmatrix} 2 & -4 \\ -1 & 3 \end{bmatrix}$, $I = \begin{bmatrix} 1 & 0 \\ 0 & 1 \end{bmatrix}$, and $N = \begin{bmatrix} -1 & 2 \\ -3 & 5 \end{bmatrix}$.

1. LI $\begin{bmatrix} 2 & -4 \\ -1 & 3 \end{bmatrix}$

2. IL $\begin{bmatrix} 2 & -4 \\ -1 & 3 \end{bmatrix}$

3. LN $\begin{bmatrix} 10 & -16 \\ -8 & 13 \end{bmatrix}$

4. Does $LN = NL$? **[no]**

Closing the Lesson

Summarize the lesson by restating each conclusion from the explorations.

1. To multiply a matrix by a scalar, multiply each entry by that scalar.

2. The product of two matrices is the matrix obtained by multiplying each row of the left matrix by each column of the right matrix.

3. The identity matrix for multiplication is the square matrix consisting of ones on the main diagonal and zeros in every other position. The product of any matrix and the identity matrix is the original matrix.

4. Multiplication of matrices is not commutative.

ASSESS

Selected Answers

Odd-numbered Exercises 7–49 and odd-numbered Extra Practice

Assignment Guide

Core level students should complete Exercises 1–51.

Core-plus level students should complete Exercises 1–51.

Error Analysis

Students sometimes have difficulty solving application matrices like those shown in Exercise 32 because they do not pay attention to the order of multiplication for matrices; that is, they do not pay attention to which matrix they put on the left before they multiply the matrices. Because the matrices in this exercise are not square, it is extremely important to keep the supplies matrix on the right when multiplying. A calculator will give an error message if the order of the matrices is wrong when attempting to multiply them.

Alternative Assessment

Using Models Have students use paper models of cars, trucks, wheels, and nails to demonstrate that they understand the multiplication of matrices shown in Exploration 1 on page 139 of the Pupil's Edition. Ask them to "undo" the matrices by showing how the entries are found from the raw data. For example, Sam makes 8 cars, each with 4 wheels and 7 nails. He also makes 15 trucks, each with 6 wheels and 9 nails. Ask them to make piles of the objects that go together in order to find the product matrix.

Portfolio Activity Have students find a real-world example of data represented by the product of two matrices. Ask them to write a report about how the matrices were generated and to include the procedures they learned for multiplying two matrices. Have them give a statement about why matrices are useful for organizing data.

RETEACH

Have students restate the procedure for multiplying two matrices as a list of steps. Then ask them to follow their steps for the exercise below. Have them compare their answer to the given solution to see if their steps are correct.

$$\begin{bmatrix} -1 & 2 & 4 \\ 3 & -2 & -5 \end{bmatrix} \bullet \begin{bmatrix} 1 \\ -2 \\ -1 \end{bmatrix} \qquad \begin{bmatrix} \mathbf{-9} \\ \mathbf{12} \end{bmatrix}$$

ENRICH

Refer to Exercise 9 on page 127 of the Pupil's Edition and form the matrix which represents the number of routes *directly* from one city to another, without passing through any other city. For example, there are 2 routes directly from A to B, and 0 routes directly from B to D. Multiply the matrix by itself. The product matrix represents the number of air routes between two cities that go through exactly one other city. How many air routes are there between city B and D that go through one other city? Which city is connected to the most other cities through one other city?

[Since

$$\begin{matrix} & A & B & C & D \\ A & 0 & 2 & 0 & 0 \\ B & 0 & 0 & 2 & 0 \\ C & 0 & 1 & 0 & 2 \\ D & 1 & 0 & 1 & 0 \end{matrix} \bullet \begin{matrix} & A & B & C & D \\ A & 0 & 2 & 0 & 0 \\ B & 0 & 0 & 2 & 0 \\ C & 0 & 1 & 0 & 2 \\ D & 1 & 0 & 1 & 0 \end{matrix} =$$

$$\begin{matrix} & A & B & C & D \\ A & 0 & 0 & 4 & 0 \\ B & 0 & 2 & 0 & 4 \\ C & 2 & 0 & 4 & 0 \\ D & 0 & 3 & 0 & 2 \end{matrix},$$

there are four air routes between cities B and D that go through exactly one other city; city B.]

Interdisciplinary Connection

Systems Analyst If you consider the matrix for Exercise 9 on page 127 of the Pupil's Edition to be a representation of how computers are connected on a network, then you will have an example of one of the jobs of a systems analyst. A systems analyst is responsible for designing a network of computers so that they are connected in a way specified by the customer.

Practice

3.3 Exploring Matrix Multiplication

In Exercises 1–5, the dimensions of two matrices are given. If the two matrices can be multiplied, indicate the dimensions of the product matrix. If they cannot, write *not possible*.

	C	D	CD
1.	2 by 3	3 by 4	________
2.	1 by 2	1 by 3	________
3.	2 by 2	2 by 3	________
4.	3 by 1	1 by 3	________
5.	2 by 1	3 by 3	________

Use the given matrices to evaluate each operation. Write *not possible* when appropriate.

$$A = \begin{bmatrix} -1 & 5 \\ 3 & -7 \end{bmatrix} \quad B = \begin{bmatrix} 4 & -8 \\ 0 & 1 \\ -6 & -2 \end{bmatrix} \quad C = \begin{bmatrix} 3 & 7 & -9 \\ 8 & -5 & 0 \end{bmatrix} \quad D = \begin{bmatrix} 2 & -4 \\ -1 & 6 \end{bmatrix}$$

6. $A + D$ ________ **7.** $4B$ ________ **8.** $2A - 3D$ ________

9. AD ________ **10.** AB ________ **11.** AC ________

12. BD ________ **13.** CB ________ **14.** BC ________

Find each product.

15. $\begin{bmatrix} 3 & -5 \\ 8 & -6 \end{bmatrix}\begin{bmatrix} 1 & 0 \\ 0 & 1 \end{bmatrix}$

16. $\begin{bmatrix} 1 & 0 \\ 0 & 1 \end{bmatrix}\begin{bmatrix} -2 & 0 \\ 12 & -9 \end{bmatrix}$

NAME ______________________ CLASS __________ DATE __________

Lesson Activity

3.3 Exploring Matrix Multiplication

Use the given matrices to find each product matrix. If the product cannot be found, write *not possible*.

$$A = \begin{bmatrix} 1 & 3 & 5 \\ 2 & 4 & 5 \end{bmatrix} \quad B = \begin{bmatrix} 1 & 4 \\ 2 & 5 \\ 3 & 6 \end{bmatrix} \quad C = \begin{bmatrix} 1 & 0 & 0 \\ 0 & 1 & 0 \\ 0 & 0 & 1 \end{bmatrix}$$

1. matrix AB = ______________

2. matrix BA = ______________

3. matrix BC = ______________

4. matrix AC = ______________

5. matrix $3A$ = ______________

6. matrix $2B$ = ______________

7. For which matrix is C the identity matrix? Why?

____

8. An identity matrix for matrix B is ______________________________.

9. Does $A \cdot B = B \cdot A$? Why or why not?

____

For Exercises 10 and 11, use $XY = \begin{bmatrix} 40 & 51 & 62 \\ 26 & 33 & 40 \end{bmatrix}$.

10. Matrix X must have ________ rows, and matrix Y must have ________ columns.

11. Create matrices X and Y that have the given product.

NAME ____________________ CLASS ____________ DATE ____________

Basic Skills Practice
Cumulative Review

Read each question and circle the best answer.

1. If $4m - 5 = -29$, what is the value of m?

A -6
B -1
C 6
D 24

2. On a farm-to-market road the maximum speed limit is 55 miles per hour, and the minimum speed is 35 miles per hour. If x represents the speed, which sentence best describes this condition?

F $35 \leq x \geq 55$
G $35 \geq x \geq 55$
H $55 \leq x \leq 35$
I $55 \geq x \geq 35$

3. The cost of placing an ad in the local paper is given by the formula $C = \$4.00 + \$0.45w$, where C is the total cost and w is the number of words. What is the cost of an ad containing 22 words?

A $4.45
B $9.90
C $13.90
D $19.80

4. If $3.4n + 8 = 31.8$, what is the value of n?

F 2
G 4
H 7
I 8

5. The average weight of a five-year-old child is 40 pounds. Most children's weight varies by 4 pounds. Which number line shows the weight range for a five-year-old?

A 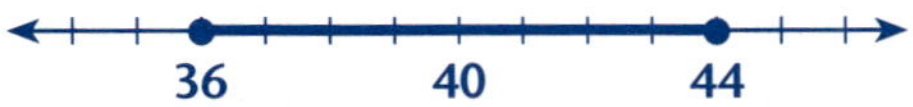

B 

C 

D 

Use the graph to answer Exercises 6 and 7.

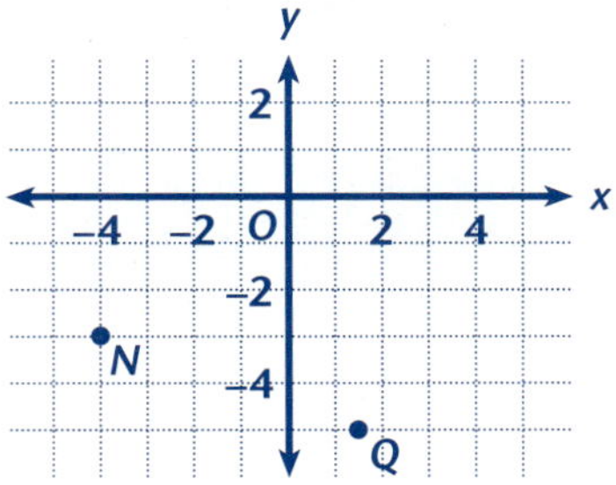

6. What is the ordered pair for point N?

F $(4, -3)$
G $(-4, -3)$
H $(4, 3)$
I $(-3, -4)$

7. What is the ordered pair for point Q?

A $(1, 5)$
B $(-1.5, -5)$
C $(1.5, -5)$
D $(1.5, 5)$

8. Which inequality or equation represents the number line below?

F $x < -3$
G $x \geq -3$
H $x = -3$
I $x > -3$

9. The average temperature in Austin, Texas, in the month of July is 85 degrees. This temperature often varies by 15 degrees. Which number line below shows this variance?

A 

B 

C 

D 

LESSON 3.4 Multiplicative Inverse of a Matrix

PREPARE

Pacing Guide
2 Days

Objectives and Overview
- Determine the inverse matrix for multiplication.
- Use technology to find the inverse of a matrix.

Students learn the definition of the inverse of a matrix and how to find the inverse by solving a system of equations. They also learn how to find the inverse of a matrix with technology.

Vocabulary
inverses
identity matrix

Assessing Prior Knowledge

Find the multiplicative inverse, or reciprocal, of each number.

1. -2 $\left[-\frac{1}{2}\right]$
2. $\frac{2}{3}$ $\left[\frac{3}{2}\right]$
3. $-\frac{4}{5}$ $\left[-\frac{5}{4}\right]$
4. $3\frac{1}{3}$ $\left[\frac{3}{10}\right]$
5. $-1\frac{3}{5}$ $\left[-\frac{5}{8}\right]$

Approach A Using the Book

Opening Discussion
Discuss the Reciprocal Property for multiplication. Does every number have a reciprocal? **[No; 0 doesn't have a reciprocal.]** What is the product of a number and its reciprocal? **[1]**

Brainstorming
Have the class start with any 2×2 matrix. Have them brainstorm how they might find the matrix whose product with the given matrix is the identity matrix.

Multiplicative Inverse of a Matrix
The multiplicative inverse of a matrix has the property that when it is multiplied by the original matrix from the right or from the left, the product is the identity matrix. Remind students of the algebraic equivalent of finding the inverse of a matrix. When they solved a linear equation of the form $ax = b$, they multiplied both sides by the reciprocal of a, giving the solution $x = b \bullet \frac{1}{a}$.

TEACH

Approach B Using Models

Opening Discussion
Discuss the Reciprocal Property for multiplication. Does every number have a reciprocal? **[No; 0 doesn't have a reciprocal.]** What is the product of a number and its reciprocal? **[1]** Can the product of two matrices ever give only ones and zeros as entries in the product matrix? **[yes]**

Brainstorming
Have the class start with any 2×2 matrix. Have them think of what matrix can be multiplied by the original matrix to get back the original matrix. **[the identity matrix]** How might they find the matrix whose product with the given matrix is the identity matrix?

Using an Algorithm
Finding the inverse of a 2×2 matrix can be done with an algorithm, or a sequence of steps. Because finding an inverse follows an algorithm, it can be programmed into a computer or graphics calculator.

Definition of *Identity Matrix*

The identity matrix can be found for any square matrix and consists of the matrix with ones along the main diagonal (which starts with the upper left entry in the top row) and zeros everywhere else. Point out that the identity matrix is like the multiplicative identity element, 1, for normal multiplication. Emphasize that the identity matrix is always a *square* matrix.

Example 1

In this example, the inverse matrix is found by multiplying the original square matrix by another square matrix with the same dimensions that has only variables for entries.

When the two matrices are multiplied, a system of equations is found that can be solved for each of the entries in the identity matrix. This method is straightforward and will help students review systems of equations. Point out that the matrix of variables representing A^{-1}, the inverse matrix for matrix A, can be multiplied from the right or from the left to get a system of equations. The product of a matrix with its inverse is the identity matrix. Emphasize that the original matrix and its inverse are both *square* matrices.

Additional Example 1

If matrix $A = \begin{bmatrix} -2 & 1 \\ 4 & 3 \end{bmatrix}$, find A^{-1}. $\begin{bmatrix} -0.3 & 0.1 \\ 0.4 & 0.2 \end{bmatrix}$

Critical Thinking, page 147

The row and column dimensions have to be the same (the matrix must be square) because the inverse matrix must be able to be multiplied by the original matrix both from the right side and from the left side. This can happen only if the matrix is square.

Example 2

This example involves a matrix that doesn't have an inverse.

continued on top of next page

The following is the algorithm for finding the inverse of matrix $M = \begin{bmatrix} A & B \\ C & D \end{bmatrix}$:

1. Store the values of the original matrix into memory locations A, B, C, and D, respectively.
2. Calculate $AD - BC$. If $AD - BC = 0$, then stop; state that this matrix has no inverse.
3. Solve the system of equations $AE + BG = 1$ and $CE + DG = 0$ for E and G.
4. Solve the system of equations $AF + BH = 0$ and $CF + DH = 1$ for F and H.
5. The inverse matrix is $\begin{bmatrix} E & F \\ G & H \end{bmatrix}$.

Ask students to study the steps in the algorithm and to verify them with an example.

Working With an Algorithm

Have students find the inverse of each of the following matrices:

A. $\begin{bmatrix} 3 & -4 \\ -2 & 4 \end{bmatrix}$ $\begin{bmatrix} 1 & 1 \\ 0.5 & 0.75 \end{bmatrix}$

B. $\begin{bmatrix} 4 & 6 \\ 2 & 1 \end{bmatrix}$ $\begin{bmatrix} -0.125 & 0.75 \\ 0.25 & -0.5 \end{bmatrix}$

An algorithm gives students a step by step procedure for finding the inverse of a matrix.

continued on bottom of next page

Approach A

Using the Book, continued

Point out to students that they may have a difficult time predicting when a matrix has no inverse. They should follow the steps shown in Example 1 on page 146 of the Pupil's Edition. If there is no solution, the corresponding system of equations has no solution. Make sure that students make the connection to the number 0, the only real number that has no multiplicative inverse.

Additional Example 2

Find the inverse of $B = \begin{bmatrix} -2 & 3 \\ 6 & -9 \end{bmatrix}$.

[There is no inverse.]

Ongoing Assessment, page 147

The entries in the second row are half of the entries in the first row. If one row is a multiple of another row, then the matrix has no inverse.

Ongoing Assessment, page 147
Try This

a. $\begin{bmatrix} \frac{1}{3} & -\frac{1}{6} \\ 1 & 0 \end{bmatrix}$, **b.** No inverse

Matrix Technology

A graphics calculator includes a quick way to find the inverse of a matrix. Enter the matrix as matrix A. Using the inverse key, [x^{-1}], type $[A]^{-1}$ to find the inverse. The message that appears on the screen when a matrix has no inverse may be ERR:SINGULAR MAT. A singular matrix is one with no inverse. Make sure students understand that unusual expressions for zero, such as 2E–13, may appear on the matrix display. Explain that a calculator does not always do operations in the same way that students do them, so 2E–13 may be as close to 0 as the calculator can get.

TEACH

Approach B

Using Models, continued

Teaching Tip

For students who are unsure of the steps of the algorithm, review the methods of solving a system of equations from Chapter 2 and how to multiply two matrices from Lesson 3.3.

Cooperative Learning

Matrix Game Have students work in groups of four to write four matrices and their inverses. One matrix can have no inverse. Ask them to cut out the matrices and their inverses (or the statement "no inverse," if appropriate) so that they are like separate puzzle pieces. Mix them up in an envelope. Then have groups switch envelopes. The groups should match the matrices in the envelope with their inverse or with the statement "no inverse." The group that matches the matrices the fastest wins.

Matrix Technology

A graphics calculator includes a quick way to find the inverse of a matrix. Enter the matrix as matrix A. Using the inverse key, [x^{-1}], type $[A]^{-1}$ to find the inverse. Point out that the message that appears on the screen when a matrix has no inverse may be ERR:SINGULAR MAT. A singular matrix is one that has no inverse. Make sure students understand that unusual expressions for zero, such as 2E–13, may appear on the matrix display. Explain that a calculator does not always do operations in the same way that students do them, so 2E–13 may be as close to 0 as the calculator can get.

Have students find the inverse of the following matrices:

A. $\begin{bmatrix} 2 & 4 \\ 1 & 3 \end{bmatrix}$ $\begin{bmatrix} \mathbf{1.5} & \mathbf{-2} \\ \mathbf{-0.5} & \mathbf{1} \end{bmatrix}$

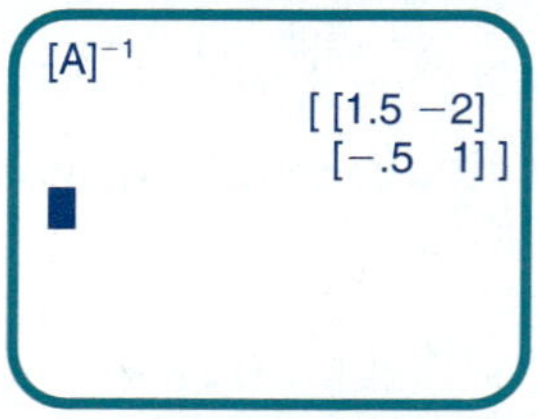

Inclusion Strategy

Enlarge each of the following matrices onto sheets of notebook paper so that they can be used like flashcards.

Matrix	Inverse
$\begin{bmatrix} 1 & 2 \\ -1 & 4 \end{bmatrix}$	$\begin{bmatrix} \frac{2}{3} & -\frac{1}{3} \\ \frac{1}{6} & \frac{1}{6} \end{bmatrix}$
$\begin{bmatrix} 1 & -2 \\ -2 & 4 \end{bmatrix}$	No inverse
$\begin{bmatrix} 2 & 1 \\ 4 & -2 \end{bmatrix}$	$\begin{bmatrix} 0.25 & 0.125 \\ 0.5 & -0.25 \end{bmatrix}$

Some matrices are the inverses of other matrices, and one matrix has no inverse. Include a flashcard that reads "No inverse." Then mix up the flashcards, and fasten them to the chalkboard. Ask students to match a matrix from a flashcard with its inverse or with the flashcard that reads "No inverse." They should find the inverse of their matrix by solving a system of equations.

Teaching Tip

One way to tell if a 2×2 matrix has an inverse is by checking its *determinant.* The matrix $\begin{bmatrix} a & b \\ c & d \end{bmatrix}$ has the determinant $ad - bc$. If $ad - bc = 0$, then the matrix has no inverse.

Closing the Lesson

Journal Entry Have students write a description of how to find the inverse of a matrix in their journals. Tell them to include sample problems and to describe how to tell when a matrix does not have an inverse.

B. $\begin{bmatrix} 1 & 3 \\ 2 & 7 \end{bmatrix}$ $\begin{bmatrix} \mathbf{7} & \mathbf{-3} \\ \mathbf{-2} & \mathbf{1} \end{bmatrix}$

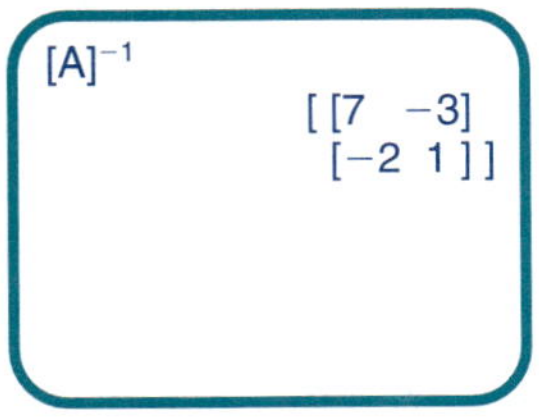

Journal Entry

Have students describe in their journal how to find the inverse of a matrix. Have them include a discussion of how to tell when a matrix does not have an inverse. Examples should be included when appropriate.

ASSESS

Selected Answers

Odd-numbered Exercises 7–39 and odd-numbered Extra Practice

Assignment Guide

Core level students should complete Exercises 1–41.

Core-plus level students should complete Exercises 1–9, 13–29, and 31–41.

Error Analysis

In Exercises 22 and 23, students are asked to make a connection between the difference of the products of the entries on the diagonals of a 2×2 matrix in a matrix that has an inverse and one that doesn't have an inverse. In general, for a 2×2 matrix, $\begin{bmatrix} a & b \\ c & d \end{bmatrix}$, $ad - bc$ is called the determinant. If the determinant of a matrix is 0, then the matrix has no inverse. For Exercises 22 and 23, ad and bc are equal (the determinant of each matrix is 0), so the matrices have no inverses.

Alternative Assessment

Another way to find the inverse of a 2×2 matrix, $\begin{bmatrix} a & b \\ c & d \end{bmatrix}$, is to switch entries a and d and take the opposite of b and c to get $\begin{bmatrix} d & -b \\ -c & a \end{bmatrix}$ and multiply each entry in this matrix by the reciprocal of $(ad - bc)$, if $(ad - bc)$ does not equal 0.

Have students use this method to find the inverse of $\begin{bmatrix} 1 & -4 \\ -1 & 8 \end{bmatrix}$. $\begin{bmatrix} 2 & 1 \\ 0.25 & 0.25 \end{bmatrix}$

RETEACH

Cooperative Learning Have students work with a partner to find the inverse of a matrix. Have each student write a matrix and then pass it to his or her partner to find the inverse. Ask pairs of students to check each other's work.

ENRICH

Finding the inverse of a 2×2 matrix can be done with an algorithm, or a sequence of steps. Because it has an algorithm, this process can be programmed into a computer or a graphics calculator. The following is the algorithm for finding the inverse of matrix $M = \begin{bmatrix} A & B \\ C & D \end{bmatrix}$:

1. Store the value of the original matrix into memory locations A, B, C, and D, respectively.
2. Find the determinant of the matrix, $AD - BC$.
3. If the determinant equals zero, then stop; state that this matrix has no inverse.
4. Find the multiplicative inverse of the determinant, $\frac{1}{AD - BC}$; call it T.
5. Multiply A, B, C, and D each by T. Store the values in E, F, G, and H, respectively.
6. The inverse is the matrix $\begin{bmatrix} H & -F \\ -G & E \end{bmatrix}$.

Ask students to study the steps in the algorithm and to use it to find the inverse of the matrix $\begin{bmatrix} 8 & 2 \\ -3 & -1 \end{bmatrix}$. Check the answer by using the matrix functions of a calculator. $\begin{bmatrix} 0.5 & 1 \\ -1.5 & -4 \end{bmatrix}$

Encourage interested students to write a graphics calculator program to match the algorithm.

NAME ______________________ CLASS __________ DATE __________

Practice

3.4 Multiplicative Inverse of a Matrix

Use the definition of inverse matrices to find the inverse matrix for each of the given matrices.

1. $\begin{bmatrix} -3 & 2 \\ -2 & 1 \end{bmatrix}$

2. $\begin{bmatrix} 1 & 3 \\ 2 & 7 \end{bmatrix}$

3. $\begin{bmatrix} 1 & 2 \\ 3 & 4 \end{bmatrix}$

4. $\begin{bmatrix} 6 & -7 \\ -4 & 5 \end{bmatrix}$

5. $\begin{bmatrix} 1 & 3 \\ 2 & 5 \end{bmatrix}$

6. $\begin{bmatrix} 1 & 6 \\ 1 & 5 \end{bmatrix}$

7. $\begin{bmatrix} -1 & 1 \\ -2 & 1 \end{bmatrix}$

8. $\begin{bmatrix} 2 & -1 \\ 3 & -2 \end{bmatrix}$

9. $\begin{bmatrix} 2 & 3 \\ 1 & 2 \end{bmatrix}$

10. $\begin{bmatrix} -3 & 11 \\ 2 & -7 \end{bmatrix}$

Does each matrix have an inverse? Use technology to find the inverse of each matrix if it exists.

11. $\begin{bmatrix} 2 & 0 \\ 3 & 5 \end{bmatrix}$

12. $\begin{bmatrix} 4 & 3 \\ 8 & 6 \end{bmatrix}$

13. $\begin{bmatrix} 2 & 1 & 6 \\ -4 & 5 & -8 \end{bmatrix}$

14. $\begin{bmatrix} -3 & 10 \\ 5 & -2 \\ 0 & 4 \end{bmatrix}$

15. $\begin{bmatrix} -3 & 4 \\ 1 & -2 \end{bmatrix}$

16. $\begin{bmatrix} 3 & 6 \\ 2 & 1 \end{bmatrix}$

NAME ______________________ CLASS __________ DATE __________

Lesson Activity

3.4 Multiplicative Inverse and Other Matrices

Fill in the blanks in the following procedures to find the multiplicative inverse for the given matrix:

$$\text{Matrix } A = \begin{bmatrix} 3 & 1 \\ 4 & 2 \end{bmatrix}$$

$$A \quad \cdot \quad A^{-1} \quad = \quad I$$

1. $\begin{bmatrix} 3 & 1 \\ 4 & 2 \end{bmatrix} \quad \begin{bmatrix} a & b \\ c & d \end{bmatrix} \quad \begin{bmatrix} \underline{\quad} & \underline{\quad} \\ \underline{\quad} & \underline{\quad} \end{bmatrix}$

2. $\begin{bmatrix} 3a + \underline{\quad} & 3b + \underline{\quad} \\ \underline{\quad}\, a + \underline{\quad} & \underline{\quad} + \underline{\quad} \end{bmatrix} = \begin{bmatrix} \underline{\quad} & \underline{\quad} \\ \underline{\quad} & \underline{\quad} \end{bmatrix}$

3. Solve for a and b.

$3a + \underline{\quad} = \underline{\quad}$ $\qquad$ $3b + \underline{\quad} = \underline{\quad}$

$\underline{\quad}\, a + \underline{\quad} = \underline{\quad}$ $\qquad$ $\underline{\quad} + \underline{\quad} = \underline{\quad}$

4. Eliminate c and d.

$-6a - \underline{\quad} = \underline{\quad}$ $\qquad$ $-6b - \underline{\quad} = \underline{\quad}$

$\underline{\quad}\, a + \underline{\quad} = \underline{\quad}$ $\qquad$ $\underline{\quad} + \underline{\quad} = \underline{\quad}$

$a = \underline{\quad}$ $\qquad$ $b = \underline{\quad}$

5. Substitute to find c and d.

$\underline{\quad} + 2c = \underline{\quad}$ $\qquad$ $\underline{\quad} + 2d = \underline{\quad}$

$c = \underline{\quad}$ $\qquad$ $d = \underline{\quad}$

6. $A^{-1} = \begin{bmatrix} \underline{\quad} & \underline{\quad} \\ \underline{\quad} & \underline{\quad} \end{bmatrix}$

Use the procedure in Steps 1–6 to find the multiplicative inverse for each matrix.

7. $\begin{bmatrix} -1 & 2 \\ 3 & -2 \end{bmatrix}$ __________ $\qquad$ **8.** $\begin{bmatrix} 2 & 4 \\ 5 & 3 \end{bmatrix}$ __________

NAME ______________________ CLASS __________ DATE __________

Basic Skills Practice

Finding Perimeter and Using the Distance Formula, Part 1

To find the perimeter, the total distance around the outside of a figure.

Example 1: Find the perimeter of the rectangle shown.
Add the length of all sides to get the total.
$P = 3\text{ ft} + 4\text{ ft} + 3\text{ ft} + 4\text{ ft}$
$P = 14\text{ ft}$

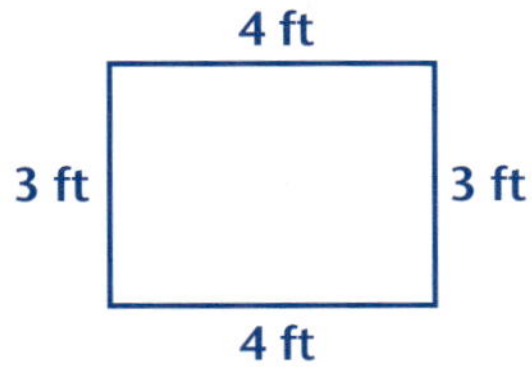

To find the distance that an object travels, multiply its rate of speed by the length of time that it travels.

Example 2: How far will a truck go in 4 hours at 60 mph?
$d = rt$ → Write the formula (d = distance, r = rate, t = time).
$d = 60(4)$ → Substitute the given values ($r = 60$, $t = 4$).
$d = 240$ → Multiply.
The truck will go 240 miles.

Find the perimeter.

1. 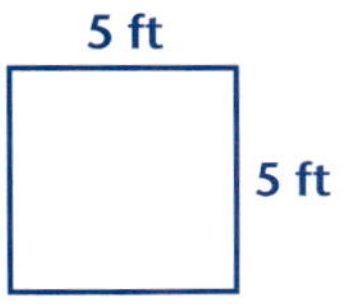

____

2. 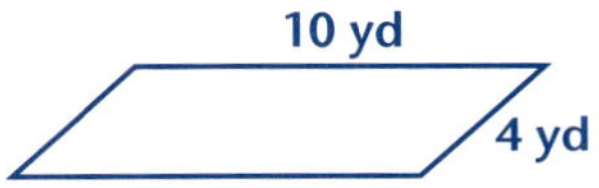

____

3. 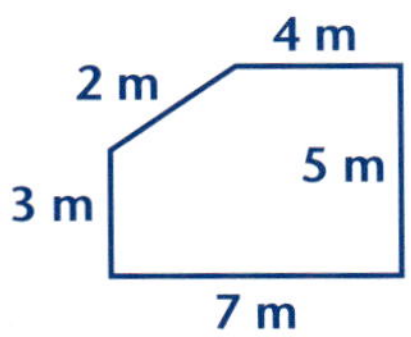

____

4. 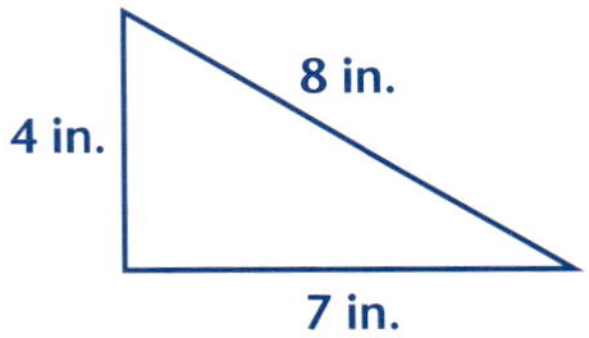

____

5. 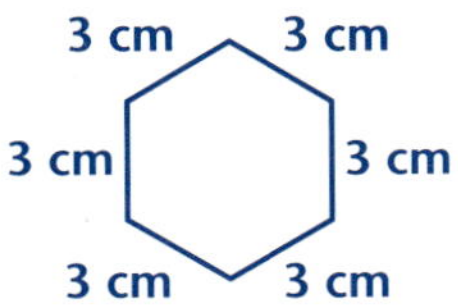

____

6. 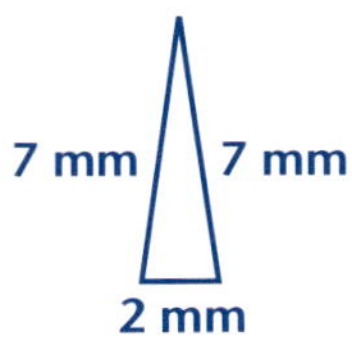

____

Find the missing amount.

7. $r = 50$ mph, $t = 6$ h, $d =$ __________ mi

8. $r = 32$ mph, $t = 4.5$ h, $d =$ __________ mi

9. $r = 475$ mph, $t =$ __________ h, $d = 2150$ mi

10. $r =$ __________ mph, $t = 1.5$ h, $d = 70.25$ mi

11. $r = 66$ mph, $t = 0.8$ h, $d =$ __________ mi

12. $r = 105$ mph, $t =$ __________ h, $d = 43$ mi

LESSON 3.5

Solving Matrix Equations

PREPARE

Pacing Guide

2 Days

Objectives and Overview

- Solve a system of equations by using matrices.
- Use technology to solve a system of equations by using matrices.

Students learn to solve a system of equations by forming a matrix equation from the coefficient matrix, the variable matrix, and the constant matrix for the original system. Then they multiply both sides of the matrix equation by the inverse of the coefficient matrix. Students also learn how to solve the system of equations with a graphics calculator.

Vocabulary

matrix equation	variable matrix
coefficient matrix	constant matrix

Assessing Prior Knowledge

Solve each equation.

1. $8x = 24$ [3]

2. $-14x = 7$ $\left[-\frac{1}{2}\right]$

3. $-5x = -36$ $\left[7\frac{1}{5}\right]$

4. $\frac{x}{7} = 4$ [28]

5. $\frac{2}{3}x = 12$ [18]

Approach A: Using the Book

Opening Discussion

Ask students to review how they solved systems of equations in Chapter 2. Point out that there is another way to solve a system of equations if matrices are used. Why would using matrices be desirable? **[You can use technology to help find the solution.]**

Brainstorming

Remind the class of the solution to the equation $ax = b$. Discuss how the solution may apply to the matrix equation $AX = B$. Do the students think the solution may look the same as that for $ax = b$?

Solving Matrix Equations

Make sure that students recall the methods learned thus far to solve systems of equations—graphing, elimination, and substitution. Go over how the system of equations is found for the consumer economics

TEACH

Approach B: Using Models

Opening Discussion

Ask students to review how they solved systems of equations in Chapter 2. Point out that a system of equations can be represented by matrices. Why would using matrices be desirable? **[You can use technology to help find the solution.]**

Brainstorming

Remind the class of the solution to the equation $ax = b$. Discuss how the solution may apply to the matrix equation $AX = B$. Do the students think the solution may look the same as that for $ax = b$?

Using a Graphics Calculator

The matrix functions on a graphics calculator can be used to solve a system of equations and are especially useful for finding the inverse of a matrix.

The inverse of a matrix may be difficult to find when the dimensions of a matrix are larger than 2×2. In order to use a graphics calculator to solve a system of example. Point out that matrices can also be used to solve a system of equations.

The Matrix Solution for Systems of Equations

Review how a system of equations can be written in matrix form by using the coefficient matrix, the variable matrix, and the constant matrix. The product matrix of the coefficient matrix and the variable matrix can then be used to form a matrix equation that can be solved by using the inverse of the coefficient matrix.

Point out that since multiplication of matrices is not commutative, it is important to multiply the constant matrix by the inverse of the coefficient matrix on the correct side, which is the left side. Both sides of the matrix equation are multiplied on the left. If students use technology to perform the multiplication, they may get an error message when they multiply on the wrong side.

Use **Transparency 14** for discussion.

Teaching Tip

You may want to stress the similarities and differences between solving an equation of the form $AX = B$ for X when A, B, and X are numbers and when A, B, and X are matrices. One similarity is the process used for both types of equations. A difference is that if A, B, and X are numbers, $A^{-1} \bullet A$ is 1 and if A, B, and X are matrices, $A^{-1} \bullet A$ is the identity matrix.

Critical Thinking, page 153

Multiplication of matrices is not commutative. The procedure is the same as solving the linear equation $ax = b$ except that A^{-1} is a matrix instead of a number, and it must be multiplied on the left.

Solving Matrix Equations With Technology

Technology, such as the inverse matrix command of a graphics calculator, can be used to solve matrix equations.

continued on top of next page

equations, the following requirements must be met: (1) the system of equations must be square, which means that the number of variables and the number of equations in the system is the same, and (2) the inverse of the coefficient matrix must exist.

The existence of the inverse can be checked easily with the calculator. The calculator will give an error message if the inverse cannot be found.

Use the following algorithm to solve a system of equations with a graphics calculator.

1. Organize the information into a system of equations.
2. Write the system of equations so that each is of the form $Ax + By = C$ or $Ax + By + Cz = D$, where A, B, C, and D are real numbers.
3. Write the coefficient matrix A, the variable matrix X, and the constant matrix B.
4. Identify the matrix equation $AX = B$.
5. Find $A^{-1} \bullet B$, if A^{-1} exists. This is the solution to a system of equations whose matrix form is $AX = B$, where A, X, and B are matrices.

Working With a Graphics Calculator

Have students solve each of the following systems of equations:

A. $\begin{cases} 3x + 2y = 4 \\ 4x - 2y = 10 \end{cases}$ **[$x = 2, y = -1$]**

B. $\begin{cases} 3x - 2y + z = -11 \\ x + 2y - 2z = 0 \\ 2x + y + z = 3 \end{cases}$ **[$x = -2, y = 4, z = 3$]**

Teaching Tip

Students need to be especially careful that they enter the matrices correctly into a graphics calculator. A common error that students make is entering the data by columns instead of by rows. Ask them to check their entries carefully before finding the inverse of the coefficient matrix. They can choose [A] on a clear screen to check what they entered. Because a calculator will give the wrong answer if the entries are not correct, checking the entries is very important.

continued on bottom of next page

Approach A

Using the Book, continued

Example 1

In this example, students solve a system of equations with technology. They first have to find the coefficient matrix A, the variable matrix X, and the constant matrix B. They then substitute the matrices into the equation $AX = B$. Point out that these steps cannot be done with technology. Technology is used only to find A^{-1} and to multiply A^{-1} by B to get the final solution. It is important to remember in the final step that if two matrices are equal, then each corresponding entry is equal. Thus, the values of x and y can be read from the product $A^{-1} \bullet B$.

Additional Example 1

Use matrices to solve $\begin{cases} 3x - 2y = 5 \\ -2x + 5y = 4 \end{cases}$.

$[x = 3; y = 2]$

Teaching Tip

Students may not understand how the inverse key on a graphics calculator is used. Point out that the inverse key returns the reciprocal of every real number except 0. However, when it is used to find the inverse of a matrix, it does not return the reciprocal of a matrix, which is not possible. It returns the inverse.

Ongoing Assessment, page 153

Try This

The coefficient matrix, variable matrix, and constant matrix become

$A = \begin{bmatrix} -1 & 1 \\ -2 & 3 \end{bmatrix}$, $X = \begin{bmatrix} x \\ y \end{bmatrix}$, and $B = \begin{bmatrix} 1 \\ 0 \end{bmatrix}$, and the

matrix equation is $\begin{bmatrix} -1 & 1 \\ -2 & 3 \end{bmatrix}\begin{bmatrix} x \\ y \end{bmatrix} = \begin{bmatrix} 1 \\ 0 \end{bmatrix}$.

The inverse matrix is $A^{-1} = \begin{bmatrix} -3 & 1 \\ -2 & 1 \end{bmatrix}$,

so the solution is $\begin{bmatrix} x \\ y \end{bmatrix} = \begin{bmatrix} -3 \\ -2 \end{bmatrix}$.

TEACH

Approach B

Using Models, continued

Solving Matrix Equations With Technology

Encourage students to use a graphics calculator to check their answers. Use the STO► key and the ALPHA key as shown below to check the solution to the system in question A on the previous page.

```
2→X
                2
-1→Y
               -1
```

```
3X+2Y
                4
4X-2Y
               10
```

The Chinese Method

A graphics calculator can also be used to simulate the Chinese method of solving a system of linear equations in matrix form. For the example given on page 155 of the Pupil's Edition, use the method shown on the screens to find the solution $y = 4$.

The matrix $[4 \;\; 5 \;\; 22]$ is entered as matrix A and the matrix $[2 \;\; 3 \;\; 13]$ is entered as matrix B. Note that instead of subtracting row 1 from row 2, row 1 is multiplied by -1 and then added to row 2.

```
*row(-1, [A], 1)→ [C]
        [[-4 -5 -22]]
*row(2, [B], 1)→ [D]
          [[4 6 26]]
```

```
[C]+[D]
            [[0 1 4]]
```

Cooperative Learning

Using a Graphics Calculator Have students work with a partner to solve each of the following systems of equations below. For question A, have one student set up the system of equations and write the coefficient matrix, the variable matrix, and the constant matrix. Ask the second student to enter the matrices into the calculator and solve the system. Ask the first student to check the solution. Then have students switch roles.

Example 2

In this example, a system of equations is given with coefficients that would make it difficult to find an inverse matrix without technology. A graphics calculator is used to solve the system.

Point out that part of the first step of organizing information is to make sure that the equations in the system are both of the form $Ax + By = C$, where A, B, and C are real numbers. This is necessary to ensure that the correct coefficient and constant matrices are found.

Using Technology

A graphics calculator can also be used to check the solution in Example 2. Use the STO► key and the ALPHA key as shown below.

```
2.25→H
                 2.25
1.50→D
                  1.5
```

```
14H+12D
                 49.5
10H+7D
                   33
```

Additional Example 2

A resort hotel offers various package options to its customers. One option includes 3 nights lodging and 5 meals for $368.25. Another option includes 6 nights lodging and 14 meals for $816.30. What is the price for each night's lodging and each meal? **[$89.50; $19.95]**

Inclusion Strategy

Modeling a System of Equations Have students work in pairs to model a system of equations with different colored chips or tiles (such as red, R, and yellow, Y). Have one student assign a value to each color of chip (for example, $5 for red and $1 for yellow), and compose a system of equations with the chips. For example, one system is $6R + 2Y = 32$ and $4R - 3Y = 17$. Then have the other student use technology to find the value of each chip.

continued on top of next page

A. On one test, m multiple-choice questions were worth 5 points each and t true-false questions were worth 3 points each. The test included a total of 105 points. On a second test, m multiple-choice questions were worth 4 points each and t true-false questions were worth 2 points each, for a total of 80 points. How many multiple-choice questions and true-false questions were on each test? **[First set up the system $\begin{cases} 5m = 3t = 105 \\ 4m = 2t = 80 \end{cases}$. This is the same as the system $AX = B$ with $A = \begin{bmatrix} 5 & 3 \\ 4 & 2 \end{bmatrix}$ and $B = \begin{bmatrix} 105 \\ 80 \end{bmatrix}$. Enter matrix A:**

```
MATRIX[A] 2×2
[5      3        ]
[4      2        ]


1, 1=5
```

Then enter B in the same way.

Calculating $A^{-1}B$, gives $X = \begin{bmatrix} 15 \\ 10 \end{bmatrix}$. So, $m = 15$ and $t = 10$.]

B. P pounds of peanuts at $1.29 a pound are mixed with C pounds of cashews at $3.39 a pound to produce a mixture that costs $8.31. The next week, P pounds of peanuts at $1.39 a pound are mixed with C pounds of cashews at $3.09 a pound to produce a mixture that costs $8.11. How many pounds of peanuts and cashews are used for the mixture each time? **[$P = 2.5$ pounds, $C = 1.5$ pounds]**

continued on bottom of next page

Approach A

Using the Book, continued

Switch roles and have the first student solve a system of equations to find the new value of the chips assigned by his or her partner.

Ongoing Assessment, page 155

Form the matrices $\begin{bmatrix} 3 & 1 & 0 \\ 2 & 3 & 7 \end{bmatrix}$.

Apply scalar multiplication, $\begin{matrix} 2 \\ 3 \end{matrix}\begin{bmatrix} 3 & 1 & 0 \\ 2 & 3 & 7 \end{bmatrix}$,

to produce $\begin{bmatrix} 6 & 2 & 0 \\ 6 & 9 & 21 \end{bmatrix}$.

Subtract row 1 from row 2 to get $[0 \quad 7 \quad 21]$.
Thus, $y = 3$.
Finally, substitute 3 for y in the first equation to get $x = -1$.

Closing the Lesson

Journal Entry Have students write the following system of equations into their journal: $3x + 8y = 36.10$ and $10x + 3y = 26.85$. Then have them list the steps necessary to solve the system of equations with technology, including the solution and a check. **[$x = 1.50$, $y = 3.95$]**

TEACH

Approach B

Using Models, continued

Journal Entry

Have students describe in their journal how to use technology to solve a system of equations by using matrices. Have them list their examples and show drawings of their calculator screens when possible. Have them include a written statement of the application problem, if one is used.

ASSESS

Selected Answers

Odd-numbered Exercises 5–23 and 27–29 and odd-numbered Extra Practice

Assignment Guide

Core level students should complete Exercises 1–34.

Core-plus level students should complete Exercises 1–34.

Error Analysis

Students who have difficulty with Exercise 19 may not realize that they need to use the matrix that they found in Exercise 18 to find the selling price of each type of sweater. Have them rewrite the results of Exercise 18 as a matrix equation of the form $AX = B$ and then use technology to solve the problem.

Alternative Assessment

The Matrix Game To involve all students in finding the solution to a system of equations, ask one student to write the equations for an application problem on the chalkboard. Then have another student give the coefficient matrix, the variable matrix, and the constant matrix. Next, have a third student rewrite the system as a matrix equation. Then ask another student to enter the matrices into a graphics calculator or other technology (preferably one for which an overhead model can be displayed). Ask the next student to show the solution given by the technology. Repeat this sequence until all students have participated.

RETEACH

Have students work with a partner to find the solution to each system of equations below. Then have them switch papers and check each other's answers.

1. $\begin{cases} 3x + 2y = 20 \\ -2x - 4y = -8 \end{cases}$ **[(8, −2)]**

2. $\begin{cases} 1.5x + 4.8y = 10.95 \\ 3.8x - 1.2y = 7.7 \end{cases}$ **[(2.5, 1.5)]**

ENRICH

The usefulness of a graphics calculator for very large systems of equations can be appreciated by solving the system of equations below.

$$\begin{cases} 3x + 4y - z + 2w = -5 \\ -2x + 3y - 4z + 4w = -27 \\ x - 6y + 3z - 5w = 27 \\ 5x - y - z - 2w = 12 \end{cases}$$

[$x = 2$, $y = -1$, $z = 3$, $w = -2$]

Encourage students to write their own system of equations and to find the solution with a graphics calculator.

Interdisciplinary Connection

Economics An economist may be responsible for solving some problems by using the method of *linear programming*. In this method, a system of linear equations is given for a problem with a complex set of conditions. Technology is then used to find the points of intersection of the linear equations. The maximum or minimum solution to the problem is located at one of these points of intersection.

NAME ______________________ CLASS ____________ DATE ____________

Practice

3.5 Solving Matrix Equations

Represent the systems of equations in matrix form, $AC = B$.

1. $\begin{cases} 6x - 3y = 15 \\ 4x + 2y = 9 \end{cases}$

2. $\begin{cases} 5y = 2x - 11 \\ x = 3y + 8 \end{cases}$

3. $\begin{cases} 2a - b + 4c = 6 \\ -a + 7b - 2c = 1 \\ 6a + 3b - c = -8 \end{cases}$

4. $\begin{cases} 8x + 7y - z = 0 \\ 5x - 4y - 3z = 13 \\ x + y + z = -4 \end{cases}$

Solve the system of equations by using matrices, if possible.

5. $\begin{cases} -3x + 2y = -13 \\ -2x + y = -8 \end{cases}$

6. $\begin{cases} x + 3y = 5 \\ 2x + 7y = 12 \end{cases}$

Match each matrix equation with its correct solution.

a. **(2,1)** **b.** **(2,−1)** **c.** **(−2,3)** **d.** **(−2,−3)**

7. $\begin{bmatrix} 4 & -3 \\ 1 & -4 \end{bmatrix} \begin{bmatrix} x \\ y \end{bmatrix} = \begin{bmatrix} 1 \\ 10 \end{bmatrix}$ ____________

8. $\begin{bmatrix} 5 & 4 \\ 2 & -1 \end{bmatrix} \begin{bmatrix} x \\ y \end{bmatrix} = \begin{bmatrix} 6 \\ 5 \end{bmatrix}$ ____________

9. $\begin{bmatrix} 2 & 3 \\ 1 & -4 \end{bmatrix} \begin{bmatrix} x \\ y \end{bmatrix} = \begin{bmatrix} 7 \\ -2 \end{bmatrix}$ ____________

10. $\begin{bmatrix} 1 & -1 \\ 3 & 6 \end{bmatrix} \begin{bmatrix} x \\ y \end{bmatrix} = \begin{bmatrix} -5 \\ 12 \end{bmatrix}$ ____________

NAME ______________________ CLASS __________ DATE __________

Lesson Activity

3.5 Solving Equations Using Matrices and Other Methods

Solve for *x* and *y* by using the four different methods suggested.

$$\begin{cases} 4x + 2y = 17 \\ 3x + 6y = 24 \end{cases}$$

1. elimination ______________________

____

2. substitution ______________________

____

3. matrices ______________________

____

4. Chinese method ______________________

____

5. Which method did you prefer to use? Explain why.

____

____

6. Motion Park is actually two activity parks in one. There is a theme park with more than 150 rides, and there is a water park with water slides and other water activities. You can buy either a combination ticket that gives you admission to both parks or a theme-park ticket that admits you only to the ride section. Two classes that visited Motion Park bought the following number of tickets: Class 1 bought 25 combination tickets and 15 theme-park tickets. The total cost was $870. Class 2 bought 10 combination tickets and 8 theme-park tickets for a total cost of $384. Use matrices to find the cost of a combination ticket and the cost of a theme-park ticket.

____

NAME ______________________ CLASS __________ DATE __________

Basic Skills Practice

Finding Perimeter and Using the Distance Formula, Part 2

Read each question and circle the best answer.

1. How many yards of fence would it take to enclose a garden with a width of 10 yards and a length of 12 yards?

A 42 yd
B 38 yd
C 44 yd
D 22 yd

2. A cargo plane flew 100 miles at 200 mph. How long did the flight take?

F 2 hr
G 1 hr
H 30 min
I 1 hr 30 min

3. A cowboy built a square pen for his cattle. Each side is 50 feet long. What is the perimeter?

A 200 ft
B 500 ft
C 250 ft
D 2500 ft

4. Each side of a certain hexagon is 6 inches long. What is the perimeter of the hexagon?

F 60 in.
G 36 in.
H 38 in.
I 40 in.

5. A bullet-train travels 300 miles in 2 hours. What is the average speed of the train?

A 100 mph
B 250 mph
C 200 mph
D 150 mph

6. Louis rides his bicycle for 10 minutes at an average speed of 30 mph. How far did he ride?

F 3 mi
G 5 mi
H 6 mi
I 8 mi

Solve each problem.

7. A football field is 30 yards wide and 100 yards long. What is the perimeter in yards? __________

8. A certain county is perfectly square. Each side is 40 miles long. What is the perimeter in miles? __________

9. An F-16 jet flew 5127.5 miles at an average speed of 1520 mph. How long did the trip take in hours? __________

10. A UFO travels at a rate of 45,500 mph. How long does it take to travel 967,350 miles in hours? __________

11. The length of a basketball court in the school gymnasium measures 28 meters and the width measures 50 meters. Find the perimeter in meters. __________

LESSON 4.1

Exploring Experimental Probability

PREPARE

Pacing Guide
2 Days

Objectives and Overview
- Find the experimental probability that an event will occur.
- Use random numbers in probability situations.

Students investigate ways in which computers and calculators can be used to generate random numbers in order to simulate probability experiments. Then they use the results of the experiments to make predications about other experiments.

Vocabulary
probability
certain event
impossible event
trials
successful event
experimental probability

Assessing Prior Knowledge

Simplify each fraction.

1. $\frac{3}{6}$ $\left[\frac{1}{2}\right]$
2. $\frac{20}{100}$ $\left[\frac{1}{5}\right]$
3. $\frac{12}{36}$ $\left[\frac{1}{3}\right]$

Approach A: Using the Book

Opening Discussion
Weather forecasters study weather patterns such as high- and low-pressure systems to help them forecast the weather for a particular area. But how do they predict the *chance* of a particular kind of weather? What does it mean when a forecaster says that there is a 20% chance of precipitation? What allows a forecaster to say that the probability of rain is 90%?

Brainstorming
Ask the students to make a list of situations that involve a probability. Examples include the chances of winning a card game, of guessing whether a basketball player will make a foul shot, and of guessing whether a child who is about to be born will be a boy or a girl.

continued on top of next page

TEACH

Approach B: Using Models

Opening Discussion
Discuss how weather forecasters study weather patterns such as high- and low-pressure systems to help them forecast the weather for a particular area. When a forecast is given, probabilities are generally used. For example, a weather forecast might state that there is a 70% chance of rain tomorrow. How is this chance predicted? What does a 70% chance of precipitation mean? What allows a forecaster to say that the probability of rain is 90%?

Brainstorming
Invite your students to brainstorm a list of everyday situations that involve finding a probability. Examples include the chances of drawing a winning card in a card game, of guessing whether a basketball player's foul shot will be successful, and of guessing whether a child who is about to be born will be a boy or a girl.

continued on bottom of next page

Approach A Using the Book, continued

Probability and Events

Your students may already have an intuitive notion of probability. They may be unsure about the meaning of the term *event* as it applies to probability. You may wish to give your students examples of an event, such as getting heads on a single toss of a coin or getting a 2 on a single roll of a number cube. Some events are certain to occur. These include getting *either* heads or tails on a single toss of a coin and tomorrow being Wednesday given that today is Tuesday. Some events are impossible. These include getting *both* heads and tails on a single toss of a coin and getting a 7 on a single roll of a six-sided number cube.

Technology

On a graphics calculator, the RAND function is often in a probability (PRB) menu. The integer (INT) function is often found in the MATH menu. Caution students against typing INT in an alpha mode instead of using the integer function on a graphics calculator.

Teaching Tip

Defining Experimental Probability

Read the formal definition of experimental probability on page 167 with your students. To make the definition easier to understand, give an example of an experiment, and identify the experiment's successful event.

Number 3 slips of paper from 1 to 3. Put the slips of paper in a hat. Tell students that you will find the experimental probability of drawing a 2. Choose a slip of paper and record the number. Put the slip of paper back into the hat. Repeat this exercise 12 times to see how many 2s are drawn.

Drawing slips of paper from the hat is the experiment. Since the exercise is repeated 12 times, there are 12 events in the experiment. An event is successful when a 2 is drawn. The experimental probability is

$\frac{\text{number of times a 2 is drawn}}{12}$.

TEACH

Approach B Using Models, continued

Defining Probability Terms

Many students may have an intuitive sense of probability. It is important, however, that they understand the terms used in probability. For example, in order to explain what an *event* is in probability, you might give these examples: getting heads on a single toss of a coin or getting an even number on a single roll of a number cube. Some events are certain to happen, such as getting either heads or tails on a single toss of a fair coin and next month being December given that today is November 11. Impossible events, on the other hand, are not possible, such as getting both a 5 and a 6 on a single roll of a number cube and tomorrow being Tuesday given that today is Friday.

Modeling a Probability Experiment

You have several choices for conducting the explorations. Students can use real coins and number cubes, or they can use slips of paper to represent the outcomes of a toss of a coin and a roll of a number cube. A third option is to use a random-number generator on a calculator or a computer to simulate the tossing of a coin or the rolling of a number cube. On a graphics calculator, the random-number generator (RAND) is often in a probability (PRB) menu.

The integer function (INT), which rounds the random decimal number *down* to the nearest integer, may be in the MATH menu. Caution students against typing INT and RAND in an alpha mode instead of using the integer and random functions. The integer and random-number functions are abbreviated with lowercase letters. The calculator may not understand the capital letters displayed in an alpha mode.

The Language of Probability

Before you give your students a formal definition of experimental probability, use an example to demonstrate the experimental probability. You could model drawing the greatest integer from the first 3 positive integers by numbering 3 slips of paper from 1 to 3. Place the slips of paper in a box. Choose a slip of

Cooperative-Learning Groups

Teacher-Guided Explorations You may wish to guide your students as they perform the experiments in the explorations. For greater student participation, you can ask one or more students to toss the coin or roll the number cube. Ask another student to record the outcome of each trial. Encourage students to share their observations of each experiment with the class. Choose students to answer the questions, and invite discussion of the answers so that as many students as possible are involved in each activity.

Student-Directed Explorations Have the class work in groups of four. Each student should perform the experiment and then answer the questions. Then all of the students in a group should combine the results of their experiments. One student should be chosen to record the group's answers to the questions. Another student should write a summary of the group's findings. Finally, one student from each group should be chosen to report his or her group's data to the class. The data from all groups should be recorded on the overhead or chalkboard.

Exploration 1

In this exploration, the students determine the experimental probability of an event from a sample space. They use their findings to examine the relationship between the probability that an event does occur and the probability that the event does not occur.

In Steps 1 and 2, students determine an experimental probability by counting the outcomes. Such an activity offers students a concrete foundation for studying what can be a difficult theoretical subject.

In Step 3, students may notice that the class results are very different from many of the individual results. The class results may more closely approximate what students might expect from many tosses of a coin, namely that a coin is just as likely to show heads as tails.

continued on top of next page

paper, and record the number. Put the slip back into the box. Repeat this exercise 12 times to see how many 3s are drawn.

Drawing a slip from a box is the event. Since the exercise is repeated 12 times, there are 12 events in the experiment. An event is successful when a 3 is drawn. The experimental probability is

$$\frac{\text{number of times a 3 is drawn}}{12}.$$

Now discuss the definition from the book.

Cooperative Learning

Before your students begin the explorations, ask them to form groups of four. Students should take turns tossing the coins or rolling the number cube. One student in each group should be chosen to record the group's responses, and another student should be chosen to write a summary of the group's activity. All students should offer input for the summary and should engage in a discussion of the questions.

Exploration 1

Students model a probability experiment involving the toss of a coin. They also examine the relationship between the probability of an event and the probability of the complement event.

Have students work in their groups and use real coins, models of coins, or a random-number generator to investigate the questions below. You can choose to write the questions on an overhead or the chalkboard, or you can include the questions on a handout.

- Use coins, models of coins, or a random-number generator to model 10 coin tosses. How many heads did you get? How many tails did you get? **[Answers will vary, but the sum of the number of heads and the number of tails should be 10.]**
- Compare the number of heads with the total number of tosses. This is the experimental probability of heads for this experiment. Compare the number of tails with the total number of tosses. This is the experimental probability of tails for this experiment.

continued on bottom of next page

Approach A

Using the Book, continued

In Step 4, students are introduced to complementary events. The probability of tossing tails is shown to be the same as the probability of not tossing heads. Since an event must either occur or not occur, the sum of the probability that the event will occur and the probability that the event will not occur is 1.

Ongoing Assessment, page 167

4. 1; 1; yes; the sums are equal. Since a toss of either heads or tails is certain to occur and one event excludes the other, the sum of the probabilities of both outcomes must be 100%, or 1.

Follow-Up Demonstration (Optional)

If you feel students would benefit from an additional demonstration, you may ask them to repeat the exploration and record their results. By combining the results of the two explorations, the sum of two complementary events can be stressed. Students should also see that by enlarging the sample space, the experimental probability of each event more closely approximates one-half.

Critical Thinking, page 167

If the experimental probability were $\frac{1}{10}$, for example, then a reasonable prediction would be 2 out of 20. If the experimental probability were $\frac{9}{10}$, then a reasonable prediction would be 18 out of 20. Therefore, a reasonable way to use the experimental probabilities is to predict that the number of heads, x, that will appear in 20 tosses is such that

$$\frac{x}{20} = \text{experimental probability.}$$

Solving for x gives

$$x = \text{experimental probability} \times 20.$$

On Their Own

A. If the experimental probability of getting heads on repeated tosses of a fair coin is $\frac{21}{50}$, what was the experimental probability of

TEACH

Approach B

Using Models, continued

[Answers will vary, but the sum of the two numbers should be 1.]

- Combine your data with another group's data. Is the experimental probability of either heads or tails for the combined data different from what it was for each group's data? How? **[Answers will vary, but the sum of the experimental probabilities for heads and tails should still be 1.]**
- Repeat the previous step by combining your data from your two groups with another combined group. Do you notice a pattern in the experimental probabilities? **[Answers will vary. Students may notice that as the data set gets larger, the experimental probabilities for both heads and tails get closer to one-half.]**
- If all the data from the entire class has not been combined, repeat the previous step until the data from your entire class have been combined. How do the experimental probabilities compare? **[Answers will vary. Students may notice that as the data set gets larger, the experimental probabilities for both heads and tails get closer to one-half.]**
- Can you use your experimental probabilities from the exploration to predict the number of times you will get heads in 20 tosses of a coin? Explain. **[For 20 tosses, you will have approximately double the number of heads that you had in 10 tosses.]**

On Their Own

A. If the experimental probability of getting heads on repeated tosses of a coin was $\frac{21}{50}$, what was the experimental probability of getting tails on the same tosses of the coin? $\left[\frac{29}{50}\right]$

B. If the experimental probability of getting tails on repeated tosses of a fair coin was $\frac{253}{500}$, what was the experimental probability of getting heads on the same tosses of the coin? $\left[\frac{247}{500}\right]$

getting tails on the same tosses of the coin? $\left[\frac{29}{50}\right]$

B. If the experimental probability of getting tails on repeated tosses of a fair coin is $\frac{253}{500}$, what was the experimental probability of getting heads on the same tosses of the coin? $\left[\frac{247}{500}\right]$

Technology

Generating Random Integers An occurrence is said to happen randomly if there is no way to tell for sure what the outcome will be. Thus, each outcome of a random experiment is equally likely. Before students generate random numbers on a calculator or on a computer, they may wish to guess the outcome. Their guesses, along with the generated numbers, may reinforce the idea of a random experiment. Explain that when computers and calculators simulate random experiments, no outcome, in a simulation, is more likely than another outcome.

Extension

The command INT(RAND*4) + 3 is examined to determine the possible numbers it generates.

Exploration 2

Students use a random-number generator to simulate rolling an ordinary six-sided cube. They record their results and use those results as a means of predicting the experimental probability for a similar experiment.

In Step 1, students are asked to make an assumption about the average number of times that any of the six numbers on the cube will show up on 10 rolls of the cube.

Step 3 allows students to demonstrate their understanding of some of the language of probability. Before students can talk about and use the laws of probability, they must understand the language.

continued on top of next page

Student Reports

One or more students from each group should report the group's findings to the class. Each reporter should copy the results of the coin tosses on the chalkboard. The reporter should also clearly state the group's observations. Questions and discussion should be invited.

Technology

Generating More Than Two Random Numbers Any number of random integers can be generated for a simulated experiment by using a graphics calculator or computer software. The command INT(k*RAND) + A will generate k random integers, beginning with A.

Students should practice generating lists of integers with the commands below.

INT(4*RAND) + 3
INT(2*RAND) + 4
INT(6*RAND) + 1

Before students begin generating the numbers, ask them to indicate the list of numbers that they believe each command could generate.

Modeling a Probability Experiment

If random-number generators are not available for the students, they can simulate the experiment in Exploration 2 with number cubes that several students have brought from home. Or the students can cut out six pieces of paper that are identical in size. Students can number the pieces of paper from 1 to 6, place them in a box, and mix the papers well. To generate a random number, a student should draw a piece of paper, record the number written on the paper, and replace the paper in the box.

Exploration 2

In this exploration, students use a random-number generator to simulate rolling an ordinary six-sided number cube. They use their results to predict the experimental probability for a similar experiment.

Have students work in their groups and use random-number generators, number cubes, or slips of paper. Each of the following questions should be answered. You may wish to write the questions on the overhead or chalkboard.

continued on bottom of next page

Approach A: Using the Book, continued

Ongoing Assessment, page 168

2. Enter the command INT(RAND • 6) + 1. Then use this command 10 times.

4. Answers will vary. Use the ratio $\frac{\text{number of times 4 appears}}{10}$.

Follow-Up Demonstration (Optional)

If you feel that students are unsure of the relationship between the experimental probability and the theoretical probability of rolling a 4 in 10 rolls of a die, you may wish to conduct the experiment several times on a calculator whose display can be projected on an overhead. Ask a student to record the results of each experiment. Then your students can estimate the number of 4s that might be expected in a similar experiment.

Closing the Lesson

Discuss the ways in which the students used counting to determine experimental probability.

TEACH

Approach B: Using Models, continued

- What numbers are marked on the faces of an ordinary six-sided number cube? If the cube were rolled six times, how many times would you expect a 4 to appear? If the cube were rolled 12 times, how many times would you expect a 4 to appear? If the cube were rolled 10 times, how many times would you expect a 4 to appear? **[1, 2, 3, 4, 5, and 6; one time; two times; between 1 and 2 times, but more often 2 times]**
- Roll an ordinary six-sided number cube 10 times, or simulate the rolls. How many times did you get a 4? **[Answers will vary.]** In this experiment, what is a successful event? How many successful events did you have? What is a trial in this experiment? How many trials did you have? **[Rolling a 4; answers will vary but should be the same as in the answer to the previous question; one roll of the number cube; 10.]**
- What fraction describes the experimental probability of your group getting a 4 in 10 rolls of an ordinary six-sided number cube? **[The numerator of each group's fraction should be the number of times that the group rolled a 4. The denominator of each group's fractions should be 10.]**
- How do you think the number of times that you get a 4 in 20 rolls of an ordinary six-sided number cube is related to the number of times you got a 4 in 10 rolls of the cube? **[The number of 4s in 20 rolls should be about twice the number of 4s in 10 rolls.]**

Student Reports

One or more students from each group should report the group's findings to the class. Each reporter should copy the results of the rolls on the chalkboard. The reporter should also clearly state the group's answers to the exploration questions. Questions and discussion of each group's results should follow each report.

Closing the Lesson

Discuss the ways in which the students used counting to determine experimental probability.

ASSESS

Selected Answers

Odd-numbered Exercises 5–49 and odd-numbered Extra Practice

Assignment Guide

Core level students should complete Exercises 1–52.

Core-plus level students should complete Exercises 1–28 and 41–52.

Error Analysis

Students who have difficulty with Exercise 2 probably do not understand the relationship between the probabilities of all the possible outcomes of an experiment. Remind students that the probability of a certain event is 1. Because the sum of the probabilities of all the outcomes is 100%, or 1, the probability of one of two possible outcomes can be found by subtracting the probability of the other outcome from 1.

Alternative Assessment

Demonstrations Define the following terms and construct an experiment to demonstrate the meaning of each term: probability, experimental probability, trial, successful event.

Portfolio Activity Write an algorithm for determining an experimental probability.

RETEACH

Ask several students to create a number line on the floor of the classroom with masking tape. The integers from 1 to 6 should be marked on the number line. Have one student generate a random integer from 1 to 6 on a calculator. Another student should write the number on the chalkboard. Then a third student should stand at the position on the number line that is directly above the integer generated. This process should be repeated 20 times. A new student should be chosen to mark the position of each number generated.

If you do not have enough students to mark the positions of 20 integers, alter the number of trials to accommodate the number of students available. When the experiment is done, ask your students to calculate the experimental probability that a student will be chosen to stand on position 1. Have them repeat the calculation for positions 2, 3, 4, 5, and 6.

ENRICH

Interdisciplinary Connection

Business If a manufacturer of circular saw blades is interested in the number of defective blades in a shipment of 300 blades, he can examine a sample of 30 blades and record the number of defective blades. There could possibly be anywhere from 0 to 30 defective blades in the sample. The number of defective blades in the sample is an example of a *random variable.* Random variables that take on integer values such as this are called *discrete* random variables. Random variables that are measured in time, weight, volume, or distance take on non-integer values and are often referred to as *continuous random variables.*

Suppose that there were 5 defective blades in a sample of size 30. What would be a reasonable estimate of the number of defective blades in the whole shipment? **[50]**

NAME ______________________ CLASS ____________ DATE ____________

Practice

4.1 Exploring Probability

The results of spinning a spinner with regions I, II, III, IV, and V are given in the table below. Use this table for Exercises 1–9.

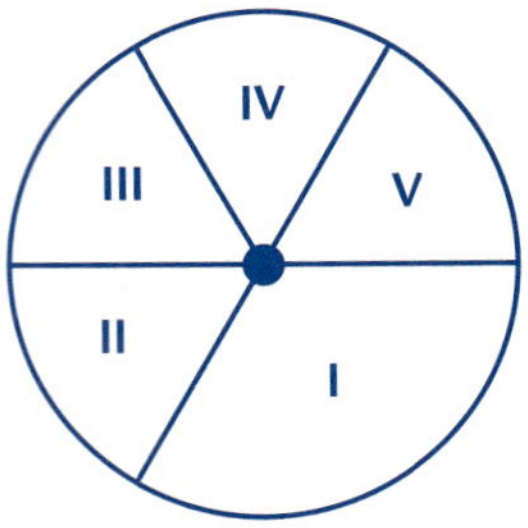

Region	I	II	III	IV	V
Frequency	34	17	15	18	16

1. Do you think it is equally likely for the spinner to land in any region of the spinner? Explain. ______________________

2. Find the experimental probability for landing in region I. ______________________

3. Find the experimental probability for landing in region II. ______________________

4. Find the experimental probability for landing in region III. ______________________

5. Find the experimental probability for landing in region IV. ______________________

6. Find the experimental probability for landing in region V. ______________________

7. Find the sum of the experimental probabilities for landing in regions I and V. Then find the sum of the experimental probabilities for landing in regions II, III, and IV. Are these sums the same? Why or why not?

____

Describe the output of each expression.

8. INT(RAND*2) ______________________

9. INT(6*RAND) ______________________

10. INT(RAND*2)+1 ______________________

11. INT(RAND*2)+5 ______________________

12. INT(RAND*10)+1 ______________________

13. INT(15*RAND)+1 ______________________

Write commands to generate the random numbers described.

14. integer values from 5 to 10 ______________________

15. integer values from 12 to 24 ______________________

NAME ______________________ CLASS __________ DATE __________

Lesson Activity
4.1 Design Your Own Games

1. Design a game of your own in which the probability of one of the events occurring is $\frac{1}{2}$. If necessary, draw a diagram to help describe your game.

2. Design a game that is different from the one in Exercise 1 in which the probability of one of the events occurring is $\frac{2}{3}$. If necessary, draw a diagram to help describe your game.

3. Design a game that is different from the ones in Exercises 1 and 2 in which one of the events has an 80% chance of occurring. If necessary, draw a diagram to help describe your game.

NAME ______________________ CLASS ____________ DATE ____________

Basic Skills Practice

Areas of Triangles, Rectangles, and Squares, Part 1

Area is the number of square units needed to cover the surface of a plane figure.

Example 1: To find the area of a rectangle multiply the length by the width.
Using the formula, $A = l \times w$, let $l = 4$ ft and $w = 3$ ft.

$$A = 4 \text{ ft} \times 3 \text{ ft}$$
$$A = 12 \text{ ft}^2$$

Example 2: To find the area of a square, square the length of one side.
Using the formula, $A = s^2$, let $s = 3$ m.

$$A = 3 \text{ m} \times 3 \text{ m}$$
$$A = 9 \text{ m}^2$$

Example 3: To find the area of a triangle multiply one-half of the base times the height.
Using the formula, $A = \frac{1}{2}bh$, let $b = 3$ ft and $h = 4$ ft.

$$A = \frac{1}{2}(3 \text{ ft})(4 \text{ ft})$$
$$A = 6 \text{ ft}^2$$

Find the area of each rectangle.

1. 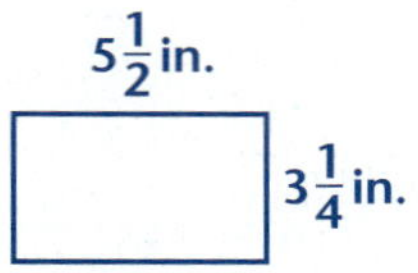

____

2. 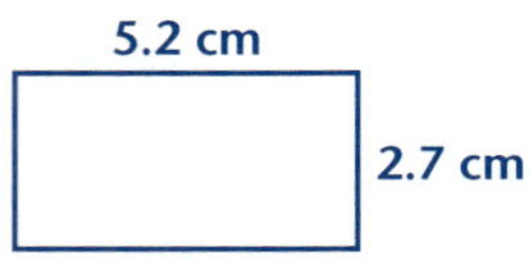

____

3. $l = 9\frac{1}{4}$ yd, $w = 3\frac{1}{2}$ yd ______________

4. $l = 4.4$ mm, $w = 2.9$ mm ______________

Find the area of each square.

5. 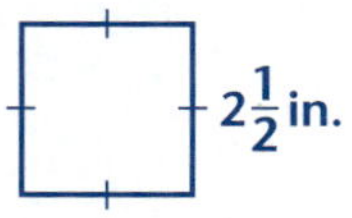

____

6. 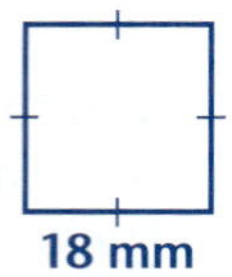

____

7. $s = 7.8$ cm ______________

8. $s = 16$ in. ______________

Find the area of each triangle.

9. 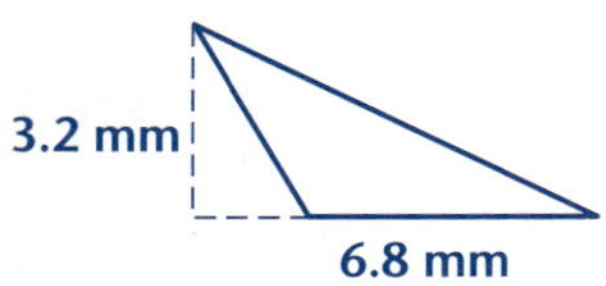

____

10. 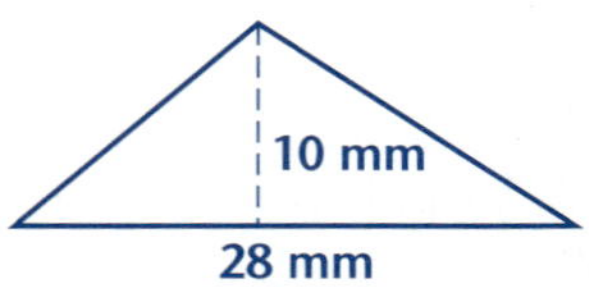

____

11. $b = 11$ cm, $h = 23$ cm ______________

12. $b = 4.5$ in., $h = 6.5$ in. ______________

LESSON 4.2 *Exploring* Simulations

PREPARE

Pacing Guide
3 Days

Objectives and Overview
- Design and perform simulations to find experimental probability.

Students examine how to design a simulation by using a spreadsheet. They plan the simulation of one trial and then perform a large number of trials. The results of the trials are used to find the experimental probability of the simulation.

Vocabulary
simulation

Assessing Prior Knowledge

Find the experimental probability of each event.

1. The experimental probability of rolling a 6 on 15 rolls of a six-sided number cube if there are 2 successful events in 15 rolls. $\left[\frac{2}{15}\right]$
2. The experimental probability of tossing tails on 20 tosses of a coin if there are 13 successful events in 20 tosses. $\left[\frac{13}{20}\right]$

Approach A Using the Book

Opening Discussion
The computer simulates Amy's 20 shots by repeating an experiment with 20 trials. How are the 20 trials in each experiment represented? In the simulation, how many times did Amy make at least 5 field goals in a row? Suppose that Amy makes 60% of her field-goal attempts. How could the simulation be modified to find the experimental probability that she will make four field goals in a row? **[Generate the integers from 1 to 5. Let 1, 2, and 3 represent successful events, and let 4 and 5 represent unsuccessful events.]**

Use **Transparency 15** for discussion of the spreadsheet on page 173 of the Pupil's Edition.

Brainstorming
Let students work in groups of 3 or 4 to brainstorm a list of five experiments that might be simulated by using a spreadsheet. Examples include finding the

continued on top of next page

TEACH

Approach B Using Models

Opening Discussion
Copy all or part of the spreadsheet from page 173 of the Pupil's Edition on the chalkboard or an overhead, making sure to include the formula. A computer spreadsheet program can be used to simulate the outcomes of attempted field goals by a player who makes 50% of his or her shots. A simulation can be designed so that one trial represents 20 attempted field goals. The trial is then repeated many times. How many trials are represented on the spreadsheet? If a 1 represents a successful field-goal attempt in the simulation, how many times did the player make at least 5 field-goals in a row?

Brainstorming
Ask students to think of situations involving experimental probabilities that might be simulated by using a spreadsheet. Examples include the following: What is the experimental probability of two sons in a family with two children? What is the

continued on bottom of next page

Approach A **Using the Book**, continued

experimental probability of giving birth to two daughters in a row in a family with three children, finding the experimental probability of rolling two 2s in a roll of two six-sided number cubes, and finding the experimental probability that 3 customers in a row will buy a pickup truck at a particular dealership if 70% of the customers who visit the dealership buy a pickup truck there.

Technology

Spreadsheets use built-in formulas called *functions* that usually take one or more values, perform an operation on the value(s), and return one or more values. The values given to a function are called *arguments*. Arguments are generally enclosed in parentheses. RAND is a spreadsheet function that returns a randomly-generated number between 0 and 1; it has no argument. Students should note, however, that they should not omit the parentheses when typing the formulas used in the explorations. The parentheses should simply be left empty.

Inclusion Strategies

If you have access to a computer or a computer lab with spreadsheet software, ask each student to simulate three trials of Amy's field-goal attempts. The students should print their results so that the class can find the experimental probability of Amy making at least four field goals in a row from the total number of trials generated by all the students.

Modeling a 50% Probability With a Coin Toss

Any random-number generator that returns exactly two values with an equal chance of occurring can be used to simulate an experiment in which there is a 50% chance of a successful event. Since tossing a coin produces one of two equally likely outcomes, heads or tails, one outcome can represent a successful event and the other outcome can represent an unsuccessful event in a simulation.

TEACH

Approach B **Using Models**, continued

experimental probability of rolling a 2 and a 6 on a roll of two six-sided number cubes? What is the experimental probability that 3 customers in a row will buy a double cheeseburger at a restaurant if 20% of the customers in the restaurant buy double cheeseburgers?

Spreadsheet Formulas

Formulas, called *functions*, that are built into a spreadsheet generally ask for one or more values to be entered. Then the spreadsheet performs an operation on the value(s) and returns one or more values. The values entered for a function are called *arguments*. Arguments are usually enclosed in parentheses. RAND, a spreadsheet function that returns a randomly generated number between 0 and 1, requires no argument. Students should note, however, that the parentheses are not omitted in the formula, INT(RAND()*2). The parentheses should simply be left empty. The *2 is necessary so that two integers, one representing a successful event and the other representing an unsuccessful event, will be generated.

Modeling With a Coin Toss

If an event in an experiment has a 50% chance of happening, any random-number generator that returns exactly two values can be used to simulate the experiment. Because tossing a coin produces one of two equally likely outcomes, heads or tails, the two events can be represented by the two outcomes. If a number cube is used, let three of the numbers represent a successful event and the other three an unsuccessful event. All other rolls of the cube should be ignored.

Cooperative Learning

Divide the class into groups of four. Before beginning the exploration, tell the class that heads will represent making a shot and tails will represent missing a shot. All groups should use the same representation. The members of each group should take turns tossing the coin until 20 trials have been completed. Each group should select one member to record the results of each trial. Another member should be chosen

Cooperative Learning

Teacher-Guided Exploration You can guide students through the exploration by recording the results of the coin tosses on the overhead. Organize the data as if it were being generated on a spreadsheet. Instead of coins, you can use slips of paper or chips.

To demonstrate how a larger number of trials influences experimental probability and to involve all students, have students take turns tossing the coin. When 20 tosses have been completed, tally the results and ask your students to calculate the experimental probability. Repeat the 20 trials again and calculate the experimental probability for these 20 trials. Then calculate the experimental probability for all 40 trials. Repeat this activity two more times.

Student-Directed Group Approach
Divide the class into groups of four. Before beginning the exploration, the class should decide which outcome, heads or tails, will represent making a shot and which outcome will represent missing a shot. All groups should use the same representation. The members of each group should take turns tossing the coin for 20 trials. Each group should select one member to record the results of each trial. Another member should be chosen to record the number of sequences of at least four heads. These members should write the number of sequences on the board so that the class can calculate the experimental probability for the total number of trials.

Exploration 1

Students use a coin or similar random-number generator to simulate a 50% chance of making a foul shot.

In Step 1, students perform the simulated trials and record the results. So that the sequences of successful events are easily observed, each group should organize its data as if it were being generated by a spreadsheet similar to the one on page 173 of the Pupil's Edition.

continued on top of next page

to record the number of sequences of at least four heads. These members should write the number of sequences on the board so that the class can calculate the experimental probability for the total number of trials. Each group should choose a member to write a summary of the group's findings.

Exploration 1

Students use a coin or a similar random-number generator to simulate a 50% chance of making a field goal.

Have students work in their groups and use a coin to simulate a 50% chance of making a field goal. Each group should toss its coin 20 times. The members should work cooperatively and take turns tossing the coin. The sequences should be recorded. Provide the students with the following instructions and questions:

- Toss a coin 20 times and record the sequence of heads and tails. How does this simulate making a field-goal? **[Heads represents a successful shot. Tails represents an unsuccessful shot.]**
- Does the sequence have at least four consecutive heads? If so, what do these heads represent? **[at least four consecutive field goals in a row]**
- Pool the results of the class. **[Results may vary.]**
- How will you use the pooled results to find the experimental probability based on the class experiment? What is the experimental probability? **[Find the number of sequences that have at least four consecutive heads. This is the numerator of the of the experimental probability. The denominator of the experimental probability is the total number of simulations conducted by the class. Experimental probabilities will vary.]**

Student Reports

One student from each group should report the results of his or her group's experiment, noting the sequences of at least four consecutive heads. If these students write the experimental

continued on bottom of next page

Approach A Using the Book, continued

In Steps 3 and 4, students pool their results and compare the results of their group's simulation with the results of the whole class.

Ongoing Assessment, page 173

4. Use the ratio

$$\frac{\left(\text{number of times there were at least 4 consecutive heads}\right)}{\left(\text{number of times the experiment was performed by the class}\right)}.$$

Follow-Up Demonstration (Optional)

If you feel that reinforcement of this exploration is necessary, you can simulate the same experiment by using the random-number generator on a graphics calculator. If you have access to a graphics calculator whose screen can be displayed on an overhead, you can call out the randomly generated integers while several students take turns recording the integers on the

Sequences of random numbers can be generated on a graphics calculator and displayed on the chalkboard for class discussion.

TEACH

Approach B Using Models, continued

probability of their group's experiment on the board, all students will be able to compare their group's probability with the class's experimental probability.

Journal Entry

Simulation Design Notes It will be helpful for students to write down in their notebook a strategy for designing a simulation. Refer to the steps for designing a simulation on page 173 of the Pupil's Edition, discuss these steps with students, and have them enter the process and how it is applied in their journal.

Exploration 2

In this exploration, students perform a simulation to find the experimental probability of rain on at least one of two days.

Have students continue to work in their groups and use slips of paper to generate sequences of random numbers. Give each group a piece of the

Students can cut out numbered slips of paper and draw them from a box to simulate the probability of rain.

same size paper. Instruct the groups to choose a member to fold the paper into sixteenths by using a series of 4 folds. A student from each of the groups should label 10 of the sections with integers from 1 to 10 inclusive. Those sections board. Generate four groups of 20 random numbers. Then calculate the experimental probability based on the number of sequences of at least four consecutive heads in all 80 trials.

Designing a Simulation

Teaching Tip Have students write in their notebook the following tips on designing a simulation:

- Before trials can be simulated, the significance of each result of the trial must be noted.
- When the probability of an event is not 50%, a random-number generator other than a coin must be used.
- Each trial after the first is simply a repetition of the first trial.
- The greater the number of trials, the closer the experimental probability will be to the theoretical probability.

On Their Own

The Homestyle Diner offers either baked haddock or fried haddock as one of its specials on Friday night. Half of the time, the special is baked haddock. The other half of the time, the special is fried haddock. Have students make a spreadsheet that simulates 20 trials of 10 consecutive Friday nights. Let 0 represent baked haddock and 1 represent fried haddock.

A. Which trials have at least three consecutive baked haddock specials? **[Answers will vary from experiment to experiment.]**

B. What is the experimental probability of the diner offering the baked haddock special on three Fridays in a row in 10 weeks? **[Use the ratio**

$$\frac{\left(\begin{array}{c}\textbf{number of trials with at least 3}\\ \textbf{consecutive baked haddock specials}\end{array}\right)}{\textbf{20}}\textbf{.]}$$

Exploration 2

In this exploration, students perform a simulation to find the experimental probability

continued on top of next page

should be cut out and placed in a box. The remaining sections will not be needed for this exploration. Provide students with the following instructions and questions:

- How can you use the 10 slips of paper to simulate a 40% chance of rain? **[Let four of the slips represent rain and the remaining six slips represent no rain.]**
- Let the slips numbered 1, 2, 3, and 4 represent rain. The other slips represent no rain. Mix the slips in the box. Then pick one slip, and record the result in a table like the one below.

Trial	1st number	2nd number	At least one number less than or equal to 4?
1	4	10	yes
2			
3			
4			
5			
. . .			

Replace the slip and mix well. Choose a second slip and record its number in the table. Indicate whether at least one number in the pair is less than or equal to 4.

- What do the pair of numbers simulate? What do your numbers indicate about rain on those two days? **[The pair of numbers simulates the weather for two days, indicating whether or not it will rain. Students' numbers will vary, but any of the digits from 1 to 4 inclusive indicate rain on that day. Any other number indicates no rain on that day.]**
- The pair of numbers you generated represents one trial in an experiment to find the experimental probability of rain on at least one of two days when the chance of rain is 40%. Perform a large number of trials, say 20, and record the results. **[Results will vary.]**
- Did any of your trials produce two numbers greater than 4? If so, which one(s)? What do the numbers indicate about rain on those

continued on bottom of next page

Approach A — Using the Book, continued

when the theoretical probability of a successful event is not 50%.

In Step 1, two methods of simulating a 40% chance of rain are explained, one method using a random number generator and the other method using slips of paper.

In Step 2, students must generate *two* numbers because this experiment simulates the weather for two days.

In Steps 4 and 5, students must identify the significance of each number generated.

Remind students that in Step 6, "at least one day" means either one or two days.

Students use the results of their experiments to find an experimental probability of having rain on at least one of two days.

Ongoing Assessment, page 174

7. $\frac{7}{10}$

Technology

If students use technology, they can also generate a sequence of random integers from 1 to 10. Have them let a number less than or equal to 4 mean that it will rain, and let a number greater than or equal to 5 mean that it will not rain.

Application: Social Engagements

Because many students can easily relate to the situation simulated in the application on page 175 of the Pupil's Edition, this activity allows the students to connect a topic in mathematics to a real-world situation.

Modeling With Number Cubes

Make sure students understand why a roll of 6 must be discounted. Even though the number cubes may generate more numbers than needed, it may be possible to model the situation by modifying the results of rolling the cube.

TEACH

Approach B — Using Models, continued

days? **[These numbers indicate that it does not rain on either day.]**

- Did any of your trials produce two numbers less than or equal to 4? If so, which one(s)? What do the numbers indicate about rain on those days? **[These numbers indicate that it rains on both days.]**
- According to your data, how many weekends had rain on at least one day in the trials? **[Answers will vary but should be the number of trials in which at least one of the numbers is 1, 2, 3, or 4.]**
- What is the experimental probability of having rain on at least one of the two days? **[Answers will vary. The numerator of each group's probability fraction should be the number indicated in their answer to the previous question. The denominator of each group's fraction should be 20. Some groups may simplify their fractions or express them as a percent.]**

Technology

If students use technology, they could also generate a sequence of random integers from 1 to 100. In this case, a number less than or equal to 40 means that it will rain. A number greater than 40 means that it will not rain.

Follow-Up Demonstration (Optional)

If you feel that reinforcement of this exploration is necessary, you can simulate the same experiment by using the random-number generator on two graphics calculators. You can ask two students to generate pairs of integers from 1 to 100 while several students take turns recording the integer pairs on the board. Generate four groups of 20 pairs of random integers. Then calculate the experimental probability based on the number of pairs that represent at least one rainy day in all 80 trials.

Modeling an Application

Because many students can easily relate to the situation simulated in the application on page 175 of the Pupil's Edition, this activity allows

Lesson Preview: Logical Connectives

In Step 3 of the application, students must recognize that the experimental probability that both friends are on time for lunch is determined by finding the trials in which the first event is successful *and* the second event is successful. In Lesson 4.4, students will learn how a database distinguishes between a situation in which both events must occur and a situation in which either one may occur. In that lesson, students will also learn how to graphically represent the difference between these two types of situations.

Critical Thinking, page 175

Situations will vary. Situations that can be simulated by a coin toss have equally likely chances of happening or not happening, such as a female baby being born. Situations that can be simulated with a roll of a number cube do not necessarily have equally likely chances of happening or not happening. Any situation with a probability of success that can be written as a simplified fraction with a denominator less than or equal to the number of faces on the cube can be simulated by using that cube.

Closing the Lesson

Discuss the steps for designing a simulation that are outlined on page 173 of the Pupil's Edition. Ask students to explain how they followed each of these steps as they completed the two explorations.

the students to connect a topic in mathematics to a real-world situation.

Have students work in pairs. Each student should assume the role of one of two people who have a lunch engagement. The probability that person A is on time is 60%. The probability that person B is on time is 80%. The students can design a simulation to find the experimental probability that both people are on time for lunch. Tell students to generate a pair of random integers from 1 to 100 by using technology. Any number less than or equal to 60 indicates that person A is on time. Any number less than or equal to 80 indicates that person B is on time. Each student should generate an integer for the role he or she has chosen. The pair of integers represents the first trial. Perform 10 trials.

Ask each pair of students to determine how many times out of 10 that they meet on time. **[The number of times will vary but should be the total number of trials in which the first number is less than or equal to 60 *and* the second number is less than or equal to 80.]**

Tell each pair of students to write the fraction that represents the experimental probability of both persons meeting on time for lunch. **[The numerator of each fraction should be the same as the number of times out of 10 that they met on time in the simulation. The denominator should be 10.]**

When Models Generate More Numbers Than Are Needed

Students could also model the application problem with two number cubes. If six-sided regular number cubes are used, a 1, 2, or 3 on the first cube can mean that person A is on time, and a 4 or 5 can mean he or she is late. A roll of a 1, 2, 3, or 4 on the second cube can mean that person B is on time, and a roll of 5 can mean he or she is late. Any roll of 6 must be ignored because it represents nothing in this simulation.

ASSESS

Selected Answers

Odd-numbered Exercises 7–31 and odd-numbered Extra Practice

Assignment Guide

Core level students should complete Exercises 1–32.

Core-plus students should complete Exercises 1–17 and 23–33.

Error Analysis

Students may have difficulty determining a model for a simulation. Explain to these students that if the probability of a successful event is not equal to 50%, then integers should be chosen to represent the probability. An event with an n% probability of success can be represented by integers from 1 to n. The unsuccessful event for the same situation can be represented by integers from $(n + 1)$ to 100.

Alternative Assessment

Simulations Have students describe a situation that can be simulated by a roll of a regular six-sided number cube. Tell them to design and describe a simulation that will give the experimental probability of the situation.

Portfolio Activity Have students write a summary describing how a toss of a coin and a roll of a number cube can both be used to simulate an experiment. The summary should include instructions on how to use the results to find the experimental probability.

RETEACH

Have students perform a simulation to find the experimental probability that more than three people in a row who are polled by telephone prefer candidate A if nine people are polled and the probability that a person polled prefers candidate A is 50%.

Choose one student to be the recorder. Divide the remaining students into two equal groups. Give each of these students a piece of paper. Half of the students should write *Candidate A* on their pieces of paper. The other students should write *Candidate B* on their papers.

Use a random-number generator to generate either a 0 or a 1. Ask the recorder to write the number on the chalkboard. If a 0 is generated, tell one of the candidate A students to stand in the front of the classroom facing the class. The student should hold up his or her paper, which reads *Candidate A*. If a 1 is generated, the same procedure should be followed by a candidate B student. Repeat this activity nine times. As numbers are generated, students should form a line across the front of the classroom.

These are the results of the first trial. When the results have been recorded, tell the students in the line to disband. Perform 10 trials.

Ask students to examine the results on the board and to count the number of trials in which at least 3 people in a row supported candidate A. Tell them to use this number to find the experimental probability that at least 3 people in a row who are polled prefer candidate A.

ENRICH

Interdisciplinary Connection

Firefighting Because those who study fires have increased understanding of how fires spread, a sophisticated computer simulation of the path of a fire can help firefighters predict a fire's progress and thereby aid in controlling the fire. The effectiveness of smoke detectors and sprinklers also aids the computer simulation in predicting the number of smoke detectors and sprinklers necessary to detect and put out a fire.

NAME ______________________ CLASS __________ DATE __________

Practice
4.2 Exploring Simulations

Suppose that the Denver Nuggets and the New York Knicks are in the basketball championship finals and each team has an equal chance of winning a given game. The finals end when one team wins four games. Use simulation of the finals for at least 10 trials.

1. Describe the three steps of the simulation.

____

____

2. Complete the chart below to give game-by-game results of each of the 10 trials. Use random numbers generated by a coin toss or technology. Use *W* for a Knicks win, and *L* for a Knicks loss.

Trial	1	2	3	4	5	6	7
1							
2							
3							
4							
5							
6							
7							
8							
9							
10							

Use your results of the basketball championship simulation to find the experimental probability of each of the following:

3. The Nuggets win the finals. ______________________

4. The Knicks win the finals in 4 games. ______________________

5. The finals last 7 games. ______________________

6. A team wins the finals after losing the first two games. ______________________

NAME ______________________ CLASS __________ DATE __________

Lesson Activity

4.2 Baseball Simulations

INT (RAND()*50 + 1)			
Trial	1st number	2nd number	3rd number
1	34	6	35
2	12	43	37
3	30	1	25
4	41	4	44
5	4	40	17
6	33	9	12
7	37	2	39
8	18	50	1
9	9	38	11
10	34	25	13

Use the simulation shown above to find the experimental probability for each of the following events. If the simulation cannot be used to simulate an event, explain why.

1. Quinn is a baseball player with a batting average of .300. This average means that he gets a hit 30% of the times he bats. What is the experimental probability that Quinn will get a hit in at least 2 of his next 3 at bats?

____

2. Roger has a batting average of .400. What is the experimental probability that he will get a hit in at least 2 of his next 3 at bats?

____

3. Jerome is a pitcher who wins 60% of his games. What is the experimental probability that he will win at least 2 of his next 3 games?

____

4. What is the experimental probability that Jerome will win all 3 of his next 3 games?

____

5. Manuel has a batting average of .280. What is the experimental probability that he will get a hit in at least 3 of his next 5 at bats?

____

NAME ______________________ CLASS ____________ DATE ____________

Basic Skills Practice

Areas of Triangles, Rectangles, and Squares, Part 2

Read each question and circle the best answer.

1. John is building an additional room that is 15 feet by 24 feet. How many ***square yards*** of carpet are needed to cover the floor?

A 39 yd^2

B 40 yd^2

C 45 yd^2

D 360 yd^2

2. The frame around a picture has a length of 20 inches and a width of 16 inches. The length of the picture inside the frame is 14 inches and the width is 10 inches. What is the area of the frame?

F 180 $in.^2$

G 320 $in.^2$

H 220 $in.^2$

I 460 $in.^2$

3. Alex plans to install wall-to-wall carpeting in his house. He needs to buy 156.25 square yards to cover a square room. What are the dimensions of the room?

A 12.5 yd × 12.5 yd

B 15.5 yd × 15.5 yd

C 17.5 yd × 17.5 yd

D 20.5 yd × 20.5 yd

4. What is the area of a rectangular house with a length of 30 feet and a width of 20 feet?

F 600 ft^2

G 500 ft^2

H 50 ft^2

I 1100 ft^2

Solve each problem.

5. Keisha plans to buy flowers to cover a triangular garden in her backyard. The dimensions of the garden are 12 feet along the bottom and 16 feet along one side. A 20 foot side completes the garden and forms a right triangle. What is the area of the garden in square feet?

6. Eric plans to install wall-to-wall carpeting in his square study. The floor is 5.5 yards wide. What is the area to be covered by the carpet in square yards? ____________

7. In the same house, Eric carpets a rectangular room that is 7.7 yards long and 6.2 yards wide. How much more area, in square yards, is covered by this carpet than by the carpet in Exercise 6? ____________

8. Calvin plans to paint the four walls and the ceiling in a guest room. Each wall is rectangular with a width of 12 feet and a height of 10 feet. The ceiling is a square with sides of 12 feet. Find the total area, in square feet, of the ceiling and the four walls. ____________

9. The area of a rectangular field is 165,000 square meters. If the length is about 480 meters, what is the approximate width, in meters, of the field? ____________

LESSON 4.3

Statistics and Probability

PREPARE

Pacing Guide

3 Days

Objectives and Overview

- Find the mean, median, and mode of a set of data.
- Make scatter plots and frequency tables from data.

Students examine measures of central tendency for a set of data. Scatter plots are introduced as a means of displaying data and determining trends. Students study frequency tables and bar graphs to determine measures of central tendency. Circles graphs are used to make predictions based on an experimental probability.

Vocabulary

mean	correlation
median	line of best fit
mode	surveys
range	populations
scatter plot	frequency table

Assessing Prior Knowledge

Each set of numbers represents a set of test grades. If each grade is equal in weight, find the average of each set of test grades. Round to the nearest whole number, if necessary.

1. 78, 82, 96, 80, 83 [84]

2. 70, 67, 67, 70, 72 [69]

3. 93, 95, 90, 90, 90 [92]

Approach A

Using the Book

Opening Discussion

Ask students if they know how a batting average is calculated. **[The total number of base hits is divided by the number of times at bat. The quotient is expressed in thousandths.]** How is this kind of average different from the average of your last five math exam grades? How is it similar to your math exam average? Do you know other ways in which an average can be expressed?

Brainstorming

Ask students to work in pairs and make up several sets of math exam grades, each containing five grades. Instruct them to choose grades so that the highest and lowest grades in half of the sets differ by 35 points, while the highest and lowest grades in the other sets differ by 7 points. Does the mean of each set seem representative of the test scores? Do you think that another number in each set is more representative? Explain. **[When the spread of scores is only 7 points,**

TEACH

Approach B

Using Models

Opening Discussion

Ask students if they can list any baseball player's batting average. Do you know how a batting average is calculated? **[The batting average is the total number of base hits divided by the number of times at bat. The quotient is expressed in thousandths.]** Would you calculate the average of your last five exam grades differently than a batting average? Explain. Do you know other ways in which an average can be expressed?

Brainstorming

Ask students to make up several sets of math exam grades, each containing five grades. They can work in pairs if they wish. Choose grades so that the highest and lowest grades in half of the sets differ by 35 points. The highest and lowest grades in the other sets should differ by 7 points. Do you think that the mean of each set is representative of the test scores in that set? Explain whether another number in the set might be more representative of the data. **[When the spread of**

the mean is representative of the score. When the spread is 35 points and four of the grades are clustered near the highest or lowest grade, for example, the fifth grade will pull the mean in its direction. This mean may not be representative of the test scores. If no such example appears in the students sets, then construct a set to illustrate this situation.]**

Defining *Descriptive Statistics*

Descriptive statistics is only one area of statistics. The importance of descriptive statistics lies in the fact that it helps people who must make decisions about a set of data to organize and describe the data. Measures such as mean, median, mode, and range and techniques such as displaying data in graphs, charts, and tables allow the data to be summarized and presented in ways that are easily understood.

Example 1

In this example, students review how to find the mean, median, mode, and range of two sets of data. There is no mode for either set of data. A summary statement compares the measures.

Ongoing Assessment, page 179

Try This

mean: 33
median: 31
mode: 29

Additional Example 1

Give the mean, median, mode, and range for the enrollment figures at Southwest Technical College and Midwest Technical College from 1990 to 1996.

	Southwest	Midwest
1990	1000	2500
1991	1100	2500
1992	1175	2775
1993	1200	3300
1994	1250	3350
1995	1225	3350
1996	1200	3200

continued on top of next page

scores is only 7 points, the mean is representative of the score. When the spread is 35 points and four of the grades are clustered near the highest or lowest grade, the fifth grade will pull the mean in its direction. This mean may not be representative of the test scores.]

Modeling Measures of Central Tendency

Using a Number Line Use a number line to mark the positions of data in order to provide a visual model for identifying measures of central tendency. Draw two number lines on the chalkboard or overhead. Label the number lines in increments of 0.01 from 0.30 to 0.40. Make sure that the tic marks on the two number lines are aligned. Give students the following data sets, which are the baseball averages of five National League and five American League players:

National League: .336, .335, .319, .330, .370
American League: .339, .329, .342, .343, .363

Mark each National League player's average above one number line and each American League player's average above the other number line using a large dot. If colored chalk is available, you can represent the players in each league with a different color of chalk.

Which number line shows the greatest range of values? **[the National League number line]** For which league is it easier to predict the mean of the averages? Explain. **[the American League, because the scores are more closely clustered]** Why is it easy to find the median of each data set on these number lines? **[The data are arranged in numerical order on the number lines.]** Does either set of data have a mode? **[no]**

Working With Number Lines

The mean, the median, and the mode are used to measure central tendency. When data are displayed using dots above a number line, the mode is easily recognized because that data point is repeated the most times. Since the median is the number in the middle position, it is also easy to find when the data are displayed on a number line. If there are two middle numbers, the median is the mean of those numbers. The mean is also easily found when a

continued on bottom of next page

Approach A — **Using the Book**, continued

[Southwest: mean = 1164, median = 1200, mode = 1200, range = 250; Midwest: mean = 2996, median = 3200, mode = 2500 and 3350, range = 850. In summary, the mean, median, and range were all higher for Midwest Technical College. The data set for Southwest Technical College has one mode and that for Midwest Technical College has two modes.]

Teaching Tip

Before discussing Example 2, you may wish to review the construction of a scatter plot and the definitions of correlation coefficient and line of best fit with your students. Remind students that the range of values for the correlation coefficient is −1 to 1. Strong negative and strong positive correlations are indicated by points that lie very close to a line and by correlation coefficients very close to either −1 or 1.

Example 2

Example 2 displays the National League batting averages data in a scatter plot. The line of best fit drawn for the data shows a moderate positive correlation. Students should remember that a line of best fit does not have to pass through any of the data points, but that the line must minimize the distance between itself and all of the data points.

Ongoing Assessment, page 179

Since the correlation coefficient is only about 0.50, the line of best fit only moderately predicts the average for a given year. Although the batting averages tended to increase as the years increased, nearly as many averages decreased between consecutive years as increased.

Additional Example 2

Make a scatter plot of the enrollment figures for each year for Midwest Technical College from the following table. Then use a calculator or

TEACH

Approach B — **Using Models**, continued

small set of data is displayed on a number line. The total number of data points can be counted and a calculator can be used to find the sum. Ask students to find the mean, median, and range of each data set above and to summarize the data by using the number lines. **[National League: mean = .338, median = .335, range = .051; American League: mean = .343, median = .342, range = .034. The mean and the median batting averages were higher in the American League. The batting averages in the National League had a greater range.]**

Teaching Tip

Students may refer to the mean as the *average*. Each of the measures of central tendency typify a set of numbers, and therefore each is a kind of average. You may want to refer more precisely to the mean as the *arithmetic mean.*

Cooperative Learning

Finding a Line of Best Fit Divide the class into groups of seven or larger. Ask several students to use masking tape to create the axes for a scatter plot of the National League batting averages on the floor of the classroom. They should label the

In this kinesthetic activity students play the role of data points.

computer to find the value of the correlation coefficient and the line of best fit.

Year	Midwest enrollment
1990	2500
1991	2500
1992	2775
1993	3300
1994	3350
1995	3350
1996	3200

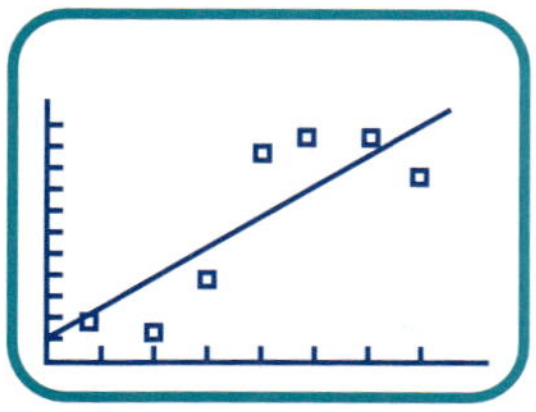

The graph shows a strong positive relationship. The correlation coefficient is about 0.86, and the line of best fit has a slope of about 156.25.

Population Samples

An important area of statistics called inferential statistics allows you to reach conclusions about a large body of data, called a *population,* by observing only a part of that data. This part is called a *sample.* Probability also allows conclusions to be made about a population based on a sample of it.

Presenting Survey Results Graphically in a Frequency Table

Remind students that when they construct a frequency table, observed values should be marked in groups of five for readability. Since the data in a frequency table are listed in numerical order, it is easy to determine the mean, median, mode, and range of the data set from the frequency table. When a mean is calculated from a frequency table, each value must be multiplied by its frequency.

continued on top of next page

vertical axis from .300 to .400 and the horizontal axis from 1989 to 1993. Five students from each group should stand in the scatter plot to mark the position of the five data points. All students should discuss the position of a line of best fit for the data, perhaps by using a string held in the air among the students. When a line is agreed on, the two remaining students should mark the line with masking tape.

Technology

Graph the scatter plot of the National League batting averages on a graphics calculator. If you have calculators available for your students, they can graph the data points and the line of best fit as well. To graph the scatter plot, the two variables must be entered in two statistical lists. A plot must be turned on, and the scatter plot option must be selected. Students must also identify which lists represent the XLIST and the YLIST on the plot. The correlation coefficient can be found by calculating the linear regression line (LinReg ($a + bx$)) for the correct XLIST and YLIST. If the linear regression equation is then copied to the Y= list, it can be graphed on the scatter plot.

Ask students to compare their model of the line of best fit with the calculator's line.

Interpreting a Correlation Coefficient

Since the correlation coefficient is only about 0.50 for the baseball data, the line of best fit only moderately matches the batting averages. Therefore, the line cannot be used with great precision to predict the average for a given year.

Inferential Statistics

Students have studied how measures of descriptive statistics, such as mean, median, and mode, help people organize and describe data. Techniques of descriptive statistics, such as displaying data in graphs, charts, and tables, allow the data to be summarized and presented in ways that are easily understood. Another area of statistics, inferential statistics, allows you to reach conclusions about a large body of data, called a *population,* by observing only a part of those data. This part is called a *sample.* When

continued on bottom of next page

Approach A Using the Book, continued

Exploration

In this exploration, students explore how a bar graph displays the information from a frequency table. Students then use either the frequency table or its bar graph to determine the experimental probability of an event and to make predictions based on the experimental probability. Circle graphs are also used to compute probabilities. Measures of central tendency are calculated.

Ongoing Assessment, page 181

6. The surveys for country music and rap music both report that 23% of the students "like" the music. However, 50% of the students felt strongly about rap music, compared with 18% for country music. Also, 41% of the students tolerate country music, while only 9% tolerate rap music. None of the measures of central tendency indicate this major difference between the surveys.

Inclusion Strategy

Conducting a Survey Have students work in groups of three and conduct surveys of their own choosing. The surveys should ask students to respond with *strongly dislike, dislike, tolerate, like,* or *strongly like* to a survey question. They should write survey questions and distribute them to their classmates. Each group should tally its results in a frequency table. The data should be presented in both a bar graph and a circle graph. Each group should report its results to the class and display its graphs.

Closing the Lesson

Journal Entry Have students ask five of their classmates to estimate their height. The students should record these heights and their own estimated height in their journal. Each student should find the mean, median, mode, and range of the heights recorded. Ask students to explain how they could use a frequency table and a bar graph to find the mean and the median of all of their classmates' heights.

TEACH

Approach B Using Models, continued

conclusions are based on a sample, probabilities are used.

Cooperative Learning

Frequency Tables, Bar Graphs, and Circle Graphs Frequency tables, bar graphs, and circle graphs can be used to model the results of surveying a population sample and to make predictions based on the sample.

Refer to the survey on page 180 of the Pupli's Edition. Have students work in groups of three and conduct similar surveys of their own choosing. The surveys should ask students to respond with *strongly dislike, dislike, tolerate, like,* or *strongly like* to a survey question. They should write out or type the survey questions and distribute them to their classmates. When the surveys are collected, each group should tally its results in a frequency table. The data should be presented in both a bar graph and a circle graph. Each group should report its results to the class and display its graphs. Make sure that all students contribute to the data collection, tally, and representation and that all students are involved in presenting their group's report to the class.

Each group's report should include the following points:

- an explanation of how they used the frequency table to construct their bar graph
- identification of any mode, if one exists, and its significance for the data set
- an estimation of the median, using the bar graph
- identification of the percent for each section of the circle graph
- the experimental probability of those who either liked or strongly liked the variable for which the survey was conducted
- their prediction of how many out of 200 randomly selected students from the school responded with either like or dislike

ASSESS

Selected Answers
Odd-numbered Exercises 9–41 and odd-numbered Extra Practice

Assignment Guide
Core level students should complete Exercises 1–45.

Core-plus level students should complete Exercises 1–18, 21–28, and 33–45.

Error Analysis
Students who answer Exercise 18 incorrectly may not understand that a range is computed from only the maximum and minimum values in the data set. Ask students to choose a data set with five data points and the high, low, and median values given in the exercise. Then tell students to add three more data points to their set. They should choose data points so that the maximum and minimum values remain the same but the median changes. Ask students to compare the ranges of the two data sets.

Alternative Assessment
Mean, Median, Mode, and Range
Ask students to write a set of instructions for calculating the mean, median, mode, and range of a data set. Then ask them how they can use a frequency table and a bar graph to determine the mean, median, mode, and range of a data set.

Portfolio Activity Ask students to draw a scatter plot of the American League batting averages given on page 178 of the Pupil's Edition and to use technology to find the correlation coefficient and a line of best fit. Ask the students to interpret the trend of the batting averages over the 5 years for which data are given.

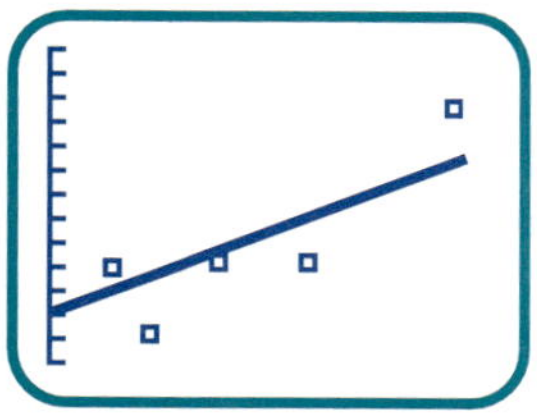

[The correlation coefficient is about 0.79, so the line of best fit predicts the average for a given year fairly well.]

RETEACH

Record each student's age in months on a chart on the chalkboard or overhead. Ask students to identify any modes and to calculate the range. Then ask each student to write down an estimate of the mean and the median of the set. How did they arrive at their estimates? Construct a frequency table and bar graph of the ages. How do the table and the graph describe the mode of the data? Use the table or the graph to calculate the mean and to find the median of the ages. Compare students' estimates.

ENRICH

Interdisciplinary Connection
Geography Since the 1960s, with the development of the electronic computer and remotely-sensed images of the Earth gathered from satellites, geographic research has employed the tools of statistics. Because large databases of geographic information are now available from remote-sensing technology, mathematically constructed models are necessary to deal with the large amounts of data collected. Geographic information systems use statistical tools to organize and model the data so that it can be used to predict future trends or location (more commonly referred to as *spatial*) patterns.

Research Activity
Ask students to prepare reports on some famous baseball players such as Joe DiMaggio, Hank Aaron, Reggie Jackson, Babe Ruth, Roger Maris, Roberto Clemente, and Stan Musial. The reports should contain statistics that contributed to the players' fame.

NAME ______________________ CLASS ____________ DATE ____________

Practice

4.3 Statistics and Probability

Given the data 85, 89, 70, 82, and 84, find

1. the mean. ______________ **2.** the median. ______________

3. the mode. ______________ **4.** the range. ______________

Given the data 14.5, 20, 19, 20, 10.4, and 6.7, find

5. the mean. ______________ **6.** the median. ______________

7. the mode. ______________ **8.** the range. ______________

Given the data 225, 146, 115, 230, 245, and 146, find

9. the mean. ______________ **10.** the median. ______________

11. the mode. ______________ **12.** the range. ______________

13. Describe how to find the median for a set of 7 data values. ______________

____

14. Describe how to find the median for a set of 8 data values. ______________

____

15. If the lowest score is 110, the median is 145, and the highest score is 182, find the range. ______________

16. If the range of the scores is 44, the median is 70, and the highest score is 92, find the lowest score. ______________

17. Construct a set of five data values with a mode of 12, a median of 19, and a mean of 21, or explain why this is impossible.

____

18. Construct a set of five data values, all positive integers, with a mode of 10, a median of 10, and a mean of 2, or explain why this is impossible.

____

NAME ______________________ CLASS ____________ DATE ____________

Lesson Activity

4.3 Extending Statistics

Use the information in the table to answer the questions below.

1992–1993 NBA Individual Statistics (Top 4 Players' Average Points per Game)					
Chicago Bulls		New York Knicks		Phoenix Suns	
Player	**Avg.**	**Player**	**Avg.**	**Player**	**Avg.**
Jordan	32.6	Ewing	24.2	Barkley	25.6
Pippen	18.6	Starks	17.5	Majerle	16.9
Grant	13.3	Smith	12.4	K. Johnson	16.1
Armstrong	12.3	Mason	10.3	Dumas	12.8

1. Which team's top 4 players had a mean point average of 19.2? ______________

2. Which team's top 4 players had a mean point average of 17.85? ______________

3. Which team's top 4 players had the greatest median point average? ______________

4. Which team's top 4 players had point averages with a range of 12.8? ______________

5. Which team's top 4 players had point averages with a range of 20.3? ______________

Exercises 6–8 refer to the team you did not use in your answers for Exercises 1–5.

6. What was the mean point average of this team's top 4 players?

____

7. What is the median point average of this team's top 4 players?

____

8. What is the range of this team's point averages? ______________

9. Which of the statistics that you calculated do you think indicates which team has the best players? Explain your answer.

____

____

____

NAME ______________________ CLASS __________ DATE __________

Basic Skills Practice
Cumulative Review

Read each question and circle the best answer.

1. Which integer is equivalent to the following expression:

$$5 \times 3^2 - 10$$

A -45
B -5
C 35
D 215

2. How many square feet of sod are needed to cover a 40-foot-by-65-foot lawn?

F 105 ft^2
G 210 ft^2
H 2400 ft^2
I 2600 ft^2

3. The distance around a square swimming pool is 48 feet. What are the dimensions of the pool?

A 12 ft by 12 ft
B 14 ft by 14 ft
C 16 ft by 16 ft
D 24 ft by 24 ft

4. Calculate the following expression:

$$12 - 4.031$$

F 4.019
G 4.021
H 7.969
I 8.031

5. Roy bought carpet for two bedrooms. One room is 9.5 ft by 11.5 ft and the other is 14 ft by 15 ft. About how many square feet of carpeting did he buy?

A 50 ft^2
B 60 ft^2
C 300 ft^2
D 319 ft^2

6. Joe saved $17.80 on a jacket he bought on sale. The regular price for the jacket was $89.00. What percent of the regular price did he save?

F 10%
G 20%
H 25%
I 30%

Solve each problem.

7. Lauren walked around her office building during her lunch hour. She walked 0.2 miles along one side of the building and 0.45 miles along another. She cut through the courtyard, which was 0.3 miles, and came back to her desk, which was another 0.25 miles. What was the total distance she walked in miles? __________

8. Brandon is preparing to plant a garden. He wants it to be 6 feet along one side and 8 feet along another. Then he wants these two sides to be connected by a 10-foot side. What is the area of the garden in square feet if the garden is a right triangle? __________

LESSON 4.4

Exploring The Addition Principle of Counting

PREPARE

Pacing Guide
3 Days

Objectives and Overview
- Draw and interpret Venn diagrams.
- Use the Addition Principle of Counting.

Students examine the difference between the connectives *and* and *or.* They see how database software counts data when *and* and *or* criteria are defined. Venn diagrams are used to model the intersection of data sets, and the Addition Principle of Counting is applied to real-world situations.

Vocabulary

database	intersection
record	union
field	disjoint
Venn diagram	Addition Principle of Counting

Assessing Prior Knowledge

These data sets are batting averages.

Team 1	.336	.348	.319	.370	.337
Team 2	.334	.350	.344	.336	.310

1. How many averages are greater than .330 and less than .350? **[6]**
2. How many players batted below .330 or above .350? **[3]**

Approach A: Using the Book

Opening Discussion

Databases use a feature called *queries* to choose a particular sample from the database. How many people who live in Texas did the database find? How many who are relatives did it find? If Meg asked the database to find all the relatives who live in Texas, how many relatives would it list? **[1 relative]** If Meg asked the database to find all the people who are relatives or who live in Texas, how many people would the database list? **[6 people]**

Inclusion Strategy

Writing Definitions Because database *records* and *fields* are loosely defined here, some students may have difficulty understanding what those terms represent.

Although formal definitions of database records and field are not necessary, students should have an

continued on top of next page

TEACH

Approach B: Using Models

Opening Discussion

Electronic databases allow large amounts of data to be organized in an efficient manner. The data is easily accessible for analyzing from a database. Discuss databases, their uses, and several examples.

Copy the database structure on the next page onto the chalkboard.

Explain to your students that you might keep a similar database of student names. One of the features of database software that makes it so powerful is its ability to choose a particular sample from the database. Such selections can be made by using *queries*. For example, you could use a query to print the names of all students whose curriculum is college prep. What would the results of this query be?

continued on bottom of next page

Approach A Using the Book, continued

awareness of the way these terms are used. By modeling the technical language on more familiar language, the students should gain sufficient understanding of the meaning of database records and fields.

Ask students what their medical record in a doctor's office might look like. The record contains their name and pieces of information about them that might be important to their doctor, such as the year of their birth, any surgeries or serious illnesses they have had, and their height and weight.

A database record of medical histories might include the same information. Just as in the doctor's office, the medical information for each person would be in a separate record. Although all the information for one patient can be stored in one file folder with the patient's name on it, in a database of medical records, each piece of information about each person is actually stored in a separate element, called a *field*. Storing the information in separate fields makes the database better able to organize and manage all its data.

Ask each student to examine the database on page 187 of the Pupil's Edition. Tell them that all the information about each person entered in the database represents a record. Tell them also that the categories Last name, First name, Street, City, State, Zip, and Relative are all fields. Let the students use this information to write their own definitions of a database record and field in their notebook. Then have several students write their definitions on the chalkboard. Any students who believe that a chalkboard definition better defines a term should copy the chalkboard definition into their notebook.

The Inclusive *Or*

The word *or* can be used two ways. First, *a* or *b* can mean *either a or b* or *both a and b*. Since this meaning includes the possibility of both *a* and *b*, this *or* is defined as the *inclusive or*. When Meg asks the database to find the people who are either Texans or relatives, she knows that the

TEACH

Approach B Using Models, continued

Student ID	Last name	First name	Curriculum	Class	Varsity sports?	JV sports?	Student council?
100	Paulsen	Kyle	vocational	freshman	☑	☐	☑
101	Cooper	Rebecca	college prep	sophomore	☐	☑	☐
102	Ling	Han	office tech	freshman	☑	☐	☐
103	Jemez	Hernando	college prep	freshman	☑	☐	☑
104	Francis	Roseanna	office tech	sophomore	☐	☑	☐
105	Powell	Yolanda	vocational	sophomore	☑	☐	☐
106	Lee	Louis	college prep	freshman	☐	☐	☐

[The results of the query of all students whose curriculum is college prep is shown below.]

Last name	First name	Curriculum	Class
Cooper	Rebecca	college prep	soph.
Jemez	Hernando	college prep	fresh.
Lee	Louis	college prep	fresh.

What results would you get if you queried the database to find all of the freshmen?

[The results of the query of all students who are freshmen is shown below.]

Last name	First name	Curriculum	Class
Paulsen	Kyle	vocational	fresh.
Ling	Han	office tech	fresh.
Jemez	Hernando	college prep	fresh.
Lee	Louis	college prep	fresh.

database will find people who are *both* Texans and relatives, as well as those people who are either relatives or Texans. Second, the word *or* can also be used to exclude the possibility of both *a* and *b* being satisfied. If a teacher schedules an exam for Tuesday or Wednesday, students do not expect to take the exam on both Tuesday and Wednesday. This use of *or* is called the *exclusive or*. The *inclusive or* is used in this lesson.

Venn Diagrams

Venn diagrams are pictorial representations of the relationship between two sets. Because these diagrams allow a visual analysis of two sets, it is possible that students will see the relationships in the database more quickly when the data are presented in a Venn diagram.

Ask students why it makes sense visually to overlap circles that represent sets with common elements. Why does it make sense visually for two disjoint sets to be represented by nonintersecting circles?

Use **Transparency 16** for discussion.

Critical Thinking, page 186

set A = set B

Cooperative Learning

Teacher-Guided Exploration You can ask students to read each question aloud and discuss solutions. Ask students how the Venn diagrams should be drawn. Then draw the diagram on the chalkboard or on an overhead. Encourage all students to discuss strategies for using the diagram to answer the questions. For greater student participation in the explorations, you can ask students to come to the chalkboard or overhead and point to the names that satisfy the conditions in each question.

Student-Directed Approach Have students work in pairs. The partners should take turns reading the questions. Both students should discuss the answers, and one student should record the answers. The other student should draw the Venn diagrams. That student should also read and, if necessary, edit

continued on top of next page

If the database were queried to find all of the freshmen whose curriculum is college prep, how many students would be listed? [2] If it were queried to list all of the students who are freshmen or whose curriculum is college prep, how many students would be listed? [5]

Constructing a Database of Your Class

Explain that each line in the database on the previous page is called a *record* and that each category is called a *field*. Work with your students to construct a database for your class that can be drawn on poster board or on the chalkboard. **Important:** For the alternate explorations to be successful, you must have at least two disjoint categories and two overlapping categories. Therefore, use categories different from those in the example, if necessary.

Modeling *And* and *Or* (Venn Diagrams)

Draw two large circles on the chalkboard. Label one circle Freshmen and the other circle College prep, or use two appropriate field labels from the database you constructed above. Ask each student to come to the chalkboard and write his or her last name in an appropriate region on the board.

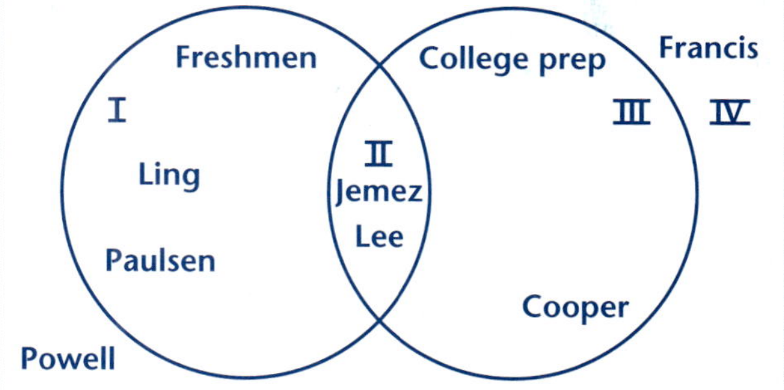

Region II, the overlapping region of the circles, represents the *intersection* of the students who are freshmen and whose curriculum is college prep. Regions I, II, and III together represent the *union*, or combined regions, of the two sets of students. These students either are freshmen or are taking a college prep curriculum.

What region represents students who are freshmen and whose curriculum is not college prep? **[Region I]**

continued on bottom of next page

Approach A: Using the Book, continued

the other student's recorded answers. Both students should initial the recorded answers.

Exploration 1

Students use a Venn diagram to demonstrate that the total number of unique elements in two disjoint sets is the sum of the elements in each set.

In Step 1, students must recognize that the two sets are disjoint and that they must be drawn as two nonintersecting circles.

In Steps 4 and 5, the exploration highlights the difference between the intersection and the union of two sets by using the logical connectives *and* and *or*. Ask students to verify their answers with the database.

Ongoing Assessment, page 187

6. They are equal.

Inclusion Strategy

Modeling And and Or (Venn Diagrams) Draw two large circles on the chalkboard. Label one circle Freshmen and the other circle Play sports. Ask each student to come to the board and write his or her last name in the appropriate region.

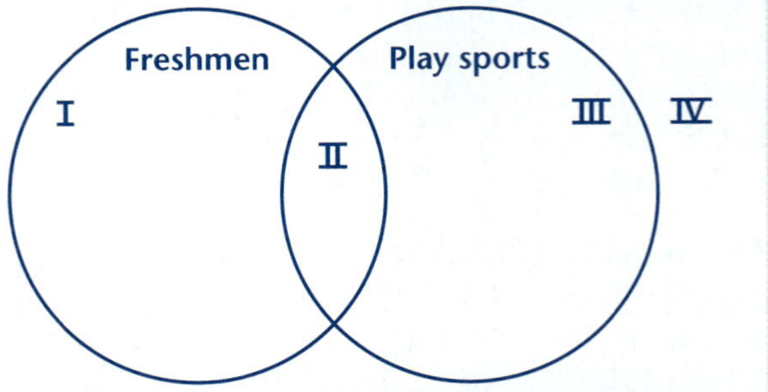

TEACH

Approach B: Using Models, continued

What region represents students whose curriculum is college prep but who are not freshmen? **[Region III]**

What region represents students whose curriculum is not college prep and who are not freshmen? **[Region IV]**

Cooperative Learning

The explorations provide excellent opportunities to involve all students in discovering the rules for finding the number of elements in the intersection and the union of two intersecting sets. Cooperatively written reports will allow all students to summarize their findings.

Alternate Exploration 1

Students use a Venn diagram to show that the total number of unique elements in two disjoint sets is the sum of the elements in each set.

Begin the exploration by asking each student to write the following statement in his or her notebook: I am a freshman or a sophomore. (Or substitute two other nonoverlapping categories.) Then ask each student to label his or her statement as either true or false.

Refer students to the database of the class that you constructed in Constructing a Database of Your Class. Modify the database so that there are freshman and sophomore fields, with responses indicated by yes or no.

Draw a Venn diagram, and label one circle Freshman and the other circle Sophomore. Ask students why the circles do not intersect.

Ask each student to write his or her name in the correct circle on the board. Tell each student to write the following in their notebook:

1. the number of freshmen
2. the number of sophomores
3. the number of people who are either freshmen or sophomores
4. the number of students who are both a freshman and a sophomore **[none]**

Region II, the overlapping region of the circles, represents the *intersection* of the students who are freshmen and who play sports. Regions I, II, and III together represent the *union* of the two sets of students. These students are either freshmen or play sports.

What region represents students who are freshmen and who do not play sports? **[Region I]**

What region represents students who play but who are not freshmen? **[Region III]**

What region represents students who do not play sports and who are not freshmen? **[Region IV]**

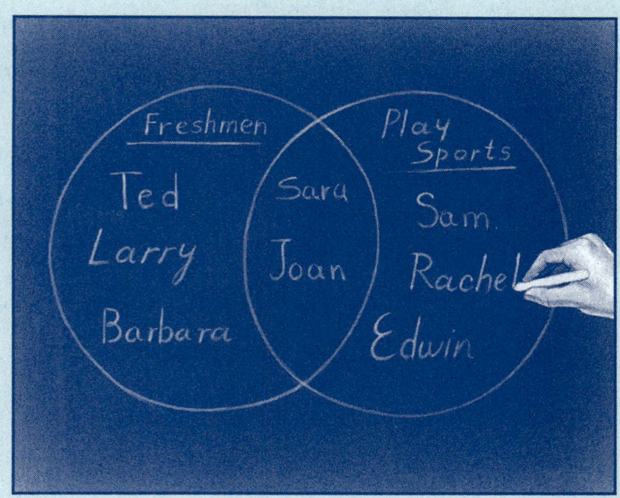

Have students write their names in the Venn diagram.

continued on top of next page

Now ask students to work in pairs to answer the following questions:

- Why is the sum of the numbers in questions 1 and 2 equal to the number in question 3? **[The sets are disjoint so the number in question 3 can be found by simply counting the freshmen then counting the sophomores and adding.]**
- Why do you think the answer to question 3 is the same as the number of students who wrote *true* for the statement "I am a freshman or a sophomore"? **[Students who wrote *true* are in either the freshman circle or the sophomore circle.]**
- Should anyone who wrote *false* for the statement "I am a freshman or a sophomore" be counted in the set of those students who are either freshmen or sophomores? **[no]** Explain. **[Anyone who wrote *false* would not be in either the freshman circle or the sophomore circle]**
- Complete the following: If the statement "x is in set A or set B" is true and sets A and B are disjoint sets, then x is in **[either in set A or set B, but not in both]**.

Student Reports

Ask the students in each pair to write a summary of their findings in the explorations. Some pairs may choose to write formal rules for finding the number of elements in the union of two disjoint sets. Call on several pairs to read their summaries to the class. Make a class list of the highlights of these summaries, which all students should then copy into their notebook.

Alternate Exploration 2

Students use a Venn diagram to examine the Addition Principle of Counting.

Refer students to the database of student names that you constructed in Constructing a Database of Your Class. Tell students to modify the structure of the database to include one field for students who play a school sport.

Draw a Venn diagram, labeling one circle College prep and the other circle Sports (or use two other overlapping categories). Ask students why the circles intersect.

continued on bottom of next page

Approach A

Using the Book, continued

On Their Own

A. Draw a Venn diagram of the last names of people living in Florida and people living in Ohio? **[Students' diagrams should show two nonintersecting circles. One circle should have the name Kang written in it. The other circle should have the names Post, Jordan, and Childress written in it.]**

B. How many people live in Florida? How many people live in Ohio? **[One person lives in Florida, and three people live in Ohio.]**

C. How many people live in either Florida *or* Ohio? **[4 people]** How many people live in Florida *and* Ohio? **[0 people]**

Exploration 2

Students use a Venn diagram to demonstrate that the number of unique elements in the union of two intersecting sets is the sum of the elements in each set minus the elements that the two sets have in common.

In Step 1, students must recognize that the sets intersect and that they must be drawn as two intersecting circles. The names of those people who live in Ohio and who are relatives must be written only once—in the area where the two circles overlap.

In Steps 4 and 5, the exploration highlights the difference between the intersection and the union of two intersecting sets by using the logical connectives *and* and *or.* Ask students to verify their answers against the database.

Steps 6 and 7 lead students to discover that the intersection has been counted twice. Therefore, it must be subtracted from the total.

Ongoing Assessment, page 187

7. $3 + 4 - 2 = 5$

TEACH

Approach B

Using Models, continued

Ask each student to write his or her name in the correct circle on the board.

Tell each student to answer the following in their notebook:

1. Count the number of college prep students. **[Answers will vary.]**
2. Count the number of students who play a sport. **[Answers will vary.]**
3. Count the total number of names in both circles. **[Answers will vary.]**

Now ask students to work in pairs to compare their answers. Each pair should discuss the following questions:

- How many students are in the intersection of the two circles? **[Answers will vary.]**
- Why is the sum of the numbers in questions 1 and 2 greater than the number in question 3? **[When the numbers in questions 1 and 2 are added, the students in the intersection are counted twice.]**
- Why do you think that the way in which you found the total number of students in both of the circles is different from the way in which you found the total number of students in both circles in Exploration 1? **[The sets in this exploration intersect. The sets in Exploration 1 were disjoint.]**

Student Reports

Ask the students in each pair to summarize their findings. The summaries should include a procedure for finding the number of elements in either of two intersecting sets. Examples should be used. One student from each pair should read the summary to the class. Other students should recommend language that would allow the procedure to be followed more easily.

Application

Copy the table from the top of page 188 of the Pupil's Edition onto the chalkboard.

Ask students how many students favor the rule. How many oppose it? For visual learners, this information may be better displayed in a Venn

Application

Addition Principle of Counting Using the survey results to find the union of two intersecting sets leads to a somewhat formal statement of the Addition Principle of Counting. This definition may be difficult for some students to understand. The important thing for them to remember is that they must add the number of elements in the two sets and then subtract any elements that the two sets have in common.

Use this example to help your students understand the definition: There are 13 boys. There are 11 students who favor the rule. 4 of the 11 students who favor the rule are boys.

Substitute the language of the example for the language of the definition:

Suppose that there are 13 boys, that 11 students favor a particular rule, and that 4 of those students have already been counted as boys. Then there are 13 + 11 − 4, or 20, students who are either boys or who favor the rule.

Application

Games In this application, the Addition Principle of Counting is applied to the game *Crazy Eights*. Remind students that the queen of hearts must not be counted because it has already been played. This is a good time to discuss with students that care must be taken when counting the number of ways that a choice can be made.

Closing the Lesson

Ask students to write the Addition Principle of Counting in their own words and to apply their definition to an example of their choosing.

diagram that shows the sex of those who oppose the rule and those who favor the rule. For example, Xs can be used to represent the boys, and Os can be used to represent the girls.

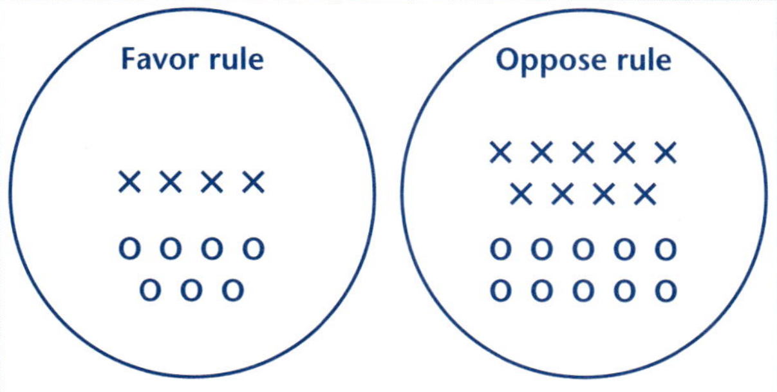

How can you use the diagram to find the number of students who favor the rule? **[Count the total number of Xs and Os in the Favor rule circle.]** How can you use the diagram to find the number of boys who favor the rule? **[Count the Xs in the Favor rule circle.]**

Tell how you can use the diagram to find the number of students who are boys or who favor the rule? **[Count the number of students in the Favor rule circle, and then add the number of boys in the Oppose rule circle; 11 + 9 = 20.]**

Tell how can you use the table to find the number of students who are boys or who favor the rule. **[Find the total number of boys, 13, and the total number of students who favor the rule, 11, and then subtract the 4 boys who were counted twice; 13 + 11 − 4 = 20.]**

Why was it not necessary to subtract when you used the diagram to find the number of students who are boys or who favor the rule? **[No boys were counted twice in this method.]**

Application

Games In this application, the Addition Principle of Counting is applied to the game *Crazy Eights*. Remind students that the queen of hearts must not be counted because it has already been played. This is a good time to discuss with students that care must be taken when counting the number of ways a choice can be made.

ASSESS

Selected Answers

Odd-numbered Exercises 9–45 and odd-numbered Extra Practice

Assignment Guide

Core level students should complete Exercises 1–46.

Core-plus level students should complete Exercises 1–20 and 31–50.

Error Analysis

Students who are unsure of the answer to Exercise 4 may not understand the relationship between the intersection and the union of two sets when the number of elements in the union of the two sets is being counted. Refer these students to the two explorations that review the difference between finding the number of elements in the union of two disjoint sets and of two intersecting sets.

Alternative Assessment

Venn Diagrams for an Old Puzzle

An old European fable tells the story of an unfortunate man who was left a widower with four small children. Sometime later, he married a widow who had three children of her own. The new husband and wife were delighted to have both a mother and a father for their children. They soon had two additional children together. Once, when writing a letter to a friend, the man explained that he now had six children of his own and his wife had five children of her own. Yet he was confused because, by counting heads, he could see that their family now had 9 children. He could not explain how $6 + 5 = 9$.

Explain the man's error in reasoning. Use a Venn diagram to demonstrate the situation.

Portfolio Activity Ask students to explain how the Addition Principle of Counting can be used to find the number of elements in the union of two disjoint sets.

RETEACH

One technique for representing relationships between two sets involves describing the four distinct regions into which two intersecting circles divide an area.

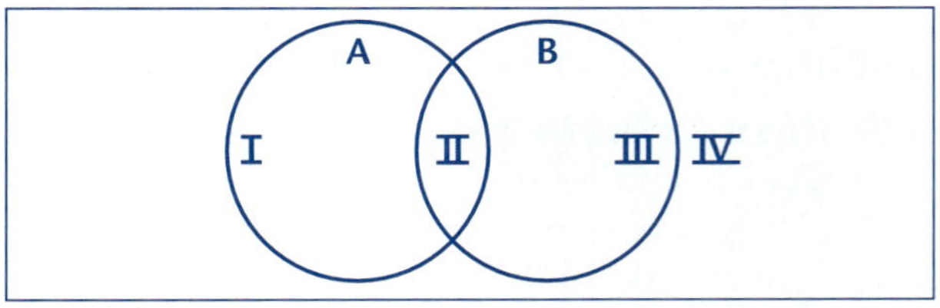

Region I represents the elements in *A,* but not in *B.*
Region II represents the elements in both *A* and in *B.*
Region III represents the elements in *B,* but not in *A.*
Region IV represents the elements in neither *A* nor *B.*

Set *A* is therefore equal to {I, II} and set *B* is equal to {II, III}. No discussion of set notation should be necessary for this approach. Students will likely recognize the braces as a grouping symbol, which is satisfactory for this lesson. This notation allows students to see that although region II is included in both sets, it should be counted only once in their union. Thus, the number of elements in the union of *A* and *B* is the number of elements in {I, II} plus the number of elements in {II, III} minus the number of elements in region II, which is the same as the Addition Principle of counting.

ENRICH

Interdisciplinary Connection

Advertising Venn diagrams can be used to model problems whose solutions are important to many businesses and industries. Suppose, for example, that a travel agency wants to know whether it should advertise at the mall. The agency knows from studies that 60% of people in the geographic area shop only at the mall, 15% shop only downtown, and 5% shop neither at the mall nor downtown.

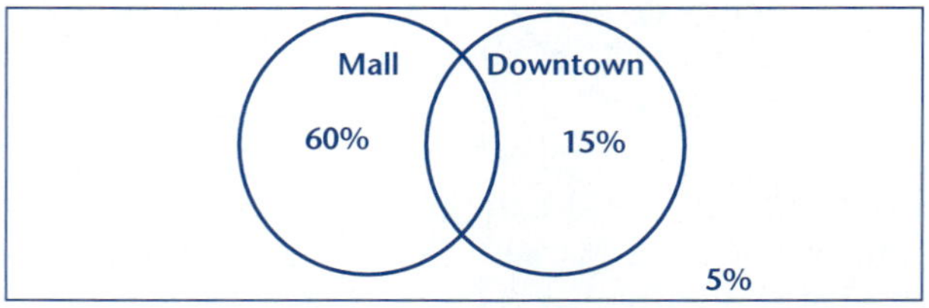

Use the diagram to find the probability that a randomly selected person in the area shops at the mall. **[*P*(shopping at the mall) = 1 − (0.15 + 0.05) = 0.8]**

NAME ______________________ CLASS ____________ DATE ____________

Practice

4.4 Exploring the Addition Principle of Counting

List the integers from 1 to 10 inclusive that are

1. odd. ______________

2. multiples of 2. ______________

3. odd AND multiples of 2. ______________

4. odd OR multiples of 2. ______________

List the integers from 1 to 20 inclusive that are

5. multiples of 4. ______________

6. multiples of 3. ______________

7. multiples of 4 AND multiples of 3. ______________

8. multiples of 4 OR multiples of 3. ______________

9. Draw a Venn diagram for multiples of 3 and multiples of 5 from 1 to 30 inclusive.

10. What numbers are both multiples of 3 and 5? ______________

11. How is this shown in the Venn diagram? ______________

A marketing representative gave supermarket customers a sample taste of a new soft drink. The results are shown in the table.

	Liked drink	Disliked drink	Total
Men	16	14	30
Women	19	10	29
Total	35	24	59

12. How many customers are men? ______________

13. How many customers disliked the new soft drink? ______________

14. How many customers are men AND disliked the soft drink? ______________

17. How many customers are men OR disliked the soft drink? ______________

NAME ______________________ CLASS ____________ DATE ____________

Lesson Activity

4.4 Exploring the Addition Principle of Counting

Last name	First name	County	M/F	Employed
Smith	Al	Montgomery	M	Y
Smith	Carol	Montgomery	F	Y
Smith	Candy	Jackson	F	N
Smith	Mike	Montgomery	M	N
Smith	Paula	Jackson	F	Y
Smith	Peter	Jackson	M	N
Smith	Rob	Montgomery	M	N
Smith	Sarah	Jackson	F	Y
Smith	Tom	Montgomery	M	N

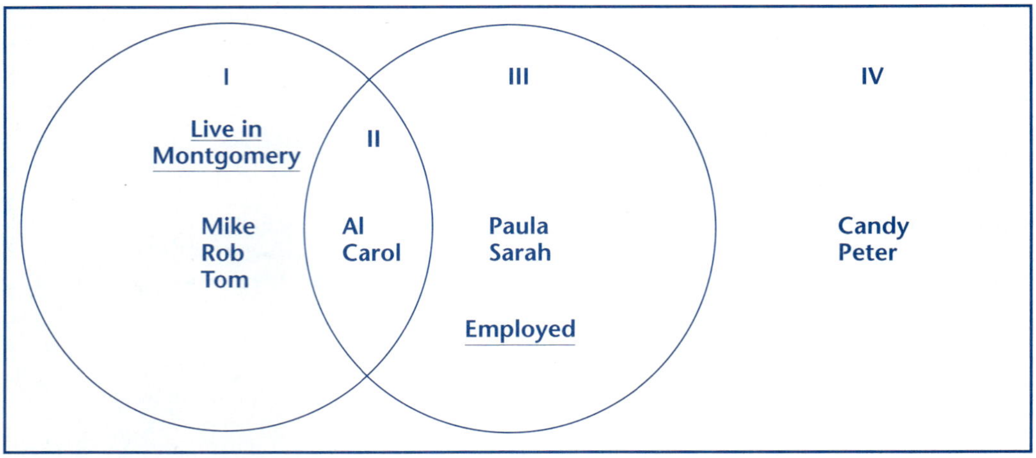

Name the region of the Venn diagram that represents each category.

1. Smiths who live in Montgomery and are not employed ______________________

2. Smiths who are not employed and do not live in Montgomery ______________________

3. Smiths who live in Montgomery and are employed ______________________

4. Smiths who do not live in Montgomery and are employed ______________________

Use the information given in the table to answer each question.

5. How many Smiths are female? ____________ **6.** How many Smiths are employed? ________

7. How many Smiths are female and employed? ______________________

8. How many Smiths are female or employed? ______________________

9. Explain how to use the Addition Principle of Counting to find out how many Smiths are living in Montgomery or are unemployed.

____

NAME ______________________ CLASS ____________ DATE ____________

Basic Skills Practice

Circumference and Area of Circles, Part 1

Circumference is the distance around a circle. To find the circumference, multiply the length of the diameter by π. Diameter is the distance across the circle and π is approximately 3.14.

Example 1: Find the circumference of the circle.

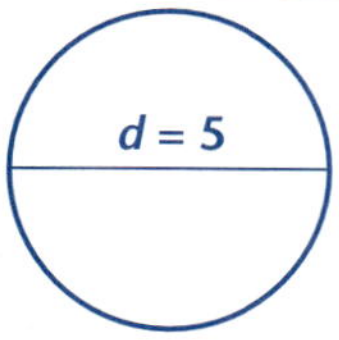

$C = \pi d$ → Write the formula.

$C \approx 3.14(5)$ → Substitute given values.

$C \approx 15.7$ → Multiply.

To find the area of a circle, multiply π by the square of the radius (r^2). Remember that the radius is half the diameter of a circle.

Example 2: Find the area of the circle.

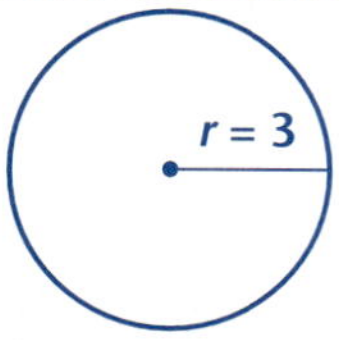

$A = \pi r^2$ → Write the formula.

$A \approx 3.14(3)(3)$ → Substitute given values.

$A \approx 28.26$ → Multiply.

Find the circumference for each circle.

1.

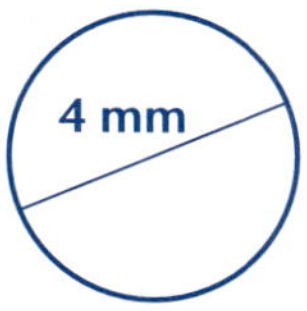

____

2.

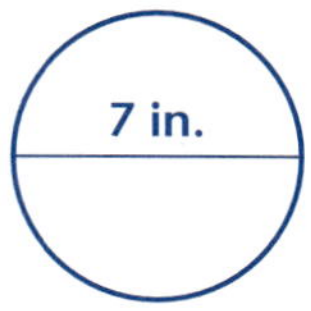

____

3.

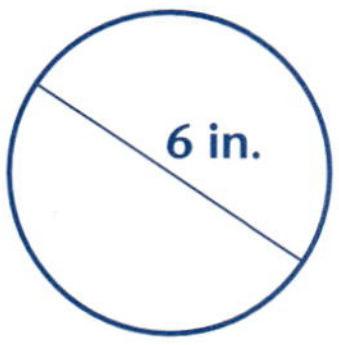

____

4.

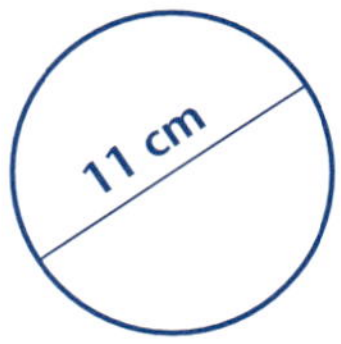

____

Find the area for each circle.

5.

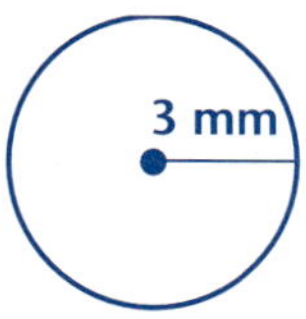

____

6.

____

7.

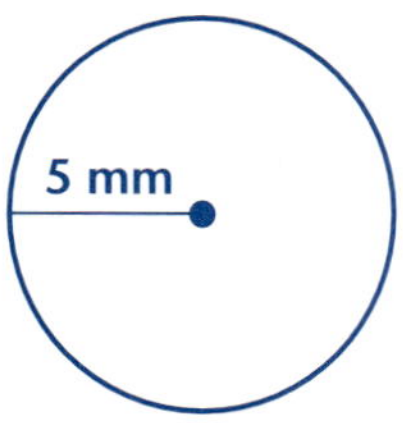

____

8.

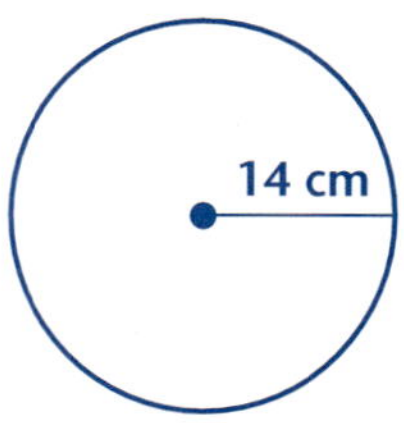

____

LESSON 4.5

Multiplication Principle of Counting

PREPARE

Pacing Guide

3 Days

Objectives and Overview

- Make a tree diagram to show possible outcomes.
- Use the Multiplication Principle of Counting to find the number of possible outcomes.

Students use tree diagrams to count the total number of ways that two tasks can be performed together. Then they apply the Multiplication Principle of Counting to the same kind of problem. Finally, they examine when to apply the Addition Principle of Counting to a situation and when to apply the Multiplication Principle of Counting.

Vocabulary

tree diagram
Multiplication Principle of Counting

Assessing Prior Knowledge

Suppose that Elizabeth's music collection includes 26 jazz CDs and 15 classical CDs. The collection also includes 37 jazz tapes and 22 classical tapes.

1. How many jazz recordings are either on CD or tape? [63]
2. How many tapes are either jazz or classical? [59]

Approach A: Using the Book

Opening Discussion

Why do you suppose that the diagram is called a tree diagram? Why is it easier to count the possible selections by using a tree diagram than it might be to count the selections by listing them?

Brainstorming

Ask each student to think of three different situations for which using a tree diagram would be a good problem-solving strategy. Examples include the number of ways that you could choose a car with two sets of options, or the number of ways a person could order a dinner from a menu that includes eight entrees and four desserts. Ask volunteers to read several of their examples.

Problem-Solving Strategy

One important problem-solving strategy involves listing all the possible solutions to the problem. Tree diagrams are a means of listing all the possible ways

TEACH

Approach B: Using Models

Opening Discussion

Ask students to name 3 required courses and 2 elective courses at your school. Refer to the tree diagram on page 192 of the Pupil's Edition, and draw a similar diagram. Ask students to compare the diagram with a stem-and-leaf plot. Both diagrams branch from a main stem. This diagram is called a tree diagram. Do you think that this name is appropriate? Explain. How does the tree diagram make it easier to count the possible pairs of course selections? Do you think it would be easier to simply list the selections and then count them?

Brainstorming

Ask each student to describe in his or her notebook three different situations for which a tree diagram would be a good problem-solving tool. Examples include how many ways you could select a car with two sets of options, and the number of ways a person could order a dinner from a menu that includes eight that two events can happen together. Make sure students see that the final branch of a tree diagram gives a *count* of the total number of ways that the events represented by all the branches can occur together. A *listing* of all the possibilities includes all the other branches of the tree diagram.

Example 1

This example uses a tree diagram to count the number of possible outcomes of two consecutive actions.

Teaching Tip

Multiplication Principle of Counting
Some students may wonder how the Multiplication Principle of Counting differs from the Addition Principle of Counting, although they recognize that two different operations are involved. Clarify the difference between the two techniques by stressing that the Multiplication Principle can be used whenever a tree diagram can be used to model a situation. Specifically, the Multiplication Principle counts the number of possible outcomes when one action that can be performed m ways is followed by a second action that can be performed n ways.

Ongoing Assessment, page 193
If there were 3 teams, then for each of the 3 teams, there would be 6 possible outcomes from the number cube. That is, $3 \cdot 6 = 18$.

Additional Example 1

There are 6 routes from town A to town B. There are 2 routes from town B to town C. Roll a number cube to select a route from town A to town B. Then flip a coin to select a route from town B to town C. How many pairings of different routes between town A and town C are possible? **[12 routes]**

Inclusion Strategy

Finding an Example Ask each student to think of an example in which a tree diagram

continued on top of next page

entrees and four desserts. Ask volunteers to read several of their examples.

Interpreting a Tree Diagram

Tree diagrams are a graphical representation of all the possible ways that two events can happen together. Students should understand that reading the final branch of a tree diagram gives only a *count* of the total number of ways that the events represented by all the branches can occur together. By reading all of the paths along the tree, all of the possible ways that two events can happen can be listed.

Cooperative Learning

Have students work with a partner. Tell them that from the two student council representatives from the freshman class, one will be chosen to attend a meeting with the principal. One of the four class officers will also be chosen to attend the same meeting. Ask students to cut six slips of paper, all the same size. Two of the slips should be labeled Student council 1 and Student council 2 and placed in a box. The other four slips should be labeled President, Vice-President, Secretary, and Treasurer and placed in a second box. Tell the students to choose the officer slips one at a time without replacement and to record the labels as branches of a tree diagram. Then the students should choose the student council slips from the box one at a time and record the labels as branches off the first officer branch. Both student council labels should then be returned to the box, drawn again, and recorded as branches off the second officer branch.

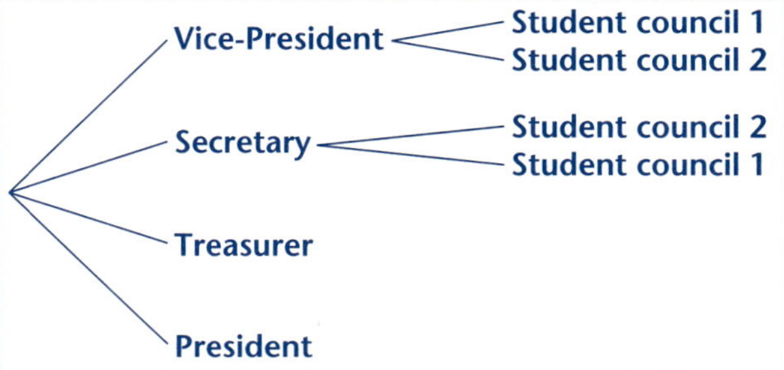

Finish constructing the tree in this manner. Ask students to answer the following questions:

continued on bottom of next page

Approach A Using the Book, continued

would be impractical for counting the number of possible outcomes for two consecutive actions. Have the students exchange papers with a partner and solve his or her partner's problem. Partners should check each other's work.

Example 2

This example shows how to apply the Multiplication Principle of Counting to a situation that involves finding the total number of outcomes of more than two consecutive actions.

Alternative Approach

If students have difficulty visualizing how 480 pairs of shoes are possible, show them how they can get an idea of the possibilities by using a partial tree diagram to determine a pattern.

Choose the first style and begin building a tree diagram. With this style you can choose 8 sizes.

Now look at the first size. This size comes in 3 different colors.

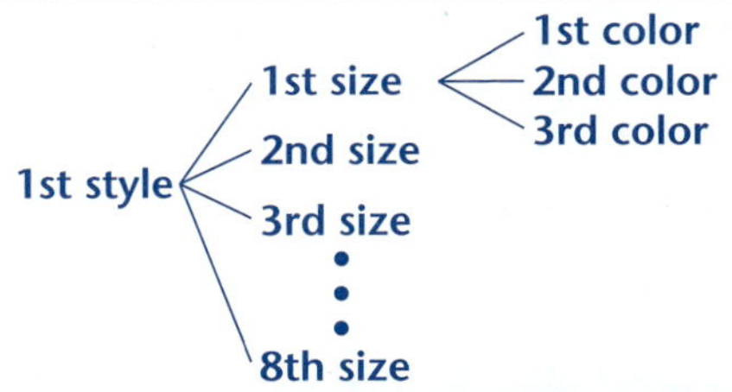

If you choose the first style and the first size, you have 3 different colors to choose from.

Since there are 8 sizes for the first style and each of the 8 sizes comes in 3 different colors, you have 8 · 3, or 24, choices of shoes in the first style. Each of these 24 combinations comes in 20 different styles.

TEACH

Approach B Using Models, continued

1. How many choices of a student council member are there to accompany each of the class officers to the meeting? [2]
2. How many possible pairs of representatives from the freshman class are there? [8]

If there were three freshman student council representatives, how many possible teams of representatives from the freshman class would there be? [12]

Using a Formula

Ask each student to think of an example in which a tree diagram would be impractical for counting the number of possible outcomes for two consecutive actions. How do the tree diagrams that students constructed suggest a formula for counting the number of possible outcomes from two consecutive actions? The Multiplication Principle of Counting suggests that when there are m ways to make a first choice and n ways to make a second choice, it follows that there are $m \bullet n$ ways to make the two choices.

Have the students exchange papers with a partner and solve his or her partner's example. Partners should check each other's work.

Modeling the Multiplication Principle of Counting

Technology Modern word processors offer people many choices. Just how many choices do they offer? Discuss a software menu from a word processor that shows several choices by drawing a picture on the chalkboard or an overhead.

Suppose that the choices for formatting on a word processor include 46 different fonts, 4 different font styles, and 16 different font sizes.

How could you use the Multiplication Principle of Counting to determine the number of possible selections of font and font style? **[There are 46 • 4, or 184, possible selections.]** Now imagine pairing each of these selections with the 16 font sizes. How could you find the total number of selections of font, font style, and font size? **[There are 184 · 16, or 2944, possible selections!]**

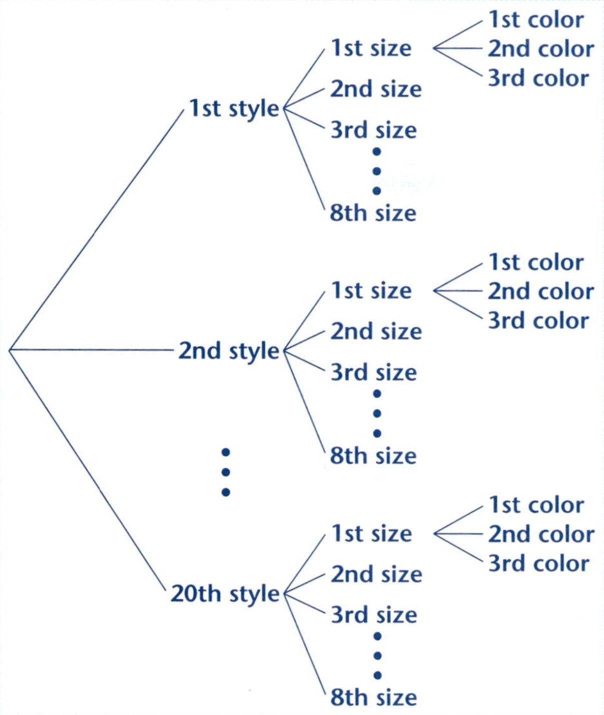

Examine the pattern on the diagram at the right. Each of the 20 sizes will have the same 24 size and color choices as the first style, so there are 20 · 24, or 480, possible pair of shoes.

Teaching Tip

Checking a Count Remind students that it is a good habit to check any solution to a problem. Whenever possible, tree diagrams should be used to verify an application of the Multiplication Principle of Counting.

Ongoing Assessment, page 194

Try This

There are 16 possible selections. The first branch shows that for each of the two kinds of pen, felt tip or ballpoint, there are four colors to choose from. Thus, there are 4 different colors of felt-tip pens and 4 different colors of ballpoint pens for a total of 8 choices of pen type and color. For each of these 8 choices, Tammy can choose one of two brands.

continued on top of next page

Application

Polygons The Multiplication Principle of Counting can also be applied to problems in geometry. For example, the principle can be used to count the number of diagonals in a hexagon. A diagonal joins two nonconsecutive vertices of a polygon. Draw a picture of a hexagon on the chalkboard. Ask each student to copy the hexagon into his or her notebook. Tell the students to label any two *consecutive* vertices.

How many nonconsecutive vertices does each of the labeled vertices have? [3]

Have students draw all possible diagonals from the two vertices.

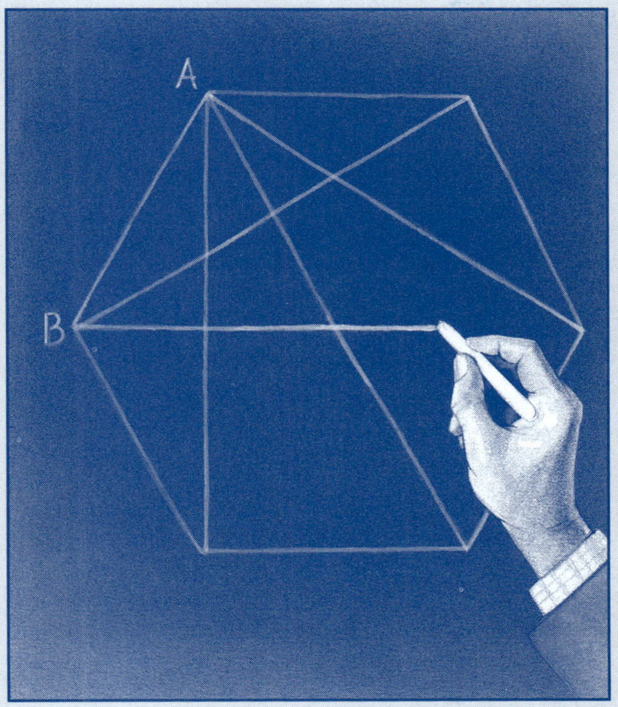

You can apply the Multiplication Principle of Counting to the diagonals of a polygon.

continued on bottom of next page

Approach A — Using the Book, continued

Therefore, she has 8 + 8, or 16, possible selections to choose from.

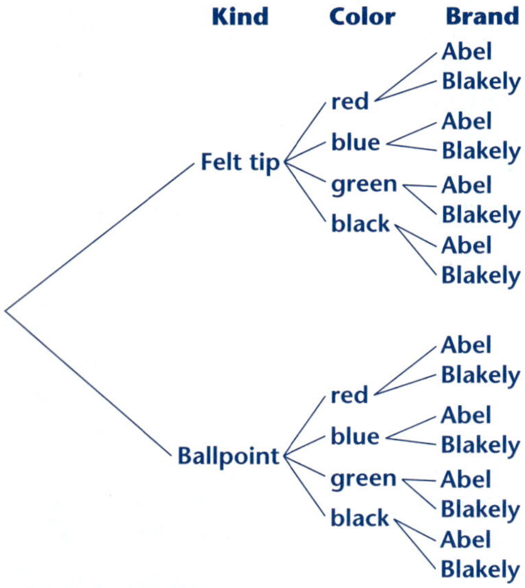

Example 3

This example shows an application of the Multiplication Principle of Counting to geometry. Make sure that students understand that they are counting the total number of ways to name the line.

Additional Example 3

Vertices *A* and *B* are two vertices of pentagon *ABCDE*. If a diagonal of the pentagon connects two nonconsecutive vertices, how many diagonals can be drawn from vertices *A* and *B*? **[4 diagonals]**

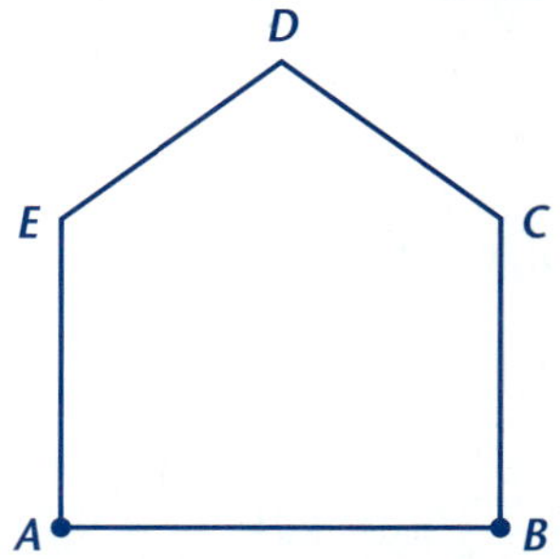

TEACH

Approach B — Using Models, continued

How could you use the Multiplication Principle of Counting to find the total number of diagonals that can be drawn from the labeled vertices? **[Since each of the labeled vertices is connected to 3 nonconsecutive vertices, there are 2 · 3, or 6, possible diagonals.]**

Problem-Solving Strategy

Reading Mathematics Students often have difficulty deciding whether to use the Multiplication Principle of Counting or the Addition Principle of Counting. Paying attention to the language of a problem and identifying key words such as *and* and *or* can help students determine which model—a Venn diagram or a tree diagram—should be used to solve a problem.

Ask students to imagine drawing one card from a regular deck of 52 playing cards. Half of the cards are red and half are black. Draw a Venn diagram to model the ways in which a red king could be drawn if there are 4 kings in a deck.

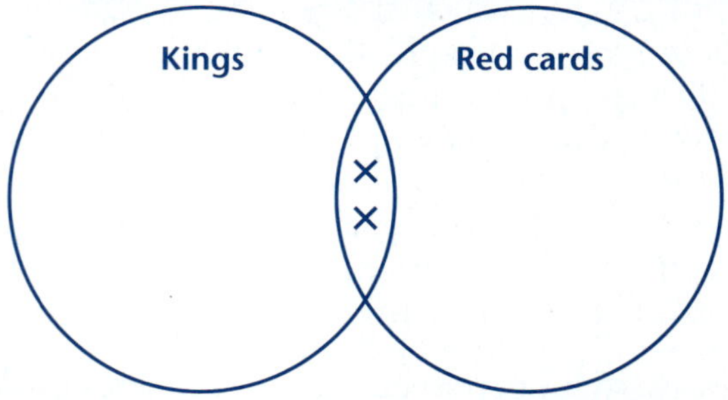

The intersection contains the 2 red kings, the king of hearts and the king of diamonds, so there are 2 ways to draw a red king.

Ask students to again imagine drawing a card from the deck. Use the same Venn diagram to model the ways in which a red card or a king can be drawn. The union of the two circles contains 26 red cards + 4 kings − 2 red kings in the intersection that have been counted twice, so there are 26 + 4 − 2, or 28, ways to draw a red card or a king.

Critical Thinking, page 195

Answers will vary. Sample answer: Two advantages of a tree diagram are that you can see the combinations and that you can also physically count the combinations. The advantage of using the Multiplication Principle of Counting is that it is easier and quicker.

Example 4

This example provides some examples that help students determine when to use the Addition Principle of Counting and when to use the Multiplication Principle of Counting.

Communicating Understanding

Journal Entry Ask students to explain why the Multiplication Principle of Counting applies to example 4, part C and why the Addition Principle of Counting applies to example 4, part B. Students should understand that the Multiplication Principle of Counting applies to situations in which one event follows another. The Addition Principle of Counting applies to situations in which either one event or another occurs.

Closing the Lesson

Ask students to complete the following statements:

A. Use the **[Multiplication]** Principle of Counting to count the number of ways that two consecutive actions can be performed.

B. Use the **[Multiplication]** Principle of Counting to find the number of ways that one action *and* another action can be performed.

C. Use the **[Addition]** Principle of Counting to find the number of ways that one set *or* another set can be combined.

Each of these situations was modeled by a Venn diagram. The key word in the second situation that signals the use of a Venn diagram is *or.* When the union of two sets is required, the Addition Principle of Counting can be used. The first situation could be reworded to ask for the ways in which a card that is both red and a king is drawn. This language clearly indicates the intersection of two sets.

If two cards are drawn from a regular deck of 52 cards, the situation may be modeled by a tree diagram. For example, if two cards are drawn and the first card is replaced before the second is drawn, the number of ways to draw a red card and then a king can be modeled by a tree diagram.

There are 26 ways to draw a red card and 4 ways to draw a king, so 4 branches of kings can be drawn from each branch of the 26 red cards. Thus, there are $26 \cdot 4$, or 104, ways to draw a red card and then a king.

Closing the Lesson

Ask students to complete the following statements:

A. Use the **[Multiplication]** Principle of Counting to count the number of ways that two consecutive actions can be performed.

B. Use the **[Multiplication]** Principle of Counting to find the number of ways that one action *and* another action can be performed.

C. Use the **[Addition]** Principle of Counting to find the number of ways that one set *or* another set can be combined.

ASSESS

Selected Answers

Odd-numbered Exercises 7–27 and odd-numbered Extra Practice

Assignment Guide

Core level students should complete Exercises 1–29.

Core-plus level students should complete Exercises 1–12 and 21–31.

Error Analysis

Some students may have difficulty distinguishing between *and* when it indicates intersection and when it indicates one action followed by another. Tell these students to determine whether the situation can be modeled by a Venn diagram or by a tree diagram. If the situation can be modeled by a tree diagram, then the Multiplication Principle of Counting should be used.

Alternative Assessment

The Principles of Counting Have students describe two situations, one that can be solved by using the Addition Principle of Counting and another that can be solved by using the Multiplication Principle of Counting. Have students exchange papers with a partner. Each student should solve his or her partner's two problems and make a formula and a diagram for each solution.

Portfolio Activity Ask students to design an experiment that involves tossing a coin. Then ask the students to design an experiment that involves rolling a number cube. Tell them to use a tree diagram to list and count the outcomes when the number-cube experiment is followed by the coin-tossing experiment.

RETEACH

Ask each student to bring in a map of your state, or ask the social studies department of your school to donate a map for each student. Choose three cities for which there are two routes between the first and second cities and three routes between the second and third cities. Ask the students to list all of the possible routes from the first city to the third city that go through the second city. Then explain to them that a tree diagram can help make counting the routes easier. Draw a tree diagram to model the problem. Trace and count the routes on the diagram. Finally, ask the students what formula they might use to find the total number of possible selections when one event is followed by another.

ENRICH

Interdisciplinary Connection

Language A tree diagram derived its name because its form is similar to that of a tree. The formal name for a tree diagram is *dendrogram,* which has the Greek root *dendro*, meaning "tree" or "tree-like." A person who studies the botanical nature of trees and other woody plants is called a *dendrologist.*

NAME ______________________ CLASS __________ DATE __________

Practice
4.5 Multiplication Principle of Counting

Jerry has 4 T-shirts, 2 pairs of jeans, and 3 pairs of shoes that can be worn together in any combination.

1. How many ways are there for Jerry to choose an outfit? ______________

2. How many ways are there to include a blue T-shirt that is one of the 4 T-shirts? ______________

3. How many ways are there to include a pair of black jeans that is one of the 2 pairs of jeans? ______________

4. How many ways are there to include a pair of brown shoes that is one of the 3 pairs of shoes? ______________

How many ways are there to arrange the letters in each of the following words?

5. cat ______________

6. math ______________

7. sport ______________

8. problem ______________

9. The Camacho's are decorating their kitchen. There are 7 different wallpaper patterns from which to choose. There are 5 borders that coordinate with each wallpaper pattern. How many combinations are possible? ______________

10. Members of the pep club are selling carnations. They are selling red, pink, or white carnations in packages of 2, 4, or 6. If all carnations in a package must be the same color, how many choices are offered? ______________

11. A popular car model comes in 5 different colors, with a standard or automatic transmission and with or without a sports package. How many different types of cars are available in this model?

____

12. How many ways are there to name the given square? ______________

13. Name the ways.

____

____

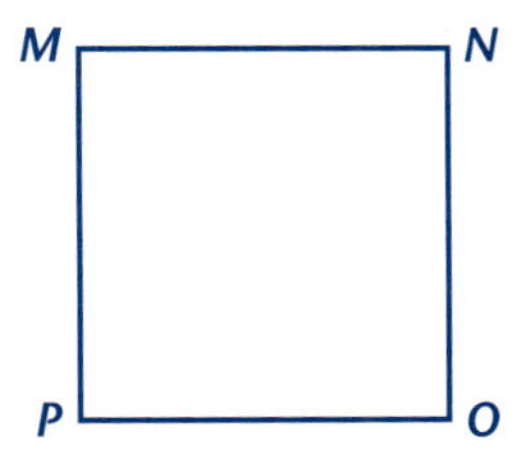

NAME ______________________ CLASS __________ DATE __________

Lesson Activity
4.5 Multiplication Principle of Counting

You have decided to buy a new car: either model A or model B. The models you like come in blue, black, red, or white with either a manual or an automatic transmission.

1. Make a tree diagram showing all of the possible car choices.

2. Explain how you could have used the Multiplication Principle of Counting to determine the number of choices.

____

____

____

____

3. The base price for either model A or model B is $9000. Both companies charge an extra $500 for a black car and an additional $750 for an automatic transmission. If you have $10,000 to spend, list the choices that you cannot afford to buy.

____

____

____

4. Suppose that you have only $9500 to spend. List the choices you cannot afford.

____

____

____

5. What is the least amount of money you would need in order to afford any of the choices listed on your tree diagram?

____

NAME ______________________ CLASS ____________ DATE ____________

Basic Skills Practice

Circumference and Area of Circles, Part 2

Read each question and circle the best answer.

1. A circular track is 300 yards across. What is its circumference?

A 390 yd
B 938 yd
C 942 yd
D 953 yd

2. A circular cooking pan has a radius of 6 inches. What is its area in square inches?

F 94.4 $in.^2$
G 113.04 $in.^2$
H 125.01 $in.^2$
I 154.38 $in.^2$

3. The face of a clock has a diameter of 14 inches. What is its area?

A 153.86 in^2
B 172.81 in^2
C 189.2 in^2
D 205.65 in^2

4. A certain castle has a circular tower. The circumference of the tower is 208 feet. What is the diameter of the tower?

F 29.2 ft
G 43.81 ft
H 58.3 ft
I 66.24 ft

5. The diameter of a large coin is 1.5 inches. What is its area in square inches?

A 1.77 $in.^2$
B 2.86 $in.^2$
C 3.07 $in.^2$
D 4.22 $in.^2$

6. The radius of a certain spherical asteroid is 6 kilometers. What is its circumference?

F 20.56 km
G 27.32 km
H 32.29 km
I 37.68 km

Solve each problem. If necessary, round answers to the nearest hundredth.

7. A round sign has a diameter of 22.5 feet. What is its area in square feet? ____________

8. Another sign has a diameter of 21.5 inches. What is its area in square inches? ____________

9. The base of a coffee mug has a diameter of 3.2 inches. What is its circumference in inches? ____________

10. The base of a paper cup has a diameter of 2.8 inches. What is its area in square inches? ____________

11. Find the circumference, in inches, of a circle that has a diameter of 9 inches. ____________

12. Find the area, in square inches, of a circle that has a diameter of 44 inches. ____________

13. Find the circumference, in inches, of a basketball with a diameter of 11 inches. ____________

LESSON 4.6 Theoretical Probability

PREPARE

Pacing Guide
3 Days

Objectives and Overview
- Find the probability that an event will occur.
- Use counting principles to find the probability that an event will occur.

Students find the theoretical probability that an equally likely event will occur by using the formula for theoretical probability. They also use counting principles to find the probability that an event will occur.

Vocabulary
event
equally likely outcomes
theoretical probability

Assessing Prior Knowledge

Solve the following:

1. If 10 tosses of a coin result in 3 heads, what is the experimental probability of tossing heads with this coin? $\left[\frac{3}{10}\right]$
2. If 20 rolls of a single number cube result in six 5s, what is the experimental probability of rolling a 5 with this number cube? $\left[\frac{6}{20}, \text{ or } \frac{3}{10}\right]$

Approach A Using the Book

Opening Discussion
Hold up a picture of a weather map. What does a weather forecaster mean when he or she says that there will be a 30% chance of rain tonight? In Lesson 4.1, what was meant by *experimental probability*? Do you think the weather forecaster's prediction is based on an experimental probability? Which chances are greater—a 4 in 20 chance of rain or a 7 in 15 chance of rain? **[a 7 in 15 chance]** Explain. **[A 4 in 20 chance is 20%, and a 7 in 15 chance is 47%.]**

Defining *Theoretical Probability*
Experimental probability is based on *the number of times that a successful event occurs in a given number of trials*. Theoretical probability is based on *the possible number of successful outcomes of an event*. Thus, experimental probability involves actual trials, while theoretical probability involves predictions of the outcome of equally likely events over a large number of trials.

TEACH

Approach B Using Models

Opening Discussion
Show your students a picture of a weather map. Ask them what they think a weather forecaster means when he or she says there will be a 30% chance of rain? Students studied experimental probability in Lesson 4.1. What does *experimental probability* mean? Do you think a weather forecaster uses experimental probability to make predictions about the weather? Which is a greater chance of rain—4 in 20 or 7 in 15? **[7 in 15]** Explain. **[A 4 in 20 chance is 20%, and a 7 in 15 chance is 47%.]**

Cooperative Learning
Modeling Theoretical Probability Because experimental probability involves a concrete experience, a probability experiment will help explain theoretical probability, which is a more abstract concept. Kinesthetic learners, particularly, will profit from this approach. Ask students to work in pairs and to toss a coin 20 times. Ask them to record their results in a frequency table and a bar graph. They should also

Inclusion Strategies

Equally Likely Outcomes You can use number cubes with different numbers of sides to reinforce the meaning of equally likely outcomes. If you have number cubes available with different numbers of sides, show them to students. Otherwise, ask students if they have played any games with number cubes that are not cubes. How many sides do the number cubes have? Tell each student to choose an integer from 1 to 6. Then ask them to determine the theoretical probabilities of rolling their integer on number cubes with 6, 8, 10, 12, and 20 sides. Copy the following chart or one similar to it onto the board, and ask several students to give the probabilities of some of the integers chosen.

Explain that the probability of rolling any integer from 1 to n on an n-sided number cube is dependent on the number of outcomes. Thus, the probability of rolling any integer from 1 to n on an n-sided number cube is $\frac{1}{n}$ because each of the outcomes for any roll is equally likely.

Number of sides	Integer chosen	Theoretical probability
6		
6		
6		
8		
8		
8		
10		
10		
10		
12		
12		
20		
20		

Example 1

The definition of theoretical probability is applied to a real-world situation.

continued on top of next page

find the experimental probability of both outcomes.

Ask students what they think the theoretical probability of heads is for tossing a coin 10 times. Then record each pair's results in a frequency table on the chalkboard. Calculate the experimental probability of the class results. As the number of experiments increases, the experimental probability should approach the theoretical probability. Are the class results close to the theoretical probability of heads?

A Weighted Coin

Show students a coin, and ask them if the probability of tossing heads is the same as the probability of *not* tossing heads? Because each event has a 1 in 2 chance, the events are equally likely and therefore have the same theoretical probability.

Show students a regular number cube, and ask them if the probability of rolling a 2 is the same as the probability of *not* rolling a 2. Because there is a 1 in 6 chance of rolling a 2 but a 5 in 6 chance of not rolling a 2, the events are not equally likely and therefore do not have the same theoretical probability.

Ask students to suppose that they had a weighted coin for which the chance of landing on one side was greater than the chance of landing on the other side. How many possible outcomes are there? Is the probability of the coin landing on each side 50%? Because the two events are not equally likely, the probabilities of the coin landing on each side are not equal.

Events With More Than One Successful Outcome

Draw six squares on the chalkboard or overhead to model the six faces of a number cube. How many faces show a number less than 3? Use colored chalk or an erasable marker to shade these faces.

1 2 3 4 5 6

What is the ratio of shaded faces to the total number of faces? Rearrange the faces to show that there is one shaded face for every 3 faces.

continued on bottom of next page

Approach A — Using the Book, continued

Students may recall seeing this or a similar database in Lesson 4.4. In that lesson, they used the database to count the number of outcomes of people living in a certain state. In this lesson, students will use the number of outcomes to find the theoretical probability of such an outcome.

Students may need reminding that when the text uses the word *probability,* it now means theoretical probability, unless otherwise noted.

Additional Example 1

Refer to Example 1 on page 199. Find the probability that an address selected at random from the database is from Florida. **[$\frac{1}{10}$, or 10%]**

Example 2

In this example, students find the probability of an outcome that involves more than one success. In the example, the total number of successes is compared with the total number of possible outcomes.

Teaching Tip

Making Connections Ask students to recall how they found the number of outcomes in the union of two disjoint sets. The total number of outcomes in the union of two disjoint sets is the sum of the outcomes in each set. In order to find the probability of an event with more than one outcome, the individual probabilities can be added. For example, to find the probability that the result of a single roll of a number cube is less than 3, you can find the sum of the probabilities of rolling a 1 *or* rolling a 2. The probability of rolling a 1 or a 2 equals the probability of rolling a 1 plus the probability of rolling a 2, or $\frac{1}{6} + \frac{1}{6} = \frac{1}{3}$.

Additional Example 2

Roll a number cube once. What is the probability that the result is 4 or greater? $\left[\frac{1}{2}\right]$

TEACH

Approach B — Using Models, continued

Thus, there is a 1 in 3 chance that a shaded face would be rolled randomly.

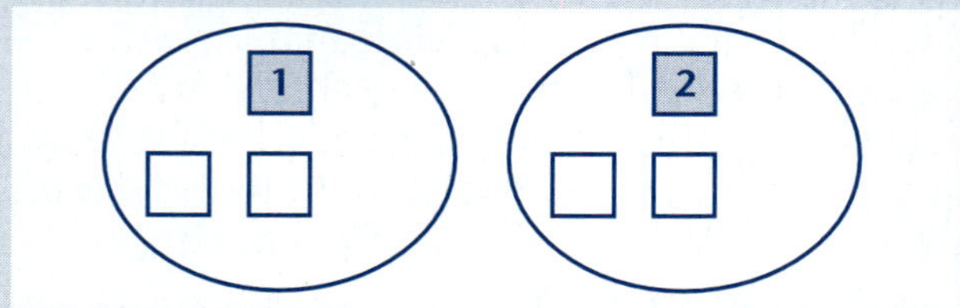

The probability of rolling a number less than 3 on a single roll of a number cube is $\frac{1}{3}$.

The Probability of Sequential Events

When two number cubes are rolled, the roll of the first cube does not affect the roll of the second cube. Thus, the two rolls can be considered sequential and independent and can be modeled by a tree diagram, as shown at right. In the tree diagram, all the shaded outcomes represent the rolls whose sum is 8.

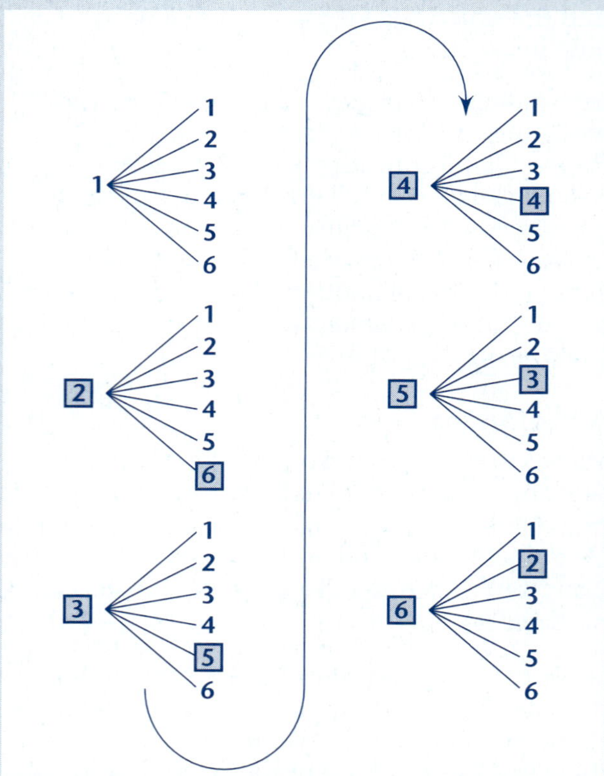

Simultaneous Events

When rolling two number cubes, the probability of rolling a pair of numbers whose sum is 5 can be calculated from the probability of one of the successful outcomes. For example, the probability of rolling the pair (1, 4) equals the probability of rolling a 1 *and* rolling a 4, or $\frac{1}{6} \cdot \frac{1}{6} = \frac{1}{36}$. Since there are 5 combinations of rolls, including rolling a 1 and rolling a 4, that give a sum of 5, the probability of rolling a sum of 5 is $4 \cdot \frac{1}{36} = \frac{4}{36}$, or $\frac{1}{9}$.

Example 3

In this example students use the Multiplication Principle of Counting to find the total number of outcomes in a set. Then they find the probability of a subset of that set.

Make sure students understand that although only 4 of the possible rolls on each cube will give successful outcomes, the probability of a sum of 5 must be based on the total number of possible outcomes, 36, because rolls of 5 and 6 on either cube are also possible.

Use **Transparency 17** for discussion.

Additional Example 3

Roll two number cubes. What is the probability that the sum is less than 5? $\left[\frac{1}{6}\right]$

Ongoing Assessment, page 200

Try This

$\left[\frac{1}{9}\right]$

Critical Thinking, page 200

Answers will vary. Sample answer: The probability that a sum of 1 is tossed on one roll of two number cubes is 0. The probability that a sum greater than or equal to 2 is rolled on one roll of two number cubes is 1. The probability of events ranges from 0 to 1.

continued on top of next page

The tree diagram shows that there are 6 • 6, or 36, possible outcomes when two number cubes are rolled. Five of those outcomes represent a successful roll, a sum of 8. The probability of rolling a sum of 8 is $\frac{5}{36}$.

Make sure students understand that although only 5 of the possible rolls on each cube will give successful outcomes, the probability of a sum of 8 must be based on a total number of possible outcome, 36, because a roll of 1 on either cube is also possible.

Equally Likely Sums

Use the tree diagram that shows the possible outcomes of rolling two number cubes. How many possible sums are there? Notice that not all of the sums are equally likely. How many ways are there to roll a sum of 5? How many ways are there to roll a sum of 12? Since there are more ways to roll a sum of 5 than a sum of 12, a sum of 5 is more likely to occur than a sum of 12. Which sum is the most likely to occur?

Journal Entry

The Range of Probabilities Answer the following questions. Then draw a number line that shows the range of values for the probability of an event. What does it mean for an event to have a 100% chance of occurring? How could you represent the probability of this event? What is the probability of rolling both a 4 and a 5 on a single roll of a number cube?

Teaching Tip

Because some situations require the application of the Addition and Multiplication Principles of Counting to find the number of possible outcomes for an event, a review of these principles may helpful. Ask students to write examples of each in their notebook. Then call on several students to read their examples aloud.

Using Counting Principles to Find Probabilities

Copy the following survey results onto the chalkboard. The table shows the results of a class survey about a new rule concerning students

continued on bottom of next page

Approach A

Using the Book, continued

Look Back/Look Ahead

Journal Entry Because it will be important for students to identify whether the outcomes in Example 4 model the Addition or Multiplication Principles of Counting, restating these principles may be helpful. Ask students to write statements of each principle in their journal. If they are unsure of the language of the principles, tell them to use the index in their text to find the statements of the principles. They can then copy the statements into their journal.

Example 4

Students apply the Multiplication and Addition Principles of Counting in order to find theoretical probabilities.

Additional Example 4

	Favor block scheduling	Oppose block scheduling	Total
Boys	12	5	17
Girls	12	8	20
Total	24	13	37

Use the results in the table to find the probability that a randomly-chosen student:

A. is a girl. $\left[\frac{20}{37}\right]$

B. favors block scheduling. $\left[\frac{24}{37}\right]$

C. is a girl *and* favors block scheduling. $\left[\frac{12}{37}\right]$

D. is a girl *or* favors block scheduling. $\left[\frac{32}{37}\right]$

Closing the Lesson

Ask students to compare experimental probability with theoretical probability. How are they similar? How are they different?

TEACH

Approach B

Using Models, continued

driving to school. A student who participated in the survey is chosen at random.

	Favor rule	Oppose rule	Total
Boys	4	9	13
Girls	7	10	17
Total	11	19	30

Use Venn diagrams to model the data in the table. The Xs represent boys and the Os girls.

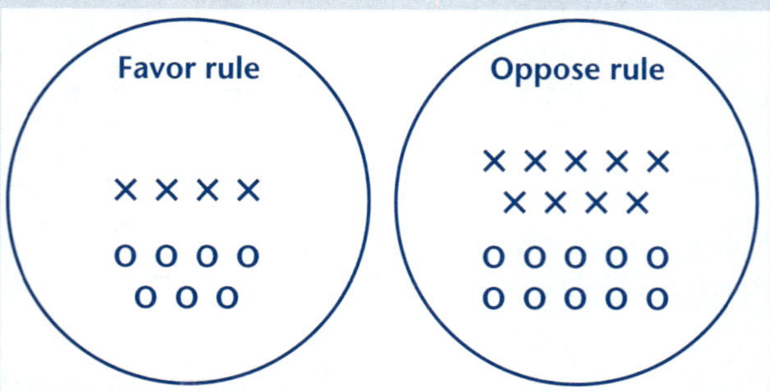

Since the Venn diagram represents disjoint sets, the number of *equally likely* possibilities can be found by adding the number of outcomes in the Favor rule circle with those in the Oppose rule circle. There are 11 + 19, or 30, students in the survey, each of which is equally likely to be chosen.

How can you use the diagram to find the probability that the student chosen favors the rule? **[Count the total number of Xs and Os in the Favor rule circle. This is the numerator of the probability. The denominator is 30.]**

The 4 Xs in the Favor rule circle represent the number of students who are boys *and* favor the rule. The probability that the student chosen is a boy who favors the rule is $\frac{4}{30}$, or $\frac{2}{15}$.

Counting the number of students in the Favor rule circle and then adding the number of remaining boys in the *Oppose* circle gives the number of students who are boys *or* favor the rule. The probability that the student chosen at random is a boy or favors the rule is $\frac{11 + 9}{30} = \frac{20}{30}$, or $\frac{2}{3}$.

ASSESS

Selected Answers

Odd-numbered Exercises 7–33 and odd-numbered Extra Practice

Assignment Guide

Core level students should complete Exercises 1–33.

Core-plus level students should complete Exercises 1–11 and 23–39.

Error Analysis

Because the terms *event* and *outcome* have been only loosely defined in this chapter, some students may confuse the two terms. Explain to these students that outcomes are the equally likely results of a random experiment, while an event is more general and can be a set of outcomes.

Alternative Assessment

Naming Triangles The vertices of a triangle can be used to name the triangle.

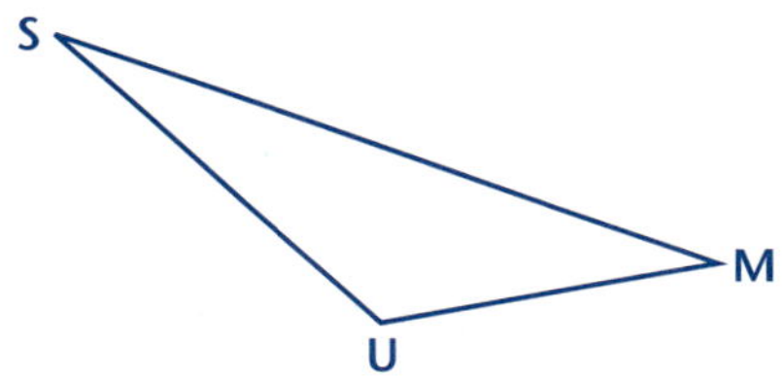

1. In how many different ways can this triangle be named? **[6 different ways]**
2. What is the probability of one of the names being drawn at random from a hat containing all the names? $\left[\frac{1}{6}\right]$
3. What is the probability that the name drawn begins with a *U*? $\left[\frac{1}{3}\right]$
4. What is the probability that the name drawn begins with a *U* or an *M*? $\left[\frac{2}{3}\right]$
5. What is the probability that the name drawn begins with a *U* and ends with an *M*? $\left[\frac{1}{6}\right]$

Portfolio Activity Suppose that there are 28 people in your class. What is the probability that each person names the triangle above differently? Explain. **[The probability is 0. This is an impossible event because there are 28 students and only 6 different ways to name the triangle.]**

RETEACH

Use the following problem to introduce students to theoretical probability. Marcia, Brian, and Ky toss a coin twice to decide who will drive to a basketball game. Marcia will drive if two heads show, Brian will drive if two tails show, and Ky will drive if one heads and one tails show. Who has the greatest chance of driving?

Model the problem with a tree diagram.

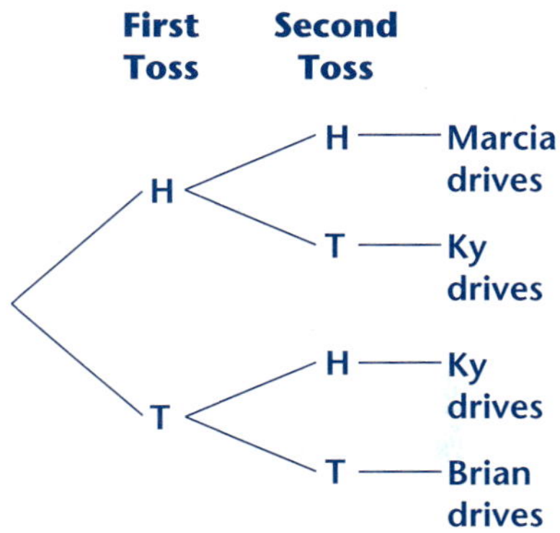

There are 4 equally likely outcomes, so the probability of each outcome is $\frac{1}{4}$. Since one of the outcomes is 2 heads, Marcia has a 1 in 4, or 25%, chance of driving. One of the outcomes is 2 tails, so Brian also has a 1 in 4, or 25%, chance of driving. Two of the outcomes have one heads and one tails, so Ky has a 2 in 4, or 50%, chance of driving. Ky has the greatest chance of driving.

ENRICH

Interdisciplinary Connection

Actuarial Science Actuaries have extensive training in mathematics, statistics, accounting, and probability. They use their knowledge to estimate the risks of events such as driving accidents and fires. Their estimations allow insurance companies to determine premium rates for their customers. For example, an actuary uses probability to determine the risk that a driver who is male and 18 years old will have an accident within a certain period of time.

NAME ______________________ CLASS __________ DATE __________

Practice

4.6 Theoretical Probability

An integer from 1 to 40 is drawn at random. Find the probability it is

1. odd. __________ **2.** even. __________

3. a multiple of 4. __________ **4.** prime. __________

5. odd AND a multiple of 5. __________ **6.** odd OR a multiple of 5. __________

The letters in the word *CONSTITUTION* are written on index cards and placed in a brown bag.

C O N S T I T U T I O N

7. A letter is selected at random. Find the probability that it is a vowel. __________

8. What is the probability of selecting the letter T? __________

9. What is the probability of selecting a consonant? __________

10. Find the probability that the name of a state chosen at random begins with the letter N.

____

11. Find the probability that the name of a state chosen at random begins with the letter B.

____

	Liked lunch	Disliked lunch	Total
Boys	16	14	30
Girls	19	10	29
Total	35	24	59

A student who was surveyed is chosen at random. What is the theoretical probability that the student

12. is a girl? __________ **13.** likes the lunch? __________

14. is a girl AND likes the lunch? __________ **15.** is a girl OR likes the lunch? __________

NAME ______________________ CLASS ____________ DATE ____________

Lesson Activity
4.6 Theoretical Probability

These games can be played at a fair.

Coin Toss
To win, the coin must land on a black square.

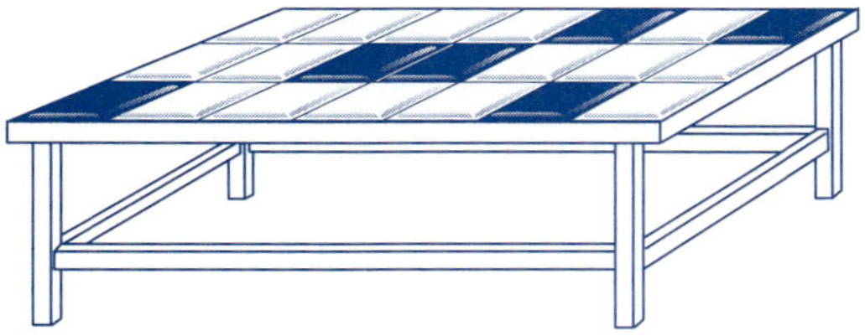

Number Spin
To win, choose three numbers. If the spinner lands on one of these numbers, you win.

Balloon Pop
Only one balloon contains a paper with a prize written on it. If you pop the balloon with the paper in it, you win.

Pick a Marble
Close your eyes and pick a marble. Any red marble is a winner.

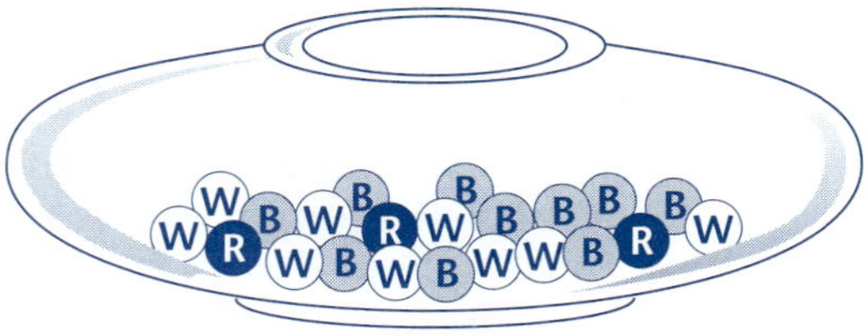

Read the rules for winning each game. Then use theoretical probability to answer each question.

1. What is the probability of winning the Coin Toss on the first try? ____________

2. What is the probability of winning the Number Spin if you predict that the spinner will land on 2, 4, or 6? ____________

3. What is the probability that you will lose at the Balloon Pop? ____________

4. Suppose that you can choose only one number for the Number Spin. With that restriction, which game do you have the highest probability of winning? What is the probability? ____________

5. What is the probability of winning at Pick a Marble? ____________

6. What is the probability that you will pick either a blue OR a white marble in Pick a Marble? ____________

7. At the Coin Toss, each time a coin lands, it is removed. Suppose that player 1 throws 1 coin, and player 2 throws 2 coins, one at a time. Compare the probabilities that each player wins. Explain your comparison.

____

NAME ________________ CLASS ________ DATE ________

Basic Skills Practice
Surface Area and Volume, Part 1

The amount of material needed to cover an object is known as its surface area. Volume is the number of nonoverlapping unit cubes that will fill the interior of a solid.

Example 1: Find the area and volume of the rectangular box.

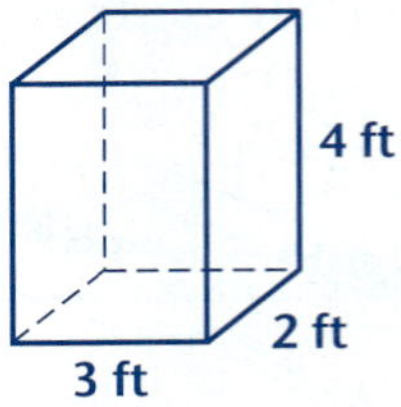

- Find the area of each face.
  Use $A = l \times w$
  Top and bottom: $3 \times 2 = 6$ ft^2 each
  Front and back: $3 \times 4 = 12$ ft^2 each
  Sides: $4 \times 2 = 8$ ft^2 each
- Add the area of each face.
  $SA = (2 \times 6 \text{ ft}^2) + (2 \times 12 \text{ ft}^2) + (2 \times 8 \text{ ft}^2) = 52 \text{ ft}^2$
- Use the formula $V = lwh$ to find the volume.
  $V = (2 \text{ ft})(3 \text{ ft})(4 \text{ ft}) = 24 \text{ ft}^3$

Example 2: Find the surface area and volume of the cube.

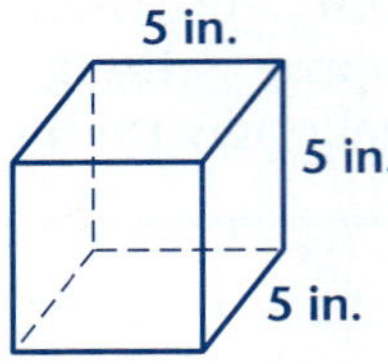

- Find the area of one face.
  Use $A = s^2$
- Since all the faces are the same, multiply the area of one face by the total number of faces (6).
  $SA = 6(25 \text{ in.}^2) = 150 \text{ in.}^2$
- Use the formula $V = s^3$ to find the volume.
  $V = (5 \text{ in.})^3 = 125 \text{ in.}^3$

Example 3: Find the lateral surface area (curved part without ends) and the volume of the cylinder.

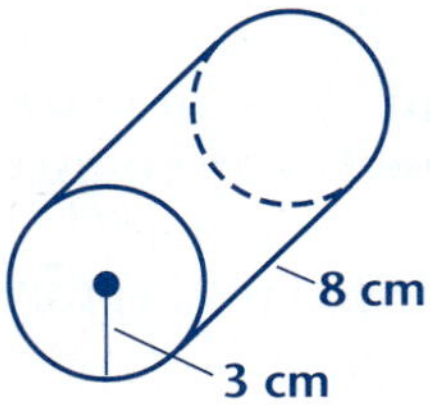

- Use $SL = 2\pi rh$.
  $SL = 2\pi(3 \text{ cm})(8 \text{ cm}) = 2\pi(24) \text{ cm}^2 = 48\pi \text{ cm}^2$
- If it is necessary to eliminate π, then use $\frac{22}{7}$.
  $SL = 48\left(\frac{22}{7}\right) \text{cm}^2 \approx 151 \text{ cm}^2$
- Use the formula, $V = \pi r^2 h$, to find the volume.
  $V = \pi(3 \text{ cm})^2(8 \text{ cm}) = \pi(9 \text{ cm}^2)(8 \text{ cm}) = 72\pi \text{ cm}^3$ or 226 cm^3

Find the surface area and volume of each solid figure.

1.

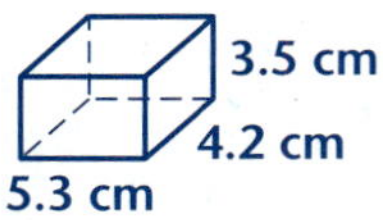

Surface area: ________ Volume: ________

2.

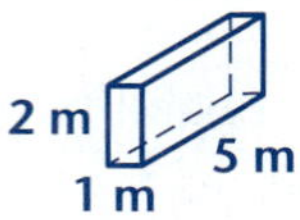

Surface area: ________ Volume: ________

3.

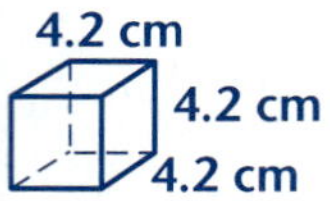

Surface area: ________ Volume: ________

4.

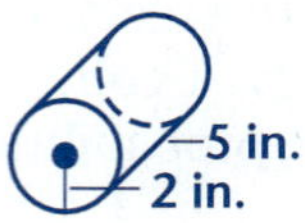

Lateral SA: ________ Volume: ________

LESSON 4.7 Independent Events

PREPARE

Pacing Guide

3 Days

Objectives and Overview

- Define and give examples of independent events.
- Find the probability that independent events will occur.

The following topics are discussed:

1. using geometric ratios to find probabilities
2. the probability of independent events
3. finding a probability by finding the complement

Vocabulary

independent events
dependent events
complement

Assessing Prior Knowledge

Find the probability of each event.

1. rolling a number greater than 2 on a single roll of a number cube $\left[\frac{2}{3}\right]$

2. rolling a sum of 3 on two consecutive rolls of a number cube $\left[\frac{1}{18}\right]$

3. getting at least 1 heads on 2 tosses of a fair coin $\left[\frac{3}{4}\right]$

Approach A: Using the Book

Opening Discussion

Some situations involve a sequence of events. Does the outcome of one roll of a number cube influence the outcome of a subsequent roll of a number cube? Explain. Suppose that you could use the letters *C, U, B,* and *E* on a license plate but that each letter could be used only once. Would the choice of the first letter affect the choice of the second letter? Why? The probability of rolling a 4 on a number cube is $\frac{1}{6}$. What is the probability of *not* rolling a 4?

Brainstorming

Ask students to suggest situations that involve two events for which the probability of the second outcome is not influenced by the first outcome. Examples include tossing heads and then tossing tails with a coin or rain occurring in New York and rain occurring in Sydney, Australia. Then ask students to suggest situations involving two events

continued on top of next page

TEACH

Approach B: Using Models

Opening Discussion

Situations that can model the Multiplication Principle of Counting involve a sequence of events. In some situations, the first outcome influences the second outcome; in other situations, the first outcome does not influence the second outcome. Do you think that the outcome of one roll of a number cube influences the outcome of a subsequent roll of a number cube? Why or why not? If you could use the letters *C, U, B,* and *E* on a license plate but each letter could be used only once, do you think that the choice of the first letter would affect the choice of the second letter? Explain. There is a 1 in 6 chance of rolling a 4 on a number cube. What is the chance of *not* rolling a 4?

Brainstorming

Ask students to work in groups of four and to make a list of situations involving two events for which the probability of the second outcome is not influenced by the first outcome. Examples include

continued on bottom of next page

Approach A Using the Book, continued

for which the probability of the second outcome *is* influenced by the first outcome. Examples include being chosen for a team after another person has been chosen (because there are fewer people from which to choose). As students make suggestions, write them on the chalkboard or overhead.

Review

Making Connections Recall from Lesson 3.1 that when data are arranged in a table of rows and columns and are enclosed by a bracket, the structure is a matrix. Like a matrix, a geometric model of probabilities may identify the outcomes of an event by row and column headings.

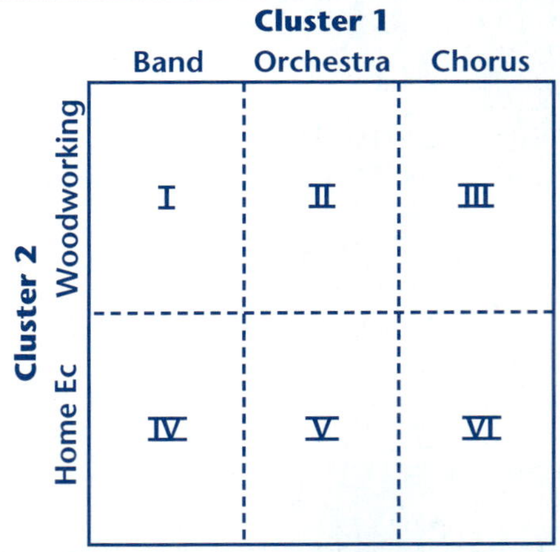

Thus, region I of the diagram above identifies the selection that includes woodworking and band. Region V identifies the selection that includes home economics and choir. Students should note that because the outcomes are equally likely, all of the regions have equal area.

TEACH

Approach B Using Models, continued

tossing a coin twice or choosing a shirt color after choosing pants of a certain color if the shirt and the pants can be the same color. Then ask the groups to brainstorm situations involving two events for which the probability of the second outcome *is* influenced by the first outcome. Examples include choosing a second member of a team after choosing the first member from a group of people or choosing a second digit for an ID number after choosing the first digit if the ID number must consist of two different digits.

Ask each group to read one of each kind of situation.

Geometry Connection: Modeling Equally Likely Events

A tree diagram can be used to model three consecutive tosses of a coin.

The diagram below shows 8 equally likely outcome and can be used to list the possible results of the three tosses.

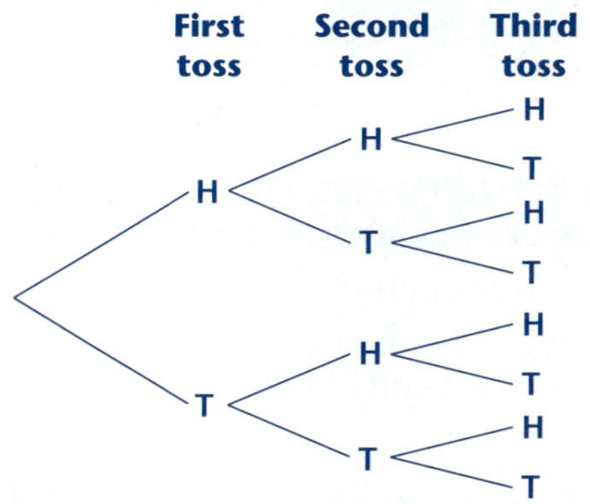

A diagram that gives a spatial representation of the results allows relationships among different kinds of outcomes to be easily determined. To make such a diagram, divide a rectangle horizontally into two equal parts to represent the equally likely results of tossing one coin. Then divide each of the resulting rectangles into four equal areas to represent the 4 equally likely outcomes of tossing a coin twice.

Example 1

The area of a planar region is used to model events that do not have equally likely outcomes. The model shows the probability of two events occurring together.

The important feature of the model is that the areas representing the outcomes are proportional to the probabilities of the outcomes.

Some students may notice that 0.48 is equal to 0.6 • 0.8. In fact, since the number of possible outcomes can be calculated by using the Multiplication Principle of Counting, the probability that Allessandro and Zita are both on time is equal to the product of the probabilities.

Use **Transparency 18** for discussion.

Ongoing Assessment, page 205

Try This

The region representing the probability that Allessandro is on time and that Zita is not has 12 squares, so the probability is $\frac{12}{100} = 0.12$, or 12%.

Additional Example 1

The probability that John shops downtown is 0.7, and the probability that Jane shops downtown is 0.4. Use an area model to find the probability that both John and Jane go shopping downtown. **[28%]**

continued on top of next page

	HH	HT	TH	TT
H	HHH	HHT	HTH	HTT
T	THH	THT	TTH	TTT

Like a matrix, each entry in the rectangle is determined by its row value and its column value. This arrangement takes less space and easily shows, for example, that 3 of 8, or 37.5%, of the results of tossing a coin three times contain exactly two heads.

Models For Outcomes That Are Not Equally Likely

Fifty students, 25 males and 25 females, were polled to see if they favored or opposed block scheduling. The results are shown in the following diagrams.

The areas represented by the circles can also be transferred to one rectangular grid. But because all of the possible outcomes for favoring or opposing block scheduling are not equally likely, equal regions of a rectangle will not model this data. Instead, a 10 × 10 square grid, which is appropriate for representing percents, can be used.

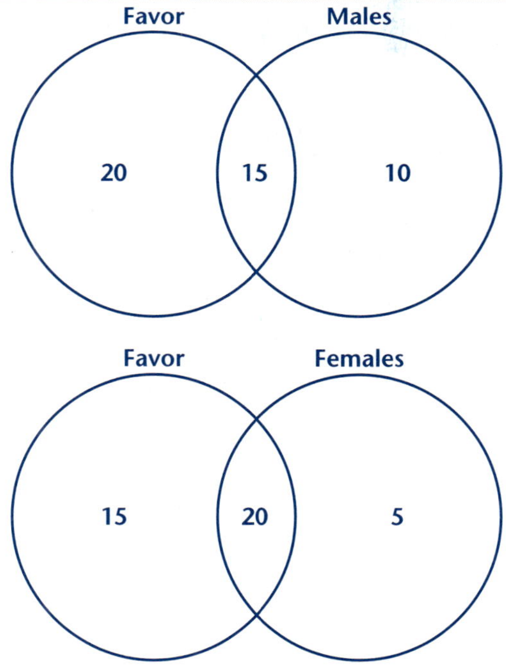

continued on bottom of next page

Approach A

Using the Book, continued

Example 2

This example applies the Multiplication Principle of Counting to a situation with two successive events.

Using Formulas

Show how the solution uses the theoretical probability formula to find the probability of drawing two black cards.

$$P = \frac{s}{n} = \frac{26 \bullet 25}{52 \bullet 51} = \frac{650}{2652}, \text{ or } \frac{25}{102}$$

Another approach involves multiplying the probability of the individual events.

P (both cards are black)

$= P(\text{first card is black}) \bullet P(\text{second card is black})$

$= \frac{26}{52} \bullet \frac{25}{51}$

$= \frac{650}{2652}, \text{ or } \frac{25}{102}$

Additional Example 2

Susan draws a card and then draws another card from a regular 52-card deck. Find the probability that the two cards drawn are both kings. In a regular deck, 4 of the cards are kings. $\left[\frac{1}{221}\right]$

Example 3

This example defines independent and dependent events. It also leads students to the discovery that for any two independent events A and B, probability(A and B) = probability(A) • probability(B).

Students should see that the Multiplication Principle of Counting can be applied to both dependent and independent events.

Teaching Tip

When applied to this situation, the formula for theoretical probability can be used to show how both the Multiplication and Addition Principles of Counting are involved.

TEACH

Approach B

Using Models, continued

Since $\frac{15}{25}$, or 60%, of the boys favor block scheduling, 60% of the grid, or 60 squares, should represent that part of the data. Similarly, $\frac{20}{25}$, or 80%, of the squares should represent the girls who favor block scheduling.

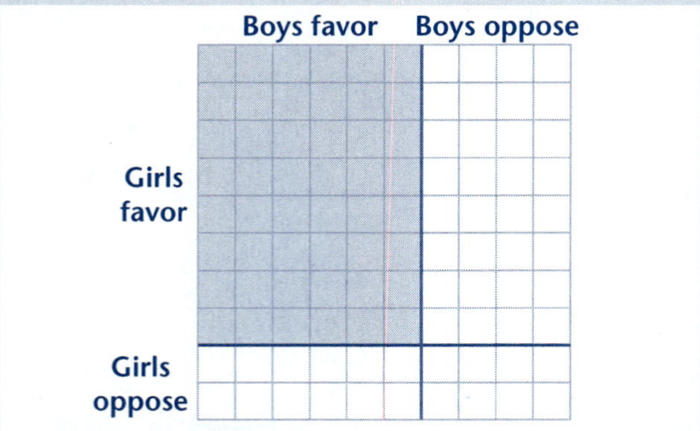

The shaded area represents the total number of males and females who favor block scheduling. It contains 8×6, or 48, out of 100 possible squares, so 48% of the students favor block scheduling.

Example 2

Making a Table Ask students to consider drawing a table to represent 2 consecutive draws of a card, without replacement, from a regular 52-card deck. A deck has two black suits and two red suits. Because each draw has so many posible results, only a partial diagram is reasonable.

Point out that 26 of the possible 52 first draws are black. Thus, if a black card is drawn without replacement, only 25 black cards remain for the second draw. The number of outcomes for the second draw is reduced and so is the probability of each outcome.

Because a table shows a sequence of events, the Multiplication Principle of Counting can be used with the formula for theoretical probability in order to determine the probability of drawing two black cards.

Three of the outcomes for a roll of a number cube are odd. The number of outcomes greater than or equal to 5 is equal to the number of outcomes that are 5 plus the number of outcomes that are 6. There are 6 • 6 total outcomes possible.

$$P = \frac{s}{n} = \frac{3 \cdot (1 + 1)}{6 \cdot 6} = \frac{1}{6}$$

Ongoing Assessments, page 206

The probabilities are multiplied.

Additional Example 3

If two number cubes are rolled, what is the probability that the first roll will show a number less than 3 and the second roll will show a number greater than 3? [$\frac{1}{9}$]

Inclusion Strategy

Using a Complement of an Event

Ask students to complete the following:

1. The probability of rolling a 4 is [$\frac{1}{6}$]. The probability of not rolling a 4 is [$\frac{5}{6}$].
2. The probability of tossing heads on a single toss of a coin is [$\frac{1}{2}$]. The probability of not tossing heads on a single toss of a coin is [$\frac{1}{2}$].
3. The probability of tossing 2 heads on 2 tosses of a coin is [$\frac{1}{4}$]. The probability of not tossing 2 heads on two tosses of a coin is [$\frac{3}{4}$].
4. What do you notice about the sum of the two probabilities in questions 1–3? **[The sum is 1.]**
5. How can you use the probability of the complement of an event to find the probability of the event? **[Subtract the probability of the complement of the event from 1.]**

continued on top of next page

	First draw	Possible second draw
Spades ♠	♠A	♠2,3,...,10,J,Q,K ♣A,2,3,...,10,J,Q,K
	♠2	♠2,3,...,10,J,Q,K ♣A,2,3,...,10,J,Q,K
	⋮	
	♣K	♠A,2,3,...,10,J,Q ♣A,2,3,...,10,J,Q,K
Clubs ♣	♣A	♠A,2,3,...,10,J,Q,K ♣2,3,...,10,J,Q,K
	♣2	♠A,2,3,...,10,J,Q,K ♣A,3,...,10,J,Q,K
	⋮	
	♣K	♠2,3,...,10,J,Q,K ♣A,2,3,...,10,J,Q

Defining Independent Events

Journal Entry Ask students to describe the difference between the three coin-tossing events that were modeled with the rectangle and the two card-drawing events that were modeled with the table.

Ask several students to read their entry. Those who correctly note that the toss of one coin did not influence the outcome of the second toss but that drawing one card without replacement did affect the drawing of the second card should be asked to read their entry again.

continued on bottom of next page

Approach A

Using the Book, continued

Example 4

In this example, the complement of an event is used to find the probability of an event.

Using Venn Diagrams

Since Example 4 describes a situation that involves two intersecting sets, a Venn diagram can be used to find the probability of the complement of an event.

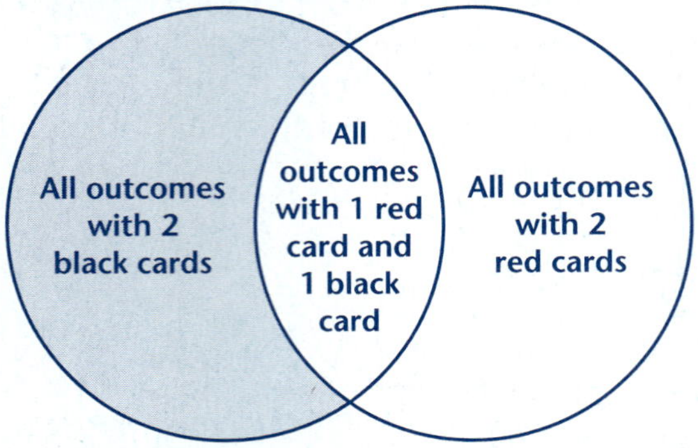

If two different cards are drawn from a deck, the outcomes that do not have at least one red card are represented by the shaded region. The probability of this region is $\frac{26 \cdot 25}{52 \cdot 51} = \frac{650}{2652}$, or $\frac{25}{102}$. The probability of the remaining area of the Venn diagram is $1 - \frac{25}{102} = \frac{77}{102}$.

Additional Example 4

A pair of number cubes is rolled 3 times. Find the probability that the cubes show the same number at least once. $\left[\frac{4}{9}\right]$

Ongoing Assessment, page 207

Try This

$\frac{35}{36}$

Critical Thinking, page 207

Since there are 26 coins and half of them, the 6 quarters and 7 dimes, have a value of at least 10 cents, the probability is $\frac{7+6}{26} = \frac{1}{2}$.

TEACH

Approach B

Using Models, continued

Tell students that the coin-tossing events should be labeled *independent events* in their journal and that the card-drawing events should be labeled *dependent events*. Then ask students to explain in their journal why these labels seem appropriate for the events.

Cooperative Learning

Multiplying Probabilities Ask students to work with a partner. Tell them to use the geometric model for the three consecutive coin tosses.

	HH	HT	TH	TT
H	HHH	HHT	HTH	HTT
T	THH	THT	TTH	TTT

Ask them to use the model to find the following probabilities:

1. the probability that the outcome of one toss of a coin is heads
2. the probability that the outcome of 2 tosses of a coin is 2 heads
3. the probability that the outcome of 3 tosses of a coin is 3 heads

How could you find the probability that only heads come up in all three tosses? Worded differently, this question asks how you could find the probability that a toss of heads is followed by a second toss of heads, which is then followed by a third toss of heads. The probability that only heads comes up is as follows:

$\frac{1 \cdot 1 \cdot 1}{2 \cdot 4 \cdot 8}$ or $\frac{1}{2} \cdot \frac{1}{4} \cdot \frac{1}{8} = \frac{1}{64}$

The probability of any sequence of n independent events can be found by multiplying the probabilities of each of the n events.

Closing the Lesson

Ask students to write an example of a situation that involves two independent events. Tell them to represent the event in two ways by using a diagram. Tell them to find the probability that both events occur. Have students also explain the difference between independent and dependent events.

Cooperative Learning: Problem-Solving Strategy

Ask students to work with a partner and develop a strategy for finding the area of the shaded region below.

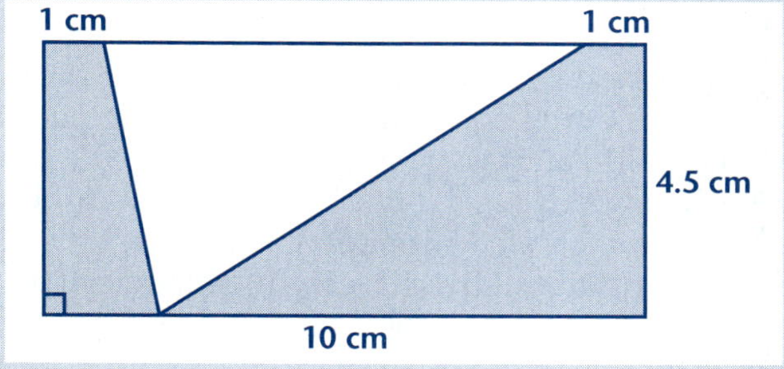

Have one student in each pair write a summary of their strategy. Have several students read their summary aloud. Students should notice that the area of the shaded region can be found by subtracting the area of the unshaded triangle from the area of the rectangle. This will give the area of the shaded region.

Point out that sometimes the easiest way to find an answer to a problem is by examining what is *not* the answer to the problem. Students use this strategy when answering a multiple-choice question. They examine the answers they think are not possible in order to narrow down the set of possible answers.

Using a Complement

Sometimes it is easiest to find the probability of an event by considering the number of ways in which the event does not occur. This strategy involves the *complement* of a set. For example, to find the probability that two rolls of a regular number cube show a sum of at least 3, finding the probability of this event by using the complement requires less counting because there is only one outcome with a sum of *less than* 3.

$P\text{ (a sum less than 3)} = \frac{1}{36}$

That leaves 35 out of 36 chances that the sum will be at least 3. Thus, the probability of rolling a sum of at least 3 is $\frac{35}{36}$.

ASSESS

Selected Answers

Odd-numbered Exercises 11–27 and 31–37 and odd-numbered Extra Practice

Assignment Guide

Core level students should complete Exercises 1–39.

Core-plus level students should complete Exercises 1–14 and 22–39.

Error Analysis

Students who have trouble calculating the probability of an event from the probability of the complement of the event may be subtracting incorrectly. A review of how to subtract a fraction from a whole number may be necessary.

Alternative Assessment

Geometry and Probability Give students examples of the kind of probability situations for which geometry is helpful when solving. When is geometry not useful for determining probabilities? Explain two other methods of determining probabilities that do no require diagrams.

RETEACH

A geometric model can be used to find the probability that a number equal to or greater than 5 shows on the red cube and that an even number greater than 3 shows on the green cube.

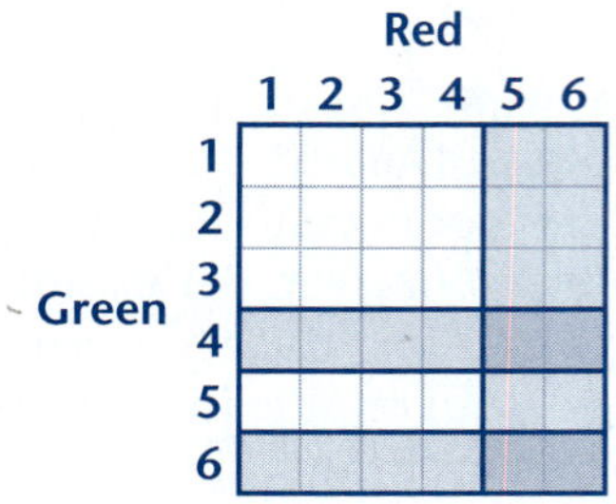

The rectangle shows 36 equally likely outcomes. The four darkly shaded squares represent rolls in which the red cube shows a number greater than 4 and the green cube shows an even number greater than 3. There are 4 out of 36 equally likely outcomes, so the probability of the event is $\frac{4}{36}$, or $\frac{1}{9}$. The formula for theoretical probability can also be used to find the probability.

Suppose that each of the probabilities is considered separately. The probability of rolling a 5 or a 6 on the red cube is $\frac{1}{3}$. The probability of rolling a 4 or a 6 on the green cube is also $\frac{1}{3}$. Notice that the product of these individual probabilities, $\frac{1}{3} \cdot \frac{1}{3}$, gives the same result, $\frac{1}{9}$. In fact, the probability of two independent events, *A* and *B*, can be calculated from the product of the probabilities of *A* and *B*.

Sometimes it is easier to find the probability of an event by first finding the probability of the complement of the event. For example, to find the probability that either cube shows a number greater than 2, most of the squares would need to be shaded and then counted. In this case, it would require less shading to find the squares that do *not* show outcomes in which either of the squares is greater than 2. There would be only 4 shaded outcomes. The probability that the rolls produce at least one cube with a number greater than 2 is then calculated as follows:

$1 - \frac{4}{36} = \frac{32}{36} = \frac{8}{9}$.

ENRICH

Interdisciplinary Connection

Biochemistry The word *complement* has several special meanings. For example, in grammar, a direct object of a verb can be called a complement. In music, an interval that, when added to another interval, completes an octave is called a complement. Biochemists study how bacterial species attack the human body and cause disease. When microorganisms enter the body, two of the body's natural defenses attempt to neutralize and destroy the organisms. One of a group of defenses includes antibodies and the *complement* system. Antibodies, which are proteins produced by special white blood cells, work together with antigens to kill, neutralize, or destroy microorganisms or to prevent them from multiplying. Another group of proteins in the body make up the complement system. These proteins, sometimes with and sometimes without the help of antibodies, interact with disease-causing microorganisms and destroy them.

NAME ______________________ CLASS ____________ DATE ____________

Practice
4.7 Independent Events

Ten cards numbered from 1 to 10 are in a bag. A card is drawn and replaced. A second card is drawn. Find the probability that

1. both are even. ____________

2. both are odd. ____________

3. both are prime. ____________

4. the first one is odd and the second is even. ____________

5. the first one is even and the second is odd. ____________

6. the first one is prime and the second is even. ____________

7. the first one is prime and the second is odd. ____________

8. the first one is even and the second is prime. ____________

9. the first one is odd and the second is prime. ____________

10. one of them is even and the other is odd (in either order). ____________

One number is selected from the list 2, 4, 6, and 8. Another is selected from the list 6, 7, and 8. Find the probability that

11. both are odd. ____________

12. they are the same. ____________

13. their sum is odd. ____________

14. the number from the second list is greater than the number from the first list. ____________

Use the area model to find the probability that

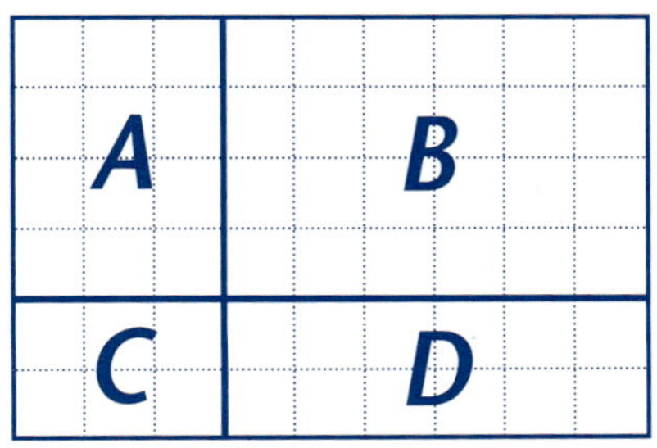

15. *B* occurs. ____________

16. *D* occurs. ____________

17. *B* occurs OR *D* occurs. ____________

NAME ______________________ CLASS ____________ DATE ____________

Lesson Activity

4.7 Independent Events

For each situation,

- **tell whether the events are dependent or independent,**
- **state the probability of the event occurring, and**
- **explain how you calculated the probability.**

1. There are 6 white socks, 8 black socks, and 4 red socks in your drawer. You pick 2 socks, one at a time, and hope that both are white.

____

____

2. There are 30 red marbles, 15 black marbles, and 5 white marbles in a bag. You pick 1 marble at a time, two times. Each time you pick a marble, you replace it before picking again. You hope to pick 1 black marble and then 1 white marble.

____

____

3. You choose 2 cards, one at a time, from a deck of 52 playing cards. You do not replace the first card. You hope to draw 2 hearts.

____

____

4. You have 5 pairs of blue jeans, 3 pairs of black pants, and 2 pairs of tan chinos hanging in your closet. Without looking, you choose one pair of pants on Monday, wear them, and throw them into the hamper. On Tuesday, you do the same. You want to wear blue jeans both days.

____

____

5. Your friend has the same numbers and colors of pants in her closet as in Exercise 4. However, on Monday, without looking, she chose a black pair of pants. She didn't want to wear black, so she replaced the pants and chose another pair without looking. This time she chose blue jeans. Explain the results of these events.

____

____

NAME ______________________ CLASS ____________ DATE ____________

Basic Skills Practice

Surface Area and Volume, Part 2

Read each question and circle the best answer.

1. A can of baked beans requires a label to cover the curved part of the can without overlapping. The radius of the can is 3 inches. The height of the can is 8 inches. Find the size of the label.

A 12π in.^{2}

B 24π in.^{2}

C 48π in.^{2}

D 72π in.^{2}

2. Al needs to paint the surfaces of a rectangular toy box. The box measures 3 feet in length and 2 feet in width. If 1 pint of paint covers 25 square feet, what else does Al need to know in order to find the number of pints that he needs?

F the weight of the box

G the number of ounces in 1 pint

H the cost of 1 ounce of paint

I the height of the box

3. James wants to know how much water his swimming pool will hold. The pool is 5 meters by 10 meters by 3 meters. How much water will the pool hold?

A 50 m^3

B 150 m^3

C 750 m^3

D 1000 m^3

4. Sara wants to cover a cube with colorful contact paper. She can cut the paper the exact size of each side of the cube. Then she can stick the paper onto each side. Find out how much contact paper she will need if one side of the cube is $1\frac{1}{2}$ ft.

F 2.25 ft^2

G 10.13 ft^2

H 13.5 ft^2

I 18 ft^2

5. Jose wants to mail a gift to his mother. The box he wants to mail it in is the shape of a cube. The cost to mail the gift depends on the amount of space it occupies. What information will be the most helpful to determine the cost to mail the gift?

A the volume of the box

B the surface area of the box

C the weight of the box

D the shape of the box

6. What is the volume of a cube-shaped fish tank whose side is equal to 2.5 feet?

F 6.25 m^3

G 7.5 m^3

H 15 m^3

I 15.625 m^3

Solve each problem.

7. A box holds 12 pieces of chalk. If the radius of each piece of chalk is 2 centimeters and the height of each piece of chalk is 7 centimeters. What is the total volume of each piece of chalk? Use $\pi = \frac{22}{7}$. ____________

8. Smith Wood Supply makes crates of many sizes and shapes. One standard crate measures 7 feet by 4 feet by 2 feet. What number would best represent the amount of gravel the crate can hold? ____________

LESSON 5.1

Functions and Relations

PREPARE

Pacing Guide

2 Days

Objectives and Overview

- Understand the definitions of *relation* and *function*.
- Use a graph to identify a function by using the vertical line test.
- Use the $f(x)$ function notation to represent and evaluate a function.

Students learn how to define the terms *relation* and *function* and learn how to identify a function by using the vertical line test. Function notation is introduced, and students use it to evaluate functions.

Vocabulary

relation	range
function	function notation
vertical line test	replacement
domain	function rule

Assessing Prior Knowledge

Graph each equation.

1. $x + y = 3$ **2.** $2x - 4 = y$

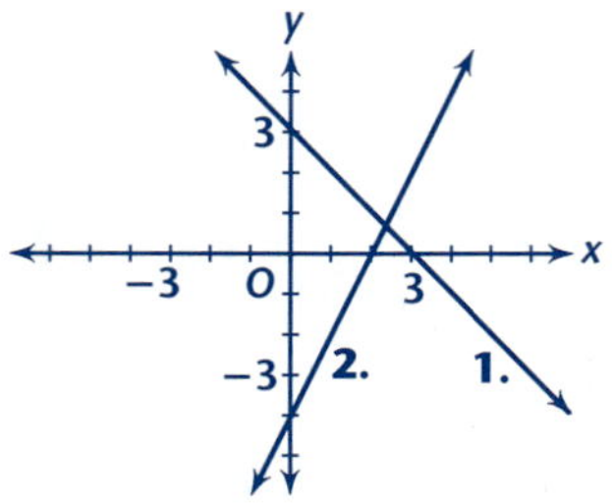

Approach A: Using the Book

Opening Discussion

How do you interpret the road atlas chart given on page 220 of the Pupil's Edition? For example, how can you tell if it is possible to go fishing in the Petrified Forest National Park? Why is this table useful?

Brainstorming

Have students brainstorm different relations that they see in the road atlas chart. Ask them to explain how they can tell how many pairs of data are in the chart. What are some of the pairs of data?

Defining *Relation* and *Function*

The road atlas chart of national parks provides a concrete way to define relation and function. Stress that a relation is *any* set of ordered pairs, while a function is a special type of relation. A function is defined as a relation for which there is exactly one second coordinate for each first coordinate.

TEACH

Approach B: Using Models

Opening Discussion

The services offered by a national park depend on which park you visit. How do you interpret the road atlas chart given on page 220 of the Pupil's Edition? For example, how can you tell if fishing is available in the Petrified Forest National Park? Why is the table given in the road atlas useful? Can the table be arranged in any other way to show how services are related to a park?

Brainstorming

Have students brainstorm different quantities that are related to each other and that can be expressed mathematically. For example, the distance that a car travels is related to its speed.

Using a Graphics Calculator

A graphics calculator can help students understand functions because it can quickly graph functions, which students can then observe and compare with the definition. Start by entering a function using the

Inclusion Strategy

Write a Function Arrange the students' desks into rows. Give each student a large sheet of paper on which the desk arrangement is drawn. Have each student take turns entering his or her gender, age, hair color, eye color, and height on an overhead copy of the arrangement. Ask students to copy the class chart from the overhead onto their paper. From this chart, name some pairings that are functions. Name some pairings that are relations. Name some pairings that are relations but not functions.

Critical Thinking, page 221

Answers may vary. For example, they may have begun the trip in a residential area and then may have had to stop for stoplights, stop signs, or traffic. They may have stopped to buy fuel a few minutes from home. Their speed levels off at 55 mph at around 19 minutes because they reached a highway.

Testing for a Function

The vertical line test is commonly used to show whether a relation is a function. If a vertical line intersects the graph of a relation more than once, then the relation is not a function. Point out that if a vertical line intersects the graph more than once, it implies that there are at least two points with the same x-coordinate. Functions do not contain two points with the same x-coordinate.

Example 1

In this example, graphs of various relations are shown, and the vertical line test is applied to each graph.

Point out that if a graph passes the vertical line test, then it is a function. If it does not pass the test, then it is not a function but is still a relation. Encourage students to think of other graphs to test with the vertical line test.

continued on top of next page

[Y=] key of the graphing menu. For example, enter $y = 2x^2 + x + 1$ as $Y_1 = 2X^2 + X + 1$, using the special [X,T,θ,n] key for convenience. Use the various capabilities of the graphics calculator to help you understand the definition of function. Press Zoom 6 or adjust the viewing window to see a graph of the function.

Working With a Graphics Calculator

Have students enter the following functions into a graphics calculator. Sketch each graph.

A. $y = x^2 - x + 2$ **B.** $y = |x - 4|$

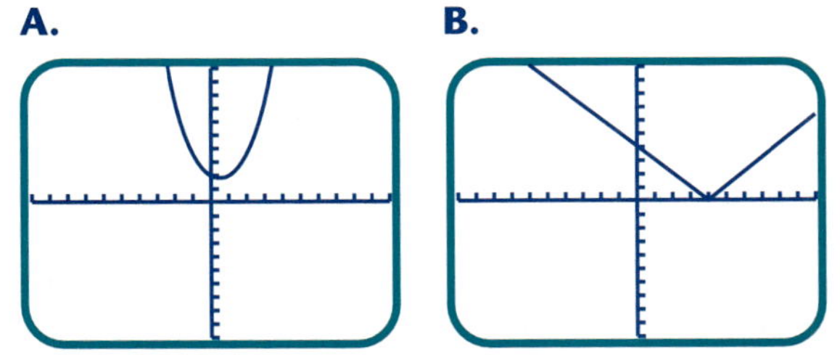

Defining *Relation* and *Function*

Use a graphics calculator to emphasize that for a function, there is exactly one second coordinate for each first coordinate. Ask students to look at the table generated for a function that has been entered by using the [Y=] key. Use the TABLE command to see that each x-value is paired with only one y-value. Below is a sample screen produced by the TABLE command for problem **B** in Working With a Graphics Calculator.

X	Y1	
−3	7	
−2	6	
−1	5	
0	4	
1	3	
2	2	
3	1	
X = 3		

Cooperative Learning

Domain and Range Divide the class into groups of four and have them use a graphics calculator to study domain and range.

Ask one student in each group to enter any function in the Y= menu of a graphics calculator. Have the second student use the

continued on bottom of next page

Approach A

Using the Book, continued

Additional Example 1

Which of these graphs represent functions?

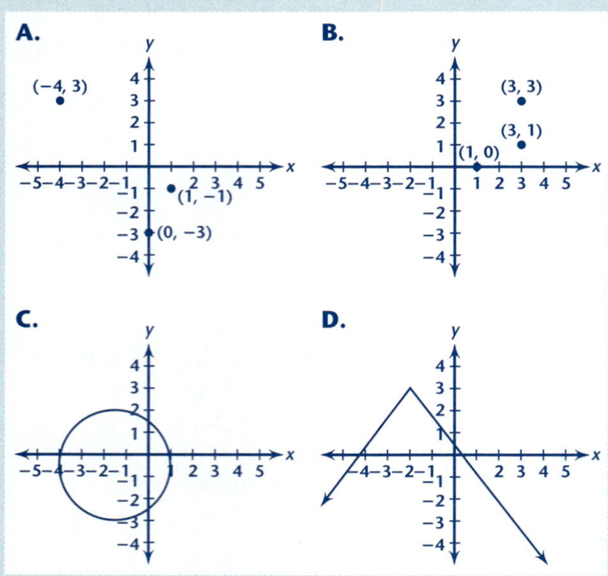

[A; D]

Definition of *Domain* and *Range*

The domain of a function is defined as the set of allowable x-values for the function, and the range of a function is defined as the set of y-values for the function. The set of allowable x-values are the only numbers that can be used with the function and are sometimes known as the "input values."

The set of values that the function produces is the range, or the "output values." Range values depend on the domain values. For example, for the function $y = x^2$, any number is in the domain, but only nonnegative numbers are in the range.

Ongoing Assessment, page 223

The range of $y = |x|$ is always greater than or equal to 0.

Function Notation

Function notation uses symbols like $f(x)$. To express the function $y = x^2$, for example, write $f(x) = x^2$. Emphasize that function notation is

TEACH

Approach B

Using Models, continued

window menu to list Xmin, the minimum domain value to be used for the graph, and Xmax, the maximum domain value to be used for the graph. Have the third student predict Ymin, the minimum range value, and Ymax, the maximum range value, that are produced by the second student's restrictions on the domain. Have the third student press the GRAPH key to show the graph and then look at the window screen to check the predicted Ymin and Ymax values. Have the fourth student write a report of the group's activities, including the function and the window values used.

Have students switch roles and work with a new function until all students have performed all four roles. Each student should discuss the contents of the report, verify the accuracy of the entries, and sign the report.

Function Notation

A graphics calculator is a good tool to explore function notation. Function notation is important because it highlights the input value for a function and the rule acting on that value.

Enter a function into the Y= menu under Y_1. Then go to the home screen and choose Y_1 from the function list (for some calculators, it is under the Y-VARS menu). Then enter the input value. For example, enter $y = 2x^2 + x + 1$ as $Y_1 = 2X^2 + X + 1$. Then type in $Y_1(3)$ on the home screen to evaluate the function at 3. For this example, $Y_1(3) = 22$.

Have students use a graphics calculator to evaluate each function below for the following domain values: −3, −2, −1, 0, 1, 2, 3.

A. $y = x^2 - x + 2$ [{14, 8, 4, 2, 2, 4, 8}]
B. $y = |x - 4|$ [{7, 6, 5, 4, 3, 2, 1}]

Teaching Tip

Function Notation and List Features

Students can use the list features to evaluate a function for a set of domain values. For problem **A** above, they should enter $Y_1 = X^2 - X + 2$ into the Y= menu. Then enter the domain values (−3, −2, −1, 0, 1, 2, 3) under L_1 of the STAT

important because it highlights which variable is dependent and which variable is independent. For this example, the function's output values depend on the value of x, so x is the independent variable and y is the dependent variable. Function notation also expresses the function with the notion of a "function rule." That is, if an input value in the domain is assigned to x, then the function rule determines the output value in the range.

Use **Transparency 19** for a discussion of functions and the notation of functions.

Teaching Tip

Make sure that students know how to read a function expressed in function notation. The expression $f(x) = x^2$ is read "f of x equals x squared." Point out that when the expression is read correctly, you are told how the function rule operates on the x-values. In this case, the rule is "square the x-values."

Example 2

In this example, a function is evaluated for a value in its domain.

Stress that the function notation $h(4)$ indicates that the value 4 should be substituted for each x in the function $h(x) = x^2 - 2x + 1$ and that the resulting expression should be simplified. Point out that the result of evaluating the function at 4 produces an ordered pair with an x-coordinate of 4; the ordered pair is (4, 9) in this case. This is a point in the coordinate plane and can be graphed.

Additional Example 2

Suppose that f is a function and $f(x) = 2x^3 + x$. The domain of the function is all positive numbers. Evaluate $f(1.5)$. **[8.25]**

Ongoing Assessment, page 224
Try This
90

continued on top of next page

EDIT screen. Then type in $Y_1(L_1)$ on the home screen. The calculator evaluates the function for every value in List 1. The calculator screen shows the following list: 14, 8, 4, 2, 2, 4, 8.

Remind students that they may need to scroll to the right to see the complete list of values. Have students use the list features to evaluate each function below for the following domain values: −3, −2, −1, 0, 1, 2, 3.

A. $y = 3x^2 - 2x - 1$ [{32, 15, 4, −1, 0, 7, 20}]
B. $y = -2x - 2$ [{4, 2, 0, −2, −4, −6, −8}]

Absolute-Value Function

The absolute-value function $g(x) = |x - a|$ represents the distance between any number and a fixed point a. The domain of g is all real numbers. Have students work with a partner to explore the range values for this function with a graphics calculator. Have one student use the STORE feature to store the value 3 for a in the calculator. Then have the student enter the function Y_1 = abs(X − A) under the Y= menu of the calculator. Have a second student use the table feature of the calculator to fill in the following table:

x	−3	−2	−1	0	1	2	3
y							

x	−3	−2	−1	0	1	2	3
y	6	5	4	3	2	1	0

Have students switch roles and make a table of values for $a = -3$. Then have them compare their tables and answer the following questions: Which table values represent the domain of $g(x)$? Which table values represent the range of $g(x)$? What are some points on the graph of $g(x)$?

continued on bottom of next page

Approach A

Using the Book, continued

The Absolute-Value Function

It is convenient to use the absolute-value function $f(x) = |x|$ to represent the distance between two points on a number line because distance is always positive. The distance between a number and a fixed point, a, is $f(x) = |x - a|$.

Example 3

In this example, the absolute-value function is used to calculate the distance between a number and a fixed point, 5.

In general, the distance between a number, x, and a fixed point, a, is $f(x) = |x - a|$. Emphasize that there are two numbers at a fixed distance from a given point on a number line.

Additional Example 3

Determine a function for the distance between any number and the fixed point −2. Find the distance from 6 to −2 and from 0 to −2.
[$f(x) = |x - (-2)|$ or $f(x) = |x + 2|$; 8 and 2]

Closing the Lesson

Journal Writing Summarize the lesson by having students write in their journal the definition of a function along with mathematical and real-world examples of functions. Ask them to describe how they can test for functions.

TEACH

Approach B

Using Models, continued

Journal Entry

Have students list the steps necessary to graph a function, make a table of values for a function, and use function notation to evaluate a function in their journal. Ask them to include sample functions that they have worked with in this lesson.

ASSESS

Selected Answers

Odd-numbered Exercises 7–61 and odd-numbered Extra Practice

Assignment Guide

Core level students should complete Exercises 1–61.

Core-plus level students should complete Exercises 1–35, 39–44, and 48–62.

Error Analysis

Students may have difficulty identifying the domain and range for the functions in Exercises 32 and 33. Remind students that the domain consists of all x-values that can be used to evaluate the function. Since students don't have the function rule for these exercises, they should look at the graph carefully and notice that there is a y-value for every value of x on the number line in both graphs, so the domain is all real numbers. Also remind students that the range consists of all the y-values that are shown on the graph or that can be shown if the graph is extended indefinitely. For Exercise 32, the range is only the number -3 because every point on the graph has a y-coordinate of -3. For Exercise 33, the range is $-1 \leq y \leq 1$ because every point of the graph has a y-coordinate between -1 and 1 inclusive.

Alternative Assessment

Function Notation Ask students to create their own "function machine" to demonstrate that they understand input values, output values, and the function rule. Have students work with a partner. Ask one student to give the rule for a function and to list the input values that can be used for the function. Ask the second student to find the output values. Then have the students switch roles. Have them repeat this activity with progressively more difficult functions until they understand how to evaluate a function for each value of the domain.

Portfolio Activity Have students find an example of a function in a newspaper or magazine and display it for class. Finding functions that are given as a function rule is difficult, so encourage them to look for tables whose entries represent functions. Ask them to write a paragraph explaining how they know that their choice is an example of a function.

RETEACH

Ask students to work with a partner to identify the domain and range of a function. Ask one student to draw a graph of any function. Have the other student verify that the graph is a function and then give the domain and range. Then ask students to switch roles and repeat the steps.

ENRICH

A graphics calculator can be used to explore many functions. Ask students to explore some other functions available on the calculator by entering any of the following: Y_1 = SIN (X), Y_1 = COS (X), Y_1 = TAN (X), Y_1 = LOG (X), Y_1 = LN (X), or $Y_1 = \sqrt{\ }$(X). Use the Zoom feature if the graph is not shown in the current window. Ask students to sketch the graph and give the domain and range of the function.

Interdisciplinary Connection

Economics An economist is responsible for finding patterns in data that relate to economics, which is the study of the production, distribution, and consumption of wealth. Patterns in data are sometimes function rules that can be useful in predicting future trends in the data.

NAME ______________________ CLASS __________ DATE __________

Practice

5.1 Functions and Relations

Which of the following relations are functions? Explain.

1. $\{(3, 5), (1, 0), (9, 5), (2, 0)\}$ ______________________

2. $\{(-1, -2), (3, 5), (1, 2), (-1, 7)\}$ ______________________

3.

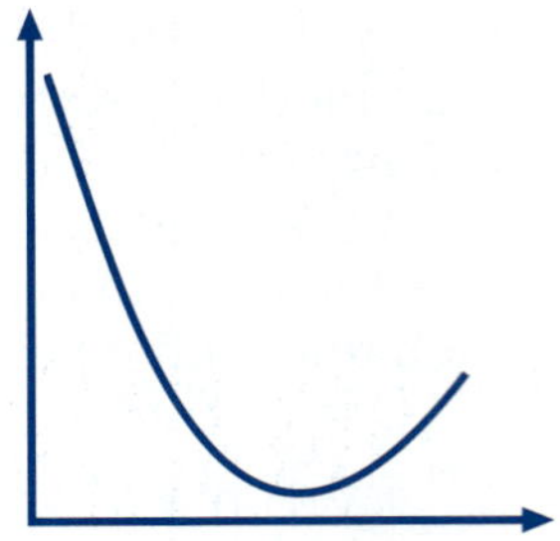

____

____

4.

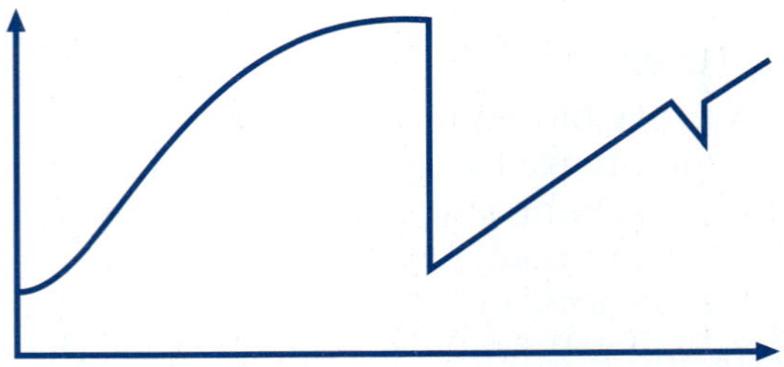

____

____

5.

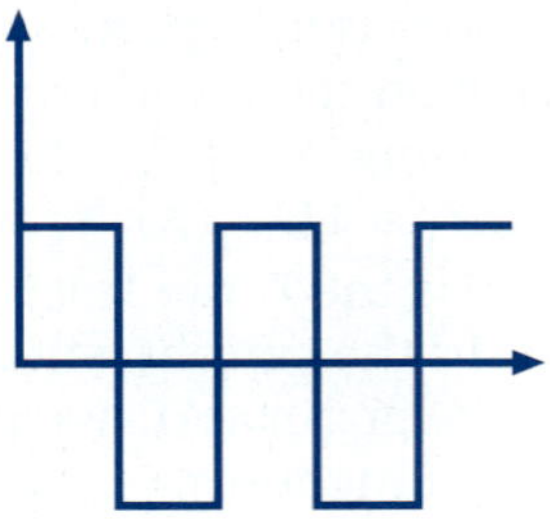

____

____

6.

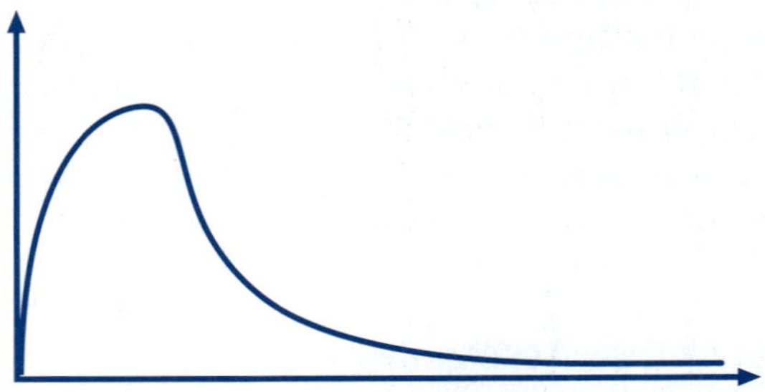

____

____

Let $f(x) = 6x$. Evaluate.

7. $f(2)$ __________ **8.** $f(-1)$ __________ **9.** $f(0.5)$ __________

10. $f(4)$ __________ **11.** $f(-5)$ __________ **12.** $f\left(\frac{1}{3}\right)$ __________

Evaluate each function when x is 3.

13. $f(x) = x^2$ __________ **14.** $g(x) = -2x$ __________

15. $f(x) = \frac{1}{x}$ __________ **16.** $h(x) = -4|x|$ __________

17. $k(x) = 2x^2 - 1$ __________ **18.** $g(x) = x^2 - 3x$ __________

NAME ______________________ CLASS ____________ DATE ____________

Lesson Activity
5.1 Going in Circles

Each of these circles has a different diameter. Trace and cut out each circle.

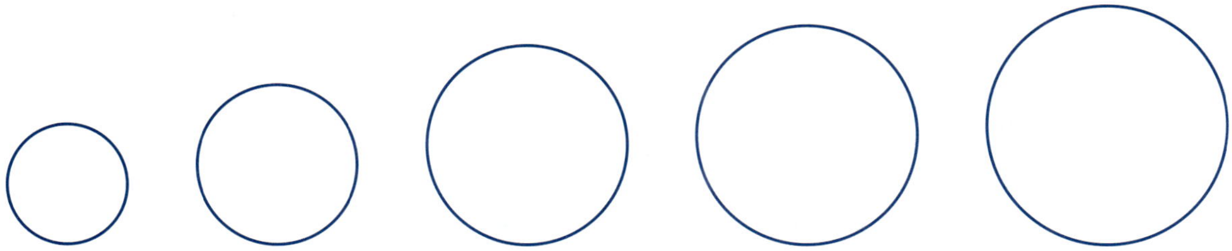

1. One at a time, place each circle on the horizontal axis of the grid provided so that the axis divides the circle into two halves. Use a string to measure the circle's circumference. As shown, lay the string parallel to the vertical axis, starting at the endpoint of the diameter. On the grid, mark the ordered pairs that correspond to the diameter and circumference of each circle. The first circle is done for you.

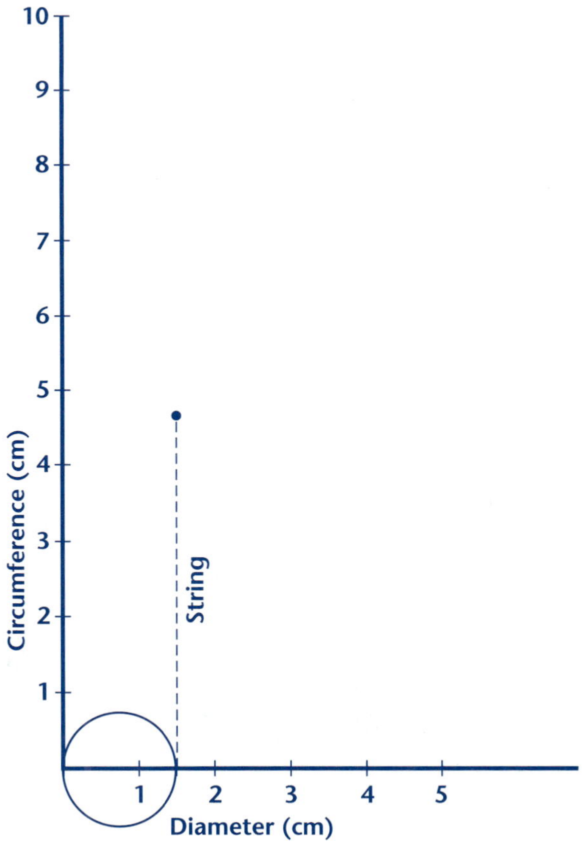

2. Examine the graph. Why is the graph a function?

____

3. Identify the domain and range.

____

4. Calculate the ratio of vertical distance to horizontal distance for each circle. Is the ratio constant?

____

5. What number does the ratio approximate?

____

6. Determine a function rule that relates circumference to diameter.

____

7. Use your function rule to find the circumference of a circle with a diameter of 3.5 cm. ______________________________

NAME ______________________ CLASS ____________ DATE ____________

Basic Skills Practice

Cumulative Review

Read each question and circle the best answer.

1. The cost of having a birthday party at a pizza restaurant is given by the formula

$$C = \$20.00 + \$2.50p,$$

where p is the number of people coming to the party. What is the cost of a pizza party with 15 people attending?

A $22.50

B $37.50

C $57.50

D $42.50

2. A crayon company wants to cover each crayon with paper that does not overlap. The radius of each crayon is 3 centimeters and the height is 8 centimeters. Find the amount of paper that it will take to cover each crayon.

F 12π cm^2

G 24π cm^2

H 48π cm^2

I 72π cm^2

3. 9.067 is between which pair of consecutive integers?

A 8 and 9

B 9 and 10

C 10 and 11

D 30 and 31

4. The diameter of a basketball hoop is 55 centimeters. A manufacturer wants to know how much metal it will take to make the rim. Find the circumference of the rim. Use 3.14 for π.

F 55 cm

G 86.4 cm

H 172.7 cm

I 345.4 cm

5. Refer to the information given in Exercise 4. About how much space within the basketball hoop will there be for a basketball player to make a basket?

A 55π cm^2

B 110π cm^2

C 756π cm^2

D 3025π cm^2

6. Elizabeth enjoys wearing ribbon in her hair. She has 6 red, 4 green-and-white striped, 2 blue, and 8 floral ribbons. What fraction of the total number of ribbons are green-and-white striped?

F $\frac{1}{5}$

G $\frac{1}{10}$

H $\frac{3}{10}$

I $\frac{2}{5}$

Solve each problem.

7. Michele gets paid $7.50 for 2 hours of baby-sitting. Friday night she baby-sat for the Andersons. They paid her $18.75. How many hours did she baby-sit? ____________

8. Unleaded gasoline costs $1.259 per gallon. You need 10.4 gallons to fill your gas tank. Rounded to the nearest hundredth, about how much does it cost to fill the tank? ____________

LESSON 5.2

Exploring Transformations

PREPARE

Pacing Guide

2 Days

Objectives and Overview

- Identify basic transformations of parent functions.
- Define stretches, reflections, and shifts as they apply to the graphs of functions.

Students explore how to identify the basic transformations of the absolute-value function, quadratic functions, and other functions. They also examine the graphs of some transformations of parent functions.

Vocabulary

vertex	parent function
symmetric	transformation
axis of symmetry	

Assessing Prior Knowledge

Graph each function.

1. $y = |x + 1|$ **2.** $y = x^2 - 1$

3. $y = 3x^2$

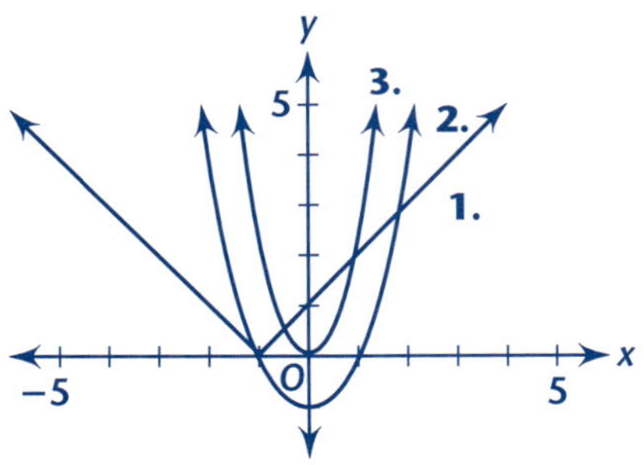

Approach A — Using the Book

Opening Discussion

Ask students if they can draw two identical function graphs in different parts of the coordinate plane. Although the graphs have different coordinates, are they related? How might the graph of $y = x^2 + 1$ be related to the graph of $y = x^2$? How could a change in the function rule be related to a change in the graph?

Brainstorming

Ask students to make a table of values for the graphs of $y = x$ and $y = x + 1$. Have them make a conjecture about how the graphs are related. Use calculator graphs of each function to check their conjectures. What might the graph of $y = x - 1$ look like?

Setting Up the Explorations

Teacher-Guided Explorations If you wish, you can guide the students through the textbook

continued on top of next page

TEACH

Approach B — Using Models

Opening Discussion

Ask students if they can draw two identical function graphs in different parts of the coordinate plane. Although the graphs have different coordinates, are they related? How could a change in the function rule be related to a change in the graph? How would adding 1 to every y-value of a table of values change the function rule?

Brainstorming

Ask students to create a table of values for the graphs of $y = x$ and $y = x + 1$. Have them make a conjecture about how the graphs are related. Use calculator graphs of each function to check their conjectures. What might the graph of $y = x - 1$ look like? What function rule would apply to the same graph if it were shifted up 5 units?

Modeling Transformations

Use graphics calculators, function-plotting software, or overhead transparencies to model the

continued on bottom of next page

Approach A

Using the Book, continued

explorations by using an overhead graphics calculator or function-plotting software. Follow each exploration step by step. Make sure that students understand how each transformation of the parent function is done and how to identify each transformation. Explain how the vertex of the graph changes as the function is transformed.

Cooperative-Learning Groups Divide the class into groups of four. Ask one student to read the steps from the textbook. Have another student write the equations for the transformed functions and draw their graphs, with input from the rest of the group. Have a third student answer the questions in the explorations and make conjectures about the graphs of transformed functions. Have the fourth student write a report detailing the group's activities, including whether the conjectures were tested. Ask each student to sign the completed report.

Symmetry and the Axis of Symmetry

Point out that several types of symmetry are possible in a graph. For example, a graph that is symmetric in the y-axis has the y-axis as its axis of symmetry, which means that you could fold the graph in half along the y-axis and the left and right halves would match exactly. Similarly, symmetry in the x-axis means that you could fold the graph in half along the x-axis and the upper and lower halves would match exactly. Point out that a graph can also have more than one type of symmetry.

Definition of *Parent Function*

A parent function is the basic function of an entire family of functions. Each member of the family of functions has the same basic characteristics as the parent function, including its graph. For example, the family of functions of the parent function $y = x$ is of the form $y = ax + k$. The graphs of the family of functions may be positioned differently in the coordinate plane, depending on how they are

TEACH

Approach B

Using Models, continued

transformation of a function. First show students how to graph the parent function and describe its basic characteristics. Also show a table of values for the parent function. Then show a transformation of the parent function and describe the similarities to the parent function, both in the function rule and in the graph. If a graphics calculator is used, use the TABLE features to show a comparison of the coordinates for the parent graph in Y_1 and for the transformation graph in Y_2 (with the same x-values). Use the models to explain how to use the parent function to graph a stretch, a reflection, or a shift.

Cooperative Learning

Before starting the explorations, divide the class into groups of four. Ask one student to read the steps from the textbook. Have another student draw the graphs used in the explorations. Have a third student make conjectures about the relationship between the parent graph and the transformed graph. Have the fourth student write a report detailing the group's activities. Ask each student to sign the completed report.

Exploration 1

Have students investigate how to transform the graph of the absolute-value function. The questions below may be listed on the chalkboard or overhead. (For ideas on guiding students, see Exploration 1 on page 229 of the Pupil's Edition.)

- How can you tell if a transformed graph is a vertical or a horizontal shift of the absolute-value parent function by comparing the function rule to $y = |x|$? **[For a vertical shift, a number will be added to or subtracted from $|x|$, such as $y = |x| + 2$. For a horizontal shift, a number will be added to or subtracted from x inside the absolute-value symbol, such as $y = |x - 2|$.]**
- How can you tell if a transformed graph is a vertical or a horizontal shift of the absolute-value parent function by examining how the function rule is applied? **[A vertical shift**

related to the parent function. Point out that every function has a parent function.

Definition of *Transformation*

A variation of a parent function is known as a transformation. Stretches, reflections, and shifts are some examples of transformations. The graph of a reflection or shift is identical to the graph of the parent function but is in a different position in the coordinate plane. More than one transformation can be applied to the parent function. For the function $y = f(x)$, a stretch is indicated by the form $y = a \cdot f(x)$ or $y = f(ax)$, a reflection through the y-axis by the form $y = f(-x)$, a reflection through the x-axis by the form $y = -f(x)$, and a horizontal shift by the form $y = f(x - a)$, where a is any number.

Exploration 1

In this exploration, students graph the parent function $y = |x|$ on a graphics calculator and then transform the graph in several different ways.

The transformed graphs appear on page 229 of the Pupil's Edition, but not in the same order as in the exploration. As students perform this exploration, emphasize that they should pay attention to how the parent graph was changed because there is a pattern to all stretches, reflections, and shifts. Once they see the pattern, they can predict the type of transformation of the parent function. See the previous section titled Definition of *Transformation* for some of the patterns.

Ongoing Assessment, page 229

5. If it has been stretched, then the sides of the graph will not be made up of the portions of the lines $y = x$ and $y = -x$ that lie above the x-axis. If it has been reflected, then it will open downward. If it has been shifted, then the vertex of the graph will not be at (0, 0).

continued on top of next page

occurs when a number is added or subtracted after the absolute value expression is evaluated; a horizontal shift occurs when a number is added to or subtracted from the x-value before the absolute value of the result is found.]

Student Reports

One or more members of each group should report the group's findings to the class. Ask them to show their graphs and describe the transformations. The reporter should state the group's conclusions clearly, including the answers to the exploration questions, and the entire group should answer questions from other groups.

Follow-Up Demonstration (Optional)

For this exploration, you may want to demonstrate a variety of transformations of the absolute-value parent function, $y = |x|$. For example, demonstrate $y = |x| + 3$, $y = 3|x|$, and $y = 3|x| + 3$. Include a discussion of the nature of each transformation and its similarity to the parent function. Ask for student input, and have them verify the transformations with a graphics calculator.

On Their Own

Describe what you must do to the function rule $y = |x|$ in order to obtain a new function rule that represents each of the following:

A. a stretch **B.** a reflection **C.** a shift

[Sample: A. Multiply $|x|$ by some number, such as 2. B. To reflect through the x-axis, let $y = -|x|$. C. Change $|x|$ to $|x - a|$, where a is some number.]

Exploration 2

Have students answer the following questions about transformations of the quadratic parent function $y = x^2$. (For ideas on guiding students, see Exploration 2 on page 230 of the Pupil's Edition.)

- How can you tell if a transformed graph is a vertical or a horizontal shift of the quadratic parent function by looking at the function rule $y = x^2$? **[For a vertical shift, a number will be added to or subtracted from x^2.**

continued on bottom of next page

Approach A

Using the Book, continued

Follow-Up Demonstration (Optional)

If you need closure for this exploration, summarize the types of transformations given in the section titled Definition of *Transformation*. Have students give examples of the various types of transformations that are different from those given in the exploration exercises.

On Their Own

Describe what you must do to the function rule $y = |x|$ in order to obtain a new function rule that represents each of the following:

A. a stretch **B.** a reflection **C.** a shift

[Sample: A. Multiply $|x|$ by some number, such as 2. B. To reflect through the x-axis, let $y = -|x|$. C. Change $|x|$ to $|x - a|$, where a is some number.]

Exploration 2

In this exploration, students draw the quadratic parent function $y = x^2$ and then transform it.

Students then compare their graphs with the graphs in Exploration 1 and determine how they are similar. Point out that the similarities arise from the type of transformation, not from the type of function. Therefore, it is important to know the transformation patterns for stretches, reflections, and shifts, which are recognizable for any function.

Ongoing Assessment, page 230

5. Stretch the graph vertically by 2, reflect it through the x-axis, and shift it vertically by 3.

Follow-Up Demonstration (Optional)

If you need closure for this exploration, restate how to transform a function by stretching, reflecting, or shifting it. Use examples from the explorations to reinforce the patterns of these transformations.

TEACH

Approach B

Using Models, continued

For example, the new function may be $y = x^2 + 2$. For a horizontal shift, a number will be added to or subtracted from the x-value before squaring, such as $y = (x - 2)^2$.]

- How can you tell if a transformation graph is a vertical or a horizontal shift of the quadratic function by examining how the function rule is applied? **[A vertical shift occurs when a number is added or subtracted after the value of x^2 is found; a horizontal shift occurs when a number is added to or subtracted from the value of x before squaring.]**
- Compare transformations of the quadratic function and the absolute-value function. Make a conjecture about similar transformations of any function. **[For any function $y = f(x)$, a vertical shift of the parent graph occurs when a number is added to or subtracted from the parent function; the transformed function has the form $y = f(x) + k$; a horizontal shift of the parent graph occurs when a number is added to or subtracted from the x-value before the function is evaluated; the transformed function has the form $y = f(x - k)$.]**

Follow-Up Demonstration (Optional)

To follow up and summarize this exploration, have students apply their conjectures to the other functions listed on page 230 of the Pupil's Edition. Use a graphics calculator to test the conjectures quickly.

On Their Own

Describe what you must do to the function rule $y = x^2$ in order to obtain a new function rule that represents each of the following:

A. a stretch **B.** a reflection **C.** a shift

[Sample: A. Multiply x^2 by some number, such as 2. B. To reflect through the x-axis, let $y = -x^2$. C. Change x^2 to $(x - a)^2$, where a is some number.]

On Their Own

Describe what you must do to the function rule $y = x^2$ in order to obtain a new function rule that represents each of the following:

A. a stretch **B.** a reflection **C.** a shift

[Sample: A. Multiply x^2 by some number, such as 2. B. To reflect through the x-axis, let $y = -x^2$. C. Change x^2 to $(x - a)^2$, where a is some number.]

Application

A common application of the quadratic function is area. For this application, the area of a square photo, including its frame, is a function of the length of the side of the square. Point out the following two ways to represent the area of the enlarged square:

1. Graph $y = x^2$ and the transformation $y = (x + 2)^2$ and note the similarities.
2. Make a table of values for $y = x^2$ and $y = (x + 2)^2$. The y-values will be the same in each table only if you shift the x-values for $y = x^2$ 2 columns to the left, as shown.

$y = x^2$

x	−3	−2	−1	0	1	2	3
y	9	4	1	0	1	4	9

$y = (x + 2)^2$

x	−3	−2	−1	0	1	2	3
y	1	0	1	4	9	16	25

Closing the Lesson

Summarize the lesson with the chart below. For each parent function, ask students to give an example of a vertical stretch, a reflection through the x-axis, and a horizontal shift.

Name	Parent function
linear	$y = x$
exponential	$y = 2^x$
quadratic	$y = x^2$
reciprocal	$y = \frac{1}{x}$
absolute value	$y = \|x\|$

Closing the Lesson

Summarize the lesson by restating each conclusion from the explorations.

1. A parent function is the most basic of a family of functions.
2. Examples of parent functions include the linear function, $y = x$; the exponential function, $y = 2^x$; the quadratic function, $y = x^2$; the reciprocal function, $y = \frac{1}{x}$; and the absolute-value function, $y = |x|$.
3. A transformation of the parent function may be a stretch, a reflection, or a shift.

ASSESS

Selected Answers

Odd-numbered Exercises 7–43 and odd-numbered Extra Practice

Assignment Guide

Core level students should complete Exercises 1–44.

Core-plus level students should complete Exercises 1–12, 14, 16–18, 20, and 22–47.

Error Analysis

Students may have difficulty with reciprocal functions. Point out that these functions follow the same pattern of stretching, reflecting, and shifting that every other function follows. Therefore, students should be able to predict that Exercise 24 is a horizontal shift.

Alternative Assessment

Using a Spreadsheet Have students use a spreadsheet to help them understand transformations. Ask them to enter x-values into column A of the spreadsheet. Use a domain of $-5 \leq x \leq 5$. Use the fill down feature of a spreadsheet to make the entries relative to the first entry. Have them enter the function in column B of the spreadsheet. Use column C for a stretch, column D for a reflection, and column E for a shift.

Compare the values generated by the spreadsheet with the values in the explorations. Change the function in column B to see a new table. Change the functions in columns C, D, or E to see new transformations.

As an extension for the spreadsheet program that has graphing capabilities, have the students compare the graphs of the transformation columns with the graph of column B.

RETEACH

Have students define parent function and describe how to transform the parent function. Then ask them to apply their description to the following exercises and to show a graph of the transformed function along with that of the parent function. Have them check their answers with a graphics calculator.

A. $y = |x - 3| + 2$ **B.** $y = (x + 2)^2 + 3$

[The parent function is the most basic of a family of functions. To graph a function of the form $f(x) = ag(x - h) + k$, stretch its parent function vertically by a units, and then translate the stretched graph horizontally by h units and vertically by k units.]

A.

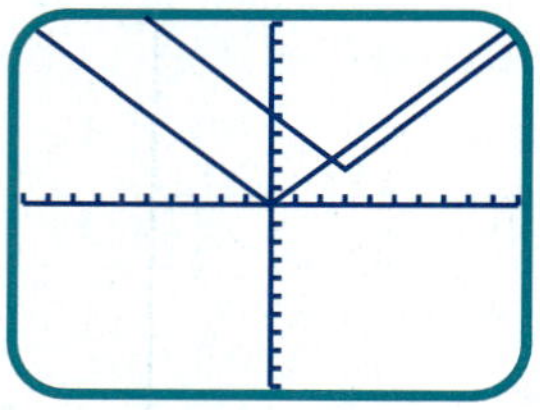

B.

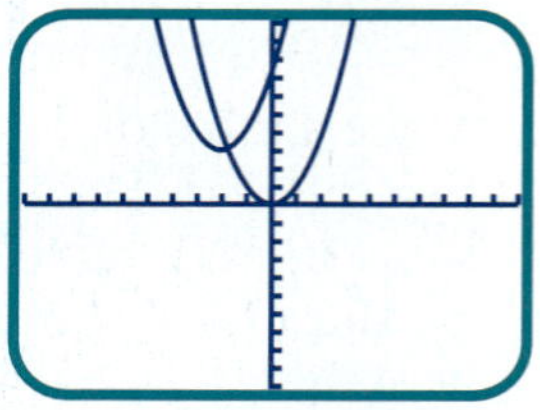

ENRICH

A graphing calculator can be used to explore the transformations of many functions. Ask students to explore the other functions available on a graphics calculator by entering the following: Y_1 = SIN (X) and Y_2 = SIN (X) + 2, or Y_1 = LOG (X) and Y_2 = 3LOG (X). Press Zoom 6 after any function if the graph is not shown in the current window. Ask students to sketch each graph and its parent graph carefully and to describe the type of transformation shown.

Interdisciplinary Connection

Computer Programming Real-world functions may be complex models with many variables. Computers can handle transformations of functions, but the computer programmer must include options for changing the function in the programs that are written.

NAME ______________________ CLASS ____________ DATE ____________

Practice
5.2 Exploring Transformations

Give the parent function for each graph. Then describe the transformation with the terms *stretch*, *reflect*, and *shift*.

1.

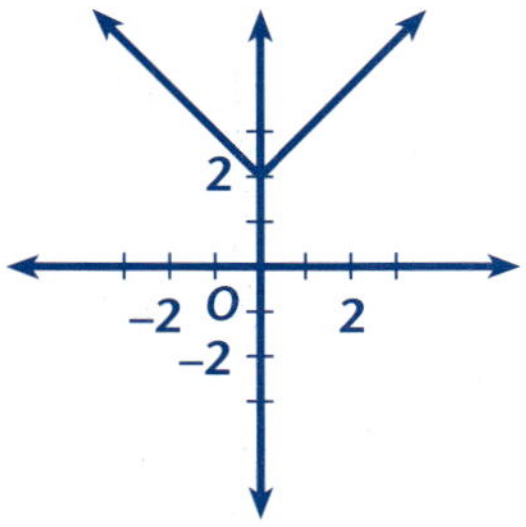

____

2.

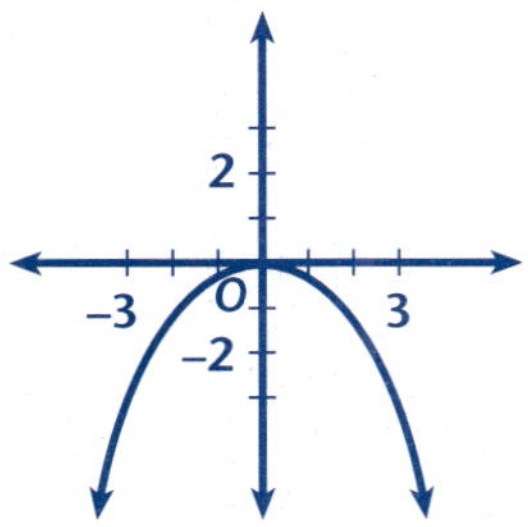

____

3.

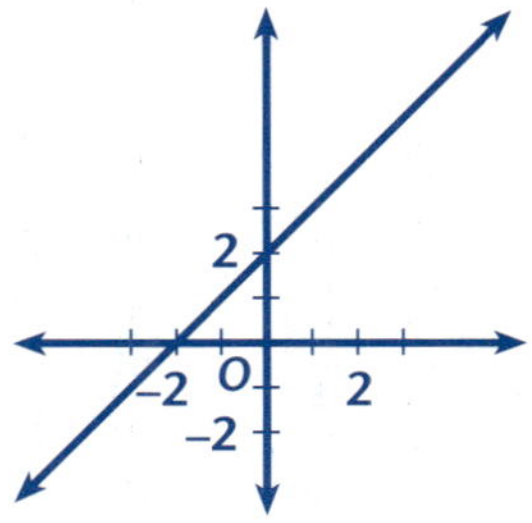

____

Graph each function. Then tell whether the graph of $y = x^2$ has been stretched, reflected, or shifted.

4. $y = x^2 - 1$

____

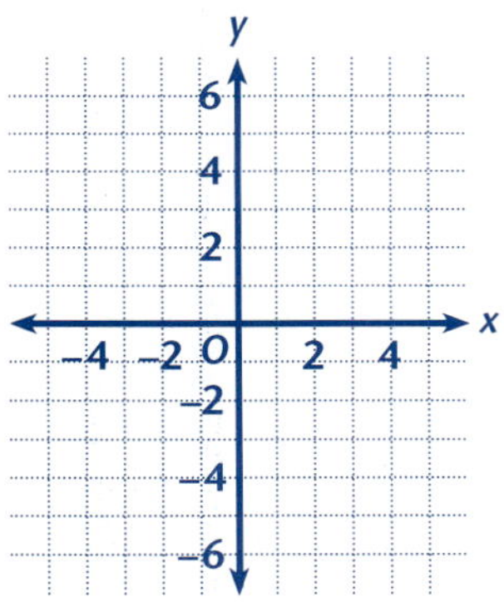

5. $y = 2x^2$

____

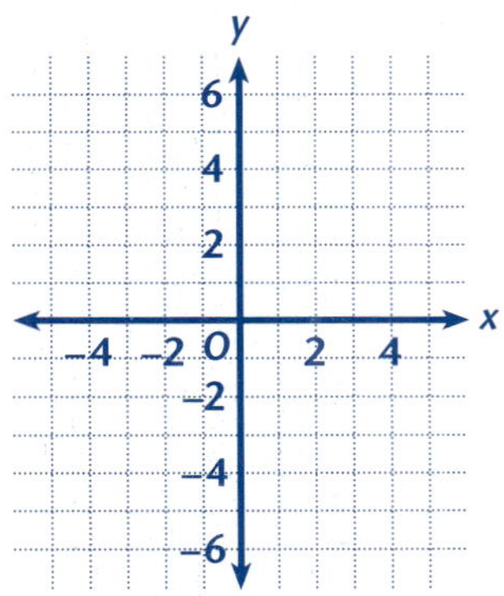

6. $y = x^2 + 2$

____

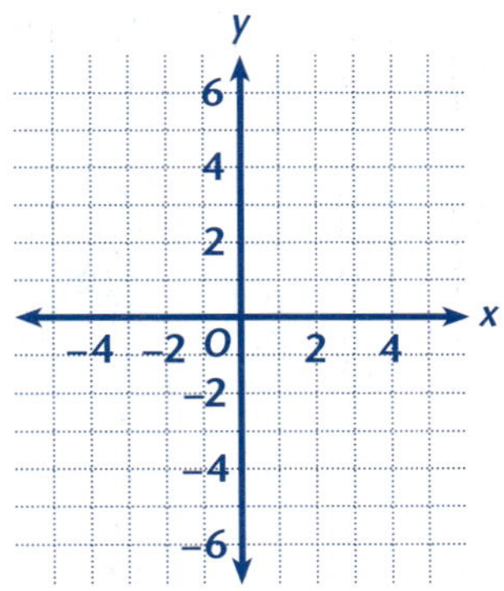

7. $y = -2x^2$

____

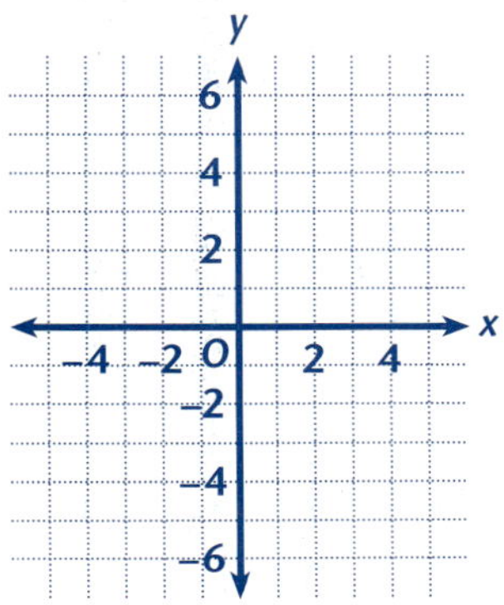

8. $y = (x + 1)^2$

____

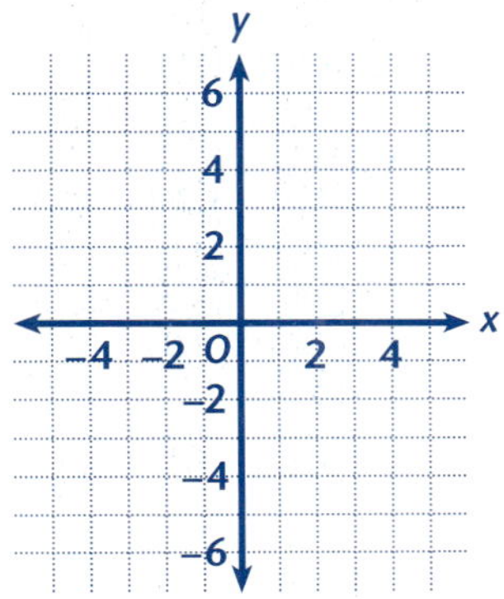

9. $y = (x - 1)^2$

____

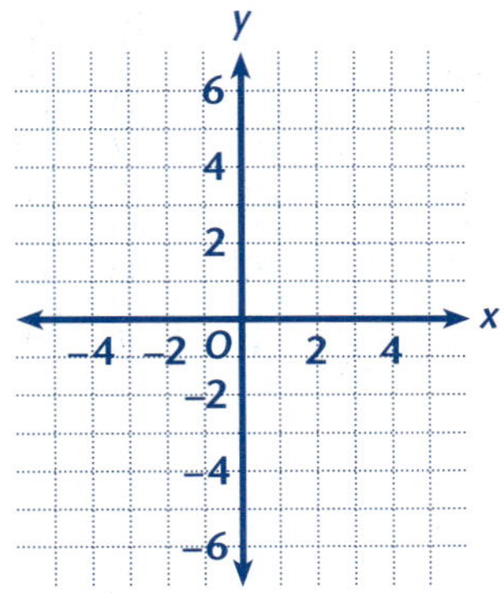

NAME ______________________ CLASS __________ DATE __________

Lesson Activity

5.2 Modeling CPI

The Consumer Price Index (CPI) is a measure of the average change in prices of goods and services over time. Currently, the CPI is based on the cost of an item or service during the fixed period from 1982–1984. The chart shows the average monthly price during 1993 compared with the average monthly price during the fixed period.

	Average monthly price—1993	Average monthly price—fixed period	Ratio (x)	CPI ($100x$)
Energy	\$104.20	\$ 99.90		
Food	\$140.90	\$ 99.40		
Shelter	\$155.70	\$ 99.10		
Medical	\$201.40	\$106.80		
Apparel	\$133.70	\$100.20		
Telephone	\$121.20	\$ 99.20		

One way to model the CPI with a function rule is to find the ratio of the average monthly price during a current year to the price during the fixed period. This ratio multiplied by 100 represents the CPI.

1. Complete the chart by finding the ratio for each category. (Round your answer to the nearest hundredth.)

2. Write a function rule for the CPI as a function of x.

____

3. Compare this function with the ratios from Exercise 1. What kind of transformation was applied? ______________________

4. Find the percent increase for apparel from the fixed period to 1993.

____

5. Find the percent increase for medical from the fixed period to 1993.

____

6. Use your answers from Exercises 4 and 5 to describe the general relationship between each ratio, x, and each percent increase.

____

7. The price of an item was \$100 during the fixed period. What is the average monthly price during 1993 if the ratio of average change in price is 1.75?

____

NAME ______________________ CLASS ____________ DATE ____________

Basic Skills Practice
Similar Figures, Part 1

Two or more figures that have the same shape but different sizes are called *similar figures*.
Triangle *ABC* and triangle *DEF* are similar figures.
This is written as $\Delta ABC \sim \Delta DEF$.

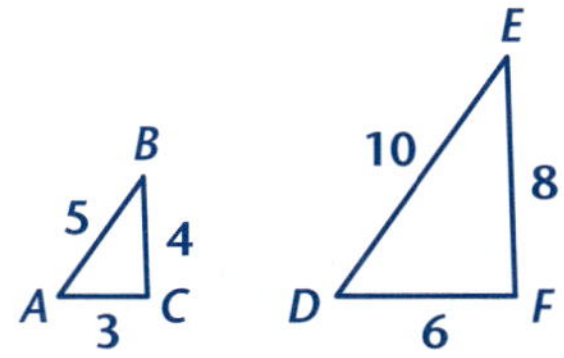

When two figures are similar,

- the corresponding angles are congruent.
  $\angle A \cong \angle D \quad \angle B \cong \angle E \quad \angle C \cong \angle F$
- the ratios of the lengths of the corresponding sides are equal.
  $\frac{AB}{DE} = \frac{5}{10} = \frac{1}{2} \qquad \frac{BC}{EF} = \frac{4}{8} = \frac{1}{2} \qquad \frac{AC}{DF} = \frac{3}{6} = \frac{1}{2}$

You can determine if two figures are similar by checking to see that they meet both conditions.

Example 1: Are rectangle *ABCD* and rectangle *EFGH* similar?

- Because both figures are rectangles, all angles are right angles, so they are all congruent.
- $\frac{AB}{EF} = \frac{2}{3} \qquad \frac{BC}{FG} = \frac{4}{6} = \frac{2}{3} \qquad \frac{CD}{GH} = \frac{2}{3} \qquad \frac{DA}{HE} = \frac{4}{6} = \frac{2}{3}$

The ratios of the corresponding sides are equal, so rectangle *ABCD* ~ rectangle *EFGH*.

You can find the length of the missing side in similar figures by using proportions.

Example 2: The two flags are similar. What is the length of *s*?

Write a proportion. → $\frac{2}{3} = \frac{4}{s}$

Solve the proportion. → $2s = 12$

$s = 6$

Determine if the pairs of figures are similar. Write *yes* or *no* in the blank spaces.

1.

____

2.

____

Each pair of figures is similar. Find the length of the missing side.

3.

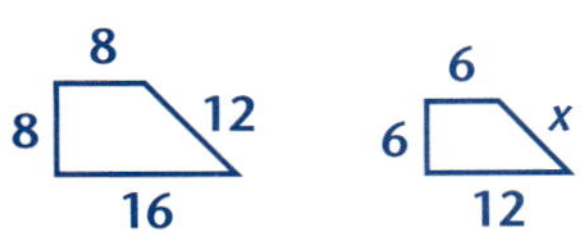

____

4.

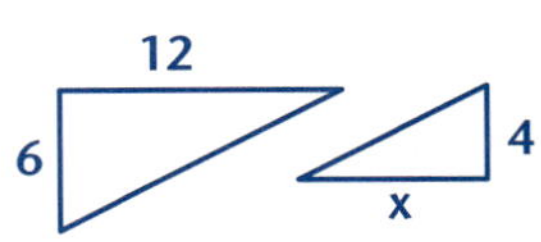

____

LESSON 5.3
Stretches

PREPARE

Pacing Guide
2 Days

Objectives and Overview
- Describe the effect of a stretch on the graph of a function when the coefficient is greater than or equal to 1 and when it is between 0 and 1.
- Identify the coefficient to determine the amount of stretch.

Students review vertical stretch from Lesson 5.2 and define scale factor. Stretches are applied to absolute-value functions, quadratic functions, and reciprocal functions, whose graphs are hyperbolas. Stretches with scale factors between 0 and 1 are examined, as well as stretches with scale factors greater than 1.

Vocabulary
scale factor
stretch
hyperbola
asymptote

Assessing Prior Knowledge

Evaluate each of the following for $f(4)$:

1.	$f(x) = x^2$	[16]
2.	$f(x) = x^2 - 12$	[4]
3.	$f(x) = 3x^2$	[48]
4.	$f(x) = 3x^2 - 12$	[36]
5.	$f(x) = 3(x^2 - 12)$	[12]
6.	$f(x) = 3(x^2 - 12)^2$	[48]

Approach A: Using the Book

Opening Discussion
Is the graph of $y = 2x^2$ stretched vertically from the graph of $y = x^2$? Review the concept of stretch from Lesson 5.2. Does a stretch make a graph lie between the parent function and the x-axis? What number values for a stretch make the graph stretch vertically so that the y-values of the transformed graph are greater than the y-values of the parent graph?

Brainstorming
Have the class start by brainstorming things that get stretched. For example, human muscles are stretched. Could stretching include being made smaller? Ask students if they think functions could be stretched.

Stretches
The falling of water over a waterfall can be modeled by the function $S = 16t^2$ because of the laws of physics, including the law of gravity. Point out that since $S = t^2$ is the parent function for the waterfall model, its graph

TEACH

Approach B: Using Models

Opening Discussion
How can you tell what the shape of the graph of $y = 2x^2$ is without graphing it? Is the graph thinner or wider than the graph of $y = x^2$? What kind of transformation from Lesson 5.2 is used? Does a stretch make a graph thinner or wider? What number values for a stretch make the graph stretch vertically? What values cause the graph to lie between the parent graph and the x-axis?

Brainstorming
Have the class start by brainstorming things that get stretched. For example, human muscles are stretched, and rubber bands are stretched. How does a stretched object look compared with the original object? Could stretching include being made smaller? Ask students if they think functions could be stretched.

Using a Graphics Calculator
A graphics calculator is very useful in the study of functions and their transformations. The functions

of height versus time will be a parabola. Students should make the quick connection that the parent graph will be stretched. The y-values will be greater for the same domain values. Ask students to conjecture the values for which a stretch will make the y-values increase or decrease. **[If the scale factor is larger than 1, the y-values will increase for the same domain values. If the scale factor is less than 1, the y-values will decrease.]**

Ongoing Assessment, page 234

256 feet

Definition of *Scale Factor*

The scale factor is defined as the coefficient of a function that makes a transformation a stretch. For the function $s = 16t^2$, 16 is the scale factor. This is an important concept because if you multiply the y-values of the parent graph by the corresponding scale factor, you can easily get the points of the transformed graph. Emphasize that when the scale factor, a, is greater than 0, $y = a \cdot f(x)$ is stretched vertically from the graph of its parent function, $y = f(x)$, by the scale factor, a.

Ongoing Assessment, page 235

In the waterfall problem, $g(5)$ represents the height of the waterfall if it takes the water 5 seconds to fall from the top of the waterfall to the bottom.

Example 1

In this example, the scale factor of a transformation is $\frac{1}{2}$, and it is used to sketch the graph of $y = \frac{1}{2}|x|$.

Point out that if a stretch has a scale factor between 0 and 1, then the stretch will decrease the y-values by that factor, and the graph will become broader (the y-values are one-half as great for the same domain values). The graph is between the parent function and the x-axis.

continued on top of next page

and their transformations are entered under the Y= menu, an appropriate window is set to see a complete graph, and then the GRAPH key is pressed to see the graphs. The power of the calculator lies in the speed with which questions about transformations can be answered.

Working With a Graphics Calculator

Have students graph each of the functions below and give the relationship between each graph and its parent graph. What is the vertex of each graph? What is the axis of symmetry of each graph?

a. $y = x^2$ **b.** $y = 2x^2$ **c.** $y = 5x^2$
d. $y = 0.5x^2$ **e.** $y = 0.05x^2$

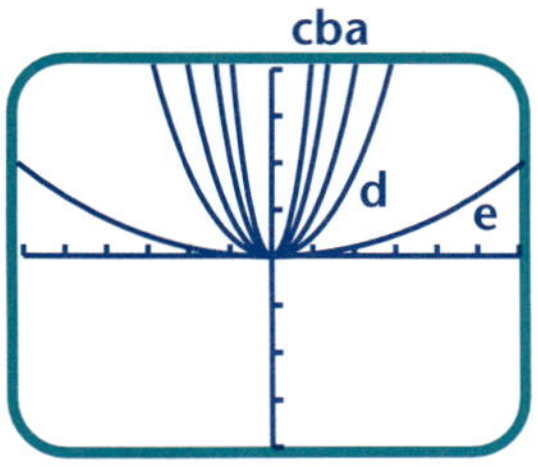

[Each graph is a vertical stretch of the parent graph $y = x^2$. Graphs b and c are stretched vertically from the parent graph, and graphs d and e are between the parent graph and the x-axis. Each has a vertex at (0, 0), and the y-axis is the axis of symmetry for each.]

The Viewing Window

The viewing window of a graphics calculator may vary for each graph, but certain elements are important to show a complete graph. First, all important features of the function must be visible on the screen. For a parabola, the vertex and points on both sides of the axis of symmetry must show.

Second, other important points should show. For more complicated functions, this may mean points where the graph turns and changes direction.

Third, it may be important to have a "square screen," which means that the calculator screen simulates normal grid paper. On some

continued on bottom of next page

Approach A Using the Book, continued

Ongoing Assessment, page 236

Try This

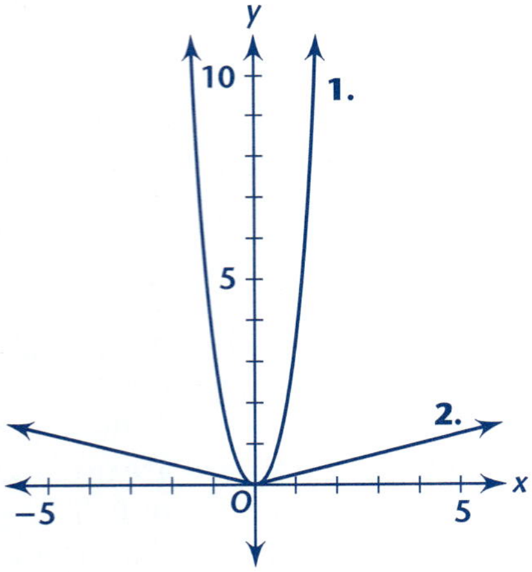

1. $f(x) = 4x^2$ **2.** $g(x) = \frac{|x|}{4}$

Additional Example 1

Use the parent function to sketch the graph of $y = \frac{1}{4}x^2$.

x	$\frac{1}{4}x^2$	y
−4	$\frac{1}{4}(-4)^2$	4
−2	$\frac{1}{4}(-2)^2$	1
0	$\frac{1}{4}(0)^2$	0
2	$\frac{1}{4}(2)^2$	1
4	$\frac{1}{4}(4)^2$	4

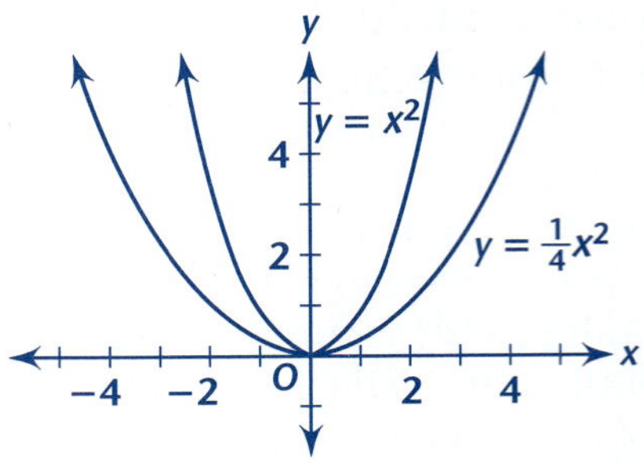

TEACH

Approach B Using Models, continued

calculators, the Zoom 5 command accomplishes this. Ask students to give the viewing window coordinates for the previous graphs. **[Sample: Xmin −6, Xmax 6, Ymin −4, and Ymax 4 give a square screen.]**

Scale Factor

A graphics calculator can be used effectively to study the scale factor between a transformed graph and its parent graph. In general, when $a > 0$, $y = a \cdot f(x)$ is stretched vertically from the graph of its parent function, $y = f(x)$, by the scale factor, a. If $0 < a < 1$, the transformed graph is between the parent function and the x-axis. Ask students to give the scale factor for the graphs in Working With a Graphics Calculator on the previous page. **[a.** 1 **b.** 2 **c.** 5 **d.** 0.5 **e.** 0.05**]**

Cooperative Learning

Table Features Have students work in groups of four to study the tables generated from transformed graphs. Ask one student to choose a parent function and list several transformed functions. Have that student enter the graphs into the Y= menu of a graphics calculator with an appropriate viewing window. Use Zoom 8 (ZInteger) to see integers for the x-coordinates.

Ask the second student to use the table command of the calculator to make a list of values for each of the functions.

Ask the third student to make conjectures, based on the table values, about the relationship between the transformed function and the parent function. Are the graphs stretched vertically, or are they between the parent function and the x-axis?

Ask all students to discuss the conjectures and to verify them with the graphs on the calculator screen. Ask the fourth student to record the group's activities in a complete report to be presented to the class.

Example 2

In this example, the graph of the reciprocal function, $y = \frac{1}{x}$, is stretched vertically by a scale factor of 6. Point out that the graph of the reciprocal function forms a curve known as a *hyperbola*. Emphasize that the reciprocal parent function does not have 0 in its domain because division by 0 is undefined.

Additional Example 2

Sketch the graph of $y = \frac{1}{3x}$. How does it compare with its parent function?

x	$\frac{1}{3}\left(\frac{1}{x}\right)$	y
$-\frac{1}{9}$	$\frac{1}{3}\left(\frac{1}{-\frac{1}{9}}\right)$	-3
$-\frac{1}{3}$	$\frac{1}{3}\left(\frac{1}{-\frac{1}{3}}\right)$	-1
$\frac{1}{3}$	$\frac{1}{3}\left(\frac{1}{\frac{1}{3}}\right)$	1
$\frac{1}{9}$	$\frac{1}{3}\left(\frac{1}{\frac{1}{9}}\right)$	3

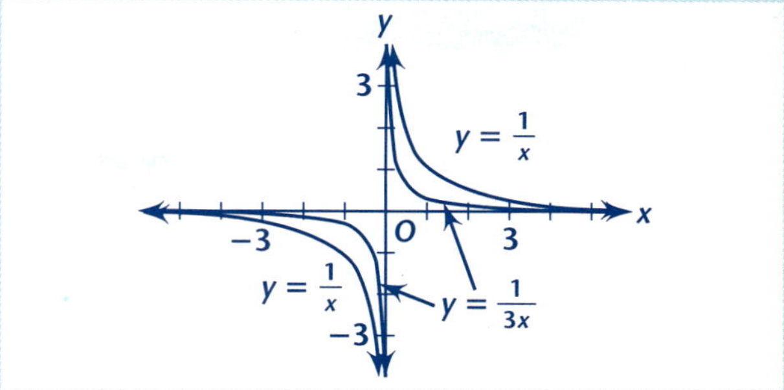

[It is stretched vertically by a factor of $\frac{1}{3}$.]

The Definition of *Hyperbola*

The hyperbola is a curve formed by the graph of the reciprocal function, $y = \frac{1}{x}$. The graph approaches the x- and y-axes but never crosses them because 0 is not in the domain of the function (division by zero is undefined). Because the graph approaches a line but never crosses it, the graph has *asymptotes*. The x- and y-axes are asymptotes for the graph of the function $y = \frac{1}{x}$. If the hyperbola is transformed by a shift, the asymptotes are also shifted by the same amount.

continued on top of next page

Hyperbolas and Asymptotes

Hyperbolas are graphs of functions of the form $y = a \cdot \frac{1}{x}$, with a scale factor of a and a parent function of $y = \frac{1}{x}$. The parent graph of the hyperbola has the x- and y-axes as asymptotes, which are lines that the graph approaches but never crosses. Have students study the graphs of $y = \frac{2}{x}$ and $y = \frac{1}{2x}$ and compare them with the parent graph. Are they vertical stretches of the parent graph? **[yes]** Do these graphs have the same asymptotes? **[yes]** What is an application of hyperbolas that students may have seen in the past? **[Sample answer: inverse variation]** Fill in the missing coordinates by tracing the graph of $y = \frac{2}{x}$: (25, ?), (100, ?), (1000, ?). **[(25, 0.08); (100, −0.02); (1000, 0.002)]** What is the greatest value of x for which the graph touches the x-axis? **[None; the graph never touches the x-axis.]**

Inclusion Strategy

Enlarge each of the following functions and their solutions on separate sheets of paper that can be used as flashcards:

Function	**Solution**
$y = \lvert x \rvert + 4$	not a stretch
$y = \frac{x^2}{4}$	stretch; scale factor of $\frac{1}{4}$
$y = \frac{1}{2x}$	stretch; scale factor of $\frac{1}{2}$

Mix up the flashcards and attach them to the chalkboard. Ask students to match a function flashcard with its solution flashcard. They should verify their answers by graphing.

Teaching Tip

Using Technology Students may notice a discrepancy between their graphs after they are stretched and a calculator graph. Point out that the standard window on a graphics calculator does not have square grids like graph paper does. The window needs to be scaled with a

continued on bottom of next page

Approach A

Using the Book, continued

Ongoing Assessment, page 237

(1, 2)

Critical Thinking, page 237

Assuming $a > 0$, in $y = ax^2$, as a increases, the graph becomes narrower, or moves *closer to* the positive y-axis. In $y = a\left(\frac{1}{x}\right)$, as a increases, the graph moves away from the origin, or moves *away from* both the x- and y-axis.

Example 3

In this example, students are asked to identify a stretch by looking at the function rule.

Point out that if a function is multiplied by a scale factor, then the transformation is a stretch, as is the case in parts **A** and **C** of this example. If a value is added to the function, then the transformation is not a stretch (it is a shift), as is the case in part **B**.

Additional Example 3

Which functions are stretches of the parent graph?

A. $y = \frac{5}{x}$ **B.** $y = \frac{1}{2}x^2$ **C.** $y = |x| - 3$

[A and B are, as can be seen in the graphs below.]

A.

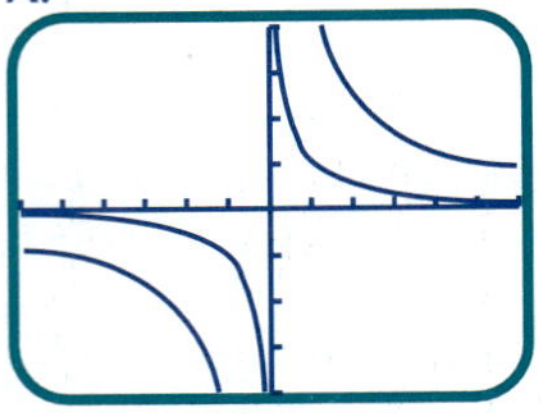

B.

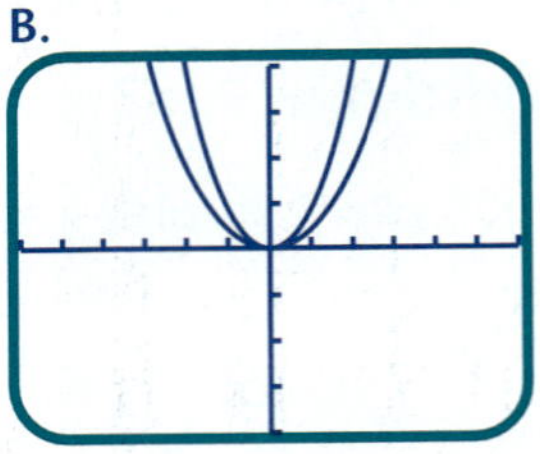

C.

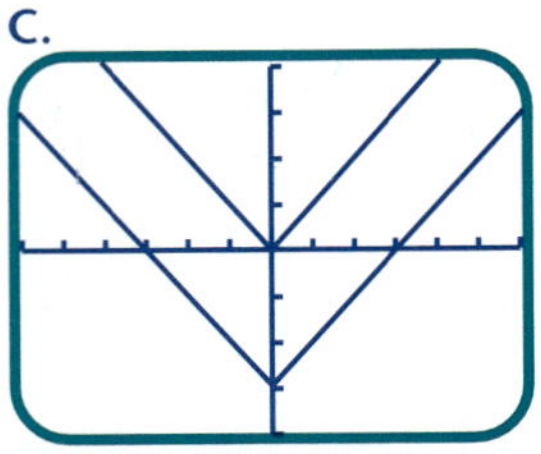

TEACH

Approach B

Using Models, continued

3 to 2 ratio for the x- and y-values. For example, Xmin $= -6$, Xmax $= 6$, Ymin $= -4$, and Ymax $= 4$ sets a square window for a graph. Any graph with this window will look similar to one done on graph paper.

Closing the Lesson

Summarize how to graph a vertical stretch with the following statements:

1. Identify the parent function. (A list is given on page 231 of the Pupil's Edition.)
2. Determine the scale factor, a.

Ask students to give the parent function and the scale factor for each of the following functions:

A. $y = 4x^2$ **B.** $y = \frac{1}{3}|x|$ **C.** $y = \frac{3}{x}$

[A. $y = x^2$; 4 **B.** $y = |x|$; $\frac{1}{3}$ **C.** $y = \frac{1}{x}$; 3**]**

ASSESS

Selected Answers

Odd-numbered Exercises 7–41 and odd-numbered Extra Practice

Assignment Guide

Core level students should complete Exercises 1–44.

Core-plus level students should complete Exercises 1–44.

Error Analysis

Exercises 22 and 23 ask students to write an equation for a graphed function. The following steps may be helpful: (1) Identify the parent function from the graph. **[quadratic]** (2) Give the function rule for the parent function. [$y = x^2$] (3) Substitute the x-coordinate of the given point into the parent function rule. **[For Exercise 22, the given point is (2, 8). Substitute 2 into $y = x^2$ to get the y-coordinate; $y = (2)^2 = 4$.]** (4) Compare the y-coordinate found in Step 3 with the y-coordinate of the given point to determine the scale factor. **[Since $8 = 2 \times 4$, the scale factor is 2.]** (5) Rewrite the function as $y = a \cdot f(x)$, with the scale factor a. [$y = 2x^2$]

Alternative Assessment

Portfolio Activity Ask students to make their own library of functions that they can refer to as they use the textbook. Have students make a chart with the following headings and then fill in the chart with function rules and their graphs as they learn them.

Function Rule: Parent	Graph	Function Rule: Stretch	Graph

RETEACH

Have students work with a partner to find which functions are stretches of the parent graph. Ask them to identify the parent graph and the scale factor. Have them check each other's work.

A. $y = \frac{|x|}{5}$ **B.** $y = 4|x|$ **C.** $y = x^2 - 4$

D. $y = \frac{x^2}{4}$ **E.** $y = \frac{1}{x} + 2$

[A. $y = |x|, \frac{1}{5}$; B. $y = |x|$, 4; D. $y = x^2, \frac{1}{4}$]

ENRICH

The stretch transformation can be applied to any function. Have students use a graphics calculator to graph each of the following parent functions. Then apply a vertical stretch of 2.

1. $y = \sqrt{8 - x^2}$ **2.** $y = -\sqrt{8 - x^2}$

In a graphics calculator, enter function **1** as $Y_1 = \sqrt{(8 - x^2)}$ and function **2** as $Y_2 = -\sqrt{(8 - x^2)}$. A vertical stretch of 2 would be entered as $Y_3 = 2\sqrt{(8 - x^2)}$ and $Y_4 = -2\sqrt{(8 - x^2)}$.

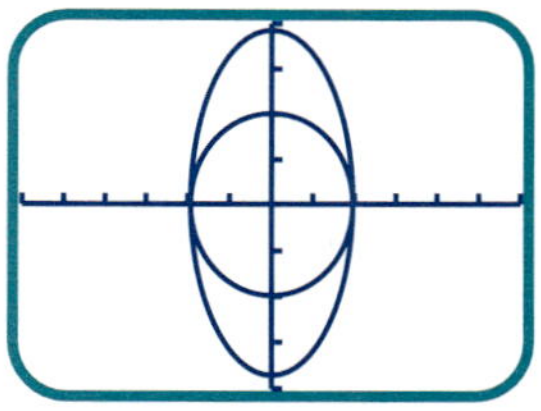

NAME ______________________ CLASS ____________ DATE ____________

Practice

5.3 Stretches

Given the function $y = 2|x|$,

1. identify the parent function. ______________

2. sketch the graphs of the function and the parent function on the axes at the right.

3. tell how the graph of the parent function was changed by the 2. ______________

____

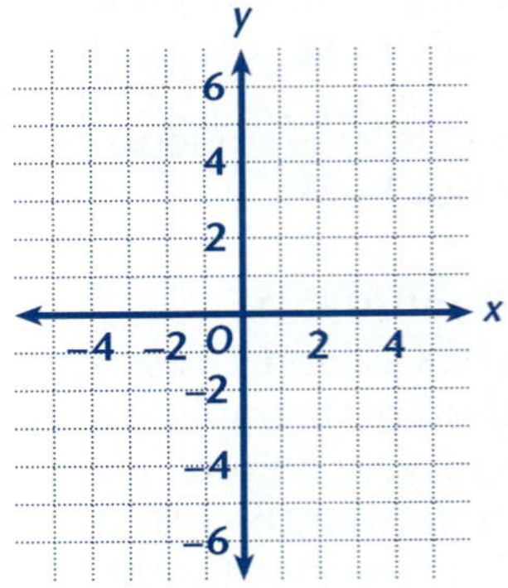

Given the function $y = \frac{3}{x}$,

4. identify the parent function. ______________

5. sketch the graphs of the function and the parent function on the axes at the right.

6. tell how the graph of the parent function was changed by the 3. ______________

____

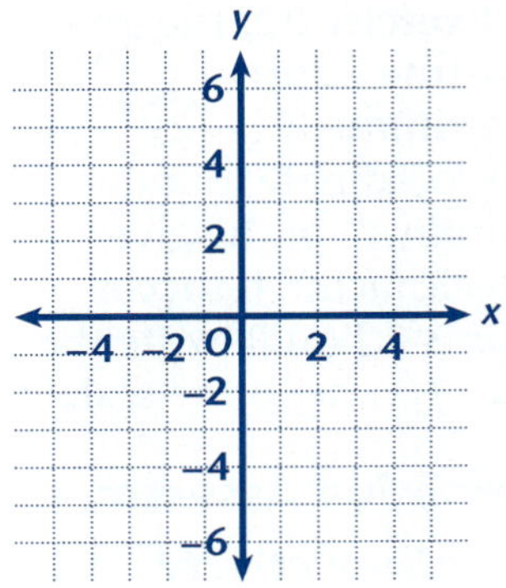

Tell whether each function results in a stretch of the parent function.

7. $y = \frac{x^2}{4}$ ______________

8. $y = \frac{8}{x}$ ______________

9. $y = |x| - \frac{1}{2}$ ______________

Write an equation for each graph.

10.

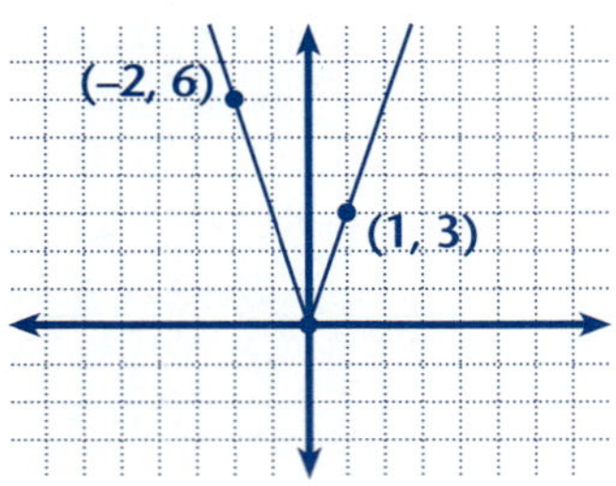

____

11.

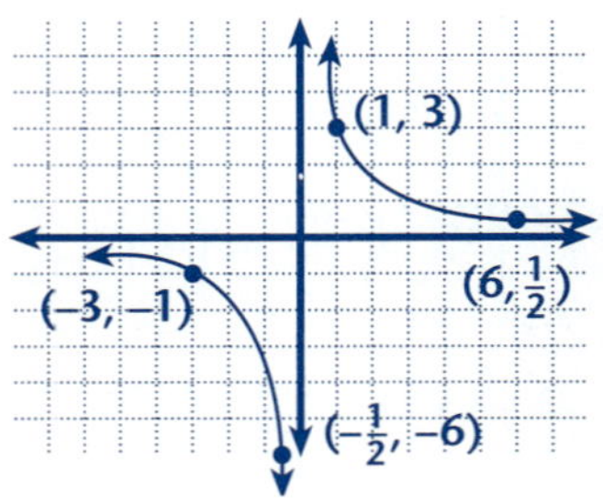

____

NAME ______________________ CLASS ____________ DATE ____________

Lesson Activity

5.3 Comparing Pricing Structures

King's Kid offers to cater a house party for a fixed fee of $50 and an additional charge of $4 for each person. For comparable service and menu items, Lady K offers to cater a house party for a fixed charge of $75 and an additional charge of $3 per person.

1. Write function rules for the two caterers to compare price and number of people.

____

2. Use a graphics calculator to graph the functions on the same set of axes. Compare the pricing structures for the two caterers. Which caterer offers the better deal?

____

To entice customers, King's Kid decides to reduce its charge per person by 25%. When Lady K learns about King's Kid new prices, they immediately reduce their charge per person by $33\frac{1}{3}$%.

3. Write function rules to model the new pricing structure for each caterer.

____

4. Graph King's Kid new pricing structure as a function of the number of people. Then graph Lady K's new pricing structure on the same axes. Which caterer offers a better deal after the changes?

____

5. Compare the points of intersection of the graphs before and after the changes in pricing.

____

6. Compare the graphs of King's Kid's pricing structure before and after the changes. Then compare the graphs of Lady L's pricing structure before and after the changes. Use transformations to explain any pattern(s) you observe.

____

____

NAME ______________________ CLASS ____________ DATE ____________

Basic Skills Practice
Similar Figures, Part 2

Read each question and circle the best answer.

In Exercises 1–4, each pair of figures is similar. Find the length of the missing side.

1.

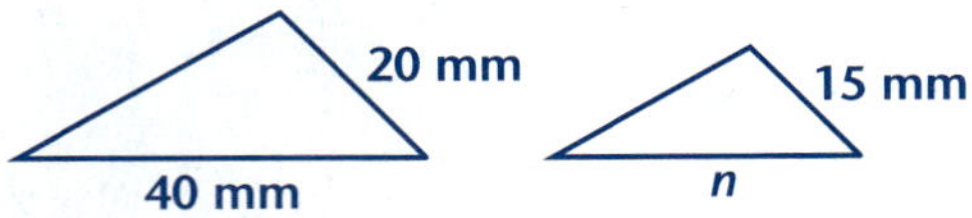

A $n = 20$ mm
B $n = 25$ mm
C $n = 30$ mm
D $n = 35$ mm

2.

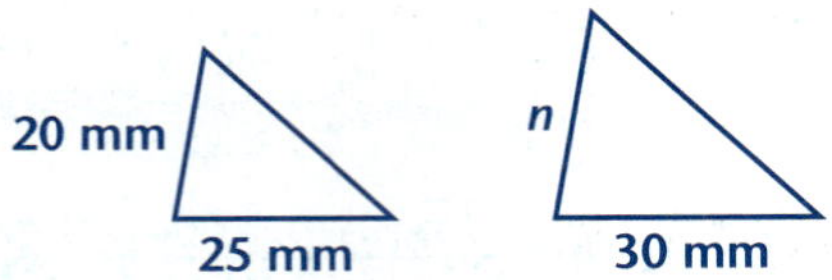

F $n = 22$ mm
G $n = 24$ mm
H $n = 26$ mm
I $n = 28$ mm

3.

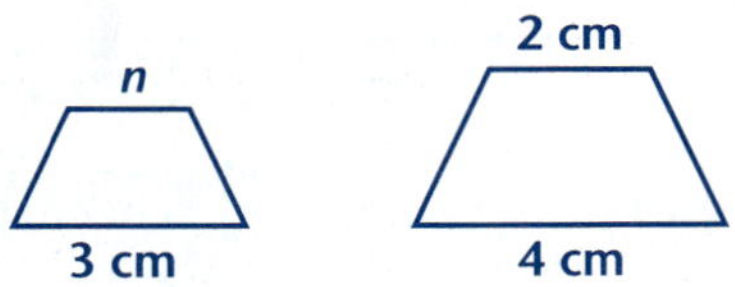

A $n = 1.0$ cm
B $n = 1.5$ cm
C $n = 2.0$ cm
D $n = 2.5$ cm

4.

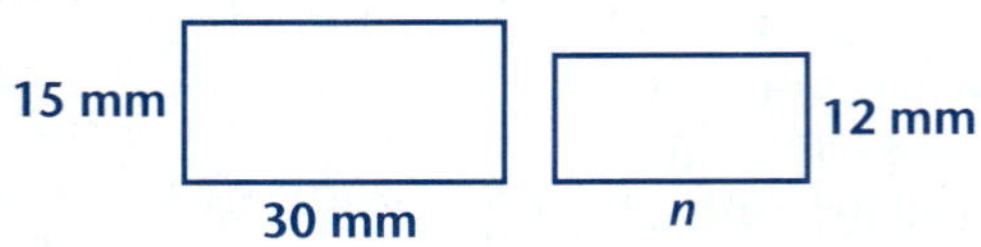

F $n = 26$ mm
G $n = 24$ mm
H $n = 22$ mm
I $n = 20$ mm

In Exercises 5–9, $\triangle ABC \sim \triangle DEF$.

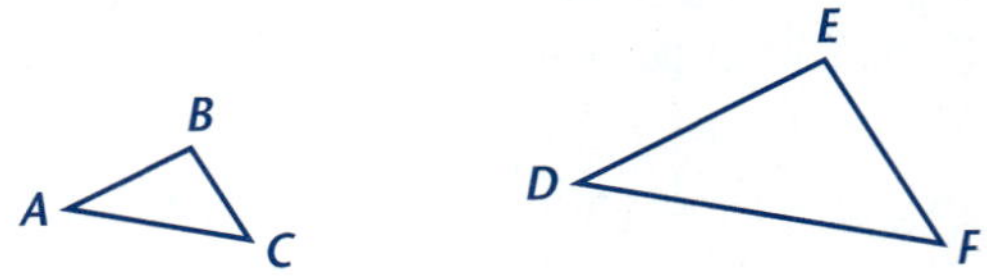

5. $\angle A \cong$ ________

A $\angle B$
B $\angle D$
C $\angle E$
D $\angle F$

6. $\angle B \cong$ ________

F $\angle F$
G $\angle E$
H $\angle D$
I $\angle B$

7. $\frac{AB}{DE} = \frac{AC}{n}$

A $n = EF$
B $n = DF$
C $n = DE$
D $n = BC$

8. Which side corresponds to $\overline{BC}$?

F $\overline{BA}$
G $\overline{CA}$
H $\overline{EF}$
I $\overline{DF}$

9. If $m\angle F = 35°$, what is $m\angle C$?

A 25°
B 30°
C 33°
D 35°

LESSON 5.4
Reflections

PREPARE

Pacing Guide
2 Days

Objectives and Overview
- Understand the connection between the reflection of a graph and the opposite or negative of a function.
- Identify the minimum or maximum value for absolute-value and quadratic functions.

Students explore reflections by comparing tables for graphs and learn how to reflect a function both vertically and horizontally. They also learn how to identify the minimum or maximum value for both absolute-value and quadratic functions.

Vocabulary
vertical reflection
horizontal reflection

Assessing Prior Knowledge

1. Graph $f(x) = x^2 + 1$.
2. Graph $g(x) = -(x^2 + 1)$.

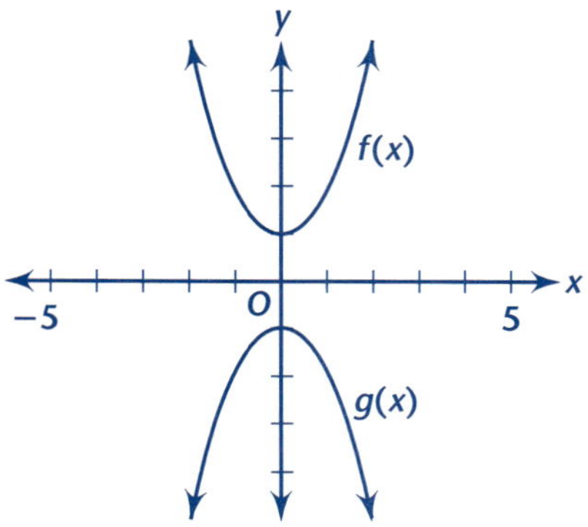

Approach A: Using the Book

Opening Discussion
Ask students to describe their image in a mirror. Are their right hands still on the right in the image? Do they look different in a mirror?

Brainstorming
Have the class practice reflecting geometric shapes through a reflection line. What are some things to consider when reflecting a geometric shape? Is the shape of the image identical to the original? Can functions be reflected also?

Exploration: A Reflection
In this exploration, students graph parent functions and then reflect them through the x-axis. Then they generalize the relationship between the graph of a function and the graph of its reflection. Point out that *all* reflections of $y = f(x)$ through the x-axis are of the form $y = -f(x)$.

continued on top of next page

TEACH

Approach B: Using Models

Opening Discussion
Ask students to describe an image in a mirror. Is the orientation of the object the same? Is the right side of the object still on the right in the image? Do objects look different in a mirror?

Brainstorming
Have the class practice reflecting objects over a vertical and a horizontal line. What are some things to consider when reflecting a geometric shape? Is the shape of the image identical to the original? Can functions be reflected also?

Using a Reflection Tool
Use reflection tools to provide a concrete foundation for doing reflections. Instruct students on how to use the tools. First graph the function, and then place the tool on the reflecting line (either axis in this case). Have them look through the reflection tool to see the reflected image of the original graph. Ask students to look at the correct

continued on bottom of next page

Approach A

Using the Book, continued

Ongoing Assessment, page 241

7. $y = -f(x)$

Teaching Tip

You may want to stress the conclusion from the exploration. For *all* functions $y = f(x)$, a reflection through the x-axis is of the form $y = -f(x)$, and a reflection through the y-axis is of the form $y = f(-x)$. The graph of a function reflected through the x-axis is found by using the same x-values and the opposite of every y-value of the original graph and then regraphing the points.

Critical Thinking, page 242

The graphs are lines through the origin, perpendicular to each other. The transformation is a reflection through the x- or y-axis. The combined graphs are the same as the graph of $y = |x|$ and its reflection through the x-axis.

Example 1

In this example, students graph functions of the form $y = -f(x)$ and then describe the parent function of each graph.

In part **B**, a vertical stretch is also applied to the parent function. Point out that if more than one transformation is applied to a parent function, then the reflection transformation is usually applied last. Emphasize that the minimum point of each parent function becomes the maximum point of the reflected image.

Additional Example 1

Describe each graph.

A. $y = -\frac{1}{x}$ **B.** $y = -\frac{x^2}{2}$

TEACH

Approach B

Using Models, continued

angle so that the figure is not distorted. Trace the reflected graph over the image given in the reflection tool.

Working With Images

Have students graph each parent function and its image reflected through both the x-axis and the y-axis by using a reflection tool.

A. $y = -x^2$ **B.** $y = -2^x$

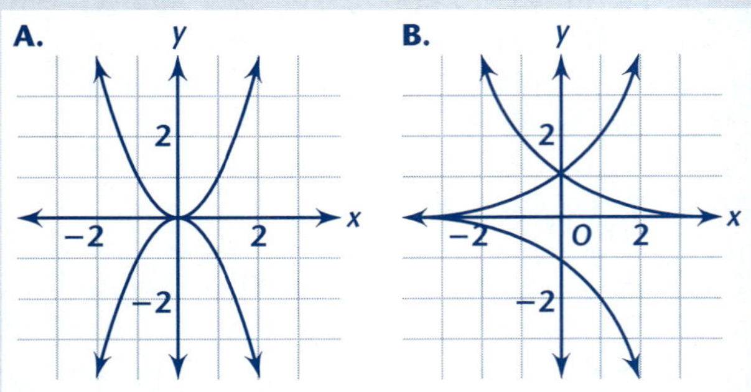

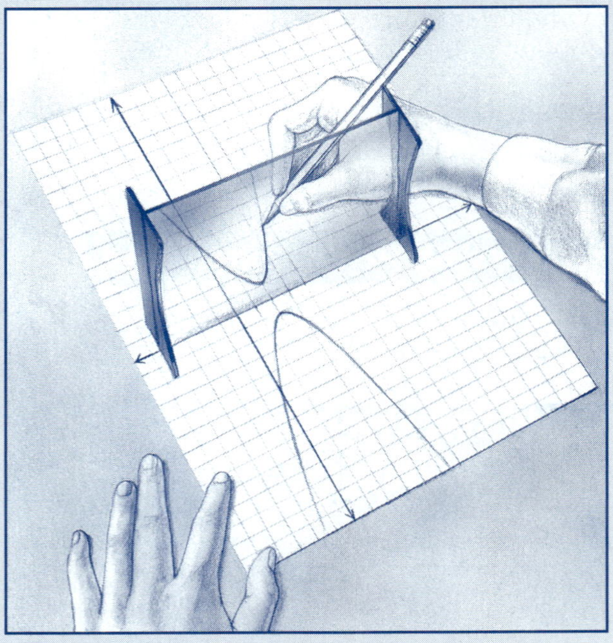

A reflection tool provides an interesting way of demonstrating reflection transformations.

[**A. The parent function is $y = \frac{1}{x}$, the reciprocal function. The y-value of every point on the graph of the parent function changes to its opposite. Thus, the graph of $y = -\frac{1}{x}$ is a reflection of the graph of $y = \frac{1}{x}$ through the x-axis.**

B. The parent function is $y = x^2$. First every point on the graph of the parent function is stretched vertically by a scale factor of $\frac{1}{2}$. The negative sign in $y = -\frac{x^2}{2}$ indicates a reflection of the graph of $y = \frac{x^2}{2}$ through the x-axis.]

Vertical Reflection

Emphasize that a vertical reflection, which is a reflection through the x-axis, occurs when each point (a, b) of the original function is replaced by the point $(a, -b)$ and then graphed. Using function notation, $y = -f(x)$ is a vertical reflection of $y = f(x)$ through the x-axis.

Use **Transparency 20** for discussion.

Horizontal Reflection

Emphasize that a horizontal reflection, a reflection through the y-axis, occurs when each point (a, b) of the original function is replaced by the point $(-a, b)$ and then graphed. Using function notation, $y = f(-x)$ is a vertical reflection of $y = f(x)$ through the y-axis.

Example 2

In this example, functions are identified as vertical reflections of parent functions.

Point out that first the parent function should be identified and a table of values should be made for the parent function, if necessary. Then the rule for the parent function should be replaced with its opposite. The new table of values should have opposite y-values for each point corresponding to a point on the parent function.

Using Technology

A graphics calculator can also be used to identify which graphs are vertical reflections of a parent

continued on top of next page

Teaching Tip

Students may notice that for problem **A** in Working With Images, the reflection of the parent function through the y-axis is the same as the original parent function. Point out that if the y-axis is the axis of symmetry for the parent function, its reflection through the y-axis gives the same graph. Have students explain why this is the case for the parent function $y = x^2$. [**A horizontal reflection replaces a with $-a$ for each point (a, b). Since $a^2 = (-a)^2$, the points are the same.**]

Vertical Reflections

Encourage students to use a reflection tool to find the coordinates of the image points for a vertical reflection through the x-axis for each of the following graphs. Have students complete the tables shown below each graph. Have them make a generalization about how each point (a, b) of the original graph is transformed.

A.

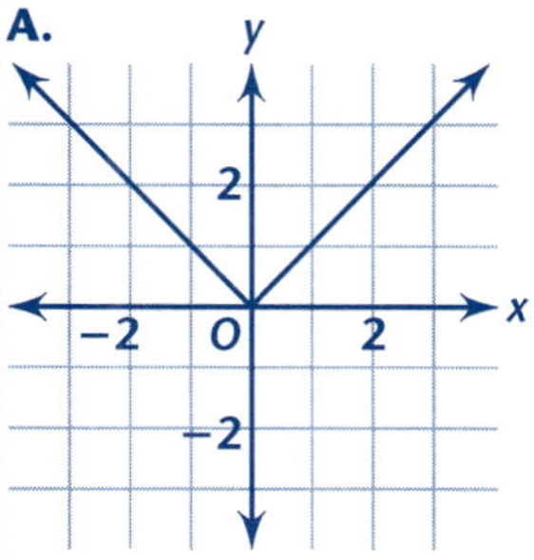

x	y_1	
−3	3	[−3]
−2	2	[−2]
−1	1	[−1]
0	0	[0]
1	1	[−1]
2	2	[−2]
3	3	[−3]

B.

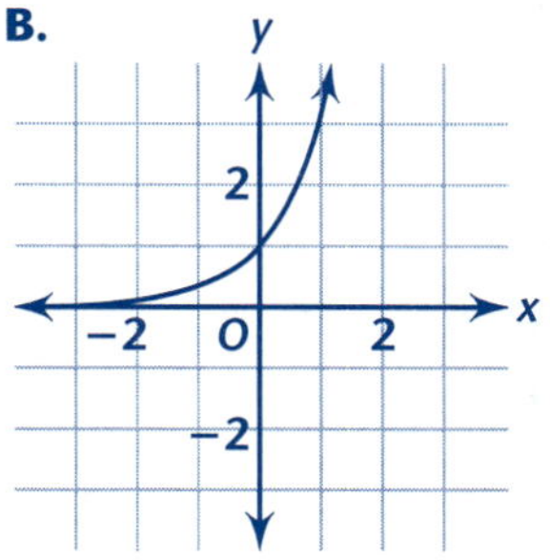

x	y_1	
−3	0.04	[−0.04]
−2	0.1	[−0.1]
−1	0.3	[−0.3]
0	1	[−1]
1	3	[−3]
2	9	[−9]
3	27	[−27]

[**Each point (a, b) is transformed into the point $(a, -b)$.**]

continued on bottom of next page

Approach A

Using the Book, continued

function. Have students first identify the parent function. Then graph both the parent function and the transformed function. Students should be able to visually determine whether one graph is a reflection of the other through the x-axis.

Additional Example 2

Identify which graphs are vertical reflections of parent functions.

A. $y = -x^2$ **B.** $y = 3^{-x}$ **C.** $y = -|x|$

[A. The values of $-x^2$ are opposite those of x^2. This is a reflection through the x-axis of the parent function $y = x^2$. B. This is not a vertical reflection because the values of 3^{-x} are not opposite those of 3^x (this is a horizontal reflection). C. The values of $y = -|x|$ are the opposite of the y-values of the function $y = |x|$. This is a reflection of $y = |x|$ through the x-axis.]

Closing the Lesson

Summarize how to graph a reflection through the x-axis:

1. Identify the parent function.
2. Identify the scale factor, and graph vertical stretches.
3. For every point (a, b), find $(a, -b)$ and graph.

Have students use these steps to graph each of the following:

A. $y = -2x$ **B.** $y = -3^x$

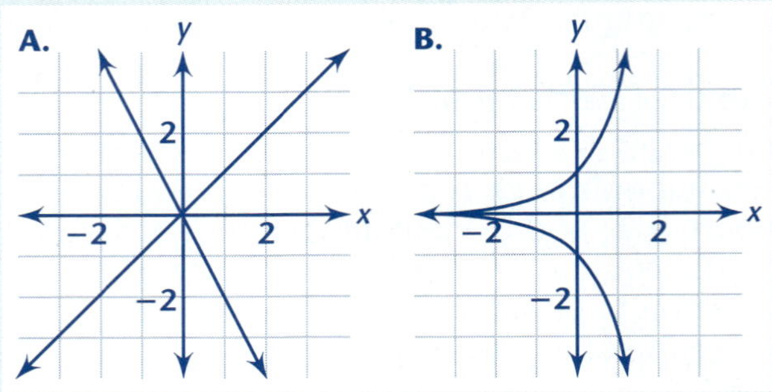

TEACH

Approach B

Using Models, continued

Cooperative Learning

Horizontal Reflections Have students work with a partner to find the horizontal reflected image through the y-axis for each graph shown below. Have them make a generalization about how each point (a, b) of the original graph is transformed.

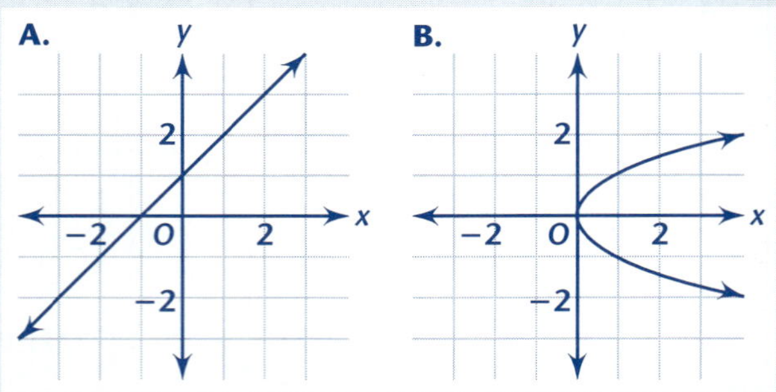

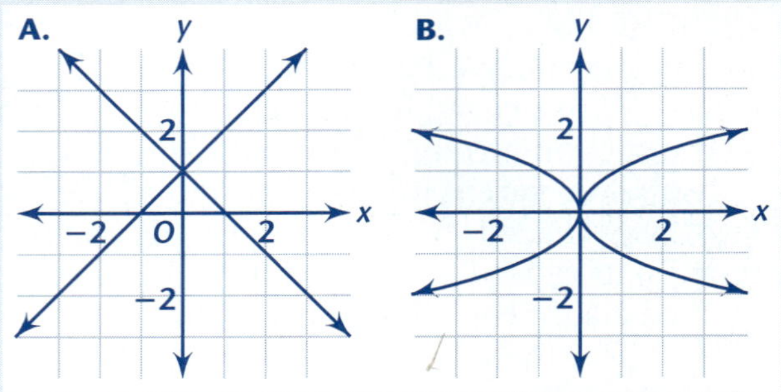

[Each point (a, b) is transformed into the point $(-a, b)$.]

Journal Entry

Have students describe how to find a vertical reflection through the x-axis and a horizontal reflection through the y-axis. Include examples and give specific changes in the function rule.

ASSESS

Selected Answers

Odd-numbered Exercises 7–31 and odd-numbered Extra Practice

Assignment Guide

Core level students should complete Exercises 1–37.

Core-plus level students should complete Exercises 1–37.

Error Analysis

Stress the difference between vertical reflections and horizontal reflections.

For Exercise 13, students may have difficulty understanding what happens to the part of the graph below the x-axis. Point out that the y coordinates of these points are negative; so when the rule for vertical reflection is applied, they become positive. So they lie above the x-axis after the reflection.

Alternative Assessment

The Reflection Game To involve all students in finding the reflection of a parent function through the x-axis, start by asking one student to write a parent function on the chalkboard. Then have the next student identify the parent function by name. Ask a third student to graph the parent function. Next have another student reflect the parent function through the x-axis. Then ask another student to give the new function rule. Ask the next student to graph five points on the reflected function. Repeat this sequence until all students have participated.

RETEACH

Have students identify which graphs are vertical reflections of parent functions.

A. $y = -x^2$ **B.** $y = 2^x - 4$

C. $y = \frac{2}{x}$ **D.** $y = -x$

E. $y = -|x|$

[A, D, E]

ENRICH

Have students draw an object of their choice in the coordinate plane and then reflect it through the x-axis. Give coordinates of key points in the original graph and corresponding points in the image graph.

Interdisciplinary Connection

Graphic Art Graphic artists need to know how to find reflected images in order to create symmetry in their art work. They may use hand tools or technology to create the reflected images.

NAME ______________________ CLASS ____________ DATE ____________

Practice

5.4 Reflections

Graph each function.

1. $y = -2x^2$

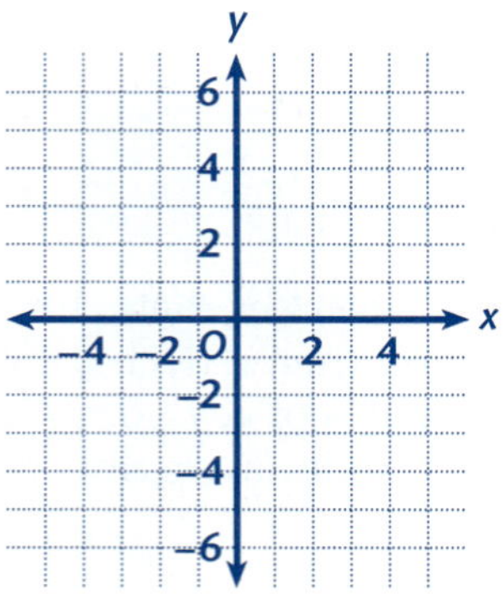

2. $y = -\frac{|x|}{2}$

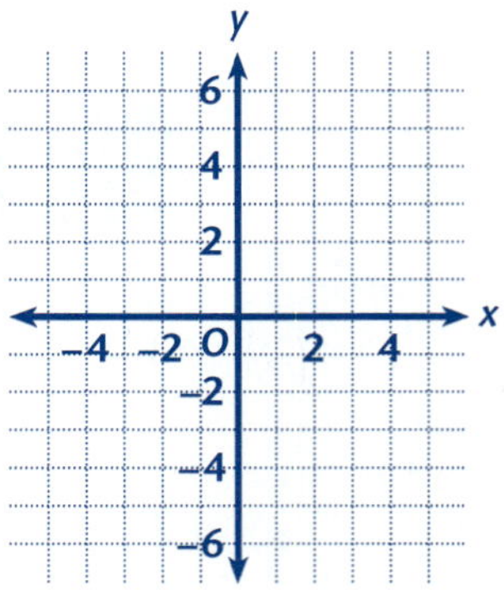

3. $y = \frac{-3}{x}$

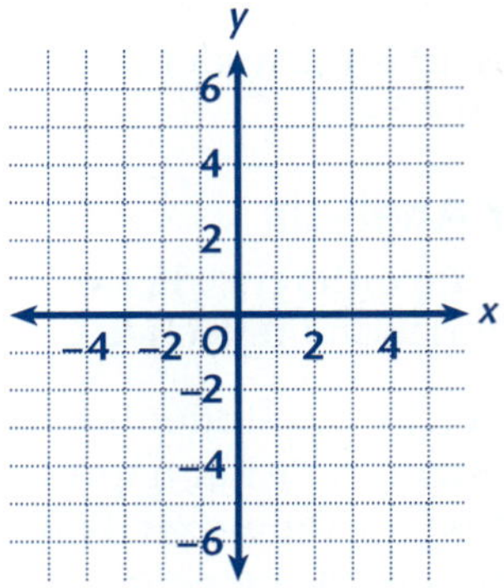

Which equations have graphs that are vertical reflections of the parent function?

4. $y = x^2 - 1$ ____________

5. $y = -2|x|$ ____________

6. $y = -\text{INT}(x)$ ____________

7. $y = \frac{1}{-x}$ ____________

8. $y = -x$ ____________

9. $y = x - 1$ ____________

Which equations have graphs that are horizontal reflections of the parent function?

10. $y = (-x) - 1$ ____________

11. $y = 2|-x + 1|$ ____________

12. $y = \text{INT}(x - 1)$ ____________

13. $y = \frac{1}{-x} - 2$ ____________

14. $y = (-x - 1)^2$ ____________

15. $y = |x - 1|$ ____________

For Exercises 16–17, find $f(2)$, and use this information to tell which kind of reflection (horizontal or vertical) is applied to the function $y = 3x + 2$.

16. $f(x) = -(3x + 2)$ ____________

17. $f(x) = 3(-x) + 2$ ____________

For Exercises 18–19, find $f(3)$, and use this information to tell which kind of reflection (horizontal or vertical) is applied to the function $y = (x - 1)^2$.

18. $f(x) = -(x - 1)^2$ ____________

19. $f(x) = (-x - 1)^2$ ____________

NAME ______________________ CLASS ____________ DATE ____________

Lesson Activity
5.4 Modeling Snowflakes

You can use reflections to create a snowflake by cutting designs in folded paper.

Fold a square sheet of paper in half. Fold in half again to form square *OPCB*. Then unfold along segment *OP*.

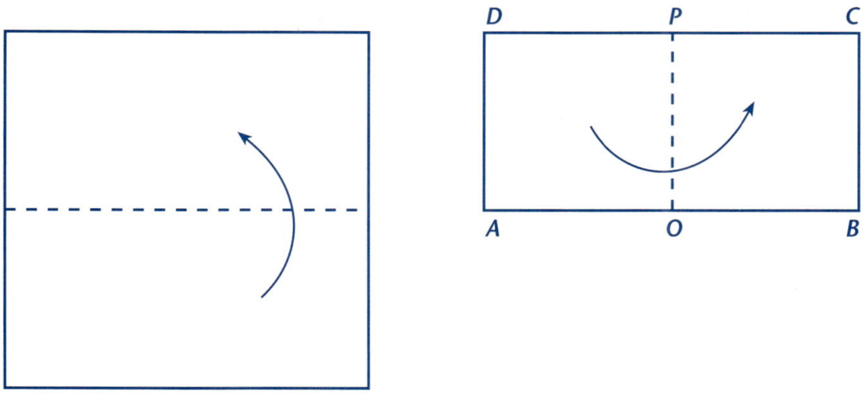

Fold square *OPCB* in half, parallel to $\overline{OP}$, to form $\overline{EF}$; then unfold $\overline{EF}$. Bring *A* onto fold $\overline{EF}$, and crease along $\overline{OG}$. Fold $\overline{OB}$ along $\overline{OA}$ so that $\overline{OB}$ lies along $\overline{OG}$. Fold the paper in half along $\overline{HO}$. Cut the folded paper along *B* to *I* so that $BI = OI$. Unfold your snowflake.

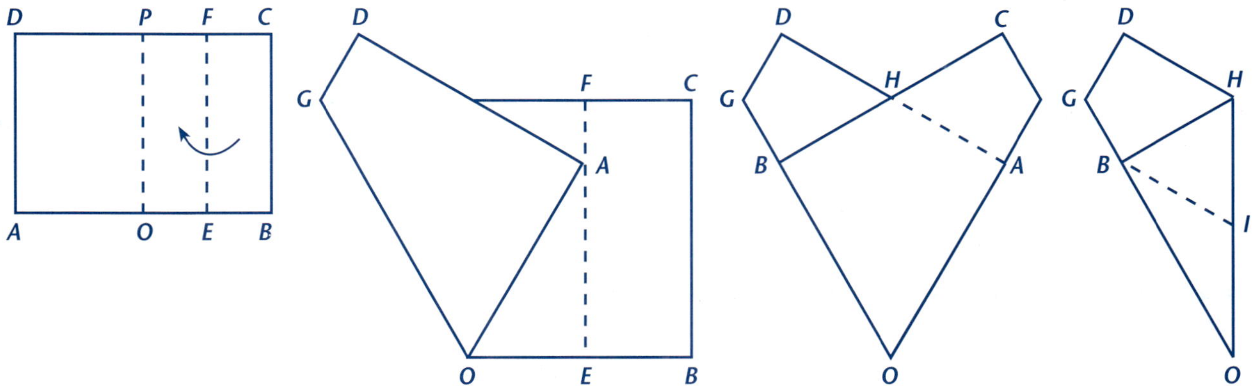

Use your snowflake to answer these questions.

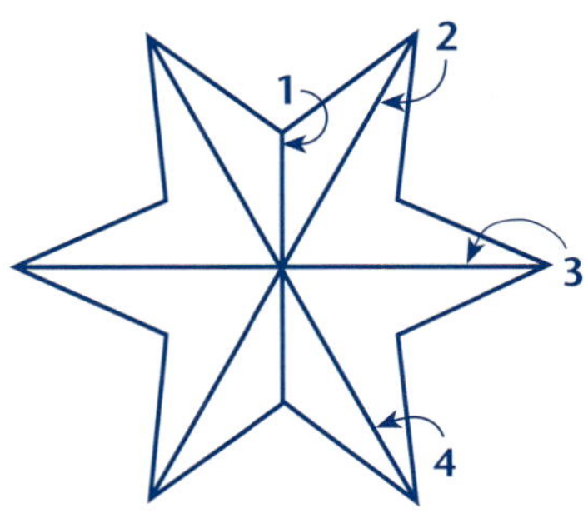

1. Number your fold lines as shown in the figure. Fold the snowflake along line 1. What do you notice about the pattern?

____

2. Unfold your snowflake and repeat the procedure for fold lines 2, 3, and 4. Describe what you notice after each fold.

____

NAME ______________________ CLASS __________ DATE __________

Basic Skills Practice
Symmetry and Rotations, Part 1

A figure has *line symmetry* if it can be folded through a line and both parts of the figure coincide or match one another. Since the parts form mirror images of one another, this is also called reflectional symmetry.

Figures may have one line of symmetry, two or more lines of symmetry, or no line of symmetry.

Example 1: Find the lines of symmetry for each figure.

a. 

b. 

c. 

a. This figure has a horizontal line of symmetry.
b. This figure has a vertical and a horizontal line of symmetry.
c. This figure has two congruent parts, but it has no line of symmetry.

Rotational symmetry is another kind of symmetry. A figure has rotational symmetry if there is a point O about which the figure can be rotated and made to coincide with its original figure. Rotational symmetries are identified by the angle of rotation required to make the figure coincide with itself. If a figure coincides with itself after being rotated 180°, the figure has *point symmetry*. If a figure only has 360° rotational symmetry, it is not considered to be symmetric because any figure will coincide with itself after being rotated in a full circle.

Example 2: Determine if each figure has rotational symmetry.

a. 

b. 

c. 

a. This figure has rotational symmetry. It will coincide with itself after being rotated 120°.
b. This figure does not have rotational symmetry. (It does have vertical line symmetry.)
c. This figure has point symmetry. It will coincide with itself after being rotated

Identify the type of symmetry, if any, in each figure.

1.  ____________

2.  ____________

3. 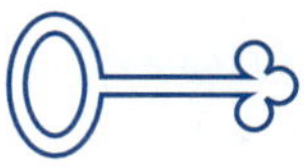 ____________

4.  ____________

5.  ____________

6.  ____________

LESSON 5.5 Translations

PREPARE

Pacing Guide

2 Days

Objectives and Overview

- Describe the effect of a translation on the graph of a function.
- Identify the relationship between the translation of a graph and the addition or subtraction of a constant to the function.

Student learn how to shift a graph vertically up or down and horizontally right or left. They also learn the function rules that correspond to translations.

Vocabulary

translation
vertical translation
horizontal translation

Assessing Prior Knowledge

Graph the function $f(x) = 2x^2$ and sketch the graph of its reflection through the x-axis.

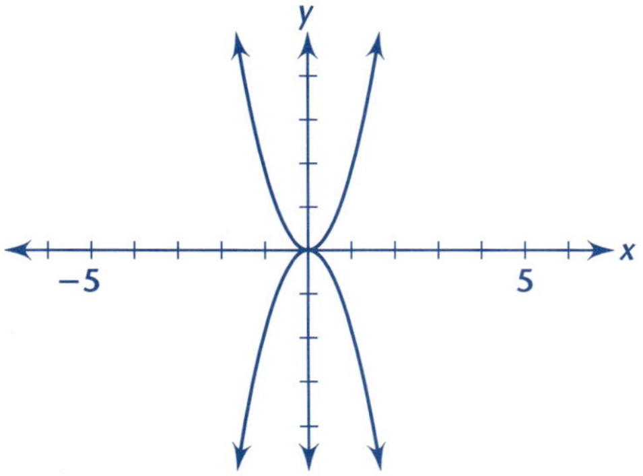

Approach A: Using the Book

Opening Discussion

Discuss the use of translations in the graphic art of Escher and in the use of tile patterns. A translation is often called a "slide." How might translations or slides be modeled by functions?

Brainstorming

Ask students for examples of objects that are created with translations. Examples include tile patterns, wallpaper patterns, and logos. Then ask students if they think functions can be translated. How might functions be translated?

Definition of *Translation*

A translation is a transformation that shifts the graph of a function in a certain direction in the coordinate plane. Only translations that shift the graph vertically or horizontally are considered in this textbook because they have a simple relationship to the original graph.

continued on top of next page

TEACH

Approach B: Using Models

Opening Discussion

Show examples of art work that include translations, or discuss the use of translations in the graphic art of Escher and in tile patterns. A translation is often called a "slide." How might translations or slides be modeled by functions?

Brainstorming

Ask students for examples of objects that are created with translations. Examples include tile patterns, wallpaper patterns, and logos. Ask them to draw some tile patterns. Then ask students if they think functions can be translated. How might functions be translated?

Definition of *Translation*

A translation is a transformation that shifts the graph of a function in a certain direction in the coordinate plane. Only translations that shift the graph vertically or horizontally are considered in

continued on bottom of next page

Approach A Using the Book, continued

Example 1

In this example, a transformed absolute-value graph is compared with the graph of its parent function, and it is determined that the parent graph was shifted horizontally 10 units to the right.

Point out that the translated image is geometrically congruent to the original graph. It is just shifted 10 units to the right. Therefore, the y-coordinates of the image graph will have the same y-coordinates as the parent graph, but the x-coordinates of corresponding points will be 10 units greater. Encourage students to compare the tables for each graph so that they can see how the points correspond.

Use **Transparency 21** for discussion.

Vertical Translation

A vertical translation occurs when the graph of a function is shifted up or down. Point out that all vertical translations of the function $y = f(x)$ are of the form $y = f(x) + k$. If k is negative, the graph is shifted downward. If k is positive, the graph is shifted upward. Emphasize that the y-values of the new function have to be k units greater than or less than the y-values for the same x-values of the parent function.

Example 2

In this example, the graph of the quadratic parent function is shifted down 3 units.

Point out that for each pair of corresponding points, the x-coordinates are the same, and the y-coordinates of the image are 3 less than the y-coordinates of the parent graph.

TEACH

Approach B Using Models, continued

this textook because they have a simple relationship to the original graph.

Using a "Translation Ruler"

A ruler or straight edge can help students find the points of a translation of a graph. The edge of a piece of paper, perhaps folded, can be used and may even be preferable because it allows students to choose their own units and mark them on it.

Instruct students to first determine the parent function and graph it. Have them highlight at least five points. Next, determine if the translation is up, down, to the right, or to the left. Then mark off the number of units to be translated on the ruler. Use the mark on the ruler to locate the translated points in the direction specified by the translation. Do this for as many points as necessary to create the image graph.

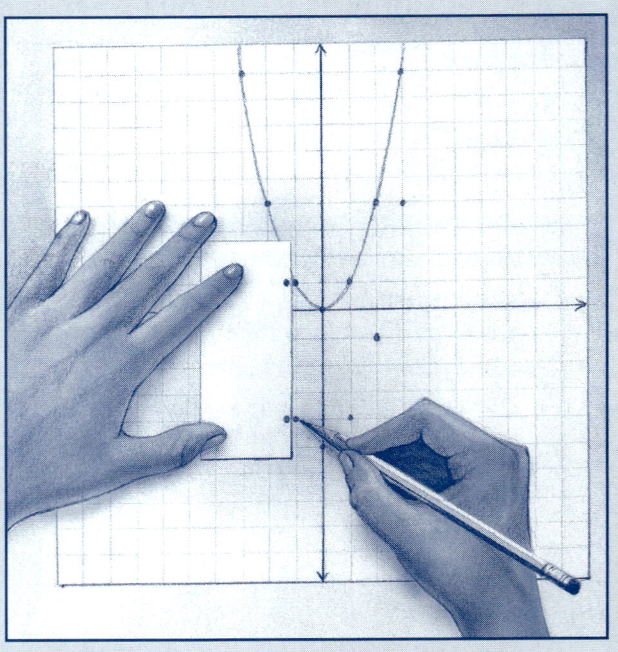

A "translation ruler" can give students tactile experiences of transformations.

Ongoing Assessment, page 247
Try This

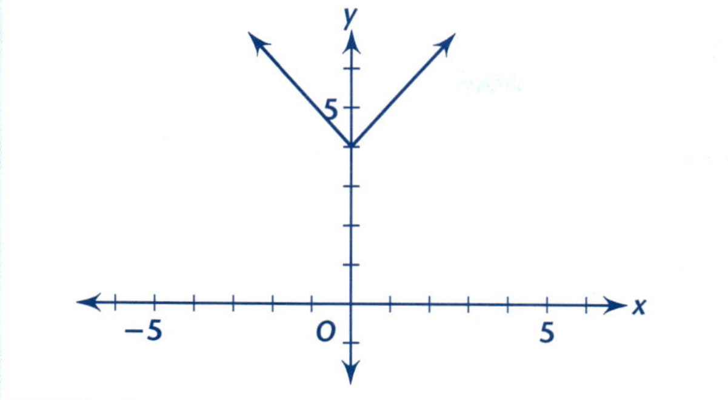

Use **Transparency 22** for discussion of the vertical translation of a parabola.

Horizontal Translation

A horizontal translation occurs when the graph of a function is shifted to the right or the left.

Point out that all horizontal translations of the function $y = f(x)$ are of the form $y = f(x - h)$. If h is negative, the graph is shifted to the left. If h is positive, the graph is shifted to the right. Point out that the minus sign between the x and the h is not the sign of h. Emphasize that the x-values of the new function have to be h units greater than or less than the x-values of the parent function in order to have the same y-values.

Example 3

In this example, the graph of the quadratic parent function $y = x^2$ is compared with the graph of $y = (x - 3)^2$.

The translation graph is shifted 3 units to the right of the parent graph. Emphasize that the order of operations in finding points is to first substitute the value of x, subtract 3, and then square it.

continued on top of next page

Working With a "Translation Ruler"

Have students graph each of the following translated functions and their parent functions:

A. $y = x + 2$ **B.** $y = x^2 - 4$

C. $y = \frac{1}{x + 2}$ **D.** $y = |x - 3|$

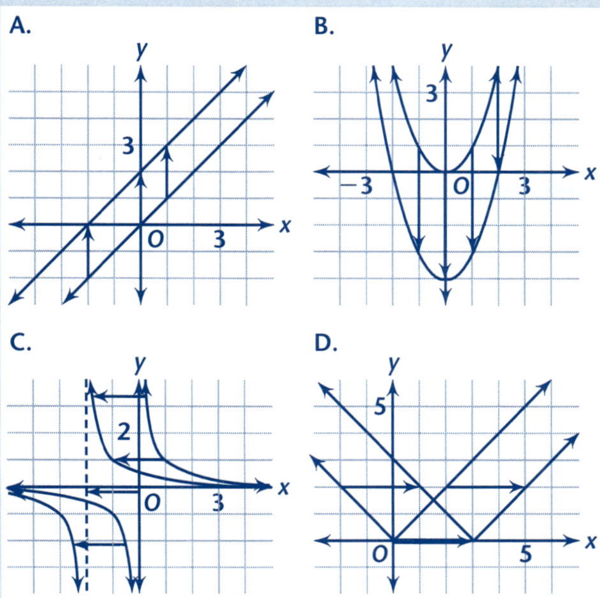

Teaching Tip

Using a physical model to help translate graphs vertically or horizontally gives a tactile connection to translations and emphasizes two geometric connections: (1) the number of units translated in each case is identical, and (2) the direction of translation is the same for each point. Have students describe the direction of the translations in Working With a "Translation Ruler." How far are the translations? **A. vertical, 2 units up; B. vertical, 4 units down; C. horizontal, 2 units left; D. horizontal, 3 units right]**

Cooperative Learning

Writing a Function Rule Help students understand the geometric connection in translations by dividing the class into groups of four. Ask one student in the group to draw a parent graph and its translated image on the same coordinate plane. Have the second student measure the magnitude of the translation and determine the direction of the translation. Ask a third student to give the function rules for the

continued on bottom of next page

Approach A Using the Book, continued

Ongoing Assessment, page 248

Try This

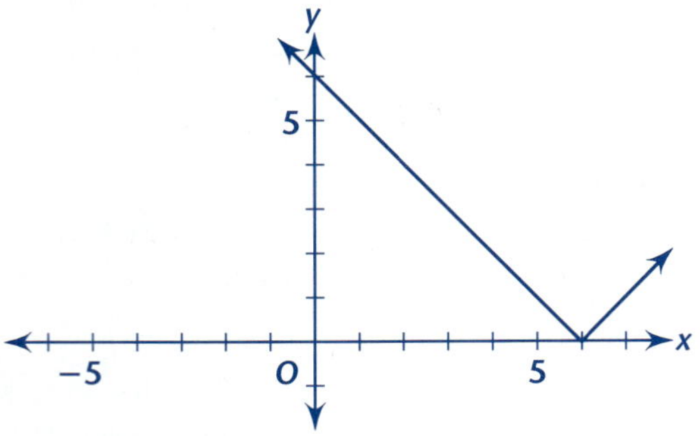

Example 4

In this example, functions are given, and the student is asked to identify which functions represent a translation of a parent graph.

Point out that only Exercise **C** is not a translation because it does not have the form of a vertical or a horizontal translation. Emphasize that in Exercise **D** the fraction bar acts like a grouping symbol, so the denominator is $(x - 10)$, which implies a horizontal translation.

Example 5

This is an example of an application of translations to a set of real-world data.

Students should notice that exactly the same amount must be added to each data value in order for it to be a translation. Emphasize that the average salary of the salespeople also increases by the same amount.

Ongoing Assessment, page 250

The vertex moves down by 5 units.

TEACH

Approach B Using Models, continued

parent graph and the translated graph. Each student should discuss his or her results. Have a fourth student record the examples used. Ask students to switch roles and repeat the activity with four more parent graphs.

Journal Entry

Have students use examples to discuss how to translate the graph of a parent function. Summarize how to graph a vertical translation and how to graph a horizontal translation. Discuss how a function rule would look if both types of translations were done to a parent function.

Closing the Lesson

Inviting Student Participation

Summarize the lesson by having a student write a parent function on the chalkboard and specify how the function should be translated. Then have a second student list the function rule for the translation. Ask another student to graph the parent function and the translation. Have successive students write parent functions and their translations until all types of translations have been reviewed.

ASSESS

Selected Answers

Odd-numbered Exercises 7–39 and odd-numbered Extra Practice

Assignment Guide

Core level students should complete Exercises 1–39.

Core-plus level students should complete Exercises 1–41.

Error Analysis

Students who have trouble with Exercises 19–26 may need help in understanding the difference between vertical and horizontal translations. Point out that point (a, b) becomes the point $(a, b + k)$ for a vertical translation of magnitude k. The point (a, b) becomes the point $(a + h, b)$ for a horizontal translation of magnitude h.

Alternative Assessment

Using Technology Ask students to use spreadsheets to help them understand vertical and horizontal translations. Ask them to put the following data on employee salaries into a spreadsheet, using consecutive numbers from 1 to 9 to identify the employees.

Employee number	Yearly salary	Holiday bonus	Total salary
1	23,000	250.00	**[23,250]**
2	25,500	250.00	**[25,750]**
3	41,000	250.00	**[41,250]**
4	18,500	250.00	**[18,750]**
5	12,800	250.00	**[13,050]**
6	29,000	250.00	**[29,250]**
7	23,000	250.00	**[23,250]**
8	23,500	250.00	**[23,750]**
9	45,000	250.00	**[45,250]**

Have students give a new spreadsheet showing the total salary for one year for each employee. Have students try three different holiday bonuses and report the change in total salary for each.

RETEACH

The following graphs represent vertical or horizontal translations of a parent graph. Give the parent graph, the type of translation, and the new function rule.

A.

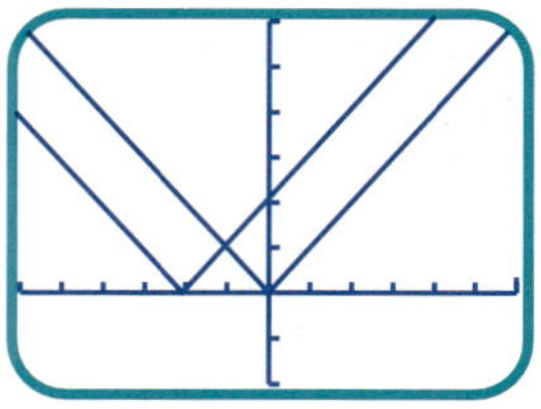

B.

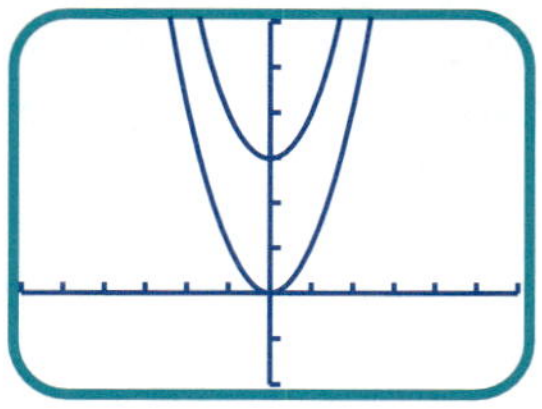

[A. $y = |x|$, horizontal, $y = |x + 2|$; B. $y = x^2$, vertical, $y = x^2 + 3$]

ENRICH

A function can also be translated in a diagonal direction. For example, if $(x, y) \rightarrow (x + 2, y + 3)$ for all points (x, y) of the graph of the function, then each point of the function moves to the right 2 units and up 3 units. Have students translate the points of $y = x^2$ by 2 units to the right and 3 units up.

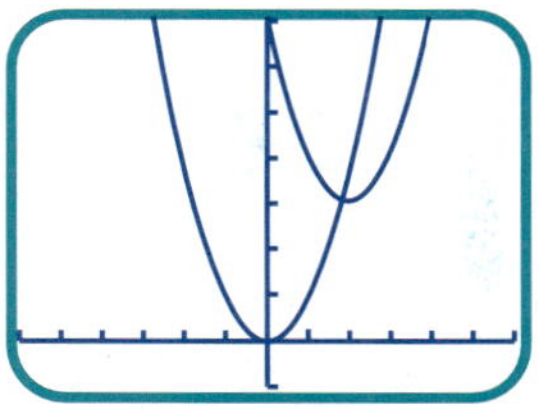

Ask students to give the new function rule.
[$y - 3 = (x - 2)^2$, or $y = (x - 2)^2 + 3$]

NAME ______________________ CLASS __________ DATE __________

Practice

5.5 Translations

Identify the parent function. Tell what type of transformation is applied to each, and then draw a carefully labeled graph. HINT: Remember to plot a few points to help you see what happens to the parent function.

1. $y = 0.5x^2$ ______________

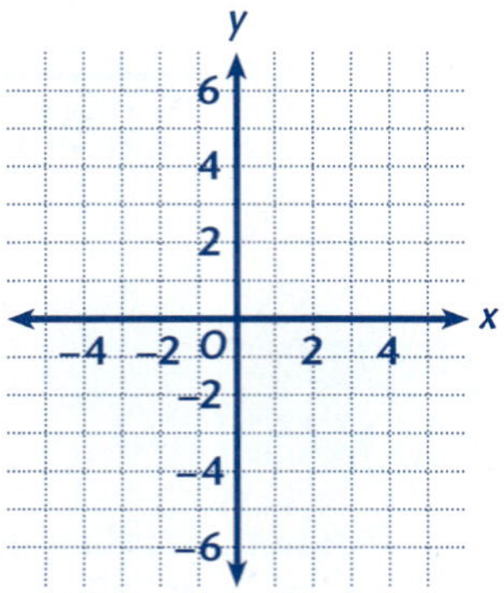

2. $y = -3|x|$ ______________

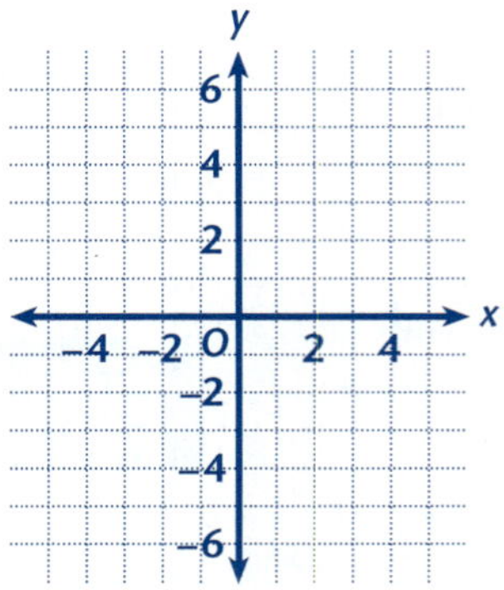

3. $y = \frac{3}{x}$ ______________

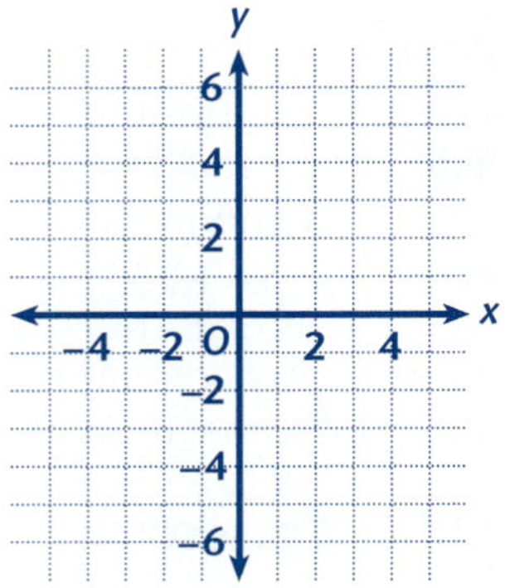

4. $y = 2x - 2$ ______________

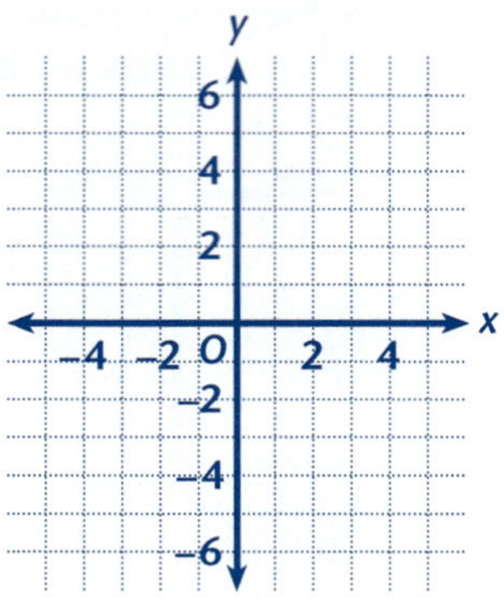

5. $y = |x - 1|$ ______________

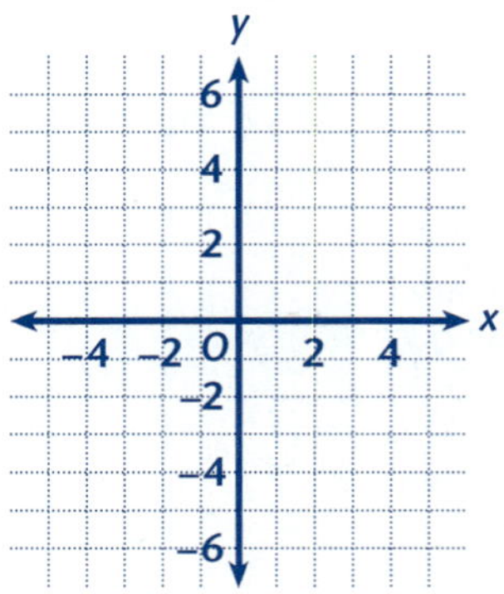

6. $y = \text{INT}(x + 2)$ ______________

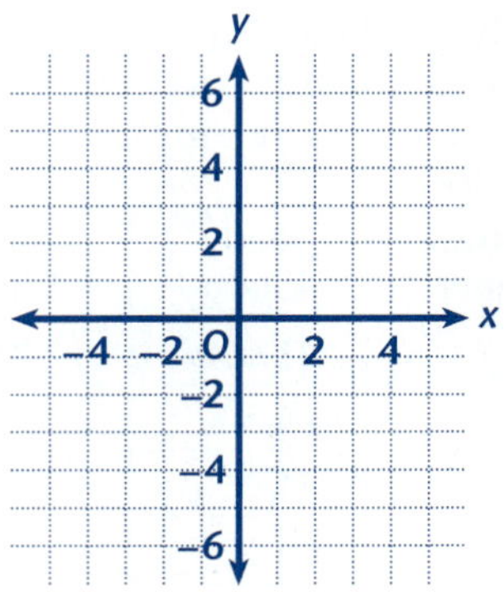

Consider the parent function $f(x) = 2^x$. Note that $f(1) = 2$, so the point (1, 2) is on the graph. In each exercise, state how the parent function is translated for the graph of each function given.

7. $f(x) = 2^x - 1$ ______________

8. $f(x) = 2^x + 4$ ______________

9. $f(x) = 2^{(x-1)}$ ______________

10. $f(x) = 2^{(x+2)}$ ______________

The point (2, 4) is on the graph of $f(x)$. Find the new coordinates of the point when each transformation is applied to the function.

11. vertical translation by 2 ______________

12. horizontal translation by 3 ______________

13. horizontal translation by −4 ______________

14. vertical translation by −5 ______________

NAME ______________________________ CLASS ____________ DATE ____________

Lesson Activity

5.5 Triangles Inside Triangles

You can use translations to demonstrate a mathematical statement about equilateral triangles. Draw two identical equilateral triangles. Select any point, *P*, inside one triangle, and draw perpendiculars from the sides of the triangle to the point, as shown.

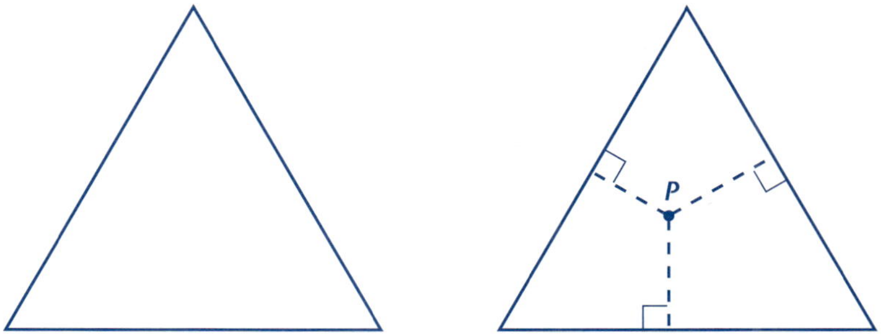

Next draw lines parallel to each of the triangle's sides through point *P*, forming the three smaller equilateral triangles shaded here. Cut out the three triangles.

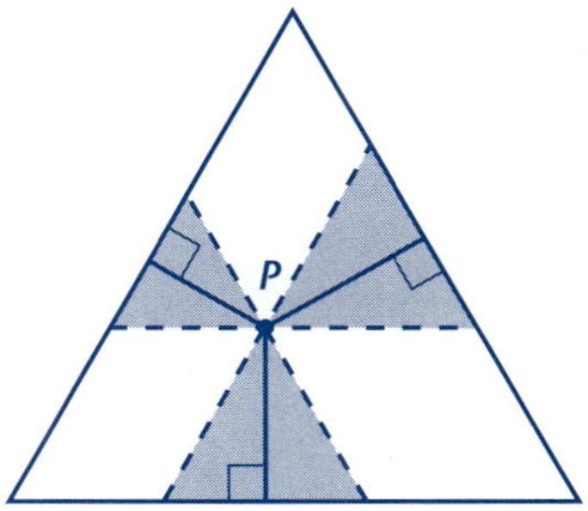

Using the remaining equilateral triangle you drew, translate one of the small triangles to a vertex of the large triangle. Translate another triangle to another vertex of the large triangle. Translate the third triangle into the space between the first two, as shown.

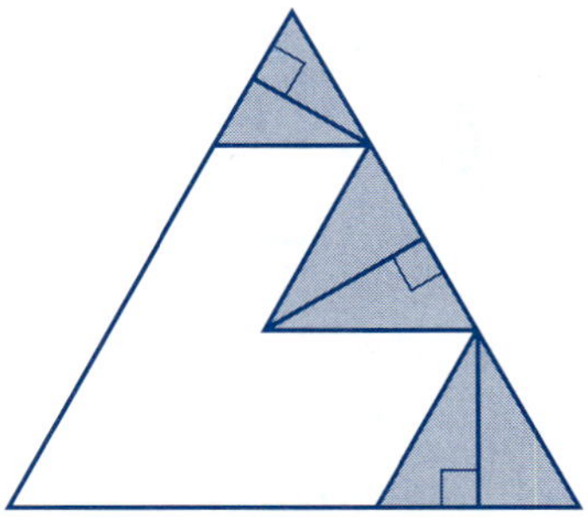

Rotate two of the small triangles, and then translate them horizontally so that all three triangles form a line as shown.

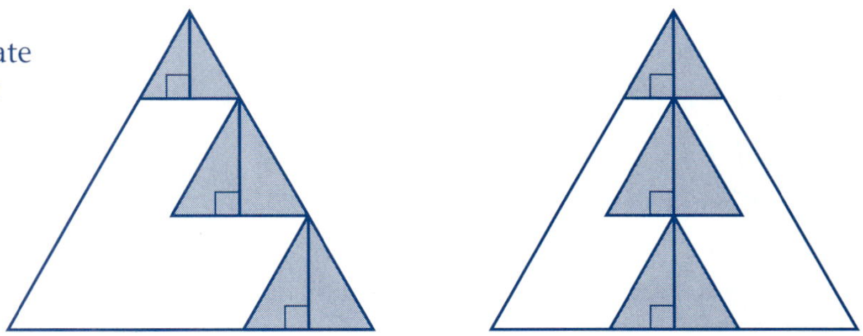

1. What geometric statement about equilateral triangles does this activity demonstrate?

____

NAME ______________________ CLASS __________ DATE __________

Basic Skills Practice

Symmetry and Rotations, Part 2

Read each question and circle the best answer.

1.

A Horizontal line of symmetry
B Vertical and horizontal lines of symmetry
C Point symmetry
D No symmetry

2.

F Vertical line of symmetry
G Horizontal line of symmetry
H Rotational symmetry
I No symmetry

3.

A Point symmetry
B Rotation symmetry
C Horizontal line of symmetry
D All of the above

4.

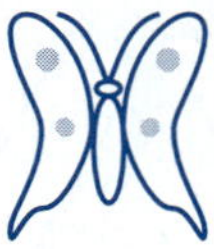

F No symmetry
G Rotational symmetry
H Horizontal line of symmetry
I Vertical line of symmetry

5.

A No symmetry
B Point symmetry
C Horizontal line of symmetry
D Vertical line of symmetry

6.

F Vertical line of symmetry
G Rotational symmetry
H Horizontal line of symmetry
I All of the above

7.

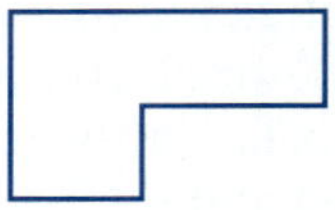

A No symmetry
B Rotational symmetry
C Horizontal line of symmetry
D Vertical line of symmetry

8.

F Vertical line of symmetry
G Horizontal line of symmetry
H Vertical and horizontal lines of symmetry
I No symmetry

LESSON 5.6 Combining Transformations

PREPARE

Pacing Guide
2 Days

Objectives and Overview
- Identify the parent function in a transformation.
- Understand the effect of order on combining transformations.

Student learn how to combine translations with stretches or reflections of a parent graph. They learn that a stretch should be done before a translation when combining transformations.

Assessing Prior Knowledge

Ask students to describe the second graph in terms of the first graph for each pair of equations.

1. $f(x) = x^2$ and $f(x) = x^2 - 3$ **[The second graph is the first graph shifted down 3 units.]**
2. $f(x) = |x|$ and $f(x) = |x + 2|$ **[The second graph is the first graph shifted 2 units to the left.]**
3. $f(x) = \frac{1}{x}$ and $f(x) = \frac{1}{x - 3}$ **[The second graph is the first graph shifted 3 units to the right.]**
4. $f(x) = x^2$ and $f(x) = (x - 2)^2$ **[The second graph is the first graph shifted 2 units to the right.]**

Approach A

Using the Book

Opening Discussion
Review the use of translations in the previous lesson. Could the translation of a function be combined with a stretch? How would the image look?

Brainstorming
Show students the figure below. Have them brainstorm which transformations may be involved in drawing it.

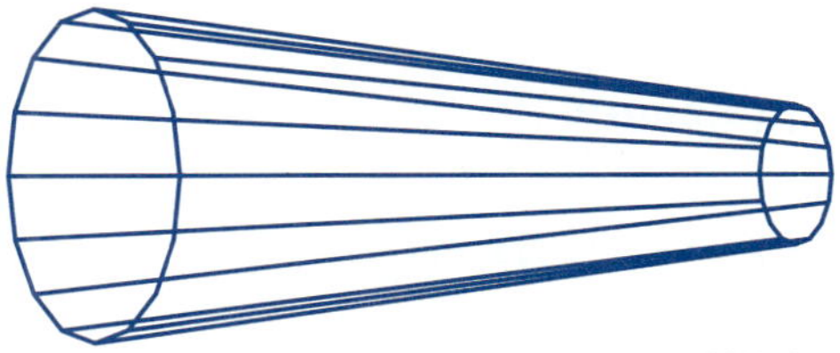

[The elliptical figure on the left is translated to the right, and the size of the figure is changed by a stretch.]

continued on top of next page

TEACH

Approach B

Using Models

Opening Discussion
Review the use of translations from the previous lesson. Could the translation of a function be combined with a stretch? How would the image look? If a function is reflected first and then translated, how would the image look?

Brainstorming
Draw the following figures in a column on the chalkboard: an ellipse, a square, a triangle, and a trapezoid. Instruct students to copy the figures onto their paper in a neat column. (Have them use grid paper if available.) Now instruct the students to draw a half-size reproduction of each figure to the right of the original. Have them brainstorm what transformations were involved in making the new figures. **[a translation to the right and a scale change]**

continued on bottom of next page

Approach A

Using the Book, continued

Combining Transformations

Emphasize that to combine transformations, first the parent function should be determined and graphed. Second, the type of transformation needs to be identified. If there is more than one transformation, identify each stretch and each translation. It is important for students to distinguish between them. Ask students to refer to Lessons 5.4 and 5.5, if necessary. Third, perform the stretch and graph the stretched function. Last, perform the translation and graph the translated function.

Use **Transparency 23** for discussion.

Ongoing Assessment, page 254

Bob and his friend must travel $\frac{50}{3}$, or $16\frac{2}{3}$, mph.

Critical Thinking, page 254

The area of the original square is multiplied by 3 and then that product is increased by 5.

Example 1

In this example, a vertical stretch of 2 units is applied to the absolute-value function, and then the stretched graph is translated 4 units to the left.

Emphasize that the translation does not need to be done after the stretch, but that it is usually easier this way. Encourage students to use graphics calculators to verify that the parent function was transformed correctly.

Additional Example 1

Sketch the graph of $y = 3(x - 2)^2$. **[The parent graph is $y = x^2$. Since the scale factor is 3, the y-values will be 3 times as great as those of the parent function. This is a vertical stretch of 3. Because 2 is subtracted from the variable, the x-values in the transformed**

TEACH

Approach B

Using Models, continued

Using Paper Models

Using a physical model to study transformations helps students make tactile connections between the final transformed graph and the original parent graph.

Ask students to draw graphs of $|x|$, $2|x|$, and $3|x|$. Have them cut out paper triangles that fit into these graphs. Then have them color the triangles with different colors and perhaps label them.

For class discussion and review, cut out a set of triangles from transparency film to use on an overhead projector.

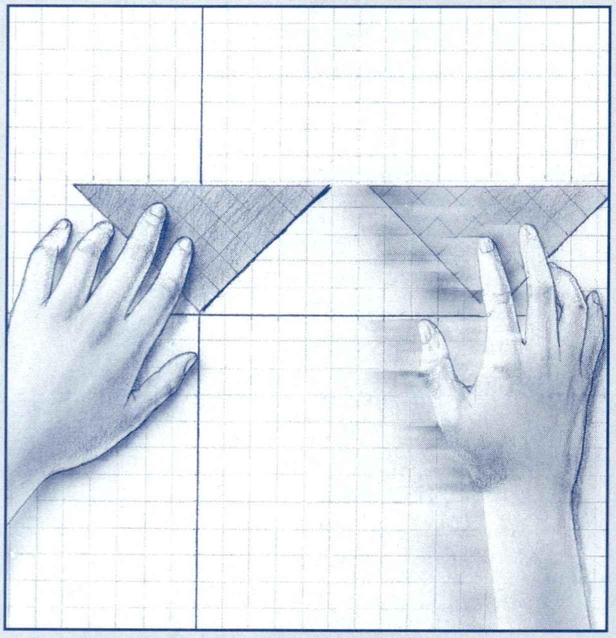

Paper models can help students understand sequences of transformations.

function will be 2 units greater in order to produce the same y-values as the stretched function. The result is a translation 2 units to the right.]

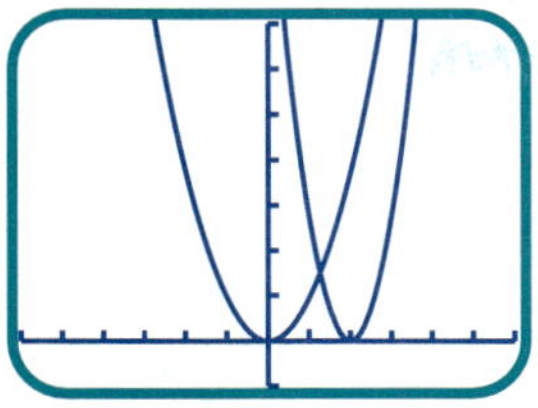

Using Technology

A graphics calculator can help students understand how to transform a function by plotting the parent graph and the transformed graph quickly. Point out the necessity of using parentheses correctly when doing a horizontal translation. For example, $y = \text{ABS}(x - 3)$ is a horizontal translation of 3 units to the right, whereas $y = \text{ABS}x - 3$ is a vertical translation 3 units down.

Ongoing Assessment, page 255
Try This

The graph is symmetric. The axis of symmetry is the vertical line $x = -4$.

Example 2

In this example, the summary statement given on page 255 of the Pupil's Edition is used to sketch the graph of the quadratic parent function with a translation 2 units down and 3 units to the left and a reflection through the x-axis.

While there is no set order for doing transformations, it is suggested that reflections be done first, stretches next (there are none for this example), and translations last. This order is easy to remember as *RST*.

Additional Example 2

Use the information from the summary on page 255 of the Pupil's Edition to sketch the graph of the function $y = -2(x - 1)^2$. **[The parent**

continued on top of next page

Working With Paper Model

Have students model the following transformations with the paper triangles:

A. $|x - 11|$ **B.** $-|x|$

C. $2|x - 5|$ **D.** $-3|x| - 3$

E. $3|x + 3| + 3$ **F.** $-2|x - 2| - 2$

[A. The $|x|$ triangle is translated by 11 units to the right. B. The $|x|$ triangle is reflected through the x-axis. C. The $2|x|$ triangle is translated 5 units to the right. D. The $3|x|$ triangle is reflected through the x-axis and shifted down 3 units. E. The $3|x|$ triangle is translated 3 units to the left and shifted 3 units up. F. The $2|x|$ triangle is reflected through the x-axis, translated 2 units to the right, and shifted 2 units down.]

From Models to Mathematics

Using the overhead, choose a triangle and position it in the plane. Then ask students to write the equation for the combined transformation modeled by the triangle. Repeat this activity for several examples.

Cooperative Learning

Writing a Function Rule Help students understand combined transformations by dividing the class into groups of four. Ask one student in the group to graph a parent function and its transformed image on the same coordinate plane. Have the second student cut out a paper model of the parent graph and count the number of separate transformations that are done to get the final image. Ask a third student to use a paper model of the transformed image to determine the amount of stretch. Ask the fourth student to give the function rules for the parent function and the combined transformation. All students should try to verify the rule. Ask students to switch roles and repeat the activity with other parent functions.

Journal Entry

Have students use examples to discuss how to do a combined transformation of the graph of a parent function. List the sequence of steps that would be followed, and tell how to check the progress of the transformation as it is being done.

continued on bottom of next page

Approach A

Using the Book, continued

function is $y = x^2$, so the graph is a parabola. Since the scale factor is negative, the parabola is reflected through the x-axis. Since there is a 2 following the negative sign, the graph is stretched vertically by a scale factor of 2. The 1 indicates a horizontal translation 1 unit to the right.]

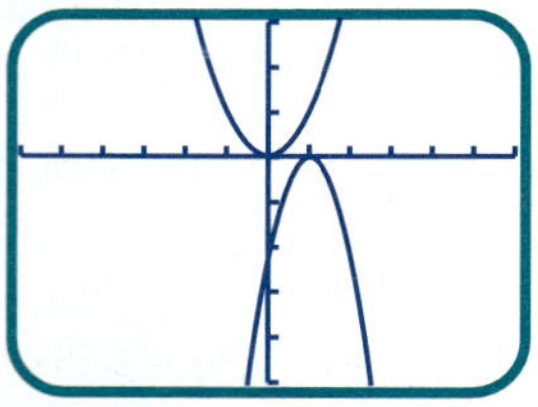

Closing the Lesson

A vertical translation occurs when the graph of a function is shifted up or down. Point out that all vertical translations of the function $y = f(x)$ are of the form $y = f(x) + k$. If k is negative, the graph is shifted downward. If k is positive, the graph is shifted upward. A horizontal translation of the function $y = f(x)$ is of the form $y = f(x - h)$. If h is positive, the graph is shifted to the right, and if h is negative, the graph is shifted to the left. A vertical stretch of the function $y = f(x)$ is of the form $y = a \cdot f(x)$, and a reflection in the x-axis of the function $y = f(x)$ is of the form $y = -f(x)$. Point out that all of these transformations can be combined into one transformation of the form $y = -a \cdot f(x - h) + k$.

TEACH

Approach B

Using Models, continued

Closing the Lesson

Inviting Student Participation

Summarize the lesson by having a student write a parent function on the chalkboard and specify how the function should be translated, stretched, or reflected. Then have a second student list the function rule for the combined transformations. Ask another student to show the graphs of the parent function and the transformed function. Have successive students write parent functions and their transformations until all students have participated.

ASSESS

Selected Answers

Odd-numbered Exercises 7–31 and odd-numbered Extra Practice

Assignment Guide

Core level students should complete Exercises 1–36.

Core-plus level students should complete Exercises 1–36.

Error Analysis

Students who have trouble with Exercises 16–23 need to understand how each transformation acts on each point of the original graph. Remind students that point (a, b) becomes point $(a, b + k)$ for a vertical translation of k. Point (a, b) becomes point $(a + h, b)$ for a horizontal translation of h. Point (a, b) becomes point $(a, -b)$ for a reflection through the x-axis, point (a, b) becomes point $(-a, b)$ for a reflection through the y-axis, and point (a, b) becomes point (a, sb) for a vertical stretch of s.

Alternative Assessment

Portfolio Activity Continue the portfolio activity started in Lesson 5.3 by asking students to add to their library of functions. Ask them to add the headings below to their chart and then to fill in the chart with the function rules and graphs as they learn them.

	Function rule	Sample graph
Translation		
Reflection		

RETEACH

The following graphs represent combined transformations of a parent graph. Give the parent function, the types of transformations, and the new function rule. Use a graphics calculator to verify the transformations.

A.

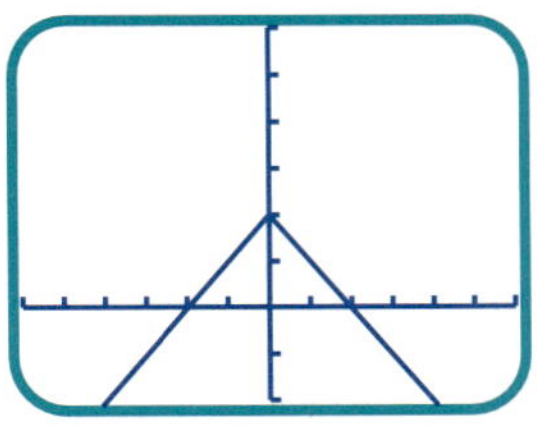

B.

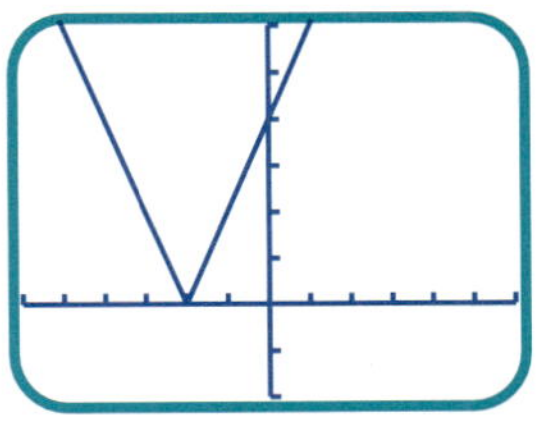

C.

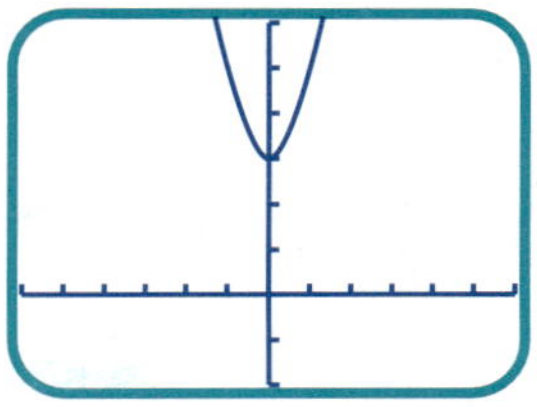

[A. $y = |x|$, reflection through the x-axis and translation 2 units up, $y = -|x| + 2$; B. $y = |x|$, vertical stretch of 2 units and horizontal translation 2 units to the left, $y = 2|x + 2|$; C. $y = x^2$, vertical stretch of 2 and translation up 3 units, $y = 2x^2 + 3$]

ENRICH

One transformation that is not done in this chapter is a rotation of 180°. For example, if $(x, y) \rightarrow (-x, -y)$ for all points (x, y) of the function, then each point of the graph of the function is rotated 180° about the origin. Another transformation is a reflection through the line $y = x$. For example, if $(x, y) \rightarrow (y, x)$ for all points (x, y) of the function, then each point of the function is reflected through the line $y = x$. Have students make a table of points for the parent function $y = x^2$, rotate the points through 180°, and then reflect the original points through the line $y = x$.

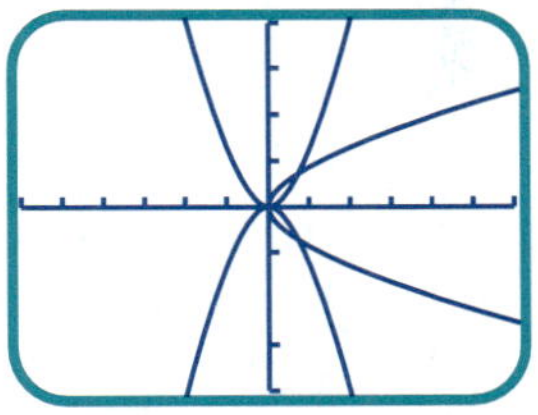

Interdisciplinary Connection

Medicine Virologists and other medical scientists use functions to help them predict the growth or decline of viruses that cause epidemics. The functions they use to model the spread of a virus often involve combined transformations.

Practice

5.6 Combining Transformations

Sketch the graph of each function.

1. $f(x) = \text{INT}(x - 2)$

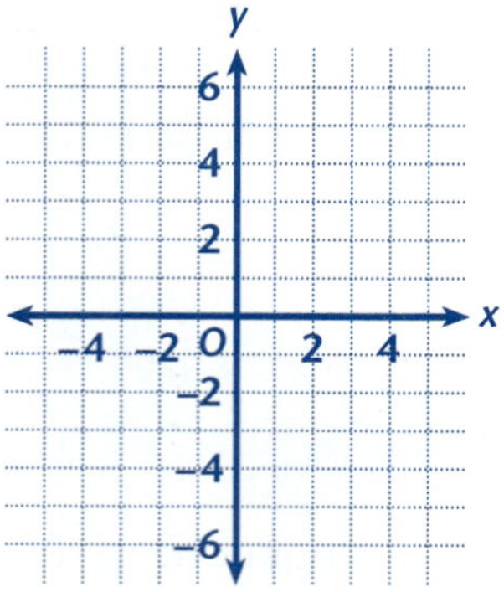

2. $g(x) = -x + 1$

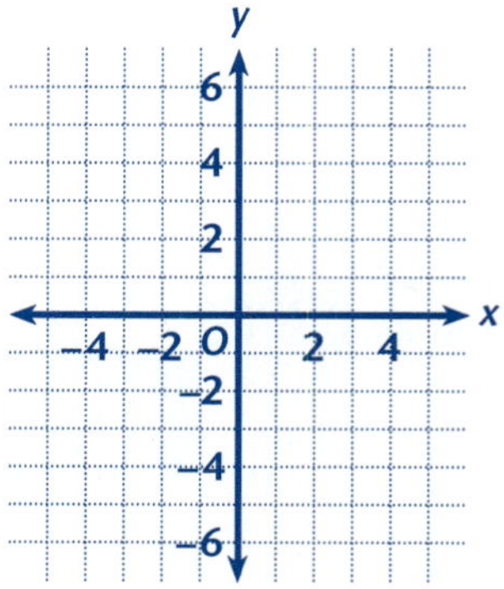

3. $h(x) = |x - 4|$

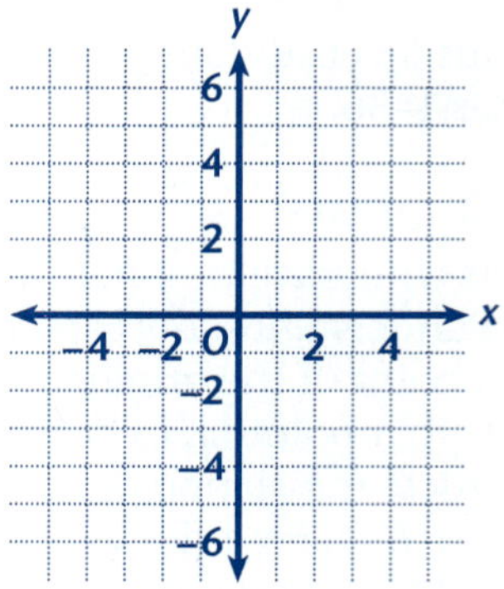

4. $p(x) = 2^x - 1$

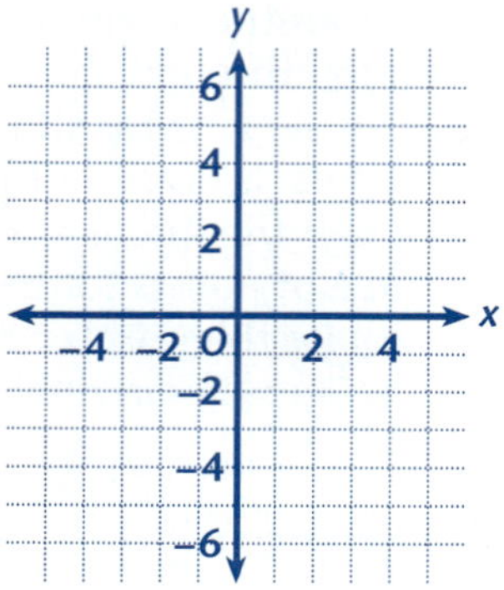

5. $k(x) = (x - 2)^2$

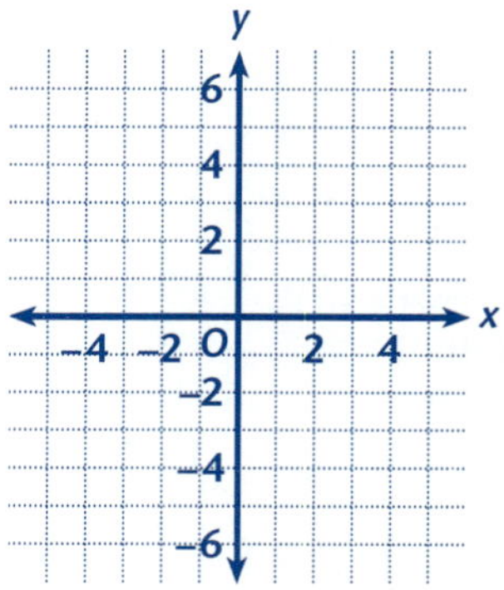

6. $h(x) = \dfrac{1}{x - 1}$

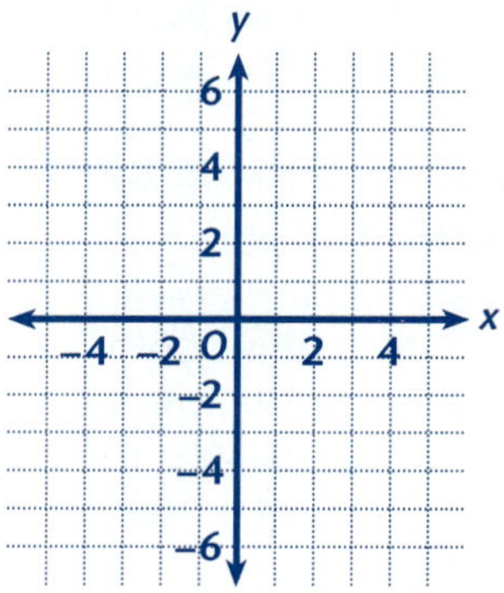

Tell what happens to the point (4, 8) on the graph of $y = 2x$ when the transformations are applied in the order given. HINT: If you are not sure, draw a sketch.

7. a vertical stretch of 2, followed by a vertical translation of 2

____

8. a vertical translation of 2, followed by a vertical stretch of 2

____

9. a horizontal translation of 5, followed by a vertical stretch of 3

____

10. a vertical stretch of 3, followed by a horizontal translation of −5

____

NAME ______________________ CLASS ____________ DATE ____________

Lesson Activity

5.6 Burning Calories

The graph shows the estimated number of Calories burned each hour when a 100-pound, a 120-pound, a 140-pound, and a 180-pound person walk at various rates.

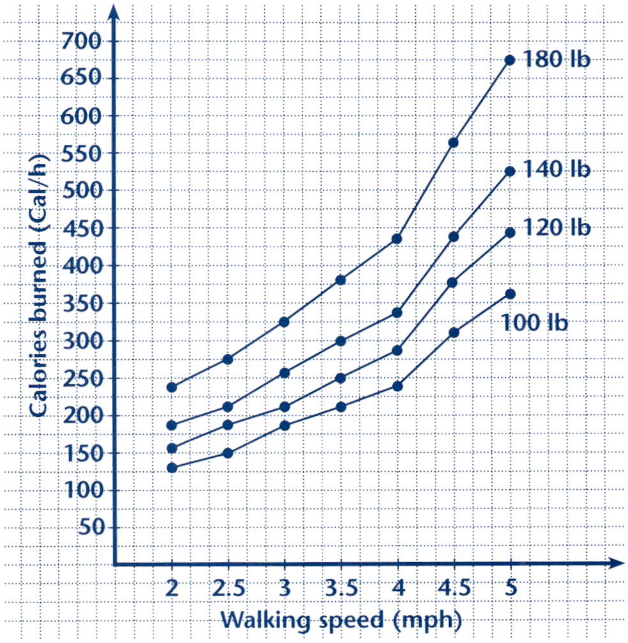

1. According to the graph, on what two factors does the number of Calories burned per hour depend?

____

2. Identify the range and domain for each person's graph. ____________________

3. According to the graph, about how fast must a 120-pound person walk to burn about 180 Calories per hour? ____________________

4. At approximately what speed must a 100-pound person walk to burn as many Calories per hour as a 140-pound person walking at 2.5 mph?

____

5. Is there a greater change in the number of Calories burned per hour when a person increases walking speed from 2 mph to 2.5 mph or from 4 mph to 4.5 mph?

____

6. Approximate the number of Calories burned per hour when a 100-pound, a 120-pound, a 140-pound, and a 180-pound person walks at 3 mph.

____

NAME ______________________________ CLASS ____________ DATE ____________

Basic Skills Practice
Cumulative Review

Read each question and circle the best answer.

1. Write a mathematical sentence for this statement: 8 subtracted from 3 times a number is 16.

A $8 - 3n = 16$
B $3n - 8 = 16$
C $16 - 3n = 8$
D $16 - 8 = 3n$

2. Find the perimeter of a rectangle with sides of 4 cm and 6 cm.

F 10 cm
G 15 cm
H 20 cm
I 24 cm

3. Given the similar figures shown, find the length of side *AB*.

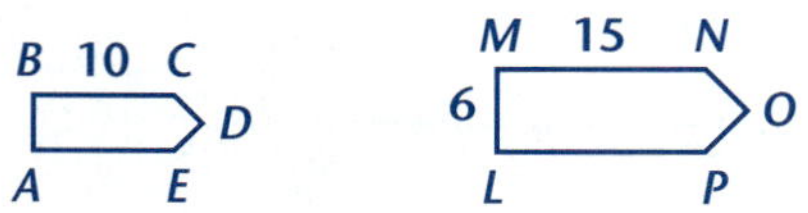

A 4 cm
B 6 cm
C 8 cm
D 10 cm

4. Identify the type of symmetry for the figure shown.

F Horizontal line of symmetry
G Vertical line of symmetry
H Point symmetry
I No symmetry

5. The formula for the area of a triangle is $\frac{1}{2}bh$, where b is the base and h is the height. Find the area of a triangle with a base of 5 cm and a height of 3 cm.

A 15 cm^2
B 7.5 cm^2
C 5 cm^2
D 3.5 cm^2

6. Which inequality describes the number line?

F $x \leq 3$
G $x \geq -2$
H $x < 6$
I $x < 0$

7. What is the volume of a cargo box with a base that measures 20 ft^2 and a height of 5 ft?

A 25 ft^3
B 50 ft^3
C 100 ft^3
D 125 ft^3

8. To solve the equation $3a + 2 = 11$, you need to—

F add 2 to both sides, and then divide both sides by 3
G subtract 2 from both sides, and then divide both sides by 3
H divide both sides by 3
I divide both sides by 11

LESSON 6.1 Exploring Exponents

PREPARE

Pacing Guide

2 Days

Objectives and Overview

- Understand the concepts of exponents and powers.
- Use the properties of exponents to simplify expressions.

Students explore different concepts relating to exponents and powers and apply the properties of exponents to simplifying expressions.

Vocabulary

customary notation
decimal notation
expanded notation
exponential notation

Assessing Prior Knowledge

Evaluate.

1. a^2 when $a = 4$ **[16]**
2. $\frac{1}{b^2}$ when $b = 3$ **$\left[\frac{1}{9}\right]$**
3. c^3d when $c = 4$ and $d = 5$ **[320]**
4. $e^4 \div f^2$ when $e = 3$ and $f = 9$ **[1]**
5. $(hj)^2$ when $h = 3$ and $j = 8$ **[576]**

Approach A Using the Book

Opening Discussion

Ask students if there is an easy way to multiply 10 by itself 100 times. Is there an easy way to express 10 multiplied by itself 100 times? Have they ever heard of the power of a number?

Brainstorming

Ask students to find a pattern in the table below. **[in the second row, each number is double the previous number]** What is a convenient way to express the second row of numbers in the table? **[multiply 2 by itself]**

1	2	3	4	5	6	7	8
2	4	8	16	32	64	128	256

If $2 \cdot 2 \cdot 2 \cdot 2 \cdot 2 = 2^5$, how could they write the numbers in the second row of the table?
[$2^1, 2^2, 2^3, 2^4, 2^5, 2^6, 2^7, 2^8$]

continued on top of next page

TEACH

Approach B Using Models

Opening Discussion

Ask students what they know about the powers of numbers. Is there an easy way to express "10 multiplied by itself 100 times?" What is a googol? **[the number 1 followed by 100 zeros, or 10 to the 100th power]**

Brainstorming

Bring an 8-feet long piece of string to class. Fold the string in half so that there are two strands and show the class. Now fold the string on itself again and count the number of strands with the class. **[4]** Explain that the original number of strands, 2, has been multiplied by 2 to produce 4 strands. Fold the string on itself again and count the number of strands with the class. **[8]** The number of strands has again doubled. Now ask the class how they would describe mathematically what has happened to the original number of strands. **[Answers will vary. Sample answer: there were 2 strands at first, then**

continued on bottom of next page

Approach A — Using the Book, continued

Setting Up the Explorations

Teacher-Guided Explorations If you wish, you can guide the students through the textbook explorations by using an overhead calculator or the chalkboard. Make sure that students understand how exponents are used to write expressions in simpler form. Explain how to read an expression with exponents.

Cooperative-Learning Groups Divide the class into groups of four. Ask one student to read the steps from the textbook. Have another student write the answers to the steps in the explorations. Have a third student make conjectures about how to simplify expressions containing exponents. Have a fourth student write a report detailing the group's activities, including whether the conjectures were tested. Ask each student to sign the completed report.

Definition of *Exponent, Power,* and *Base*

Discuss exponents and powers with students. Explain that exponents provide a simple way to express a number with many digits. For example, by using an exponent, $10 \cdot 10 \cdot 10 \cdot 10 \cdot 10$ can be written as 10^5. The exponent is 5, and it represents 5 factors of 10 in the expression $10 \cdot 10 \cdot 10 \cdot 10 \cdot 10$. The number 10 is called the *base*. When an expression with exponents is read, it is customary to use the word *power.* When reading the expression above, it is customary to say "10 to the 5th power" or "10 raised to the power of 5." Numbers written with exponents are sometimes simply called powers. Emphasize that exponents are used for convenience and that the exponent is the number of repeated factors of a number.

Definition of *Customary Notation, Expanded Notation,* and *Exponential Notation*

Customary notation, or *decimal notation,* and *expanded notation* are introduced here.

TEACH

Approach B — Using Models, continued

the number doubled, then doubled again. This can be written a $2 \cdot 2 \cdot 2$.]

Explain that the expression $2 \cdot 2 \cdot 2$ is written as 2^3. In this notation, 3 is an exponent.

Exponents and Powers

Ask the class how they could write $2 \cdot 2 \cdot 2 \cdot 2$. [$2^4$] How would they write $2 \cdot 2 \cdot 2 \cdot 2 \cdot 2 \cdot 2 \cdot 2 \cdot 2 \cdot 2 \cdot 2$? [2^{10}] What does 2^7 mean? [$2 \cdot 2 \cdot 2 \cdot 2 \cdot 2 \cdot 2 \cdot 2$]

Discuss the definition at the top of page 267.

Ask students to find a pattern in the table below. **[in the second row, each number is double the previous number]** What is a convenient way to express the second row of numbers in the table? **[multiply 2 by itself]**

1	2	3	4	5	6	7	8
2	4	8	16	32	64	128	256

If $2 \cdot 2 \cdot 2 \cdot 2 \cdot 2 = 2^5$, how could they write the numbers in the second row of the table? [$2^1, 2^2, 2^3, 2^4, 2^5, 2^6, 2^7, 2^8$]

Customary, Decimal, and Expanded Notation

Cooperative Learning Discuss the three notations for a number on page 267 of the Pupil's Edition.

Have students work in groups of three. One student will be the customary notation expert, one student the expanded notation expert, and the last student the exponential notation expert for the following exercise:

Have each student make up a 4-digit number and write it in the notation of his or her respective expertise. Then have them share their number with the other two members of the group. Each student must read the numbers and write them in the notation for which he or she is responsible.

Repeat the exercise two more times, giving each student the opportunity to be an expert in each of the three notations.

Point out that expanded notation shows the value of each digit. *Exponential notation,* on the other hand, shows the powers of 10 that are inherent in the place value of the digit. Emphasize that all of these forms are equivalent.

Ongoing Assessment, page 267

$1 \cdot 10^4 + 5 \cdot 10^3 + 2 \cdot 10^2 + 0 \cdot 10^1 + 8$

On Their Own

A. Write 5723 in expanded notation.

B. Write $7 \cdot 1000 + 4 \cdot 100 + 5 \cdot 10 + 9$ in customary notation.

C. Write 8917 in exponential notation.

D. Write $6 \cdot 10^4 + 4 \cdot 10^3 + 7 \cdot 10^2 + 3 \cdot 10^1 + 1$ in customary notation.

[A. $5 \cdot 1000 + 7 \cdot 100 + 2 \cdot 10 + 3$; B. 7459; C. $8 \cdot 10^3 + 9 \cdot 10^2 + 1 \cdot 10^1 + 7$; D. 64,731]

Exploration 1

In this exploration, students discover how to multiply two exponential expressions with the same base.

Tell students to look for patterns in the multiplications so that the usefulness of exponents will become apparent. Check their generalization in Step 5. Make sure they understand that a is the base and can represent any number. Because some students have a difficult time connecting algebraic expressions to numeric ones, do several examples of the rule in Step 5, $a^m \cdot a^n = a^{m+n}$. Emphasize that $a^m + a^n \neq a^{m+n}$ by showing the counterexample $2^3 + 2^4 = 8 + 16 = 24$ and $2^7 = 128$.

Use **Transparency 24** for discussion or review.

Ongoing Assessment, page 267

5. $a^m \cdot a^n = a^{m+n}$

continued on top of next page

Cooperative-Learning Groups

Before starting the explorations, divide the class into groups of four. Ask one student to read the steps from the textbook. Have another student make the tables and do the calculations for the explorations. Have a third student make conjectures about the properties of exponents. Have the fourth student write a report detailing the group's activities. Ask each student to sign the completed report.

Exploration 1

Have students work in their groups to complete Exploration 1. Then have students further investigate the product of two powers with the same base by answering the following questions. The questions may be listed on the chalkboard or the overhead.

- Is $2^3 \cdot 3^4 = 6^7$ true? Make a conjecture from your answer. **[no; in general, $a^m \cdot b^n \neq (ab)^{m+n}$]**
- Can the base of an exponential expression be less than 1? **[yes, as long as it is positive]**
- Explain why you add the exponents when you multiply two powers of the same base. **[Because each power represents a number of factors; if you wrote out each power the long way and counted the factors, it would be the same as adding the number of factors]**

Student Reports

One or more members of each group should report their group's findings to the class. Ask them to show their calculations and discuss their conjectures. The reporter should state the group's conclusions clearly, including the answers to the exploration questions, and the entire group should answer questions from other groups.

On Their Own

Simplify each expression and write the answer in decimal form.

A. $2^7 \cdot 2^5$ **B.** $10^6 \cdot 10^3$ **C.** $2 \cdot 2^8$

[A. $2^{12} = 4096$; B. $10^9 = 1{,}000{,}000{,}000$; C. $2^9 = 512$]

continued on bottom of next page

Approach A

Using the Book, continued

On Their Own

Simplify each expression.

A. $2^7 \cdot 2^5$ **B.** $10^6 \cdot 10^3$ **C.** $2 \cdot 2^8$

[**A.** 2^{12}; **B.** 10^9; **C.** 2^9]

Exploration 2

In this exploration, students discover that to divide two exponential expressions with the same base, the exponent of the denominator is subtracted from the exponent of the numerator.

Emphasize that the bases must be the same.

Ongoing Assessment, top of page 268

4. a^{m-n}

6. No; the bases of numbers expressed in exponential form must be the same in order to combine the exponents by dividing powers.

On Their Own

Simplify each expression.

A. $\frac{10^5}{10^2}$ **B.** $\frac{5^6}{5^2}$ **C.** $\frac{3^8}{3}$

[**A.** $10^3 = 1000$; **B.** $5^4 = 625$; **C.** $3^7 = 2187$]

Critical Thinking, page 268

$(ab)^n$ means that ab is a factor n times. By rearranging the factors, a is a factor n times and b is a factor n times. Hence $(ab)^n = a^nb^n$. Similarly $\left(\frac{a}{b}\right)^n$ means $\frac{a}{b}$ is a factor n times, so a is a factor in the numerator n times and b is a factor in the denominator n times. Hence, $\left(\frac{a}{b}\right)^n = \frac{a^n}{b^n}$.

Exploration 3

In this exploration, students discover how to find the power of the power of a number, or $(a^m)^n = a^{mn}$, for positive integers m and n and the positive number a.

Emphasize that the power of a power is a short way to write repeated factors, each of which is a

TEACH

Approach B

Using Models, continued

Exploration 2

Have students work in their groups to complete Exploration 2. Then have students answer the following questions about dividing powers:

- Is $\frac{6^4}{3^2} = 2^2$ true? Make a conjecture about your result. **[no; in general, $\frac{a^m}{b^n} \neq \left(\frac{a}{b}\right)^{m-n}$]**
- Explain why subtraction is used in the rule to divide powers, $\frac{a^m}{a^n} = a^{m-n}$. **[If both the numerator and denominator were written in customary notation, the division would reduce the number of factors of *a* in the numerator by *n* factors, so the number of factors left would be *m* − *n*.]**

On Their Own

Simplify each expression and write the answer in decimal form.

A. $\frac{10^5}{10^2}$ **B.** $\frac{5^6}{5^2}$ **C.** $\frac{3^8}{3}$

[**A.** $10^3 = 1000$; **B.** $5^4 = 625$; **C.** $3^7 = 2187$]

Exploration 3

Have students work in their groups to complete Exploration 3. Then have students answer the following questions about the power of a power:

- Is $3^3 \cdot 3^2 = (3^3)^2$ true? Make a conjecture about your result. **[no; in general, $a^m \cdot a^n \neq (a^m)^n$]**
- Use your conjecture to explain why the power of a power is not the same as multiplying powers. **[$(a^m)^n$ is really $a^m \cdot a^m \cdot \ldots \cdot a^m$, *n* times, so there are *n* factors of a^m. If *m* is repeated *n* times, there are *mn* factors of *a*. This is usually not the same as *m* + *n*.]**

On Their Own

Simplify each expression and write the answer in decimal form.

A. $(3^4)^2$ **B.** $(10^2)^3$ **C.** $(2^2)^5$

[**A.** $3^8 = 6561$; **B.** $10^6 = 1{,}000{,}000$; **C.** $2^{10} = 1024$

Application

The following physical model can be used to help students visualize the rapid growth of bacteria explored in the application on page 269 of the Pupil's Edition.

power. This is the only rule in which you multiply powers because multiplication is repeated addition. Show students that $(a^m)^n = a^{mn}$ is not the same as $a^m \cdot a^n = a^{m+n}$ for $a = 2$, $m = 3$, and $n = 4$.

Ongoing Assessment, bottom of page 268

5. $(a^m)^n = a^{mn}$

On Their Own

Simplify each expression and write the answer in decimal notation.

A. $(3^4)^2$ **B.** $(10^2)^3$ **C.** $(2^2)^5$

[A. $3^8 = 6561$; B. $10^6 = 1{,}000{,}000$; C. $2^{10} = 1024$]

Application

A common application of exponential functions is exponential growth. If students have difficulty understanding how there could be 8000 bacteria in 3 hours, show the expanded form $1000 \cdot 2 \cdot 2 \cdot 2 = 8000$. There are 3 factors of 2 because the bacteria doubles every hour for 3 hours.

Teaching Tip: Extension

Have students experiment with the exponent key on their calculator to determine exactly how to enter an exponential expression. Stress the use of parentheses as grouping symbols for computations involving fractions. Point out that answers may be displayed in "E" form. For example, 10^{10} may be displayed as 1E10. That means the answer is a 1 with 10 zeros after it.

Ongoing Assessment, page 269

a. True **b.** False; $10^3 + 10^4 \neq 10^7$

Critical Thinking, page 269

Exponents can be added only when the powers are being multiplied and when the bases are the same.

Closing the Lesson

Summarize the lesson by reviewing the rules for multiplying powers, dividing powers, and finding the power of a power.

Have students take a piece of notebook paper and fold it in half. Students should record the number of regions and fold the paper again. Have them record the number of regions. Have students continue the process and keep a running total at each step. The number of regions doubles with each fold.

Now, letting one region represent 1000 bacteria, have them verify that there will be 8000 bacteria at 3 P.M. **[Fold the piece of paper 3 times. This produces 8 regions, each representing 1000 bacteria. Thus, there will be 8000 bacteria.]** How many bacteria will there be at 4 P.M.? **[Fold the paper once more to produce 16 regions, which represent 16,000 bacteria.]**

Extension

Cooperative Learning After demonstrating the calculator keys, keystrokes, and examples from the Extension, give students the following instructions for an activity that they can complete with a partner:

1. Think of three quotients of powers and write them on notebook paper.
2. Exchange papers with your partner.
3. Simplify your partner's quotients.
4. Work with your partner to check the answers and make corrections if needed.

Check students' work and discuss any common errors with the class.

Closing the Lesson

Summarize the lesson by reviewing the rules for multiplying powers, dividing powers, and finding the power of a power.

ASSESS

Selected Answers

Odd-numbered Exercises 11–63 and odd-numbered Extra Practice

Assignment Guide

Core level students should complete Exercises 1–71.

Core-plus level students should complete Exercises 1–71.

Error Analysis

Students may have difficulty with Exercises 35–40. Point out that they should analyze each problem carefully to see if one of the properties they learned in this lesson applies. Suggest that they state the property, use it for the exercise, and then check the answer with a calculator.

Alternative Assessment

Using a Spreadsheet Have students use a spreadsheet to help them understand the properties of exponents learned in this lesson. Ask them to enter the exponent 1–6 into column A of the spreadsheet. Use the fill down feature to make the entries relative to the first entry. Have them enter the base into column B of the spreadsheet. If this column is also made relative, then only the first number needs to be changed to explore other bases. Use column C to multiply powers, column D to divide powers, and column E to find the power of a power.

Compare the values generated by the spreadsheet with the values found in the explorations. Change the number in column B to see a new table. Change the formula in column C, D, or E to see different calculations. Encourage students to predict the value in the cell before it is entered.

RETEACH

Have students review the rules they learned in the explorations. Then ask them to do the following exercises. Have them show the answer in exponential notation and then simplify. They should also check their answers with a calculator.

A. $2^4 \cdot 2^5$ **B.** $\frac{2^9}{2^3}$ **C.** $(2^5)^4$ **D.** $\frac{2^5 + 2^4}{2^3}$

[A. $2^9 = 512$; B. $2^6 = 64$; C. $2^{20} = 1{,}048{,}576$; D. 6]

ENRICH

Graphics Calculators and Powers of Two A graphing calculator can be used to explore the powers of a number if an exponential function with that number as a base is graphed. For example, to find powers of 2, graph Y = 2^X. Then use ZOOM INTEGER (ZInteger) to show integer values of 2 to the power *x*. Look at the table or the graph to see positive and negative powers of 2. Ask students to experiment with similar exponential graphs to help them make the connection to future lessons.

Interdisciplinary Connection

Biology Real-world applications of exponential growth are common in biology. Finding the *growth factor* (the base of the number written in exponential notation) for a biological specimen may take many years of scientific experimentation.

NAME ______________________ CLASS ____________ DATE ____________

Practice

6.1 Exploring Exponents

Change each to customary notation.

1. 2^4 ________ **2.** 10^7 ________ **3.** 4^3 ________ **4.** 10^6 ________

5. 6^3 ________ **6.** 5^3 ________ **7.** 1^9 ________ **8.** 4^4 ________

Evaluate each of the following:

9. $(3^4)^2$ ________ **10.** $(7^2)^2$ ________ **11.** $(10^4)^3$ ________

12. $\frac{5^3}{5^2}$ ________ **13.** $\frac{7^5}{7^3}$ ________ **14.** $3^2 + \frac{3}{3}$ ________

15. $2^4 \cdot \frac{2^2}{2^3}$ ________ **16.** $\frac{3^2 \cdot 3^2}{3^3}$ ________ **17.** $\frac{10^4 \cdot 10^5}{10^2}$ ________

18. $m^4 \cdot m^5$ ________ **19.** $f^4 \cdot f^4$ ________ **20.** $r^9 \cdot r^2$ ________

21. $(a^4)^2$ ________ **22.** $(b^3)^7$ ________ **23.** $\frac{x^3}{x^2}$ ________

24. $m^4 \cdot \frac{m^2}{m^3}$ ________ **25.** $y^2 + \frac{y}{y}$ ________ **26.** $m^3 \cdot m^5$ ________

27. $(r^3)^2 \cdot r^4$ ________ **28.** $q^4 \cdot (q^3)^5$ ________ **29.** $(t^3)^3 \cdot (t^2)^4$ ________

30. $\frac{d^7}{d^3}$ ________ **31.** $\frac{w^5(w^2)}{w^3}$ ________ **32.** $\frac{(w^4)^3}{w^2}$ ________

Some of the following can be expressed as a single power of 10; others cannot. If possible, write the given expression as a single power of 10. Then evaluate each expression.

33. $10 \cdot 10^2$ ________ **34.** $10 + 10^2$ ________

35. $10^2 \cdot 10^4$ ________ **36.** $10^7 \div 10^5$ ________

37. $10^7 - 10^5$ ________ **38.** $(10^2)^4$ ________

39. $10^2 + 10^4$ ________ **40.** 10^{4+2} ________

41. $\frac{10^9}{10^2}$ ________ **42.** $\frac{10^{10} - 10^3}{10^5}$ ________

NAME ______________________ CLASS ____________ DATE ____________

Lesson Activity
6.1 Modeling Properties of Exponents

To model some properties that are used for computing with exponents, begin by cutting out a 1 in. × 2 in. rectangle from grid paper. Label the rectangle with its area, "2." Place the rectangle on grid paper, trace around it, and then slide it up and trace again, as shown.

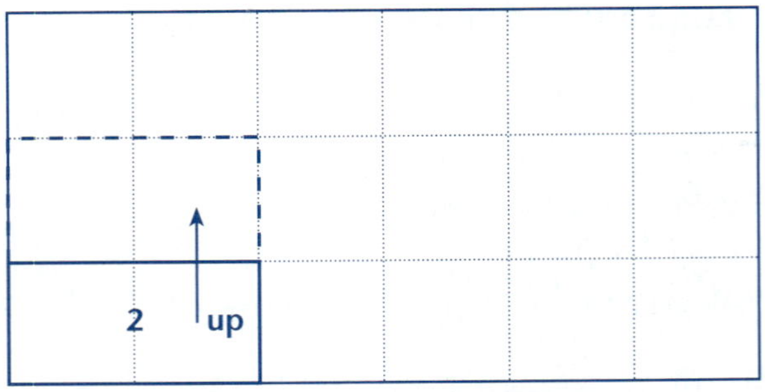

Cut out and label this piece "2 × 2." Place the 2 × 2 piece on grid paper. Trace around it, slide it right, and trace again. Label this piece "2 × (2 × 2)," and cut it out.

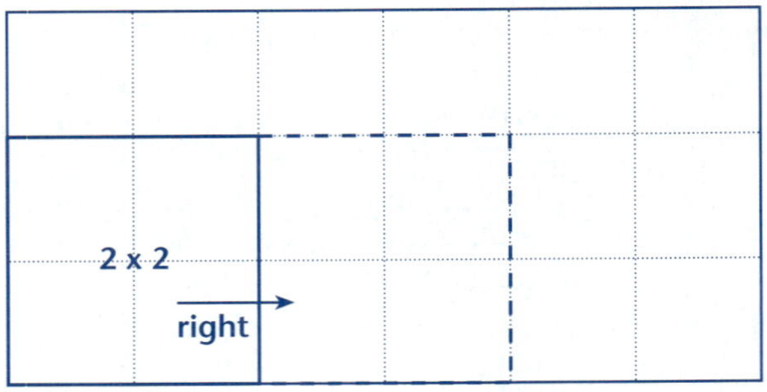

Place the 2 × (2 × 2) piece on grid paper, and trace around it. Slide it either up or right, and trace again. Label this piece "2 × (2 × 2 × 2)," and cut it out.

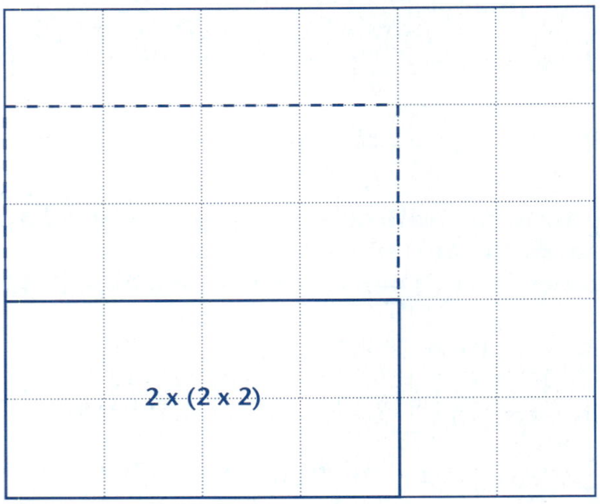

Place the pieces on a flat surface in order from left to right.

1. Describe the geometric shapes as you look at them from left to right.

____

____

2. Write the exponential name of each shape in order from left to right.

____

3. What is the relationship between the exponent and the number of factors of 2 used to describe the area of each shape?

____

4. What geometric shape do the even exponents describe? the odd exponents?

____

NAME ______________________ CLASS __________ DATE __________

Basic Skills Practice

Reflections and Translations, Part 1

A *reflection* is a special type of transformation that involves flipping a figure over a line called the line of reflection. Line *l* is a line of reflection. The original figure, ΔABC, is called the pre-image. The figure resulting from the transformation, ΔDEF, is called the image. Pre-images and images have the same size and shape, and are always congruent.

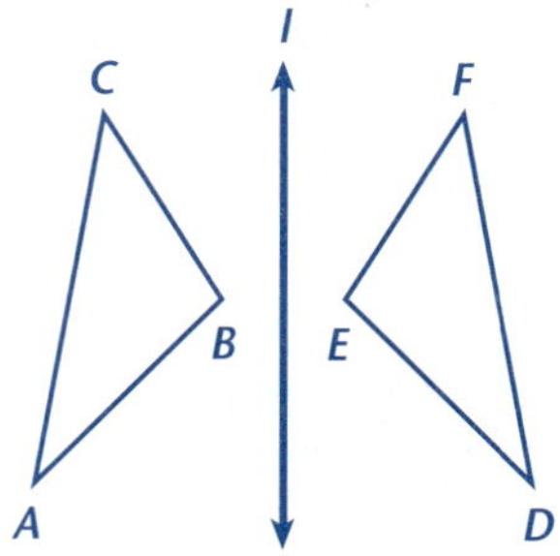

A *translation* is a transformation that moves all the points in a figure a fixed distance in a given direction. A translation retains the same size and shape of the original figure, so the original figure and the translation image are always congruent. Translations are also called *slides*.

Example: Translate ΔABC 6 units to the right and 1 unit up, and label the translation image ΔXYZ.

$A(1, 1) \rightarrow X(1 + 6, 1 + 1)$ or $X(7, 2)$
$B(6, 1) \rightarrow Y(6 + 6, 1 + 1)$ or $Y(12, 2)$
$C(1, 5) \rightarrow Z(1 + 6, 5 + 1)$ or $Z(7, 6)$

The graph shows that ΔXYZ is congruent to ΔABC.

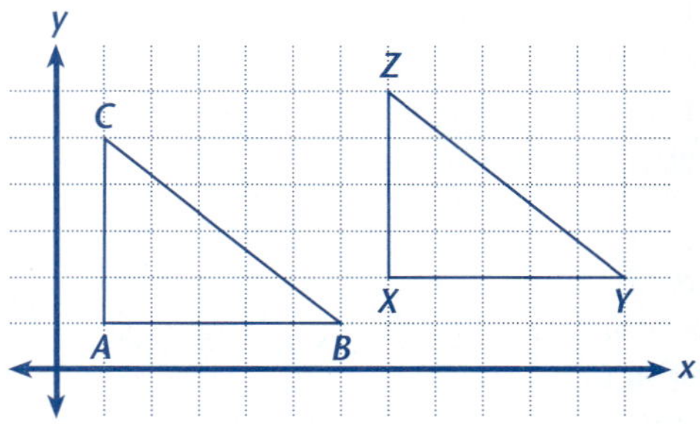

The *y*-axis is the line of reflection in the diagram shown.

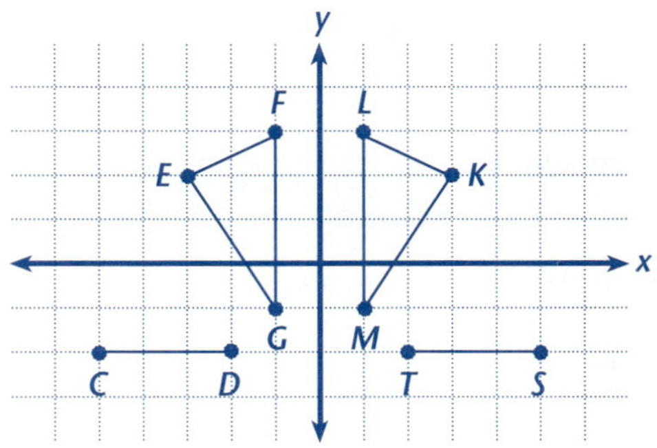

1. Give the coordinates of the reflection of point *E*. ______

2. What is the reflected image of ΔEFG? ______

3. Give the coordinates of the image point for *C*. ______

4. What is the reflected image of segment *CD*? ______

Find the translation image of each of the figures whose vertices are provided.

5. $A(2, 2)$, $B(7, 2)$, $C(7, 5)$, $D(2, 5)$; slide 2 units right and 3 units up ______

6. $A(-3, 2)$, $B(-8, 2)$, $C(-8, 4)$, $D(3, 4)$; slide 2 units left and 1 unit up ______

7. $A(3, -2)$, $B(6, 2)$, $C(6, -4)$, $D(3, -4)$; slide 3 units right and 2 units down ______

8. $A(-3, 1)$, $B(-1, 3)$, $C(0, 0)$; slide 2 units right and 1 unit down ______

LESSON 6.2

Multiplying and Dividing Monomials

PREPARE

Pacing Guide
2 Days

Objectives and Overview
- Understand the structure and concept of a monomial.
- Use the properties of exponents to simplify expressions.

Students learn how to find the product and quotient of monomial expressions that contain exponents. They learn the Product-of-Powers Property, the Power-of-a-Power Property, the Power-of-a-Product Property, and the Quotient-of-Powers Property and apply them to monomial expressions.

Vocabulary
monomial

Assessing Prior Knowledge

Evaluate.

1.	$(-2)^2$	[4]
2.	$(-2)^3$	[−8]
3.	$(2)(3)(4)$	[24]
4.	$(2)(-3)(4)$	[−24]
5.	$(-2)(-3)(-4)$	[−24]
6.	$(-2)(3)(-4)$	[24]

Approach A: Using the Book

Opening Discussion
How do you multiply two powers? How do you divide two powers? How do you find the power of a power?

Brainstorming
Have students brainstorm ways to simplify the following expression: $(2 \cdot 2 \cdot 2 \cdot 2 \cdot 2 \cdot 3 \cdot 3 \cdot 3 \cdot 3 \cdot 3 \cdot 3 \cdot 3)(2 \cdot 2 \cdot 2)(3 \cdot 3 \cdot 3 \cdot 3 \cdot 3)$. **[After commuting the 2s, there is a sequence of 5 + 3, or 8, 2s and a sequence of 7 + 5, or 12, 3s. The compact notation agreed on by mathematicians is $2^8 \cdot 3^{12}$.]** How can the exponent key on a calculator be used to simplify the product quickly?

Product-of-Powers Property
Discuss the Product-of-Powers Property with students. Point out how the rule captures the pattern seen above.

TEACH

Approach B: Using Models

Opening Discussion
How do you multiply two powers? How do you divide two powers? How do you find the power of a power? What do you do if the expressions are more complicated?

Brainstorming
Have students brainstorm ways to simplify the following expression: $(2 \cdot 2 \cdot 2 \cdot 2 \cdot 2 \cdot 3 \cdot 3 \cdot 3 \cdot 3 \cdot 3 \cdot 3 \cdot 3)(2 \cdot 2 \cdot 2)(3 \cdot 3 \cdot 3 \cdot 3 \cdot 3)$. **[After commuting the 2s, there is a sequence of 5 + 3, or 8, 2s and a sequence of 7 + 5, or 12, 3s. The compact notation agreed on by mathematicians is $2^8 \cdot 3^{12}$.]** How can the exponent key on a calculator be used to simplify the product quickly?

Product-of-Powers Property
Discuss the Product-of-Powers Property with students. Point out how the rule captures the pattern seen above.

Defining *Monomial*

The textbook defines *monomial* as an algebraic expression that is either a constant, a variable, or a product of a constant and one or more variables. The key to distinguishing a monomial from a binomial or other polynomial is that there are no arithmetic operators like + or − in a monomial expression. A monomial expression can also contain powers of variables with positive integer exponents.

Example 1

In this example, the Product-of-Powers Property is used to simplify the product of monomials.

Point out that in Lesson 6.1, the property for the product of exponential expressions with the same base was given. In this lesson, the same property is used repeatedly with more complex expressions.

Additional Example 1

Find the products.

A. $(4xy)(-3x^2y^4)$ $\quad$ **$[-12x^3y^5]$**

B. $(-3g^2h^2)(-gh^3)(2h^3k^4)$ $\quad$ **$[6g^3h^8k^4]$**

Critical Thinking, page 274

Since the bases are the same in each problem, add the exponents.

a. $x^{3c} \cdot x^{4c} = x^{(3c + 4c)} = x^{7c}$

b. $(x + c)^d \cdot (x + c)^{2d} = (x + c)^{(d + 2d)} = (x + c)^{3d}$

Teaching Tip

Make sure that students understand the difference between the Product-of-Powers Property and the Power-of-a-Power Property. Point out that the Power-of-a-Power Property is really just a short way to use the Product-of-Powers Property. For example, use the Product-of-Powers Property to rewrite $(4^2)^3$ as $(4^2)(4^2)(4^2) = 4^{2+2+2} = 4^6$.

continued on top of next page

Defining *Monomial*

The textbook defines *monomial* as an algebraic expression that is either a constant, a variable, or a product of a constant and one or more variables. The key to distinguishing a monomial from a binomial or other polynomial is that there are no arithmetic operators like + or − in a monomial expression. A monomial expression can also contain powers of variables with positive integer exponents.

Discuss this definition with the class and give examples.

Cooperative Learning

Have students work in groups of four to complete the following activities. Bring blank index cards to class and distribute them to the students.

Alternate Example 1

After discussing Example 1 from the Pupil's Edition with the class, instruct the students to write a monomial of their own on their index card. If necessary, remind students of the definition of monomial. For variety, encourage them to include two letters from those used in the example and to use their favorite numbers for the coefficient and powers.

Now tell the students to work with another member of their group to complete the following exercise:

1. On notebook paper, write your monomial.
2. Write your partner's monomial beside your monomial.
3. Working with your partner, find the product by using the Product-of-Powers Property.

Students should repeat this exercise until each student has worked with each of the other members of the group.

Alternate Examples 2 and 3

After explaining the Power-of-a-Power Property and the Power-of-a-Product Property in the Pupil's Edition and presenting Examples 2 and 3, have each student pick a power and write "Raised to the power of _____" on the back of their index card, with the space filled in with their power.

continued on bottom of next page

Approach A: Using the Book, continued

Example 2

In this example, the Power-of-a-Power Property is used to simplify a numeric and an algebraic expression.

Encourage students to write out the expression by using the Product-of-Powers Property (see the Teaching Tip on the previous page) so that they make the connections between the properties. Encourage them to visualize the differences between expressions that involve the different properties.

Additional Example 2

Simplify.

A. $(4^3)^2$ **[$4^6 = 4096$]**

B. $(y^5)^3$ **[y^{15}]**

Example 3

In this example, the base of the expression is a monomial.

Point out that the monomial is a factor of the answer that occurs the same number of times as the power of the monomial, so the expression $(4x)^3$ equals $4x \cdot 4x \cdot 4x$, which can be simplified by using the Product-of-a-Power Property.

Additional Example 3

Simplify the expression $(x^2y^4)^3$. **[x^6y^{12}]**

Example 4

In this example, the Power-of-a-Power Property is used to simplify more complicated monomial expressions.

The example shows that the Power-of-a-Product Property can be used to simplify expressions with numerical coefficients as well as variables. If the coefficient is within the parentheses grouping the monomial term, then the Power-of-a-Power Property applies to it also.

TEACH

Approach B: Using Models, continued

Now have the students work with each of the other students in their group to complete the following exercise:

1. On notebook paper, write your partner's monomial.
2. Raise your partner's monomial to your power.
3. Work with your partner to find the result using the Power-of-a-Power Property.

As students are completing these exercises, check their work and remind them of the definitions and properties, if necessary.

Alternate Example 4

Use Example 4 to reinforce how powers of a monomial are simplified and to discuss what happens if a negative monomial is raised to a power. Have students make their monomials negative and repeat the exercise above with at least one partner from their group.

Alternate Example 5

After discussing the Quotient-of-Powers Property and presenting Example 5, have students complete the following exercise with a partner:

1. On notebook paper, write your partner's monomial power.
2. Find a new monomial with powers that are smaller than those of your partner's monomial.
3. Now divide your partner's monomial by your new monomial and record the result.

Check student's work and remind them of the definitions and properties, if necessary.

Working With a Scientific Calculator

Have students evaluate each of the following pairs of expressions by using a scientific calculator. Are the pairs equal?

A. $(-3x^2y^3)(-4xy^2)$ and $12x^3y^6$ for $x = 4$ and $y = 5$ **[not equal]**

B. $(-3x^2y^4)^3$ and $-27x^6y^{12}$ for $x = 3$ and $y = -2$ **[equal]**

C. $\frac{8s^2t^4}{4st^2}$ and $2st^2$ for $s = -4$ and $t = 3$ **[equal]**

Additional Example 4

Simplify each expression.

A. $(4x^2y^5)^3$ **B.** $(-y)^4$ **C.** $(-y)^3$ **D.** $(-4x)^5$

[**A.** $64x^6y^{15}$; **B.** y^4; **C.** $-y^3$; **D.** $-1024x^5$]

Critical Thinking, page 275

The expression $(-x)^m$ is positive whether x is positive or negative as long as m is even because an even power of any number is positive. When m is odd and x is negative, then $(-x)^m$ is positive because $-x$ is positive. When m is odd and x is positive, then $(-x)^m$ is negative because an odd power of a negative number is negative.

Example 5

In this example, the Quotient-of-Powers Property is used to simplify the quotient of two monomial expressions.

Suggest to students that they simplify the quotient of the coefficients first and then simplify the quotient of the variable terms.

Remind them that the base must be the same for the Quotient-of-Powers Property to apply.

Additional Example 5

Simplify.

A. $\frac{-6x^3y^4}{3x^2y}$ **B.** $\frac{s^5t^3u}{s^2t^3v}$

[**A.** $-2xy^3$; **B.** $\frac{s^3u}{v}$]

Ongoing Assessment, page 276

Try This

a. $-3ab$

b. $-\frac{5}{9}t^2$

Closing the Lesson

Journal Writing Summarize the lesson by having students write in their journal each property learned in this lesson, along with examples.

Teaching Tip

Students may need to be reminded that powers of variables should be positive integers. They may lose sight of this if they use graphics calculators because a calculator will show an answer even if negative exponents are used. Ask students to pay careful attention to the restrictions on each property as they are simplifying expressions.

Using a Scientific Calculator

A scientific calculator can help students understand the power properties because the exponent key can be used to verify how the properties apply. For example, to verify that $(-t)^5 = -t^5$, substitute any number for t. If $t = 3$, enter $(-3)^5$ in the calculator (don't forget the parentheses) to get -243. Then enter -3^5 in the calculator to get the same result.

Journal Entry

Have students describe the similarities and differences in simplifying the following pair of expressions: $(-4xy^2)^3$ and $(-4xy^2)(-4xy^2)^2$.

ASSESS

Selected Answers

Odd-numbered Exercises 7–65 and odd-numbered Extra Practice

Assignment Guide

Core level students should complete Exercises 1–66.

Core-plus level students should complete Exercises 1–12, 15–19, 23–31, and 36–70.

Error Analysis

Remind students of the importance of parentheses when simplifying an expression. Point out that if a number is inside the parentheses when an expression is raised to a power, then that number is also raised to the power, according to the Power-of-a-Power Property.

Use **Transparency 25** for discussion of Exercises 63–66.

Alternative Assessment

Simplification Machine Ask students to create their own "simplification machine" to demonstrate that they understand the power properties. Have students work with a partner. Ask one student to give a power expression to simplify, along with the property to use. Ask the second student to find the output values of the machine by using the property. Then have them check the answer with a calculator. If the answer is not correct, have the first student change the property and try again. If the answer is correct, ask them to switch roles. Have them repeat this pattern with progressively more difficult expressions until they understand how to apply all of the power properties.

RETEACH

Ask students to match each expression with its simplified form.

1. $(3xy^2)(4x^3)$	**A.** $16x^4y^{12}$
2. $(-xy^3)(-x^3y^2)$	**B.** $12x^4y^2$
3. $(-2xy^3)^4$	**C.** $16x^3y$
4. $\frac{32x^4y^3}{2xy^2}$	**D.** x^4y^5
5. $(-4x^2y^{10})^2$	**E.** $16x^4y^{20}$

[**1.** B; **2.** D; **3.** A; **4.** C; **5.** E]

ENRICH

Ask students to explain why $(x^2 + y^3)^2 \neq x^4y^6$. **[It is sufficient to find one counterexample. For example, if $x = 3$ and $y = 4$, then $(3^2 + 4^3)^2 = 5329$ and $3^4 \cdot 4^6 = 331{,}776$.]**

Interdisciplinary Connection

Statistics A statistician has to deal with the powers of numbers when certain financial tables are created. For example, if interest is left to compound in an account, it will grow exponentially.

NAME ______________________ CLASS ____________ DATE ____________

Practice

6.2 Multiplying and Dividing Monomials

Find each product.

1. $(3a^2)(4b^3)$ ____________

2. $(10x)(3x^2)$ ____________

3. $(-4p^3)(2pq)$ ____________

4. $(-4c^2)(-2c^3)$ ____________

5. $(-a^4)(-ab^3)$ ____________

6. $(xy^3)(x^2y^4)$ ____________

7. $(36a^2b^3)(0.5xb^4)$ ____________

8. $(1.2x^2y^2)(0.1x^3y^3)$ ____________

9. $(-8p^3q^2)(3pq)^2$ ____________

10. $(2cd^2)^3(3c^3d^3)$ ____________

Find each quotient.

11. $\frac{8a^6}{2a^3}$ ____________

12. $\frac{-4x^3}{2x}$ ____________

13. $\frac{-r^3}{7r^2}$ ____________

14. $\frac{-2y^5}{12y^3}$ ____________

15. $\frac{s^3t^5}{2st}$ ____________

16. $\frac{-9m^7n^4}{-3m^4n^3}$ ____________

17. $\frac{2u^2v^5}{0.2u^2v^2}$ ____________

18. $\frac{0.4r^3st}{0.1st^3}$ ____________

Simplify each of the following:

19. $(3x^3)^2$ ____________

20. $(-2a^4)^2$ ____________

21. $(ab^2)^4$ ____________

22. $10(2x^2y^3)^2$ ____________

23. $(xy^2)(x^3y^2)^4$ ____________

24. $-2(5n^2p^2)^3$ ____________

25. $(a^2b^2)^3(a^3)^4$ ____________

26. $(-3pq^2)^3(p^2q)^4$ ____________

27. $\left(\frac{a}{b}\right)^3$ ____________

28. $\left(\frac{3w}{y^4}\right)^2$ ____________

29. $\left(\frac{4p^3}{3q^2}\right)^2$ ____________

30. $\left(\frac{-5x^5}{2x^3}\right)^3$ ____________

NAME ______________________ CLASS ____________ DATE ____________

Lesson Activity

6.2 Modeling Number Patterns

Use what you know about powers to formulate a number pattern involving consecutive integers.

Begin by arranging centimeter blocks to form a 1 cm × 2 cm × 3 cm rectangular prism. Arrange another set of blocks to form a 2 cm × 1 cm × 1 cm rectangular prism. Combine and rearrange the two rectangular prisms to form a cube.

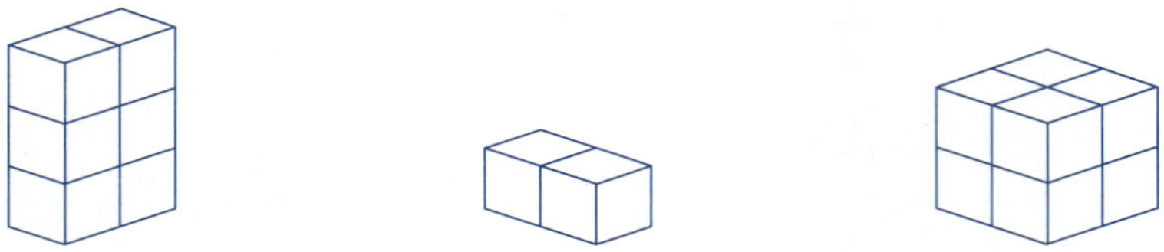

Next arrange centimeter blocks to form one 2 cm × 3 cm × 4 cm rectangular prism and one 3 cm × 1 cm × 1 cm rectangular prism. Use the two rectangular prisms to form another cube.

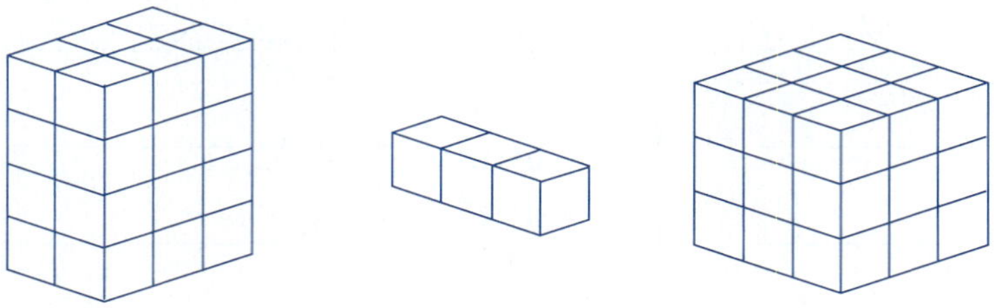

Finally form a 3 cm × 4 cm × 5 cm rectangular prism and a 4 cm × 1 cm × 1 cm rectangular prism using centimeter blocks. Then use the two rectangular prisms to form another cube.

1. Write the dimensions of each cube. Then use exponents to express the number of centimeter blocks in each cube you formed. ______________________

____

2. What are the next structures in the pattern? ______________________

3. Use consecutive positive integers to write a number sentence that relates the dimensions of each rectangular prism to the total number of blocks in each cube.

____

4. Let x represent the smallest dimension of a rectangular prism. Write a rule, using variables, to describe the pattern you developed in Exercises 1–3.

____

NAME ______________________ CLASS ____________ DATE ____________

Basic Skills Practice
Reflections and Translations, Part 2

Read each question and circle the best answer.

1. In the figure, Y is the reflected image of X. What is the measure of the angle formed by the line of reflection and the segment connecting points X and Y?

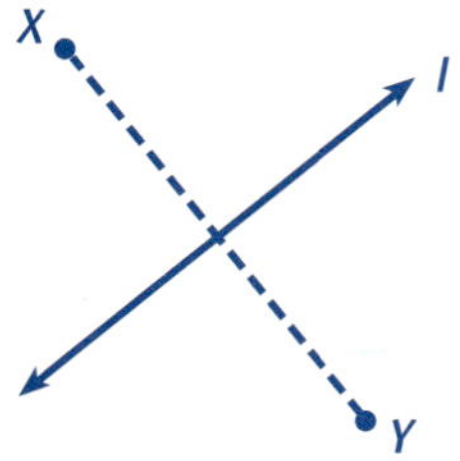

A 55°
B 60°
C 90°
D 180°

2. A line segment with endpoints at (2, 2) and (2, −2) is reflected over the y-axis. The endpoints of the reflected image are—

F (−2, 2) and (−2, −2)
G (0, 2) and (0, −2)
H (−1, 2) and (−1, −2)
I (0, −1) and (0, 1)

3. The drawing shows an example of—

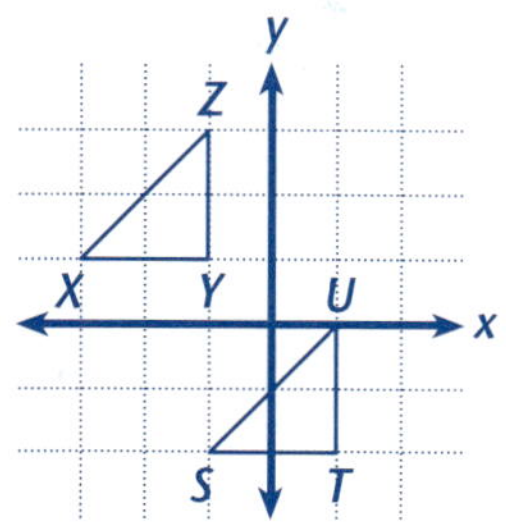

A a translation
B a reflection
C a rotation
D an inflection

4. Suppose ΔABC is translated 3 units down and 1 unit to the left to create ΔDEF. The vertices of ΔABC are (1, 2), (4, 3), and (3, 1), respectively. What are the coordinates of the vertices of the image ΔDEF?

F $D(1, 2)$, $E(4, 3)$, $F(3, 1)$
G $D(0, 1)$, $E(2, 1)$, $F(3, 1)$
H $D(0, -1)$, $E(3, 0)$, $F(2, -2)$
I $D(2, -2)$, $E(3, 0)$, $F(0, -1)$

5. Rectangle $ABCD$ is translated, forming image rectangle $FGHI$. What best describes the relationship between rectangle $ABCD$ and rectangle $FGHI$?

A They are similar.
B They are congruent.
C They are mirror images.
D They are dilations.

6. A line segment with endpoints at (3, 1) and (2, 4) is reflected over the y-axis. The endpoints of the reflected image are—

F (−3, 1) and (−2, 4)
G (1, 3) and (2, −4)
H (−1, 3) and (−2, −4)
I (3, −1) and (2, 4)

7. The drawing shows an example of—

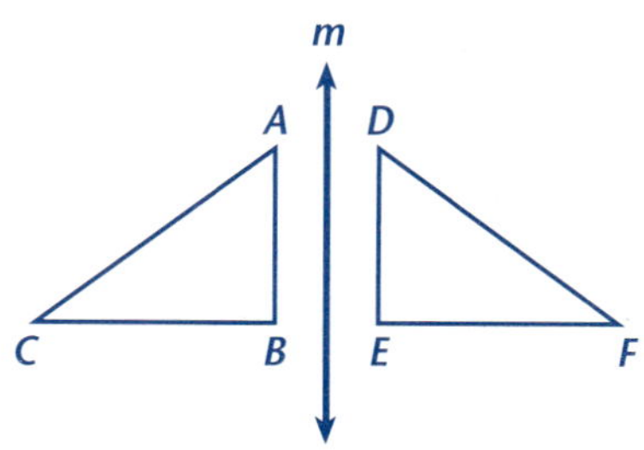

A a translation
B a reflection
C a rotation
D an inflection

LESSON 6.3

Negative and Zero Exponents

PREPARE

Pacing Guide

2 Days

Objectives and Overview

- Understand the concept of negative and zero exponents.
- Simplify expressions containing negative and zero exponents.

Monomials with negative or zero exponents are defined, and expressions involving negative or zero exponents are simplified. The power properties are then restated to include all integer exponents.

Vocabulary

zero exponent

Assessing Prior Knowledge

Evaluate.

1.	$5 - 9$	$[-4]$
2.	$-2 - 4$	$[-6]$
3.	$-9 - (-9)$	$[0]$
4.	$2 - (-4)$	$[6]$

Approach A: Using the Book

Opening Discussion

Students may have accidentally discovered that if the power of an expression was entered as a negative number in a calculator, they still got an answer. Explain that the meaning of this number and the meaning of zero exponents will be found in this lesson.

Brainstorming

What might negative exponents mean in an expression? What might a zero exponent mean?

Ongoing Assessment, page 279

On most calculators 10^{-1} can be evaluated in two ways.

a. 10 [x^{-1}] [ENTER]

b. 10 [^] -1 [=]

Allow students to become familiar with how to use these keys on their calculator.

TEACH

Approach B: Using Models

Opening Discussion

Students may have accidentally discovered when they were using calculators that if the power of an expression was entered as a negative number, they still got an answer. What expression goes with the answer? Why? What might negative exponents mean in an expression? How could a conjecture about negative exponents be verified?

Brainstorming

Have the class enter 5 into a scientific calculator and then press the reciprocal key, [x^{-1}]. What is the result? What did the reciprocal key do? Have students brainstorm the meaning of raising a number to the power of -1. What about raising a number to a different negative power?

Negative Exponents

Negative exponents are introduced by considering the quotient of two numbers with the same base and a higher power in the denominator, $\frac{10^3}{10^4}$. Applying the Quotient-of-Powers Property is a natural way to introduce negative powers. Point out that $\frac{1}{10} = \frac{10^3}{10^4} = 10^{-1}$, which means that 10 to the negative 1 is *not* a negative number; it is a reciprocal. Extend this concept to include other negative powers. The negative power of a number is not a negative number; it is a reciprocal.

Example 1

In this example, expressions with negative powers are simplified.

In part **A**, the Product-of-Powers Property is used with negative powers. In part **B**, the Quotient-of-Powers Property is used with negative powers. In part **C**, no properties apply. Point out the importance of checking for the same base in part **C**.

Additional Example 1

Simplify each expression.

A. $3^{-2} \cdot 3^4$ **B.** $\frac{2^4}{2^{-3}}$ **C.** $3^4 \cdot 4^2$

[**A.** 9; **B.** 128; **C.** 1296]

Critical Thinking, page 280

$(-2)^3 = -8$; $(-2)^{-3} = -\frac{1}{8}$

The numbers are reciprocals of each other.

On Their Own

Simplify each expression.

A. $2^{-4}2^5$ **B.** $\frac{3^{-2}}{3^{-4}}$ **C.** $2^3 \cdot 3^{-1}$

[**A.** 2; **B.** 9; **C.** $\frac{8}{3}$]

Exploration: Defining x^0

In this exploration, the definition of the zero power of a number is discovered. x^0 is the quotient of any number with itself, or 1.

continued on top of next page

Cooperative-Learning Groups

Have students work in groups of four to complete the following activities. Bring blank index cards to class and distribute them to the students.

Example 1

After demonstrating Example 1 to the students, continue the activities from Lessons 6.1 and 6.2 as follows: Assign a number to each group. Then instruct the students to write that number raised to a positive or negative number. Have each student work with a partner from the same group and perform the following tasks:

1. On notebook paper, multiply your number by your partner's number and simplify.
2. Divide your partner's number by your number and simplify.
3. Exchange papers and check your partner's work.

Next, have each student work with a partner from *another* group to multiply their numbers and simplify the result with a calculator.

Check students' work and discuss common errors with the class.

Alternate Exploration: Defining x^0

The following activity uses power tables to motivate the definitions of this lesson. Once the students have completed this activity, have them work in their groups to complete the exploration in the Pupil's Edition.

Using a Power Table A table showing the powers of a number may help students to see how negative powers are shortcuts for reciprocals and may suggest the definition of x^0. A table showing sequential powers of 2, for example, is shown on the next page.

continued on bottom of next page

Approach A — Using the Book, continued

Ongoing Assessment, page 280

4. $x^0 = 1$ for any nonzero number x

Example 2

In this example, expressions with powers are simplified.

In part **A**, the definition of a zero exponent is used. Explain how to use of the Product-of-Powers Property to get the zero exponent.

Additional Example 2

Simplify each expression.

A. $a^{-3}a^3$ **B.** $-4s^{-3}$ **C.** $\frac{y^4}{x^{-2}}$

[**A.** 1; **B.** $\frac{-4}{s^3}$; **C.** y^4x^2]

Critical Thinking, page 281

$$\left(\frac{a}{b}\right)^{-n} = \frac{a^{-n}}{b^{-n}} = \frac{\frac{1}{a^n}}{\frac{1}{b^n}} = \frac{b^n}{a^n} = \left(\frac{b}{a}\right)^n$$

On Their Own

Simplify each expression.

A. $3x^{-4}x^4$ **B.** $\frac{x^{-2}x^2y^3}{y^{-3}}$

[**A.** 3; **B.** y^6]

Teaching Tip

Using Technology Use a scientific calculator to verify the properties of positive, negative, or zero exponents. Caution students that the

TEACH

Approach B — Using Models, continued

$2^4 = 16$
$2^3 = 8$
$2^2 = 4$
$2^1 = 2$
$2^0 = 1$
$2^{-1} = \frac{1}{2}$
$2^{-2} = \frac{1}{4}$
$2^{-3} = \frac{1}{8}$
$2^{-4} = \frac{1}{16}$

Use the table to point out the decreasing sequence of powers of 2, the pattern in the answers (one-half of the previous answer), and the use of negative powers as reciprocals.

Working With a Power Table Have students make a table of powers for each of the following:

A. 3^{-4} to 3^4

B. 10^{-4} to 10^4

[A.]

$3^4 = 81$
$3^3 = 27$
$3^2 = 9$
$3^1 = 3$
$3^0 = 1$
$3^{-1} = \frac{1}{3}$
$3^{-2} = \frac{1}{9}$
$3^{-3} = \frac{1}{27}$
$3^{-4} = \frac{1}{81}$

[B.]

$10^4 = 10{,}000$
$10^3 = 1000$
$10^2 = 100$
$10^1 = 10$
$10^0 = 1$
$10^{-1} = \frac{1}{10}$
$10^{-2} = \frac{1}{100}$
$10^{-3} = \frac{1}{1000}$
$10^{-4} = \frac{1}{10{,}000}$

reciprocal of a power may be a decimal, which may complicate the simplification. Below, 2^2, 2^{-2}, and 5^0 have been calculated.

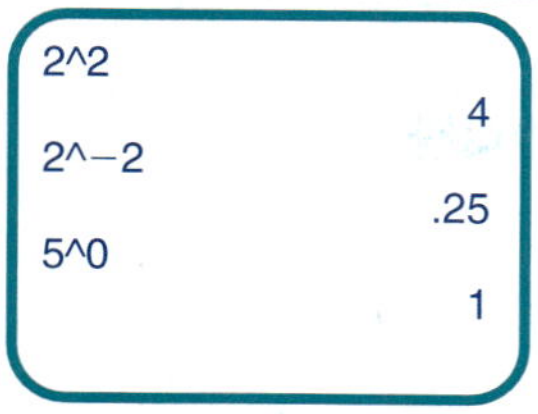

Inclusion Strategy

Enlarge each of the following expressions and solutions on pieces of paper that can be used as flashcards. Ask students to give answers without negative or zero exponents.

Expression	Solution
$x^{-4} \cdot x^3$	$\frac{1}{x}$
$\frac{x^{-3}}{x^4}$	$\frac{1}{x^7}$
$(x^{-4})^3$	$\frac{1}{x^{12}}$

Mix up the flashcards and fasten them to the chalkboard. Ask students to match an expression from a flashcard with its solution. Encourage students to make up their own flashcards.

Closing the Lesson

Use **Transparency 26** to summarize the power properties.

Defining x^0 From the power table above, the definition of x^0 is suggested by the position of the zero power in the table. x^0 has to be 1 in order for the sequence of numbers in the table to be continued. Ask students to evaluate each of the following by inspection or to make a table, if necessary:

A. y^0 **B.** $(xy)^0$ **C.** $(2 \cdot 3)^0$ **D.** $(10^{-4})^0$
[A. 1; **B.** 1; **C.** 1; **D.** 1**]**

Making Tables on a Calculator Have students work in their groups to study the table generated by an exponential function. Ask one student to enter Y = 2^X into the Y= menu of a graphics calculator. Use the table set feature (TblSet) to set TblMin = −4 and △Tbl = 1. Ask a second student to use the table command of the calculator to make a list of table values for the powers of 2, including negative powers. Ask a third student to make conjectures, based on the table values, about the relationship between 2^0 and the other powers of 2. Ask all students to discuss the table values, make conjectures, and verify them by checking with a calculator. Now have students switch roles and make tables for the powers of 3, 10, and 1. Ask the fourth student to record the group's activities and write a complete report to be presented to the class.

Example 2

Cooperative-Learning Group After demonstrating Example 2, have the students work with a partner to simplify the following expressions:

A. a^7a^{-7} **B.** $x^{-2}y$ **C.** $\frac{r^{-2}s^5}{s^4t^{-3}}$

[A. 1; **B.** $\frac{y}{x^2}$; **C.** $\frac{st^3}{r^2}$**]**

Closing the Lesson

Brainstorm with the class to write the properties of powers on the chalkboard (compare the Summary of Power Properties at the end of the lesson). Have students record the table in their journal.

ASSESS

Selected Answers

Odd-numbered Exercises 7–45 and odd-numbered Extra Practice

Assignment Guide

Core level students should complete Exercises 1–46.

Core-plus level students should complete Exercises 1–48.

Error Analysis

In Exercises 23 and 24, students simplify fractions containing several monomials. You may want to suggest that students completely simplify the numerator first, then simplify the denominator, and finally simplify the fraction. There may be less chance for error if a pattern is used to simplify expressions.

Use **Transparency 27** to discuss Exercises 47 and 48, which examine Fermat numbers.

Alternative Assessment

Using a Spreadsheet Ask students to use a spreadsheet to show tables for 2^x, 5^x, and 10^x. Make sure that they include negative and zero exponents.

RETEACH

Have students simplify each expression, giving answers without negative or zero exponents.

A. $r^3 \cdot r^{-3}$ **B.** $\frac{r^2}{r^3}$ **C.** $(r^3)^{-2}$

D. $(r^{-3})^{-2}$ **E.** $(r^{-3}s)^{-2}$

[A. 1; B. $\frac{1}{r}$; C. $\frac{1}{r^6}$; D. r^6; E. $\frac{r^6}{s^2}$]

ENRICH

A *perfect number* is a number whose integral factors other than itself add up to the number. All perfect numbers are of the form $2^{p-1}(2^p - 1)$, where p is a prime number and $(2^p - 1)$ is a prime number. Ask students to find the first five perfect numbers. **[6; 28; 496; 8128; 33,550,336]**

NAME ______________________ CLASS ________ DATE ________

Practice

6.3 Negative and Zero Exponents

Evaluate each expression.

1. 2^{-4} ______ **2.** $(-2)^4$ ______ **3.** -2^4 ______ **4.** -2^{-4} ______

5. $(-2)^{-4}$ ______ **6.** 3^{-2} ______ **7.** 10^{-4} ______ **8.** -7^2 ______

9. $(-7)^2$ ______ **10.** $(-10)^3$ ______ **11.** $(-8)^{-2}$ ______ **12.** $(-3)^{-3}$ ______

Write each of the following without negative or zero exponents:

13. x^{-1} ______ **14.** y^0 ______

15. a^{-3} ______ **16.** $2n^{-2}$ ______

17. $(x^2y^3)^0$ ______ **18.** $-x^{-2}x^{-3}$ ______

19. $a^{-1}b^{-2}c^3$ ______ **20.** $-c^3d^{-4}$ ______

21. m^5m^{-2} ______ **22.** $k^{-4}k^7$ ______

23. $r^{-2}s^{-2}t^4$ ______ **24.** $-a^3a^{-6}a^{-5}$ ______

25. $\frac{5}{x^{-1}}$ ______ **26.** $\frac{7^0}{(xy)^0}$ ______

27. $\frac{y^{-6}}{y^2}$ ______ **28.** $\frac{rt^{-4}}{t^3}$ ______

29. $\frac{w^{-8}}{w^{-2}}$ ______ **30.** $\frac{3q^{-5}}{q^{-5}}$ ______

31. $\frac{-2a^{-4}}{a^2b}$ ______ **32.** $\frac{5q^{-4}}{q^2s^{-2}}$ ______

33. $\frac{(2a^3)(3a^5)}{a^{-2}}$ ______ **34.** $\frac{(7d^{-6})(-4d^{-2})}{2d^{-4}}$ ______

35. $\frac{x^{-4}x^3}{x^{-3}x^4}$ ______ **36.** $\frac{-z^3z^{-6}}{-3z^2}$ ______

NAME ______________________ CLASS ____________ DATE ____________

Lesson Activity

6.3 Modeling Negative and Zero Exponents

To model negative and zero exponents, cut out a 1 in. × 2 in. rectangle from grid paper. Label the piece with its area, "2." Place this rectangle on grid paper, and then fold the right half over the left half, as shown. Trace around the folded piece, cut out, and label this piece "1."

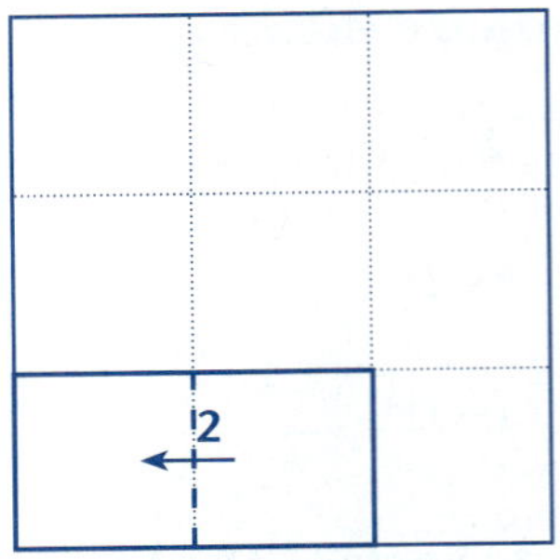

Place this piece on grid paper, fold the top half over the bottom half, trace around it, and cut it out. Label this piece "$\frac{1}{2} \times 1$."

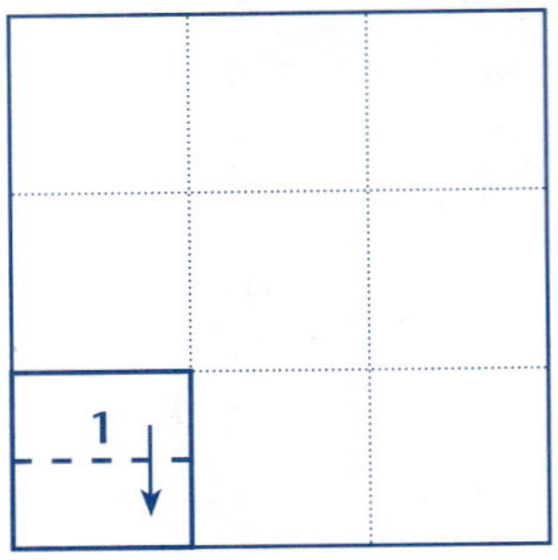

Place the $\frac{1}{2} \times 1$ piece on grid paper, fold the right half over the left half, trace, and cut out this piece.

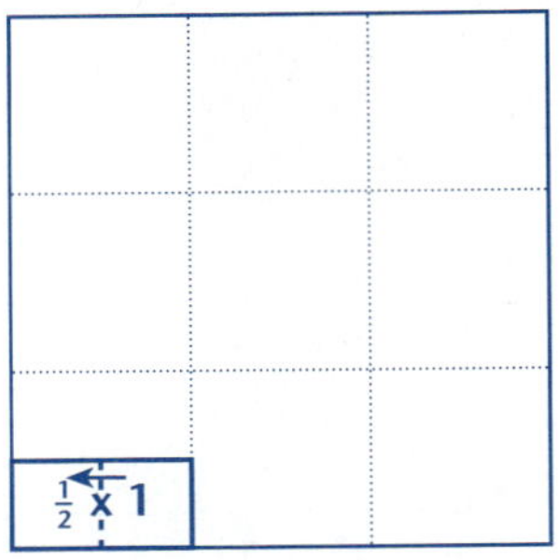

1. How would you label this piece? ____________________

2. Place your sequence of cutouts on a flat surface in order from left to right. Reading from left to right, are the areas increasing or decreasing?

____

3. Are the factors of 2 located in the numerator or the denominator in the two smallest cutouts? ____________________

4. If a negative sign indicates that the factors of 2 are in the denominator, write an exponential name for each of your pieces.

____

5. Describe the patterns revealed by the construction of these shapes.

____

NAME ______________________________ CLASS ______________ DATE ______________

Basic Skills Practice

Scale and Dilations, Part 1

A dilation is a transformation that changes the size of a figure while preserving the orientation, shape, and measures of the angles in the figure. Dilations refer to both enlargements and reductions in the size of the original figure. A scale factor or multiplier determines the amount of enlargement or reduction of the original figure. A dilation is a reduction if the scale factor is less than 1, and it is an enlargement if it is greater than 1.

Example 1: Transform segment AB using its endpoints $A(4, 0)$ and $B(2, 3)$ and a scale factor of $\frac{1}{2}$.

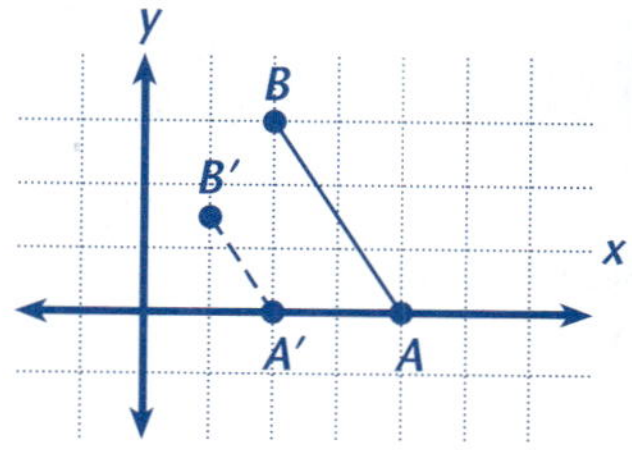

Multiply both coordinates of each point.

$A(4, 0) \rightarrow A'\left(\frac{1}{2} \times 4, \frac{1}{2} \times 0\right)$ or $A'(2, 0)$

$B(2, 3) \rightarrow B'\left(\frac{1}{2} \times 2, \frac{1}{2} \times 3\right)$ or $B'\left(1, 1\frac{1}{2}\right)$

The transformation maps every point in segment AB to segment $A'B'$.

Example 2: What is the scale factor used to transform ΔPQR to the image $\Delta P'Q'R'$?

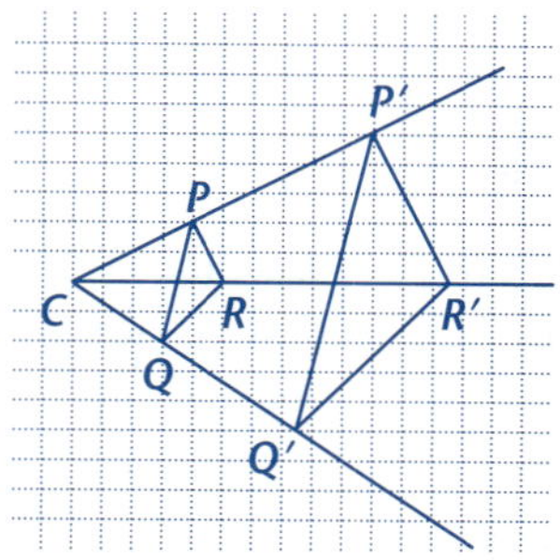

Every dilation has a point known as the center of dilation, which in this case is point C. $\Delta P'Q'R'$ is an enlargement of ΔPQR, so the scale factor must be greater than 1. The distance from C to P' is $\frac{5}{2}$ times greater than the distance from C to P (the same is true for the distances from C to Q' and C to Q, and from C to R' and C to R), so we call this an enlargement with a scale factor of $\frac{5}{2}$.

A square with sides 3 centimeters long undergoes a dilation with a scale factor of 2.5.

1. How long is each side of the new image? ______________

2. What is the surface area of the new image? ______________

A triangle with a base of 3 centimeters and a height of 4 centimeters undergoes a dilation with a scale factor of 2.

3. What is the area of the original triangle? ______________

4. What is the area of the new image? ______________

5. If the scale factor of a dilation is n, what is the ratio of the area of the image to the area of the pre-image? ______________

LESSON 6.4
Scientific Notation

PREPARE

Pacing Guide
2 Days

Objectives and Overview
- Recognize the need for special notation to facilitate computation with very large and very small numbers.
- Perform computations involving scientific notation, with and without a calculator.

Students explore the use of scientific notation in astronomy and physics. They learn how to perform computations with numbers in scientific notation and how to use scientific notation with a calculator.

Vocabulary
scientific notation
first factor
second factor

Assessing Prior Knowledge

Simplify.

1.	$20 \cdot 100$	[2000]
2.	$2.64 \cdot 10^2$	[264]
3.	$10^3 \cdot 10^{-3}$	[1]
4.	$200 \cdot 10^{-3}$	[0.2]
5.	$1.36 \cdot 10^{-4}$	[0.000136]

Approach A — Using the Book

Opening Discussion
Ask students if they know what the age of the universe is. **[between 13 and 18 billion years]** What are some of the distances discussed in science-fiction movies? How small is an atom? Discuss the idea that some things are so big or so small that they need special notation to be expressed easily.

Brainstorming
Have the class brainstorm objects that can be measured in terms of powers of 10.

Definition of *Scientific Notation*
Point out that scientific notation is very specific. A number is expressed in scientific notation if the first factor is a number between 1 and 10 and the second factor is a positive, negative, or zero power of 10. The number 6.1×10^{-3} is written correctly in scientific notation. The number 0.61×10^{-2} is not.

TEACH

Approach B — Using Models

Opening Discussion
Discuss the circumference of Earth in meters **[40,075,161 meters]** and the distance to the sun in meters **[149,600,000,000 meters]**. What are some of the distances discussed in science-fiction movies? How small is an atom? Discuss the idea that some things are so big or so small that they need special notation to be expressed easily.

Brainstorming
Have the class brainstorm objects that can be measured in terms of powers of 10. List objects measured in metric units in order to make the size comparisons meaningful.

Example 1

In this example, a number in customary notation is expressed in scientific notation.

Point out that if the decimal point is moved to the left to convert a number in customary notation to scientific notation, then the power of 10 is positive and equal to the same number of places that it takes to relocate the decimal. If the decimal point is moved to the right, then the power of 10 is negative.

Additional Example 1

The Earth is estimated to be about 15,000,000,000 years old. This number is read as "15 billion." Write this number in scientific notation. **[1.5×10^{10}]**

Ongoing Assessment, top of page 287

Try This

8.75×10^5

Example 2

This example shows two different ways to do a computation with large numbers.

One way is to do long division. The other way is to express the number in scientific notation and then use the properties of exponents. Emphasize that both methods are equivalent but that using the properties of exponents can save time.

Ongoing Assessment, page 287

About 685 million

Using Technology

A calculator can also be used to perform the division in this example. Enter the first number as 6 × 10^15 [÷] 1000, where × is the multiplication key. The advantage to this method is that the number remains in scientific notation and the properties of exponents are not used.

continued on top of next page

Students can use a meter stick to explore scientific notation.

Using a Meter Stick

A meter stick can be used to help students see the size difference between objects measured in terms of powers of 10. Have students begin by comparing measurements of different objects in millimeters. For example, they can compare the approximate measure of a pencil point (1 mm), a fingernail (10 mm), the height of two books (100 mm), and the height of a desk (1000 mm).

Next, have students think about how to convert the values to centimeters by using powers of 10. For example, the measure of the pencil point is converted as follows:

$$1 \text{ mm} = \frac{1}{10} \text{ cm} = 0.1 \text{ cm} = 10^{-1} \text{ cm}$$

Next, have students complete the following table. Remind students that 1 decimeter (dm) = 10 cm.

	mm	cm	dm	m
1 mm =	1	10^{-1}	[10^{-2}]	[10^{-3}]
1 cm =	[10^{1}]	1	[10^{-1}]	[10^{-2}]
1 dm =	[10^{2}]	[10^{1}]	1	[10^{-1}]
1 m =	[10^{3}]	[10^{2}]	[10^{1}]	1

continued on bottom of next page

Approach A

Using the Book, continued

Additional Example 2

In science-fiction movies, spaceships travel at "warp speed." If warp 1 is about 1×10^{12} meters per hour (the speed of light) and the universe is about 10^{27} meters wide, how long would it take a spaceship in a science-fiction movie to travel across the universe?

[$\frac{1 \times 10^{27}}{1 \times 10^{12}} = 1 \times 10^{15} = 10^{15}$, or 1000 trillion hours]

Ongoing Assessment, bottom of page 287

Try This

3×10^{24}

Teaching Tip

Negative Exponents You may want to review the concept that a negative exponent is another way of representing the reciprocal of a number, not the opposite of a number. Negative exponents are especially useful in expressing very small quantities. They are usually written as a number in scientific notation with a negative power of 10. For example, an atom has a diameter of about 1×10^{-10} m.

Example 3

In this example, a very small number, the wavelength of an X ray, is written in scientific notation.

Point out that if a number is very small (positive, but less that 1), the decimal point needs to be moved to the right in order to express the number in scientific notation. If the decimal point is moved to the right, then the power of 10 is negative and is equal to the number of places that it took to relocate the decimal. Remind students that a negative power means a reciprocal, not a negative number.

Additional Example 3

An atom has a diameter of about 0.0000000001 meters. Write this number in scientific notation. **[1×10^{-10} m]**

TEACH

Approach B

Using Models, continued

Have students discuss the pattern in the table. Remind them that $1 = 10^0$. **[Answers will vary, but students should observe the decreasing powers of 10 from left to right in the table.]**

Have students use single-digit numbers to write equivalent expressions for different units. Show students that

$1 \text{ m} = 10^3 \text{ mm}$, or 1×10^3 mm,
so
$2 \text{ m} = 2 \times 10^3 \text{ mm}$

Point out that the expressions on the right in each case are in **scientific notation.**

Definition of *Scientific Notation*

Point out that scientific notation is very specific. A number is expressed in scientific notation if the first factor is a number between 1 and 10 and the second factor is a positive, negative, or zero power of 10. The number 6.1×10^{-3} is written correctly in scientific notation. The number 0.61×10^{-2} is not.

Explain that to put 0.61×10^{-2} into correct form, you must *multiply* 0.61 by 10 to get 6.1. Therefore, to keep the value of the expression the same, the power of 10 (that is, 10^{-2}) must be *divided* by 10. The result is 10^{-3}. Have students provide examples for which they divide the first factor and multiply the second.

Multiplying in Scientific Notation

Explain to students that to multiply two numbers in scientific notation, they multiply the first numbers of each expression together and then add the exponents of 10 to find the appropriate power of 10 for the result.

Students can investigate the multiplication rule by considering 2-by-4 rectangles with various units of measure. For example, the lengths of the sides of the rectangle below are expressed in both meters and millimeters. Ask students to find the area of the figure on the next page in terms of each unit.

Ongoing Assessment, page 288

Try This

4.02×10^{-8}

Critical Thinking, page 288

12×10^{-2}
$= 1.2 \times 10^{1} \times 10^{-2}$
$= 1.2 \times 10^{-1}$

0.12×10^{-2}
$= 1.2 \times 10^{-1} \times 10^{-2}$
$= 1.2 \times 10^{-3}$

Example 4

In this example, students compare the sizes of two numbers written as powers of 10.

Point out that the easiest way to compare two powers of 10 is to use the Quotient Property of exponents.

Using Technology

A calculator can also be used to perform the division in this example. Enter the calculation as 10 × 10^(−2) [÷] 10 × 10^(−10), where × is the multiplication key.

Additional Example 4

How many times wider is an atom with a diameter of 10^{-10} m than an electron with a diameter of 10^{-18} m? **[10^8 times wider]**

Example 5

This example shows how to find the product and the quotient of two numbers written in scientific notation.

Point out that for part **A**, the Product of Powers Property is used to add the exponents and that for part **B**, the Quotient of Powers Property is used to subtract the exponents. Use the Commutative Property for addition to regroup the factors so that these numbers are separate from the powers of 10.

continued on top of next page

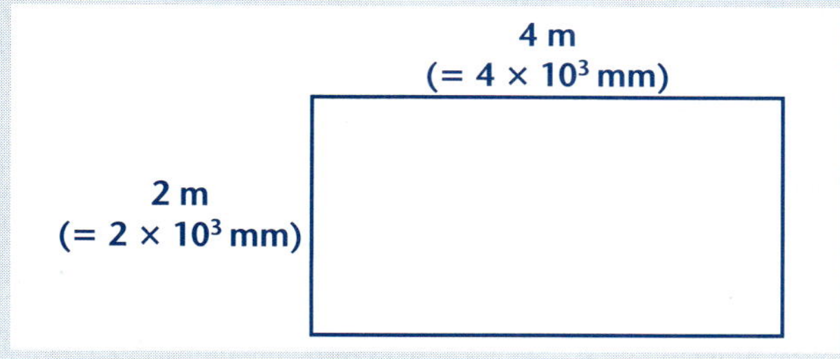

[8 m^2 and 8×10^6 mm^2; note that in scientific notation the first number is 8×10^0 m^2.]

The following example illustrates the use of negative exponents. Have the students find the area in terms of each unit of measure.

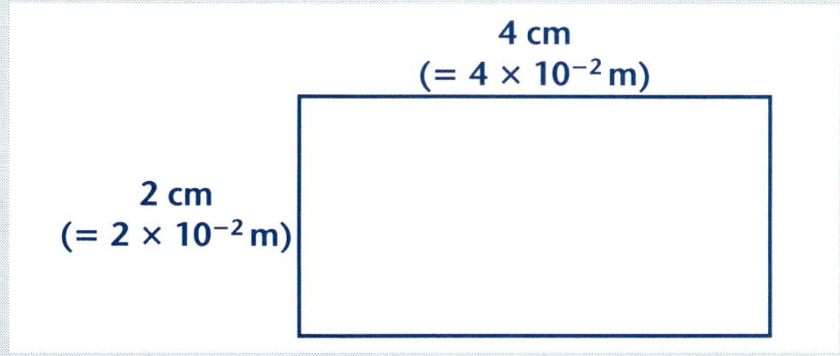

[8 cm^2 and 8×10^{-4} m^2; note that in scientific notation the first number is 8×10^0 cm^2.]

Students should be encouraged to check their work in terms of more familiar notation. For example, for the first rectangle, the area in millimeters is

$$A = 2000 \text{ mm} \times 4000 \text{ mm} = 8{,}000{,}000 \text{ mm}^2.$$

They should see that this result agrees with their answer in scientific notation. Encourage students to use different combinations of units to extend these exercises.

Dividing in Scientific Notation

Explain to students that to divide two numbers in scientific notation they divide the first numbers in each expression (as in conventional notation) and then subtract the exponents of 10 to find the appropriate power of 10 for the result. For example:

$$(4 \times 10^5) \div (2 \times 10^2) = 2 \times 10^3$$

continued on bottom of next page

Approach A

Using the Book, continued

Additional Example 5

Express the product and quotient in scientific notation.

A. $(5 \times 10^4)(3 \times 10^{-6})$ **B.** $\frac{5 \times 10^{-2}}{4 \times 10^3}$

[A. 1.5×10^{-1}; B. 1.25×10^{-5}]

Example 6

This example shows that a calculator with an EE key or an EXP key can also be used to do computations with scientific notation.

Ask students to experiment with their calculator to learn how to do these computations.

Additional Example 6

The speed of light is about 2.98×10^5 kilometers per second. If the sun is about 5.900×10^9 kilometers from the planet Pluto, how long does it take sunlight to reach Pluto? **[about 19,800 seconds]**

Inclusion Strategy

Using Calculators Have students work in pairs to practice doing calculator computations in scientific notation. Have each student make up a computation problem for the other student to do. Ask them to check each other's work. Do as many problems as necessary for students to become comfortable with the special calculator keys.

Closing the Lesson

Summarize how to do each of the following:

1. Convert a number in customary notation to scientific notation.
2. Multiply two numbers in scientific notation.
3. Divide two numbers in scientific notation.

Have students simplify each of the following. Express the answers in scientific notation.

A. $(2.3 \times 10^4)(1.8 \times 10^{-6})$ **B.** $\frac{5.1 \times 10^{-6}}{2 \times 10^{-4}}$

[A. 4.14×10^{-2}; B. 2.55×10^{-2}]

TEACH

Approach B

Using Models, continued

The first numbers were divided ($4 \div 2 = 2$), and the exponents were subtracted ($5 - 2 = 3$).

Students can use their rectangles to check the rules for division.

Closing the Lesson

Summarize how to do each of the following:

1. Convert a number in customary notation to scientific notation.
2. Multiply two numbers in scientific notation.
3. Divide two numbers in scientific notation.

Have students simplify each of the following. Express the answers in scientific notation.

A. $(2.3 \times 10^4)(1.8 \times 10^{-6})$ **B.** $\frac{5.1 \times 10^{-6}}{2 \times 10^{-4}}$

[A. 4.14×10^{-2}; B. 2.55×10^{-2}]

ASSESS

Selected Answers

Odd-numbered Exercises 7–73 and odd-numbered Extra Practice

Assignment Guide

Core level students should complete Exercises 1–77.

Core-plus level students should complete Exercises 1–7, 12–17, 22–28, and 35–77.

Error Analysis

Students may have difficulty with significant digits in Exercises 49–56. Point out that significant digits are used to designate a level of accuracy—the more digits that are used (even if they are zeros), the more accurate the measurement.

Alternative Assessment

The Scientific Notation Game To involve all students in computations with scientific numbers, have one student write a number in scientific notation on the chalkboard. Then have the next student write another number in scientific notation on the chalkboard. Ask a third student to show the product of these two numbers. Have another student show the quotient of these two numbers. Repeat this sequence until all students have participated.

RETEACH

Have students work with a partner and explain how to simplify each expression. Have them check each other's work.

A. $(8.3 \times 10^6)(7.1 \times 10^4)$

B. $(3.1 \times 10^{-5})(2.1 \times 10^4)$

C. $\frac{8 \times 10^7}{2 \times 10^4}$

D. $\frac{(1.4 \times 10^3)(3.2 \times 10^4)}{2.2 \times 10^4}$

[A. 5.893×10^{11}; B. 6.51×10^{-1}; C. 4×10^3; D. 2.04×10^3]

ENRICH

Ask students to use the Internet to search for information on distances related to the solar system. For example, they can search on the keyword *solar system*. Ask them to write a report on what they find and to list the distances in scientific notation.

Interdisciplinary Connection

Science Biologists use scientific notation to express the size of human cells. Physicists depend on scientific notation to express the vast distances in space, the fast speed of light, and the very tiny size of particles. Chemists use scientific notation to express the number of molecules in a sample of a substance.

NAME ______________________ CLASS __________ DATE __________

Practice

6.4 Scientific Notation

Write each number in scientific notation.

1. 200,000 ______________

2. 500,000,000 ______________

3. 210,000 ______________

4. 40,000 ______________

5. 6,500,000,000 ______________

6. 840,000,000 ______________

7. 466,000,000,000 ______________

8. 37,500 ______________

9. 0.20 ______________

10. 0.0008 ______________

11. 0.000076 ______________

12. 0.00645 ______________

13. 0.00000362 ______________

14. 0.000978 ______________

Write each number in customary notation.

15. 5×10^3 ______________

16. 6×10^5 ______________

17. 9×10^4 ______________

18. 5.6×10^7 ______________

19. 8.56×10^{10} ______________

20. 3.56×10^8 ______________

21. 6×10^{-2} ______________

22. 6.7×10^{-5} ______________

23. 8.76×10^{-6} ______________

24. 9.22×10^{-8} ______________

Perform each computation. Write your answers in scientific notation.

25. $(4 \times 10^2)(3 \times 10^2)$ ______________

26. $(6 \times 10^4)(2 \times 10^5)$ ______________

27. $(2 \times 10^3) + (4 \times 10^3)$ ______________

28. $(7 \times 10^7) + (5 \times 10^7)$ ______________

29. $(7 \times 10^5) - (2 \times 10^5)$ ______________

30. $(9 \times 10^6) - (2 \times 10^6)$ ______________

31. $\dfrac{12 \times 10^4}{2 \times 10^2}$ ______________

32. $\dfrac{6 \times 10^6}{2 \times 10^3}$ ______________

33. $\dfrac{(2 \times 10^3)(4 \times 10^2)}{4 \times 10^4}$ ______________

34. $\dfrac{(8 \times 10^{10})(6 \times 10^2)}{2 \times 10^8}$ ______________

NAME ______________________ CLASS __________ DATE __________

Lesson Activity
6.4 Fermi's Method

Enrico Fermi (1901–1954), a famous American physicist, used a series of estimates to solve many problems that were almost impossible to solve by exact measurement. For example, suppose that you wanted to determine how many kernels of corn there are in a cubic meter of popped corn. To use Fermi's method to solve this problem, start by estimating the number of kernels in a tablespoon of unpopped corn.

Picture a tablespoon and imagine putting unpopped kernels of corn in it.

1. What do you think is the least number of kernels that will fit in the spoon? What do you think is the greatest number of kernels that will fit?

____

2. To estimate the number of unpopped kernels that will fit in a tablespoon, take the square root of the product of your estimates in Exercise 1. What is your result? ______________

3. Suppose that it takes 2 tablespoons of unpopped kernels to make a liter of popcorn. If one cubic meter equals 1000 liters, how many kernels are there in a cubic meter of popped corn? Write your answer in scientific notation.

____

Another way you can estimate the number of kernels in a cubic meter of popped corn is by using the formula

$$\text{Number of kernels} = \frac{\text{number of cm}^3 \text{ in a cubic meter}}{\text{volume, in cm}^3\text{, of a piece of popped corn}}.$$

4. If 10^3 cm^3 = 1 L and 10^3 L = 1 m^3, find the number of cubic centimeters in 1 cubic meter. ______________

5. Estimate the measure of the diameter of a piece of popped corn in centimeters. ______________

6. Find the volume of a piece of popped corn by cubing your estimate of the diameter. ______________

7. Use the formula to calculate the number of kernels in a cubic meter of popped corn. Express your answer in scientific notation. ______________

8. Compare the two methods of estimation.

____

NAME ______________________ CLASS ____________ DATE ____________

Basic Skills Practice
Scale and Dilations, Part 2

Read each question and circle the best answer.

1. Find the coordinates of the image points for $A(2, 2)$ and $B(6, 6)$ under a dilation with a scale factor 0.25. The origin is the center of dilation.

A $A(0.5, 0.5)$, $B(1.5, 1.5)$

B $A(1, 0.5)$, $B(1.5, 1.5))$

C $A(-0.5, 0.5)$, $B(1.5, -1.5)$

D $A(2, 2)$, $B(6, 4)$

2. The dotted line figure is a dilation of the solid line figure. Identify whether it is an enlargement or a reduction and find the scale factor.

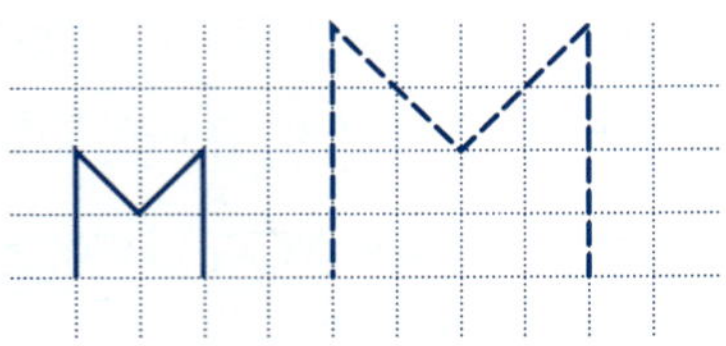

F enlargement, scale factor of 3

G reduction, scale factor of $\frac{1}{3}$

H enlargement, scale factor of 2

I reduction, scale factor of $\frac{1}{2}$

3. The dotted line figure is a dilation of the solid line figure. Identify whether it is an enlargement or a reduction and find the scale factor.

A reduction, scale factor of $\frac{1}{2}$

B enlargement, scale factor of 2

C reduction, scale factor of $\frac{1}{4}$

D enlargement, scale factor of 4

4. Fanny is drawing blueprints for a new house. On the blueprint, the entry hall measures 5 inches by 3.5 inches. If the scale used for the drawing is 0.5 in. = 1 ft, What are the dimensions of the actual entry hall?

F 5 ft by 3.5 ft

G 2.5 ft by 1.75 ft

H 10 ft by 7 ft

I 8 ft by 9 ft

5. Diana is going to enlarge a family picture. The original print is 4 inches by 6 inches. She is going to use a scale factor of 10. What are the dimensions of the enlargement?

A 14 in. by 16 in.

B 40 in. by 60 in.

C 4 ft by 6 ft

D 40 ft by 60 ft

6. Using a copy machine, a drawing of a tower that was 24 inches high was reduced to a tower that measured 8 inches high. What was the scale factor used for the reduction?

F $\frac{1}{2}$

G $\frac{1}{3}$

H $\frac{1}{4}$

I $\frac{1}{8}$

7. Given a triangle with vertices at $A(2, 2)$, $B(3, -1)$, and $C(1, 0)$, find the vertices of the image of ΔABC, using the origin as the center of dilation and a scale factor of 2.

A $A(2, 4)$, $B(4, 2)$, $C(3, 6)$

B $A(2, 1)$, $B(-1, 3)$, $C(0, 1)$

C $A(4, 4)$, $B(6, -2)$, $C(2, 0)$

D $A(4, 4)$, $B(-2, 6)$, $C(0, 2)$

LESSON 6.5
Exponential Functions

PREPARE

Pacing Guide
2 Days

Objectives and Overview
- Understand the exponential function and how it is used.
- Identify the graph of the exponential function for different bases.

Students learn how to apply exponential function models to problems of population growth. They also study the graph of an exponential function.

Assessing Prior Knowledge

Use a calculator to find each of the following:

1.	3^3	[27]
2.	3^5	[243]
3.	3^7	[2187]
4.	3^9	[19,683]

TEACH

Approach A: Using the Book

Opening Discussion
Discuss the application of exponential functions to model problems about population growth, finance, and science. How do cells grow? **[They increase their number by dividing.]**

Brainstorming
Ask students to explain how interest is earned on an account. What happens if the interest is left in the account? **[The account increases more quickly.]**

Ongoing Assessment, page 293

The figure for one year was divided by 365 and rounded to the nearest thousandth.

Example 1
In this example, a population table representing the world's population is given, and students make

continued on top of next page

Approach B: Using Models

Opening Discussion
"If you were to hide a dollar a day in a secret place, the amount of growth per year would always be the same. What kind of function models this kind of growth? **[a linear function]** Some growth, like the growth of population of the world, is not linear. In fact, world population growth is sometimes referred to as an explosion. How is it different from linear growth?" **[The amount of growth each year keeps increasing.]**

Brainstorming
Ask students to explain how interest is earned on a bank account. What happens if the interest is left in the account? **[For a fixed principal, the amount of interest per year increases.]** What happens if the interest is taken out? **[The amount of interest per year stays the same.]** How is this like population growth?

continued on bottom of next page

Approach A Using the Book, continued

estimates of future population levels. Point out that the values in the right column are derived by multiplying the previous value in the column by 1.017, or 100% + 1.7%. Emphasize that there is a pattern in the numbers that can be generalized in a formula (the general growth formula given in the next column).

Additional Example 1

Certain bacteria are known to increase at the rate of 8% per day. A bacterial culture starts with 500,000 bacteria.

A. Estimate the number of bacteria on the 10th day. **[about 1,079,462 bacteria]**

B. If the bacterial growth continues at the same rate, how many bacterial will be added to the culture in each second of the 10th day? **[about 13 bacteria per second]**

	A	B
1	**Day**	**# Bacteria**
2	0	500,000
3	1	540,000
4	2	583,200
5	3	629,856
6	4	680,244.48
7	5	734,664.0384
8	6	793,437.16147
9	7	856,912.13439
10	8	925,465.10514
11	9	999,502.31355
12	10	1,079,462.4986

General Growth Formula

The general formula for exponential growth, $P = A(1 + r)^t$, can be used as a model for many applications of exponential growth and decay. For example, if A is a bank account balance and r is the interest rate, the formula represents the growth of an account with an annual interest

TEACH

Approach B Using Models, continued

Growth of a Savings Account

Cooperative-Learning Groups Have students create a table for the amount of money in a savings account if no withdrawals or deposits are made. Have them agree on a starting amount, which is to be deposited on January 1, 2000, and an interest rate. Help them fill in the first few amounts as shown, if necessary.

Date	Interest earned at 3% per year	Total amount in account
Jan. 1, 2000	$0.00	$1000.00
Jan. 1, 2001	$30.00	$1030.00
Jan. 1, 2002	$30.90	$1060.90
Jan. 1, 2003	$31.83	$1092.73
Jan. 1, 2004	$32.79	$1125.51
Jan. 1, 2005	$33.77	$1159.27
Jan. 1, 2006	$34.78	$1194.05
Jan. 1, 2007	$35.82	$1229.87
Jan. 1, 2008	$36.90	$1266.77
Jan. 1, 2009	$38.00	$1304.77
Jan. 1, 2010	$39.14	$1343.92

Graphing an Exponential Function

Have students graph the amounts in the savings account over the 10-year period. A sample graph for the data in the table is shown on the next page. Ask students to describe the shape of the curve.

payment. Another use of the model is with a negative rate of growth. This represents a decay. For example, a population may be decreasing at a rate of 2% per year. In this case, r would equal -0.02 in the formula. Students are introduced to decay examples in the next lesson.

Example 2

In this example, the general growth formula is used to estimate the population of India, which has a known growth rate of 1.9%.

Point out that the formula yields the total population after t years. To find out the increase in population, the original population must be subtracted.

Additional Example 2

When Christopher was born, his parents put $1000 into a savings account, and it has grown at a yearly rate of 4.5%. By how much did the account increase in 18 years? **[$1208.48]**

Ongoing Assessment, page 295

Try This

above 19 million people

Example 3

In this example, $y = 2^x$ is graphed. The y-values increase as the x-values increase. Make sure students see that the x-axis is an asymptote for the graph.

Functions of the form $y = b^x$ are called exponential functions because they represent the values of a base, b, raised to a variable power, x. Point out that $b > 0$. If $0 < b < 1$, then the function values decrease as x increases. If $b > 1$, then the function values increase as x increases.

Additional Example 3

Make a table of values. Graph and describe the function $f(x) = 5^x$. Show why it is a function.

continued on top of next page

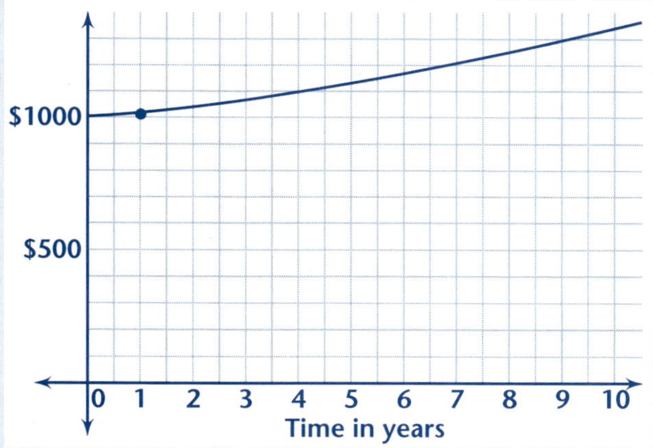

[The slope of the curve—the rate of growth of the account—keeps increasing. Students may notice that this fact is also shown by the increasing numbers in the second column of their table.]

The General Growth Formula

Ask students how they filled in the amounts in their table. Can they think of an easy way to fill in the right-hand column? Is there a pattern?

Call students' attention to the first two amounts in the right-hand column. They should notice that the second amount is $(1 + r)$ times the first amount, where r is the interest rate. Do they think the pattern continues? Have them check, using their calculators, to see that it does.

Explain to the students that since each new entry in the third column is produced by multiplying by $(1 + r)$, the entries in rows 2–4 of the third column are given by the following formulas, respectively:

$$\$1000(1 + r) = \$1000(1 + r)^1$$
$$\$1000(1 + r)(1 + r) = \$1000(1 + r)^2$$
$$\$1000(1 + r)(1 + r)(1 + r) = \$1000(1 + r)^3$$

Students should be able to see that for a starting amount A, the amount P in the account after t years can be found by the following formula:

$$P = A(1 + r)^t.$$

Have students check their results by the using the formula. For the data above, the final amount in the account is found as follows:

$$\$1000(1 + 0.03)^{10} = \$1343.92$$

continued on bottom of next page

Approach A
Using the Book, continued

x	−2	−1	0	1	2
y	0.04	0.2	1	5	25

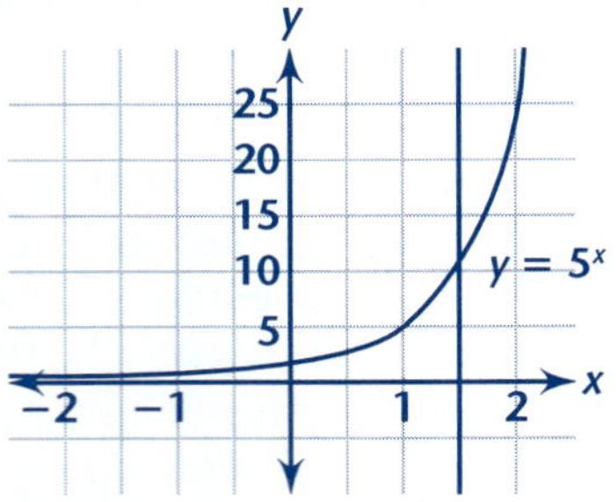

[The vertical line test shows that $f(x) = 5^x$ is a function. The line $x = 1.5$ is shown on the graph.]

Example 4
In this example, the graphs of $y = 3^x$ and $y = 4^x$ are graphed on the same coordinate plane and compared.

Point out that each graph has the x-axis as an asymptote and that the graphs intersect at the point (0, 1) because that point represents the zero power of each function.

Additional Example 4
Graph $y = 5^x$ and $y = 10^x$ on the same axes. Explain how they are similar and how they are different.

x	−2	−1	0	1	2
5^x	0.04	0.2	1	5	25
10^x	0.01	0.1	1	10	100

TEACH

Approach B
Using Models, continued

The Exponential Function
The basic exponential function consists of a positive constant, known as the **base,** raised to a **variable power:**

$$f(x) = a^x$$

Examples include 2^x, 3^x, 4^x, 1.03^x, 10^x, etc. Ask the students the following questions:

- Do you think a basic exponential function will increase, decrease, fluctuate, etc., as x increases? **[For bases greater than 1, the function constantly increases.]**
- What is true of all basic exponential functions at the point where $x = 0$? **[They all have the value 1.]**
- Do you think the value of a basic exponential function can ever be 0? **[No; the value of the function is always positive.]**
- Is there an upper limit to the function? **[No; for bases greater than 1, the value of the function gets very large as x increases.]**
- Describe the shape that all basic exponential functions have in common. Without calculating any points, draw a sketch of a basic exponential function. **[For bases greater than 1, the function starts out close to x-axis and increases slowly until it reaches the point (0, 1). After that, it increases more rapidly and is without limit. A sample sketch is shown.]**

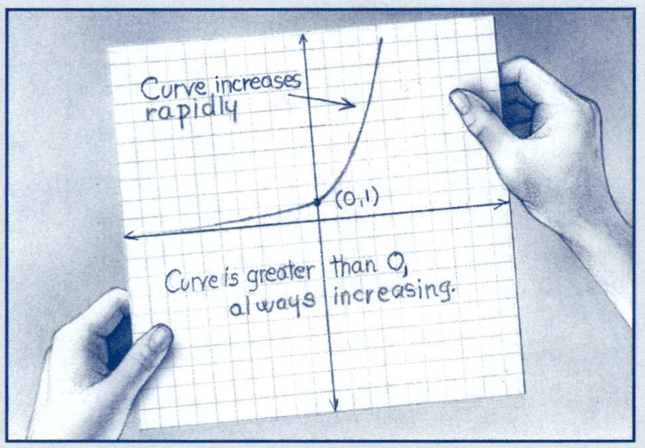

A revealing sketch of an exponential curve can be made without calculating any values.

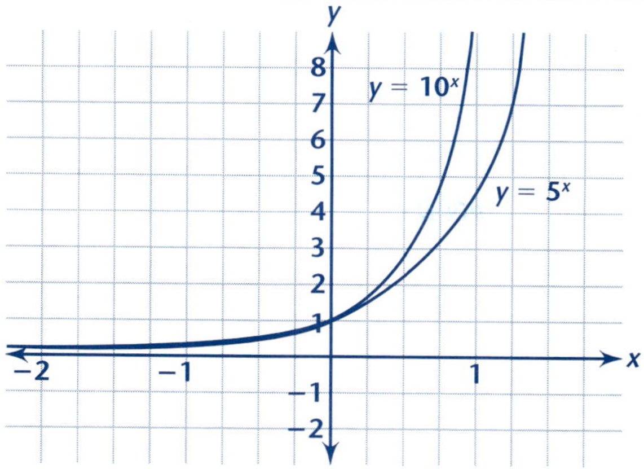

[The graphs are similar in that both rise from left to right, both cross the y-axis at $y = 1$, and both get closer to the x-axis as x decreases. They are different in that the graph of $y = 10^x$ rises faster than the graph of $y = 5^x$ for values of x greater than 0.]

Ongoing Assessment, page 296

For the graphs of $f(x) = a^x$, for $a > 0$, the graph rises increasingly faster as a increases.

Closing the Lesson

Inviting Student Participation

Summarize the lesson by having a student write an exponential function of the form $y = b^x$ on the chalkboard. Have another student list a table of values for the function. Have a third student show its graph (this may be done with a graphics calculator). Have a fourth student give a new function. Continue with this pattern until all students have participated.

Graphing the Basic Exponential Functions

A table of values is a good visual aide for understanding exponential growth. For example, to compare the graphs of $y = 2^x$, $y = 5^x$, and $y = 10^x$, the table below will be helpful. Students should easily see that (0, 1) is a common point.

x	−2	−1	0	1	2
2^x	0.25	0.5	1	2	4
5^x	0.04	0.2	1	5	25
10^x	0.01	0.1	1	10	100

Transforming the Basic Exponential Functions

Have students make a table of values for each exponential function. Explain how the graphs compare with the parent function in problem **A**.

A. $y = 2^x$ **B.** $y = 2^x + 2$

C. $y = 2^{x+2}$ **D.** $y = 2(2^x)$

	−3	−2	−1	0	1	2	3
$y = 2^x$	0.125	0.25	0.5	1	2	4	8
$y = 2^x + 2$	2.125	2.25	2.5	3	4	6	10
$y = 2^{x+2}$	0.5	1	2	4	8	16	32
$y = 2(2^x)$	0.25	0.5	1	2	4	8	16

[The y-values in the graph for problem B are all 2 units higher that the y-values in the graph for the parent function in problem A. This is a vertical shift of 2 units. The y-values in the graph for problem C are all 2 units to the left of the y-values in the graph for part A for each corresponding x-value. This is a horizontal shift of 2 units to the left. In problem D, the entire function is stretched vertically by a factor of 2. The function crosses the y-axis at (0, 2).]

Closing the Lesson

Ask students how to explain their graphs of the interest growth in terms of a basic exponential function. **[The equation has a base of 1.03, with a stretch factor of 1000; that is, $1000(1.03^x)$. Notice that the graph passes through the point (0, 1000).]**

ASSESS

Selected Answers

Odd-numbered Exercises 7–33 and 37–43 and odd-numbered Extra Practice

Assignment Guide

Core level students should complete Exercises 1–43.

Core-plus level students should complete Exercises 1–46.

Error Analysis

Students who have trouble with Exercises 15–18 may need to review translation concepts from Chapter 5. Point out that point (a, b) becomes the point $(a, b + 3)$ for a vertical translation of magnitude 3 in Exercise 15. Point (a, b) becomes the point $(a + (-3), b)$ for a horizontal translation of magnitude -3 in Exercise 16. Exercise 17 is a vertical stretch, and Exercise 18 is a horizontal stretch.

Alternative Assessment

Using Technology Ask students to use a graphics calculator to help them understand the similarities and differences in the graphs of exponential functions. Enter the functions in the Y= menu and then use the table features to see the coordinates. Compare the coordinates. Have students sketch the table screen from a graphics calculator for the graphs of $y = 2^x$, $y = 3^x$, and $y = 4^x$. Ask them to compare the coordinates.

X	Y1	Y2
−3	.125	.03704
−2	.25	.11111
−1	.5	.33333
0	1	1
1	2	3
2	4	9
3	8	27

X = 3

X	Y2	Y3
−3	.03704	.01563
−2	.11111	.0625
−1	.33333	.25
0	1	1
1	3	4
2	9	16
3	27	64

Y3 = 64

[Each table (and graph) contains the point (0, 1). The higher the base, the higher the corresponding y-value for each x-coordinate. All of the y-values increase as the x-values increase.]

RETEACH

The population of China was about 1,203,097,268 in 1995 and was growing at a rate of 1.04%. Have students estimate the population of China in the following years. Use a graph or a table.

A. 1996

B. 2000

C. 2005

D. 2010

[A. 1,215,609,480; B. 1,266,973,200; C. 1,334,240,490; D. 1,405,079,197]

ENRICH

Ask students to find population data on a country of their choice. Almanacs and the CIA World Factbook on the Internet are good sources. Copy the population data for that country, and then show a graph of the data. Does the graph look like the graph of an exponential function? Draw a smooth curve for the graph. What does the point on the y-axis represent? Can the graph be used to make a projection of data for the future?

NAME ______________________ CLASS ____________ DATE ____________

Practice

6.5 Exponential Functions

Graph each of the following:

1. $y = 7^x$

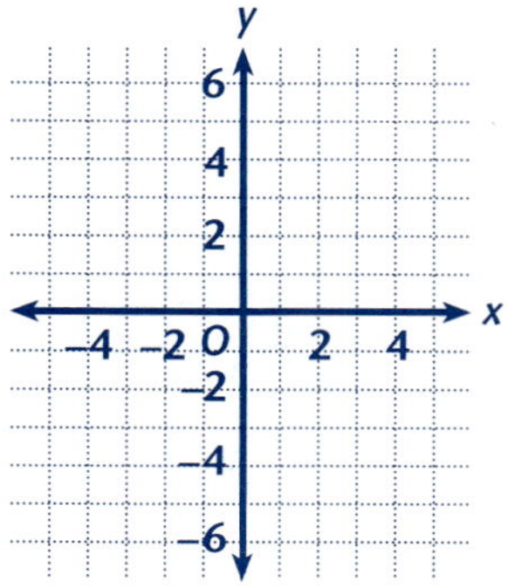

2. $y = 0.1^x$

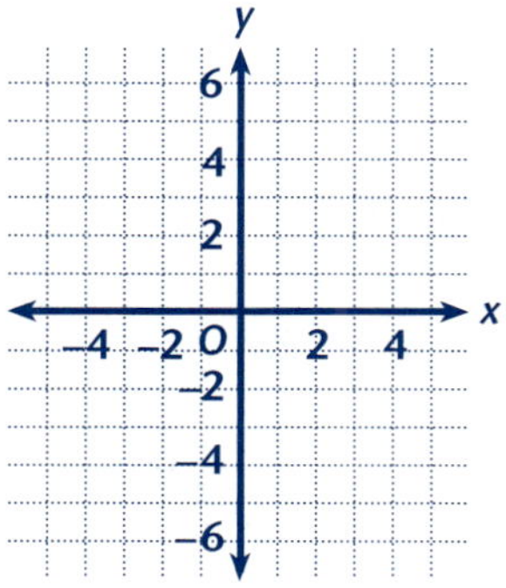

3. $y = 1.5^x$

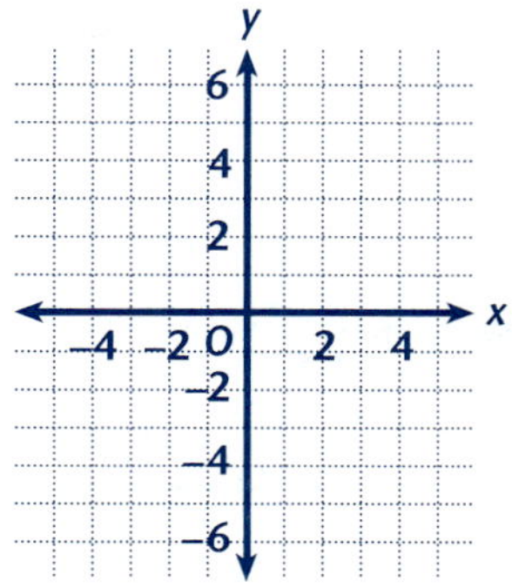

Graph each function. In each case, describe the effect of 1 in the transformation of the parent function.

4. $y = 2^x + 1$ ____________

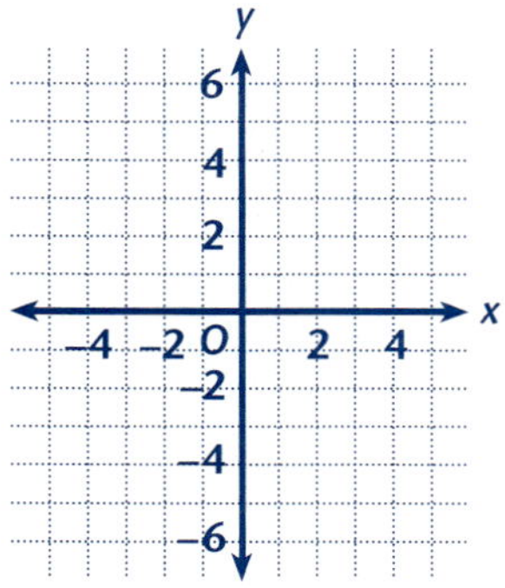

5. $y = 2^{x+1}$ ____________

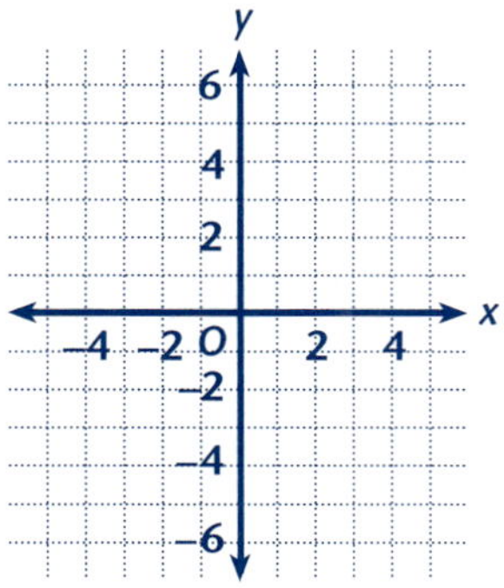

6. $y = 1 \cdot 2^x$ ____________

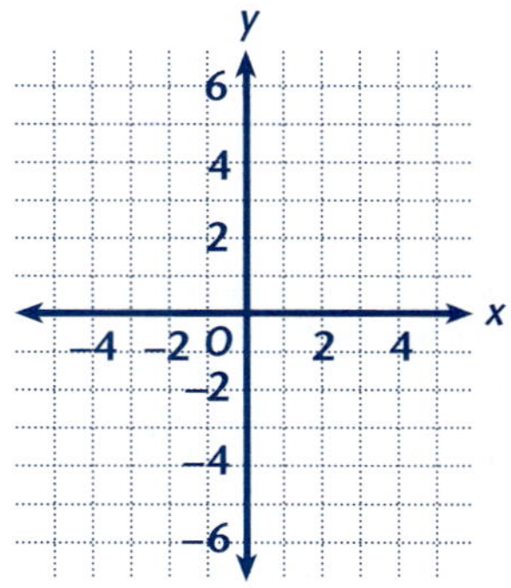

If $f(x) = \left(\frac{1}{2}\right)^x$, find the following values for the function:

7. $f(2)$ ________ **8.** $f(0)$ ________ **9.** $f(-1)$ ________ **10.** $f(-2)$ ________

11. $f(-3)$ ________ **12.** $f(-4)$ ________ **13.** $f(4)$ ________ **14.** $f(3)$ ________

The population of a city is about 200,000 and is growing at a rate of about 0.6% per year.

15. What multiplier is used to find the new population each year? ____________

16. Estimate the population 4 years from now. ____________

NAME ______________________________ CLASS ______________ DATE ______________

Lesson Activity

6.5 Comparing Exponential Functions

Answer the questions below for each of these six exponential functions:

$y = 3^x$ $\quad y = \left(\frac{1}{3}\right)^x$

$y = 4(3)^x$ $\quad y = 4\left(\frac{1}{3}\right)^x$

$y = 8(3)^x$ $\quad y = 8\left(\frac{1}{3}\right)^x$

1. Complete the table of values.

x	3^x	$4(3)^x$	$8(3)^x$	$\left(\frac{1}{3}\right)^x$	$4\left(\frac{1}{3}\right)^x$	$8\left(\frac{1}{3}\right)^x$
−3						
−2						
−1						
0						
1						
2						
3						

2. For which functions do the y-values always increase as the x-values increase? ______________

3. For which functions do the y-values decrease as the x-values increase?

____

4. Which functions get closer and closer to the x-axis as x takes on smaller values?

____

5. What function values are common for at least two of the functions?

____

6. Describe the effect of the 4 on the transformation of $y = \left(\frac{1}{3}\right)^x$. Then describe the effect of the 8.

____

7. Describe the effect of the 4 on the transformation of $y = 3^x$. Then describe the effect of the 8.

____

NAME ____________________ CLASS ____________ DATE ____________

Basic Skills Practice
Cumulative Review

Read each question and circle the best answer.

1. Pam mixes 2 quarts of window cleaner concentrate with 10 quarts of water. How many ***gallons*** of window cleaner does this make?

A 3 gal
B 4 gal
C 12 gal
D 20 gal

2. A student, 5 feet tall, casts a shadow 2 feet long. Find the height of a building that casts a shadow of 52 feet at the same time.

F 21 ft
G 52 ft
H 104 ft
I 130 ft

3. A stockholder received 4.575 shares as a dividend. Prior to the dividend, he owned 360.45 shares of the stock. How many shares of the stock does the stockholder now own?

A 364.925 shares
B 365.025 shares
C 375.025 shares
D 405.20 shares

4. In a scale drawing, a 280-ft tower is drawn 7 inches high. A building next to it is drawn 2 inches high. How tall is the actual building?

F 140 ft
G 120 ft
H 100 ft
I 80 ft

5. Find the circumference of the circle. (Use $\pi = 3.14$)

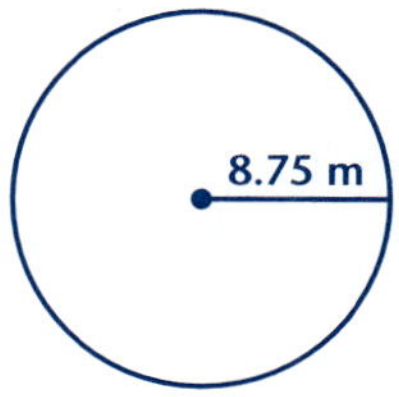

A 13.738 m
B 27.475 m
C 27.575 m
D 54.95 m

6. Sally, age 43, has two sons named John and Rob. John is 7 years older than Rob. The sum of the sons' ages is less than Sally's age. Which inequality could be used to find J, the maximum age John could be?

F $43 > J + 7 - J$
G $2J - 7 < 43$
H $43 - J > J + 7$
I $J + 7 < 43$

7. If a pipe that is 12 feet 9 inches long is cut into 30 sections of equal length, how long is each section?

A 0.67 in.
B 4.3 in.
C 5.1 in.
D 6.7 in.

LESSON 6.6

Applications of Exponential Functions

PREPARE

Pacing Guide

2 Days

Objectives and Overview

- Use exponential functions to model applications.
- Use exponential functions to solve problems about growth and decay in different contexts.

Students extend the concepts they learned in Lesson 6.5 to include exponential functions that represent decay models and to find the value of an investment several years ago (a negative exponent).

Assessing Prior Knowledge

Use a calculator to find each of the following:

1. $3 \cdot 0.2^2$ [0.12]

2. $3 \cdot 0.2^3$ [0.024]

3. $3 \cdot 0.2^4$ [0.0048]

Approach A

Using the Book

Opening Discussion

Carbon-14 is a radioactive form of carbon that occurs naturally in the atmosphere. Ask students how they think carbon-14 gets into the bodies of living organisms. **[Living plants absorb carbon-14 directly from the air, and animals get it, indirectly, from plants.]** When a plant or animal dies, it no longer absorbs carbon-14 from the environment. Carbon-14 spontaneously decays at a steady rate such that after about 5700 years only half of the original amount remains.

Brainstorming

By measuring the amount of carbon-14 in a sample of a once-living object, scientists can approximate the time that has past since it died. Ask students to think of objects that scientists might test by this method. **[Examples include bones, teeth, baskets, cloth, grain, wood from structures or campfires, etc.]**

TEACH

Approach B

Using Models

Opening Discussion

Carbon-14 is a radioactive form of carbon that occurs naturally in the atmosphere. Ask students how they think carbon-14 gets into the bodies of living organisms. **[Living plants absorb carbon-14 directly from the air, and animals get it, indirectly, from plants.]** When a plant or animal dies, it no longer absorbs carbon-14 from the environment. Carbon-14 spontaneously decays at a steady rate such that after about 5700 years only half of the original amount remains.

Brainstorming

By measuring the amount of carbon-14 in a sample of a once-living object, scientists can approximate the time that has past since it died. Ask students to think of objects that scientists might test by this method. **[Examples include bones, teeth, baskets, cloth, grain, wood from structures or campfires, etc.]**

Carbon-14 Dating

In a living thing, a certain small percent of the carbon in its makeup is expected to be carbon-14. After it dies, however, the amount of carbon-14 begins to reduce by half every 5700 years. This fact is used to approximate the age of an object.

Use **Transparency 28** to discuss carbon-14 dating.

Example 1

In this example, a table for the powers of $\frac{1}{2}$ is used to estimate the age of an object with carbon-14 dating.

Point out that real carbon-14 dating involves precise instruments that give closer estimates.

Additional Example 1

Use the table on page 300 of the Pupil's Edition to estimate the age of an object that has 40% of its original carbon-14. Is it $\frac{1}{4}$ as old as the object in Example 1? **[6500 years; no]**

Example 2

In this example, the general growth formula from Lesson 6.5 is used to find out how much money is needed now in order to have a certain down payment in 3 years.

Point out that this is one of the more important uses of the general growth formula because it allows you to look at an investment plan in terms of the final amount of the investment. Point out that this formula assumes an annual compounding period. A different formula exists if interest is compounded at a different time.

Additional Example 2

Suppose that your parents want to save $10,000 for your college education and that you will graduate in 4 years. The current interest rate is 4% compounded annually. How much will you need to invest in a savings account to have $10,000 in 4 years? **[$8548.04]**

continued on top of next page

Carbon-14 Dating

In a living thing, a certain small percent of the carbon in its makeup is expected to be carbon-14. After it dies, however, the amount of carbon-14 begins to reduce by half every 5700 years. This fact is used to approximate the age of an object.

Use **Transparency 28** to discuss carbon-14 dating.

Using Paper Models

Using paper models can help students understand how rapidly exponential growth or decay can change the population or amount of a substance. For example, if students make a paper model of an object and then cut away half of it at specified, equal intervals, they can get a sense for how quickly the amount of the object is reduced. For carbon-14, the specified interval is 5700 years. For the classroom experiment, the interval might be 1 minute.

Working With a Paper Model

Cooperative-Learning Groups Students should conduct the following experiment with a partner:

Have students cut out squares from graph paper that measure 32 grids on each side. Let the entire square represent the original amount of carbon-14 in a living organism. Have students remove, by cutting, half of the squares per minute (1 minute = 1 half-life). Have them determine how long after the organism dies it will be before only one square is left. **[5 half-lives, or 28,500 years]**

Next, have students construct a table of the percent of carbon-14 remaining for intervals of 1 half-life.

Half-lives elapsed	0	1	2	3	4	5
Percent of carbon-14 remaining	100	50	25	12.5	6.25	3.125

Ask students to use the table to determine the age of the artifact modeled by the paper square

continued on bottom of next page

Approach A — Using the Book, continued

Example 3

In this example, the general growth formula from Lesson 6.5 is used with a negative exponent to find out the value of an investment 4 years ago.

Point out that this is another important use of negative exponents because it allows you to look backward into an investment plan.

Additional Example 3

The value of a certain stock has been growing at a rate of 2% per year for 6 years. If the stock is now worth $2500, what was the value of the investment 6 years ago? **[$2219.93]**

Ongoing Assessment, page 301

Try This

about $152,765

Example 4

In this example, an exponential decay model is used for a probability application.

Point out that the logic of this problem is that Ginny gets the first step right and the second step right and the third step right, and so on. A sequence that involves this use of "and" necessitates a multiplication of the probability for each step. Therefore, the decay model applies because the probability for each step is less than 1.

Additional Example 4

Suppose that a high-tech ball bounces to 95% of its original height on every bounce. It is dropped from an initial height of 6 feet.

A. What will be the height after the second bounce? **[5.415 ft]**

B. What will be the height after 10 bounces? **[3.59 ft]**

C. After how many bounces will the height be 1 foot? **[35 bounces]**

TEACH

Approach B — Using Models, continued

if 20% of the original carbon-14 remains. **[between 2 and 3 half-lives; that is, between 11,400 and 17,100 years]**

Have students graph the results from the table and use the graph to improve their estimate of the elapsed time.

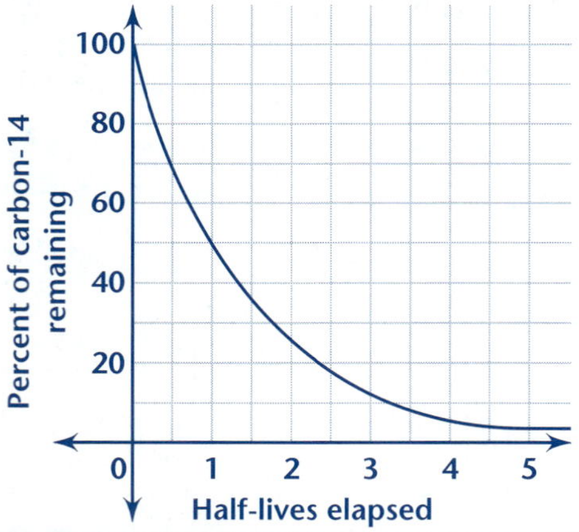

[about 2.3 half-lives, or about 13,110 years]

Ask students the following questions:

- Can they determine the equation for the curve they graphed? If they need a hint, tell them that it is an exponential function with a base less than 1. $\left[f(x) = \left(\frac{1}{2}\right)^x\right]$
- How is this curve different from the exponential curves they studied in Lesson 6.5? How is it similar? **[It is decreasing instead of increasing, but like the ones in Lesson 6.5, it passes through the point (0, 1). It is, in fact, a reflection through the *y*-axis of the graph of $f(x) = 2^x$.]**
- Next, ask students if they can relate this equation to the general growth equation, $P = A(1 + r)^t$. **[In this case, *r* is the negative value −0.5, *t* is the elapsed time in half-lives, and *A* is 1.]**
- If basic exponential functions with bases greater than one *increase,* while those with bases less than one *decrease,* what would they expect for an in-between case? **[When the base is 1, the function is a horizontal line passing through (0, 1).]**

Critical Thinking, page 301
over 15 steps

Example 5

In this example, the graphing and table methods are used to find the population of a city *x* years ago.

Point out that a graphics calculator makes these methods useful because it is efficient and fast. Use the Zoom Integer function to trace in even increments of one year.

Additional Example 5

The deer population in a certain region of the United States was estimated at 460 in 1995 and was decreasing at the rate of 2% per year. When will the number of deer fall below 300? Solve the problem by the following methods:

A. with a graph **B.** with a table

A. [about 21.1 years]

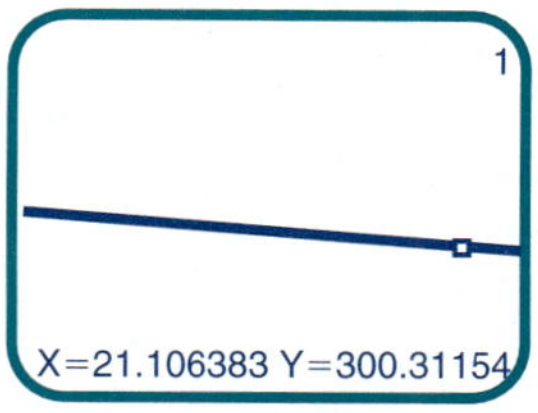

B.

x	$460(0.98)^x$	x	$460(0.98)^x$	x	$460(0.98)^x$
0	460	10	376	19	313
2	442	12	361	20	307
4	424	14	347	21	301
6	407	16	333	22	295
8	391	18	320	23	289

Example 6

In this example, translated graphs of the parent function $y = 2^x$ are compared.

continued on top of next page

Solving the Growth Equation When *t* Is Given

With a firm understanding of exponentials based on models, students should have little difficulty understanding the procedures in Examples 2 and 3 of the textbook. You can use those directly, or you can modify the previous modeling exercise. For example, you can have them check their graphic answer to the question by asking the following question:

What percent of the original substance will be left after 2.3 half-lives have elapsed?
[$(0.5)^{2.3} = 0.203 \ldots$, or about 20.3%, so there is good agreement with the graphic result.]

Probability and Exponential Functions

Example 4 in the textbook deals with the probability of getting a certain number of events of a given probability in a row. It is easy to set up a classroom experiment to confirm such results. For example, give each student a 12-sided number "cube" and ask them what the probability is of getting three rolls in a row with values of 3 or greater and of 4 or greater.
$\left[\left(\frac{5}{6}\right)^3 = 0.58 \ldots, \left(\frac{5}{6}\right)^4 = 0.48 \ldots\right]$

Have students roll their cubes 3 and 4 (or more) times and report their results to the class. Calculate the percent of the students who got "successful" results and compare these with the predicted probability.

Graphing Exponential Functions

Have students graph the exponential functions given in Example 5 in the textbook and explain in their own words why the original parent function is shifted *down* in one case and to the *right* in the other.

As an additional exercise, have students graph the basic exponential function, $f(x) = a^x$, for values of a greater than 1, less than 1, and equal to 1. The results should be memorable and instructive.

continued on bottom of next page

Approach A

Using the Book, continued

Remind students that they learned how to translate functions vertically (problem **A** below) and horizontally (problem **B** below) in Chapter 5. Emphasize that all of the graphs have the same shape and are just repositioned in the coordinate plane.

Additional Example 6

Compare the graphs of the transformations with the graph of the parent function $y = 3^x$.

A. $y = 3^{x+1}$ **B.** $y = 3^x + 1$

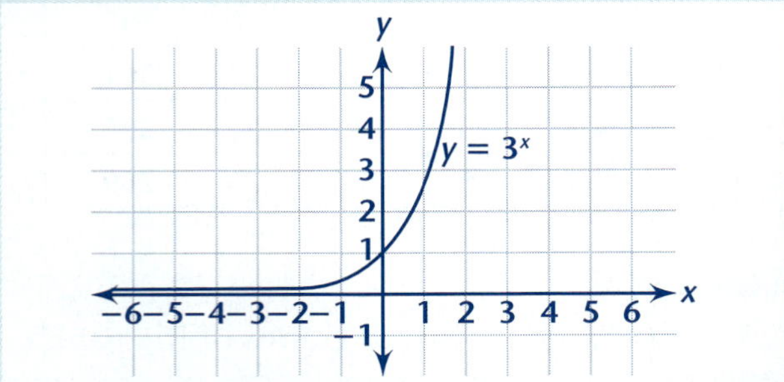

[parent function $y = 3^x$]

A.

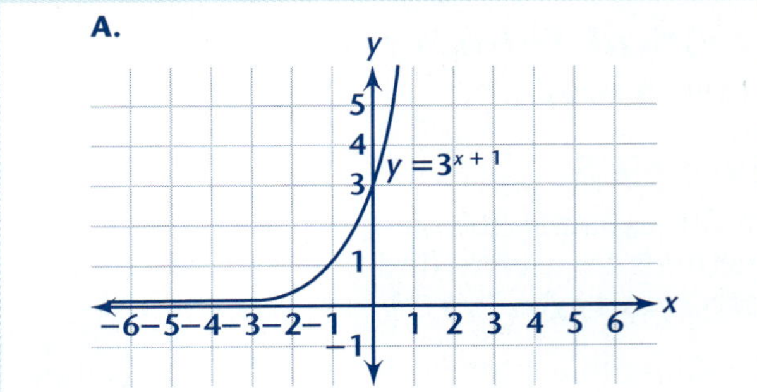

B.

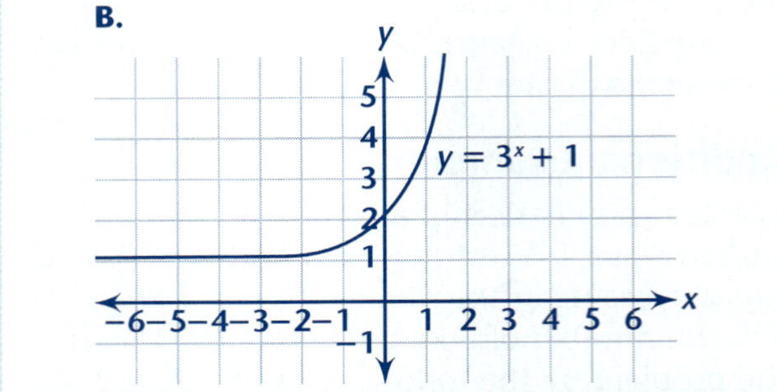

[A. The graph of $y = 3^{x+1}$ is shifted to the left 1 unit from the graph of the parent function, $y = 3^x$. B. The graph of $y = 3^x + 1$ is shifted up 1 unit from the graph of the parent function, $y = 3^x$.]

TEACH

Approach B

Using Models, continued

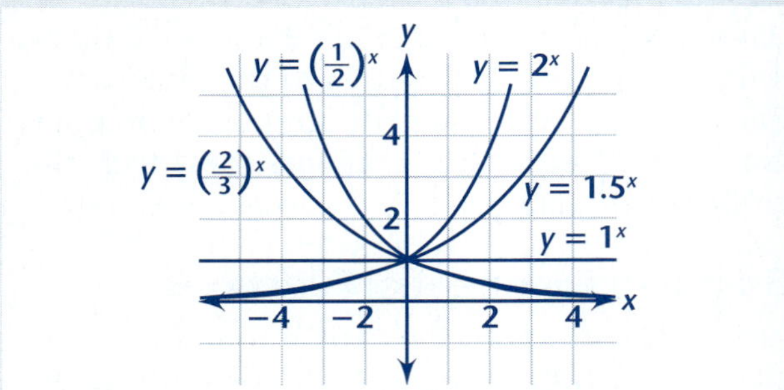

Teaching Tip

A physical model gives students a tactile connection to how exponential decay works. Encourage students to experiment with their squares and different decay factors.

Cooperative Learning

Ask students to work in groups of four to devise their own decay experiment. As a group, they should decide on what decay factors to test, how many squares of the object to start with, and how long to conduct the experiment. Also ask them to include an experiment in which the squares represent the current amount of money in an account (play money can also be used here) and they are asked to determine how much money was invested 5 years ago (at a rate they determine) to get to the current balance. Ask students to share the responsibilities and to submit a complete report.

Journal Entry

Have students describe how an exponential decay pattern can be used to write an exponential function of the form $P = A(1 - r)^t$, where P is current amount or population, A is the starting amount, r is the decay rate, and t is the length of time. Encourage them to include an example in their description.

ASSESS

Selected Answers

Odd-numbered Exercises 7–41 and odd-numbered Extra Practice

Assignment Guide

Core level students should complete Exercises 1–43.

Core-plus level students should complete Exercises 1–43.

Error Analysis

Students who have trouble with Exercises 13–20 may need to pay closer attention to whether exponential growth or exponential decay is described. If an investment is growing, then they should use $P = A(1 + r)^t$, where P is current amount or population, A is the starting amount, r is the growth rate, and t is the length of time. If an investment is decreasing, then use a negative value for r in the same formula.

Alternative Assessment

Exponential Modeling Ask students to enter each of the following points into the STAT EDIT mode of a graphics calculator: (0, 1000), (1, 512), (2, 250), and (3, 125). The x-values go into List 1 and the y-values go into List 2. Then choose STAT CALC ExpReg to find a model of the form $y = a \cdot b^x$, where a is the initial amount of the substance, b is the growth or decay rate, and y is the amount of substance remaining after x time periods. The results are shown below. This is close to a half-life model because b rounds to 0.5.

```
ExpReg
 y=a*b^x
 a=1009.531751
 b=.4988155787
 r = −.9999186178
```

RETEACH

Have students create a table of values for the amount of carbon-14 left in a fossil that originally contained 100 grams of carbon-14 and that is believed to be 57,000 years old.

	A	B
1	Number of 5700-year periods	Amount of carbon-14 left (grams)
2	0	100
3	1	50
4	2	25
5	3	12.5
6	4	6.25
7	5	3.125
8	6	1.5625
9	7	0.78125
10	8	0.390625
11	9	0.1953125
12	10	0.09765625

ENRICH

Have students do research on carbon-14 dating and write a report.

Interdisciplinary Connection

Biology Specialized fields of biology are concerned with finding the half-life of once-living organisms so that the history of the planet can be better understood. Chemistry and biochemistry are involved in experiments that are conducted to determine half-life.

NAME ______________________ CLASS __________ DATE __________

Practice

6.6 Applications of Exponential Functions

The percent of female drivers in the United States has risen steadily from 1964 to 1992. The percent increase is modeled by $y = 40(1.03)^x$, where x is the number of years beginning with $x = 0$ in 1964, and y is the percent increase.

1. Use the model to estimate the percent increase of female drivers in the United States in 1975. ______________________

2. Estimate the percent increase of female drivers in the United States in the year 2000. ______________________

The 'Mendelssohn' Stradivarius violin was estimated to be worth approximately $1,700,000 in 1990. The violin is expected to increase in value by approximately 7.5% each year.

3. Use a calculator to complete the table.

Number of years	0	1	2	3	4	5
Value (in $)						

4. Find a formula to model the growth in value of the violin. ______________________

5. Graph this function by using a graphics calculator.

6. Estimate the value of the violin in the year 2010. ______________________

The International Telecommunication Union estimated 130,000,000 telephone subscribers in the United States in 1991. The number of telephone subscribers is estimated to increase 8% each year.

7. Find a formula to model this growth in subscribers. ______________________

8. Use the formula to estimate the number of subscribers in the year 2001. ______________________

9. Graph this function by using a graphics calculator.

10. Estimate the number of subscribers in the year 2011. ______________________

NAME ____________________ CLASS __________ DATE __________

Lesson Activity

6.6 Musical Scales and Exponential Functions

Frets are the ridges on the neck of a guitar. The placement of the frets is related to the notes of the musical scale.

1. Draw a horizontal line through the top fret, representing the note E. Draw a vertical line through the corresponding E on the right end of the horizontal axis. Extend the lines until they meet. Mark the point of intersection. Repeat for the notes G, A# (A sharp), C# (C sharp), E, and G. Connect the points by extending the part of the graph already drawn.

2. Describe the graph.

____

____

____

3. Explain how this application can be tied to the topic of exponential functions.

____

____

4. Develop at least two application problems for exponential functions. Write them in the space provided and exchange them with a classmate.

____

____

____

____

____

NAME ______________________ CLASS ____________ DATE ____________

Basic Skills Practice
Cumulative Review

Read each question and circle the best answer.

1. Two tanks hold a total of 45 gallons of water. One tank holds 6 gallons more than twice the amount in the other. How much water is in the smaller tank?

A 13 gal
B 15 gal
C 19 gal
D 11 gal

2. What is the length of side *AB*?

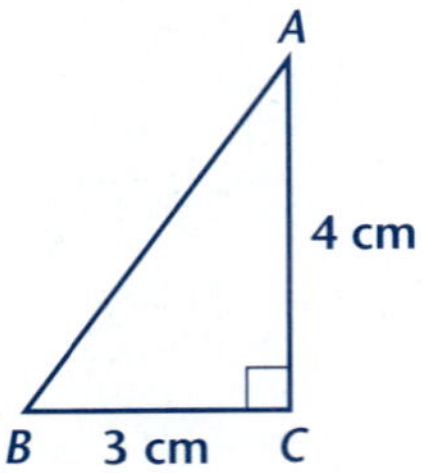

F 4 cm
G 3.5 cm
H 5 cm
I 7 cm

3. What is $\frac{2}{5}$ expressed as a decimal?

A 0.04
B 0.2
C 0.25
D 0.4

4. Find the mode for the following heights: 98 cm, 115 cm, 150 cm, 107 cm, 115 cm, 140 cm, 133 cm.

F 98 cm
G 115 cm
H 123 cm
I 858 cm

5. An international traveler is checking 7 pieces of luggage having a combined weight of 183.4 kilograms. What is the mean (average) weight of the pieces of luggage?

A 26.2 kg
B 26.7 kg
C 27.31 kg
D 34.7 kg

6. What is the surface area of a cube that has edges of 3 feet?

F 9 ft^2
G 18 ft^2
H 27 ft^2
I 54 ft^2

7. A pipe made of material that is 0.15 millimeters thick has an outside diameter of 3.575 millimeters. What is the inside diameter of the pipe?

A 3.275 mm
B 3.425 mm
C 3.545 mm
D 3.572 mm

8. Susan wants to rent a garden plot near her apartment. Which is the best estimate of the area of a rectangular plot that is 38 feet wide and 62 feet long?

F 1500 ft^2
G 2000 ft^2
H 2400 ft^2
I 2500 ft^2

LESSON 7.1

Exploring Polynomial Functions

PREPARE

Pacing Guide

3 Days

Objectives and Overview

- Evaluate polynomial functions.
- Apply surface area and volume functions.

Students solve volume and surface area problems by exploring the behavior of functions. They also examine an identity.

Vocabulary

volume
surface area
lateral faces
polynomial function
monomial
identity

Assessing Prior Knowledge

Fill in the chart below for each value of x.

$y = -0.5x - 1$			
	x	y	(x, y)
1.	3	[−2.5]	[(3, −2.5)]
2.	−10	[4]	[(−10, 4)]
3.	5	[−3.5]	[(5, −3.5)]

Approach A: Using the Book

Opening Discussion

Hold up a box or a picture of a box, or draw a box on the chalkboard. How would you find the volume of the box? On what variables does the volume depend? How could you find the surface area of the box? Why might a company want to minimize the surface area of the box?

Brainstorming

Ask students to list several reasons why a manufacturer might want to change the volume or surface area of a product's packaging. Examples include the following: to decrease the cost of the packaging, to make the package size and shape more visually appealing, to make the package easier to stack or store, to make the package easier to hold, or to make the product available in a variety of sizes.

continued on top of next page

TEACH

Approach B: Using Models

Opening Discussion

Hold up a box or a picture of a box, or draw a box on the chalkboard. Ask students how they would find the volume of the box. Which variables determine the volume of the box? What do you think is meant by the surface area of the box? How could you find the surface area of the box? Why do you think a company would want to minimize the surface area of the box?

Brainstorming

Ask students to think of several reasons why a manufacturer might want to change the volume or surface area of its product's packaging. Examples include the following: to decrease the cost of the packaging, to make the package size and shape more visually appealing, to make the package easier to stack or store, to make the package easier to hold, or to make the product available in a variety of sizes.

continued on bottom of next page

Approach A: Using the Book, continued

Volume as a Function of Width

By keeping the height of a box constant and expressing the length of the box in terms of the width, the volume of the box can be expressed as a function of the square of the width. Since volume varies directly as x^2, students are able to examine how changes in the width of a box affect its volume.

Teaching Tip

Interpreting a Formula Students may need reminding that in the formula $V = Bh$, B represents the *area* of the base, not its length.

Technology

A spreadsheet can be used to display values of the independent and dependent variables and has an advantage over a calculator when a large number of values must be displayed. For example, the fill series command can be used to rapidly enter values for the width of a box. Then a formula can be entered for the volume of the box and copied into the adjacent cells. In a spreadsheet formula, an asterisk (*) indicates multiplication, and a caret (^) indicates an exponent.

Cooperative Learning

Teacher-Guided Exploration You can guide the exploration by calling on different students to answer the questions. If you have a graphics calculator that displays on an overhead, you may find it helpful to direct the exploration by highlighting the appropriate values in the calculator table. Encourage students to offer any observations they have about how width affects volume and surface area.

Student-Directed Approach Divide the class into groups of four. Each group should select a member to read the questions. Another student should be chosen to record the group's answers. A third student should write a summary of the group's answers, and the fourth student

TEACH

Approach B: Using Models, continued

Using a Table

Tell your students that a box has dimensions of 20 cm long × 10 cm wide × 30 cm high. Tell them that a manufacturer wants to lower the cost of producing the box by changing the package size. Assume that the height of the box will remain the same and that the ratio of length to the width will remain 2 to 1. How can the length be determined from the width? Fill in the chart below to indicate how the volume of each box will be found.

Width (cm)	Length (cm)	Volume (cm^3)
5	$5 \cdot 2$	$5 \cdot 5 \cdot 2 \cdot 30 = 5^2 \cdot 60 = 1500$
6	$[6 \cdot 2]$	$[6 \cdot 6 \cdot 2 \cdot 30 = 6^2 \cdot 60 = 2160]$
7	$[7 \cdot 2]$	$[7^2 \cdot 60 = 2940]$
8	$[8 \cdot 2]$	$[8^2 \cdot 60 = 3840]$

By keeping the height of the box constant and expressing the length of the box in terms of the width, the volume of the box can be expressed as a function of the square of the width. Since volume varies directly as x^2, students are able to examine how changes in the width of a box affect its volume.

Teaching Tip

Abbreviating a Formula Since the area of the base, B, of the box is equal to the product of the length of the base, l, and the width of the base, w, the formula for the volume of a box, $V = lwh$, can be expressed as $V = Bh$. Caution students that B represents the *area* of the base, not the length.

Technology

A spreadsheet can be used to display values of the independent and dependent variables and has an advantage over a calculator when a large number of values must be displayed. Refer to Using the Book for an explanation of how a spreadsheet can be used.

should report the group's results to the class. Students should take turns answering the questions, and all students should be involving in a discussion of the answers. If graphics calculators are available, students may want to enter the equations and examine the results in the table.

Exploration

In this exploration, students discover formulas for the volume and the surface area of a box.

The exploration is divided into two parts. In the first part, students explore the effect that changing the width of a box has on its volume. In the second part, the students explore the effect that changing the width of a box has on the surface area of the box.

Part I In Step 3, an expression in one variable is written for the volume. Students should see that by leaving the height of the box constant and expressing the length of the base as a function of the width, changes in volume can be examined as the width is varied.

Step 5 highlights one way that technology can help to solve a problem efficiently. Students should see that if the step function in the table is set to increments smaller than 0.1, the calculator will more closely approximate 5400.

Part II Ask students to write down the procedure they followed in Step 1. They can then refer to this procedure as they complete Step 3.

When students have completed Step 6, refer them to their answer from Part I, Step 5. Ask them what new width they might choose for the box in order to accommodate their findings from these two steps.

In Step 7, students may need reminding that the fraction $\frac{\text{original volume} - \text{new volume}}{\text{original volume}}$ will give the percent reduction in volume. Similarly, the fraction $\frac{\text{original surface area} - \text{new surface area}}{\text{original surface area}}$ will give the percent reduction in surface area. Ask students if they think the reductions are a good compromise.

continued on top of next page

Modeling Volume (Graphics Calculator)

Use a graphics calculator or draw a picture on the chalkboard to demonstrate the calculator displays. Show students how a calculator can be used to determine the width of a box for a given volume when the length of the base is dependent on the width and the height is constant. Enter the volume function into the calculator, and then display the table, showing the ordered pairs for width and volume.

Cooperative Learning

Before you begin the exploration, divide the class into groups of four. If graphics calculators are available, give one to each student, if possible; otherwise, give one to each group. Explain to students that everyone in the group should be involved in the discussion of the questions you will ask. One member of the group should be selected to record the group's answers, and another should write a summary of the group's results. A third student should be assigned to read the report to the class. If each group has only one calculator, the fourth student in the group should manipulate the calculator.

Exploration

In this two-part exploration, students first explore the effect that changing the width of a box has on its volume. Then students explore the effect that changing the width has on the surface area of the box.

Guide the students through the explorations with the following instructions and questions:

Part I: Volume

- Ask students to find the volume of a box with these dimensions: 20 cm long × 10 cm wide × 30 cm high. **[6000 cm^3]**
- Tell students to find the new volume of the box if the volume of the box is reduced by 10%. **[5400 cm^3]**
- Tell students to assume that the company would like to reduce the volume of the box by 10%. Ask students to recall the table of

continued on bottom of next page

Approach A

Using the Book, continued

Student Reports

The designated student from each group should read the group's report to the class. Students giving the report should draw charts on the chalkboard or use a graphics calculator to demonstrate their group's reasoning. Students should be invited to ask questions of each reporter. Any member may assist the group's reporter in answering questions.

Follow-Up Demonstration (Optional)

If you think that the concepts demonstrated in the exploration are not clear to some of your students, you may want to guide them through another exploration. Use the box from the exploration and vary the height while keeping the length constant. The volume can then be expressed as the function $V(x) = \frac{20}{3}x^2$. Find $V(30)$ to determine the original volume. A height of 28.4 cm gives a volume of about 5400 cm^3.

On Their Own

The dimensions of a box are 15 inches × 45 inches × 40 inches. The function $V(x) = 120x^2$, where x is the length and 40 is the height, can be used to represent the volume of the box.

A. Find $V(15)$. **[27,000]** What does this number represent? **[the volume of the box in cubic inches when the length is 15 inches]**

B. If the volume of the box is reduced by 10%, what is the new volume? **[24,300 cubic inches]**

C. Examine the graphics calculator table to find the length of the base that will reduce the volume of the original box by about 10%. **[about 14.3 inches]**

TEACH

Approach B

Using Models, continued

volumes they generated for a given width. If the width of the box is x, what is the length? **[2x]** Complete the following:

$V(x) = [\underline{2x}] \cdot [\underline{x}] \cdot 30 = 60x^2$

- Why do you think the exponent is 2 even though volume is expressed in *cubic* units? **[The height is constant.]**
- Enter the function $Y_1 = 60X^2$ into the calculator. Set the starting value of the table at 9.4 and set the increment at 0.1. Display the table. (Compare with the calculator display on page 317 of the Pupil's Edition.)
- What does X represent in the table? **[the width of the box as defined by the function $V(x) = 60x^2$]** What does Y_1 represent in the table? **[the volume of the box]** Find $V(10)$ in the table. **[6000]** What does this number represent in terms of the box? **[the original volume, when the width was 10 cm]**
- What value of Y_1 is closest to 5400? **[5415]** What is the corresponding value of X? **[9.5]** What does this value represent in terms of the volume of the box? **[the width that can be used to reduce the original volume of the box by approximately 10%]**

Part II: Surface Area Tell students that the *surface area* of a box is the total area of its *faces*. That is, the surface area is the total area of all the faces of the box, or the area of the material used to make the box.

- How could you find the area of the box in Part I? **[Double the product of the width and length. Double the product of the width and the height. Double the product of the length and the height. Find the sum of the three products.]** What is the surface area of the box? **[2200 cm^3]**
- Find the new surface area if the original surface area is reduced by 10%. **[1980 cm^3]**
- You can use the procedure for your calculations in the first step of Part II to find a surface area function.

$$\begin{aligned} S(x) &= 2(2x \cdot x) + 2(2x \cdot 30) + 2(x \cdot 30) \\ &= 4x^2 + 120x + 60x \\ &= 4x^2 + 180x \end{aligned}$$

X	Y_1	
14.2	24,197	
14.3	24,539	
14.4	24,883	
14.5	25,230	
14.6	25,579	
14.7	25,931	
14.8	26,285	

X = 14.8

Application: Business

Marketing Strategy In this application, the radius of the base of a cylinder is held constant and the height is allowed to vary, so the volume is a function of the height.

Students should see that the techniques used to solve this problem and the problems in the exploration are the same: only one variable is allowed to vary, while the other values are held constant. Enter the resulting equation into a graphics calculator. Find the desired value in the Y column of the calculator table. The corresponding value in the X column is the desired height.

Technology

Before the function values can be viewed in a calculator table, the starting value, or initial x-value, should be set and the increment for x should be defined.

Defining *Polynomial Functions*

Students studied the equations of straight lines in Lesson 2.1 and learned the slope-intercept form of a linear equation. A relationship between two variables, defined by the equation $y = mx + b$, is a polynomial function. In this function, the monomial mx is the product of a constant and a variable. The monomial b is a constant.

The volume function that was studied in the exploration, $V(x) = 60x^2$, consists of one monomial, $60x^2$. The surface area function, $S(x) = 4x^2 + 180x$, consists of two monomials, $4x^2$ and $180x$.

continued on top of next page

What do the expressions in the parentheses represent? **[the areas of the three lateral faces]** Why is each of these values doubled? **[Each face has an opposite face with the same area.]**

- Enter the function $Y_1 = 4X^2 + 180X$ into the graphics calculator. Set the initial value of the table to 9.7 and set the increment to 0.1. (Compare with the calculator display on page 317 of the Pupil's Edition.)
- Compute $S(10)$ for $S(x) = 4x^2 + 180x$. Then view the table. Find $S(10)$. Are the numbers the same? **[yes]**
- Scroll up through the table to find the width of the base that can be used to reduce the original surface area of the box by about 10%. **[between 9.1 and 9.2 cm]** (Compare with the calculator display on page 317 of the Pupil's Edition.)

If a width of 9.5 gives about a 10% reduction in the volume of the box and a width of 9.1 gives about a 10% reduction in the surface area of the box, what do you think is a good width for the new box? Calculate the percent reduction in volume and in surface area for your width. Is each reduction about 10%?

Student Reports

The designated student from each group should read the group's report to the class. Students giving the report should draw charts on the chalkboard or use a graphics calculator to demonstrate their group's reasoning. Students should be invited to ask questions of each reporter. Any member may assist the group's reporter in answering questions.

On Their Own

The dimensions of a box are 15 inches × 45 inches × 40 inches. The function $V(x) = 120x^2$, where x is the length of the base and 40 is the height of the box, can be used to represent the volume of the box.

A. Find $V(15)$. **[27,000]** What does this number represent? **[the volume of the box in cubic inches when the length is 15 inches]**

B. If the volume of the box is reduced by 10%, what is the new volume? **[24,300 cubic inches]**

C. Examine the graphics calculator table to find what width of the base can reduce the

continued on bottom of next page

Approach A — Using the Book, continued

Notation

The notation $V(x) = 60x^2$ indicates that the relationship between x and V is a function and that the value of the variable V is dependent on the variable x. Consequently, V is called the *dependent* variable, and x is called the *independent* variable.

Journal Entry

Ask students to describe how the volume and surface area functions that they used in the exploration are different from a function that describes a straight line.

Inclusion Strategy

Portfolio Activity Have students refer to their textbooks from other courses to find examples of functional relationships. If the students cannot find any such relationships in their own textbooks, tell them to ask their other teachers if they might use copies of other textbooks. Or you can invite them to look through subsequent lessons of their math textbook for examples.

Extension

Explaining Identities In this extension, identity is defined, and it is shown that possible identities can be investigated with a graphics calculator. You can demonstrate that two equations are equivalent by graphing them on a graphics calculator. Show students that although two functions are defined, one for Y_1 and the other for Y_2, only one line is graphed. This is true because $(x + 2)^3$ is another form of $x^3 + 6x^2 + 12x + 8$.

TEACH

Approach B — Using Models, continued

volume of the original box by about 10%. **[about 14.3 inches]**

X	Y_1
14.2	24,197
14.3	24,539
14.4	24,883
14.5	25,230
14.6	25,579
14.7	25,931
14.8	26,285

X = 14.8

Follow-Up Demonstration (Optional)

If you think that the concepts demonstrated in the exploration are not clear to some of your students, you may want to guide them through another exploration. Use the box from the exploration and vary the height while keeping the length constant. The volume can then be expressed as the function $V(x) = \frac{20}{3}x^2$. Find $V(30)$ to determine the original volume. A height of 28.4 cm gives a volume of about 5400 cm^3.

Business Application: Volume of a Cylinder

Hold up a can or a picture of a can, or draw a picture of a can on the chalkboard.

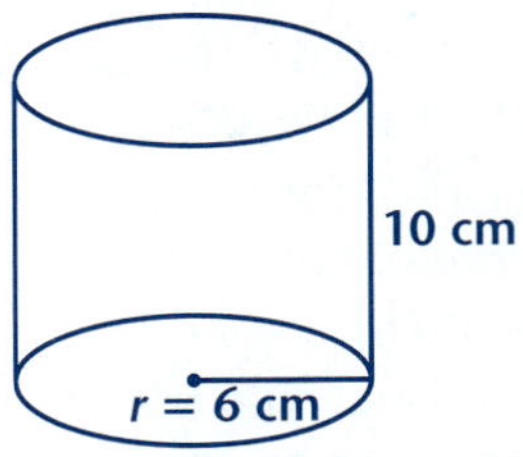

Like a box, the volume of a cylinder is $V = Bh$. Since the base of a cylinder is a circle, the volume of a cylinder can also be expressed as $V = \pi r^2 h$. The volume of the cylinder pictured can therefore be expressed as $V = 360\pi$, or approximately 1131 cm^3.

Ongoing Assessment, page 319

Try This

Enter $Y_1 = 2\pi X^2 + 20\pi X$ and $Y_2 = 2\pi X(X + 10)$. Then generate a table:

X	Y1	Y2
−3	−131.9	−131.9
−2	−100.5	−100.5
−1	−56.55	−56.55
0	0	0
1	69.115	69.115
2	150.8	150.8
3	245.04	245.04

$Y1 = 2\pi X^2 + 20\pi X$

Because the values agree, it appears that the functions are equivalent.

Tell students to assume that the manufacturer of the can wants to keep the size of the base of the can constant while increasing its volume by 20%. Consequently the volume of the can will be dependent on the height of the can. The function $V = 36\pi h$ represents this relationship. Ask students what the new volume of the can will be. **[1131 + 20% · 1131 = 1131 + 226.2, or 1357.2 cubic centimeters]**

Enter the function $V = 36\pi h$ into a graphics calculator. Then display the table with initial value of X = 11.7 and an increment of 0.1. If the students have calculators, instruct them to do the same. If no calculators are available, copy the table of values from page 318 of the Pupil's Edition onto the chalkboard.

Scroll down and highlight X = 12. The corresponding value of Y_1 is 1357.2, so the height that increases the volume of the can by approximately 20% is 12 centimeters.

Defining *Polynomial Functions*

Discuss the definition of polynomial function on page 318 of the Pupil's Edition. Visual learners will benefit from the following diagram and example of a polynomial:

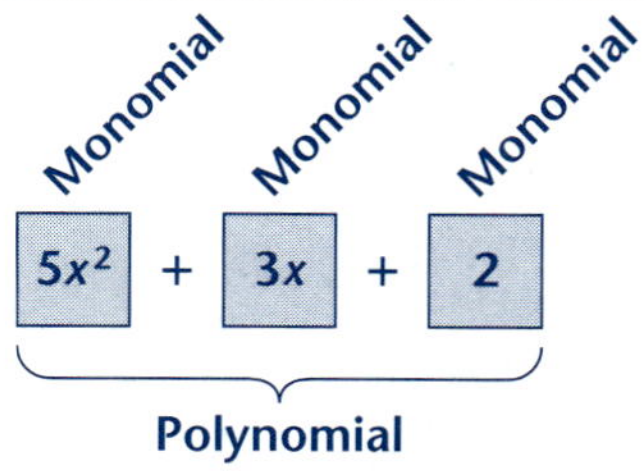

Extension

Cooperative Learning Have students work in pairs to graph the functions $y = (x + 2)^3$ and $y = x^3 + 6x^2 + 12x + 8$ on graph paper for x-values of −3, −2, −1, . . ., 2. Then have them compare the graphs.

Next have them make a table of values like the one on page 319 of the Pupil's Edition and compare the values.

ASSESS

Selected Answers

Odd-numbered Exercises 5–39 and odd-numbered Extra Practice

Assignment Guide

Core level students should complete Exercises 1–39.

Core-plus level students should complete Exercises 1–16 and 24–42.

Error Analysis

Students who have trouble with Exercises 8–19 may be evaluating the expressions incorrectly. You may wish to review the order of operations with your students before they begin these exercises. Students can also check their answers by entering each pair of equations in a graphics calculator and viewing the table of values.

Alternative Assessment

Changing Volume

a. Find the volume of a rectangular solid with a length of 5 cm, width of 15 cm, and height of 8 cm. **[600 cm^3]**

b. What is the new volume of the solid if the length is doubled and the ratio of length to height remains constant? **[2400 cm^3]**

c. By what percent did the volume increase? **[300%]**

Portfolio Activity Have students write a summary of their group's exploration activities. Their summary should include the reasoning and procedures that the group used, charts of the results, and a clear statement of the group's conclusions.

RETEACH

Ask students to recall the slope-intercept form of a linear equation. Choose an equation such as $y = 2x - 1$ and construct a table of values.

$y = 2x + 1$	
x	y
−3	−7
−2	−5
−1	−3
0	−1
1	1
2	3
3	5

Filling in the table requires you to choose values of x in order to find corresponding values of y. Every value of x has a unique corresponding y-value. For this reason, y is said to be a *function* of x. Because the value of y depends on the value chosen for x, y is called the *dependent* variable. Any variable that stands by itself on one side of the equation is the dependent variable. Thus, in the equation $V = 30x^2$, V is the dependent variable.

The language of mathematics contains a convention that allows an equation to show the relationship between two variables. For example, the equation $y = 2x + 1$ can be written as $f(x) = 2x + 1$. This notation shows that y is dependent on the value of x or that y is a function of x.

ENRICH

Interdisciplinary Connection

Personal Finance The number of dollars per month that a car costs is a function of the number of miles the car is driven per month and can therefore be modeled by a linear function. For example, suppose that the monthly cost of driving m miles can be predicted by the equation $c = 0.2m + 110$, where c is the cost in dollars and m is the number of miles driven per month.

Number of miles driven per month, m	Equation	Cost per month, c
200	$c = 0.2(200) + 110$	\$150
500	$c = 0.2(500) + 110$	\$210
1000	$c = 0.2(500) + 110$	\$310

NAME ______________________ CLASS __________ DATE __________

Practice

7.1 Exploring Polynomial Functions

Give the surface area and volume functions for each geometric solid.

1. a cube with an edge that is 3 centimeters long

____

2. a cube with an edge that is x inches long

____

3. a rectangular solid with a base length of 5 inches and a base width of 3 inches ______________________________

4. a rectangular solid with a base length of 4.5 centimeters and a base width of 3.2 centimeters ______________________________

5. a cylinder with a height of 7 inches and a radius of x inches

____

6. a cylinder with a height of 1.3 meters and a diameter of x inches

____

7. a cone with a height of 8 feet and a radius of x feet

____

8. a cone with a height of 16.4 centimeters and a diameter of x centimeters ______________________________

Show that each equation is true by substituting x-values of −1, 0, and 1.

9. $x^2 - x - 12 = (x - 4)(x + 3)$

10. $x^2 + 7x + 10 = (x + 5)(x + 2)$

11. $x^2 - 6x + 9 = (x - 3)^2$

12. $x^2 - 36 = (x + 6)(x - 6)$

NAME ______________________ CLASS ____________ DATE ____________

Lesson Activity

7.1 Write Your Own Functions

For this activity, you will need a piece of string, a ruler, and several different sized balls.

1. Use your piece of string and ruler to find the circumference of a great circle of each ball. Then find the radius of each ball. Record your measurements in the table below.

Ball	Circumference of great circle	Radius of sphere	Surface area $S(r) = 4\pi r^2$	Volume $V(r) = \frac{4\pi r^3}{3}$
1				
2				
3				
4				
5				
6				
7				
8				
9				
10				

2. Complete the table by recording the value of the surface area and volume functions for each ball.

NAME ______________________ CLASS __________ DATE __________

Basic Skills Practice

Using Scientific Notation, Part 1

Scientific notation is a condensed way of expressing numbers. Very large or very small numbers often have many zeros that make them difficult to read and to use for calculations. A number written in scientific notation is written with two factors, a number between 1 and 10, but not including 10, and a power of 10.

Example 1: Write 3,270,000 in scientific notation.

3.27 → Write the first factor (a number between 1 and 10).

10^6 → Write the power of 10.

$3{,}270{,}000 = 3.27 \times 10^6$

The exponent is the number of places that the decimal point moved.

- If the original number is greater than 10, then the exponent is positive.
- If the original number is less than 1 then the exponent is negative.
- If the original number is between 1 and 10, then the exponent is 0.

Example 2: Write 0.0068 in scientific notation.

6.8 → Write the first factor.

10^{-3} → Write the power of 10.

$0.0068 = 6.8 \times 10^{-3}$

Write the missing exponent.

1. $345 = 3.45 \times 10^?$ ______

2. $5432 = 5.432 \times 10^?$ ______

3. $500{,}000 = 5 \times 10^?$ ______

4. $0.00038 = 3.8 \times 10^?$ ______

5. $0.21 = 2.1 \times 10^?$ ______

6. $42{,}000 = 4.2 \times 10^?$ ______

7. $854 = 8.54 \times 10^?$ ______

8. $0.0298 = 2.98 \times 10^?$ ______

9. $78 = 7.8 \times 10^?$ ______

10. $0.0000098 = 9.8 \times 10^?$ ______

Write in scientific notation.

11. 46 ______

12. 7900 ______

13. 800 ______

14. 0.000067 ______

15. 6,500,000 ______

16. 0.62 ______

17. 0.37 ______

18. 5276 ______

19. 4 ______

20. 856 ______

LESSON 7.2

Adding and Subtracting Polynomials

PREPARE

Pacing Guide
3 Days

Objectives and Overview
- Add and subtract polynomials.

This lesson introduces students to the terminology associated with polynomials. Algebra tiles are then used to model the addition and subtraction of polynomials.

Vocabulary
polynomial
degree of a polynomial in one variable
standard form of a polynomial
descending order of a polynomial
vertical form
horizontal form

Assessing Prior Knowledge

Simplify.

1.	$a + 2a$	$[3a]$
2.	$8b - 3b$	$[5b]$
3.	$3cd + 7cd$	$[10cd]$
4.	$2e - (3f + 5e) + 3f$	$[-3e]$

Approach A: Using the Book

Opening Discussion
Begin the lesson by discussing the difference between a constant and a variable. How does the value of 5 differ from the value of $5x$? How are the monomials $5x$ and $5x^2$ alike? How are they different? Is the equation $5x^2 = 5x$ true? Explain.

Communicating Mathematics
Any language is best communicated when both the vocabulary and the structure of the language are understood. Similarly, mathematics is best communicated when its language and structure are understood and used correctly.

Although the number of vocabulary words related to polynomials may intimidate students, stress that once they recognize and are able to use the vocabulary, they will better understand discussions of polynomials. Equally important, they will be better able to communicate their own questions about polynomials.

TEACH

Approach B: Using Models

Opening Discussion
Ask students to explain the difference between a constant and a variable. Does the value of 5 differ from the value of $5x$? Explain. Explain how the monomials $5x$ and $5x^2$ are alike. How are they different? Is the equation $5x^2 = 5x$ true or false? Explain.

Naming Polynomials
Make a chart of prefixes on the chalkboard or overhead as shown here.

Mono-	Bi-	Tri-
[monocle]	[bifocals]	[triangle]
[monologue]	[bicycle]	[triathlon]
etc.	etc.	etc.

Ask students to fill in the chart with as many words with these prefixes as they can think of. Have volunteers read their words and give a definition. Review the meaning of the term *monomial* with your

Ongoing Assessment, top of page 323

The degree is 4.

Journal Entry

Ask students why $3x + 5x^2 = 5x^2 + 3x$, but $3x - 5x^2 \neq 5x^2 - 3x$. Ask them to write an equivalent expression for $3x - 5x^2$ in which the first term is the quadratic term.

Teaching Tip

Some students may be confused by the names *quadratic* and *quartic,* thinking that both names have a root meaning "four" or "fourth." In fact, the word *quadratic* comes from a Latin word meaning "having been made square." A square, of course, does have four sides, but it is the *squaring,* not the number of sides, that gives a polynomial of degree 2 its name.

Defining *Standard Form of a Polynomial*

Have students refer to the polynomial chart on page 323. Which of the polynomials do you think is in *descending* order? Explain. Although the order of terms in a polynomial doesn't matter, the terms of a polynomial in *standard form* are given in descending order from left to right.

Ongoing Assessment, middle of page 323

$5x^2 + x^3$ in standard form is $x^3 + 5x^2$.

Use **Transparency 29** to discuss or review the polynomial expression table.

Making Algebra Tiles

You can create algebra tiles to use for a demonstration of the lesson's concepts from six sheets each of two different colors of overhead transparency film. Cut two of the sheets of each color into 1-inch squares to represent constants. Cut two other sheets of each color into 1-inch $\times$ 3-inch rectangles to represent x-terms, and cut the other two sheets of each color into 3-inch squares to represent x^2-terms.

continued on top of next page

students. Any polynomial with just one term can be called a monomial. Ask your students to guess what the names of polynomials with two terms or three terms might be.

Degree of a Polynomial

When students think of a degree, they may think of a unit of measure used on a temperature scale, the unit of measure used to designate the size of an angle, or a title of achievement awarded by an educational institution.

The *degree of a polynomial* in one variable is the exponent with the greatest value of any of the polynomial's terms. Thus, the degree of $9 - 4x^2$ is 2, and the degree of $2a^2 + 5a^4 - 3a^3 + 1$ is 4. If a polynomial has no variable, its degree is zero.

Discuss the table on page 323 of the Pupil's Edition.

Quadratic Polynomials

A Venn diagram can be used to demonstrate that a polynomial of degree n need not have any specific number of terms. A polynomial of degree 2, for example, might be a monomial, a binomial, or a trinomial. Polynomials of degree 2 are called *quadratic* polynomials.

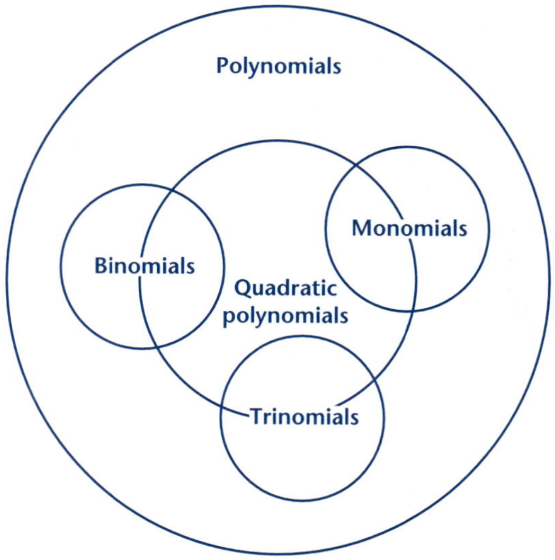

Have students identify each of the polynomials in the diagram above by degree.

continued on bottom of next page

Approach A

Using the Book, continued

Choose one color to represent negative units and the other color to represent positive units. You may wish to draw a plus sign (+) on the positive tiles and a negative sign (−) on the negative tiles.

Inclusion Strategies

Using Models Manipulate algebra tiles on an overhead as you study the examples with the students. You may wish to encourage more participation by asking several students to manipulate the tiles on the overhead.

Adding Polynomials

When algebra tiles are used to represent adding polynomials in vertical form, tiles of the same size are grouped in the same columns. Then the groups of tiles in each column are added.

When algebra tiles are used to represent the addition of polynomials in horizontal form, the tiles must be regrouped before they can be summed. For example, to model $(x^2 + 2x) + (2x^2 - x)$, first model each of the expressions. Then regroup the x^2-tiles and the x-tiles. Add both groups of tiles.

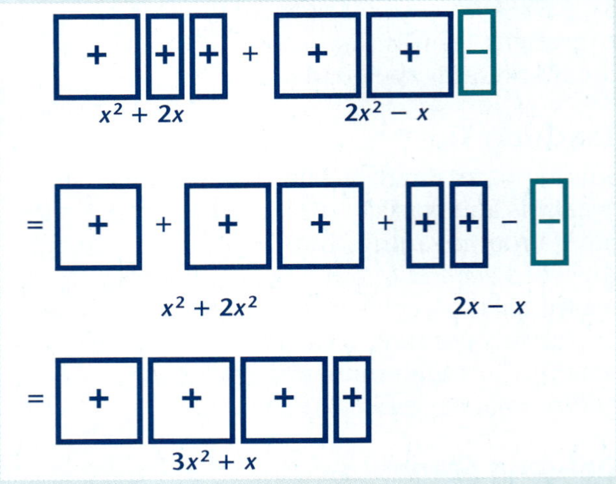

TEACH

Approach B

Using Models, continued

Teaching Tip

Some students may be confused by the names *quadratic* and *quartic,* thinking that both names have a root meaning "four" or "fourth." In fact, the word *quadratic* comes from a Latin word meaning "having been made square." A square, of course, does have four sides, but it is the *squaring,* not the number of sides, that gives a polynomial of degree 2 its name.

Defining *Standard Form of a Polynomial*

What do you think it means to say that a polynomial is in *descending* order? Explain. Although the order of terms in a polynomial doesn't matter, the terms of a polynomial in *standard form* are given in descending order from left to right.

Making Algebra Tiles

You can create algebra tiles to demonstrate the addition of polynomials from six sheets each of two different colors of overhead transparency film. Cut two sheets of each color into 1-inch squares to represent constants. Cut two other sheets of each color into 1-inch × 3-inch rectangles to represent x-terms, and cut the other two sheets of each color into 3-inch squares to represent x^2-terms. Choose one color to represent negative units and the other color to represent positive units. You may wish to draw a plus sign (+) on the positive tiles and a negative sign (−) on the negative tiles.

Modeling Polynomials With Algebra Tiles

Show students how algebra tiles can be used to model polynomial addition. Refer to the modeling of $(x^2 + x + 1) + (2x^2 + 3x + 2)$ and $(2x^2 - 3x + 5) + (4x^2 + 7x - 2)$ on page 324 of the Pupil's Edition. Invite students to manipulate the tiles on an overhead. If the tiles are not available, you can draw squares and rectangles on the board to represent the tiles. Make sure that the squares for the x^2-tiles are larger than those for the constants.

When algebra tiles are used to represent the addition of polynomials in vertical form, tiles of

Example 1

In this example, algebra tiles are used to model the addition of polynomials in a vertical format.

Remind students that a neutral pair of tiles represents a sum of zero. The object of using tiles to add polynomials is to identify as many pairs of neutral tiles as possible. When polynomials are added vertically, the Commutative and Associative Properties allow regrouping.

Additional Example 1

Use algebra tiles to model $(3x^2 + 5x - 1) + (2x^2 - 6x - 2)$. Then use a horizontal form to simplify the sum. $[5x^2 - x - 3]$

Ongoing Assessment, page 325

Try This

$x^2 - x + 4$

Example 2

The rules for adding polynomials are used to find the perimeter of polygons. The sum of the polynomials is checked by substituting a value for the variable and evaluating the indicated sum of the polynomials as well as the simplified sum.

Students should notice that when a polynomial is missing any term, that position in the column is left empty.

Teaching Tip

Some students may find it helpful to add zeros as place holders where necessary when they are adding polynomials in a vertical format. Students may also find it helpful to indicate a coefficient of 1.

Additional Example 2

Find the perimeter of each polygon. Check the addition by substituting a value for each variable.

continued on top of next page

the same size are grouped in the same columns. Then the groups of tiles in each column are added. When algebra tiles are used to represent the addition of polynomials in horizontal form, the tiles must be regrouped before they can be added. For example, to model $(x^2 + 2x) + (2x^2 - x)$, first model each of the expressions. Then regroup the x^2-tiles and the x-tiles. Add both groups of tiles.

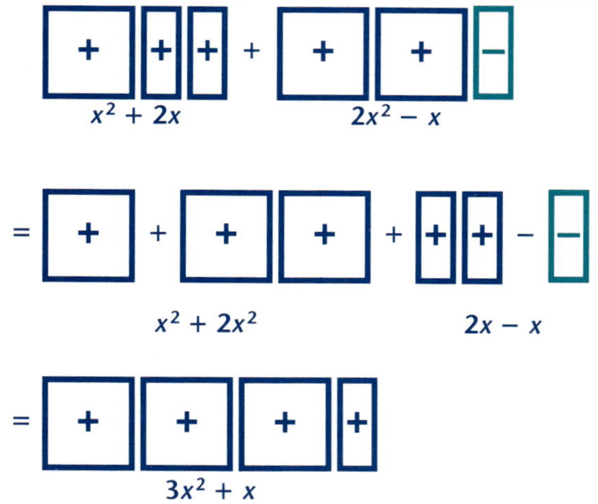

Remind students that a neutral pair of tiles represents a sum of zero. The object of using tiles to add polynomials is to identify as many pairs of neutral tiles as possible. When polynomials are added vertically, the Commutative and Associative Properties allow regrouping.

Example 1

Cooperative Learning Have students play the part of algebra tiles for adding and subtracting $(2x^2 - 3x + 5)$ and $(4x^2 + 7x - 2)$.

Begin by having six students cut out 8-inch square pieces of paper. Each of these students should write a large $+X^2$ on one side of the paper and a large $-X^2$ on the other side of the paper. Ten other students should cut out 8-inch $\times$ 4-inch rectangles of paper. These rectangles should be marked on one side with a large $+X$ and on the other side with a large $-X$. Seven more students should cut out 4-inch $\times$ 4-inch square pieces of paper and mark one side with a large $+1$ and the other with a large -1.

continued on bottom of next page

Approach A Using the Book, continued

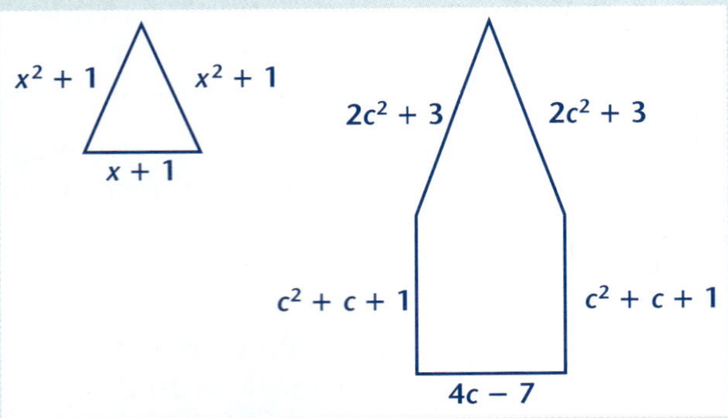

$[2x^2 + x + 3]$ $[6c^2 + 6c + 1]$

Critical Thinking, page 325

By using the Commutative and Assosicative Properties, terms are rearranged so that like terms are together, thus making mental math much easier.

Technology

You can use a graphics calculator to check the sums of polynomials. For example, to check $(a^3 + 3a + 1) + (2a^3 + a + 3) = 3a^3 + 4a + 4$, build a table of values for each side of the equation. Enter the left side of the equation for Y_1 and the right side of the equation for Y_2. Compare the Y_1 and Y_2 columns. The corresponding values should be equal.

X	Y1	Y2
−3	−89	−89
−2	−28	−28
−1	−3	−3
0	4	4
1	11	11
2	36	36
3	97	97

X = 1

Cooperative Learning

Subtracting Polynomials You may wish to first review how an expression such as $5 - 3$ can be rewritten as $5 + (-3)$ by using the definition of subtraction.

TEACH

Approach B Using Models, continued

1. Ask 2 students with x^2-tiles to stand in a group at the front of the room, holding up the positive sides of their tiles. Ask 3 students with x-tiles to stand in a group next to the x^2-tiles, holding up the negative side of their tiles. Ask 5 students with 1-tiles to stand in a third group at the front of the room, holding up the positive side of their tiles.
2. Four students with x^2-tiles should join the other x^2-tiles at the front of the room, holding up the positive sides of their tiles. Seven students with x-tiles should join the other x-tiles, holding up the positive sides of their tiles. Two students with 1-tiles should join the other 1-tiles, holding up the negative side of their tiles.

Ask as many neutral pairs of tiles as possible to sit down in their seats. Six positive x^2-tiles, four negative x-tiles, and three positive 1-tiles should remain, modeling the following sum:

$$(2x^2 - 3x + 5) + (4x^2 + 7x - 2) = 6x^2 - 4x + 3$$

Example 2

Cooperative Learning Refer to the polygons in Example 2 on page 325 of the Pupil's Edition. Copy these figures onto the chalkboard and label the sides. Divide the class into groups of four. Ask two students in each group to use algebra tiles or drawings of algebra tiles to find a simplified expression for the perimeter of the rectangle. Have the remaining students do the same for the triangle. When the students are done, ask them to exchange papers. Each pair of students should then simplify the polynomials algebraically to check their teammates' work.

Technology

You can use a graphics calculator to check the sums of polynomials. For example, to check $(a^3 + 3a + 1) + (2a^3 + a + 3) = 3a^3 + 4a + 4$, build a table of values for each side of the equation. Enter the left side of the equation for Y_1 and the right side of the equation for Y_2. Compare the Y_1 and Y_2 columns. The corresponding values should be equal.

Have students work in pairs. Ask one person in each pair to evaluate the expression $(2x^2 + x) - (3x^2 - 2x)$ for $x = 2$. The other member of each pair should evaluate $(2x^2 + x) + (-3x^2 + 2x)$ for $x = 2$. Ask students to compare answers. Why are the answers the same?

Example 3

Two polynomials are subtracted in both a vertical form and a horizontal form. In the vertical form, a variable term with a coefficient of 0 is used as a placeholder for the linear term.

Definition of *Subtraction*

When the definition of subtraction is applied to polynomials, the opposite of *each term* of the polynomial being subtracted must be taken.

When a number a is to be subtracted from a number b, the symbolic notation is $b - a$. When converted to a vertical format, the notation is

$$\begin{array}{r} b \\ \underline{-a} \end{array}.$$

Additional Example 3

Subtract $7x^2 - 2x + 3$ from $10x^2 + 5$.
[$3x^2 + 2x + 2$]

Critical Thinking, page 326

Our number system has a base of 10, so $1 = 10^0$, $10 = 10^1$, $100 = 10^2$, $1000 = 10^3$ and so on. Thus, our numbers can be represented as polynomials by letting $x = 10$ when numbers are written in expanded form.

Closing the Lesson

Journal Entry To close the lesson and highlight the procedure for adding and subtracting polynomials, ask students to explain the difference between adding $3x^2 + 2x - 1$ to $5x^2 - x + 4$ and subtracting $3x^2 + 2x - 1$ from $5x^2 - x + 4$. Ask students to find the sum and the difference.

Subtracting Polynomials

When the definition of subtraction is applied to polynomials, the opposite of *each term* of the polynomial being subtracted must be taken. If students use a vertical format to subtract two polynomials, the opposite of each term in the bottom polynomial must be taken. When a horizontal format is used to subtract polynomials, the opposite of each term in the polynomial on the right must be taken.

Subtraction Notation

When a number a is to be subtracted from a number b, the symbolic notation is $b - a$. When converted to vertical format, the notation is

$$\begin{array}{r} b \\ \underline{-a} \end{array}.$$

Example 3

Present Example 1 to the students again and instruct them in the vertical form and horizontal form for subtraction.

Cooperative Learning Have students play the part of algebra tiles again to model the subtraction of the polynomials from Example 1 as follows: Repeat Steps 1 and 2 of the addition activity, but tell all students taking part in Step 2 to *turn their tiles over.*

Ask as many neutral pairs of tiles as possible to sit down in their seats. Two negative x^2-tiles, 10 negative x-tiles, and 7 positive 1-tiles should remain, modeling the following difference:

$$(2x^2 - 3x + 5) - (4x^2 + 7x - 2) = -2x^2 - 10x + 7.$$

Relate this activity to the vertical form and horizontal form by pointing out that like terms are subtracted.

Closing the Lesson

Journal Entry To close the lesson and highlight the procedure for adding and subtracting polynomials, ask students to explain the difference between adding $3x^2 + 2x - 1$ to $5x^2 - x + 4$ and subtracting $3x^2 + 2x - 1$ from $5x^2 - x + 4$. Ask students to find the sum and the difference.

ASSESS

Selected Answers

Odd-numbered Exercises 7–53 and odd-numbered Extra Practice

Assignment Guide

Core level students should complete Exercises 1–53.

Core-plus level students should complete Exercises 1–26, 31–37, and 40–54.

Error Analysis

When students add or subtract polynomials, particularly when they use a horizontal format, they may combine terms that have the same variable but not the same exponent. For example, a common error is $5x + 3x^2 = 8x^3$. Point out to these students that because $5x$ and $3x$ can be represented by the same kind of tile, $5x$ and $3x$ can be added. The sum is 8 x-tiles, or $8x$. The terms $5x$ and $3x^2$, however, are represented by different-sized tiles. Since the tiles cannot be combined, $5x$ and $3x^2$ cannot be added. Recommend to students that they check all sums and differences of polynomials by evaluating the polynomials for some value of the variable.

Alternative Assessment

Using Horizontal Form Add or subtract as indicated, using a horizontal form. Identify the definition or property that justifies each step.

1. $(r^2 + 5r - 3) + (3r^2 - 2r + 8)$

2. $(r^2 + 5r - 3) - (3r^2 - 2r + 8)$

Portfolio Activity Write two procedures, one for adding polynomials and the other for subtracting polynomials. Write the procedures for either vertical form or horizontal form.

RETEACH

Calculate $(7x^2 + 2x + 4) + (5x^2 - x + 2)$ and $(7x^2 + 2x + 4) - (5x^2 - x + 2)$ in vertical and horizontal form. **[$12x^2 + x + 6$, $2x^2 + 3x + 2$]**

ENRICH

Forestry One of a forester's jobs is to decide how much lumber can be obtained from a wood lot. Lumber is measured in board feet, which is actually a cubic measure equal to 1 foot × 1 foot × 1 inch. Since this measure is a cubic measure, foresters can often use a cubic equation to predict the number of board feet of lumber that can be obtained from a tree with a given diameter. A cubic equation has a polynomial of degree 3, or a cubic polynomial.

NAME ______________________ CLASS ____________ DATE ____________

Practice

7.2 Adding and Subtracting Polynomials

1. Write the polynomials and the sum modeled by the tiles.

____

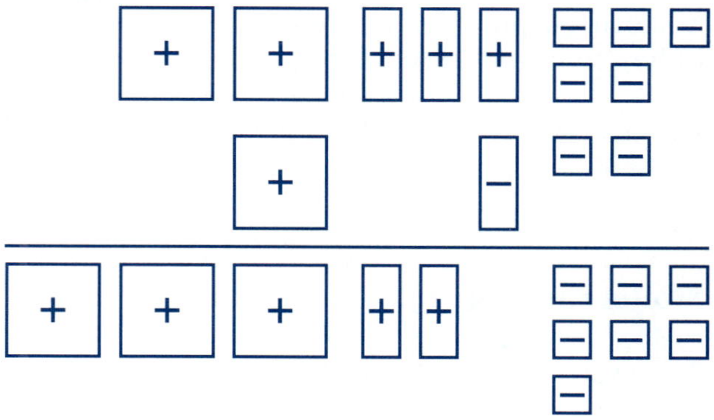

Rewrite each polynomial in standard form.

2. $2 - 8c + c^2$ ____________

3. $10 + m^3$ ____________

4. $5n^2 + 2n + 3n^4 - n$ ____________

5. $6x - 2x^3 + 3x^2 + 1$ ____________

Write the degree of each polynomial.

6. $125 - d^3$ ____________

7. $2s$ ____________

8. $9a^3 - a^2 + 3a^6 - a$ ____________

9. $4y^2 + 3y^5 - 5 + 8y^3$ ____________

Use vertical form to add.

10. $3f^2 - f + 7f$ and $2f^2 - 3f - 4$ ____________

11. $6 + c$ and $3c^3 + c^2 - 12c - 6$ ____________

Use horizontal form to add.

12. $4q + 2q^3 - 1$ and $1 - 2q + 6q^2$ ____________

13. $3y^3 - 3y^2 + 6y$ and $2y^3 - 6y - 3$ ____________

Use vertical form to subtract.

14. $4m^2$ from $3m^3 - 6m^2 + m - 1$ ____________

15. $3n - 1$ from $2n^2 - 1$ ____________

Use horizontal form to subtract.

16. $10a^2 + 2a - 8 - (10a^2 + 8)$ ____________

17. $7x^3 - 3x + 2 - (x^2 + 2)$ ____________

Lesson Activity
7.2 Modeling Perimeters

You can use pipe cleaners of length *a*, tongue depressors or ice cream sticks of length *b*, and unit tiles to model the perimeters of geometric figures.

Suppose that you want to model a figure with a perimeter of $a + 2b + 1$. One way is to arrange the pieces as shown.

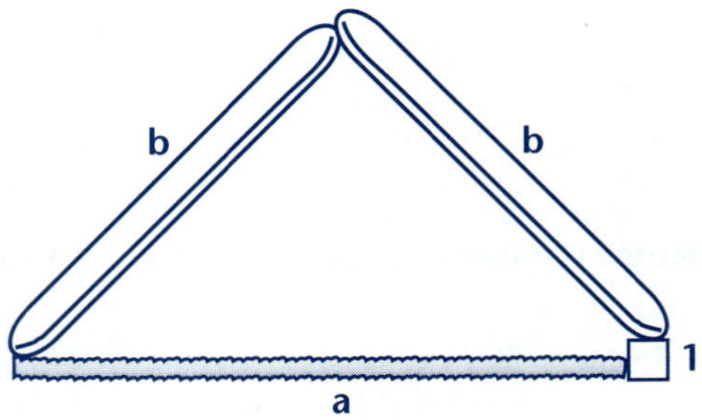

Model each figure.

1. a square with a perimeter of $8 + 4a$
2. a hexagon with a perimeter of $6b + 6$
3. a rectangle with a perimeter of $2a + 4b + 2$
4. a pentagon with a perimeter of $a + 2b + 10$

Write an expression for the perimeter of each figure. Use the Distributive Property to write an equivalent expression, if possible.

5. 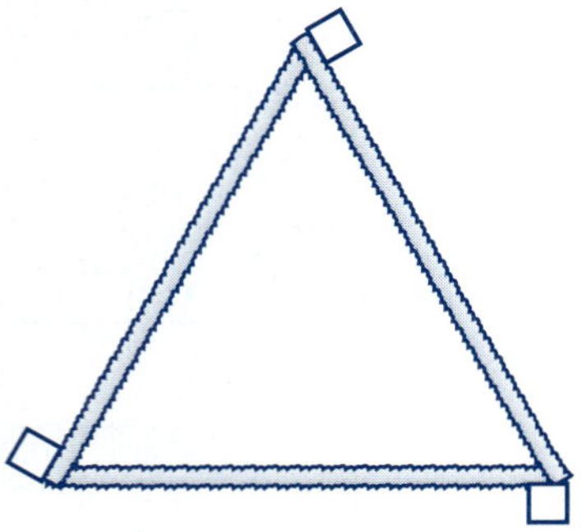

____

6. 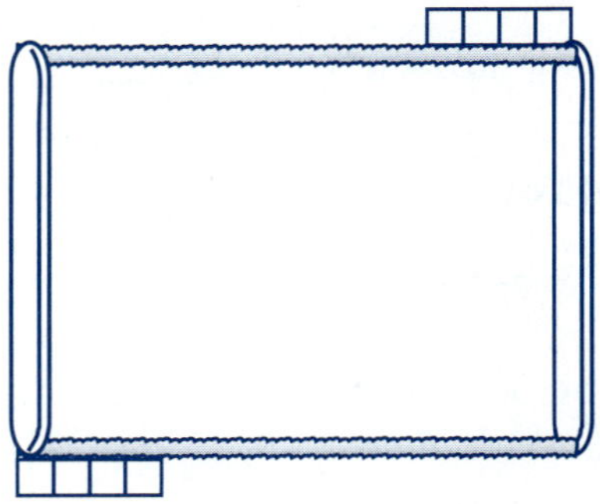

____

7. 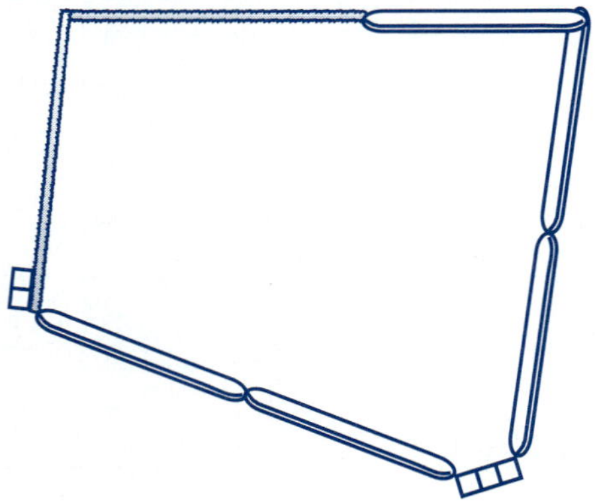

____

8. 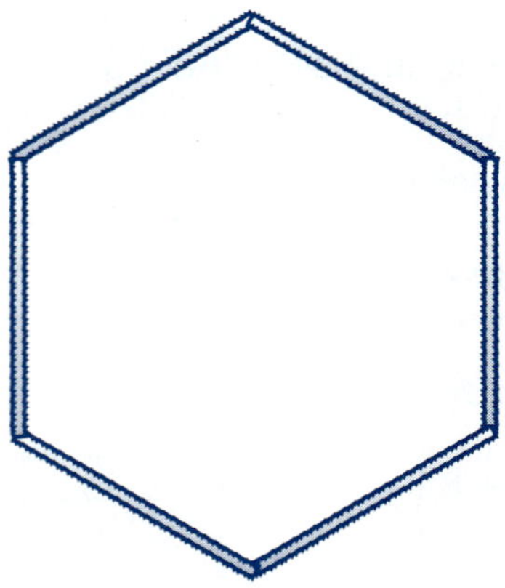

____

NAME ______________________________ CLASS ______________ DATE ______________

Basic Skills Practice
Using Scientific Notation, Part 2

Read each question and circle the best answer.

1. The distance of a galaxy from the Earth is expressed as 8.557×10^{10} miles. Which is another way to express this amount?

A 75,530,000 mi
B 855,7000,000 mi
C 8,557,000,000 mi
D 85,570,000,000 mi

2. The diameter of a particle is 1.2×10^{-6} millimeters. Express in standard notation.

F 0.00000012 mm
G 0.0000012 mm
H 0.00012 mm
I 0.0012 mm

3. The number of grains of sand on a section of a beach is 6.85×10^{9}. Express in standard notation.

A 6,850,000,000
B 685,000,000
C 68,500,000
D 685,000

4. The thickness of a soap bubble is about 0.00002 meters. What is this in scientific notation?

F 0.2×10^{-4} m
G 2×10^{-5} m
H 0.2×10^{-6} m
I 20×10^{-6} m

5. A convention attracted 1.17×10^{4} left-handed people. Express in standard notation.

A 117
B 1170
C 11,700
D 1,170,000

6. A collector owns 11,231 baseball cards. Express in scientific notation.

F 1.1231×10^{-4}
G 11.231×10^{-3}
H 1.1231×10^{2}
I 1.1231×10^{4}

7. A chemical compound contains 0.0428 milligrams of phosphate. Express in scientific notation.

A 4.28×10^{-3} mg
B 4.28×10^{-2} mg
C 4.28×10^{-1} mg
D 4.28×10^{3} mg

8. The distance from the Earth to the moon is about 239,000 miles. How is this measurement expressed in scientific notation?

F 2.39×10^{6} mi
G 2.39×10^{5} mi
H 23.9×10^{5} mi
I 2×10^{6} mi

Solve each problem.

9. There were 4.544×10^{4} fans at a concert. Express in standard notation. ______________

10. A jet flew at an altitude of 1.4×10^{4} feet. Express in standard notation. ______________

LESSON 7.3

Exploring Multiplication Models

PREPARE

Pacing Guide
3 Days

Objectives and Overview
- Use the Distributive Property to find the product of a monomial and a binomial.
- Use tiles to model the multiplication of binomials.

Algebra tiles are used to model the multiplication of binomials. Finding the zeros of two binomial factors is also discussed.

Vocabulary
zeros

Assessing Prior Knowledge

Simplify by using the Distributive Property.

1.	$3(x + 4)$	[$3x + 12$]
2.	$5(2x + 7)$	[$10x + 35$]
3.	$-2(8x - 1)$	[$-16x + 2$]
4.	$6x(x - 5)$	[$6x^2 - 30x$]

Approach A: Using the Book

Opening Discussion
Draw two groups of algebra tiles on the chalkboard. Each group should have 1 x-tile and 1 1-tile.

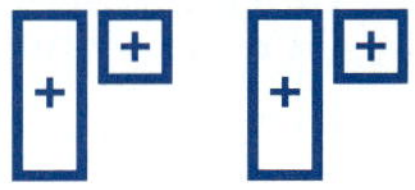

What binomial does each group represent? [$x + 1$] How many x-tiles are there? How many 1-tiles are there? Complete this statement: $2(x + 1) =$ [$\underline{2x}$] + [$\underline{2}$].

Making Algebra Tiles
For instructions on making algebra tiles, see earlier lessons.

Cooperative Learning
Teacher-Guided Exploration If you choose to guide your students through the explorations, place

TEACH

Approach B: Using Models

Opening Discussion
Draw two groups of algebra tiles on the chalkboard. Each group of tiles should have 1 x-tile and 1 1-tile.

What binomial is modeled by each group of tiles? [$x + 1$] Count the number of x-tiles. Count the 1-tiles. Complete this statement: $2(x + 1) =$ [$\underline{2x}$] + [$\underline{2}$].

Making Algebra Tiles
Refer to previous lessons for instructions on making algebra tiles.

algebra tiles as they are arranged in the steps of the exploration. (You may use drawings of algebra tiles instead.) Use a pointer to show the product of each pair of factors.

When a calculator demonstration is required, having students enter expressions in their own calculators will allow them to better connect the expression with the results they see on the calculator display.

Student-Directed Approach Divide the class into groups of four to complete each exploration. Students should take turns answering questions, and a discussion should follow each answer. One member of each group should be chosen to take notes on the group's activities. Another member should write a report of each exploration from the notes. Each student should sign the completed report.

Exploration 1

In this exploration, students study diagrams of algebra tiles to discover the product of two binomials, each of which is a sum.

In Step 1, the tiles demonstrate that the product of a binomial and a constant is a binomial. In the matrix, each product tile takes its width from the row tile and its length from the column tile.

Make sure students recognize that the factor 2 must be represented by 2 individual 1-tiles.

The same factor, $x + 1$, is used in Steps 1 and 2.

In Step 3, the placement of row and column tiles is important. The common factor, $x + 1$, is used just once so that it can be shared by both of the other factors, x and 2, which occur in different rows.

In Steps 5 and 7, students are asked to generalize their findings and to apply them to two different products.

continued on top of next page

Modeling Multiplication of Binomials

Using transparent tiles and an overhead projector, show students how to model the multiplication of two binomials of the form $(a + b)(c + d)$.

Place the tiles representing the terms of the first binomial as the row headings of a matrix and place the tiles representing the terms of the second binomial as the column headings of the matrix. Multiply each row tile by each column tile. Draw a tile to represent each product. Each product tile takes its height from the row tile and its width from the column tile. Place those product tiles in the appropriate positions in the matrix. Write the algebraic expression for all of the product tiles. This is the product of the binomials. (Compare page 329.)

Cooperative Learning

Divide the class into groups of four. Give each group a set of algebra tiles. If tiles are not available, instruct students to draw the necessary tiles. Each group should assign one student to read the questions. A second student should manipulate or draw the tiles. A third student should record the answers, and the fourth student in each group should write the report and read it to the class. All group members should be involved in a discussion of the results of each step in the exploration. All group members should sign the group's report.

Exploration 1

Students use algebra tiles to find the product of two binomials, each of which is a sum.

Have students work in their groups and use algebra tiles or drawings to explore the following questions, which you may write on the overhead or on a chalkboard:

- Use algebra tiles to show how you could multiply two binomials of the form $(a + b)(c + d)$, such as $(x + 3)(2x + 5)$. **[Students should set up a matrix with tiles that represent the terms of one binomial as the column headings. The row headings should be tiles that represent the terms of the other binomial. The entries in the matrix should be tiles representing the**

continued on bottom of next page

Approach A Using the Book, continued

Algebras tiles help students understand multiplication of polynomials.

When checks are required, remind students that they can use graphics calculators in two ways to check their results: to perform the calculations and to build a table of values that can be examined.

Ongoing Assessment, top of page 330

3. $x^2 + 3x + 2$

Ongoing Assessment, middle of page 330

7. Arrange the tiles in a matrix so that the column tiles represent $x + 4$. Thus, the column tiles should be 1 positive x-tile and 4 positive 1-tiles. The row tiles should represent $2x + 2$, so there should be 4 rows in the matrix. Two of the row tiles should be positive x-tiles, and two should be positive 1-tiles. Multiply each row tile by each column tile. Count the total number of tiles in the resulting matrix entries. You can check your answer by evaluating it for some value of x, such as 10. Then evaluate $(2x + 2)(x + 4)$ for the same value of x. The answers should be equal.

TEACH

Approach B Using Models, continued

product of each row tile with each column tile.]

- What is the product of your binomials? How did you find the product? **[The product of $(a + b)(c + d)$ is $ac + bc + ad + bd$. The product can be found by adding the tiles that are the products of the row and column tiles.]**
- Evaluate your two binomials for $x = 10$. Evaluate the product of your two binomials for $x = 10$. Explain the results. **[If the binomials were multiplied correctly, the two numbers should be equal.]**

Student Reports

One or more members from two groups can demonstrate their group's tile arrangement on the chalkboard. Members from selected groups should read their findings to the class.

On Their Own

Have students use algebra tiles or drawings to model the following multiplication problems:

A. $(2x + 1)(x + 3)$ **[$2x^2 + 6x + 3$]**

B. $(x + 2)(x + 4)$ **[$x^2 + 6x + 8$]**

Exploration 2

Students investigate the results of multiplying two binomials when at least one binomial is a difference.

Use algebra tiles to model the product $(x + 2)(x - 3)$. Then use the model to pose the following questions to students, who should answer the questions with their group. (Compare with the model at the bottom of page 330 in the Pupil's Edition.)

- Ask students how the result of this multiplication model is different from the result of the model they used in Exploration 1. **[This model involves the removal of neutral pairs to find the product of the binomial.]**
- Why does this model have neutral pairs of tiles? **[Some of the x-tiles are the result of**

Follow-Up Demonstration (Optional)

If you feel students need another demonstration, use algebra tiles to model $(x + 3)(2x + 1)$. Ask students how the tiles should be arranged, how to use the tiles, and how to interpret the results.

On Their Own

Have students use algebra tiles or drawings to model the following multiplication problems:

A. $(2x + 1)(x + 3)$ **[$2x^2 + 6x + 3$]**

B. $(x + 2)(x + 4)$ **[$x^2 + 6x + 8$]**

Exploration 2

Whether students work in groups or are guided through the exploration by you, they should see the results of multiplying two binomials when at least one binomial is a difference.

In Step 3 students should recognize that because the multiplication produced both positive and negative tiles, some neutral pairs must have been formed. Ask students why neutral pairs did not occur in Exploration 1.

In Step 7, students are introduced to the zeros of a polynomial function. The Zero Product Property of real numbers states that if either of the factors is 0, then the product of the factors is 0. Identifying the zeros, therefore, involves identifying the values that will make either factor equal to 0.

Ongoing Assessment, bottom of page 330

5.

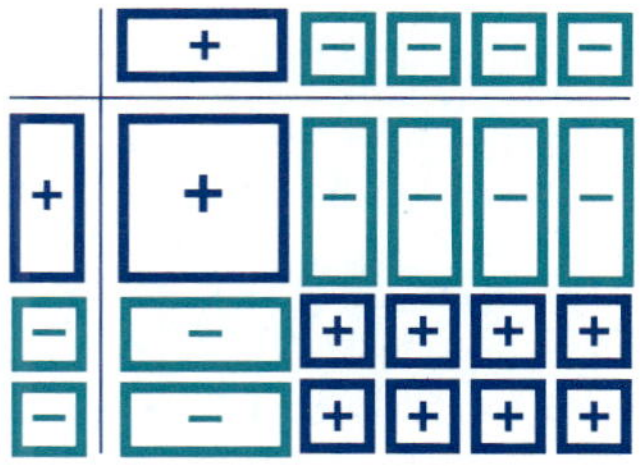

So $(x - 2)(x - 4) = x^2 - 6x + 8$.

continued on top of next page

multiplying a positive x-tile and a negative 1-tile and are therefore negative. Other x-tiles are the result of multiplying a positive x-tile and a positive 1-tile and are therefore positive.]

Now model $(x - 2)(x - 4)$ with algebra tiles.

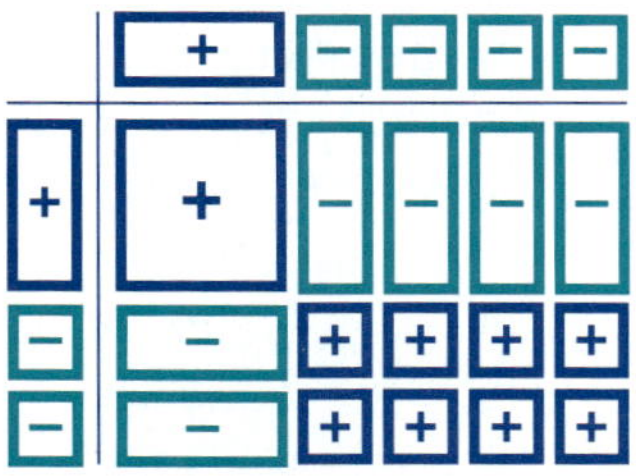

- Ask students if any neutral pairs of tiles must be removed. Why? **[No neutral pairs must be removed because all of the x-tiles in the product are the result of multiplying a positive x-tile and a negative 1-tile and are therefore all negative. All of the 1-tiles in the product are the result of multiplying two negative 1-tiles and are therefore all positive.]**
- What is the product $(x - 2)(x - 4)$? **[$x^2 - 6x + 8$]**
- When will a product of two binomials require the removal of neutral pairs of tiles? **[when the signs in the binomials are different]**
- If you evaluate $(x - 2)(x - 4)$ and $x^2 - 6x + 8$ for the same value of x, what results do you expect? **[You should get the same number for each expression.]**
- Evaluate $(x - 2)(x - 4)$ for $x = 2$. Then evaluate $(x - 2)(x - 4)$ for $x = 4$. What are the results? Why? **[0; 0; $2 - 2 = 0$ and $4 - 4 = 0$; the product of 0 and any number is 0.]**

Use a graphics calculator to build a table of values for $y = (x - 2)(x - 4)$, or construct a table for integral values of x from -5 to 5.

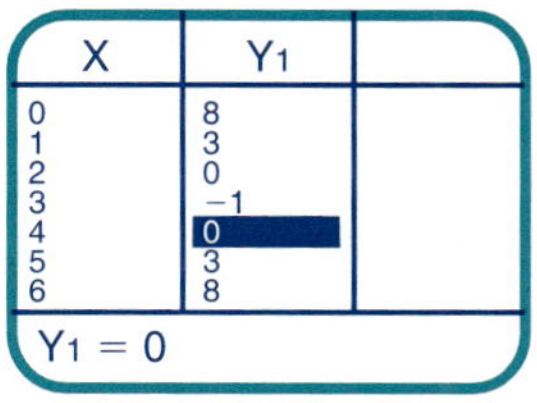

continued on bottom of next page

Approach A

Using the Book, continued

Ongoing Assessment, top of page 331

7. 5, −3; when either 5 or −3 is substituted for x in the expression $(x - 5)(x + 3)$, the product is zero.

Journal Entry

Ask students to explain why no neutral pairs were formed in Step 6 of Exploration 2 even though both positive and negative tiles were used.

Extension

Interpreting a Table of Values The chart shows that for values of x from −6 to −4, the y-values decrease and the graph crosses the x-axis at $x = -5$. Between what two integer values does the chart show that the values of y begin to increase as the values of x increase? **[−3 to 0]** As the curve changes direction, it crosses the x-axis again at $x = -2$.

Exploration 3

Students explore the special products $(x + y)^2$ and $(x + y)(x - y)$.

Part 1 Ask students how the terms of the product are related to the terms in each of the factors. **[The first term of the product is the product of the first two terms of the binomials. The last term of the product is the product of the last two terms of the binomials.]**

TEACH

Approach B

Using Models, continued

- For what values of x does $y = 0$? **[$x = 2$ and $x = 4$]**

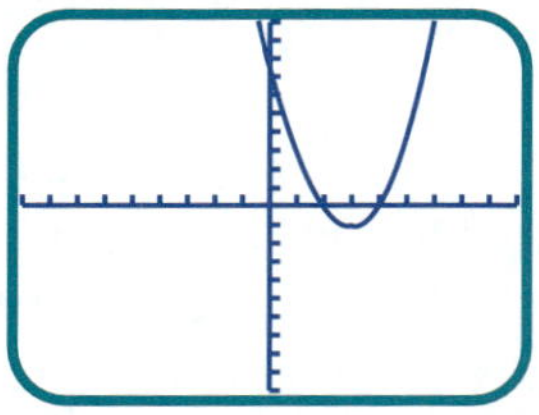

Using a standard viewing window, graph the function $y = (x - 2)(x - 4)$, or plot the points from your table and estimate the curve of the graph.

- At what two points does the graph appear to cross the x-axis? **[(2, 0) and (4, 0)]**
- Tell students that the values 2 and 4 are called the *zeros* of the product $(x - 2)(x - 4)$. Ask the students what they think the zeros of the product $(x - 5)(x + 3)$ are. **[5 and −3]** Have each group explain its choice of numbers. **[When either 5 or −3 is substituted for x in the expression $(x - 5)(x + 3)$, the product is zero.]**

Student Reports

Call on one or more members from two groups to demonstrate their group's tile arrangement on the chalkboard. Ask members from selected groups to read their findings to the class. Invite all students to comment on the tile models or on any of the reports.

Exploration 3

Students explore the special products $(x + y)^2$ and $(x + y)(x - y)$.

Ask students to continue to work with their group to complete the following exercises:

Part I

- Model the product $(x + 2)(x - 2)$. (Compare with the model on page 331 of the Pupil's Edition.) How many x-tiles remain when you remove neutral pairs? **[No x-tiles remain.]**

Ongoing Assessment, bottom of page 331

4. Square x. Then square 5. Write the product as a difference: $x^2 - 25$.

Part II In Step 3 students are asked to generalize the results of Steps 1 and 2. The model of $(x + 3)(x + 3)$ should demonstrate that the coefficient of the linear term in the product is the result of doubling, not squaring, the 3. Since $2 \cdot 2 = 2^2$, students may not recognize from the product of $(x + 2)(x - 2)$ that the coefficient of the linear term and the constant term of $(x + y)^2$ are not necessarily the same.

Ongoing Assessment, page 332

4. Square x. Double the product of 5 and x. Square 5. Write the product as the sum of three terms: $x^2 + 10x + 25$.

Closing the Lesson

Highlight the similarities and differences in products such as $(x + 4)(x - 4)$, $(x + 4)(x + 4)$, and $(x - 4)(x - 4)$. Ask students to find all three products as well as the zeros of each product. **[$(x + 4)(x - 4) = x^2 - 16$, zeros: -4 and 4; $(x + 4)(x + 4) = x^2 + 8x + 16$, zero: -4; $(x - 4)(x - 4) = x^2 - 8x + 16$, zero: 4]**

- What is the product $(x + 2)(x - 2)$? **[$x^2 - 4$]**
- Add one 1-tile to each binomial in the model. What multiplication do the tiles now model? **[$(x + 3)(x - 3)$]**
- How many x-tiles remain when you remove all of the neutral pairs of x-tiles from the new model? What is the product $(x + 3)(x - 3)$? **[No x-tiles remain; $x^2 - 9$.]**
- How many x-tiles do you think will remain in a model of the product $(x + 5)(x - 5)$ if all of the neutral pairs of x-tiles are removed? What is the product of $(x + 5)(x - 5)$? **[No x-tiles will remain; $x^2 - 25$.]**

Part II Show your students how algebra tiles can be used to model the product $(x + 2)(x + 2)$. (Refer to Part II of Exploration 3 on page 332 of the Pupil's Edition.)

Ask students how many x-tiles are in the product. **[four x-tiles]** How many 1-tiles are in the product? **[four 1-tiles]**

Student Reports

Call on one or more members from two groups who have not given a demonstration. Ask them to demonstrate their group's tile arrangement for Exploration 3 on the chalkboard. Ask a member of each group to read the group's findings to the class. Make sure these students can be heard by every student in the class. Invite all students to comment on the tile models or on any of the reports.

ASSESS

Selected Answers

Odd-numbered Exercises 5–73 and odd-numbered Extra Practice

Assignment Guide

Core level students should complete Exercises 1–74.

Core-plus level students should complete Exercises 1–14, 24–38, and 47–78.

Error Analysis

If students have trouble getting the correct sign of a zero, explain that any zero can be found by setting the binomial factor to zero and solving the resulting equation for the variable.

Alternative Assessment

Writing Multiplication Problems Have students work with a partner. Each student should write four multiplication problems with two binomial factors. Tell students that some of the binomials should be sums and others should be differences. Have the students exchange papers and find the products of their partner's binomials by using tiles or mental calculations.

When students are done, they should return their paper to the author for checking. Students should check their partner's products by evaluating the expressions for some value of x or by examining a table of values.

Portfolio Activity Have students draw an algebra-tile model of the product of two binomials. Tell them to explain how they can use the model to find the product of their binomials, and ask them to state the product. Then have students find the zeros of the product. Ask your students to explain why these numbers are zeros of the product of their binomials.

RETEACH

Compare the multiplication of two binomials with the multiplication of two whole numbers. Use $(x + 2)(2x + 3)$ and $(12)(23)$ as examples.

Begin by demonstrating the product of 12 and 23.

$$\begin{array}{r} 12 \\ \underline{\times\ 23} \end{array} \quad = \quad \begin{array}{rr} & 10 + 2 \\ & \underline{\times\ 20 + 3} \\ & 30 + 6 \\ \underline{200 + 40} & \\ 200 + 70 + 6 & \\ = \; 276 & \end{array}$$

Then show how a similar method can be used to multiply $(x + 2)(2x + 3)$. Write the binomials in vertical form, and multiply each term of the top binomial by each term of the bottom binomial.

$$\begin{array}{r} x + 2 \\ \underline{\times\ 2x + 3} \\ 3x + 6 \\ \underline{2x^2 + 4x \quad\quad} \\ 2x^2 + 7x + 6 \end{array}$$

ENRICH

Interdisciplinary Connection

Statistics When you studied the zeros of the product of two binomials in this lesson, you saw that the graph of an equation such as $y = (x + 5)(x + 2)$ looks like a curve rather than a straight line. Sometimes a curve like this is a better model than a straight line for making predictions. Such a curve is called a quadratic regression.

NAME ______________________ CLASS ____________ DATE ____________

Practice

7.3 Exploring Multiplication Models

Write the factors and the product for each model.

1.

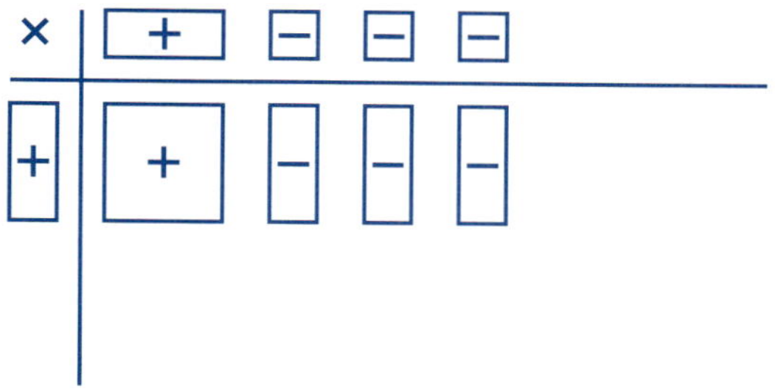

____

2.

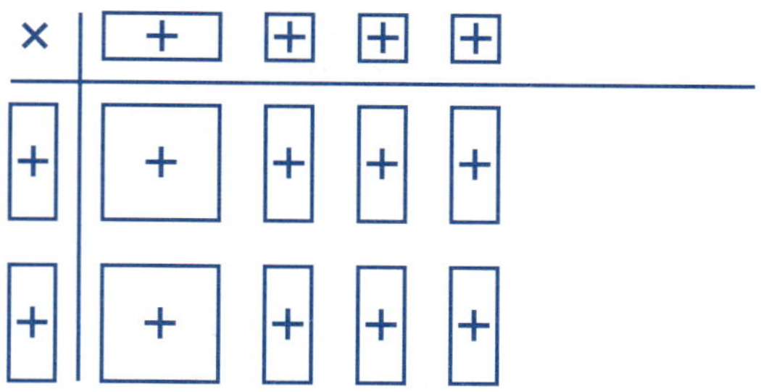

____

Use the Distributive Property to find each product.

3. $6(x + 2)$ ______________________

4. $7(x - 2)$ ______________________

5. $9(2x - 3)$ ______________________

6. $-2(3x + 4)$ ______________________

7. $-4(x - 5)$ ______________________

8. $10(2x - 2)$ ______________________

9. $x(x + 2)$ ______________________

10. $x(8 - x)$ ______________________

11. $-x(x + 3)$ ______________________

12. $6x(x - 3)$ ______________________

13. $-2x(x + 2)$ ______________________

14. $-4x(-x + 5)$ ______________________

Model each product with tiles. Check by substituting values for *x*.

15. $x(x + 4)$ ______________________

16. $2x(2x + 1)$ ______________________

Identify the zeroes for the products in the exercises below.

17. $(x + 3)(x - 2)$ ______________________

18. $(x - 5)(x - 7)$ ______________________

19. $(x - 1)(x + 1)$ ______________________

20. $(x + 12)(x - 2)$ ______________________

21. $(x + 9)(x + 9)$ ______________________

22. $(x - 2)(x + 4)$ ______________________

23. $(2x + 1)(x - 3)$ ______________________

24. $(3x + 2)(x + 3)$ ______________________

NAME ______________________ CLASS __________ DATE __________

Lesson Activity
7.3 Find the Factors

State two factors whose product can be modeled with the tiles described.

1. 1 positive x^2-tile, 3 positive x-tiles, 2 positive 1-tiles ____________________

2. 1 positive x^2-tile, 2 positive x-tiles, 1 positive 1-tile ____________________

3. 1 positive x^2-tile, 1 positive x-tile, 1 negative x-tile, 1 negative 1-tile ____________________

4. 1 positive x^2-tile, 2 positive x-tiles, 2 negative x-tiles, 4 negative 1-tiles

____

5. 2 positive x^2-tiles, 3 positive x-tiles, 1 positive 1-tile ____________________

6. 2 positive x^2-tiles, 5 negative x-tiles, 3 positive 1-tiles ____________________

7. 1 positive x^2-tile, 3 negative x-tiles, 10 negative 1-tiles ____________________

8. 1 positive x^2-tile, 2 positive x-tiles, 8 negative 1-tiles ____________________

9. 1 positive x^2-tile, 4 negative x-tiles, 4 positive 1-tiles ____________________

10. 2 positive x^2-tiles, 1 negative x-tile, 10 negative 1-tiles ____________________

11. 1 positive x^2-tile, 3 negative x-tiles, 18 negative 1-tiles ____________________

12. 2 positive x^2-tiles, 5 positive x-tiles, 7 negative 1-tiles ____________________

13. 1 positive x^2-tile, 6 positive x-tiles, 9 positive 1-tiles ____________________

14. 1 positive x^2-tile, 5 positive x-tiles, 6 negative 1-tiles ____________________

15. 1 positive x^2-tile, 5 negative x-tiles, 5 positive x-tiles, 25 negative 1-tiles

____

NAME ______________________ CLASS ______ DATE ______

Basic Skills Practice

Effects of Dimensional Changes, Part 1

Charlie works for a company that builds cardboard boxes. Their most popular box is a cube whose sides measure 2 feet, whose surface area is 24 square feet, and whose volume is 8 cubic feet. Charlie wants to design a box with dimensions twice as big.

Example 1: Find the surface area of the new box.
Use the formula for the surface area of a cube, $S = 6s^2$, to find the surface area of the new box.

$S = 6(2 \times 2)^2$ Multiply the dimension of the sides by 2.
$S = 6(4)^2$
$S = 6(16)$
$S = 96\text{ ft}^2$

Example 2: Find the volume of the new box.
Use the formula for the volume of a cube, $V = s^3$, to find the volume of the new box.

$V = (2 \times 2)^3$ Multiply the dimension of the sides by 2.
$V = (4)^3$
$V = 64\text{ ft}^3$

Use the same method to find the dimensional changes for rectangular prisms and cylinders.

Solve each problem.

1. A rectangular prism has sides of 3 inches, 4 inches, and 5 inches. What are the new dimensions of the prism if the sides are reduced by one-half? ________
2. A cylinder has a radius of 1 meter and a height of 2.5 meters. What is the volume of a cylinder with dimensions three times larger? ________
3. A cube has sides that measure 6 inches. What is the surface area of the cube if its dimensions are reduced by one-third? ________
4. A cylinder has a radius of 12 inches and a height of 4 inches. What is the volume of the cylinder if its dimensions are reduced by one-fourth? ________
5. A rectangular prism has sides of 3 inches, 4 inches, and 5 inches. What is the volume of the prism if its dimensions are doubled? ________

Travis makes planters in the shape of cubes. He usually makes planters with sides that measure 6 inches, but he now wants to make a planter with sides that measure 4 inches.

6. What is the volume of the planter Travis usually makes? ________
7. What is the volume of the planter Travis wants to make? ________

LESSON 7.4

Exploring Multiplication of Binomials

PREPARE

Pacing Guide

2 Days

Objectives and Overview

- Use the Distributive Property to multiply binomials.
- Use the FOIL method to multiply binomials.

The Distributive Property and the FOIL method are introduced as ways to multiply two binomials.

Vocabulary

FOIL

Assessing Prior Knowledge

Multiply. Write each answer in standard form.

1. $x(x^2)$ $[x^3]$

2. $x(x - 3)$ $[x^2 - 3x]$

3. $-2(x + 5)$ $[-2x - 10]$

Approach A: Using the Book

Opening Discussion

Draw an algebra-tile model of the product $(x + 3)(x + 5)$ and circle the groups of x-tiles whose sum represents the linear term. You may also use algebra tiles on the overhead.

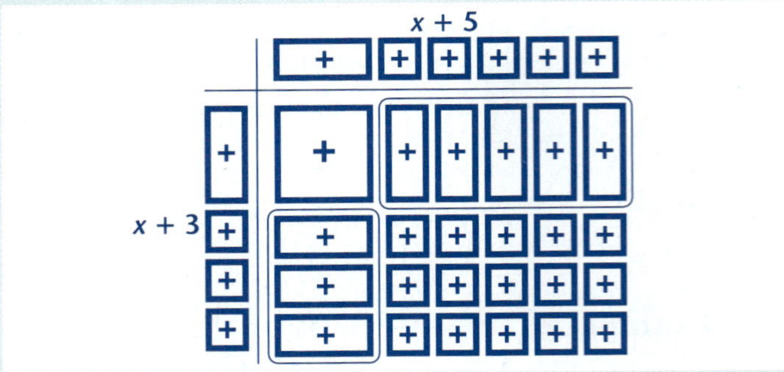

What are the factors of the x^2-tile? What multiplication resulted in each group of x-tiles? What are the factors of the group of 1-tiles?

TEACH

Approach B: Using Models

Opening Discussion

Draw an algebra-tile model of the product $(x + 3)(x + 5)$ and circle the groups of x-tiles whose sum represents the linear term. You may also use algebra tiles on the overhead.

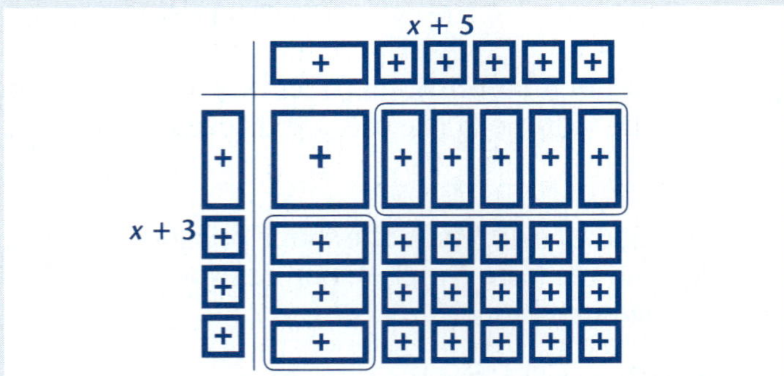

What are the factors of the x^2-tile? Name the factors of each group of x-tiles. What are the factors of the group of 1-tiles?

Ongoing Assessment, top of page 335

5. Use algebra tiles to build a rectangle from $x + 4$ and $x + 5$. The rectangle will consist of 1 x^2-tile, 9 x-tiles, and 20 1-tiles.

Teaching Tip

Modeling the Distributive Property With Algebra Tiles Ask students to circle the part of the algebra-tile model that represents $x(x + 3)$. Then ask them to circle the part of the model that represents $2(x + 3)$. The tile model multiplies each row term by each column term. When the Distributive Property is used, the same operation *distributes* $(x + 3)$ to each term of $(x + 2)$.

Ongoing Assessment, bottom of page 335

$$\begin{aligned}(x + 4)(x + 5) &= x(x + 5) + 4(x + 5)\\ &= x^2 + 5x + 4x + 20\\ &= x^2 + 9x + 20\end{aligned}$$

Inclusion Strategy

Preparing a Portfolio Activity Ask students to write down their answers to Exploration 2 on page 336. Tell them to phrase their answers to Step 4 carefully because they can use this answer to complete Step 6. Tell the students that they will use their answers later in a portfolio activity.

Ongoing Assessment, page 336

6. $x^2 + (a + b)x + ab$ or $x^2 + ax + bx + ab$

Journal Entry

Ask students to copy the FOIL model on page 336 into their journal, or ask them to make up their own model. Drawing a model of this shortcut, even if the model is copied, will help reinforce this strategy for multiplying binomials.

Use **Transparency 30** to discuss or review the FOIL method.

continued on top of next page

Exploration 1

Using Algebra Tiles Use algebra tiles to provide a familiar and concrete model that shows how to use the Distributive Property to multiply two binomials. Model the product $(x + 3)(x + 1)$ for your students.

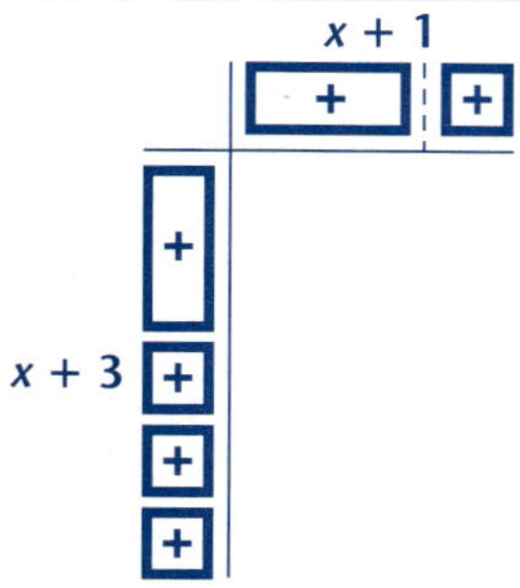

The dashed line separates two sets of tiles, showing that $(x + 3)$ is distributed to each term of $(x + 1)$:

$$(x + 3)(x + 1) = x(x + 3) + 1(x + 3)$$

The resulting tiles show that each term of $(x + 3)$ is distributed to each term of $(x + 1)$, or $x(x + 3) + 1(x + 3) = x^2 + 3x + x + 3$.

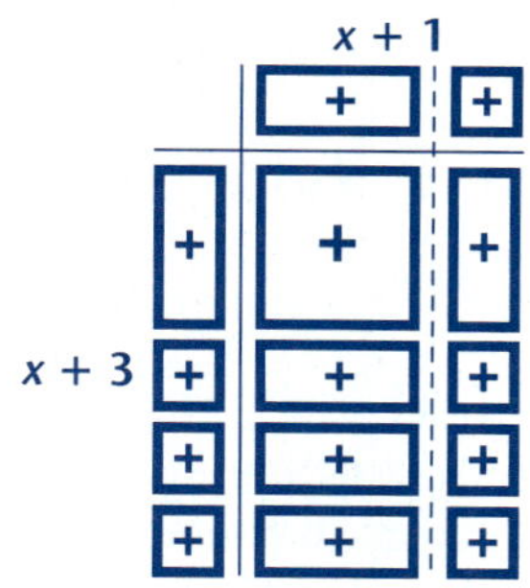

Finally, counting like tiles represents combining like terms.

$$x^2 + 3x + x + 3 = x^2 + 4x + 3$$

Exploration 2: Multiplying Binomials

Draw three identical algebra-tile models of $(x + 2)(x + 3) = x^2 + 5x + 6$. Write the equation below each model. Use the three models to identify the factors of each term of the trinomial by shading the tiles and the tile factors that represent each term.

Ask students which tile in the first model represents the first term of the trinomial. Ask

continued on bottom of next page

Approach A — Using the Book, continued

Example 1

Students use the FOIL method to multiply binomials.

Looking for a Pattern

Ask your students to examine the solutions in Example 1 in order to determine a rule for establishing the sign of the linear term.

Additional Example 1

Use the FOIL method to find each product.

A. $(2x + 1)(5x + 3)$ $[\mathbf{10x^2 + 11x + 3}]$

B. $(4b - 7)(2b + 3)$ $[\mathbf{8b^2 - 2b - 21}]$

C. $(3r - 5)(3r + 5)$ $[\mathbf{9r^2 - 25}]$

Technology

Students can store a value of x in their graphics calculator and then use that value to check the product of two binomials. Alpha mode is used to store a value in X. If 50 is stored in X, the expressions $(3x + 1)(3x - 2)$ and $9x^2 - 3x - 2$ can be entered by using the X,T,θ,n key if the calculator is in function mode. The expressions will be evaluated for $x = 50$. If the two expressions are equivalent, the same value will be returned for both expressions.

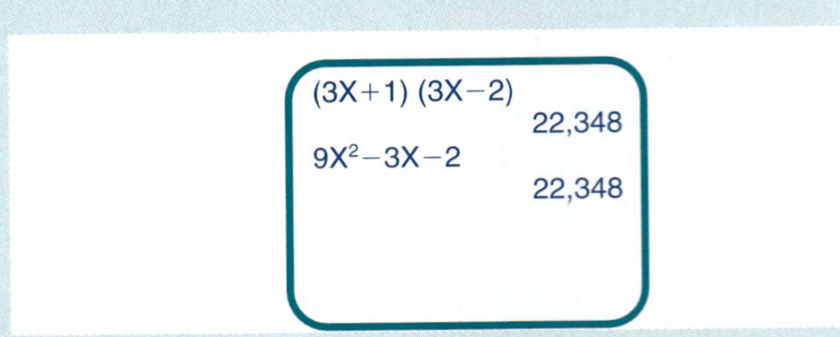

TEACH

Approach B — Using Models, continued

them to also identify the factors of this tile. Shade all three tiles. Then circle the 2 xs in the factors and the x^2-term in the product.

In the second model, circle the two sets of x-tiles in the product model. Call on a student to identify the factors of the set of 2 x-tiles. Shade these 2 x-tiles and their factors. Then ask a student to identify the factors of the set of 3 x-tiles. Use a different shading for these 3 x-tiles and their factors. Finish this model by circling the corresponding pairs of terms in the binomial factors and the middle term of the trinomial.

Finally, ask your students how they could use shading in the third model to indicate the third term of the trinomial and its factors. Shade the 6 1-tiles in the product model and their factors. Then circle the two last terms in each binomial and the third term of the trinomial.

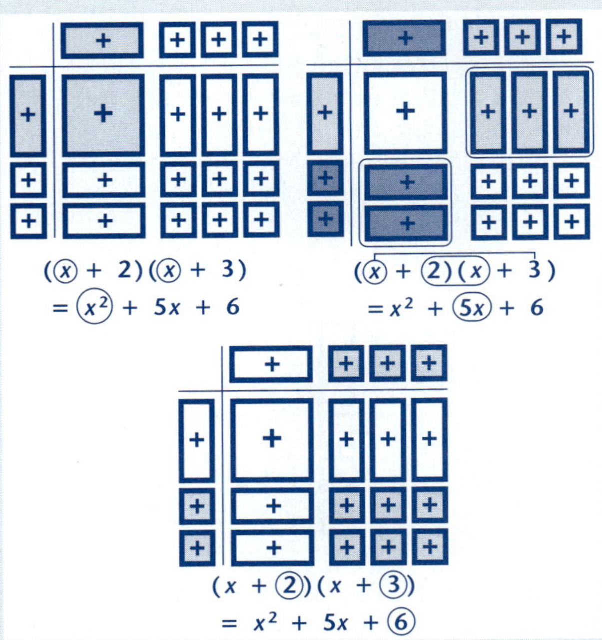

Journal Entry

Ask students to explain how this diagram can be used as a set of instructions for multiplying two binomials.

Critical Thinking, page 337

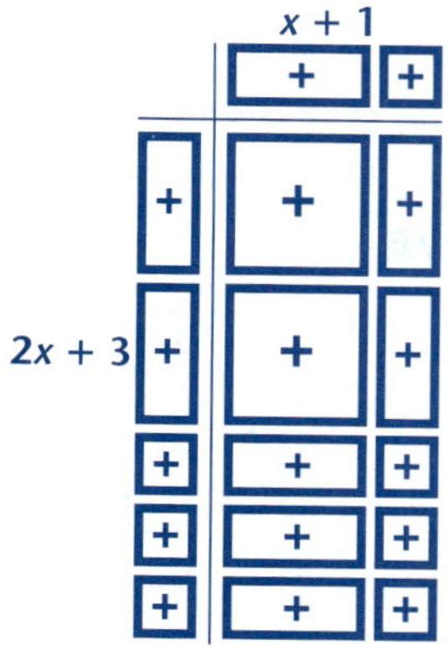

Distributive Property:

$(2x + 3)(x + 1) = 2x(x + 1) + 3(x + 1)$
$= 2x^2 + 2x + 3x + 3$
$= 2x^2 + 5x + 3$

FOIL method:

$(2x + 3)(x + 1) = 2x^2 + (2x + 3x) + 3$
$= 2x^2 + 5x + 3$

Example 2

Students apply the procedures for multiplying binomials to a geometry problem.

Encourage students to check their answers by evaluating the expression for $x = 12$ in order to determine if the area of the original mat minus the area of the cut out mat equals the area of the mat board.

Additional Example 2

A brick path surrounds a flower garden. If the area of the path is 60 square feet, find the dimensions of the garden with its brick path. **[20 feet × 12 feet]**

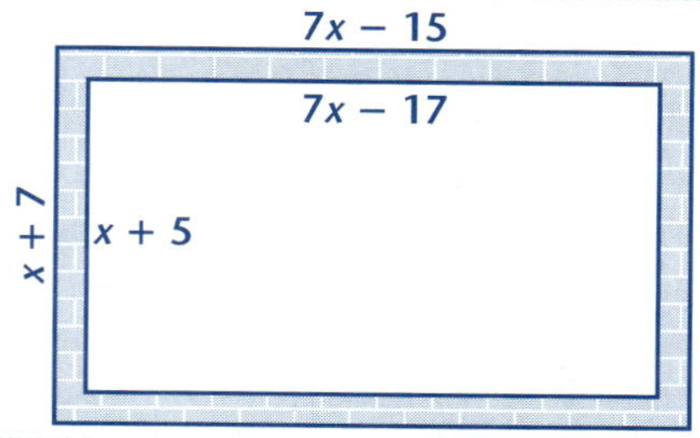

continued on top of next page

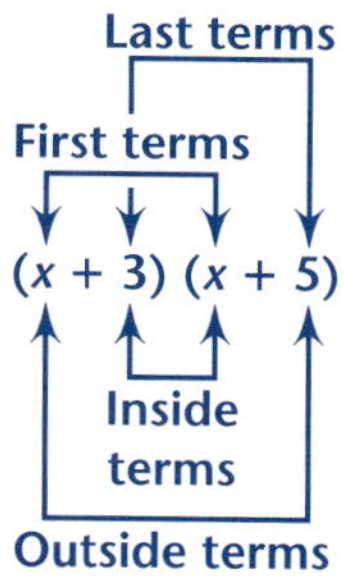

Teaching Tip

Using an Acronym To help students remember what FOIL stands for you can have them copy this diagram in their notebooks.

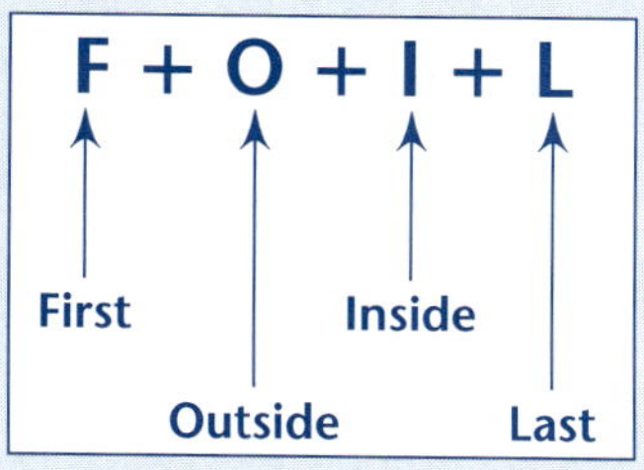

Point out to students that the letters representing the first and last terms of a trinomial are in the first and last positions of the acronym. The letters that represent the middle term of a trinomial are in the middle two positions of the acronym.

Example 1

Cooperative Learning Have students work with a partner. Ask each student to choose two linear binomials and find the products of his or her binomials by using the FOIL method. Partners should then exchange papers and check each other's work by using the Distributive Property.

Alternate Example 2

Application: Area of a Rectangle

Ask students to recall how to find the area of a rectangle. Remind students that the area of a rectangle is given in square units. Draw two rectangles on the board. Tell students that the rectangles represent rooms whose floors will be tiled. Label the dimensions of one room as

continued on bottom of next page

Approach A

Using the Book, continued

Example 3

In this example, a binomial is squared in order to find a genetic probability.

Use **Transparency 31** to discuss the Punnett square, a connection to probability that shows the possible results of crossing two flowers.

Notation

Students have thus far expressed the square of a binomial, $(a + b)$, as $(a + b)(a + b)$. A simpler expression for this product is $(a + b)^2$. Point out to students, however, that $(a + b)^2 \neq a^2 + b^2$.

Additional Example 3

Show how squaring a binomial models the probability that a toss of two coins gives a 50% probability of one heads and one tails. **[When binomials are multiplied, each term of the first binomial is distributed and multiplied by each term of the second binomial. The same pattern appears when listing the possible results of tossing two coins.**

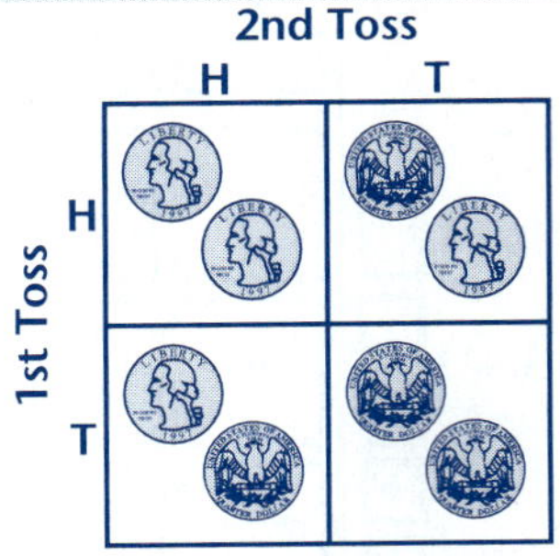

$(H + T)^2 = HH + HT + TH + TT = H^2 + 2HT + T^2$ Since there are four possibilities and two are HT, the probability of one heads and one tails is 50%.]

Closing the Lesson

Ask students to use either the Distributive Property or the FOIL method to simplify $(x + 3)(x + 3)$ and $(x - 3)(x + 3)$. Have students exchange papers and check their partner's answers either by evaluating the expressions for a value of x or by building a table of values.

TEACH

Approach B

Using Models, continued

$7x - 15$ and $x + 7$. Label the dimensions of the other room as $7x - 17$ and $x + 5$.

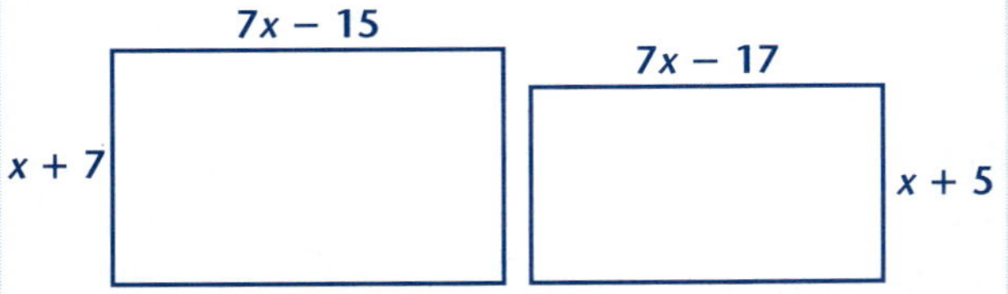

Ask half your students to find an expression for the area of the larger floor. **[$7x^2 + 34x - 105$]** The other students should find an expression for the area of the smaller floor. **[$7x^2 + 18x - 85$]**

If the difference in areas is 60 square feet, how many feet of square tiles will be needed to tile the larger floor? **[240 square feet]**

Example 3

In demonstrating Example 3, use the following tree diagram to help students understand the different possible combinations of genes.

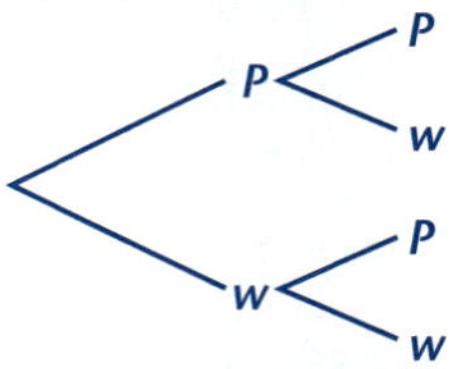

Closing the Lesson

Ask students to use either the Distributive Property or the FOIL method to simplify $(x + 3)(x + 3)$ and $(x - 3)(x + 3)$. Have students exchange papers and check their partner's answers either by evaluating the expressions for a value of x or by building a table of values.

ASSESS

Selected Answers

Odd-numbered Exercises 5–43 and odd-numbered Extra Practice

Assignment Guide

Core level students should complete Exercises 1–44.

Core-plus level students should complete Exercises 1–9 and 17–47.

Error Analysis

Some students have trouble with the FOIL method because they are unsure whether the terms referred to are terms of the individual binomials or one of the four terms of the two binomials considered together. Remind students that first and last terms are calculated from the first and last terms of each binomial. The middle term is calculated from the outside and inside terms of all four terms of the two binomials.

Alternative Assessment

FOIL, Algebra Tiles, and the Distributive Property Describe how the results of the FOIL method are related to an algebra-tile model of the product of two binomials. How is the FOIL method related to the Distributive Property?

Portfolio Activity Ask students to write a set of instructions for multiplying a binomial by using the FOIL method. An example should be included. Then students should use the same example to help them write a second set of rules for multiplying a binomial by using the Distributive Property.

RETEACH

Divide the class into four groups. Assign one of the following products to each group:

1. $(x + 5)(x + 6)$

2. $(x + 5)(x - 6)$

3. $(x - 5)(x + 6)$

4. $(x - 5)(x - 6)$

Each group should multiply its binomials by using the Distributive Property. One member of each group should write the group's steps on the chalkboard. Examine the results as a class. How can the first term of the product for each binomial pair be easily determined? How can you quickly determine the last term in the product? Why is the middle term of the product determined differently? Which products determine the middle term of the trinomial?

Use students' answers to formulate a rule for using the FOIL method to find the product of two linear binomials.

Examine the product of a binomial in the form $(a + b)(a - b)$.

ENRICH

Interdisciplinary Connection

Genetics The field of biology that studies the way in which offspring inherit biological traits from their parents is call genetics. In 1865, an Austrian monk named Gregor Mendel, working in his garden at a monastery, crossed pea plants and found that when he crossed a tall pea plant with a short pea plant, he got a tall pea plant. But when he crossed these second-generation tall pea plants with short pea plants, some short pea plants resulted. Mendel reasoned that something within the pea plants controlled their height. He called these influences *factors*. Today we call these factors *genes* and credit Mendel with laying the groundwork for modern genetics.

NAME ______________________ CLASS ____________ DATE ____________

Practice

7.4 Multiplying Binomials

1. Write the factors and the product modeled by the tiles.

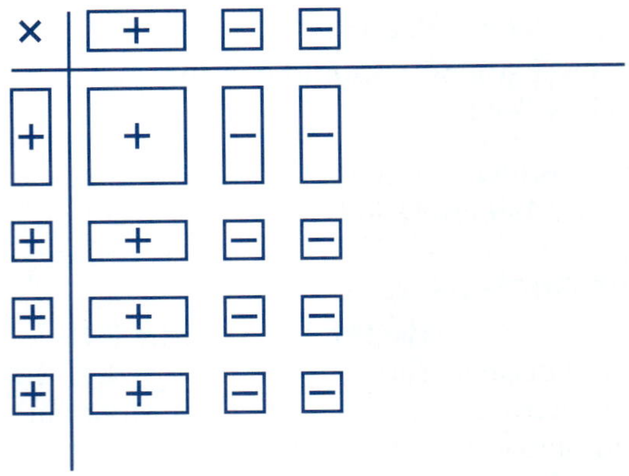

Use the Distributive Property to find each product.

2. $(a + 4)(a + 5)$ ______________

3. $(2n - 3)(n - 2)$ ______________

4. $(t + 6)(2t - 3)$ ______________

5. $(4y - 1)(2y + 3)$ ______________

6. $\left(\frac{1}{4}c - 3\right)(4c - 1)$ ______________

7. $(x + 0.2)(x - 0.8)$ ______________

Use the FOIL method to find each product.

8. $(r - 1)(r - 2)$ ______________

9. $(a - 7)(a - 3)$ ______________

10. $(m - 4)(m - 4)$ ______________

11. $(w + 0.5)(w - 0.4)$ ______________

12. $\left(s + \frac{1}{3}\right)(2s - 9)$ ______________

13. $\left(y - \frac{3}{4}\right)(y + 4)$ ______________

14. $(3c - 5d)(5c + 3d)$ ______________

15. $(3m + 2)(3m + 4)$ ______________

16. $(3x + 2y)(3x - 2y)$ ______________

17. $(2m + 5)(2m + 5)$ ______________

18. $(2y - 1)(3y - 1)$ ______________

19. $(2x + 1)(5x + 2)$ ______________

20. $(7s + 1)(7s + 1)$ ______________

21. $(x^2 + 1)(x^2 + 2)$ ______________

22. $(m^2 + n)(m^2 + n)$ ______________

23. $(4d - 2c)(2d + c)$ ______________

24. $(0.2x + 2)(0.1x - 5)$ ______________

25. $(2.3y - 5)(1.4y + 1)$ ______________

NAME ______________________ CLASS ____________ DATE ____________

Lesson Activity

7.4 Patterns in Binomial Products

Euclid used geometric models to solve problems involving binomials. So can you.

1. Use a square sheet of paper to model $(a + b)^2$. Choose a point, P, on the top edge. At point P, form a vertical crease parallel to the edge, as shown. Label the longer and shorter segments a and b. Then fold the upper right-hand corner over to the crease to locate point Q, as shown. Next make a horizontal crease through Q, as shown. Draw and explain how this models $(a + b)^2$.

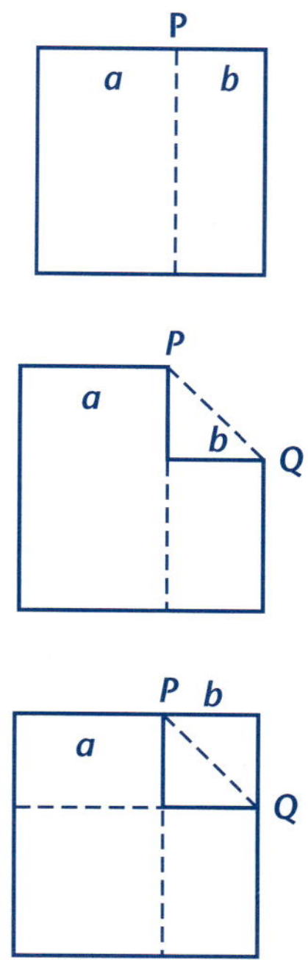

____

____

2. Use a square sheet of paper to create a model of $(a - b)^2$. Hint: Use a square of side a, and remove rectangular sections. Draw and explain how the model shows that $(a - b)^2 = a^2 - 2ab + b^2$.

____

____

3. Using construction paper of two different colors, cut out two identical squares of side a and two identical squares of side b. Rearrange the pieces to form a square, as shown. Draw and explain how you can use this model to find a simplified expression for $(a + b)^2 + (a - b)^2$.

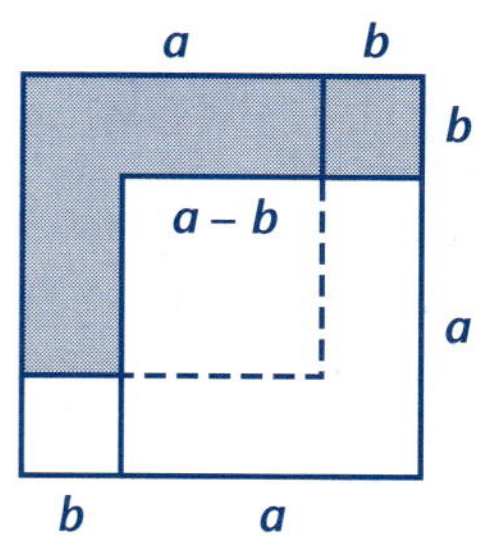

____

____

NAME ______________________ CLASS ____________ DATE ____________

Basic Skills Practice

Effects of Dimensional Changes, Part 2

Read each question and circle the best answer.

1. A cube has a surface area of 24 square inches. What is the area of the cube if each side is doubled?

A 54 $in.^2$
B 96 $in.^2$
C 30 $in.^2$
D 12 $in.^2$

2. How much more volume does a cube with sides measuring 5 inches have than a cube with sides measuring 3 inches?

F 20 $in.^3$
G 98 $in.^3$
H 150 $in.^3$
I 61 $in.^3$

3. Joy is in charge of buying containers for the school's recycling center. Of the two cylindrical containers that she can choose, one container has dimensions that are twice as big as the other container. If the smaller cylinder has a height of 2 meters and a radius of 1 meter, what is the volume of the larger cylinder?

A 12π m^3
B 16π m^3
C 24π m^3
D 40π m^3

4. How much more volume does the larger cylinder have than the smaller cylinder?

F 16π m^3
G 2π m^3
H 14π m^3
I 8π m^3

5. A rectangular prism has dimensions of 2 inches, 4 inches, and 6 inches. What is the surface area of the prism if the dimensions are multiplied by 1.5?

A 118 $in.^2$
B 108 $in.^2$
C 198 $in.^2$
D 218 $in.^2$

6. What is the volume of the larger prism?

F 162 $in.^3$
G 118 $in.^3$
H 198 $in.^3$
I 218 $in.^3$

7. How much more volume does the larger prism hold than the smaller prism?

A 44 $in.^3$
B 114 $in.^3$
C 88 $in.^3$
D 108 $in.^3$

Solve each problem.

8. The ratio of the sides of two rectangular prisms is 1:3. What is the ratio of their surface areas? ____________

9. What is the ratio of their volumes? ____________

LESSON 7.5 Common Factors

PREPARE

Pacing Guide
2 Days

Objectives and Overview
- Factor a polynomial by removing the greatest common factor.
- Factor a polynomial by removing a binomial factor.

Students factor polynomials by factoring out common monomial or binomial factors.

Vocabulary
prime numbers
composite numbers
factor
common monomial factor
prime polynomial
common binomial factor

Assessing Prior Knowledge

Write the prime factorization of each number.

1.	18	$[2 \cdot 3^2]$
2.	20	$[2^2 \cdot 5]$
3.	15	$[3 \cdot 5]$
4.	42	$[2 \cdot 3 \cdot 7]$
5.	400	$[2^4 \cdot 5^2]$

Approach A: Using the Book

Opening Discussion
Write the fraction $\frac{9}{12}$ on the chalkboard. Is this fraction in simplest form? Explain. What is the greatest common factor of the numerator and the denominator? What do you think is the simplest form of the fraction $\frac{9x}{12x}$?

Brainstorming
Write the expression $64x^3$ on the chalkboard or overhead. Ask students to make a list of factor pairs whose product is $64x^3$. Examples include $8 \cdot 8x^3$, $64x \cdot x^2$, and $4x^2 \cdot 16x$.

Exploration: Prime Numbers
This exploration provides students with an opportunity to work with a geometric model of prime and composite factors. Each rectangle represents a pair of factors because the area of the rectangle is the product of its length and width.

continued on top of next page

TEACH

Approach B: Using Models

Opening Discussion
Begin the lesson by writing the fraction $\frac{9}{12}$ on the chalkboard. Ask students if the fraction is in simplest form. What is its simplest form? By what common factor did you divide both the numerator and the denominator? How would you reduce the fraction $\frac{9x}{12x}$ to simplest form?

Brainstorming
Tell students that in this lesson they will be asked to find common factors. Then write the expression $64x^3$ on the chalkboard or overhead. Ask students to make a list of factor pairs whose product is $64x^3$. Examples include $8 \cdot 8x^3$, $64x \cdot x^2$, and $4x^2 \cdot 16x$.

Cooperative Learning
Modeling Prime Numbers Divide the class into groups of four. Give each group a geoboard or a set of toothpicks. Show students how they can use

continued on bottom of next page

Approach A
Using the Book, continued

Ongoing Assessment, top of page 341

4. It has only one distinct factor, while prime numbers have exactly two distinct factors; 1 can be factored into any number of 1s, but they are not distinct.

5. A prime number has exactly two distinct factors, one and the number itself.

Use **Transparency 32** for discussion or review.

Cooperative Learning

Journal Entry It is difficult to learn and understand a language unless a person practices speaking the language. Listening carefully to the language as it is spoken is also an important tool for learning a language. Ask students to summarize the results of the exploration, incorporating all of the boldface vocabulary terms introduced on page 341. Their summary should include examples.

Have students form groups of three and quietly read their summary to each other. Students who are listening should offer constructive criticism and ask questions of those who are reading.

Ongoing Assessment, middle of page 341

The prime factors of 24 are 2 and 3 because $24 = 2^3 \cdot 3$.

Teaching Tip

Some students may suggest that $3(x^2 + 4x)$ also represents a factorization of $3x^2 + 12x$. Encourage these students to share their observations with the class. Generally, when they are asked to factor, a complete factorization is expected. Because the terms of the binomial factor have a common factor of x, the complete factorization is $3(x^2 + 4x) = 3 \cdot x(x + 4) = 3x(x + 4)$.

TEACH

Approach B
Using Models, continued

their geoboards or toothpicks to model a 2-by-2 rectangle and a 1-by-4 rectangle.

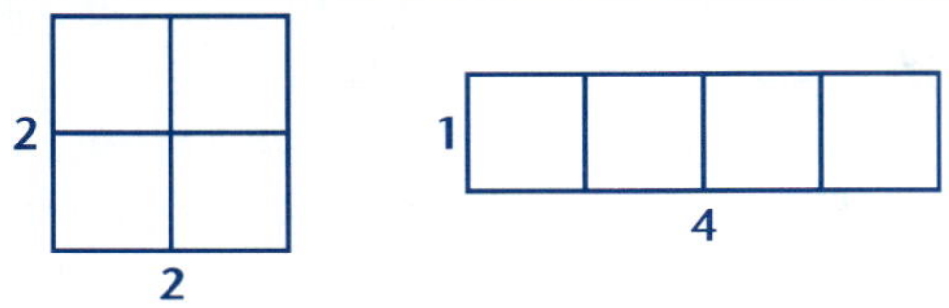

What is the area of each rectangle? Is there any other way to construct the rectangle so that its area will still be 4 square units? (A 4-by-1 rectangle is considered the same as a 1-by-4 rectangle.)

Tell students to use their modeling tools to form all possible rectangles whose areas are the integers from 2 to 13, inclusive. As they build the rectangles, they should fill in the following chart, which shows how many different rectangles can be built for a given area.

Area in square units	Number of rectangles	Dimension of the rectangles
2	1	2-by-1
3	1	3-by-1
4	2	4-by-1, 2-by-2
5	?	?
⋮	⋮	⋮
13	?	?

Learning the Language

Use the results in the chart to introduce the lesson's vocabulary.

Which areas can be represented by exactly one rectangle? These numbers are called *prime numbers*. The dimensions of the rectangle are called the *factors* of the prime number. Thus, every prime number has exactly two factors, itself and 1.

Example 1

Students practice factoring out the greatest common monomial factor from a polynomial.

Students should check to see that the polynomial is factored correctly and that the GCF has been factored out.

Additional Example 1

Factor each polynomial.

A. $4y^3 - 16y^4$ **[$4y^3(1 - 4y)$]**

B. $6p + 15p^2 - 9pq$ **[$3p(2 + 5p - 3q)$]**

C. $8r^4 + 17s^4$ **[$8r^4 + 17s^4$ is prime.]**

Ongoing Assessment, bottom of page 341

The results will be the same.

Critical Thinking, page 341

$2x + 3y$ has no polynomial factors with integral coefficients except itself and 1.

Example 2

Students practice identifying and factoring out a common binomial factor.

A strategy for factoring a binomial is developed, based on factoring a simpler polynomial. A substitution is made so that the polynomial models the polynomials for which students have factored out a common monomial factor.

Students should recognize the geometric model of the factored form as being the same as the model of the product of two binomials.

Additional Example 2

Factor $a(b - 3) + c(b - 3)$. **[$(a + c)(b - 3)$]**

Ongoing Assessment, middle of page 342

Try This

$(a + 6)(x - 3)$

continued on top of next page

The other areas can be represented by rectangles with dimensions other than 1 and the area number. These numbers are called *composite numbers*. How are the dimensions of these rectangles related to the composite numbers?

What is the only even prime number?

Monomial Factors

Write the fraction $\frac{3}{12}$ on the chalkboard. Ask students to list the factors of 3 and of 12. Which factor would they use to reduce $\frac{3}{12}$ to simplest form? Write $\frac{3}{12}$ as a reduced fraction. Ask students what factor other than 3 they could use to reduce $\frac{3x^2}{12x}$. Since $3x^2$ and $12x$ both have factors of $3x$, $\frac{3x^2}{12x} = \frac{x}{4}$. Apply this reasoning to the sum $3x^2 + 12x$. Since $3x$ is a factor of both terms, $3x^2 + 12x = 3x(x + 4)$. $3x$ is called a *common monomial factor.* To *factor* a polynomial means to express it as the product of factors whose coefficients are usually integers.

When a polynomial, such as $4a + 3b$, has no polynomial factors with integral coefficients except itself and 1, it is a *prime polynomial* with respect to the integers.

Journal Entry

Factoring and the Distributive Property Tell students to factor each polynomial by factoring the greatest common factor from each term.

A. $12rs + 18r^2s$ **[$6rs\ (2 + 3r)$]**

B. $3z^3 - 18z^2 + 24z$ **[$3z\ (z - 2)(z - 4)$]**

C. $47a^2 - 48b^2$ **[prime]**

Ask students to explain the relationship between the factored forms of the polynomials and the Distributive Property.

Teaching Tip

To reinforce the idea of a prime polynomial, ask students whether the statement $6\left(\frac{1}{3}x + \frac{1}{2}y\right) = 2x + 3y$ is true or false. Tell them that although

continued on bottom of next page

Approach A

Using the Book, continued

Example 3

This example demonstrates the regrouping of polynomial terms in order to factor them.

Students should be able to give a reason for each step.

- What property of real numbers allows the terms to be regrouped?
- What are the common monomial factors of the two terms in each binomial?
- What is the common factor in the expression $x(x + 1) + 2(x + 1)$?

Additional Example 3

Factor $mn + mp + 5n + 5p$. $[\mathbf{(m + 5)(n + p)}]$

Ongoing Assessment, bottom of page 342

Try This

$(a + b)(x + y)$

Closing the Lesson

Write the following factorization on the board:

$2ax + 2ay + max + may$
$= a(2x + 2y + mx + my)$
$= a\,[(2x + 2y) + (mx + my)]$
$= a\,[2(x + y) + m(x + y)]$
$= a(2 + m)(x + y)$

Have students identify the steps in which

1. a common monomial is factored out.
2. the Associative Property for Addition is used.
3. a common binomial is factored out.

TEACH

Approach B

Using Models, continued

the statement is true according to the Distributive Property, $6\left(\frac{1}{3}x + \frac{1}{2}y\right)$ is not a factored form of $2x + 3y$ over the set of integers. The binomial $2x + 3y$ is considered prime because $2x + 3y$ does not have any polynomial factors with *integral* coefficients except itself and 1.

Modeling Binomial Factors: Algebra Tiles

Ask students to recall this algebra-tile model.

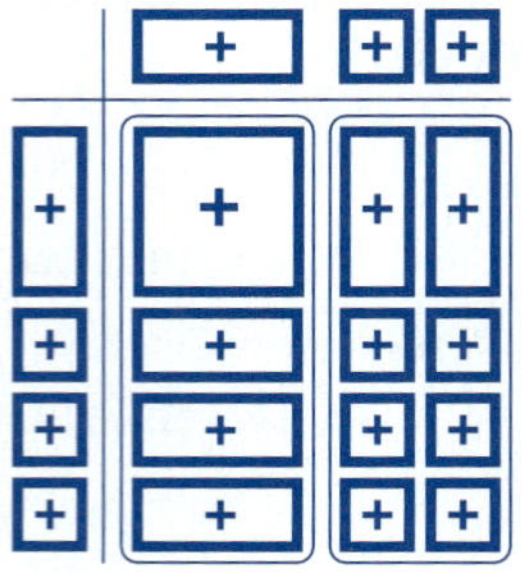

What expression represents the sum of the two circled sets of tiles? $[\mathbf{x(x + 3) + 2(x + 3)}]$ This expression represents the fact that $(x + 3)$ is distributed over each term of $(x + 2)$. What are the two factors of $x(x + 3)$? What are the two factors of $2(x + 3)$? By factoring out the common binomial factor, $x(x + 3) + 2(x + 3) = (x + 3)(x + 2)$. How is $(x + 3)(x + 2)$ related to the algebra-tile model?

Closing the Lesson

Write the following factorization on the board:

$2ax + 2ay + max + may$
$= a(2x + 2y + mx + my)$
$= a\,[(2x + 2y) + (mx + my)]$
$= a\,[2(x + y) + m(x + y)]$
$= a(2 + m)(x + y)$

Have students identify the steps in which

A. a common monomial is factored out.
B. the Associative Property for Addition is used.
C. a common binomial is factored out.

ASSESS

Selected Answers

Odd-numbered Exercises 9–55 and odd-numbered Extra Practice

Assignment Guide

Core level students should complete Exercises 1–56.

Core-plus level students should complete Exercises 1–18 and 26–57.

Error Analysis

If students have trouble with Exercises 11–22, they may not be factoring out the greatest common monomial factor. For these students, checking their answers with the Distributive Property will not show that they have not completely factored the polynomial. Tell students to check the remaining factors after they have factored out a monomial factor. If either factor can be factored again, the GCF was not chosen as the common factor the first time.

Alternative Assessment

Justifying Steps in a Calculation Ask students to give the justification for each step of the following factorization:

$12m^2n - 2m^3 + 18mn - 3m^2$
$= m(12mn - 2m^2 + 18n - 3m)$
[Distributive Property]

$= m[(12mn - 2m^2) + (18n - 3m)]$
[Associative Property for Addition]

$= m[2m(6n - m) + 3(6n - m)]$
[Distributive Property]

$= m(6n - m)(2m + 3)$ **[Distributive Property]**

Portfolio Activity Ask students to draw sketches of a right pyramid and a rectangular solid. Tell them to show how the formulas for the surface area of each figure can be written in factored form.

RETEACH

Hold up a cylinder or a picture of a cylinder, or draw a picture of a cylinder on the chalkboard.

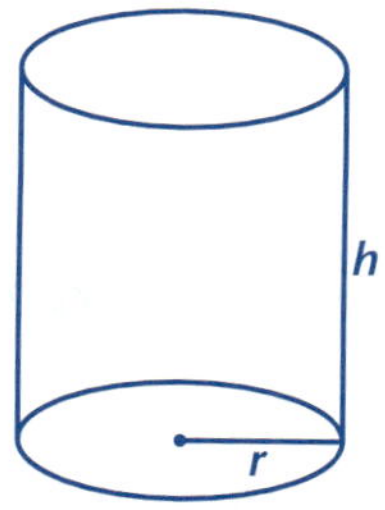

Tell students that the manufacturer of a can would like to decrease the expense of making the can by decreasing its surface area. The surface area of a cylinder such as this can be found by using the formula $A = 2\pi r^2 + 2\pi rh$.

Since both terms on the right side of the formula are multiplied by $2\pi r$, the Distributive Property can be used to *undo* the formula.

$A = 2\pi r^2 + 2\pi rh = 2\pi r(r + h)$

If the manufacturer needs to find the surface area of several cans, why do you think he might prefer to use the formula $A = 2\pi r(r + h)$?

Because $2\pi r$ is a factor of both terms in the original formula, it can be factored out and multiplied by each term just once. To check whether a polynomial has been factored correctly, apply the Distributive Property to the factored form. The resulting polynomial should be the original, unfactored polynomial.

ENRICH

Interdisciplinary Connection

Electronics The electrical resistance of a wire at a temperature, T, can be found by using the expression $R + aRT - aRT_0$, where T_0 is a reference temperature, R is the resistance of the wire at T_0, and a is the *temperature coefficient of resistance*. The expression for the resistance can be written in factored form by factoring out a common monomial factor twice.

$R + aRT - aRT_0 = R(1 + aT - aT_0) =$
$R[1 + a(T - T_0)]$

NAME ____________________ CLASS ____________ DATE ____________

Practice

7.5 Common Factors

Identify each polynomial as prime or composite.

1. $9x^2 + 24x + 15$ ____________

2. $9a^2 + 24a$ ____________

3. $9m^2 + 16$ ____________

4. $x^2 + 5x - 6$ ____________

5. $2x^2 + 2x + 4$ ____________

6. $4x^2 + 25y^2$ ____________

Factor each polynomial by removing the GCF.

7. $3x^2 - 9x$ ____________

8. $6r^3 + 3r^2 - 7r$ ____________

9. $6w^4 + 3w^2 - 9$ ____________

10. $3q^2 + 6q^5 + 9q$ ____________

11. $18b^3 - 36b^2 - 9$ ____________

12. $3v^6 - 9v^4 + 147v^2$ ____________

13. $xy - 2x^2y + xy^3$ ____________

14. $10wz^2 - 5wz + 15w^2z$ ____________

Write each as the product of two binomials.

15. $y(y + 1) + 2(y + 1)$ ____________

16. $3(c + d) - a(c + d)$ ____________

17. $2(x + w) - z(x + w)$ ____________

18. $6(x - 2) + y(x - 2)$ ____________

19. $a(a - b) - b(a - b)$ ____________

20. $3x(x + 4) - 2(x + 4)$ ____________

21. $4(x - 2y) + x(x - 2y)$ ____________

22. $z(4 - w) + y(4 - w)$ ____________

23. $7x(x - 1) + (x - 1)$ ____________

24. $mn(r + 1) - (r + 1)$ ____________

Factor.

25. $4ax - bx + 4ay - by$ ____________

26. $2pq^2 + 4pq - 2q - 4$ ____________

27. $3a + 3 - a^2 - a$ ____________

28. $6m^3 + 4m - 9m^2 - 6$ ____________

29. $3m - 12 + m^3 - 4m^2$ ____________

30. $ab^2 + 5a - 6b^2 - 30$ ____________

31. $2cd - c - 6d + 3$ ____________

32. $4 - 2x + 6 - 3x$ ____________

33. $2xz + 2yz + x + y$ ____________

34. $mx + 5m + nx + 5n$ ____________

35. $xy + 2y + x + 2$ ____________

36. $3x^2 + 3xy - 2xy - 2y^2$ ____________

NAME ______________________ CLASS ____________ DATE ____________

Lesson Activity

7.5 A Common Factor Pattern

Use greatest common factors to discover the number of unit squares intersected by a diagonal of a rectangle formed entirely of unit squares.

1. For each rectangle in the diagram, count the number of unit squares intersected by the diagonal. Then calculate the greatest common factor of the sides. Record your results in a table like the one shown below. The results for rectangles with lengths of 1 and 2 units have been entered for you.

l	*w*	GCF (*l*, *w*)	Unit squares intersected (*n*)
1	1	1	1
2	1	1	2
2	2	2	2
3	1		
3	2		
3	3		
.	.		
.	.		
.	.		

2. Use a sheet of paper to continue the pattern by drawing all of the rectangles with lengths of 6 and 7 units. Record your results in a table.

3. For which rectangles does the number of squares intersected equal one less than the sum of the lengths of the sides? What is the GCF of the length and width for these rectangles?

____

4. For which rectangles does the number of squares intersected equal two less than the sum of the lengths of the sides?

____

5. Write a formula in which the greatest common factor is used to describe the pattern or relationship between the sum of the lengths of the sides of a rectangle and the number of squares intersected by a diagonal of the rectangle.

____

Basic Skills Practice
Cumulative Review

Read each question and circle the best answer.

1. A cube has sides that measure 6 inches. By how much would its volume increase if the dimensions were multiplied by 1.5?

A 513 $in.^3$
B 216 $in.^3$
C 529 $in.^3$
D 100 $in.^3$

2. By how much would the surface area increase for the cube in Exercise 1?

F 118 $in.^2$
G 270 $in.^2$
H 486 $in.^2$
I 529 $in.^2$

3. The dimensions of a rectangular prism are 6 inches by 12 inches by 15 inches. What would the dimensions be if they were reduced by $\frac{1}{3}$?

A 3 in. × 6 in. × 9 in.
B 2 in. × 4 in. × 5 in.
C 4 in. × 8 in. × 10 in.
D 3 in. × 9 in. × 12 in.

4. By how much would the volume decrease for the rectangular prism in Exercise 3?

F 240 $in.^3$
G 360 $in.^3$
H 540 $in.^3$
I 760 $in.^3$

5. Find the median for the following scores: 9, 4, 8, 9, 10, 11, 4, 0, and 6.

A 6.7
B 61
C 9
D 8

6. The radius of a cylinder is 4 inches. The height is 24 inches. What would the volume be if its dimensions were reduced by three-fourths?

F 6π $in.^3$
G 12π $in.^3$
H 20π $in.^3$
I 24π $in.^3$

7. What would be the resulting cylinder's lateral surface area in Exercise 6?

A 6π $in.^2$
B 9π $in.^2$
C 12π $in.^2$
D 24π $in.^2$

8. Given a square with vertices at $A(2, 1)$, $B(2, 4)$, $C(5, 4)$, and $D(5, 1)$, find the vertices of the image of $ABCD$ by using the origin as the center of dilation and a scale factor of 2.

F $A(2, 1)$, $B(2, 4)$, $C(5, 4)$, $D(5, 1)$
G $A(4, 2)$, $B(4, 8)$, $C(10, 8)$, $D(10, 2)$
H $A(4, 1)$, $B(4, 4)$, $C(10, 4)$, $D(10, 1)$
I $A(2, 2)$, $B(2, 8)$, $C(5, 8)$, $D(5, 2)$

Solve each problem.

9. The dimensions of a rectangular prism are 15 inches by 20 inches by 25 inches.

What is the volume, in cubic inches, if the dimensions were multiplied by $\frac{1}{5}$? ____________

LESSON 7.6

Factoring Special Polynomials

PREPARE

Pacing Guide
2 Days

Objectives and Overview
- Factor perfect-square trinomials.
- Factor the difference of two squares.

Students use algebra tiles to explore methods for factoring perfect-square trinomials and the difference of two squares.

Vocabulary
perfect-square trinomial
difference of two squares

Assessing Prior Knowledge

Decide whether each sentence is true or false. If it is false, change the sentence to make it true.

1. The product of $(a + b)(a + b)$ is $a^2 + b^2$. **[False; the product is $a^2 + 2ab + b^2$.]**
2. The product of $(a + b)(a - b)$ is $a^2 - b^2$. **[true]**

Approach A: Using the Book

Opening Discussion

Begin with a drawing of a 2-by-2 square. What is the area of this square? Suppose that one unit is added to each dimension of the square.

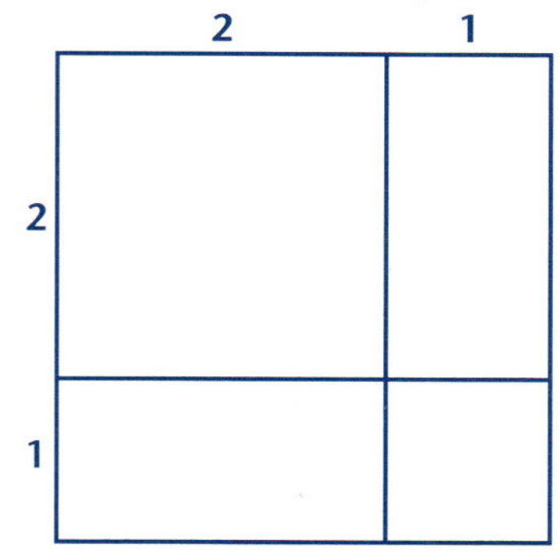

What is the area of the new square? **[4 + 2 · 2 + 1 = 9]**

continued on top of next page

TEACH

Approach B: Using Models

Opening Discussion

Begin by drawing a 2-by-2 square on the chalkboard or overhead. Ask students to find the area of the square. Then add one unit to each dimension of the square. What is the area of the new square? **[4 + 2 · 2 + 1 = 9]**

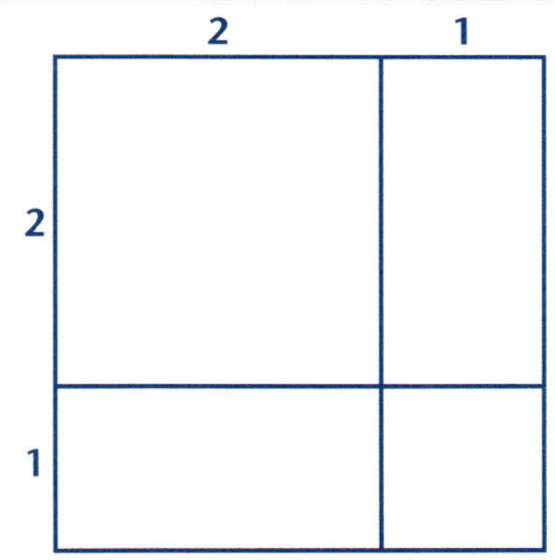

continued on bottom of next page

Approach A

Using the Book, continued

Perfect Squares

The procedure for finding the middle term of a perfect-square trinomial is presented without defining square root. The definition can be avoided by asking students what expressions have been squared to get the first and last terms of a perfect-square trinomial. Doubling the product of these expressions will give the middle term.

You may wish to explain that the middle term of a perfect-square trinomial is the product of the square roots of the first and last terms of the trinomial.

Use **Transparency 33** for discussion or review of the square of a binomial.

Ongoing Assessment, page 345

$$(2x + 6)^2 = (2x)^2 + (2)(2x)(6) + 6^2 = 4x^2 + 24x + 36$$

$$(2x - 6)^2 = (2x)^2 + (2)(2x)(-6) + 6^2 = 4x^2 - 24x + 36$$

The middle terms have opposite signs.

Example 1

Students are asked to identify whether trinomials are perfect-square trinomials.

As students learn to recognize perfect-square trinomials, it is helpful to ask them to verify that a trinomial is a perfect-square trinomial by confirming that each term of the trinomial satisfies the definition of a perfect-square trinomial. For example, students would state that $4x^2$ is the square of $2x$, that 25 is the square of 5, and that $2(2x \cdot 5) = 20x$. Therefore, $4x^2 + 20x + 25$ is a perfect-square trinomial.

TEACH

Approach B

Using Models, continued

Suppose that another unit is added to each dimension of the square. What is the area of this square?

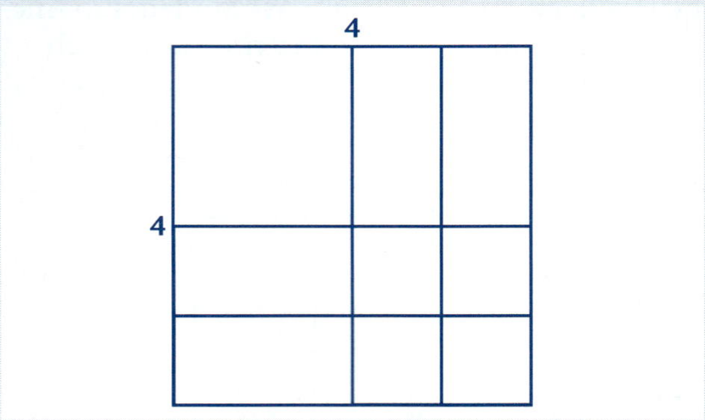

Because 4, 9, and 16 are the result of squaring an integer, they are called *perfect squares*. Find the next three perfect squares.

Modeling Perfect-Square Trinomials: Square Numbers

Numbers that can be represented by geometric forms are called *figurate numbers*. Square numbers can be arranged in a square array of dots. Draw the first four square numbers on the chalkboard or overhead.

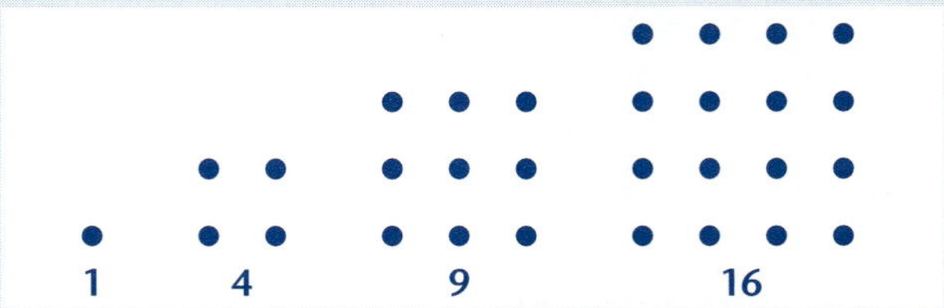

What numbers do the first four figures represent? What do you think the next square number is? The numbers 1, 4, 9, and 25 are called *perfect squares*. How are perfect squares formed?

Square numbers can be used to model special quadratic trinomials. Consider the following arrangements of square numbers:

Additional Example 1

Examine each expression. Is the expression a perfect-square trinomial? If not, why not?

A. $x^2 + 14x + 49$ **[yes]**

B. $x^2 - 12xy + 36y^2$ **[yes]**

C. $4a^2 - 18a + 81$ **[No; $4a^2$ and 81 are perfect squares, but the middle term is not equal to $2(2a)(9)$.]**

D. $25r^2 + 30r + 3$ **[No; 3 is not a perfect square.]**

E. $16m^2 + 16m + 16$ **[yes]**

F. $x^2 + 20x - 100$ **[No; x^2 and 100 are perfect squares, but the sign in front of 100 is negative, not positive.]**

Inclusion Strategy

Team Approach Ask students to work with a partner to write a flowchart for factoring perfect-square trinomials. The chart should include a decision branch for determining the operation symbols (+ or −) in the factored binomials. Have one student in each pair make up a perfect-square trinomial. The other student should use the flowchart to factor the trinomial. After factoring, students should decide if their flowchart needs revising.

Example 2

Students are asked to factor trinomials and to check their factoring by multiplying.

Additional Example 2

Factor each expression.

A. $a^2 + 4ab + 4b^2$ **[$(a + 2b)^2$]**

B. $36x^2 - 12x + 1$ **[$(6x - 1)^2$]**

C. $25 - 30y + 9y^2$ **[$(5 - 3y)^2$]**

D. $100m^2 + 140m + 49$ **[$(10m + 7)^2$]**

Critical Thinking, page 346

$x^2 + 6x + 9 = (x + 3)^2$

So $x + 3 = 5$.

Therefore, $x = 2$.

continued on top of next page

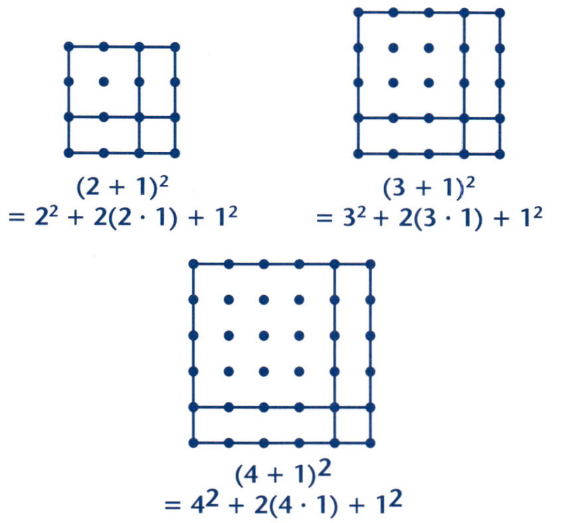

The area of each square can be found by adding the areas of 2 squares and 2 rectangles. More generally:

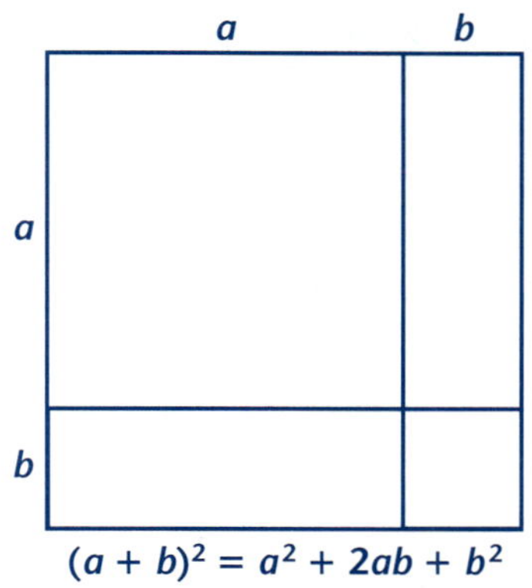

A trinomial of the form $a^2 + 2ab + b^2$ is called a perfect-square trinomial and has the factors $(a + b)(a + b)$ or $(a + b)^2$.

A Geometric Model for $(a - b)^2$

Drawing a model for $(a - b)^2$ requires subtracting 2 $a \times b$ rectangles from an $a \times a$ square.

continued on bottom of next page

Approach A — Using the Book, continued

Journal Entry

Preview As preparation for the next topic, ask students to multiply $(x + 3)(x - 3)$ in their journal. Tell them to use the product to write the answers to these questions.

- What happens to the x-terms in the product?
- What will happen to the x-terms in the product $(x - 2)(x + 2)$?
- Is the product of two binomials always a trinomial?

Example 3

In this example, students practice identifying binomials that are the difference of two squares.

Additional Example 3

Examine each expression. Is the expression a difference of two squares? If not, why not?

A. $81x^2 - 16$ **[yes]**

B. $4y^2 + 49$ **[No; $4y^2 + 49$ is not a difference; it is a sum.]**

C $48 - 25a^2$ **[No; 48 is not a perfect square.]**

D. $a^4 - b^4$ **[yes]**

E. $64 - 5a^4$ **[No; $5a^4$ is not a perfect square.]**

Example 4

This example allows students to practice factoring the difference of two squares.

In Part D, factoring the difference of two squares does not completely factor the original binomial. One of the resulting binomials is also the difference of two squares and must be factored. Remind students that to factor usually means to factor completely.

Additional Example 4

Factor each expression.

A. $x^2 - 49$ **[$(x + 7)(x - 7)$]**

B. $25m^2 - 36$ **[$(5m - 6)(5m + 6)$]**

C. $9r^2 - 100s^2$ **[$(3r - 10s)(3r + 10s)$]**

D. $16a^4 - b^4$ **[$(2a - b)(2a + b)(4a^2 + b^2)$]**

TEACH

Approach B — Using Models, continued

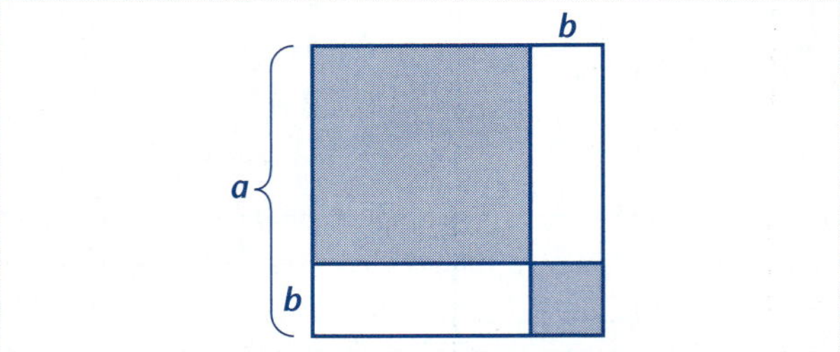

To find the shaded area, $(a - b)^2$, each of the rectangles with area ab must be subtracted from the area of the large square, a^2. Because subtracting the area of both rectangles results in the area of the smaller square, b^2, being subtracted twice, b^2 must be added once.

$$(a - b)^2 = a^2 - ab - ab + b^2 = a^2 - 2ab + b^2$$

A trinomial of the form $a^2 - 2ab + b^2$ is also a perfect-square trinomial and, as the model illustrates, has the factors $(a - b)(a - b)$ or $(a - b)^2$.

Adapting the Model to the Difference of Two Squares

You can use the models of square numbers to explore an expression for the area that remains when a square of side b is removed from a square of side a. The shaded area can be found by adding the area of two rectangles.

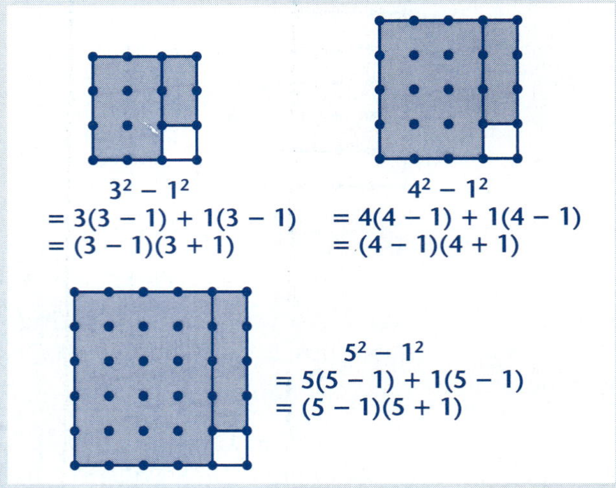

Ongoing Assessment, page 347

Try This

$(5w + 9)(5w - 9)$

Critical Thinking, page 348

$a^8 - b^8$ can be written as $(a^4)^2 - (a^4)^2$, which is a difference of squares. The rule for factoring a difference of squares then gives $a^8 - b^8 = (a^4 + b^4)(a^4 - b^4)$. But $(a^4 - b^4) = (a^2)^2 - (b^2)^2$, which is a difference of squares and equals $(a^2 + b^2)(a^2 - b^2)$ when factored. Factoring $(a^2 - b^2)$ which is yet another difference of squares, gives $(a + b)(a - b)$. Therefore, $(a^8 - b^8) = (a^4 + b^4)(a^2 + b^2)(a+ b)(a - b)$, which consists of four factors.

Example 5

Students apply the procedure for factoring the difference of two squares to the multiplication of two integers.

This example reminds students who may have forgotten that variables represent numbers. Thus, rules for algebra are rules for numbers.

Since our base-10 numbers are simplified polynomials, the rules for factoring the difference of two squares can be used to multiply special pairs of base-10 numbers.

Additional Example 5

Find each product by using the difference of two squares.

A. $39 \cdot 41$ [1599]
B. $53 \cdot 47$ [2491]
C. $12 \cdot 8$ [96]

Technology

Students should recall from Lesson 7.3 that the binomial factors of a quadratic trinomial will give the zeros of the trinomial. The points where the graph crosses the *x*-axis are the zeros of the binomial factors.

Students can verify the zeros from the CALC menu of a graphics calculator by calculating *y*-values for given *x*-values of a quadratic trinomial. When the zeros are entered as the *x*-values, the *y*-values should both be 0.

More generally:

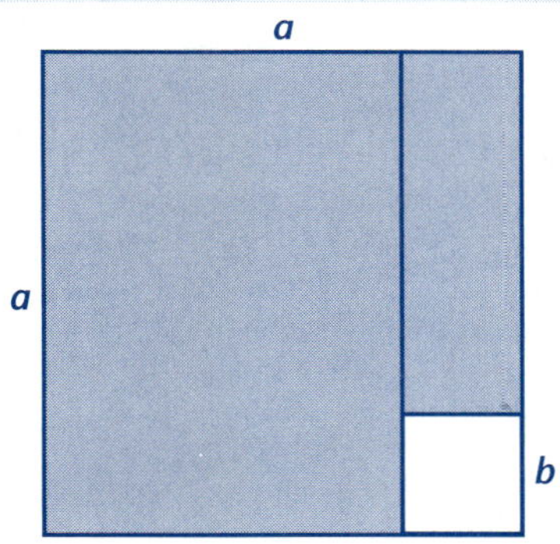

$$a^2 - b^2$$
$$= a(a - b) + b(a - b)$$
$$= (a + b)(a - b)$$

A binomial of the form $a^2 - b^2$ is called the difference of two squares and has the factors $(a + b)(a - b)$.

Application: Finding the Product of Two Integers

Ask students to recall the square array that modeled 4^2. The area of the shaded region was found by subtracting $4^2 - 1^2 = (4 - 1)(4 + 1)$. Since $(4 - 1)(4 + 1) = 3 \cdot 5$, the product of 3 and 5 can be found by using the formula for factoring the difference of two squares.

Ask students to complete the following:

$31 \cdot 29 = ([\underline{30}] + 1)([\underline{30}] - 1) = [\underline{30}]^2 - [\underline{1}]^2 = 900 - 1 = 899$

How could you find $17 \cdot 13$ by using the difference of two squares?

ASSESS

Selected Answers

Odd-numbered Exercises 5–63 and odd-numbered Extra Practice

Assignment Guide

Core level students should complete Exercises 1–63.

Core-plus level students should complete Exercises 1–36 and 42–64.

Error Analysis

Make sure that students do not incorrectly factor the sum of two squares. If you have students who factor $a^2 + b^2$ as $(a + b)(a - b)$, ask them to substitute values for a and b in both expressions. The results are almost always different.

Alternative Assessment

Application Frank Harrington announced that he would pay each employee a bonus of $1 for each of his employees. Since Frank has 31 employees, he owes a total of (31×31) dollars in bonuses to his employees. Explain how Frank can use the formula $(a + b)^2 = a^2 + 2ab + b^2$ to find the total amount that he owes his employees.

Portfolio Activity

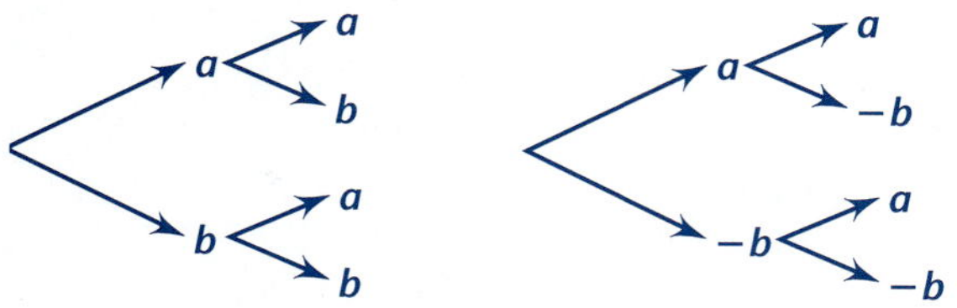

Explain how the diagrams model the method for squaring a binomial. **[Using the distrubutive property, $(a + b)^2 = a(a + b) + b(a + b) = aa + ab + ab + bb$. The diagram on the left models the multiplications involved. Follow the paths. $(a - b)^2$ is similar.]**

Flowchart Activity Have students work in pairs to build a flowchart for factoring the difference of two squares. The flowchart should show that for all real numbers a and b, $a^2 - b^2 = (a + b)(a - b)$. Ask each group to write four binomials, only two of which are the difference of two squares. Then have each pair of students exchange papers with another pair of students. Each group of two pairs should decide whether each binomial is the difference of two squares. If not, they should explain why not. If so, they should use their flowchart to factor the binomials. Students should then return their paper to the authors for correcting.

RETEACH

Draw or make algebra-tile models of the products $(x + 3)(x + 3)$, $(x - 3)(x - 3)$, and $(x + 3)(x - 3)$. Show only the product tiles. Do not show the factor tiles.

Ask students to identify the factors of each product. Which drawing is an example of $a^2 + 2ab + b^2 = (a + b)^2$? Which drawing models $a^2 - 2ab + b^2 = (a - b)^2$? Use the drawings to factor each of the following:

1. $a^2 + 6a + 9$

2. $4x^2 - 20x + 25$

3. $81m^2 - 49n^2$

ENRICH

Physics If an object moves in a straight line with constant acceleration, the distance the object has moved can be found by taking a kind of average of the squares of the beginning and ending velocities of the object. The expression for this displacement is $\frac{v_f^2 - v_0^2}{2a}$, where v_f represents the ending velocity, v_0 represents the beginning velocity, and a represents the acceleration. By factoring the difference of two squares, this expression becomes $\frac{(v_f - v_0)(v_f + v_0)}{2a}$.

NAME ______________________ CLASS ____________ DATE ____________

Practice

7.6 Factoring Special Polynomials

Use the generalization of a perfect-square trinomial or the difference of two squares to find each product.

1. $(x - 1)^2$ ____________

2. $(2x + 3)^2$ ____________

3. $(a + 5)^2$ ____________

4. $(3g + 1)^2$ ____________

5. $(2a - 5)^2$ ____________

6. $(x - y)^2$ ____________

7. $(x + 2y)^2$ ____________

8. $(3m - n)^2$ ____________

9. $(x - 3)(x + 3)$ ____________

10. $(4y + 1)(4y - 1)$ ____________

11. $(5x + 2)(5x - 2)$ ____________

12. $(6r - s)(6r + s)$ ____________

Factor each polynomial completely.

13. $x^2 - 25$ ____________

14. $y^2 - 49$ ____________

15. $2x^2 - 2$ ____________

16. $4y^2 - 16$ ____________

17. $16y^2 - 25$ ____________

18. $121 - x^2$ ____________

19. $9 - 16y^2$ ____________

20. $x^2 - 6x + 9$ ____________

21. $x^2 + 10x + 25$ ____________

22. $y^2 - 16y + 64$ ____________

23. $z^2 - 20z + 100$ ____________

24. $64x^2 - 48xy + 9y^2$ ____________

25. $25 - 10a + a^2$ ____________

26. $4m^2 + 4m + 1$ ____________

27. $81r^2s^2 - 100t^4q^4$ ____________

28. $16x^2 + 24x + 9$ ____________

29. $x^2 + 2xy + y^2$ ____________

30. $4a^2 + 4ab + b^2$ ____________

31. $y^2 - x^2$ ____________

32. $25a^2 - 1$ ____________

33. $1 - x^2y^4$ ____________

34. $144c^2 - 120cd + 25d^2$ ____________

35. $6a^2 - 216b^2$ ____________

36. $49x - xy^2$ ____________

37. $12x^4 - 12$ ____________

38. $b^4 - 2b^2 + b^2 - 2$ ____________

39. $4c^2 - 24c + 36$ ____________

40. $4m^4n^2 + 4m^3n^2 + m^2n^2$ ____________

NAME ______________________ CLASS ____________ DATE ____________

Lesson Activity
7.6 Modeling the Difference of Two Cubes

Let a be the edge of a cube, and let b be less than a. Then you can model $a^3 - b^3$ by removing a cube with edge b from the cube with edge a.

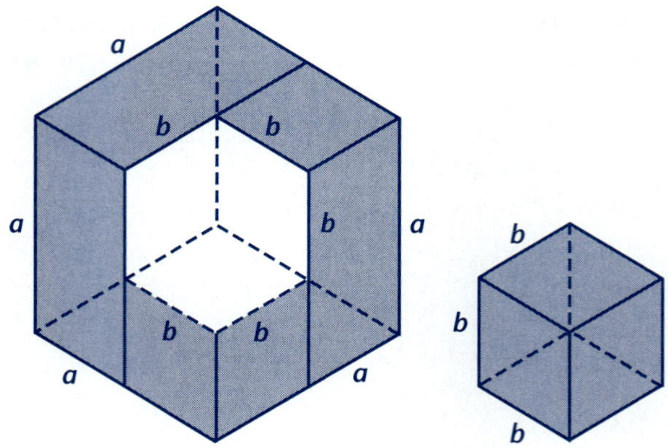

To find the factors of $a^3 - b^3$, consider the three remaining regions of the cube.

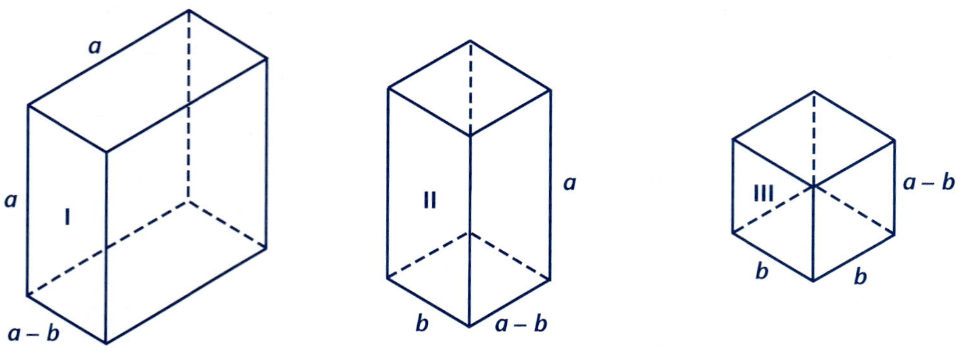

1. Find the volume of region I. ______________________

2. Find the volume of region II. ______________________

3. Find the volume of region III. ______________________

4. What is the total volume of regions I, II, and III? ______________________

5. Simplify your answer to Exercise 4. ______________________

6. Explain how you know that the expressions that you wrote in Exercise 5 are the factors of $a^3 - b^3$.

____

NAME ______________________ CLASS ____________ DATE ____________

Basic Skills Practice

Counting Outcomes and Tree Diagrams, Part 1

A sample space is a listing of all of the possible outcomes in a given situation. You can use a tree diagram to show all of the possible outcomes. Each "branch" represents a choice or possible outcome.

Example 1: A private high school offers its students a variety of uniform choices to choose from. Students can wear khaki or navy blue pants. The shirt can be white or light blue, and the tie can be black or blue. Make a diagram to show all of the possible uniforms.

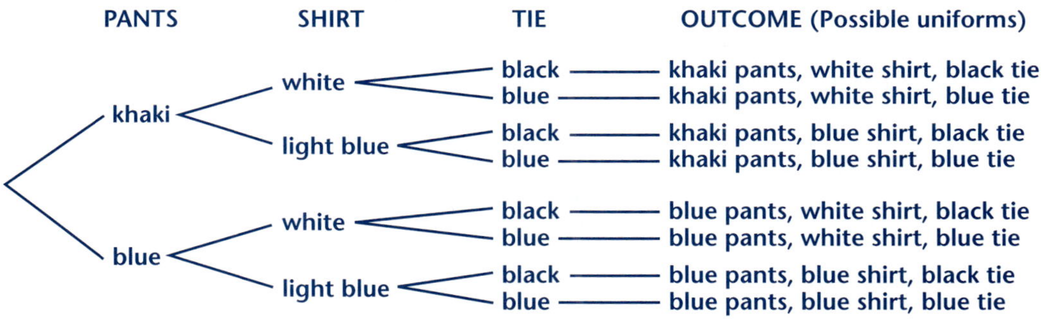

There are 8 possible uniforms.

To find the total number of possible outcomes of a series of events, multiply the number of ways each event can occur.

Example 2: A store issues identification codes to its employees. The codes start with a letter and are followed by a two-digit number. How many different codes are possible?

letter (A to Z)		digit (0 to 9)		digit (0 to 9)
26	×	10	×	10

2600 different codes are possible.

1. The lunch special at Stuart's Cafe includes soup, salad, and dessert. Complete the tree diagram to show all of the possible lunch special combinations.

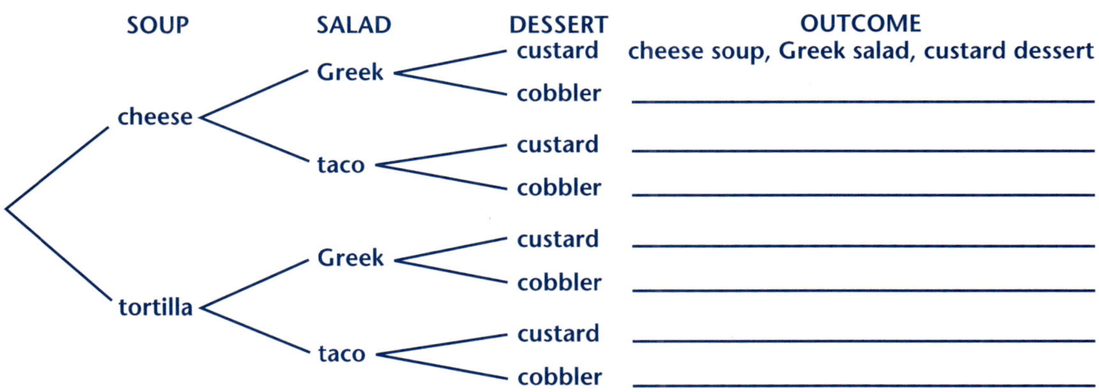

2. How many possible lunch specials are there? ____________

LESSON 7.7 Factoring Trinomials

PREPARE

Pacing Guide
3 Days

Objectives and Overview
- Use algebra tiles and guess-and-check to determine factors of a trinomial.
- Use signs and number patterns to determine the factors of a trinomial.

Students learn a method for factoring trinomials whose x^2-term has a coefficient of 1.

Vocabulary
irreducible polynomial

Assessing Prior Knowledge

Factor each polynomial.

1. $x^2 - 6x + 9$ $[(x - 3)^2]$
2. $6x^2 - 18x$ $[6x(x - 3)]$
3. $16x^2 + 40x + 25$ $[(4x + 5)^2]$
4. $x^2 - 144$ $[(x + 12)(x - 12)]$
5. $36x^2 + 12x + 1$ $[(6x + 1)^2]$
6. $x^4 - 81$ $[(x^2 + 9)(x + 3)(x - 3)]$

Approach A: Using the Book

Opening Discussion
Is $x^2 + 5x + 6$ a perfect-square trinomial? Explain. How many x-tiles are there in a tile model of $x^2 + 5x + 6$? How many 1-tiles are there? How many 1-tiles are there in the tile model of $x^2 - x - 6$? What are the possible factors of the 1-tiles?

Using Geometry to Find Factors
A rectangle with a given area has a finite number of possible integral dimensions and thus is an excellent model for factoring a trinomial.

By first laying out the tiles that model $x^2 + 6x + 8$, students are able to see that there are a finite number of tile arrangements to test in a multiplication grid. The physical arrangement of the tiles produces a rectangle whose dimensions are $x + 2$ and $x + 4$.

TEACH

Approach B: Using Models

Opening Discussion
Begin the lesson by writing the trinomial $x^2 + 5x + 6$ on the chalkboard or overhead. Ask students if the trinomial is a perfect-square trinomial. Why not? Ask students how many x-tiles would be in a tile model of $x^2 + 5x + 6$. How many 1-tiles would there be? What are the possible factors of the x-tiles? How many 1-tiles would there be in a tile model of $x^2 - x - 6$? What are the possible factors of the 1-tiles?

Factoring With Algebra Tiles
Ask students to compare the following three algebra-tile models:

Guess-and-Check With Tiles

Journal Entry Although some students may recognize that there are a great deal of possible arrangements of all 11 tiles that represent $x^2 + 2x - 8$, most students will have no trouble placing the x^2-tile. To verify that students understand that the placement of the remaining tiles must model the linear and constant terms of the trinomial, ask them to answer these questions in their journal.

- Why must the arrangement of the 1-tiles be rectangular in shape?
- Why must the area of the 1-tiles be 8?
- Why must additional x-tiles be added to the arrangement?

Does adding x-tiles to the model change the value of the trinomial product? Explain.

Ask several students to share their entries with the rest of the class. Allow students to comment on their own journal entries or on those that have been read aloud.

Ongoing Assessment, page 352

Try This

a. To make a rectangle, you have to add two neutral pairs of x-tiles. Then the rectangle has dimensions of $x - 4$ and $x + 2$. Check: $(x - 4)(x + 2) = x^2 - 2x - 8$

b. To make a rectangle, arrange the tiles so that the dimensions are $x - 2$ and $x - 4$. Since $(-)(-) = (+)$, the units are positive. Check: $(x - 2)(x - 4) = x^2 - 6x + 8$

Problem-Solving Strategy

Guess-and-Check Guessing and checking will be less haphazard for students if there is a strategy to their guesses. Ask students to compare the sum of the factors of the constant term with the coefficient of the linear term in examples **A–D** on page 353. Students should see that the sum of the factors chosen for the constant must be equal to the coefficient of the linear term.

continued on top of next page

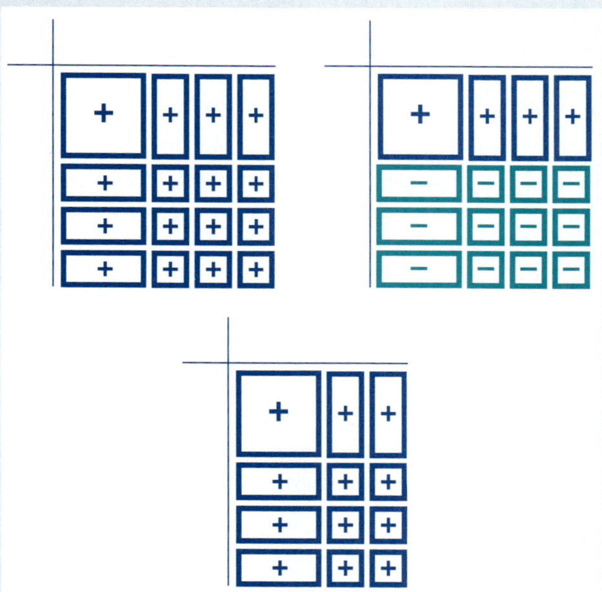

For which of the models is the arrangement of 1-tiles *not* a square?

Which model is not the difference of two squares or a perfect-square trinomial? What trinomial is represented by this model? What are its binomial factors?

Cooperative Learning

Factoring Trinomials Ask students to work with a partner. Tell them to use algebra tiles or drawings of tiles to show which tiles are needed to represent $x^2 + 2x - 8$.

Tell students to begin modeling $x^2 + 2x - 8$ as the product of two binomials by placing the x^2-tile in the matrix. Then arrange the 8 negative 1-tiles in a 1-by-8 rectangle.

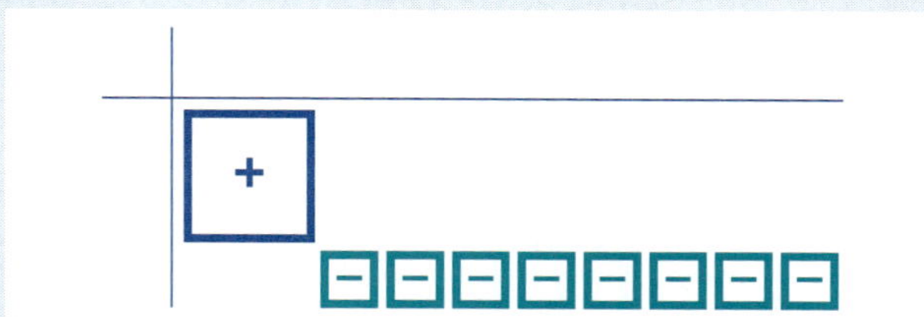

continued on bottom of next page

Approach A: Using the Book, continued

Ongoing Assessment, page 353

3. The other pairs of factors (1 and 12, and 2 and 6) will not work because their sum is not 1. Furthermore, using −4 and 3 will not work because their sum is −1.

Example

Students are asked to factor quadratic trinomials by using the guess-and-check method.

Remind students to check their answers by multiplying.

Additional Example

Factor each trinomial.

A. $x^2 + 3x + 1$ $[(x + 2)(x + 1)]$

B. $x^2 - 6x + 16$ $[(x - 8)(x + 2)]$

C. $x^2 - 10x + 21$ $[(x - 3)(x - 7)]$

D. $x^2 + 2x - 15$ $[(x - 3)(x + 5)]$

Technology

Encourage students to check their answers by multiplying. They can use graphics calculators to check their factoring by graphing both the trinomial and its binomial factors. If a trinomial has been factored correctly, the calculator will display only one graph.

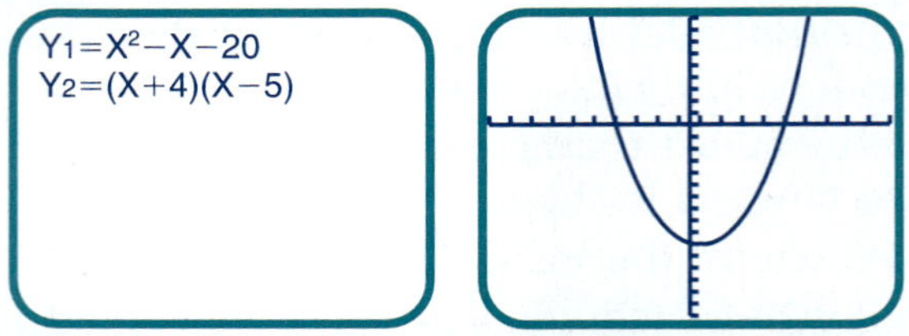

If the trinomial has been factored incorrectly, two graphs will be displayed.

TEACH

Approach B: Using Models, continued

Will this arrangement allow the two remaining x-tiles to complete a rectangle?

Have students arrange the 1-tiles in a 2-by-4 rectangle. There are two possible arrangements.

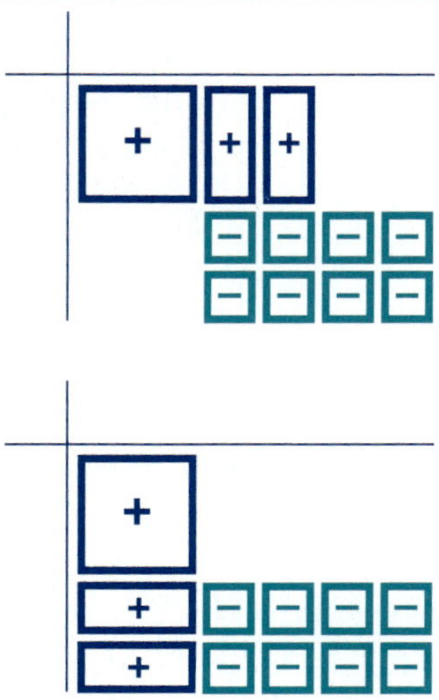

How many neutral pairs of x-tiles are needed to complete the rectangle?

What are the factor tiles for this model?

Complete this statement:
$x^2 + 2x - 8 = (x\ [\underline{-2}])(x\ [\underline{+4}])$

Inclusion Strategy

Ask students to test arrangements of algebra tiles to find factors of $x^2 - 3x - 10$ and $x^2 - 7x + 10$. Tell students to check their arrangements by multiplying. When students are sure that they have the correct factors, tell them to copy one tile model into their portfolio and explain how it models the factorization of the trinomial.

Inclusion Strategy

Guess-and-Check Hand out $8\frac{1}{2} \times 11$ pieces of paper to 16 students. Ask each of these students to draw one of the symbols listed below on their piece of paper.

Each of these is needed once: 1, 2, 3, 4, 6, 12.

Each of these is needed twice: (,), x, +, −.

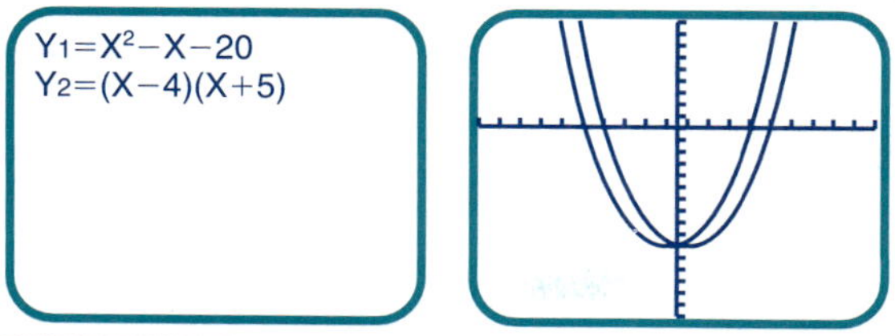

Critical Thinking, page 354

Common monomial factors should always be factored out first.

Inclusion Strategy

Each polynomial in the example on page 353 is a quadratic polynomial and can be written in the form $ax^2 + bx + c$, where a is the coefficient of the quadratic term, b is the coefficient of the linear term, and c is the constant term. The values of a, b, and c can be used to determine if a polynomial is prime. Ask students to fill in the following chart:

Polynomial	a	b	c	$b^2 - 4ac$
$x^2 - x - 20$	1	-1	-20	$1 - 4(1)(-20) = 81$
$x^2 - 10x + 16$				
$x^2 + 4x - 21$				
$x^2 + 9x + 18$				

Now have them decide if each polynomial is a perfect square and if it is factorable or prime.

continued on top of next page

Write the trinomial $x^2 + x - 12$ on the chalkboard in large characters. Tell your students that they will model the factors of this trinomial. Ask the four students holding the two pairs of parentheses to position themselves in front of the classroom. Then the two students holding the *x*s should position themselves appropriately.

What is the sign of the 12 in the trinomial? What should be the signs in the binomials? Ask two students to position themselves between the pairs of parentheses, one holding a + sign and the other holding a − sign. Then ask the two students holding the 1 and the 12 to position themselves.

Ask the remaining students if this set of factors gives the correct middle term for $x^2 + x - 12$. **[No; the middle term is incorrect.]**

Tell the students holding the 1 and the 12 to switch positions. Does this set of factors give the correct middle term? **[no]**

Have students continue choosing positions for different factor pairs of 12 until they find the correct binomial factors for $x^2 + x - 12$. **[$(x + 4)(x - 3)$]**

Repeat this activity with different students to find the factors of $x^2 - 7x + 12$. **[$(x - 4)(x - 3)$]**

A kinesthetic model for factoring trinomials

continued on bottom of next page

Approach A: Using the Book, continued

Closing the Lesson

Ask students to compare the difference between the methods they could use to factor each of the following:

1. $4x^2 + 12x + 9$
2. $16x^2 - 25$
3. $x^2 + 2x - 63$

TEACH

Approach B: Using Models, continued

Journal Entry

In preparation for Exercises 26–37, ask students to factor a common monomial factor from $3x^3 - 24x^2 + 45x$. What are the binomial factors of the resulting trinomial?

Defining *Irreducible Polynomials*

Ask students to reduce $\frac{8}{12}$ to simplest terms. What common factor did they divide by? The fraction $\frac{2}{3}$ cannot be reduced because 2 and 3 have no common factor other than 1. These numbers, 2 and 3, are said to be *relatively prime.*

The polynomial $x^2 - 6x - 8$ has no integral binomial factors because no factors of -8 have a sum of -6. We say that $x^2 - 6x - 8$ is a *prime polynomial* or that it is *irreducible* over the set of integers.

Closing the Lesson

Ask students to compare the difference between the methods they could use to factor each of the following:

1. $4x^2 + 12x + 9$
2. $16x^2 - 25$
3. $x^2 + 2x - 63$

ASSESS

Selected Answers

Odd-numbered Exercises 9–45 and odd-numbered Extra Practice

Assignment Guide

Core level students should complete Exercises 1–46.

Core-plus level students should complete Exercises 1–22 and 30–48.

Error Analysis

Students often factor trinomials incorrectly because they reverse the signs in the binomials. Tell students to be sure that they check the sign of the coefficient of the middle term.

Alternative Assessment

Factoring Trinomials Have students factor the following trinomials:

1. $x^2 - 15x + 56$

2. $x^2 + 3x - 18$

3. $x^2 + 11x + 18$

Tell your students to check each answer by substituting the zeros of the polynomial in the trinomial. The result of each substitution should be a value of zero.

Portfolio Activity Ask students to summarize the methods they have studied for factoring polynomials. Their summaries should include discussions of perfect-square trinomials, the difference of two squares, and trinomials that are not perfect-square trinomials.

RETEACH

Ask students to examine the following factors and trinomials:

$(x + 2)(x + 4) = x^2 + 6x + 8$
$(x - 2)(x - 4) = x^2 - 6x + 8$
$(x - 2)(x + 4) = x^2 + 2x - 8$
$(x + 2)(x - 4) = x^2 - 2x - 8$

Because the coefficient of x^2 is 1 in each trinomial, the linear term of each trinomial is the sum of the factors of the constant term. Consequently, factoring each of these trinomials involves finding the factors of the constant term whose sum is the coefficient of the linear term.

A table can help students find the correct factors.

Trinomial	Constant term	Factors of constant term	Sum of factors
$x^2 + 2x - 8$	−8	1, −8	−7
		−1, 8	7
		−2, 4	2 ✔
		2, −4	−2

Thus, $x^2 + 2x - 8 = (x - 2)(x + 4)$.

ENRICH

Interdisciplinary Connection

History You have seen that the zeros of a polynomial such as $x^2 - 3x + 2$ show where the graph of the equation $y = x^2 - 3x + 2$ crosses the x-axis. Since $x^2 - 3x + 2 = (x - 1)(x - 2)$, the zeros of the polynomial are 1 and 2. Thus, the graph of $y = x^2 - 3x + 2$ crosses the x-axis at $x = 1$ and $x = 2$. The value of y at $x = 1$ and $x = 2$ is 0.

The set of axes on which we graph an equation to find its zeros is called a coordinate plane. The coordinate plane is often referred to as the *Cartesian plane* in honor of the sixteenth century mathematician and logician Rene Descartes. Coordinates on the Cartesian plane, such as (1, 0) and (2, 0), are called Cartesian coordinates.

NAME ______________________ CLASS ____________ DATE ____________

Practice

7.7 Factoring Trinomials

1. Complete the model. Then write the factors and their product.

____

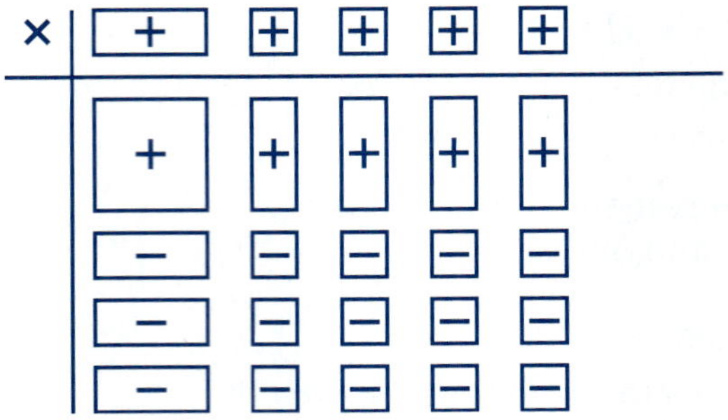

Write each trinomial in factored form.

2. $x^2 + x - 30$ __________ **3.** $m^2 + 9m + 20$ __________ **4.** $c^2 - c - 72$ __________

5. $d^2 - 7d + 12$ __________ **6.** $y^2 + y - 156$ __________ **7.** $f^2 - 2f - 48$ __________

For each polynomial, write all of the factor pairs of the third term, and then circle the pair that would successfully factor the polynomial.

8. $n^2 - 8n + 15$ ______________ **9.** $t^2 - 121$ ______________

10. $s^2 + 5s + 4$ ______________ **11.** $q^2 - 2q - 35$ ______________

Write each trinomial as a product of its factors. Use factoring patterns, graphing, or algebra tiles to assist you in your work.

12. $g^2 - 3g - 40$ ______________ **13.** $h^2 + 6h - 40$ ______________

14. $j^2 + 22j + 40$ ______________ **15.** $k^2 - 39k - 40$ ______________

16. $x^2 - x - 12$ ______________ **17.** $y^2 - 7y - 18$ ______________

18. $a^2 - 9a + 14$ ______________ **19.** $x^2 - 5x - 6$ ______________

20. $x^2 - 8x + 15$ ______________ **21.** $p^2 + 18p + 45$ ______________

NAME ______________________________ CLASS ____________ DATE ____________

Lesson Activity

7.7 How Many Arrangements?

Using the algebra tiles shown, how many different rectangular arrangements can you form to represent the trinomial $x^2 + 3x + 2$ if reflections and rotations are not included?

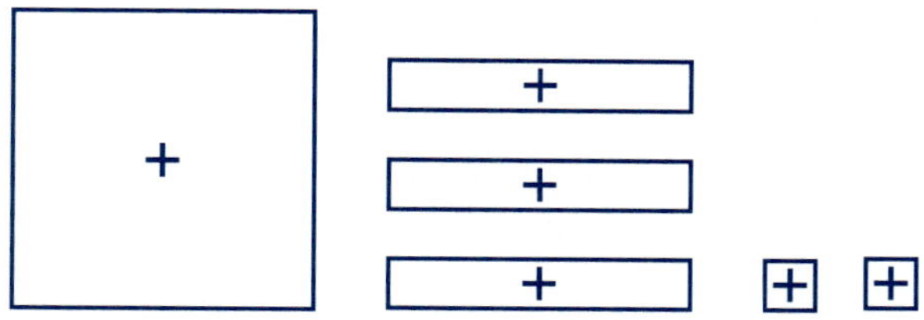

1. Keeping the x^2-tile fixed, arrange the two unit tiles and three x-tiles in different ways to form the rectangles. Draw your results on the rectangles provided. One rectangular arrangement has been drawn for you.

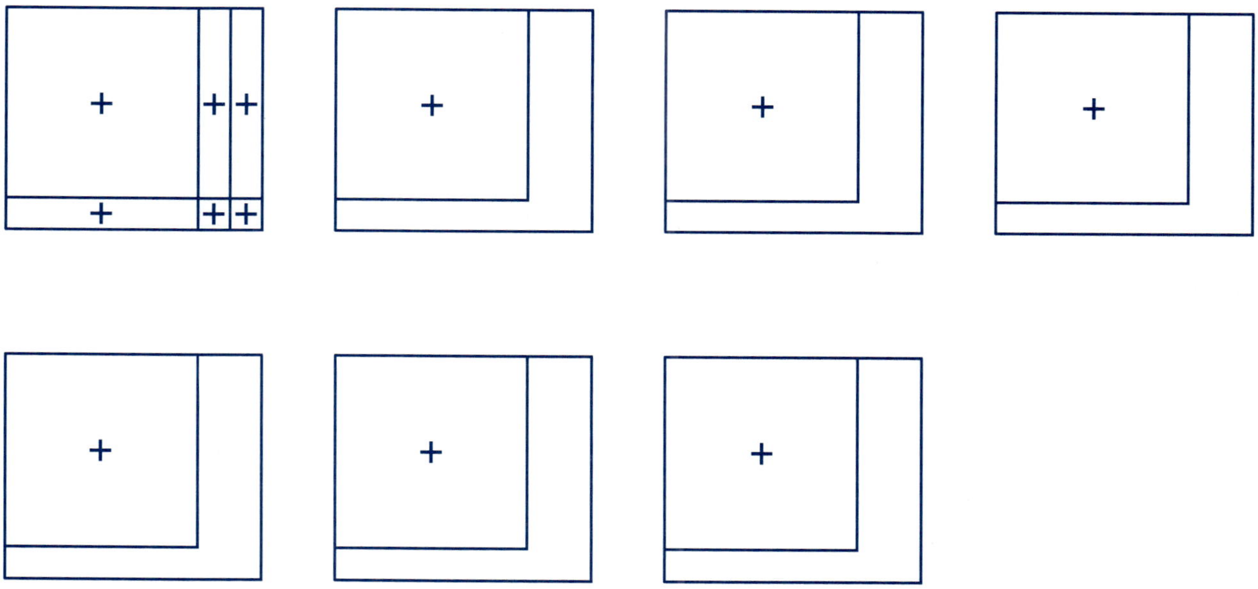

2. Next, place the x^2-tile in as many different ways as possible, and use the rest of the tiles to complete the picture. How many additional rectangular arrangements can you make to represent the trinomial $x^2 + 3x + 2$ if reflections and rotations are not included?

____

3. Find the area of each figure. In terms of factors, what do all of the arrangements have in common?

____

NAME ______________________ CLASS ____________ DATE ____________

Basic Skills Practice
Counting Outcomes and Tree Diagrams, Part 2

Read each question and circle the best answer.

The frozen yogurt shop is having a back-to-school special. You can choose from three flavors of yogurt, two kinds of hot sauces, and three kinds of dry toppings to make a sundae for the special price of $1.75. Refer to the tree diagram to answer Exercises 1–6.

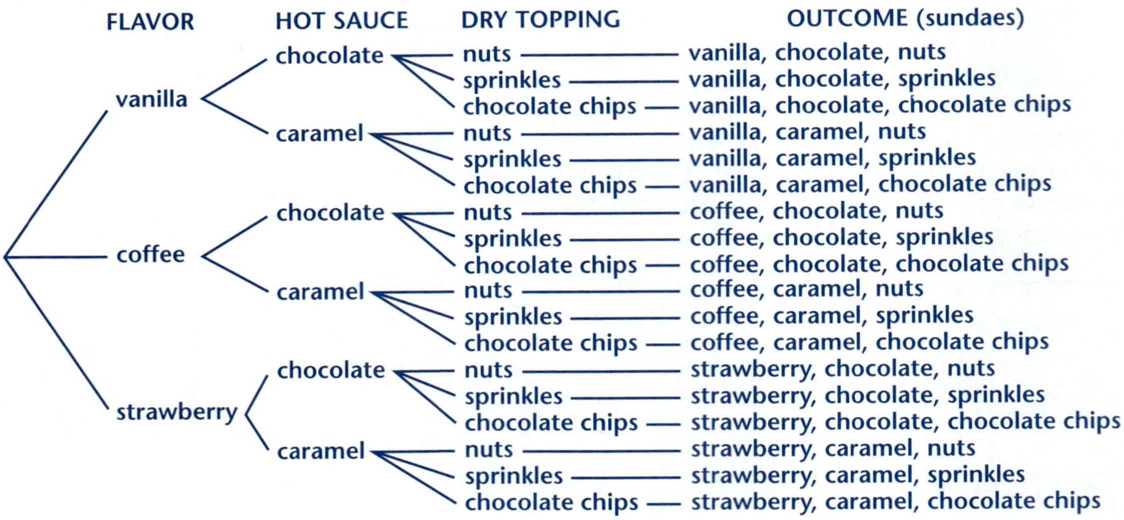

1. How many sundaes include coffee-flavored yogurt?

A 2
B 4
C 6
D 8

2. How many sundaes have sprinkles as a dry topping?

F 7
G 6
H 5
I 4

3. How many sundaes do ***not*** include nuts?

A 9
B 10
C 12
D 14

4. How many sundaes include vanilla-flavored yogurt with caramel sauce?

F 1
G 3
H 6
I 9

5. How many sundaes that include chocolate sauce can be made?

A 9
B 7
C 6
D 3

6. How many sundaes do ***not*** include vanilla-flavored yogurt?

F 10
G 12
H 15
I 18

LESSON 8.1 *Exploring* Parabolas

PREPARE

Pacing Guide
2 Days

Objectives and Overview
- Explore the parabola as the graph of a quadratic function.
- Examine graphs of quadratic functions to determine the vertex, axis of symmetry, and zeros.

Students explore the parabola as the graph of a quadratic function, as a transformation of the parent quadratic function, and as the graph of a polynomial. They also explain the relationship between the second differences in a table and the resulting graph.

Vocabulary
zero of a function

Assessing Prior Knowledge
Graph each equation.

1. $y = 2x$ **2.** $y = 2x - 3$
3. $y = 2x + 4$

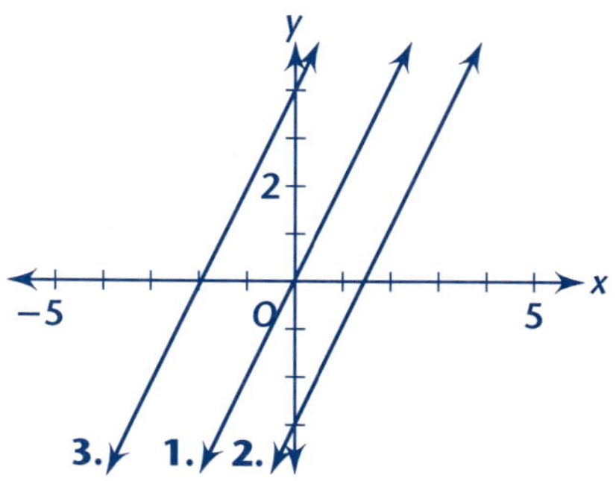

Approach A Using the Book

Opening Discussion
Ask students to recall the graphs of quadratic functions from Chapter 5. What are some characteristics of the graph of the quadratic parent function, $y = x^2$, and the graphs of transformed functions?

Brainstorming
Ask students to think of real-world phenomena in which parabolic shapes appear. Examples are satellite dishes, the stream of water coming from a water fountain, and the path of a football.

Setting Up the Explorations
Teacher-Guided Explorations If you wish, you can guide students through the textbook explorations by using an overhead graphics calculator or function-plotting software. Follow each exploration step by step. Make sure that students understand how each transformation of the parent

continued on top of next page

TEACH

Approach B Using Models

Opening Discussion
Ask students to recall how to graph a quadratic function from Chapter 5. Have them give some characteristics of the graph of the quadratic parent function, $y = x^2$, and the graphs of transformed functions.

Brainstorming
Ask students to think of real-world phenomena in which parabolic shapes appear. Examples are satellite dishes, the stream of water coming from a water fountain, the path of a football, and the flight of a rocket.

Modeling Quadratic Functions
Using a graphics calculator or an overhead projector, show students that a quadratic function of the form $y = a(x - h)^2 + k$ can also be written in the form $y = ax^2 + bx + c$. Graph both quadratics and show that they are identical if the correct values of a, b, c, h, and k are chosen. Also show that the quadratic

continued on bottom of next page

Approach A Using the Book, continued

function is done and how to identify each transformation. Explain how to locate the zeros of the function and to identify the vertex and the minimum or maximum value of the function.

Cooperative-Learning Groups Divide the class into groups of four. Ask one student to read the steps from the textbook. Have another student write the steps used in the explorations. Have a third student answer the questions and compare the graphs in the explorations. Have the fourth student write a report detailing the group's activities, including whether the conjectures were tested. Ask each student to sign the completed report.

Definition of *Constant Difference*

From a table of values for a function of the form $y = f(x)$, finding the constant differences helps determine the type of function represented in the table. First find the differences in the y-values in the table. These are known as the first differences.

If all of these differences are the same (constant), it indicates that the original function is linear. If the differences are not the same, find the differences of these first differences, which are known as the second differences. If the second differences are constant, it indicates that the original function is quadratic. Continue with this pattern until constant differences are found.

Teaching Tip

You may want to review the combined transformations of the quadratic function $y = x^2$ from Chapter 5. A combined transformation is of the form $y = -a(x - h)^2 + k$, where a is the scale factor of a stretch, h is the magnitude of a horizontal translation (shift), k is the magnitude of a vertical translation (shift), and the negative sign in front represents a reflection through the x-axis. Point out that the vertex of the transformed graph is (h, k) and that the axis of symmetry is the vertical line $x = h$.

TEACH

Approach B Using Models, continued

function in vertex form, $y = a(x - h)^2 + k$, has a vertex of (h, k), axis of symmetry of $x = h$, and minimum value of k.

Cooperative Learning

Cooperative-Learning Groups Before starting the explorations, divide the class into groups of four. Ask one student to read the steps from the textbook. Have another student draw the graphs in the explorations. Have a third student identify the zeros, the vertex, and the axis of symmetry. Have the fourth student write a report detailing the group's activities. Ask each student to sign the completed report.

Exploration 1

Have students investigate constant differences by answering the following questions, which may be listed on the chalkboard or overhead. (For ideas on guiding the students, see Exploration 1 on page 364 of the Pupil's Edition of the textbook.)

- Predict the type of function represented by a table in which the first differences are constant. **[linear]**
- Predict the type of graph represented by the table of values below.

x	0	1	2	3	4	5	6
y	−4	−3	0	5	12	21	32

[parabola]

Student Reports

One or more members of each group should report their group's findings to the class. Ask them to describe the patterns they found. The reporter should state the group's conclusions clearly, including answers to the exploration questions.

Follow-Up Demonstration (Optional)

For this exploration, you may want to give students another table of values for a quadratic function. Have students find the first and second differences in the following table and tell what kind of graph is generated by plotting the points.

Exploration 1

In this exploration, a table of values for a function is shown and students are asked to find the second differences in the y-values.

Point out that if the second differences are constant, the function represented by the values in the table is quadratic.

Ongoing Assessment, page 364

5. A parabola. If the second differences are constant, then the function is a quadratic function.

Follow-Up Demonstration (Optional)

If you need closure for this exploration, ask students to verify that the curve they drew is a parabola by graphing $y = x^2$ in the same coordinate plane. Are the graphs similar? Is one a transformation of the other?

On Their Own

The tables show some points on the graph of the function $y = f(x)$. Describe each kind of graph.

A.

x	0	1	2	3	4	5	6
y	1	0	1	4	9	16	25

B.

x	0	1	2	3	4	5	6
y	−6	−4	−2	0	2	4	6

[A. parabola; B. line]

Exploration 2

In this exploration, students identify the vertex of $y = 2(x - 3)^2 - 8$ by inspection and then find more points on the graph. They identify the graph as a parabola and compare it with the graph they generated in Exploration 1.

Point out that since the shape of the two graphs is the same, one graph is a transformation (a translation) of the other and that they are both transformations of the parent graph. Point out that the quadratic function $y = 2(x - 3)^2 - 8$ is written in this form so that the vertex can be easily identified.

continued on top of next page

x	1	2	3	4	5	6
y	3	5	11	21	35	53

[first differences: 2, 6, 10, 14, 18; second differences: 4, 4, 4, 4; parabola]

On Their Own

The tables below show some points on the graphs of two different functions, $y = f(x)$. Describe the graph of each.

A.

x	0	1	2	3	4	5	6
y	1	0	1	4	9	16	25

B.

x	0	1	2	3	4	5	6
y	−6	−4	−2	0	2	4	6

[A. parabola; B. line]

Exploration 2

Have students tell how to read the vertex from the equation of a quadratic function in the form $y = a(x - h)^2 + k$. (For ideas on guiding the students, see Exploration 2 on page 365 of the Pupil's Edition of the textbook.)

- Simplify the expression $y = 2(x - 3)^2 - 8$ (see Chapter 7) and graph the function. What is the vertical stretch of the function? **[2]**
- What is the advantage of expressing a quadratic function in the form $y = 2(x - 3)^2 - 8$? **[The vertex and minimum value are easy to read.]**
- What are some ways to find the vertex of a quadratic function that is not expressed in this form? **[Factor the function to get the vertex form, or graph the function and zoom and trace to find the maximum or minimum point.]**
- What are the zeros, axis of symmetry, and minimum value of the function? **[zeros: 1 and 5; axis of symmetry: $x = 3$; minimum value: −8]**

continued on bottom of next page

Approach A Using the Book, continued

Follow-Up Demonstration (Optional)

If you need closure for this exploration, review how to transform a function by stretching, reflecting, or shifting. Point out that (h, k) is the vertex of a quadratic function written in the form $y = a(x - h)^2 + k$. Use examples from the explorations to reinforce the patterns in these transformations.

Ongoing Assessment, top of page 365

4. A parabola; the graphs are the same.

Ongoing Assessment, middle of page 365

There are *no zeros* if the parabola does not cross the x-axis, *one zero* if it touches the x-axis at one point, and *two zeros* if it intersects the x-axis at two points.

On Their Own

Give the vertex of each quadratic function.

A. $y = -2(x + 3)^2 - 4$ **B.** $y = 2(x - 1)^2 + 6$

[A. (−3, −4); B. (1, 6)]

Definition of a Zero of a Function

A zero of a function is the x-value of any point of the function that is on the x-axis. A function can have no zeros or as many zeros as the highest power in the function. (This is proved in later mathematics courses.) Emphasize that at a zero of the function, the function has the value of zero, or $f(a) = 0$, where a is the zero.

Exploration 3

In this exploration, a quadratic function is graphed. Students are asked to find the zeros, the average of the zeros, the vertex, the axis of symmetry, and the minimum value of the graph. The graph is then compared with the graph in Exploration 2. They are identical.

TEACH

Approach B Using Models, continued

Follow-Up Demonstration (Optional)

To follow up this exploration, ask students to write an equation for the quadratic function with a vertex of (−1, 4) and a stretch of 3. **[$y = 3(x + 1)^2 + 4$]**

On Their Own

Give the vertex of each quadratic function.

A. $y = -2(x + 3)^2 - 4$ **B.** $y = 2(x - 1)^2 + 6$

[A. (−3, −4); B. (1, 6)]

Exploration 3

Have students find special quantities related to a quadratic function of the form $y = ax^2 + bx + c$. (For ideas on guiding the students, see Exploration 3 on page 365 of the Pupil's Edition of the textbook.)

- How is the vertical stretch in $y = 2x^2 - 12x + 10$ related to the vertical stretch in $y = 2(x - 3)^2 - 8$? **[Both are 2.]**
- Find $f(0)$. Is the answer a zero of the function? **[No; it is the y-intercept and is not a zero of the function unless $f(0) = 0$.]**
- Why is the average of the two zeros equal to h, the x-coordinate of the vertex? **[The parabola has symmetry, so the average of the zeros will be the x-coordinate of the vertex.]**

Follow-Up Demonstration (Optional)

To follow up this exploration, ask students to find the zeros, vertex, and axis of symmetry for $y = 3x^2 + 6x + 7$. **[There are no zeros; the vertex is (−1, 4); and the axis of symmetry is $x = -1$.]**

On Their Own

Give the vertex and zeros of each quadratic function.

A. $y = -2x^2 - 12x - 22$ **B.** $y = 2x^2 - 4x + 2$

[A. (−3, −4), no zeros; B. (1, 0), zero: $x = 1$]

Point out that the equation of the graph in Exploration 2 is given in a different form, the vertex form, compared with the equation of the graph in Exploration 3.

Ongoing Assessment, bottom of page 365

7. -8

Application

In this application, students use a parabola to model the path of a small rocket.

Use **Transparency 34** for discussion or review.

Critical Thinking, page 365

The zeros of the function are the beginning and ending times of the rocket's flight.

Closing the Lesson

Summarize the lesson with the chart below.

Quadratic function	$y = a(x - h)^2 + k$
Vertex	(h, k)
Axis of symmetry	$x = h$
Maximum (minimum) value	k

Closing the Lesson

Summarize the lesson by restating each conclusion from the explorations.

A quadratic function, $y = a(x - h)^2 + k$ (which can also be written in the form $y = ax^2 + bx + c$), has a vertex of (h, k), axis of symmetry of $x = h$, and minimum value of k.

ASSESS

Selected Answers

Odd-numbered Exercises 7–45 and odd numbered Extra Practice

Assignment Guide

Core level students should complete Exercises 1–45.

Core-plus level students should complete Exercises 1–11, 14–17, 20–26, and 30–46.

Error Analysis

Students may have difficulty finding the zeros in Exercises 20–28. Remind students of the steps used to find zeros by graphing: (1) graph the function; (2) guess the points where the graph crosses the x-axis; (3) check the guesses by substituting into the original equation; (4) if $y = 0$, then the x-value is a zero, and if $y \neq 0$, use the zero feature (this may be under the CALCULATE menu) of the calculator to find the zeros.

Alternative Assessment

Using Factoring Another way to find the zeros of a polynomial is to factor the quadratic. For example, since $x^2 + 8x - 9 = (x + 9)(x - 1)$, -9 and 1 are values that make the product of the binomials equal to zero. Guess-and-check can be used here. Ask students to try this technique in Exercises 20–28 and to check their answers by graphing.

RETEACH

Inviting Student Participation Ask one student to write a quadratic function in the form $y = a(x - h)^2 + k$ on the chalkboard. Ask the next student to give the vertex of the function. Then ask a third student to give the axis of symmetry. Have a fourth student find the zeros of the function, and have the next student rewrite the function as a polynomial of the form $f(x) = 0$. Have the next student give the roots of the polynomial. Continue with this pattern until all students have participated.

ENRICH

Finding a Model for a Quadratic Function

Use a graphics calculator or function-plotting software to show students how to find the equation of a quadratic function from the points listed in a table. First show students how to enter the table into List 1 and List 2 of a calculator (using STAT mode). Then choose QuadReg from the STAT CALC menu to get the values of a, b, and c for the quadratic equation in the form $y = ax^2 + bx + c$. Point out that they will convert this form to vertex form in Lesson 8.3.

Interdisciplinary Connection

Physics Projectile motion is described by physicists as the motion of objects thrown or launched into the air. The paths of these objects assume the shape of a parabola. The general equation for parabolic motion is $h(t) = -16t^2 + v_0t + h_0$, where h is the height in feet of the object at time t (given in seconds), v_0 is the initial velocity of the object, and h_0 is the initial height of the object.

NAME ______________________ CLASS ____________ DATE ____________

Practice

8.1 Exploring Parabolas

Compare the graph of each function with the graph of $y = x^2$. Describe the transformation on the parent function.

1. $y = 2(x + 5)^2$ ______________________

2. $y = (x - 1)^2 + 5$ ______________________

3. $y = -3(x + 1)^2$ ______________________

4. $y = (x + 6)^2$ ______________________

5. $y = -(x + 3)^2 - 2$ ______________________

6. $y = \frac{1}{2}(x - 4)^2$ ______________________

Determine the vertex and axis of symmetry for each function, and then sketch a graph from the information.

7. $y = -(x - 5)^2$

____

____

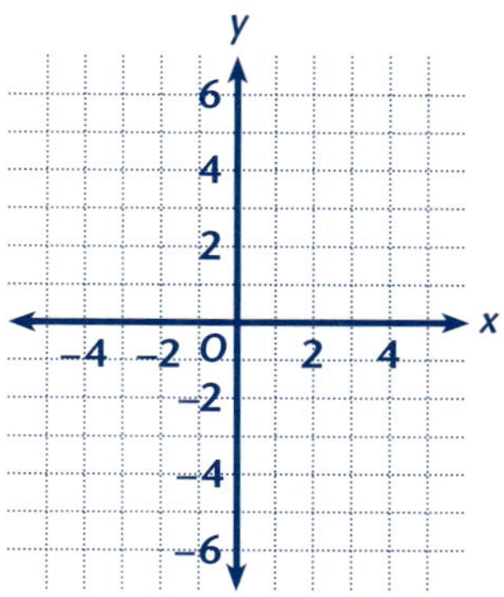

8. $y = (x - 1)^2 + 2$

____

____

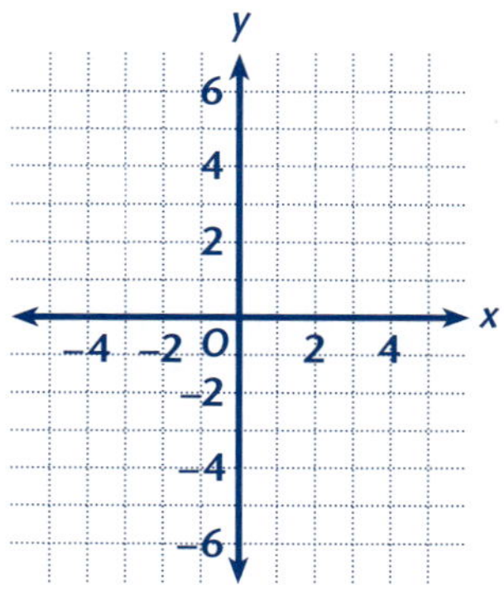

9. $y = 2(x - 3)^2$

____

____

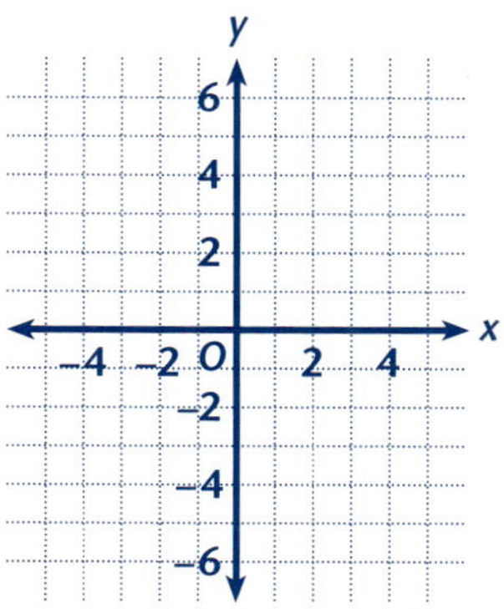

10. $y = 3(x + 2)^2 - 1$

____

____

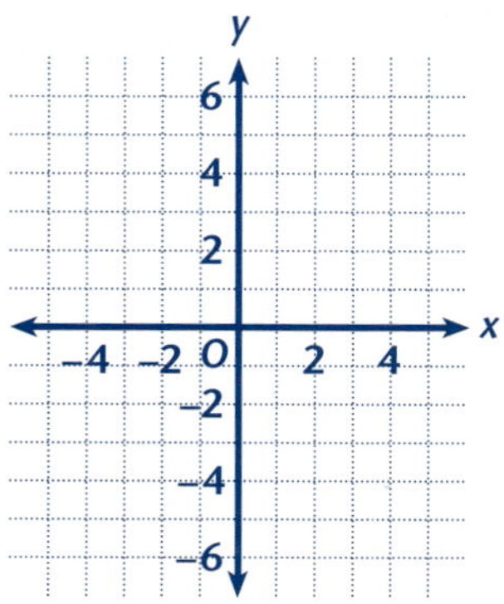

Find the zeros of each polynomial by graphing.

11. $x^2 - 2x - 15$ ______________

12. $x^2 + 2x - 15$ ______________

13. $x^2 + 8x + 15$ ______________

14. $x^2 + 5x - 14$ ______________

15. $x^2 - 11x + 30$ ______________

16. $x^2 - 3x - 4$ ______________

NAME ______________________ CLASS __________ DATE __________

Lesson Activity

8.1 Modeling a Quadratic Function

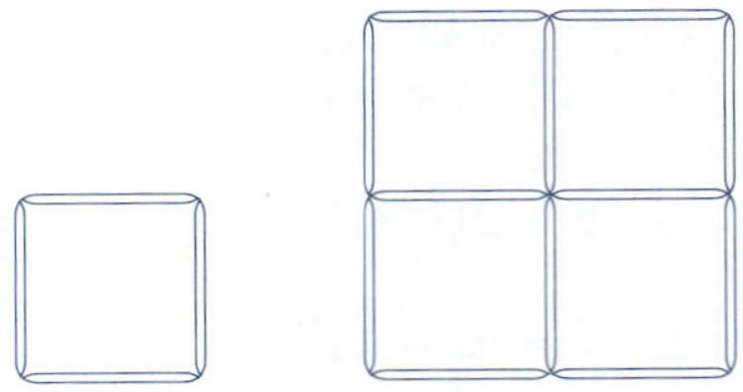

In this activity you will find a mathematical model for the number of toothpicks needed to construct a square of any size by using 1×1 squares made of toothpicks.

To begin, arrange the toothpicks to form a 1×1 square.

1. How many toothpicks are needed to form a 1×1 square?

____

Next, add toothpicks to form a 2×2 square. One way to count the number of toothpicks for a given square is to analyze the way in which the total number of toothpicks changes. Count only the new toothpicks added, and then add this number to the previous square's total.

2. How many new toothpicks were added to form a 2×2 square? __________

3. What is the total number of toothpicks for the 2×2 square? __________

4. How many new toothpicks must be added to a 2×2 square to form a 3×3 square? __________

5. What is the total number of toothpicks in a 3×3 square? __________

6. How many new toothpicks must be added to a 3×3 square to form a 4×4 square? __________

7. What is the total number of toothpicks in a 4×4 square? __________

8. Place your data in a table like the one shown. Look for a pattern. Predict the number of toothpicks needed to build a 5×5 square according to your pattern.

____

Number on a side	Total number in square
0	
1	
2	
3	
4	
5	

9. Work backward to fill in the rest of the table. Use your pattern to determine the number of new toothpicks needed to go from a 0×0 square to a 1×1 square. __________

10. Calculate the first and second differences for the values in your table. What do you notice about the second differences?

____

NAME ______________________ CLASS __________ DATE __________

Basic Skills Practice

Probability, Part 1

The probability of an event is the likelihood that the event will occur. When an event is certain to occur, it has a probability of 1. When an event is certain not to occur, it has a probability of 0. When an event is equally likely to occur or not occur, it has a probability of $\frac{1}{2}$, or 0.5.

The probability (P) that an outcome will occur is the ratio of the number of successful outcomes to the number of possible outcomes:

$$P = \frac{\text{successful outcomes}}{\text{possible outcomes}}$$

When you toss a coin, there are two possible outcomes, heads or tails.

The chances of tossing tails are $\frac{\text{successful outcomes}}{\text{possible outcomes}} = \frac{\text{tails}}{\text{heads or tails}} = \frac{1}{2}$.

Example 1: When rolling a number cube, what is the probability of getting a 3?

successful outcomes: 3

possible outcomes: 1, 2, 3, 4, 5, 6

$$P = \frac{\text{number of successful outcomes}}{\text{number of possible outcomes}} = \frac{1}{6}$$

Example 2: When rolling a number cube, what is the probability of getting an even number?

successful outcomes: 2, 4, 6

possible outcomes: 1, 2, 3, 4, 5, 6

$$P = \frac{3}{6} = \frac{1}{2}$$

Exercises 1–4 refer to one roll of a number cube.

1. What is the probability of getting an even number? __________

2. What is the probability of getting a 3 or a 5? __________

3. What is the probability of getting a multiple of 3? __________

4. What is the probability of getting 7? __________

Exercises 5–8 refer to a bag filled with 3 yellow balls, 4 green balls, 2 blue balls, and 1 orange ball.

5. What is the probability of picking a red ball out of the bag? __________

6. What is the probability of picking a blue or yellow ball? __________

7. What is the probability of picking an orange ball? __________

8. What is the probability of picking a green ball? __________

9. What is the probability of picking a green or orange ball? __________

LESSON 8.2

Solving Equations of the Form $x^2 = k$

PREPARE

Pacing Guide

2 Days

Objectives and Overview

- Use different methods to find the value of a function.
- Use square roots to solve quadratic equations.

Students learn how to find exact solutions to the quadratic equation $x^2 = k$ by algebraic methods. This includes the square-root method.

Assessing Prior Knowledge	
Find each product.	
1. $8 \cdot 8$	[64]
2. $17 \cdot 17$	[289]
Factor each polynomial.	
3. $x^2 - 4$	$[(x + 2)(x - 2)]$
4. $x^2 + 4x + 4$	$[(x + 2)(x + 2)]$
5. $x^2 + 3x + 2$	$[(x + 1)(x + 2)]$
6. $x^2 + x - 2$	$[(x + 2)(x - 1)]$

Approach A

Using the Book

Opening Discussion

Discuss how quadratic functions apply to the law of uniformly accelerated motion, whose equation is of the form $h(t) = -16t^2 + h_0$, where h is the height in feet, t is the time in seconds, and h_0 is the initial height. Ask students how to calculate the height at 2 seconds. How could you estimate how long it takes for an object to drop to the ground from the top of a tower?

Brainstorming

Have students brainstorm different ways to apply the law of uniformly accelerated motion. Have them estimate how long it takes an object to fall to the ground if its initial height is 185 ft. Then graph the function $h(t) = -16t^2 + 185$ and check the solution.

Tabular Method

A table of values for a function is useful for two reasons: (1) it gives some points of the function that

TEACH

Approach B

Using Models

Opening Discussion

Discuss how quadratic functions apply to the law of uniformly accelerated motion, which is a function of the form $h(t) = -16t^2 + h_0$, where h is the height in feet, t is the time in seconds, and h_0 is the initial height. Ask students how this law applies to an object being dropped from a tall building. How could you estimate how long it takes for an object to drop to the ground from the top of a tower?

Brainstorming

Have students brainstorm different ways to apply the law of uniformly accelerated motion. Have them estimate how long it takes an object to fall to the ground if its initial height is 185 feet. How could they tell whether it takes 10 times as long as it takes for an object to drop from a height of 18.5 feet? Graph the function $h(t) = -16t^2 + 185$ and $h(t) = -16t^2 + 18.5$ and check the solutions.

can be graphed, and (2) it may provide a visual clue about where the zeros of the function are located. If the y-values change sign, then a zero of the function exists between the corresponding x-values (known in later mathematics courses as the location principle).

Ongoing Assessment, page 368
Make another table for values of t from 3.40 to 3.50.

Graphing Method
Point out that the graphing method is useful for approximating the x-value of the point where the height is 0. Students can zoom in on the point where the graph crosses the x-axis and then trace the coordinates. The root feature in the CALCULATE menu can also be used to approximate the answer to the nearest millionth.

First the function is entered under the Y= menu and graphed. Then the root feature is chosen from the CALCULATE menu. The cursor must be placed to the left of the root (the zero) for the lower bound and to the right of the root for the upper bound. Press ENTER to see a very close approximation. Emphasize that for some applications, a negative value for a zero of a function may not be appropriate.

Finding Square Roots
Finding the exact solution to a quadratic equation of the form $0 = -16t^2 + k$ involves finding square roots. Point out that in this book, the square root of a number refers to the positive root, although there are two square roots for any given positive number. For example, $\sqrt{9} = 3$ although 3 squared and -3 squared both equal 9. If the negative root is desired, it must be specified by putting a negative sign before the square root symbol, $-\sqrt{9}$.

continued on top of next page

Using a Motion Detector
This activity enables students to see the connection between the motion of a falling body and a quadratic equation. It can be conducted as a classroom demonstration, or it can be done in groups if the required technology is available for the number of groups you have.

Place the motion detector on the floor. Have one student prepare to drop a ball (or other object) in the direction of the detector, with another student in position to catch it. At a cue from a third student, who is managing the calculator-based laboratory and the calculator, the first student should drop the ball. The calculator will record the position of the falling object at regular intervals.

NOTE: You will need to refer to the manufacturer's instructions for programming the calculator before you set up the experiment.

You can use a calculator and a motion detector to demonstrate the equation of a falling body.

continued on bottom of next page

Approach A

Using the Book, continued

Example 1

This example proves that the square root of a fraction is the quotient of the square roots of the numerator and the denominator.

This example is useful because it shows two ways to find the square root of a fraction:
(1) find the quotient of the numerator and denominator, and then take the square root;
(2) find the square root of the numerator and of the denominator, and then find the quotient.

Additional Example 1

Show that $\sqrt{\frac{25}{16}} = \frac{\sqrt{25}}{\sqrt{16}}$.

[Simplify the square roots under the radical signs:

$\sqrt{\frac{25}{16}} = \frac{5}{4}$ because $\frac{5}{4} \cdot \frac{5}{4} = \frac{25}{16}$

$\frac{\sqrt{25}}{\sqrt{16}} = \frac{5}{4}$ because $5 \cdot 5 = 25$ and $4 \cdot 4 = 16$

Since $\sqrt{\frac{25}{16}} = \frac{5}{4}$ and $\frac{\sqrt{25}}{\sqrt{16}} = \frac{5}{4}$, $\sqrt{\frac{25}{16}} = \frac{\sqrt{25}}{\sqrt{16}}$, by substitution.]

Example 2

This example shows how to solve a quadratic equation of the form $x^2 = k$ by finding the positive and negative square roots of k, or $\sqrt{k}$ and $-\sqrt{k}$.

Point out that k must be positive or 0 in order for this equation to have a solution. Explain that if k were negative, it would mean that the square of a number is negative, which is not possible for real numbers. Have students verify this by attempting to find $\sqrt{-9}$ with a calculator. They should get an error message on most calculators.

Additional Example 2

Solve the following equations:

A. $x^2 = \frac{25}{16}$ **B.** $x^2 = 1.21$ **C.** $x^2 = 150$

[A. $\frac{5}{4}$ and $-\frac{5}{4}$; B. 1.1 and -1.1;
C. approximately 12.25 and -12.25]

TEACH

Approach B

Using Models, continued

The graph will contain points (horizontal lines) that need to be removed before proceeding. Then students can use the STAT CALC QuadREG L3, L4 function (or other appropriate function, depending on the calculator used) to find a quadratic equation for the ball. A typical example might be $-15.20x^2 + 5.3$, where 15.20 is an approximation of the accepted coefficient, -16.

Graphics Calculator Method

A very close approximation of the exact length of time that it takes for an object to fall to the ground can be found by using the root function of a graphics calculator. For example, for $h(t) = -16t^2 + 80$, enter $Y_1 = -16X^2 + 80$. Then choose "zero" under the CALCULATE menu. The calculator will graph the function. Make sure that the zeros are visible on the screen. Use the arrow keys to move the cursor to the left side of the positive zero and press ENTER. Then move the cursor to the right side of the zero and press ENTER. Press ENTER again to get 2.236068, or about 2.24.

Have students use the root function on a calculator to find the approximate length of time it takes for an object to fall to the ground for each function below.

A. $h(t) = -16t^2 + 40$ **B.** $h(t) = -16t^2 + 140$

[A. 1.58; B. 2.96]

Teaching Tip

Function Notation and List Features

Students can use the list features of a graphics calculator to evaluate the uniformly accelerated motion function for a set of domain values. For problem **A** above, they should enter $Y_1 = -16X^2 + 40$ into the Y= menu. Then enter the domain values {0 1 2 3 4 5} under L_1 of the STAT EDIT screen.

Algebraic Method

The algebraic method reiterates the content of Example 2. The algebraic solutions to a quadratic equation of the form $x^2 = k$ are the positive and negative square roots of k, or $\sqrt{k}$ and $-\sqrt{k}$. Point out that a calculator solution is an approximation of the square root unless k is a perfect square.

Ongoing Assessment, page 370

The negative value is not a solution because time cannot be negative.

Critical Thinking, page 370

The zeros of $h(t)$ are the values of t that make $h(t) = 0$ true. Since $h(t) = -16t^2 + 185$, solving $0 = -16t^2 + 185$ gives the zeros of h.

Example 3

In this example, students learn how to solve quadratic equations of the form $(x - h)^2 - k = 0$, or $(x - h)^2 = k$.

Remind students how to read $\pm$ notation: ± 3 is a short way to show two solutions, $+3$ and -3. Also emphasize that you take the square root of both sides first and then add h to each side.

Additional Example 3

Solve the following equations:

A. $(s - 5)^2 - 16 = 0$ **B.** $(y + 4)^2 = 34$

[A. 5 $\pm$ 4, or 9 and 1; B. $(-4) \pm \sqrt{34}$, or approximately 1.83 and -9.83]

Coordinate Geometry

Stress the geometric connection between the solutions to an equation such as $(x - 5)^2 - 16 = 0$ and the zeros of the related quadratic function $y = (x - 5)^2 - 16$. The solutions to the equation are the zeros of the function, so the parabola crosses the x-axis at $x = 9$ and $x = 1$ (see Additional Example 3 above).

Use **Transparency 35** for discussion or review of the intercepts of parabolas.

continued on top of next page

Next type in $Y_1(L_1)$ on the home screen. The calculator evaluates the function at every value in List 1. The screen shows the following list: {40 24 −24 −104 −216 −360}. Type $Y_1(L_1)$ STORE L_2 to store the values in List 2. Look at the lists under the STAT EDIT menu. Then compare List 1 with List 2 in order to determine that a zero of the function occurs between 1 second and 2 seconds (the sign of the y-values changed).

Enter the new domain values {1.0 1.1 1.2 1.3 1.4 1.5 1.6 1.7 1.8 1.9 2.0} for List 1, and repeat the steps. Read the answer to the nearest hundredth, 1.58. Have students use the list features to find how many seconds it takes for an object to fall to the ground from a height of 200 feet.
[3.54 seconds]

Algebraic Method

The exact length of time it takes for an object to fall to the ground can be found by the algebraic method. For the model of uniformly accelerated motion, $h(t) = -16t^2 + k$, the exact solutions are $\pm\sqrt{\frac{k}{16}}$, although the negative root is not considered a solution. Ask students to find the exact length of time it takes for an object to fall to the ground from each height below.

A. 144 feet **B.** 180 feet
C. 200 feet **D.** 95 feet

[A. 3 seconds; B. $\sqrt{11.25}$ seconds; C. $\sqrt{12.5}$ seconds; D. $\sqrt{5.9375}$ seconds]

continued on bottom of next page

Approach A

Using the Book, continued

Closing the Lesson

Journal Writing Summarize the lesson by having students list the steps for solving a quadratic equation of the form $x^2 = k$ by using square roots. Ask students to include examples that have exact solutions and examples that have approximate solutions.

TEACH

Approach B

Using Models, continued

Journal Entry

Have students discuss each of the following ways to approximate the length of time that it takes for an object to fall to the ground from a tower: tabular method, graphics calculator method, and algebraic method. Ask them to include examples and solutions.

ASSESS

Selected Answers

Odd-numbered Exercises 9–65 and odd-numbered Extra Practice

Assignment Guide

Core level students should complete Exercises 1–67.

Core-plus level students should complete Exercises 1–27, 38–47, and 49–67.

Error Analysis

Students may be confused about when to give both square roots as solutions in Exercises 18–33. Point out that both the positive and negative square roots are solutions to an equation of the form $x^2 = k$ unless the equation is associated with an application that has only a positive solution.

Alternative Assessment

Portfolio Activity Have students graph the function $y = (x - 3)^2 - 8$ in the coordinate plane. Ask them to write a paragraph in which they list all of the characteristics of functions of this form that they have learned so far. Make sure that they include the vertex, axis of symmetry, name of the graph, minimum value, zeros, and how to find the solutions to the related quadratic equation $(x - 3)^2 - 8 = 0$. Because finding the solutions is the focus of this lesson, have them elaborate on the different methods of finding solutions.

RETEACH

Ask students to work with a partner to solve each equation below. The partners should discuss which method is the best to use for each. Then one student should solve the equations, and the other student should check the solutions.

A. $x^2 = 36$ **B.** $(x - 2)^2 = 81$

C. $-16t^2 + 96 = 0$ **D.** $-16t^2 + 110 = 0$

[**A.** ± 6; **B. 11 and** -7; **C.** $\pm\sqrt{6}$; **D.** $\pm\sqrt{6.875}$]

ENRICH

Report Ask students to write a report on the life of Italian astronomer Galileo Galilei. Have them include his contributions to mathematics, physics, and astronomy. How was his work related to the law of uniformly accelerated motion discussed in this lesson?

Interdisciplinary Connection

Physics The law of uniformly accelerated motion is just one of many laws that describe how objects behave when they are in motion. Others include Newton's laws of motion, laws of projectile motion, and laws of motion in a circle.

NAME ______________________ CLASS ____________ DATE ____________

Practice

8.2 Solving Equations of the Form $x^2 = k$

Find each square root. Round answers to the nearest hundredth.

1. $\sqrt{1}$ ________ **2.** $\sqrt{81}$ ________ **3.** $\sqrt{16}$ ________ **4.** $\sqrt{225}$ ________

5. $\sqrt{30}$ ________ **6.** $\sqrt{18}$ ________ **7.** $\sqrt{110}$ ________ **8.** $\sqrt{55}$ ________

Solve each equation. Round answers to the nearest hundredth.

9. $x^2 = 16$ ____________ **10.** $x^2 = 900$ ____________

11. $x^2 = 75$ ____________ **12.** $x^2 = \frac{1}{4}$ ____________

13. $x^2 = \frac{4}{49}$ ____________ **14.** $x^2 = \frac{9}{25}$ ____________

15. $(x - 2)^2 = 16$ ____________ **16.** $(x + 2)^2 = 16$ ____________

17. $x^2 - 4 = 0$ ____________ **18.** $4 = (x + 3)^2$ ____________

19. $-(x + 3)^2 + 4 = 0$ ____________ **20.** $(x - 2)^2 - 36 = 0$ ____________

21. $(x + 6)^2 = 3$ ____________ **22.** $(x - 2)^2 = 14$ ____________

Find the vertex, axis of symmetry, and zeros of each function. Sketch a graph.

23. $f(x) = (x - 1)^2 - 1$ ____________ **24.** $g(x) = (x + 2)^2 - 4$ ____________

NAME ______________________ CLASS __________ DATE __________

Lesson Activity

8.2 Finding Square Roots Geometrically

In this activity you will use base ten blocks to find positive perfect square roots.

To find the square root of 1225, arrange 9 10 cm × 10 cm blocks, two groups of 15 10 cm × 1 cm rods, and 25 1 cm × 1 cm units to form a square.

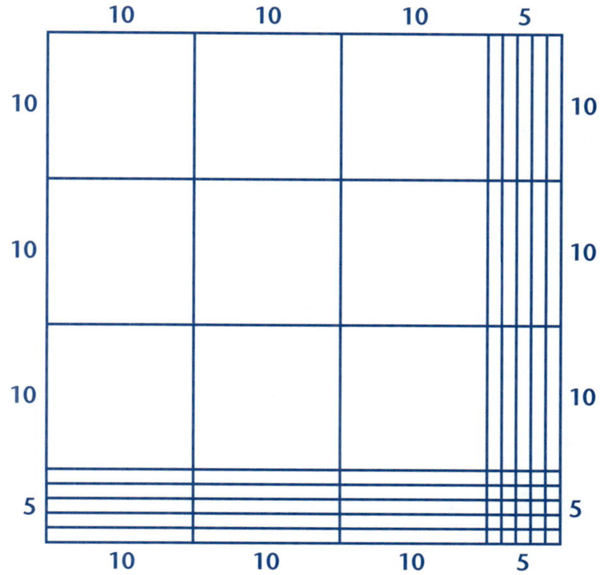

1. What is the length of the side of the resulting square?

____

2. What is the relationship between the length of the side of the square and the square root of 1225?

____

3. Why do you think this procedure can be used to find positive perfect square roots?

____

4. Find the square root of 1764 by using base ten blocks, and describe how you arranged the blocks.

____

____

Sometimes it is necessary to make a picture as a model. Suppose that you want to find the square root of 42.25. First form a 6 × 6 square. Split one rod in half lengthwise, and place the two $\frac{1}{2} \times 6$ rods on two sides of the square. Then add a $\frac{1}{2} \times \frac{1}{2}$ square to complete the large square with an area of 42.25 square units. Since there are 6.5 units on a side, the square root of 42.25 is 6.5.

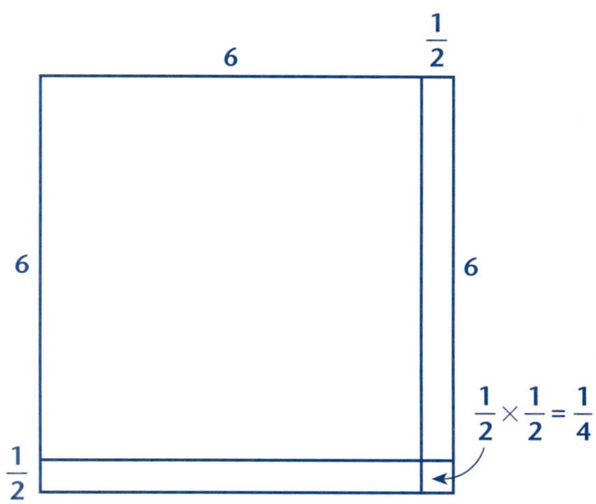

5. Find the square root of 29.16 by using a picture model.

____

NAME ______________________________ CLASS ____________ DATE ____________

Basic Skills Practice
Probability, Part 2

Read each question and circle the best answer.

Use the spinner below to answer Exercises 1–5.

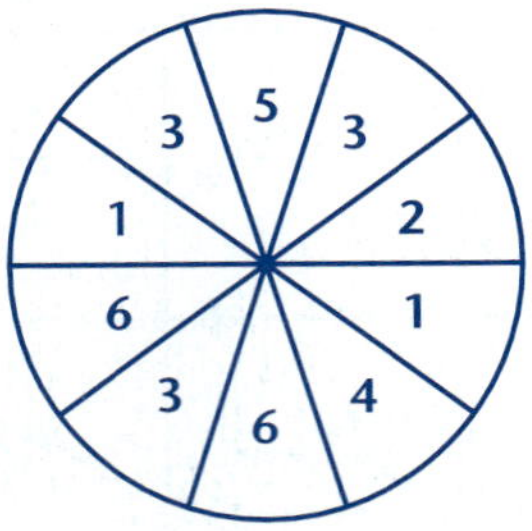

1. Find the probability of spinning an odd number.

A $\frac{1}{5}$

B $\frac{2}{5}$

C $\frac{3}{5}$

D $\frac{4}{5}$

2. Find the probability of spinning a 5 or a 6.

F $\frac{1}{10}$

G $\frac{2}{10}$

H $\frac{3}{10}$

I $\frac{7}{10}$

3. Find the probability of spinning a 1.

A $\frac{4}{5}$

B $\frac{3}{5}$

C $\frac{2}{5}$

D $\frac{1}{5}$

4. Find the odds of spinning a 3 or a 4.

F $\frac{1}{10}$

G $\frac{4}{10}$

H $\frac{2}{10}$

I $\frac{5}{10}$

5. Find the odds of spinning an even number.

A $\frac{1}{10}$

B $\frac{3}{10}$

C $\frac{4}{10}$

D $\frac{7}{10}$

6. A letter from the word GUARANTEE is picked at random. What are the odds of picking a vowel?

F $\frac{7}{9}$

G $\frac{5}{9}$

H $\frac{4}{9}$

I $\frac{2}{9}$

7. What is the probability of getting an odd number in one roll of a number cube?

A $\frac{1}{10}$

B $\frac{3}{10}$

C $\frac{1}{2}$

D $\frac{7}{10}$

LESSON 8.3 Completing the Square

PREPARE

Pacing Guide
2 Days

Objectives and Overview
- Complete the square.
- Rewrite a quadratic function in the form $y = (x - h)^2 + k$.
- Identify the vertex.
- Find the minimum value of a quadratic function from the vertex.

Students convert a parabola from general form, $y = x^2 + bx + c$, to the vertex form, $y = (x - h)^2 + k$, by completing the square. They use the vertex form to identify the vertex and the minimum or maximum value of the function.

Vocabulary
completed square
minimum

Assessing Prior Knowledge

Factor each of the following:

1.	$x^2 - 4x + 4$	$[(x - 2)^2]$
2.	$x^2 + 2x + 1$	$[(x + 1)^2]$
3.	$x^2 - 12x + 36$	$[(x - 6)^2]$
4.	$x^2 + 20x + 100$	$[(x + 10)^2]$

Approach A: Using the Book

Opening Discussion
Suppose that a baseball is hit during a game. When would it be important to find the maximum height of the baseball? How long would it take to reach that height?

Brainstorming
Have the class brainstorm situations for which it is important to find the maximum height of the path of an object. For example, in sports, when a ball has to clear a net, fence, or goal post, the maximum height is important. Ask students to name sports in which height is important.

Completing the Square
A geometric connection to completing the square can be seen with algebra tiles. Have students use algebra tiles to practice completing the square with known factors such as $(x + 3)$ and $(x + 3)$. Make sure students understand the connection between the

continued on top of next page

TEACH

Approach B: Using Models

Opening Discussion
Suppose that a baseball is hit during a game. When would it be important to find the maximum height of the baseball? How long would it take to reach that height? What is the starting height of the ball?

Brainstorming
Have the class brainstorm situations for which it is important to find the maximum height of the path of an object. For example, in sports, when a ball has to clear a net, a fence, or a goal post, the maximum height is important. How can you find the maximum value for a quadratic function?

Using Algebra Tiles
Algebra tiles help students see that a completed square of a binomial in the form $x^2 + bx$ consists of 1 square tile, or 1 x^2-term, bx-tiles, or 1 bx-term, and $\left(\frac{b}{2}\right)^2$ 1-tiles, or 1 term equal to $\left(\frac{b}{2}\right)^2$. Place half

continued on bottom of next page

Approach A — Using the Book, continued

length of the side of the completed square and the factor, as illustrated by the last diagram on page 374 of the Pupil's Edition. Emphasize that the completed square is the *square of the factor.*

Ongoing Assessment, page 374

When the nine 1-tiles are added, the figure is a square.

Exploration: How Many Ones?

In this exploration, students learn that the number of 1-tiles necessary to complete the square is the square of one-half the number of x-tiles. Have students repeat the exploration with a different number of x-tiles if they need more practice.

Ongoing Assessment, page 375

6. You will need 1369 1-tiles. Divide 74 by 2, and square the result.

Example 1

This is an example of completing a square in which half of the number of x-tiles is a fraction. Since it is not possible to take half of a tile, a picture is used and the same model is applied.

Ask students to generalize the method of completing the square from this example. **[To complete the square of $x^2 + bx$, add $\left(\frac{b}{2}\right)^2$.]**

Additional Example 1

Complete the square for $x^2 + 9x$.

[$x^2 + 9x + 20\frac{1}{4}$, or $\left(x + \frac{9}{2}\right)^2$]

Example 2

In this example, the algebraic technique for completing the square is given.

TEACH

Approach B — Using Models, continued

of the x-tiles along each side of a square tile. Then count the number of tiles needed to complete the square. If algebra tiles are not available, draw the tiles and the square pattern on a piece of grid paper. The factored form of the completed square is then $\left(x + \frac{b}{2}\right)^2$.

Working With Algebra Tiles

Have students use algebra tiles to complete the square for each binomial below. Then write the completed square in factored form.

A. $x^2 + 4x$ **B.** $x^2 + 2x$ **C.** $x^2 + 8x$

A.

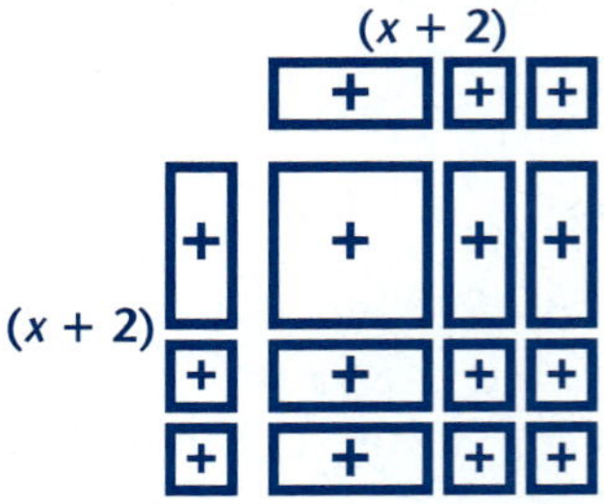

[$(x + 2)^2$]

B.

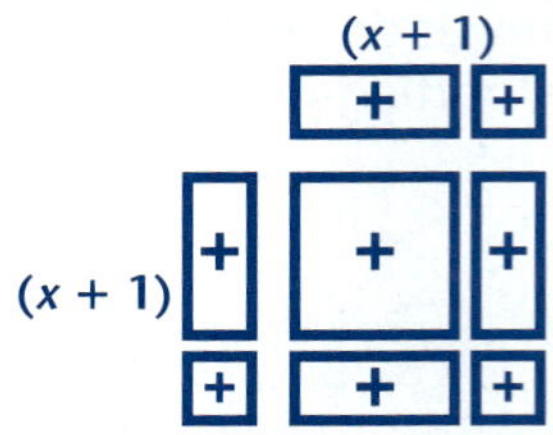

[$(x + 1)^2$]

C.

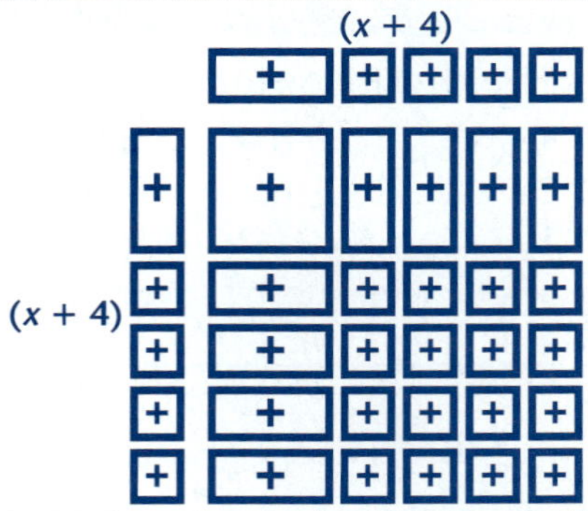

[$(x + 4)^2$]

How Many Ones?

The number of 1-tiles used to complete the square for $x^2 + bx$ is the square of one-half of

Point out that this technique requires the coefficient of the squared term to be 1. Take half of the coefficient of the x-term, square it, and add it to the other terms. Emphasize that this technique is useful because it allows a quadratic function to be written in the form $y = (x - h)^2 + k$.

Ongoing Assessment, top of page 376

Divide the coefficient of x by 2 and square the result.

Ongoing Assessment, middle of page 376

because this is the same as adding zero to one side of the equation

Additional Example 2

Complete the square for $x^2 - 16x$.
[$x^2 - 16x + 64$, or $(x - 8)^2$]

Complete the Square for a Quadratic Function

The technique of completing the square can now be applied to a quadratic function of the form $y = x^2 + bx$. Complete the square as usual to get $y = x^2 + bx + \left(\frac{b}{2}\right)^2$, but now subtract the same term from the right side of this equality to keep both sides equal. The result is $y = x^2 + bx + \left(\frac{b}{2}\right)^2 - \left(\frac{b}{2}\right)^2$. Group the terms and factor to get $y = \left[x + \left(\frac{b}{2}\right)\right]^2 - \left(\frac{b}{2}\right)^2$. The equation is now of the form $y = (x - h)^2 - \left(\frac{b}{2}\right)^2$, or $y = (x - h)^2 + k$, and the vertex and maximum or minimum value can be found easily.

Example 3

In this example, students are asked to complete the square for a quadratic function and then use the resulting vertex form to give the minimum value of the function.

Point out that completing the square gives the vertex of the parabola and that the minimum or maximum can be read from the vertex.

continued on top of next page

the coefficient of x. Identify b in each exercise above and verify that the number of 1-tiles is the square of one-half the coefficient of x.

[A. $b = 4$, $\left(\frac{b}{2}\right)^2 = \left(\frac{4}{2}\right)^2 = 4$; this is the same as the number of 1-tiles used.
B. $b = 2$; $\left(\frac{b}{2}\right)^2 = \left(\frac{2}{2}\right)^2 = 1$; this is the same as the number of 1-tiles used.
C. $b = 8$, $\left(\frac{b}{2}\right)^2 = \left(\frac{8}{2}\right)^2 = 16$; this is the same as the number of 1-tiles used.]

Paper Tiles

If algebra tiles cannot be used, as is the case for $x^2 + 3x$, then paper tiles can be used to represent algebra tiles. Students should still follow the same procedure as described above but should cut a paper tile in half so that $1\frac{1}{2}$ tiles can represent one-half of the coefficient of the x-term, or $\frac{b}{2}$. Ask students to factor each binomial below by completing the square.

A. $x^2 + 7x$ **B.** $x^2 + x$
[A. $\left(x + 3\frac{1}{2}\right)^2$; **B.** $\left(x + \frac{1}{2}\right)^2$**]**

Moving from the Concrete to the Abstract

Once students understand the process of completing the square by using algebra tiles, they should be able to understand the traditional algebraic approach more readily. You may want to use the examples in the textbook to explain the process.

As an alternative, you may want to present the process as a step-by-step procedure first, using the procedure that follows. Be sure students understand that the quadratic function must be in the form $y = x^2 + bx + c$.

1. Find the square of half of the value of b.
2. Add and subtract the square from the quadratic expression.
3. Group the terms.
4. Write the quadratic in the form $y = (x - h)^2 + k$.

Ask students to complete the square and give the vertex for each of the following:

A. $y = x^2 + 12x + 12$ **B.** $y = x^2 - 18x - 4$
[A. $y = (x + 6)^2 - 24$; $(-6, -24)$
B. $y = (x - 9)^2 - 85$; $(9, -85)$**]**

continued on bottom of next page

Approach A

Using the Book, continued

Additional Example 3

Find the minimum value of the function $f(x) = x^2 + 15x$. $\left[-\frac{15}{2}\right]$

Example 4

In this example, students complete the square for a quadratic function of the form $f(x) = x^2 + bx + c$.

Point out that the method is the same as that for the previous example, except that c is added to the $\left(\frac{b}{2}\right)^2$ term that is subtracted. The vertex and minimum are derived from the answer in the same way.

Additional Example 4

A. Rewrite $y = x^2 + 14x + 10$ in the form $y = (x - h)^2 + k$.

B. Find the vertex of the parabola.

[A. $y = (x + 7)^2 - 39$; **B.** $(-7, -39)$**]**

Ongoing Assessment, page 377

Try This

(3, 2)

Closing the Lesson

Summarize how to complete the square for a quadratic function of the form $y = x^2 + bx + c$.

1. Find the square of half of the value of b.
2. Add and subtract the square from the quadratic.
3. Group the terms.
4. Write the quadratic function in the form $y = (x - h)^2 + k$.

Ask students to complete the square and give the vertex for each of the following:

A. $y = x^2 + 12x + 12$ **B.** $y = x^2 - 18x - 4$

[A. $y = (x + 6)^2 - 24$; $(-6, -24)$;

B. $y = (x - 9)^2 - 85$; $(9, -85)$**]**

TEACH

Approach B

Using Models, continued

Have the students discuss the relationship between the result they achieve by using the steps given and by using the algebra tiles.

Cooperative Learning

Vertex Form Have students work in groups of four to convert a quadratic function of the form $y = x^2 + bx + c$ to vertex form, $y = (x - h)^2 + k$. Ask the first student to choose a function of the form $y = x^2 + bx + c$. Have the second student find the number that needs to be added and subtracted from the function to complete the square. Ask the third student to group the terms and factor the function until it is in the form $y = (x - h)^2 + k$. Ask the fourth student to identify the vertex and minimum value. Have that student record the problem and the sequence of steps used to solve it. Then have the students switch roles and repeat the exercise until each student has done all of the steps.

Journal Entry

Have students describe in their journal how to find the minimum or maximum value of a quadratic function. Have them include the steps, as well as examples to demonstrate the steps.

ASSESS

Selected Answers

Odd-numbered Exercises 9–59 and odd-numbered Extra Practice

Assignment Guide

Core level students should complete Exercises 1–60.

Core-plus level students should complete Exercises 1–17, 23–26, and 29–62.

Error Analysis

Exercises 45–52 use physics models that have a maximum value instead of a minimum value. Ask students if they can predict when a quadratic function will have a maximum value instead of a minimum value. **[when the coefficient of the squared term is negative]** Have them verify their prediction with a graphics calculator.

Alternative Assessment

Using Technology An alternative way to find the maximum or minimum value of a quadratic function is to use special features of a graphics calculator. For example, to find the minimum value of $y = x^2 - 4x + 8$, enter the function into the Y= menu as $Y = X^2 - 4X + 8$.

Next choose the minimum feature under the CALCULATE menu. The calculator will ask for the lower bound. Move the cursor to the left of the minimum value and press ENTER. Move the cursor to the right of the minimum value and press ENTER. Now press ENTER again. The calculator will give the exact minimum or a very close approximation. Follow the same steps to find a maximum value. For the function named above, an estimate of the minimum value is 4. Ask students to use this procedure to find the minimum or maximum value of the following functions:

A. $y = x^2 + 6x + 3$ **B.** $y = -x^2 - 4x - 1$

C. $y = x^2 + 3x - 2$

[**A.** −6; **B.** 3; **C.** −4.25]

RETEACH

Have students work with a partner to rewrite each function below in the form $y = (x - h)^2 + k$. Find the minimum value for each. Have them check each other's work.

A. $y = x^2 + 6x + 18$ **[$(x + 3)^2 + 9$; 9]**

B. $y = x^2 + 12x - 12$ **[$(x + 6)^2 - 48$; −48]**

C. $y = x^2 - 3x - 2$ **[$\left(x - \frac{3}{2}\right)^2 - 4\frac{1}{4}$; $-4\frac{1}{4}$]**

ENRICH

Ask students to explain how algebra tiles can be used to verify that $y = x^2 + 4$ and $y = (x + 2)^2$ are different functions. **[Sample: First use algebra tiles to show that $x^2 + 4 \neq (x + 2)^2$, as shown below.**

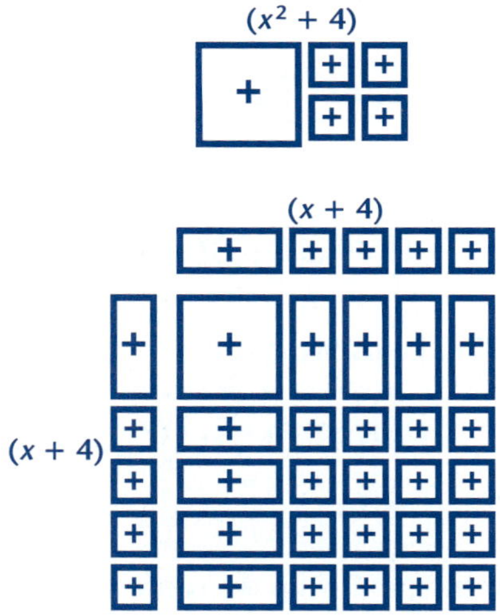

Then, since $x^2 + 4 \neq (x + 2)^2$, the related quadratic functions must be different.]

NAME ______________________ CLASS ____________ DATE ____________

Practice

8.3 Completing the Square

1. Write the product shown by the tiles in the form $x^2 + bx + c$ and in the form $(x + \underline{?})^2$.

____

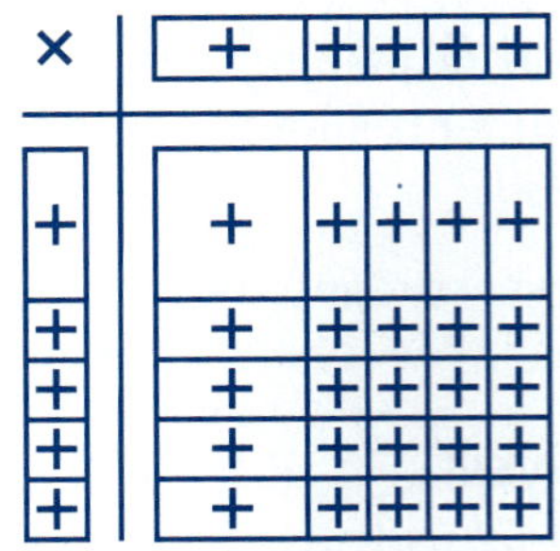

Complete the square. Write your answer in the form $x^2 + bx + c$ and in the form $(x + \underline{?})^2$.

2. $x^2 + 8x$ ______________________

3. $x^2 - 8x$ ______________________

4. $x^2 + 11x$ ______________________

5. $x^2 - 11x$ ______________________

Rewrite each function in the form $y = (x - h)^2 + k$.

6. $y = x^2 - 4x + 4 - 4$ ______________________

7. $y = x^2 + 4x + 4 - 4$ ______________________

8. $y = x^2 + 18x + 81 - 81$ ______________________

9. $y = x^2 - 18x + 81 - 81$ ______________________

Refer to the equation $y = x^2 + 6x$ for Exercises 10–12.

10. Complete the square and rewrite in the form $y = (x - h)^2 + k$. ______________________

11. Find the vertex. ______________________

12. Find the maximum or minimum value. ______________________

Find the vertex of each quadratic function.

13. $f(x) = x^2 - 3$ ______________________

14. $f(x) = x^2 - 2$ ______________________

15. $f(x) = x^2 + 6$ ______________________

16. $f(x) = x^2 + 7$ ______________________

17. $y = x^2 + 2x - 8$ ______________________

18. $y = x^2 - 4x + 3$ ______________________

NAME ______________________ CLASS ____________ DATE ____________

Lesson Activity

8.3 A Family of Parabolas

What happens to the vertex of the parabola represented by $y = x^2 + bx$ when you change the value of b?

1. Find the vertex of the parabola $y = x^2 + bx$ by completing the square to rewrite the equation in the form $y = (x - h)^2 + k$. Find the coordinates of the vertex for b-values of −4, −3, −2, −1, 0, 1, 2, and 3. Record your results in the table.

	Vertex	
b	x-coordinate	y-coordinate
−4		
−3		
−2		
−1		
0		
1		
2		
3		

2. Find the first and second differences for the y-coordinates. What do you notice about the second differences?

____

3. Graph all of the points in your table on a sheet of graph paper. Connect the points with a smooth curve. What kind of curve did you draw?

____

4. Compare your graph to the graph of $y = x^2$. Then find an equation for your graph. Test your equation by substituting into it each value of the x-coordinate of the vertex in your table. If your equation does not predict the y-coordinate of the vertex in your table, adjust the equation until it does, and repeat the calculations.

____

5. What can you say about what happens to the vertex of any parabola of the form $y = x^2 + bx$ when you change the value of b?

____

NAME ______________________ CLASS __________ DATE __________

Basic Skills Practice

Cumulative Review

Read each question and circle the best answer.

1. The Corner Diner offers a "design your own lunch" special with soup, salad, and dessert. If there are 2 soups, 3 salads, and 4 desserts, how many different lunch specials can be designed?

A 9

B 10

C 20

D 24

2. Mr. Baehr said he would give a surprise test one day next week. What is the probability of him giving the test on Monday?

F $\frac{1}{2}$

G $\frac{1}{3}$

H $\frac{1}{4}$

I $\frac{1}{5}$

3. What is 0.0037 expressed in scientific notation?

A 3.7×10^{-2}

B 3.7×10^{-3}

C 37×10^{-4}

D 37×10^{-5}

4. Given ΔFGH with vertices at $F(-2, 2)$, $G(4, 2)$, and $H(0, -2)$, what are the vertices of its image under a dilation with its center at the origin and a scale factor of $\frac{3}{2}$?

F $F(-3, 3)$, $G(6, 3)$, $H(0, -3)$

G $F(3, 3)$, $G(6, 2)$, $H(0, -3)$

H $F(-3, 3)$, $G(6, 3)$, $H(1, -3)$

I $F(-5, 3)$, $G(6, 2)$, $H(0, -3)$

5. What relationship best describes ΔABC and its translation, $\Delta A'B'C'$?

A They are similar triangles.

B They are congruent triangles.

C They are parallel triangles.

D They are rotated triangles.

6. In one roll of a number cube, what is the probability of rolling a 5 or a 3?

F $\frac{1}{2}$

G $\frac{1}{3}$

H $\frac{1}{4}$

I $\frac{1}{6}$

Solve each problem.

7. What is the missing exponent in $0.0000087 = 8.7 \times 10^{-?}$ ______________

8. Given a bag with the names of 10 male students and 12 female students, what is the probability of reaching into the bag and picking a female student's name? ______________

9. What is the probability of getting an even number on one roll of a six-sided number cube? ______________

LESSON 8.4 Solving Equations of the Form $x^2 + bx + c = 0$

PREPARE

Pacing Guide
2 Days

Objectives and Overview
- Solve quadratic equations by completing the square or by factoring.

Students learn how to solve quadratic equations by completing the square or by factoring. They may use the Zero Product Property to solve a factored quadratic function or use graphics calculator technology to find the point where a quadratic function intersects a line.

Vocabulary
Zero Product Property

Assessing Prior Knowledge

Factor each of the following:

1. $x^2 + 5x + 6$ $[(x + 3)(x + 2)]$
2. $x^2 + x - 6$ $[(x + 3)(x - 2)]$
3. $x^2 - x - 6$ $[(x - 3)(x + 2)]$
4. $x^2 - 4x + 4$ $[(x - 2)^2]$
5. $x^2 + 2x + 1$ $[(x + 1)^2]$

Approach A: Using the Book

Opening Discussion
Ask students to recall how to solve equations of the form $x^2 = k$. How might these equations be related to solving equations of the form $(x - h)^2 = k$?

Brainstorming
Have the class use a graphics calculator to graph $y = x^2 - 8x + 7$ and $y = (x - 4)^2 - 9$. How are these two graphs related? **[They are identical.]** What does this mean about the zeros of each function? How are the zeros related to the x-intercepts?

Solving Equations of the Form $x^2 + bx + c = 0$
The vertex of a quadratic function in the form $y = x^2 + bx + c$ can be found by completing the square, as was shown in Lesson 8.3. Point out that if $y = 0$, the equation $x^2 + bx + c = 0$ can then be solved by using the form $x^2 = k$, which was learned

continued on top of next page

TEACH

Approach B: Using Models

Opening Discussion
Ask students to recall how to solve equations of the form $x^2 = k$. How might these equations be related to solving equations of the form $(x - h)^2 = k$? If a y-value of a quadratic equation is known, how can you find the x-value that is paired with it?

Brainstorming
Have the class use a graphics calculator to graph $y = x^2 - 8x + 7$ and $y = (x - 4)^2 - 9$. How are these two graphs related? **[They are identical.]** What does that mean about the zeros of each function? How are the zeros related to the x-intercepts? How can you check? How would you find the points where a line intersects either of these two graphs?

Using a Graphics Calculator
A graphics calculator can provide a simple way to find the intersection of a line with a parabola. If the

continued on bottom of next page

Approach A **Using the Book**, continued

previously, where k is derived from the vertex form, $(x - h)^2 + k$. Ask students to verify this for $y = x^2 - 8x + 7$, the example given in the textbook on page 382.

Critical Thinking, page 383

The zeros of a function are the x-values of the points where the value of the function is zero. The solutions to the equation are also the points where the related function has y-values of zero.

Example 1

In this example, a quadratic equation is solved, and the zeros of a function are found.

Point out that first the expression is written in the form $(x - h)^2 + k$ by completing the square. Then $(x - h)^2 = -k$ is solved by the method learned in Lesson 8.2. The zeros of the function are also found by completing the square and solving in the same way.

Ongoing Assessment, page 383

Try This

$x = 5$ or $x = 2$

Zero Product Property

Emphasize that the Zero Product Property is used to solve quadratic equations that can be factored mentally or by trial and error. The Zero Product Property states that if the product of two factors equals 0, then one or both of the factors must be zero. That means that each factor can be set to zero and solved.

Example 2

In this example, a quadratic equation is solved by completing the square and by factoring.

Students should notice that if a quadratic equation is factorable, the factoring method may be the easiest way to solve it.

TEACH

Approach B **Using Models**, continued

calculator does not have an intersect feature, zoom and trace to find the intersection points. If the calculator does have an intersect feature, choose it from the CALCULATE menu, and scroll the cursor until it is near the intersection point before giving the calculator the lower and upper bounds. Hit ENTER to see the intersection point at the bottom of the screen. Choose the intersect feature again and scroll to the other intersection point to find its coordinates.

Working With a Graphics Calculator

Have students find the intersection of each line with the graph of the quadratic function.

A. $y = 2x - 2$, $y = x^2 + x - 4$ **B.** $y = 7x - 6$, $y = x^2 + 5x - 6$

[**A.** (2, 2), (1, −4); **B.** (0, −6), (2, 8)]

Teaching Tip

If the line is the x-axis, then the intersection of the line with the parabola gives the x-intercepts of the parabola. To use the intersect features to find the x-intercepts of a quadratic function, intersect the line $y = 0$ with the function. For example, in exercise **B** of Working With a Graphics Calculator, the points of intersection of the line $y = 0$ and the quadratic function are $(-6, 0)$ and $(1, 0)$, which means that -6 and 1 are the x-intercepts of the graph of $y = x^2 + 5x - 6$.

Solving Equations of the Form $x^2 + bx + c = 0$

Intersecting a line with a parabola can also provide an alternative way to solve equations of the form $x^2 + bx + c = 0$. Use $y = 0$ as the line and use the related quadratic function $y = x^2 + bx + c$. Then follow the procedure in the Teaching Tip above. Interpret the x-intercepts as the solutions to the equation. Ask students to solve each equation below.

A. $x^2 + 2x - 8 = 0$

B. $x^2 + 5x + 6 = 0$

C. $x^2 + 5x + 4 = 0$

[**A.** −4, 2; **B.** −3, −2; **C.** −4, −1]

Ongoing Assessment, page 384

Most students will choose the method that makes the most sense to them. Explain that either method can be used.

Additional Example 2

Solve the quadratic equation $x^2 - 2x = 15$ by the methods indicated.

A. completing the square

B. factoring

[A. $x^2 - 2x = 15$
$x^2 - 2x + 1^2 = 15 + 1$
$(x - 1)^2 = 16$
$x - 1 = 4$ or $x - 1 = -4$
$x = 5$ or $x = -3$

B. $x^2 - 2x = 15$
$x^2 - 2x - 15 = 0$
$(x - 5)(x + 3) = 0$
$x - 5 = 0$ or $x + 3 = 0$
$x = 5$ or $x = -3$

The solutions are 5 and −3.]

Example 3

In this example, a graphics calculator is used to show students how to find the *x*-value of a quadratic function when the *y*-value is −5. Both $f(x) = x^2 + 3x - 15$ and $f(x) = -5$ are graphed, and then their points of intersection are found.

Point out that this method could also be used to find the *x*-intercepts. Graph $f(x) = x^2 + 3x - 15$ and $f(x) = 0$. Although students cannot see the graph of $f(x) = 0$ because it is the *x*-axis (already shown), it allows students to use the intersect feature of the calculator to find the intercepts.

Use **Transparency 36** for discussion or review.

Ongoing Assessment, page 385

Try This

$x = 2$ or $x = 6$

Example 4

In this example, squares are cut from the corners of a square piece of cardboard, which is then folded into an open box. The volume of the box is used to find the optimum size for the piece of cardboard. The size of the cardboard is calculated by solving a quadratic equation.

continued on top of next page

Hands-on experience with the box problem can be a valuable aid to understanding.

Cooperative Learning

Have partners experiment with cutting out squares from the corners of a larger square piece of cardboard. Have them measure the original square and the dimensions of the box that can be formed after the corner squares are removed. Ask them to generalize their procedure by giving the related quadratic equation for the size of the piece of cardboard that should be used for a given volume. Use the intersect features of a graphics calculator to find the solutions to the quadratic equation.

Journal Entry

Have students describe how to solve a quadratic equation of the form $x^2 + bx + c = 0$ by each of the following methods:

1. Complete the square to get the form $(x - h)^2 + k$, and solve by using the $x^2 = k$ method.
2. Find the zeros of the related quadratic function, $y = x^2 + bx + c$.
3. Factor and use the Zero Product Property.

continued on bottom of next page

Approach A

Using the Book, continued

Teaching Tip

If students don't understand how to derive the volume line, $8 = (x - 4)(x - 4)(2)$, in the solution, you may want to use a paper model. Demonstrate that if a side of the square is x inches long and you remove 2 inches from each side, the length of each side is then $x - 2(2)$, or $x - 4$. Also show students that after the cardboard is folded into a box, the height is 2.

Example 5

In this example, a line intersects a parabola, and the points of intersection are found.

Point out that if a graphics calculator is used, the points of intersection can be found by using the zoom and trace features or by using the intersect features.

Additional Example 5

Solve the system. $\begin{cases} x^2 - 7x + 10 = y \\ y = x - 2 \end{cases}$

First substitute $x - 2$ (from the second equation) for y in the first equation and simplify.

$x^2 - 7x + 10 = x - 2$
$x^2 - 8x + 12 = 0$

Then solve the new quadratic equation by factoring.

$(x - 6)(x - 2) = 0$
$x - 6 = 0$ or $x - 2 = 0$
$x = 6$ or $x = 2$

Substitute each value of x into the equation $y = x - 2$.

$y = x - 2$	$y = x - 2$
$y = 6 - 2$	$y = 2 - 2$
$y = 4$	$y = 0$

The graphs intersect at (6, 4) and (2, 0).

Ongoing Assessment, page 386
Try This
(3, −2) and (−1, 6)

TEACH

Approach B

Using Models, continued

4. Find the points of intersection of the related quadratic function with the line $y = 0$.

Closing the Lesson

Use the equation $x^2 - 5x + 6 = 0$ to review the contents of the lesson:

1. Solve the quadratic equation. **[solutions: 3, 2]**
2. Find the x-intercepts of the related quadratic function, $y = x^2 - 5x + 6$. **[3, 2]**
3. Find the zeros of the related quadratic function. **[3, 2]**
4. Find the points where the related quadratic function intersects the line $y = 0$. **[(3, 0) and (2, 0)]**

ASSESS

Selected Answers

Odd-numbered Exercises 9–63 and odd-numbered Extra Practice

Assignment Guide

Core level students should complete Exercises 1–63.

Core-plus level students should complete Exercises 1–17 and 24–66.

Error Analysis

Students may have difficulty setting up Exercise 46. Remind students that consecutive even integers are two units apart, so if x is the first integer, then $x + 2$ is the next integer. If their product is 224, then an equation that can be solved is $x(x + 2) = 224$. This can be solved by guess-and-check or by getting 0 on one side and solving the quadratic equation that results.

Alternative Assessment

Solving a Quadratic Function by Factoring To help students understand how factoring can help solve a quadratic function, have them reverse the solution steps and create their own quadratic functions. For example, ask them to choose any two factors, such as $(x - 4)$ and $(x + 2)$. Then multiply the factors to get a trinomial. Set the trinomial equal to 0 to get a quadratic equation. Students should understand the connection after doing several of these exercises.

RETEACH

Have students work in groups of four to solve the quadratic equations below by the following methods:

1. Complete the square to get the form $(x - h)^2 + k$, and solve by using the $x^2 = k$ method.
2. Find the zeros of the related quadratic function, $y = x^2 + bx + c$.
3. Factor and use the Zero Product Property.
4. Find the points of intersection of the graph of the quadratic function and the line $y = 0$.

A. $y = x^2 + 2x - 3$ **B.** $y = x^2 - 4x + 3$

C. $y = x^2 + 6x - 7$ **D.** $y = x^2 + 7x + 6$

[A. −3, 1; B. 3, 1; C. −7, 1; D. −6, −1]

ENRICH

An open box can also be created by cutting squares from each corner of a rectangular piece of cardboard and folding up the sides. Ask students to suppose that the rectangle has dimensions of x and $2x$ and that a 1-inch square is cut from each corner of the rectangle. If the resulting volume is 24 cubic inches, have students find the dimensions of the original rectangular piece. **[10 inches by 5 inches]**

NAME ______________________ CLASS __________ DATE __________

Practice

8.4 Solving Equations of the Form $x^2 + bx + c = 0$

Find the *x*-intercepts, or zeros, of each function. Graph to check.

1. $y = x^2 - 13x + 42$ __________
2. $y = x^2 + 13x + 42$ __________
3. $y = x^2 - x - 42$ __________
4. $y = x^2 + 7x + 12$ __________
5. $y = x^2 - x - 12$ __________
6. $y = x^2 + x - 12$ __________

Solve each equation by factoring.

7. $x^2 - 4x - 60 = 0$ __________
8. $x^2 - 10x + 21 = 0$ __________
9. $x^2 + 5x + 4 = 0$ __________
10. $x^2 + 2x - 24 = 0$ __________
11. $x^2 - x - 6 = 0$ __________
12. $x^2 - 5x + 4 = 0$ __________
13. $x^2 + 11x + 30 = 0$ __________
14. $x^2 + 2x - 8 = 0$ __________
15. $x^2 + 6x + 9 = 0$ __________
16. $x^2 - 2x + 1 = 0$ __________
17. $x^2 + x - 30 = 0$ __________
18. $x^2 + 5x - 14 = 0$ __________

Solve each equation by completing the square.

19. $x^2 + 5x + 6 = 0$ __________
20. $x^2 + 3x + 2 = 0$ __________
21. $x^2 + x - 12 = 0$ __________
22. $x^2 + 7x + 10 = 0$ __________
23. $x^2 - 2x = 3$ __________
24. $x^2 + 4x = 5$ __________
25. $x^2 + 6 = 7x$ __________
26. $x^2 + 24 = 11x$ __________

Find the point or points where the graphs intersect. Graph to check.

27. $\begin{cases} y = 9 \\ y = x^2 \end{cases}$ __________
28. $\begin{cases} y = -x - 5 \\ y = x^2 + 3x - 10 \end{cases}$ __________
29. $\begin{cases} y = 7x - 35 \\ y = x^2 - 3x - 10 \end{cases}$ __________
30. $\begin{cases} y = x - 2 \\ y = x^2 - 7x + 10 \end{cases}$ __________

NAME ______________________________ CLASS ________ DATE ________

Lesson Activity

8.4 Why Factors Solve Quadratic Equations

You know that solutions to the equation $-x^2 - 6x - 5 = 0$ can be found by factoring and by using the Zero Product Property. Why are the zeros of the function $f(x) = -x^2 - 6x - 5$ and the solutions to the given equation the same?

1. What are the factors of $-x^2 - 6x - 5 = 0$? ______________________________

2. Find the x-intercepts of the function $f(x) = -x^2 - 6x - 5$.

____

3. Write the expression $-x^2 - 6x - 5$ in the form $(x - h)^2 + k$ by completing the square.

____

4. On the grid provided, graph the parabola in the form $(x - h)^2 + k$ and the lines that represent the factors you found in Exercise 1.

5. Find the x-intercept of each linear function.

____

6. Compare the x-intercepts of the parabola with the x-intercept of each linear function. Describe the relationship.

____

____

7. How are the x-intercepts related to the zeros of a function?

____

8. Why are the zeros of the function and the solutions to the corresponding quadratic equation the same?

____

NAME ______________________ CLASS ____________ DATE ____________

Basic Skills Practice
Cumulative Review

Read each question and circle the best answer.

1. Write a mathematical sentence for "a number tripled and decreased by 17."

A $3x - 17$
B $17 - 3x$
C $3x + 17$
D $17 + 3x$

2. The soccer coach is 5 years older than twice the age of the youngest team member. Write the coach's age as an algebraic expression.

F $5 - 2x$
G $2x + 5$
H $5x$
I $5x + 2$

3. The area of a square field is 400 square feet. What is the length of each side?

A 100 ft
B 40 ft
C 20 ft
D 10 ft

4. Which number sentence matches the number line?

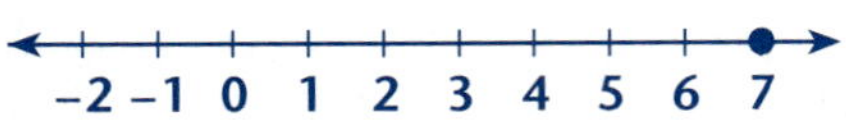

F $x \le 7$
G $x \ge -2$
H $x = 7$
I $x < -2$

5. Find the volume of a cylinder with a radius of 5 feet and a height of 8 feet. Use 3.14 for π.

A 628 ft^3
B 528 ft^3
C 125.6 ft^3
D 31.4 ft^3

6. A storage bin is 693 square feet. It is 9 feet long and 11 feet wide. How tall is the bin?

F 5 ft
G 7 ft
H 9 ft
I 11 ft

7. How many outfits can be made with 2 color choices for pants, 3 color choices for shirts, and 3 color choices for vests?

A 8
B 9
C 12
D 18

8. The Corner Paint Shop sells 4 brands of house paint. Each brand has 20 colors to choose from. How many different choices of paint does the paint shop sell?

F 16
G 24
H 40
I 80

Solve each problem.

9. A rectangular barn has a width of 10 feet, a length of 12 feet, and a height of 9 feet. Find the volume of the barn. ____________

10. What is 15.39 rounded to the nearest tenth? ____________

LESSON 8.5 The Quadratic Formula

PREPARE

Pacing Guide
2 Days

Objectives and Overview
- Use the quadratic formula to find solutions to quadratic equations.
- Use the quadratic formula to find the zeros of a quadratic function.
- Evaluate the discriminant to determine the number of real roots and to determine if the quadratic equation can be factored over the integers.

Students use the quadratic formula to find the solutions to a quadratic equation and to find the zeros of a quadratic function. They use the discriminant of the quadratic function to explore the number of solutions to the quadratic equation and to determine whether the equation can be factored.

Vocabulary
discriminant

Assessing Prior Knowledge

Solve by factoring. If not factorable, write NF.

1. $x^2 - 5x - 6$	$[(x - 6)(x + 1)]$
2. $x^2 + 5x - 6$	$[(x + 6)(x - 1)]$
3. $x^2 + 5x + 6$	$[(x + 3)(x + 2)]$
4. $x^2 - 5x + 16$	[NF]

Approach A Using the Book

Opening Discussion
Ask students for some examples of formulas they know. Ask them if they remember any patterns when they completed the square in Lesson 8.3. Could there be a formula for solving quadratic equations?

Brainstorming
Ask students to brainstorm examples of quadratic equations that cannot be factored. Shouldn't these quadratic equations also have solutions? What is a way to find these solutions?

Deriving the Quadratic Formula
The quadratic formula is derived by completing the square for a quadratic equation of the form $ax^2 + bx + c = 0$, where a, b, and c are constants. Point out that the $x^2 = k$ method of solving a quadratic equation is used in the derivation of the formula.

continued on top of next page

TEACH

Approach B Using Models

Opening Discussion
Ask students for some examples of formulas that they know. What is the advantage of using a formula? Ask them if they remember any patterns when they completed the square in Lesson 8.3. Could there be a formula for solving quadratic equations?

Brainstorming
Ask students to brainstorm examples of quadratic equations that cannot be factored. Shouldn't these equations also have solutions? What is a way to find these solutions? Is the graph of a quadratic equation that is not factorable still a parabola? Does it touch the x-axis?

Using the Quadratic Formula
A formula is really an algebraic model that can be applied when certain conditions in a problem are true. The quadratic formula can be applied whenever a quadratic equation of the form

continued on bottom of next page

Approach A

Using the Book, continued

Emphasize that $\frac{-b \pm \sqrt{b^2 - 4ac}}{2a}$ is really a shortened notation for two solutions, $x = \frac{-b + \sqrt{b^2 - 4ac}}{2a}$ and $x = \frac{-b - \sqrt{b^2 - 4ac}}{2a}$.

Use **Transparency 37** for discussion or review.

Example 1

In this example, solutions to a quadratic equation are found by ordering the terms so that the equation is in standard form and then applying the quadratic formula directly.

Point out that it is easy to use the wrong values for a, b, and c when the quadratic equation is not in standard form, so this is something to watch for.

Additional Example 1

Use the quadratic formula to solve $x^2 + 7 - 8x = 0$. [1, 7]

Ongoing Assessment, page 392

Try This

$x = -4$ or $x = 3$

Critical Thinking, page 392

$-\frac{1}{2}$, 4; answers may vary, but most students will choose the quadratic equation; no, sometimes a quadratic equation cannot be factored easily.

Example 2

In this example, the quadratic formula is used to find the zeros of a quadratic function. It is important to point out the connection between the zeros of a quadratic function and the solutions to the related quadratic equation.

Point out that the quadratic equation in this example does not have integer solutions, so it cannot be factored.

TEACH

Approach B

Using Models, continued

$ax^2 + bx + c = 0$ is to be solved. That makes the formula $x = \frac{-b \pm \sqrt{b^2 - 4ac}}{2a}$ a model to find the x-intercepts of the related quadratic function, $f(x) = ax^2 + bx + c$, to find the zeros of the related quadratic function, and to find how $ax^2 + bx + c$ can be factored. From the formula we can also get a model for other important aspects of the quadratic function: (1) the axis of symmetry is $x = -\frac{b}{2a}$; (2) the vertex is $\left[-\frac{b}{2a}, f\left(-\frac{b}{2a}\right)\right]$; and (3) the sum of the roots is $-\frac{b}{a}$. These models have useful applications as well.

Working With the Quadratic Formula

A. Solve $2x^2 + 2x - 12 = 0$ by using the quadratic formula. **[−3, 2]**

B. Find the zeros of $f(x) = 3x^2 - 9x - 12$. **[−1, 4]**

C. Find the x-intercepts of $f(x) = -x^2 + 9x + 8$. **[−1, −8]**

D. Find the vertex of $f(x) = 2x^2 + 5x - 12$. **[(1.5, −4)]**

E. Find the sum of the roots in **C.** **[−9]**

F. Find the axis of symmetry for the quadratic function in **B**. **[$x = \frac{3}{2}$]**

G. Factor $x^2 - 2x - 3$. **[$(x + 1)(x - 3)$]**

Teaching Tip

Factoring To find the factors of the quadratic expression in **B** at the left, write it in the form $(x - F1)(x - F2)$, where F1 and F2 are the zeros, in this case $(x + 1)(x - 4)$. Multiply the two binomials together to get $x^2 - 3x - 4$. This is not the original expression, but notice that if every term is multiplied by 3, then it is the original quadratic expression. Therefore, the factorization is $3(x + 1)(x - 4)$, or $(3x + 3)(x - 4)$.

Cooperative Learning

Help students understand the nature of a quadratic function by asking them to work in groups of seven. Give each group the following quadratic functions and tasks, and have each member of the group do a different task

Additional Example 2

Find the zeros of $g(x) = 4x^2 - 2x - 3$. **[about −0.65 and about 1.15]**

Ongoing Assessment, page 392

Write the equation as $3n^2 - 2n - 2 = 0$, and use the quadratic formula with $a = 3$, $b = -2$, and $c = -2$. The approximate solutions are −0.55 and 1.21.

Exploration: The Discriminant

This exploration helps students make the connection between the value of the discriminant and the number of x-intercepts of the quadratic function being graphed.

Point out to students that if they know the number of solutions ahead of time, they can predict when the quadratic formula will yield no real solutions, one solution, or two solutions. If there is one solution, then the quadratic equation is factorable, and factoring can be used.

Ongoing Assessment, page 393

2. If the value of the discriminant is greater than zero, there are two solutions. If the value of the discriminant is zero, there is one solution (or two solutions that are the same). If the value of the discriminant is negative, there are no real solutions because there are no zeros of the equation—the graph does not cross the x-axis.

Use **Transparency 38** for discussion or review.

Example 3

In this example, students find the value of the discriminant in parts **A** and **B**. In part **A,** the value of the discriminant is negative. Point out that a negative discriminant value indicates that there are no real solutions. The graph is still a parabola, but it does not intersect the x-axis. In part **B,** since the value of the discriminant is 0, $\sqrt{b^2 - 4ac}$ also equals zero, so there is one

continued on top of next page

independently and then compare answers. Then have students switch roles and work on the next quadratic function.

Quadratic Functions	Solutions
1. $f(x) = x^2 - 5x - 6$	[−1, 6]
2. $f(x) = x^2 + 3x - 4$	[1, −4]
3. $f(x) = 2x^2 - 2x - 24$	[−3, 4]
4. $f(x) = 3x^2 - 15x + 18$	[2, 3]
5. $f(x) = 4x^2 - 4$	[1, −1]
6. $f(x) = -x^2 + 5x - 6$	[3, 2]
7. $f(x) = -2x^2 - 2x + 4$	[−2, 1]

Tasks:

A. Solve the quadratic function by using the quadratic formula.
B. Find the zeros of the function.
C. Find the x-intercepts of the function.
D. Find the vertex of the function.
E. Find the sum of the zeros for the function.
F. Find the axis of symmetry for the function.
G. Factor the quadratic function.

Journal Entry

Have students use examples to discuss how to use the discriminant to determine the number of real solutions to a quadratic equation. Ask them to use a different quadratic function as an example for each possibility.

continued on bottom of next page

Approach A

Using the Book, continued

solution, $-\frac{b}{2a}$. In part **C**, the equation is not of degree 2, so it is not a quadratic equation. The quadratic formula does not apply.

Additional Example 3

What does the discriminant tell you about the solution to each of the following equations?

A. $2x^2 - 4x + 1 = 0$

B. $2x^4 + 3x = 3$

C. $4x^2 + 36 = 24x$

[A. Find $b^2 - 4ac$, where $a = 2$, $b = -4$, and $c = 1$. The discriminant is $(-4)^2 - 4 \cdot 2 \cdot 1 = 8$. When the value of the discriminant is positive, there are two x-intercepts, so $2x^2 - 4x + 1 = 0$ has two solutions.
B. The equation is not quadratic. There is no discriminant.
C. Rewrite the equation in standard form to find a, b, and c. If $4x^2 - 24x + 36 = 0$, then $a = 4$, $b = -24$, and $c = 36$. The value of the discriminant is $(24)^2 - 4 \cdot 4 \cdot 36 = 576 - 576 = 0$. When the value of the discriminant is 0, there is exactly one real solution, so $4x^2 + 36 = 24x$ has one real solution.]

Example 4

The purpose of this example is to show how the discriminant can indicate whether the quadratic equation is factorable. If the value of the discriminant is a perfect square, then $\sqrt{b^2 - 4ac}$ can be simplified further to give two solutions. These can lead to the factors of the quadratic.

Additional Example 4

Factor $18x^2 - 27x - 5$. **[$(6x + 1)(3x - 5)$]**

TEACH

Approach B

Using Models, continued

Closing the Lesson

Inviting Student Participation

Summarize the lesson by having a student write a quadratic equation on the chalkboard. Then have a second student write the equation in standard form and list the values of a, b, and c. Ask another student to find the value of the discriminant and to predict the number of real solutions. If there are solutions, ask another student to write the quadratic formula and to substitute the values of a, b, and c. Ask the next student to solve the formula. Have the sixth student put the answers back into the original equation to check the solutions. Repeat this pattern until all students have participated.

ASSESS

Selected Answers

Odd-numbered Exercises 7–59 and odd-numbered Extra Practice

Assignment Guide

Core level students should complete Exercises 1–59.

Core-plus level students should complete Exercises 1–17, 22–32, 37–42, and 46–60.

Error Analysis

Remind students that for Exercise 46 the maximum value of a quadratic function is the y-coordinate of the vertex and that the number of cases of napkins that will give the maximum value is the x-coordinate of the vertex, $\frac{-b}{2a}$.

Alternative Assessment

Using Technology Ask students to use the program below or one similar to it to solve a quadratic equation by using the quadratic formula.

```
Program:QUAD
:Input "A=",A
:Input "B=",B
:Input "C=",C
:B^2 − 4AC → D
:If D>0
:Then
:Output(5,4,(−B+√D)/(2A))
:Output(7,4,(−B−√D)/(2A))
:Else: If D=0
:Then
:Output(5,1,−B/(2A))
:Else
:Disp "NO SOLUTION"
:End
```

Portfolio Activity Ask students to solve the quadratic equation $3x^2 + 4x - 4 = 0$ in three ways—by factoring, by using the quadratic formula, and by completing the square. Describe the steps followed for each method. **[$\frac{2}{3}$, −2]**

RETEACH

Find the value of the discriminant of each equation, and give the number of solutions. Then use the quadratic formula to solve each equation that has real solutions.

1. $x^2 + 5x - 6 = 0$ **[49; 2 solutions; 1, −6]**

2. $2x^2 + 2x - 24 = 0$ **[196; 2 solutions; 3, −4]**

3. $x^2 + 10x + 25 = 0$ **[0; 1 solution; −5]**

4. $4x^2 + 4x + 4 = 0$ **[no real solutions]**

ENRICH

The vertex of a quadratic function in the form $f(x) = ax^2 + bx + c$ is $\left(\frac{-b}{2a}, \frac{4ac - b^2}{4a}\right)$. Find the vertex for each quadratic function in the Reteach exercises above.

[1. $\left(\frac{-5}{2}, \frac{-49}{4}\right)$; **2.** $\left(\frac{-1}{2}, -24\frac{1}{2}\right)$; **3.** $(-5, 0)$; **4.** $\left(\frac{-1}{2}, 3\right)$**]**

NAME ______________________ CLASS ____________ DATE ____________

Practice

8.5 The Quadratic Formula

Identify *a*, *b*, and *c* for each quadratic equation.

1. $x^2 + 3x + 2 = 0$ ______________

2. $x^2 - 2x + 1 = 0$ ______________

3. $x^2 + 7x - 3 = 0$ ______________

4. $x^2 - 5x + 3 = 0$ ______________

5. $3x^2 - 4 = 0$ ______________

6. $2x^2 + 15x = 0$ ______________

7. $10x^2 + 1 + 6x = 0$ ______________

8. $7x + x^2 - 2 = 0$ ______________

9. $x^2 = 36$ ______________

10. $x^2 + 36 = 12x$ ______________

11. $x^2 - 6x = 0$ ______________

12. $3x^2 = 5x$ ______________

Find the value of the discriminant and determine the number of solutions for each equation.

13. $4x^2 + 3x + 1 = 0$ ______________

14. $x^2 + 5x + 1 = 0$ ______________

15. $x^2 + 2x - 6 = 0$ ______________

16. $2x^2 + 2x + 4 = 0$ ______________

17. $x^2 - 5x + 1 = 0$ ______________

18. $7 - x^2 + 6x = 0$ ______________

19. $x^2 - 4x + 4 = 0$ ______________

20. $3 - 4x + 2x^2 = 0$ ______________

Use the quadratic formula to solve each equation. Give answers to the nearest hundredth. Check by substitution.

21. $3x^2 - 4x - 2 = 0$ ______________

22. $2x^2 - 5x - 4 = 0$ ______________

23. $3x^2 - 8 = 0$ ______________

24. $6x^2 - 5x + 1 = 0$ ______________

25. $x^2 - 8x + 8 = 0$ ______________

26. $x^2 - 5 = 0$ ______________

Choose any method to solve each quadratic equation. Give answers to the nearest hundredth.

27. $x^2 - x - 2 = 0$ ______________

28. $x^2 - 7x + 12 = 0$ ______________

29. $x^2 + 3x + 2 = 0$ ______________

30. $x^2 - 25 = 0$ ______________

31. $3x^2 + 2x - 5 = 0$ ______________

32. $x^2 + 3x + 1 = 0$ ______________

33. $2x^2 - 10x + 8 = 0$ ______________

34. $x^2 + 5x + 4 = 0$ ______________

NAME ______________________ CLASS ____________ DATE ____________

Lesson Activity

8.5 Another Family of Parabolas

What happens to the vertex of the parabola represented by $y = ax^2 + bx + c$ when the value of a is changed and the values of b and c remain fixed?

1. Find the vertex of the parabola represented by $y = ax^2 + 4x - 5$ by completing the square and rewriting the equation in the form $y = (x - h)^2 + k$. Find the coordinates of the vertex for a-values of -4, -2, -1, 2, 3, and 4. Record your results in the table.

	Vertex	
a	x-coordinate	y-coordinate
-4		
-2		
-1		
2		
3		
4		

2. Plot all of the points in your table on a sheet of graph paper. Connect the points. What did you draw?

____

3. Write an equation that represents your graph. Test your equation by substituting into it each value of the x-coordinate of the vertex in your table. If your equation does not predict the y-coordinate of the vertex, adjust your equation until it does.

____

4. What can you say about what happens to the vertex of any parabola represented by $y = ax^2 + 4x - 5$ when a varies and b and c remain fixed?

____

5. Repeat the activity for the parabola represented by $y = ax^2 - 6x + 9$.

____

6. Describe what happens to the vertex of a parabola represented by $f(x) = ax^2 + bx + c$ when a varies and b and c are kept constant.

____

NAME ______________________________ CLASS ______________ DATE ______________

Basic Skills Practice

Cumulative Review

Read each question and circle the best answer.

1. Find the equation that matches the number line.

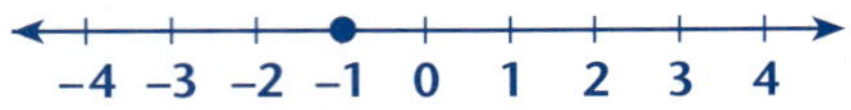

A $x + 5 = 5$

B $x - 2 = -3$

C $2x + 2 = 8$

D $16 - 8 = 3x$

2. Find the inequality that matches the number line.

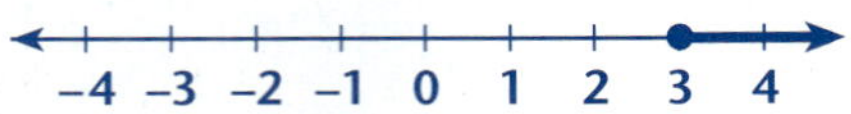

F $6n + 4 \leq 10$

G $3n \geq 9$

H $n \leq -6$

I $n > 0$

3. What is the volume of a cylinder with a radius of 5 inches and a height of 8 inches?

A 150π in.^{3}

B 200π in.^{3}

C 500π in.^{3}

D 1500π in.^{3}

4. Find the graph that matches the inequality.

$$2x + 3 < 5$$

F
G
H
I

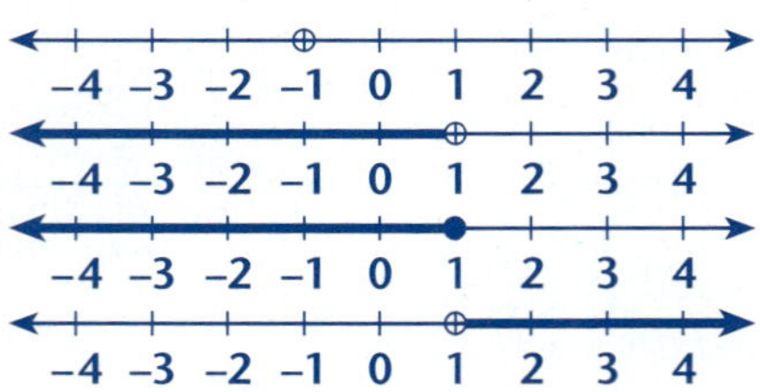

5. Triangle *ABC* is similar to triangle *DEF*. What is the value of x?

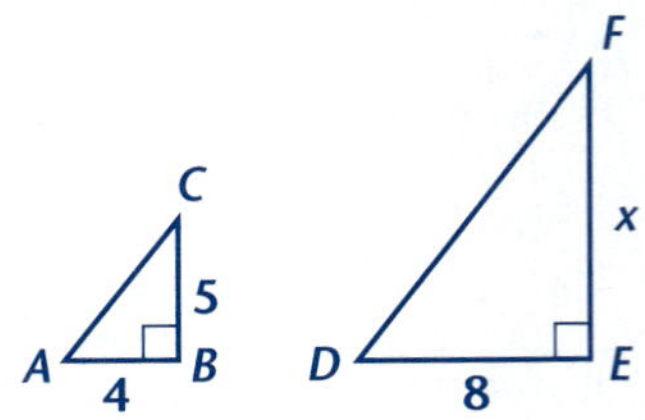

A 12

B 10

C 9

D 6

6. Joy's Ice Cream Parlor has 13 flavors of ice cream, 3 hot sauces, and 4 dry toppings. Using 1 flavor of ice cream, 1 hot sauce, and 1 dry topping, how many different sundaes can you make?

F 20

G 25

H 120

I 156

7. Susan is pregnant with twins. What is the probability that she will have one boy and one girl?

A 0

B $\frac{1}{3}$

C $\frac{1}{2}$

D $\frac{2}{3}$

LESSON 8.6

Graphing Quadratic Inequalities

PREPARE

Pacing Guide
2 Days

Objectives and Overview
- Solve and graph quadratic inequalities, and test solution regions.

Students learn how to graph quadratic inequalities of the form $ax^2 + bx + c > 0$ and of the form $y < ax^2 + bx + c$. They also test solution regions.

Vocabulary
set of points outside the parabola

Assessing Prior Knowledge

Solve each equation.

1. $x^2 - 2x - 15 = 0$
   [$x = -3$ **and** $x = 5$]
2. $3x^2 - 4x + 4 = 0$ [**no real roots**]

Approach A: Using the Book

Opening Discussion
What are some words that are used with inequalities? [**less than, greater than**] If a quadratic equation applies to a situation, when might a quadratic inequality apply? An example is a quadratic inequality that applies to heights less than a certain maximum height.

Brainstorming
Ask students to list some points from the graph in exercise **1** of Assessing Prior Knowledge at the left. What are the coordinates of some points above the graph? What are the coordinates of some points below the graph? What sentence could be written to express the coordinates of the points that are above the graph?

continued on top of next page

TEACH

Approach B: Using Models

Opening Discussion
Use the phrase *less than* or *greater than* in a sentence. When might less than or greater than be used with a quadratic expression? This is called a quadratic inequality. For example, suppose that a quadratic expression models how high a baseball is thrown. A quadratic inequality might apply to all the heights less than the maximum height.

Brainstorming
Ask students to list some points from the graph in exercise **1** of Assessing Prior Knowledge at the left. What are the coordinates of some points above the graph? What are the coordinates of some points below the graph? Write a sentence describing these points.

Using Relational Operators
Relational operators on a graphics calculator are the operators used to graph inequality statements. On some calculators, they are located in the TEST

continued on bottom of next page

Approach A — Using the Book, continued

Exploration: Production and Profit

A quadratic inequality is used as a model for production and profit. Point out that an inequality is created when the value of the quadratic expression is compared with 0 by asking whether it is greater than or less than 0.

Ongoing Assessment, page 397

4. 5000 watches per week

5. $4000 per week

Example 1

In this example, a quadratic inequality is solved by two methods. In one method, a graphics calculator is used to graph a quadratic function, and then the trace features of the calculator are used to see where values of the function are greater than 0. In the other method, the Zero Product Property is used to plot the solutions on a number line.

Emphasize the importance of testing points in each interval of the number line for the second method. Remind students that if the endpoints of an interval are not solutions, then they should be graphed with an open circle.

Additional Example 1

Solve $x^2 + 3x - 4 < 0$. [$x > -4$ **and** $x < 1$]

Calculator method:

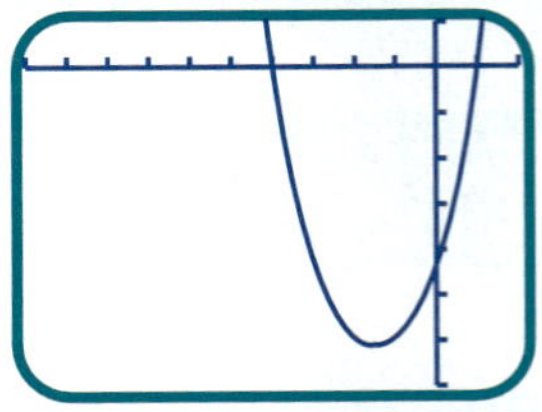

Number line method:

TEACH

Approach B — Using Models, continued

menu. The operators are =, ≠, >, ≥, <, and ≤. For example, suppose that you are asked to solve $x^2 - 2x - 3 > 0$. Enter $Y1 = x^2 - 2x - 3 > 0$ under the Y= menu. Choose Dot MODE. Press GRAPH. What you see is the graph of the solution on a number line, using the line $y = 1$ as the number line.

Explain that the way the operators work is that if an x-value, say a, is a solution, then y takes on a value of 1, and the calculator graphs $(a, 1)$. If a is not a solution, then y takes on a value of 0, and the calculator graphs $(a, 0)$. The result simulates the number line solution described in the Pupil's Edition on page 398. The resulting graph can be traced like any other graph to determine the endpoints.

Working With Relational Operators

Have students solve each inequality with a number line graph on a calculator.

A. $2x^2 - x - 3 < 0$ **B.** $3x^2 - 4x - 4 > 0$

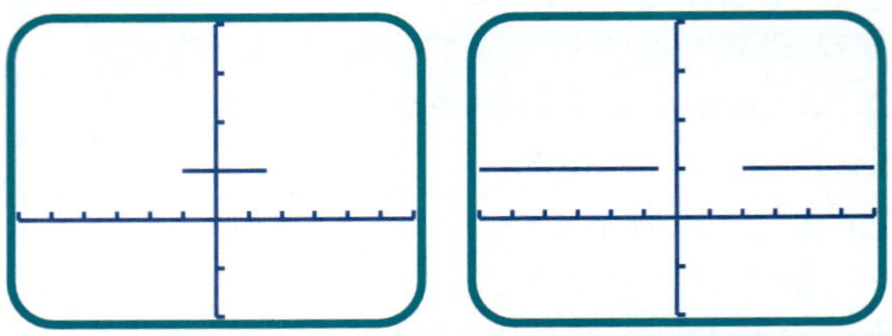

[**A.** $x > -1$ ***and*** $x < 1\frac{1}{2}$ **B.** $x < -\frac{2}{3}$ ***or*** $x > 2$]

Teaching Tip

Point out that the endpoints of the number line graph may be difficult to determine by zooming and tracing unless the endpoints are integers. Stress the importance of testing the endpoints in the original inequality to see if they are solutions. If they are not solutions, then show the endpoints as open on the graph. If the endpoints are solutions, the graph should have solid dots as its endpoints. To find the exact coordinates of the endpoints, enter the related function (not the inequality) in the Y= menu of the graphics calculator, and use the root key.

Ongoing Assessment, page 398

Try This

$x \leq -4$ or $x \geq 3$

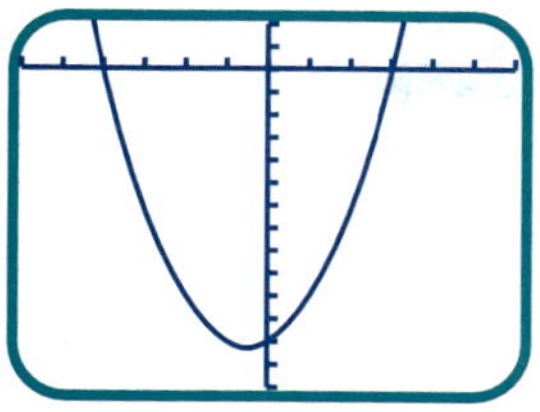

Check students' graphs.

Critical Thinking, page 398

Test the interval between the two boundary points. If the number satisfies the inequality, shade that area; if not, shade outside the boundary points.

Example 2

This example emphasizes that a parabola divides the plane into three regions: the points of the parabola, the points outside the parabola, and the points inside the parabola.

Stress that points in each of the three regions should be tested to determine if they are in the solution region.

continued on top of next page

Cooperative Learning

Have students work in pairs to solve the following quadratic inequalities. Ask one student to solve an inequality by graphing, and ask the other to solve it by using the *x*-axis as a number line. Then have students switch roles.

1. $x^2 - 5x - 6 > 0$ [$x < -1$ ***or*** $x > 6$]

2. $x^2 + 3x - 4 < 0$ [$x > -4$ ***and*** $x < 1$]

3. $2x^2 - 2x - 24 > 0$ [$x < -3$ ***or*** $x > 4$]

4. $3x^2 - 15x + 18 < 0$ [$x > 2$ ***and*** $x < 3$]

Students can work in pairs to study quadratic inequalities.

continued on bottom of next page

Approach A — Using the Book, continued

Additional Example 2

Graph $y \leq 2x^2 - 3x + 1$.

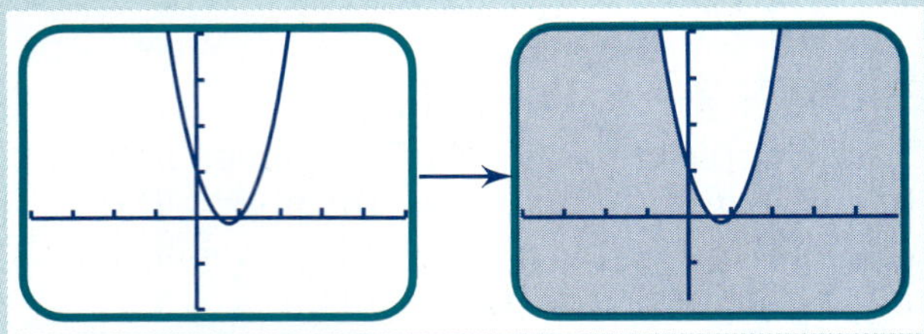

Example 3

In this example, a parabola is used to model a light reflector, and the interior of the reflector is shaded. Two inequalities are given to correspond to the shading, and both are tested to see which one applies.

Point out that it takes only one solution point to conclude that the entire region containing the solution point is the solution region.

Additional Example 3

The cross-section of a satellite dish is described by the equation $y = 0.05x^2 + 28$. Radio waves bounce off the sides of the dish in such a way that they are concentrated at the focus of the parabola. Determine which of the inequalities indicates the region where the focus is located.

A. $y > 0.05x^2 + 28$ **B.** $y < 0.05x^2 + 28$

[A. $y > 0.05x^2 + 28$
Test (0, 10).
$10 > 0.05(0)^2 + 28$
$10 > 28$ **No**

B. $y < 0.05x^2 + 28$
Test (0, 10).
$10 < 0.05(0)^2 + 28$
$10 < 28$ **Yes**

The inequality $y < 0.05x^2 + 28$ **contains the focus of the parabola.]**

TEACH

Approach B — Using Models, continued

Journal Entry

Have students graph any quadratic equation and discuss how to shade the graph according to the type of inequality that is needed. For example, have them discuss the graph of $y < -2x^2 + 3$, the graph of $y = -2x^2 + 3$, and the graph of $y > -2x^2 + 3$.

ASSESS

Selected Answers

Odd-numbered Exercises 9–43 and odd-numbered Extra Practice

Assignment Guide

Core level students should complete Exercise 1–43.

Core-plus level students should complete Exercises 1–16, 21–31, and 36–44.

Error Analysis

For Exercises 21–32, remind students to pay attention to whether the boundary parabola is part of the solution region. If the inequality includes $\leq$ or $\geq$, then the boundary parabola is part of the solution and should be solid. If the inequality includes $<$ or $>$, then the boundary parabola is not part of the solution and should be dashed.

Alternative Assessment

Cooperative Learning Ask students to work in groups of four to create their own group portfolio of quadratic inequality graphs. Have them list an inequality of their choice and describe the steps necessary to solve the inequality if it is greater than 0 and if it is less than 0. Then graph y-values greater than the inequality and y-values less than the inequality. Have each student contribute an inequality.

RETEACH

The graphs in the next column represent the solution regions for the following quadratic equation and inequalities. Match each graph with the corresponding equation or inequality.

A. $y \geq 3x^2 - 2x - 3$

B. $y = 3x^2 - 2x - 3$

C. $y \leq 3x^2 - 2x - 3$

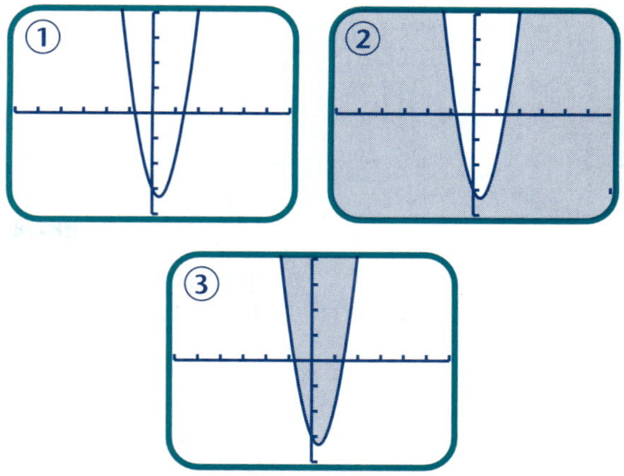

[A. 3; B. 1; C. 2]

ENRICH

The following is another exploration that students can do to help them understand quadratic inequalities:

The stopping distance of a car is the number of feet it takes for the car to stop after the brakes have been applied. A quadratic model for stopping distance is $D(s) = s + \frac{s^2}{20}$, where s is the speed of the car in miles per hour and D is the stopping distance.

1. At what speed is the stopping distance 0? **[0]**
2. For what values of s is the value of this function greater than 0? **[for all values of s greater than 0, or for any speed (because speed is always positive)]**
3. What is the minimum value of the function? Does this minimum value make sense in this problem? What minimum value does make sense? **[-5; no, it doesn't make sense because distance is always positive; 0 is the minimum value that makes sense.]**
4. At what speed is the stopping distance 100 feet? **[about 35.8 mph]**

NAME ______________________ CLASS ____________ DATE ____________

Practice

8.6 Graphing Quadratic Inequalities

Solve each quadratic inequality by using the Zero Product Property. Graph the solution on a number line.

1. $x^2 - 3x - 40 \geq 0$ ____________

2. $x^2 + 9x + 14 < 0$ ____________

3. $x^2 + 2x + 1 > 0$ ____________

4. $x^2 - x - 30 < 0$ ____________

5. $x^2 - x - 12 > 0$ ____________

6. $x^2 + 3x + 2 > 0$ ____________

Graph each quadratic inequality on the grid provided. Shade the solution region.

7. $y > x^2 - 4$

8. $y \geq -x^2 + 4$

9. $y < x^2 + 4x$

10. $y \leq x^2 + 4x + 4$

NAME ______________________ CLASS __________ DATE __________

Lesson Activity

8.6 Modeling Quadratic Inequalities

When a telephone company designs a transmission line, a few of the many different components that must be considered are the sag of the wire conductor, the tension in the line, and the span for the length of the line. The quadratic function $s = \frac{1.8678x^2}{70720}$ describes the relationship between the sag (s) at the center of the span and the span length (x) in adverse weather conditions for a certain type of wire, when sag and span length are measured in feet.

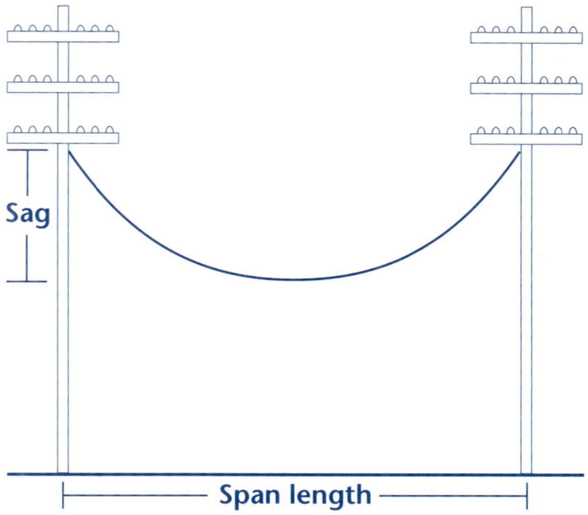

1. Complete the table of values. Round the sag to the nearest tenth.

Span length	Sag
0	
82.0	
125.2	
198.5	
246.0	
302.1	
328.0	
384.1	

2. If the span is 246.0 feet, what is the sag? ______________________

3. For what span lengths is the sag greater than 1 foot? ______________________

4. For what span length is the sag 0 feet? ______________________

5. Suppose safety standards require that an 18-foot distance be maintained between the wire conductor and the ground and that 10% of the pole's height plus 2 feet must be underground. What is the minimum pole height that could be used for a span of 198.5 feet?

____

NAME ________________ CLASS __________ DATE __________

Basic Skills Practice
Cumulative Review

Read each question and circle the best answer.

1. What happens to the area of a circle when the radius is doubled in size?

A The area remains the same.
B The area doubles.
C The area triples.
D The area quadruples.

2. What happens to the volume of a cylinder when the height is halved?

F The volume remains the same.
G The volume is one-half of what it was before.
H The volume is double what it was before.
I The volume is one-third of what it was before.

3. To solve the equation $5a + 2 = 17$, you need to—

A add 2 to both sides and then divide both sides by 3
B subtract 2 from both sides and then divide both sides by 5
C divide both sides by 5
D divide both sides by 17

4. According to the scale on a map, 1 inch represents 5 miles. The distance between the museum and the airport is 3.5 inches on the map. What is the actual distance in miles?

F 15 mi
G 17.5 mi
H 19.5 mi
I 21 mi

5. The ratio of the size of an object in a drawing to the size of the actual object is called a(n)—

A scale
B reduction
C enlargement
D rotation

6. Amalia is carpeting two bedrooms with the same type of carpet. One room is 8 feet by 10 feet, and the other is 9 feet by 12.5 feet. How many square feet of carpeting does she need?

F 190.5 ft^2
G 192.5 ft^2
H 205 ft^2
I 215 ft^2

7. Given similar triangles *ABC* and *DEF*, what is true about $\angle A$ and $\angle D$?

A They are different.
B They are complementary.
C They are congruent.
D They are similar.

8. The figures below represent—

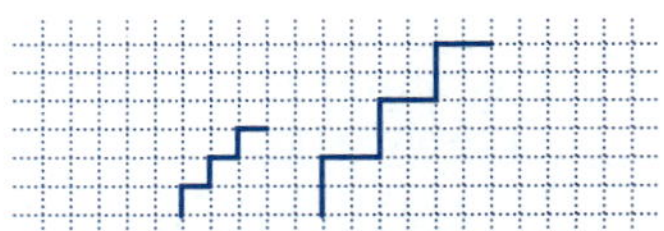

F a rotation
G a translation
H a reflection
I a dilation

Solve.

9. What is the probability of getting a 7 on one roll of a six-sided number cube? __________

LESSON 9.1

Exploring Square-Root Functions

PREPARE

Pacing Guide
3 Days

Objectives and Overview
- Find or estimate square roots.
- Explore transformations of the square-root function by using graphs.
- Identify positive and negative square roots.

The positive square root of a perfect square is introduced with an area problem. Students use guess-and-check to estimate positive square roots. Irrational and rational numbers are defined and placed in the context of the real number system. The square-root function is introduced, and its transformations are explored.

Vocabulary

perfect square	real number
rational number	dense set
irrational number	principal square root
radical sign	

Assessing Prior Knowledge

Find each product. Round answers to the nearest hundredth.

1.	5^2	[25.00]
2.	3.1^2	[9.61]
3.	8.88^2	[78.85]
4.	10.01^2	[100.20]

Approach A: Using the Book

Opening Discussion
Use the tile example to recall the concept of square root, which was first introduced in Lesson 8.2. Students should understand that the length of the side is the square root of the area of the square. Point out that 6 is the positive square root of 36. What other number is a square root of 36? [**−6**]

Brainstorming
Have students give areas of other squares that could be made with tiles. What are the side lengths of these squares? Examples include the following: area of 81 with side length of 9, area of 49 with side length of 7, and area of 9 with side length of 3. Explain that these areas are **perfect squares** because they have square roots that are integer-valued.

Ongoing Assessment, page 412
1, 4, 9, 16, 25, 36, 49, 64, 81, 100

continued on top of next page

TEACH

Approach B: Using Models

Opening Discussion
Give each student 36 unit algebra tiles. If tiles are not available, use small squares of paper, or use squares of transparency film on an overhead projector. Ask students to form the pieces into a larger square. What is the area? [**36**] What is the length of the side? Students should understand that the length of the side is the square root of the area of the square. Point out that 6 is the positive square root of 36. What other number is a square root of 36? [**−6**]

Brainstorming
Have students give areas of other squares that could be made with tiles. What are the side lengths of these squares? Examples include the following: area of 81 with side length of 9, area of 49 with side length of 7, and area of 9 with side length of 3. Explain that these areas are **perfect squares** because they have square roots that are integer-valued.

continued on bottom of next page

Approach A

Using the Book, continued

Setting Up the Exploration

Teacher-Guided Explorations If you wish, you can guide the students through the textbook explorations. For Exploration 2, you can draw the graphs on an overhead projector or have students draw them on the chalkboard. At each stage of the exploration, invite student input and encourage discussion. Students should keep a record of the responses in their notebook.

Cooperative-Learning Groups Divide the class into groups of four. In Exploration 1, students should take turns guessing and checking the answers. In Exploration 2, students should take turns graphing the functions and writing the comparisons. Each member should be responsible either for writing the group summary or for reporting the results of one exploration. All group members should be involved in the discussion of the activity as the exploration proceeds. Each student should sign the completed report.

Exploration 1

A guess-and-check strategy is used to estimate the positive square root of 12. The procedure developed is then used to find $\sqrt{20}$. Remind students that $\sqrt{12}$ is the positive number that is squared to get 12.

Ongoing Assessment, page 413

7. Procedures may vary. Sample: $4 \cdot 4 = 16$ and $5 \cdot 5 = 25$, so $\sqrt{20}$ is between 4 and 5.

Try 4.5: $4.5^2 = 20.25$.
Try 4.4: $4.4^2 \approx 19.36$.
Try 4.47: $4.47^2 \approx 19.98$.
Try 4.48: $4.48^2 \approx 20.07$.

$\sqrt{20}$ is between 4.47 and 4.48. Rounded to two decimal places, $\sqrt{20} \approx 4.47$.

TEACH

Approach B

Using Models, continued

Cooperative Learning

Before starting the explorations, divide the class into groups of four. In Exploration 1, students should take turns guessing and checking the answers. In Exploration 2, students should take turns making the tables, graphing the functions, and writing the comparisons. Each member should be responsible either for writing the group summary or for reporting the results of one exploration. All group members should be involved in the discussion of the activity as the exploration proceeds. Each student should sign the completed report.

Alternate Exploration 1

Have students work in their groups. Give each group a cardboard cutout of a square with an area of 12 square units. Base the units on one of the algebra tiles or on the paper squares described earlier. (Each side of the square should be $\sqrt{12}$, or about 3.5, units long.) Each group should investigate the following questions, which you may write on the overhead or chalkboard. (For ideas on guiding the students, see Exploration 1 on page 413 of the Pupil's Edition of the textbook.)

- Using the algebra tiles or paper squares as models, find squares whose areas are as close as possible to but less than 12 and as close as possible to but greater than 12. **[squares with areas of 9 and 16]** What are the side lengths of these squares? **[3 and 4]** What does this say about $\sqrt{12}$? **[It must be between 3 and 4.]**
- Guess 3.5. Square this to get 12.25. This is too large. Try 3.4. **[11.56]** Continue this process to find $\sqrt{12}$ to two decimal places. **[3.46]**
- Write a procedure for estimating the positive square root of a number. Use this procedure to find $\sqrt{20}$. **[Find the two perfect squares on either side of the number. Start with a number that is 0.5 more than the square root of the smaller perfect square. Square this number, and compare the result with the number whose square root you are trying to find. If it is too large, decrease your estimate and try again. If it is too small, increase your estimate and repeat. $\sqrt{20} \approx 4.47$]**

On Their Own

Have students find $\sqrt{42}$ to the nearest hundredth. **[6.48]**

Irrational Numbers

Read or have a volunteer read the section on irrational numbers. Students need to understand the difference between terminating, repeating, and nonrepeating decimals. Stress that the radical sign is used to designate only the positive square root of a number. Have students enter some rational and irrational numbers into their calculator to see if it rounds or truncates the decimals.

Inclusion Strategies

Language Help students understand and remember the definition of rational and irrational numbers by explaining the origin of these terms. *Rational* in this context means that the number can be written as a **ratio,** or fraction, of integers. *Irrational* means that the number cannot be written as a ratio of integers.

Real Numbers

Review the definition of real numbers and have students study the diagram on page 414. You may want to have them copy the diagram and place examples of each type of number in the appropriate area. Demonstrate the density property of real numbers by having students give two numbers and then find a number between them.

Use **Transparency 39** for discussion or review of real numbers.

Ongoing Assessment, top of page 414

Between any two rational numbers there is another rational number. However, between any two integers, say 1 and 2, there may not always be another integer.

The Principal Square-Root Function

The principal square root is the positive square root. You can work with the students to make a table of values for this function. What happens when a negative number is substituted for x?

continued on top of next page

Student Reports

Choose one group to report their findings for this exploration. The student who is responsible for reporting for that group should come to the front of the room and show the class the steps that the group used to find $\sqrt{20}$.

Irrational Numbers

Point out that **rational numbers** are numbers that can be represented by terminating or repeating decimals and that **irrational numbers** are numbers whose decimal part never repeats or terminates. Students need to understand the difference between terminating, repeating, and nonrepeating decimals. Square roots are often irrational numbers. Demonstrate how to find a square root with a calculator. Stress that the radical sign is used to designate only the positive square root of a number. Have students enter some rational and irrational numbers into their calculator to see if it rounds or truncates the decimals. (Compare page 414.)

Inclusion Strategies

Language Help students understand and remember the definition of rational and irrational numbers by explaining the origin of these terms. *Rational* in this context means that the number can be written as a **ratio,** or fraction, of integers. *Irrational* means that the number cannot be written as a ratio of integers.

Real Numbers

Explain that real numbers consist of all rational and irrational numbers. Ask students to name other number systems that they have studied. Some examples are integers, whole numbers, and natural numbers. Organize these systems into a diagram. You can have them copy the diagram and place examples of each type of number in the appropriate area. (Compare page 414.)

The Principal Square-Root Function

The principal square root is the positive square root. You can work with students to make a table of values and a graph of this function,

continued on bottom of next page

Approach A

Using the Book, continued

Explain. **[The function cannot be evaluated for negative numbers because there is no real number that gives a negative number when squared.]**

Ongoing Assessment, bottom of page 414

The domain and range of the principal square-root function is the set of real numbers greater than or equal to zero.

Critical Thinking, page 414

The positive part of $y = x^2$ corresponds to the positive part of $y = \sqrt{x}$, but it has been reflected through the line $y = x$.

Exploration 2

Transformations of the parent function $f(x) = \sqrt{x}$ are explored. The goal is for students to be able to recognize translations, scale changes, and reflections of the principal square-root function by examining an equation. The graphs can be made by hand or with graphing technology. If using graphing technology, warn students to use parentheses with square roots that contain more than one term. Using different colors to sketch the graphs will help students distinguish them.

Use **Transparency 40** for discussion or review.

Ongoing Assessment, page 415

9. Adding a constant to the principal square-root function shifts the graph up if the constant is positive and down if the constant is negative. Subtracting a constant within the square root sign shifts the graph right if the constant is positive and left if the constant is negative. Multiplying the radical by a positive constant results in a vertical stretch. Multiplying the radical by a negative constant results in a vertical stretch and a reflection through the x-axis.

TEACH

Approach B

Using Models, continued

$f(x) = \sqrt{x}$. What happens when a negative number is substituted for x? Explain. **[The function cannot be evaluated for negative numbers because there is no real number that gives a negative number when squared.]** What are the domain and range of this function? **[The domain and range of the principal square-root function is the set of real numbers greater than or equal to zero.]** (Compare page 414.)

Alternate Exploration 2

Transformations of the parent function $f(x) = \sqrt{x}$ are explored. The goal is for students to be able to recognize translations, scale changes, and reflections of the principal square-root function by examining an equation. Each group needs at least one graphics calculator and an erasable graph. (For instructions on how to make erasable graphs, see Using Models for Lesson 2.1.)

The group should use the table features of the calculator to create a table of values for each function for x-values from -10 to 10. The calculator will display an error message for some of the y-values. This is because the corresponding x-values are not in the domain of the function. The student producing the table should read the integer-valued points for another group member to plot on the erasable graph. In this fashion, a quick sketch of the graph can be made. Warn students to use parentheses when entering square roots that contain more than one term. Using different colors to sketch the graphs will help students distinguish them.

- Graph several functions of the form $g(x) = \sqrt{x - h}$. Compare each graph with the graph of $f(x) = \sqrt{x}$. What transformation results when $f(x)$ is changed to $g(x) = \sqrt{x - h}$? **[Subtracting a constant within the square-root sign shifts the graph to the right if the constant is positive and to the left if the constant is negative.]**

On Their Own

Have students describe the transformations of the graph of $f(x) = \sqrt{x}$ that would produce the graph of each equation.

A. $h(x) = \sqrt{x - 4}$ **[horizontal translation of 4 units to the right]**

B. $j(x) = -\frac{1}{2}\sqrt{x}$ **[vertical stretch by a factor of $\frac{1}{2}$ and reflection through the x-axis]**

Multiple Transformations

Use the graphs at the bottom of page 409 and the Extension to discuss a combination of transformations applied to the principal square-root function. Note that in the second graph and in the Extension the reflection occurs before the vertical translation.

Ongoing Assessment, page 415

g: a stretch of 2, a vertical shift of 4 units up, and a horizontal shift of 3 units to the right

h: a stretch of -2, a vertical shift of -3, and a horizontal shift of 4 to the left

Application

Read or have a student volunteer read this problem, which illustrates how the principal square-root function can be used to describe the distance to the horizon.

Critical Thinking, page 416

There is no real number that will produce a negative number when squared.

- Graph several functions of the form $h(x) = \sqrt{x} + b$. Compare each graph with the graph of $f(x) = \sqrt{x}$. What does this transformation do to the graph of the parent function? **[Adding a constant to the square-root function shifts the graph up if the constant is positive and down if the constant is negative.]**
- Graph several functions of the form $k(x) = c\sqrt{x}$, where c is positive. Compare each graph with the graph of $f(x) = \sqrt{x}$. What does this transformation do to the graph of the parent function? **[Multiplying the radical by a positive constant results in a vertical stretch.]**
- Graph several functions of the form $m(x) = d\sqrt{x}$, where d is negative. Compare each graph with the graph of $f(x) = \sqrt{x}$. What transformation does this do to the graph of the parent function? **[Multiplying the radical by a negative constant results in a vertical stretch and a reflection through the x-axis.]**

Student Reports

Ask the reporters from several of the groups to read their comparison of the parent function and one of the four transformations. Each reporter should come to the chalkboard and illustrate his or her group's findings with a quick sketch of the graphs.

Multiple Transformations

Use the functions $g(x) = 2\sqrt{x - 3} + 4$ and $h(x) = -2\sqrt{x + 4} - 3$ to demonstrate what happens when a combination of transformations is applied to the principal square-root function. First describe the transformations, and then show how each one affects the graph. Note that in the second graph the reflection occurs before the vertical translation. (Compare page 415.)

ASSESS

Selected Answers

Odd-numbered Exercises 7–55 and odd-numbered Extra Practice

Assignment Guide

Core level students should complete Exercises 1–56.

Core-plus level students should complete Exercises 1–6 and 9–60.

Error Analysis

Watch for students who forget to use parentheses in the expression for a radical when using technology. In Exercises 27 and 29, students must remember to reflect the graph before translating it vertically.

Alternative Assessment

Posters Have each student pick a function of the form $y = a\sqrt{x + b} + c$. They should illustrate and describe the different transformations of the function $y = \sqrt{x}$ that produce the graph of the chosen function. (See the graphs at the bottom of page 415 of the Pupil's Edition.) This work can be done on poster board for display in the classroom.

Portfolio Activity Ask students to review their work for Exercises 22 and 23. The function created by using the first and third columns of the tables is $f(x) = \sqrt{x^2}$. Have students use a graphics calculator to make a table of values for this function and to verify that this table is a combination of the ones made in the exercises. Then have students graph the function. Does it look familiar? What function is the same as $f(x)$? [$g(x) = |x|$]

RETEACH

From Graphs to Equations

Give students a sheet of paper showing about 10 graphs of transformed square-root functions. Have them work with a partner to write an equation for each one. Remind them to pick a point from the graph to test in their equation. Also have them write the domain and range for each function.

ENRICH

Interdisciplinary Connection

History Pythagoras is generally credited with the discovery of irrational numbers. Have students research to find out more about Pythagoras and his discovery. This research can be written in report form.

The Density Property of Real Numbers

Have students develop a method for producing a real number that is between any two given real numbers. The easiest way to do this is to write the two numbers in decimal form. Eventually, there will be a digit that is different for each of these numbers. Create a new number that has the digits that are common to the two original numbers, with the next to last digit being the smaller of the first two different digits and with the last number being one more than the last digit in the smaller of the two original numbers. For example, consider the two numbers $\frac{1}{301}$ and $\frac{1}{302}$. Write each as a decimal 0.003322259 . . . and 0.003311258 . . . (Note: Although it is not clear from these truncated decimals, these decimals eventually begin to repeat because they are rational numbers.) The fifth digits to the right of the decimal point are the first two that are different. 1 is the smaller of the two digits. The next digit after the 1 in the smaller number is also a 1. Make a new number with a fifth digit of 1 and a sixth digit of 2. This number, 0.003312, is between the two given numbers.

Practice

9.1 Exploring Square-Root Functions

Estimate each square root. If the square root is irrational, find the value to the nearest hundredth.

1. $\sqrt{841}$ __________
2. $-\sqrt{289}$ __________
3. $-\sqrt{1440}$ __________
4. $\sqrt{1.44}$ __________
5. $\sqrt{0.0081}$ __________
6. $\sqrt{\frac{100}{169}}$ __________

Identify the transformation(s) and describe the effect on the parent function $y = \sqrt{x}$.

7. $y = \sqrt{x} - 3$ __________
8. $y = 2\sqrt{x}$ __________
9. $y = -\sqrt{x - 4}$ __________
10. $6\sqrt{x - 3}$ __________
11. $y = \sqrt{x + 5} + 3$ __________
12. $y = 2.5 + 5.75\sqrt{x}$ __________

Write the equation of the function graphed.

13.
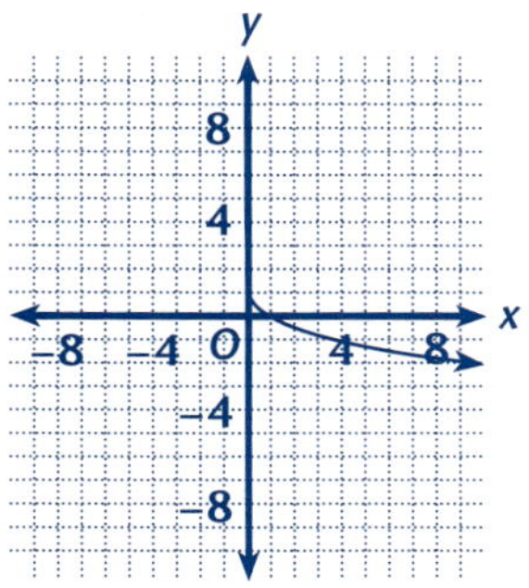
____

14.
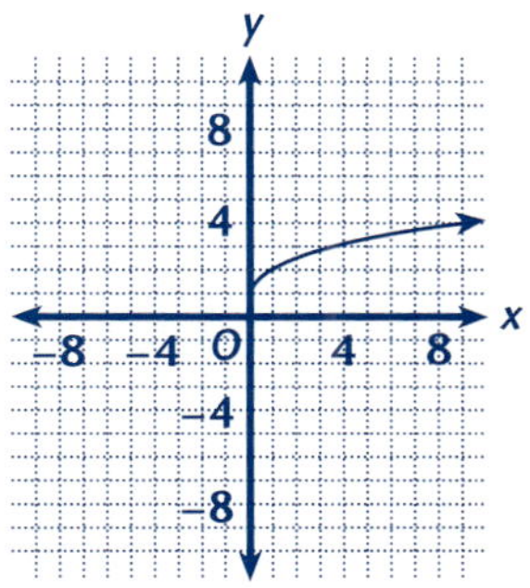
____

15.
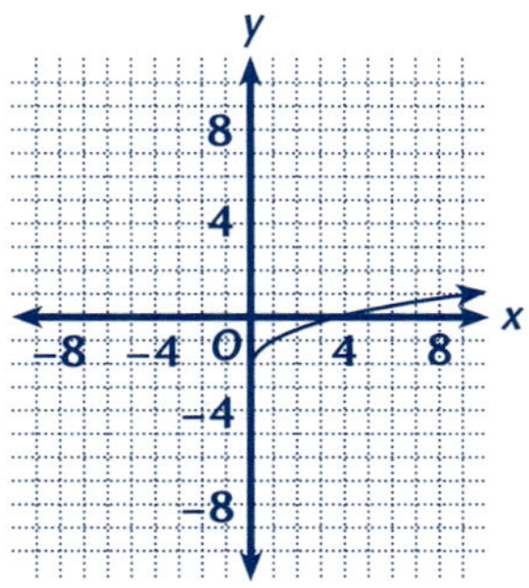
____

16.
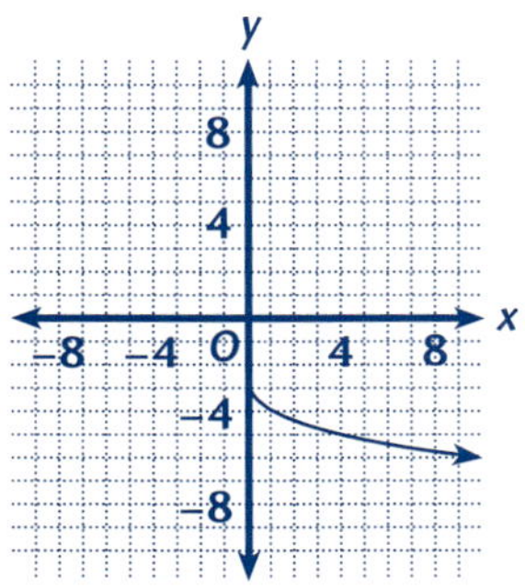
____

17.
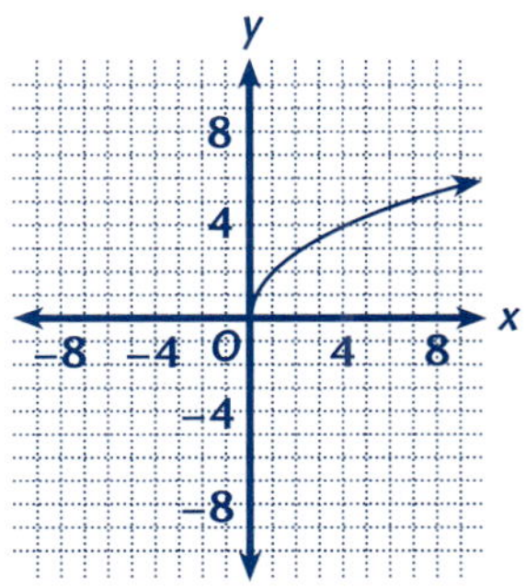
____

18.
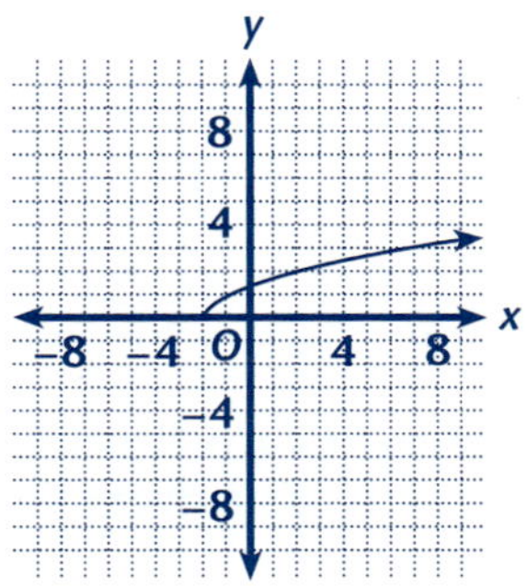
____

NAME ______________________ CLASS __________ DATE __________

Lesson Activity

9.1 Approximating Square Roots

Heron of Alexandria devised a method for calculating square roots by using sequential approximations. Follow these steps to see how to use Heron's method to approximate $\sqrt{42}$ to the nearest thousandth.

Step 1 Choose any two factors of 42. For example, choose 3 and 14.

Step 2 Find the average of the factors. $\frac{(3 + 14)}{2} = 8.5$

Step 3 Divide 42 by the average of the factors. $\frac{42}{8.5} = 4.941176471$

Step 4 Find the average of the two new factors. $\frac{(8.5 + 4.941176471)}{2} = 6.720588235$

Step 5 Divide 42 by the result. $\frac{42}{6.720588235} = 6.249452954$

Step 6 Find the new average. $\frac{(6.720588235 + 6.249452954)}{2} = 6.485020595$

Step 7 Divide 42 by the result. $\frac{42}{6.485020595} = 6.476463627$

Step 8 Find the new average. $\frac{(6.485020595 + 6.476463627)}{2} = 6.48074211$

Step 9 Divide 42 by the result. $\frac{42}{6.480742111} = 6.480739286$

When the two factors are within one thousandth of each other, you can round off to the nearest thousandth. The square root of 42 is approximately 6.481.

1. What do you notice about the two factors of 42 used to find the average in each successive step?

____

Use Heron's method to approximate the square root of the following numbers to the nearest thousandth:

2. $\sqrt{65}$ ____________________ **3.** $\sqrt{108}$ ____________________

4. Approximate $\sqrt{42}$ by using two negative factors as your initial factors. What happens to your approximation?

____

NAME ______________________ CLASS ____________ DATE ____________

Basic Skills Practice

Interpreting Bar and Circle Graphs, Part 1

Graphs provide a visual representation of data. A bar graph is a useful tool to compare data. A double bar graph can be used to compare two sets of data. Bar graphs should have a title and a scale.

The horizontal bar graph at right shows the cost of long distance calls to 4 different countries.

Example 1: Which country is the most expensive to call?
To find the most expensive country, look for the longest bar which is Israel.

The double bar graph at right shows the favorite types of food for 2 different age groups.

Example 2: Which food is equally popular with both groups?
To find the food that is equally popular, look for bars that have the same height which is chicken.

A circle graph is useful to show data as part of a whole. A circle graph should have a title, and each portion of the circle should be labeled and quantified as a part of a whole (as a fraction or as a percent).

The circle graph at right shows the distribution of Parker's expenses.

Example 3: What fraction of Parker's expenses go toward books?
To find the answer, look for the portion of the circle that is labeled *Books,* and read the fraction that is written for it, $\frac{1}{8}$.

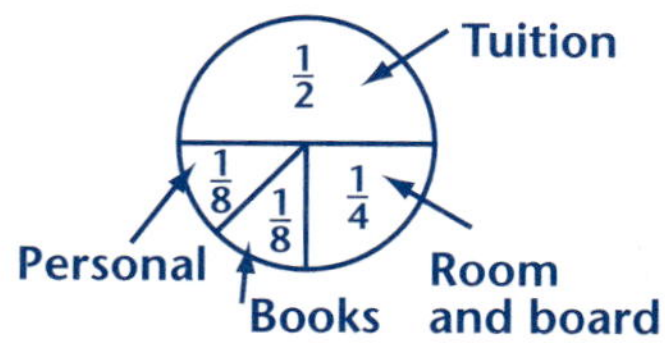

The school cafeteria conducted a survey to see how students felt about 4 different kinds of food. The results are shown in the bar graph on the right.

Use the bar graph to answer Exercises 1–5.

1. How many students like hamburger? ____________

2. How many students like chicken? ____________

3. How many more students like turkey than fish? ____________

4. How many students answered the survey? ____________

5. Which food item was the favorite among the students surveyed? ____________

LESSON 9.2 Operations With Radicals

PREPARE

Pacing Guide
3 Days

Objectives and Overview
- Define square root, and write square roots in simplest radical form.
- Examine mathematical operations with radicals.

The formal definition of square root is given and used to simplify radical expressions. Simplest radical form is defined, and the Multiplication and Division Properties of Square Roots are used to write radical expressions in simplest form. The addition, subtraction, and multiplication of radicals is also examined.

Vocabulary
radical sign
radicand
simplest radical form
Multiplication Property of Square Roots
Division Property of Square Roots
rationalizing the denominator

Assessing Prior Knowledge

Simplify.

1.	$\frac{a^4b^3}{a^3b^5}$	$\left[\frac{a}{b^2} \text{ or } ab^{-2}\right]$
2.	$(-5c^2)^2$	$[25c^4]$
3.	$-5(c^2)^2$	$[-5c^4]$
4.	$m^3n^6 \cdot m^4n^2$	$[m^7n^8]$

Approach A

Using the Book

Opening Discussion
In the last lesson, radicals were used to answer questions involving area. Radicals are also used for applications in statistics and physics. Sometimes radicals involve numbers, and other times they may involve expressions. Special rules are needed to simplify, add, subtract, multiply, and divide radical expressions. Some of these rules are similar to the special rules learned to add, subtract, and multiply polynomials.

Brainstorming
Ask students to describe the special rules used to add, subtract and multiply polynomial expressions. **[To add and subtract polynomials, you add like terms. To multiply monomials, you multiply the constants and add the exponents of factors with the same base. To multiply larger polynomials, use the Distributive Property.]**

TEACH

Approach B

Using Models

Opening Discussion
In the last lesson, radicals were used to answer questions involving area. Radicals are also used for applications in statistics and physics. Sometimes radicals involve numbers, and other times they may involve expressions. Special rules are needed to simplify, add, subtract, multiply, and divide radical expressions. Some of these rules are similar to the special rules learned to add, subtract, and multiply polynomials.

Brainstorming
Ask students to describe the special rules used to add, subtract and multiply polynomial expressions. **[To add and subtract polynomials, you add like terms. To multiply monomials, you multiply the constants and add the exponents of factors with the same base. To multiply larger polynomials, use the Distributive Property.]**

Ongoing Assessment, page 420

If x is not restricted, x can be positive or negative. However, x^2 will still be positive.

Definition of *Square Root*

This is a formal statement of the concepts that students used in the previous lesson. Point out that this definition will be used to simplify expressions such as $\sqrt{3}\sqrt{3}$ and $\sqrt{x}\sqrt{x}$.

Use **Transparency 41** for discussion or review.

Example 1

The definition of square root is used to simplify square roots with perfect-square radicands. A calculator is used to approximate a radical whose radicand is not a perfect square.

Additional Example 1

Find each square root.

A. $\sqrt{36}$ **[6]**

B. $-\sqrt{121}$ **[−11]**

C. $\sqrt{13}$ **[about 3.61]**

Ongoing Assessment, page 421

Try This

a. 14

b. −15

Simplifying Radicals and Radical Expressions

Define simplest radical form. This is a standard form that helps mathematicians and scientists communicate with each other. This is similar to the rules for writing fractions in lowest terms. Have students copy the rules for simplest radical form in their notebook or journal for later reference. Next to rule 1, have them list the first 10 perfect squares.

Teaching Tip

Encourage students to keep a list of the rules for simplest radical form in a handy location. They can then refer to it while simplifying radicals. After completing each problem, students should go

continued on top of next page

Using a Spreadsheet

Creating a Model Set up a spreadsheet whose first two columns, A and B, each have about 25 nonnegative integers in them. These columns represent the different test values for the variables A and B. Columns D and E will be used for formulas involving A and B. The spreadsheet will be used to discover formulas that are true and to disprove those that are false.

Using the Model Enter the formula = SQRT(A1)*SQRT(A1) in cell C1 and copy it by using the fill down command. What mathematical expression does this formula represent? [$\sqrt{A}\sqrt{A}$] Compare the three columns. What do you notice? [**The entries in columns A and C are equal.**] What property have you discovered? [$\sqrt{A}\sqrt{A} = A$] Repeat this process for $-\sqrt{A} \cdot -\sqrt{A}$.

Definition of *Square Root*

Show **Transparency 41,** which gives the definition of square root. This is a formal statement of the properties that were discovered with the spreadsheet. Write the definition in words. For example, the square root of a is the number you multiply by itself to get a. Use this verbal definition to simplify some radicals, such as $\sqrt{25} = 5$: the number you square to get 25 is 5.

Point out that the definition of square root will be used later to simplify expressions such as $\sqrt{3}\sqrt{3}$ and $\sqrt{x}\sqrt{x}$ to get 3 and x, respectively. Recall the term *radical sign,* and introduce the term *radicand* to name the expression under the radical sign.

Discovering the Properties of Square Roots

Can you add, subtract, multiply, or divide square roots by adding, subtracting, multiplying or dividing radicands? This activity allows students to discover the Multiplication and Division Properties of Square Roots. They should also see that similar properties for addition and subtraction do not exist. Have students work in small groups at a computer, or do this activity as a class demonstration with an overhead and computer. Set up the spreadsheet as described in Creating a Model on the

continued on bottom of next page

Approach A

Using the Book, continued

through the rules to make sure that the simplified expression meets each condition.

Example 2

Three numerical expressions involving radicals are written in simplest radical form. The Multiplication and Division Properties of Square Roots are used. These properties are formally defined next.

Additional Example 2

Simplify.

A. $\sqrt{18}$ $[3\sqrt{2}]$

B. $\frac{\sqrt{125}}{5}$ $[\sqrt{5}]$

C. $\frac{\sqrt{80}}{\sqrt{4}}$ $[2\sqrt{5}]$

The Multiplication Property of Square Roots

Write this property on the board for students to write in their notebook or journal. Also summarize the property in words: the square root of a product is equal to the product of square roots. Point out that this property is used both ways. In part **A** of Example 2, it was used to write $\sqrt{2^2 \cdot 3}$ as $\sqrt{2^2}\sqrt{3}$. In Example 6, the property will be used to multiply two radicals. Demonstrate how the Multiplication Property is used to simplify radicals such as $\sqrt{a^4b^5}$ by rewriting it as a product of even powers.

Example 3

The Multiplication Property of Square Roots is used to simplify square roots that contain a product of factors. Stress that students should look for the largest perfect-square factors and for even-powered variable expressions.

Additional Example 3

Simplify.

A. $\sqrt{32}$ $[4\sqrt{2}]$

B. $\sqrt{3600}$ $[60]$

C. $\sqrt{a^8b^6}$ $[a^4b^3]$

TEACH

Approach B

Using Models, continued

previous page. Start by entering the formula = SQRT(A1) + SQRT(B1) in cell C1 and the formula = SQRT(A1 + B1) in cell D1. Copy both formulas down. Do students see any patterns? What have they discovered?

In general, $\sqrt{A} + \sqrt{B} \neq \sqrt{A + B}$. That is, you cannot add radicals by adding their radicands. What about multiplication? Alter columns C and D to see if $\sqrt{A}\sqrt{B} = \sqrt{AB}$. Students should see right away that the two columns match. They have discovered the Multiplication Property of Square Roots. Repeat this process to demonstrate the Division Property of Square Roots and to show that no such property exists for subtraction.

The Multiplication and Division Properties of Square Roots

Discuss the Multiplication and Division Properties of Square Roots on page 422 of the Pupil's Edition. These are the properties just discovered. These properties are used to simplify radicals. When simplifying radicals, the form most frequently used in algebra is called **simplest radical form.** Discuss the rules for simplest radical form on page 421 of the Pupil's Edition. Have students list the first 10 perfect squares for future reference.

Simplifying Radicals

Demonstrate how different types of radicals can be changed to simplest radical form. Point out places where the Multiplication and Division Properties of Square Roots are used. Students should take detailed notes of the solution for each example. Some good examples are as follows:

A.	$\sqrt{18}$	$[3\sqrt{2}]$	**B.**	$\frac{\sqrt{125}}{5}$	$[\sqrt{5}]$
C.	$\frac{\sqrt{80}}{\sqrt{4}}$	$[2\sqrt{5}]$	**D.**	$\sqrt{32}$	$[4\sqrt{2}]$
E.	$\sqrt{3600}$	$[60]$	**F.**	$\sqrt{a^8b^7}$	$[a^4b^3]$
G.	$\sqrt{\frac{49}{81}}$	$\left[\frac{7}{9}\right]$	**H.**	$\sqrt{\frac{11}{25}}$	$\left[\frac{\sqrt{11}}{5}\right]$

Ongoing Assessment, page 422
Try This
$6mn^2\sqrt{2n}$ for $m \geq 0$ and $n \geq 0$

The Division Property of Square Roots

Write this property on the board for students to write in their notebook or journal. Also summarize the property in words: the square root of a quotient is equal to the quotient of square roots. Point out that this property is used both ways. In part **C** of Example 2, it was used to write $\frac{\sqrt{18}}{\sqrt{2}}$ as $\sqrt{\frac{18}{2}}$. In Example 4, it will be used to write a single radical as the quotient of two radicals.

Example 4

The Division Property of Square Roots is used to simplify the radical of a quotient. Note that the quotient should be reduced to lowest terms before applying the property. Summarize the procedure for **rationalizing the denominator.**

Additional Example 4

Simplify.

A. $\sqrt{\frac{49}{81}}$ $\left[\frac{7}{9}\right]$

B. $\sqrt{\frac{11}{25}}$ $\left[\frac{\sqrt{11}}{5}\right]$

C. $\sqrt{\frac{36}{7}}$ $\left[\frac{6\sqrt{7}}{7}\right]$

D. $\sqrt{\frac{x^5y^2}{z^4}}$ **[For x, y, and $z \geq 0$,** $\sqrt{\frac{x^5y^2}{z^4}} = \frac{x^2y}{z^2}\sqrt{x}$**]**

Operations and Radical Expressions

Have a student volunteer read this section. Have students choose two other numbers to verify that $\sqrt{a+b} = \sqrt{a} + \sqrt{b}$ is a false statement. Students should understand that for a property like this to be true, it must be true for all values of a and b. Finding one counterexample is enough to show that the property is false. Point out that addition of radicals is very similar to addition of polynomials: you can add only like terms (those with the same radicand).

continued on top of next page

I. $\sqrt{\frac{36}{7}}$ $\left[\frac{6\sqrt{7}}{7}\right]$ **J.** $\sqrt{\frac{x^5y^2}{z^4}}$

[For x, y, and $z \geq 0$, $\sqrt{\frac{x^5y^2}{z^4}} = \frac{x^2y}{z^2}\sqrt{x}$**]**

For problem **I**, you will need to multiply by $\frac{\sqrt{7}}{\sqrt{7}}$ to clear the radical from the denominator. Explain that this process is called **rationalizing the denominator.**

Cooperative Learning

Have students work in groups of three. Give each group a sheet with many radical expressions similar to those above. Each student should pick one of the three rules for simplest radical form. The group should work together on each problem. Each student is responsible for ensuring that the answer satisfies his or her rule. After several problems, students should pick a new rule.

Operations and Radical Expressions

Addition and Subtraction Use the spreadsheet program again, and set it up as described in Creating a Model. What is $\sqrt{A} + \sqrt{A}$? How many square roots of A are there? Could the expression equal $2\sqrt{A}$? Enter formulas for columns C and D to test this theory. It is true. What about $\sqrt{A} + 2\sqrt{A}$? Continue trying new addition problems in which the radicands are equal. Ask students to write a procedure for adding radical expressions with equal radicands. Then repeat this process for subtraction. Students should see the parallel between adding and subtracting radicals and adding and subtracting monomials. Only like terms can be combined, and these are combined by adding or subtracting the coefficients. Students should practice with examples such as the following:

A. $7\sqrt{10} - 10\sqrt{10}$ $[-3\sqrt{10}]$

B. $3 + 5\sqrt{13} - 1 + \sqrt{13}$ $[2 + 6\sqrt{13}]$

C. $-4\sqrt{7} + 3\sqrt{5} + 2\sqrt{5} - 9\sqrt{7}$ $[-13\sqrt{7} + 5\sqrt{5}]$

D. $c\sqrt{dx} - 2a\sqrt{dx}$ $[(c - 2a)\sqrt{dx}]$

continued on bottom of next page

Approach A — Using the Book, continued

Example 5

In this example, radical expressions are simplified by combining terms that have the same radicand.

Additional Example 5

A. $7\sqrt{10} - 10\sqrt{10}$ $[-3\sqrt{10}]$

B. $3 + 5\sqrt{13} - 1 + \sqrt{13}$ $[2 + 6\sqrt{13}]$

C. $-4\sqrt{7} + 3\sqrt{5} + 2\sqrt{5} - 9\sqrt{7}$
$[-13\sqrt{7} + 5\sqrt{5}]$

D. $c\sqrt{dx} - 2a\sqrt{dx}$ $[(c - 2a)\sqrt{dx}]$

Critical Thinking, page 424

It is not true in general because it is not true for every *a* and *b*. In fact, the condition is true only when *b* is 0 or when *a* equals *b*.

Example 6

In this example, radical expressions involving multiplication are simplified.

Stress the different properties that are used in each solution. Note the similarity of part **D** to the multiplication of binomials.

Additional Example 6

Simplify.

A. $(3\sqrt{6})^2$ $[54]$

B. $\sqrt{5}\sqrt{15}$ $[5\sqrt{3}]$

C. $\sqrt{3}(1 - \sqrt{2})$ $[\sqrt{3} - \sqrt{6}]$

D. $(2 + \sqrt{5})(3 - \sqrt{5})$ $[-19 + \sqrt{5}]$

Ongoing Assessment, page 424

19

TEACH

Approach B — Using Models, continued

Multiplication

When multiplying radical expressions, the Multiplication Property of Square Roots can be used to multiply within the radical, such as in the problem $\sqrt{5}\sqrt{15} = \sqrt{75}$. Note that this should now be simplified as $\sqrt{75} = 5\sqrt{3}$. When squaring a radical expression, you can multiply within the radical and simplify, or you can use the definition of square root, which states that when the number is positive, the square of its square root is the number itself. Thus, $(\sqrt{8})^2 = 8$, for example. Ask students to simplify $(4\sqrt{5})^2$. **[80]**

The Distributive Property is used to multiply expressions such as $\sqrt{3}(1 - \sqrt{2})$. $[\sqrt{3} - \sqrt{6}]$ Differences and sums are multiplied in the same fashion as are binomials:

$$(2 + \sqrt{5})(3 - \sqrt{5}) = 1 + \sqrt{5}$$

ASSESS

Selected Answers

Odd-numbered Exercises 7–65 and odd-numbered Extra Practice

Assignment Guide

Core level students should complete Exercises 1–66.

Core-plus level students should complete Exercises 1–6, 9–18, 21–42, and 45–71.

Error Analysis

Students may confuse the different properties or strategies that are needed for each type of problem. Encourage these students to find a similar problem in the examples and to follow its solution.

Alternative Assessment

Student Examples Ask students to give an example with a worked-out solution of a radical expression that is simplified by using the Multiplication Property of Square Roots and of one that is simplified by using the Division Property of Square Roots. They should indicate in their solutions where the properties are used.

Portfolio Activity Have students look through the homework to find as many different kinds of problems as possible. They should include these problems in their portfolio along with a description of the method used to solve each one.

RETEACH

Demonstrate for students how radical expressions can be simplified by finding the prime factorization of the radicand and by grouping like factors in pairs. Have students cross out each pair of like factors inside the radical and replace the pair with a single factor outside the radical. Any numbers or variables that cannot be paired stay under the radical in the simplified expression. For example, to simplify the expression $\sqrt{90x^2y^5}$, first write the prime factorization of the radicand, $\sqrt{2 \cdot 3 \cdot 3 \cdot 5 \cdot x \cdot x \cdot y \cdot y \cdot y \cdot y \cdot y}$. You can group together a pair of 3s, a pair of *x*s, and two pairs of *y*s. This means that a 3, an *x*, and two *y*s will be factors in front of the radical, while the remaining 2, 5, and *y* will stay under the radical. The solution is $3xy^2\sqrt{10y}$. Have students work with a partner to make up and simplify radical expressions by using this method.

ENRICH

Interdisciplinary Connection

Construction The length of the diagonal of a square, d, is related to the length of the side, s, by the formula $d = \sqrt{2}s$. A builder wants to put the window shown below into a house.

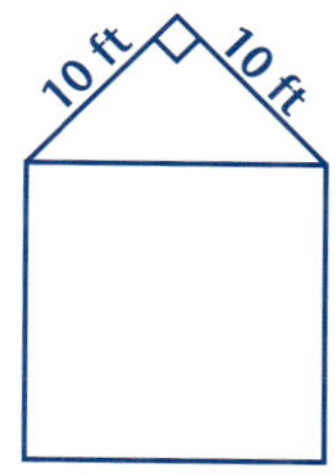

Calculate the perimeter and area of this window. **[$(20 + 30\sqrt{2})$ feet; 250 square feet]**

Factoring Over the Real Numbers

Some polynomials that could not previously be factored can be factored if we allow not only rational coefficients and constant terms, but also real coefficients and constant terms. For example, the polynomial $x^2 - 3$ is factored as $(x + \sqrt{3})(x - \sqrt{3})$. Have students multiply these factors to see that this is the correct factorization. Then ask students to find other polynomials that can be factored over the real numbers.

NAME ________________ CLASS ________ DATE ________

Practice

9.2 Operations With Radicals

Simplify each radical expression by factoring.

1. $\sqrt{144}$ ________________ **2.** $\sqrt{72}$ ________________

3. $\sqrt{288}$ ________________ **4.** $\sqrt{4000}$ ________________

5. $\sqrt{3264}$ ________________ **6.** $\sqrt{8775}$ ________________

Express each radical expression in simplest radical form.

7. $\sqrt{8}\sqrt{14}$ ________________ **8.** $\sqrt{3}\sqrt{27}$ ________________

9. $\sqrt{12}\sqrt{6}$ ________________ **10.** $\frac{\sqrt{72}}{\sqrt{6}}$ ________________

11. $\frac{\sqrt{162}}{\sqrt{3}}$ ________________ **12.** $\frac{\sqrt{500}}{\sqrt{50}}$ ________________

Simplify each radical expression. Assume that all variables are nonnegative and that all denominators are nonzero.

13. $\sqrt{m^6n}$ ________________ **14.** $\sqrt{a^8b^6}$ ________________

15. $\sqrt{\frac{x^3}{y^6}}$ ________________ **16.** $\sqrt{\frac{x^5}{y^9}}$ ________________

If possible, perform each indicated operation and simplify your answer.

17. $\sqrt{12} + 3\sqrt{3}$ ________________ **18.** $3\sqrt{27} - 5\sqrt{3}$ ________________

19. $\frac{\sqrt{1} + \sqrt{25}}{\sqrt{2}}$ ________________ **20.** $(5 + \sqrt{5}) + (2 - \sqrt{3})$ ________________

Simplify each radical expression.

21. $(2\sqrt{3})^2$ ________________ **22.** $2(\sqrt{3} + \sqrt{12})$ ________________

23. $(\sqrt{12} + 2)(\sqrt{12} - 2)$ ________________ **24.** $\sqrt{3}(2 + \sqrt{12})$ ________________

25. $(\sqrt{3} + 2)^2$ ________________ **26.** $(\sqrt{3} + \sqrt{12})^2$ ________________

Lesson Activity

9.2 Modeling Square Roots

Heron's method of approximating square roots, as discussed in Lesson Activity 9.1, can be used as a basis for a computer program to find square roots. In programming, a process of sequential approximations such as this is called iteration. When a function undergoes iteration, the initial input is called the seed, and the sequence of outputs created by iterating (repeating) the function is called an orbit. For example, the iteration of

$$f(x) = \frac{x + \frac{21}{x}}{2}$$

can be used to find an approximation of $\sqrt{21}$. Start by choosing any number for the seed. Then substitute the seed for x to find the first orbit.

Suppose you choose 4 as the seed. The first orbit is $\frac{4 + \frac{21}{4}}{2} = 4.625$.

1. Use the first orbit as your new seed, and find the second orbit.

____

2. Use the second orbit as your new seed, and find the third orbit.

____

3. Repeat the process two more times. What is your final orbit?

____

4. Describe any patterns you notice. How is the final orbit related to $\sqrt{21}$?

____

5. Find the first four orbits of $g(x) = \frac{x + \frac{16}{x}}{2}$ by choosing an initial seed that is less than zero. What happens to the orbits?

____

6. Find the first four orbits of $h(x) = \frac{x + \frac{7}{x}}{2}$ by starting with 2 as the initial seed. Explain the relationship between the orbits and the square root of 7.

____

NAME ______________________ CLASS ____________ DATE ____________

Basic Skills Practice

Interpreting Bar and Circle Graphs, Part 2

Read each question and circle the best answer.

Use the bar graph to answer Exercises 1–4.

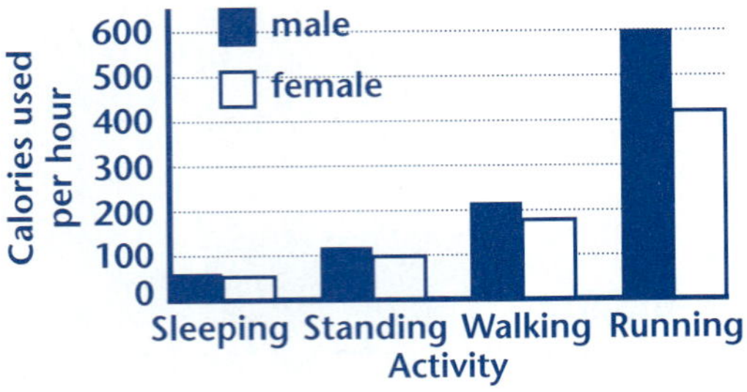

1. How many calories does a female use during 1 hour of standing?

A 200
B 100
C 150
D 50

2. How many calories does a male use by walking for 1 hour?

F 180
G 220
H 440
I 600

3. How many more calories does a male use than a female during 1 hour of running?

A 100
B 150
C 200
D 300

4. Which activity uses the fewest calories in 1 hour?

F walking
G standing
H sleeping
I running

Use the bar graph to answer Exercises 5–8.

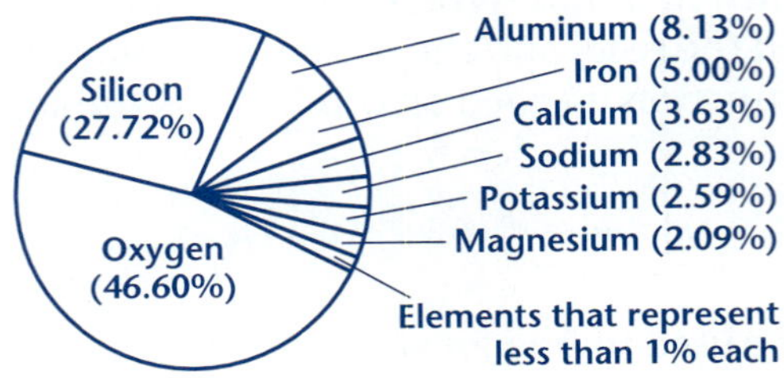

5. Which element makes up almost half of the Earth's crust?

A iron
B oxygen
C silicon
D aluminum

6. What percent of the Earth's crust is made up of sodium?

F 2.83%
G 3.63%
H 27.72%
I 30.52%

7. Oxygen and silicon make up 74.32% of the Earth's crust. What percent do the rest of the elements represent?

A 35.68%
B 25.68%
C 15%
D 2%

8. What percent of the Earth's crust is made up of elements that are present in less than 1%?

F 1%
G 1.41%
H 3%
I 4.3%

LESSON 9.3

Solving Radical Equations

PREPARE

Pacing Guide

3 Days

Objectives and Overview

- Solve equations containing radicals.
- Solve equations by using radicals.

Equations that contain radicals are solved by using graphs and by squaring both sides of the equation. Quadratic equations are solved by using square roots.

Assessing Prior Knowledge

Simplify.

1.	$(3\sqrt{5})^2$	$[45]$
2.	$\sqrt{9} \cdot \sqrt{8}$	$[6\sqrt{2}]$
3.	$7\sqrt{5} - 9\sqrt{5}$	$[-2\sqrt{5}]$
4.	$\sqrt{\frac{120}{16}}$	$\left[\frac{\sqrt{30}}{2}\right]$
5.	$\frac{\sqrt{9}}{\sqrt{8}}$	$\left[\frac{3\sqrt{2}}{4}\right]$

Approach A

Using the Book

Opening Discussion

Equations involving radicals are often used in science to model natural phenomena. In this lesson, students will see examples involving the swing of a pendulum, kinetic energy, and the balance of a beam on a fulcrum. Begin with the pendulum. A simple pendulum can be made from a string with a washer attached to one end. Demonstrate what is meant by a full swing of the pendulum.

Modeling the Swing of a Pendulum

Read or have a student volunteer read the description of the procedure used to develop a model for the time required for a full swing of a pendulum with a given length. To check that $a = \frac{1}{5}$ works for the data, have students enter the data into the statistical registers of a graphics calculator. Then enter the expression $Y_1 = (1/5)\sqrt{\ }(X)$ in the Y= list.

continued on top of next page

TEACH

Approach B

Using Models

Opening Discussion

What natural phenomena can be modeled by a square-root function? How is the model used to better understand the phenomena? In this lesson, students will see examples of square-root functions that model pendulums, kinetic energy, and the weights that balance a beam on a fulcrum. Students will also learn techniques to solve the equations for various values in order to find out more about the situation.

Using a Square-Root Function to Model the Motion of a Pendulum

The objective of this activity is to produce a set of real-world data that can be modeled by a square-root function. The data produced is the time required for a full swing of a pendulum of various lengths. Explain what a pendulum is and illustrate a full swing (one complete back-and-forth motion).

continued on bottom of next page

Approach A — Using the Book, continued

Graph both the plot and the function together to see how well the model fits the data. If graphics calculators are not available, you can substitute the *l*-values from the data table into the equation $t = \frac{1}{5}\sqrt{l}$ and compare these *t*-values with the actual values in the table. To see how close the science class's data is to the theoretical value given by the formula, rewrite the formula as $t = \frac{2\pi}{\sqrt{g}}\sqrt{l}$ and then replace *g* with 980. The constant $\frac{2\pi}{\sqrt{980}} \approx 0.201$ is very close to the class's constant of $\frac{1}{5}$, or 0.2.

Use **Transparency 42** for discussion of the graph on page 428 of the Pupil's Edition.

Ongoing Assessment, page 428

3 seconds

Example 1

In part A, the formula that relates the swing time and length of a pendulum is used to find the swing time of a pendulum with a given length. This is done by using substitution and a calculator. In part B, the same formula is used to find the length of a pendulum with a given swing time. In this case, after substituting the value for *t*, the formula must be solved for *l*. The definition of square root is used to eliminate the radical.

Teaching Tip

Technology To minimize the rounding error that occurs when using a calculator, encourage students to solve for the desired variable before performing any calculations, as was done for *l* in part B of Example 1. They should then enter the complete expression and evaluate it in one step. For part B of Example 1, enter $[3\sqrt{980}/(2\pi)]^2$.

TEACH

Approach B — Using Models, continued

You will need at least six strings of different lengths, each with a washer tied to the end. These will function as the pendulum. You will also need a meter stick, something to hang the string from, and a stopwatch.

Hang one of the strings. Have student volunteers come to the front of the room to measure the string in centimeters, swing the pendulum, and time a full swing in seconds. Have another student make a table on the chalkboard or overhead that includes the length of the string in centimeters, *l*, in the first column and the time in seconds, *t*, that it takes for the pendulum to make one full swing in the second column. Ask another group to come up and collect data for the next string length. Continue this process until all six string lengths have been used.

Creating a Model Use a graphics calculator with the overhead to make a scatter plot of the data with *l* on the horizontal axis and *t* on the vertical axis. Include (0, 0) as the point representing the time that it takes for a pendulum of 0 centimeters to complete a swing. What function resembles the graph? **[the square-root function]** How has the principal square-root function been altered? **[It has been stretched.]** Thus, an equation for this phenomenon is $l = a\sqrt{t}$, where *a* is the scale factor of the stretch. Use guess-and-check to find an approximation for *a*. Begin with $a = 0.1$. Graph the function $y = 0.1\sqrt{x}$ as a scatter plot. How well does it fit the data points? Continue to guess-and-check values for *a* until a reasonably good match has been found.

Using the Model Show students how to use the model to predict the swing time for pendulums of other lengths. How can it be used to predict the length of a pendulum whose swing time is not one of those collected? Substitute a value for *t* and demonstrate how to solve the equation for *l* by using the definition of square root.

Additional Example 1

The time in seconds that it takes an object near Earth's surface to fall d feet is given by the equation $t = \frac{1}{4}\sqrt{d}$.

A. Find the time it takes for an object to fall 368 feet. **[about 4.8 seconds]**

B. Find the distance that the object falls in 2.4 seconds. **[about 92.2 feet]**

Example 2

An equation involving a radical is solved by using a graphics calculator and then by using algebra to square both sides of the equation and eliminate the radical. An appropriate window on the graphics calculator is one in which the x-coordinates are multiples of 0.1. Use the settings Xmin = −9.4, Xmax = 9.4, Ymin = −6.2, and Ymax = 6.2. Note also that parentheses must be used to enter $\sqrt{x + 2}$.

Additional Example 2

Solve $\sqrt{x - 4} = 5$ by the methods below.

A. graphically **[29]**

B. algebraically **[29]**

Ongoing Assessment, page 430

Try This

9.5

Ongoing Assessment, top of page 430

$x \geq -2$; the radicand must be positive. The domain can be seen from the graph in the solution to part A of Example 2.

Example 3

An equation containing a radical expression and another variable term is solved. The radical is eliminated by squaring both sides of the equation. Remind students that the radical expression should be isolated on one side of the equation before squaring. This method may

continued on top of next page

The Theoretical Model

The equation $t = 2\pi\sqrt{\frac{l}{g}}$ is the theoretical model for the swing time in seconds of a pendulum whose length is l centimeters. The constant g is the acceleration due to gravity (980 centimeters per second squared). Show the class how to simplify this formula by using 980 for g so that they can compare the constant with the value found experimentally. Ask students to find the time that it takes a 75-centimeter pendulum to make a full swing and the length of a pendulum that takes 3 seconds to make a full swing. Point out that for the second problem, they will need to isolate $\sqrt{l}$ before squaring. (Compare Example 1 on page 429.)

Cooperative Learning

This activity requires the separation of the class into two different groups: expert groups and study groups. Initially, divide the class into four groups; each of these groups should work to become experts at one of the four types of equations that involve radicals. The class is then reorganized into study groups containing at least one expert on each type of problem. Every student in the study group is responsible for teaching the other group members how to solve the type of equation on which they are experts.

Teaching Tip

To provide motivation and assessment for this activity, explain that after the activity you will give a quiz on the four types of equations. When grading the quiz, figure the class's average score on each type of equation. Award bonus points to the expert group whose equation has the highest average. Then calculate each study group's average quiz score. Award bonus points to the study group that has the highest average.

Four Types of Equations

Assign one of the following types of equations to each of the expert groups. Give the group a handout that includes one example of this type of problem worked out in detail and several examples with answers for students to practice. Explain that the goal of this part of the activity is for all members to become an expert at their

continued on bottom of next page

Approach A Using the Book, continued

produce an **extraneous solution,** which is a solution to the equation that is formed by squaring both sides and that is not a solution to the original equation. For this reason, it is imperative that students check all solutions to radical equations. Note that the number of solutions can also be seen by graphing each side of the equation and counting the intersections.

Additional Example 3

Solve $\sqrt{2x + 7} = x + 2$. [1]

Ongoing Assessment, bottom of page 430
The graphs intersect at only one point.

Solving Equations by Using Radicals

In the previous examples, an equation containing a radical expression was solved by squaring each side. Here, equations containing squared expressions are solved by taking the square root of each side. The simplest of these problems was seen previously in Chapter 8. Remind students that they must include the positive and negative square roots.

Example 4

Quadratic equations are solved by using square roots. Point out that in each case, the term that was being solved for was isolated on one side of the equation before the square roots were taken. Stress that this method leads to two answers. Extraneous answers are not possible in this type of problem. However, the negative answer may need to be eliminated in application problems that involve only positive quantities such as distance or length. Note that answers are given in simplest radical form.

TEACH

Approach B Using Models, continued

group's type of equation. Later, they will help other students learn how to solve this type of equation.

After a specified amount of time, have students meet with their study groups. Each member should be given time to explain and discuss the procedure for solving his or her type of equation. They should help the other group members with a practice problem and answer any questions.

Here is a description of each type of problem, including the procedure typically used to solve each, and an example. For other examples, see pages 429–432 of the Pupil's Edition.

1. Equations With a Radical on One Side and a Constant on the Other

These problems are of the form $\sqrt{x + a} = b$. Students should know how to solve these equations by using a graphics calculator. To do this, they enter the functions $y = \sqrt{x + a}$ and $y = b$. The solution is the x-coordinate of the intersection. They also need to know how to solve these by using algebra. To do this, they should square each side of the equation, which eliminates the radical, and then solve for x.

Example: $\sqrt{x - 4} = 5$ [29]

2. Equations With a Radical on One Side and a Linear Expression on the Other

These problems are of the form $\sqrt{x + a} = x + b$. Students need to know how to solve these types of equations by using algebra. To do this, they should square each side of the equation, which eliminates the radical, and then solve for x. When squaring the right-hand side, they must square a binomial. The equation is then written in standard form and factored. Note that this procedure can produce an **extraneous solution;** this is a solution to the equation that is formed by squaring both sides and that is not a solution to the original equation. For this reason, it is imperative that students check all solutions to radical equations. Note that the number of solutions can also be found by

Additional Example 4

Solve the following equations for x:

A. $x^2 = 75$ $[x = \pm 5\sqrt{3}]$

B. $x^2 = 13^2 - 12^2$ $[x = \pm 5]$

C. $x^2 = y^2 - z^2$ $[x = \pm\sqrt{y^2 - z^2}]$

Example 5

The formula for the area of a circle is used to find the radius of a circle for a given area. Note that the negative solution to the equation is discarded.

Additional Example 5

The area of a circular stained-glass window is 336 square feet. Find the radius of the window. **[about 10.3 feet]**

Example 6

Radicals are used to solve a quadratic function that is a perfect-square binomial on one side. You might also want to have students use the quadratic formula to verify the solution.

Additional Example 6

Solve the equation $x^2 + 14x + 49 = 81$. **[−16, 2]**

Ongoing Assessment, page 432

$10^2 - 8(10) + 16 \stackrel{?}{=} 36$

$36 = 36$ True

$(-2)^2 - 8(-2) + 16 \stackrel{?}{=} 36$

$36 = 36$ True

Critical Thinking, page 432

The graph of $y = x^2 - 8x + 16$ is a parabola translated 4 units to the right. It intersects the graph of $y = 36$ at the points $(-2, 36)$ and $(10, 36)$.

graphing each side of the equation and counting the points of intersection.

Example: $\sqrt{2x + 7} = x + 2$ [1]

3. Simple Quadratic Equations That Are Solved by Using Radicals

These problems are of the form $ax^2 = d$. After dividing by a to isolate the variable, this problem is solved by using the generalization learned in Chapter 8, namely, that for $k \geq 0$, if $x^2 = k$, then $x = \pm k$. An example of a literal equation should be included. Also include an application problem that involves finding the radius of a circle with a given area.

Examples:

A. $x^2 = 75$ $[x = \pm 5\sqrt{3}]$

B. $x^2 = 13^2 - 12^2$ $[x = \pm 5]$

C. $x^2 = y^2 - z^2$ $[x = \pm\sqrt{y^2 - z^2}]$

D. The area of a circular stained-glass window is 336 square feet. Find the radius of the window. **[about 10.3 feet]**

4. Quadratic Equations With Perfect-Square Binomials

In this type of problem, one side of the equation is a perfect-square binomial. The equation is solved by rewriting it in the form $(x + a)^2 = k$ and then by using the generalization described in (3) to conclude that $x + a = \pm\sqrt{k}$. Finally, the equation is solved for x.

Example: $x^2 + 14x + 49 = 81$ **[−16, 2]**

Closing the Lesson

Work with the class to write on the chalkboard the names of the four types of equations that were solved. For each type of problem, write the characteristics under the corresponding name.

ASSESS

Selected Answers

Odd-numbered Exercises 7–55 and odd-numbered Extra Practice

Assignment Guide

Core level students should complete Exercises 1–55.

Core-plus level students should complete Exercises 1–11, 14–30, and 32–56.

Error Analysis

Students often forget to check and eliminate extraneous solutions when solving radical equations. In order to avoid this problem, you can suggest that they check all of their solutions by using graphics calculators. When using radicals to solve quadratics, it is common to forget the solution that contains the negative square root. Remind students that every positive number has both a positive and a negative square root.

Alternative Assessment

Posters Ask students to make a poster illustrating either how to solve a radical equation or how to use radicals to solve a quadratic equation. The poster should include a description of the type of equation, an explanation of the procedure used to solve the equation, and a worked-out example. Display these posters to remind all students of these problems.

Portfolio Activity Have students include one example like Exercise 17 in their portfolio. Then change the problem to $\sqrt{x+3} = \sqrt{x} + 1$. Have students square both sides, simplify, isolate the remaining radical term, and square again to find the solution. [**The solution is 1.**] Remind students to check all solutions.

RETEACH

Introduce the idea of inverse operations. For example, adding and subtracting are inverse operations. If you add a number to an expression and then subtract it, you are left with the original expression. One inverse operation undoes the effect of the other. Similarly, multiplication and division are inverse operations. Squaring and taking the square root are also inverse operations when restricted to positive numbers. (When negative numbers are included, squaring may produce numbers that are not really solutions; checking all answers will solve this problem.) When square roots are taken to undo a square, both the negative and positive square roots must be included.

Give students some additional practice at solving these types of equations. Encourage them to remember that to undo squaring, you take the square root and that to undo a square root, you square.

ENRICH

Interdisciplinary Connection

Physics Exercises 43–45 introduce students to the relationships involved when a beam with weights on each end is balanced on a fulcrum. You can model this process by using a triangular prism for the fulcrum and a small, lightweight board for the beam. You will also need a set of small weights, which you might be able to borrow from a science classroom. Pick out an object whose weight is unknown, and find its weight by using this process. Then weigh the object to determine its true weight.

How Many Solutions?

Ask students to investigate the number of solutions that are possible for a radical equation of the form $\sqrt{x+a} = x + b$. To do this, have students graph both sides of an equation of this form. One graph is a square-root graph and the other is a line. In how many ways can these graphs intersect? They can intersect twice, once, or not at all. These correspond to the number of solutions for this type of problem, either two, one, or none. Challenge students to find one example of an equation of this type for each possible number of solutions.

NAME ______________________ CLASS __________ DATE __________

Practice

9.3 Solving Radical Equations

Solve each equation algebraically. Be sure to check your solution.

1. $\sqrt{x - 7} = 3$ ______________

2. $\sqrt{7 - x} = 3$ ______________

3. $\sqrt{-x - 7} = 3$ ______________

4. $\sqrt{x - 3} = x$ ______________

5. $\sqrt{2x + 3} = \sqrt{x + 7}$ ______________

6. $\sqrt{2x + 3} = 5$ ______________

7. $\sqrt{5 + x} = 3$ ______________

8. $\sqrt{11 - 10x} = 9$ ______________

9. $\sqrt{3x - 2} = -1$ ______________

10. $\sqrt{x - 2} = 3$ ______________

11. $\sqrt{2x + 1} = x - 1$ ______________

12. $\sqrt{2x + 15} = x$ ______________

Use graphing technology or sketch the graphs on your own graph paper to solve each equation, if possible.

13. $4\sqrt{x} = -x - 3$ ______________

14. $\sqrt{x} + 0.1 = 1.0$ ______________

Solve each equation, if possible. If not possible, explain why.

15. $5x^2 = 125$ ______________

16. $x^2 - 1 = 120$ ______________

17. $4x^2 = 120$ ______________

18. $\sqrt{x} = -49$ ______________

19. $x^2 + 169 = 0$ ______________

20. $\sqrt{x} + 20 = 5$ ______________

21. $\sqrt{2x} = 50$ ______________

22. $\sqrt{x - 1} + 6 = 2$ ______________

23. $\sqrt{x + 4} = 1$ ______________

Lesson Activity
9.3 Finding the Significance of a Nonsolution

When you solve $\sqrt{x+2} = x$ algebraically, the possible solutions are $x = 2$ and $x = -1$. When you check each possible solution by substituting it into the original equation, 2 is a solution and -1 is not a solution. What is the significance of $x = -1$?

Consider what happens when you square both sides of the equation and simplify:

$$\left(\sqrt{x+2}\right)^2 = x^2$$
$$x + 2 = x^2$$

Square both sides and simplify each equation.

1. $\left(-\sqrt{x+2}\right)^2 = (-x)^2$ ______________

2. $\left(-\sqrt{x+2}\right)^2 = x^2$ ______________

3. $\left(\sqrt{x+2}\right)^2 = (-x)^2$ ______________

4. Compare the results. What do you notice?

____

Use a graphics calculator to graph each of the following equations. Enter the left side of the equation as Y1, and enter the right side of the equation as Y2. How many points of intersection does each graph have?

5. $\sqrt{x+2} = x$ ______________

6. $-\sqrt{x+2} = -x$ ______________

7. $-\sqrt{x+2} = x$ ______________

8. $\sqrt{x+2} = -x$ ______________

Find the *x*-value of the points of intersection for each of the graphs in Exercises 5–8.

9. Exercise 5 ______________

10. Exercise 6 ______________

11. Exercise 7 ______________

12. Exercise 8 ______________

13. Compare the *x*-value of the intersection points for each graph with the possible algebraic solutions $x = 2$ or $x = -1$. Why is $x = 2$ the only solution to the original equation?

____

14. For which equation(s) in Exercises 5–8 is $x = -1$ a solution? ______________

15. What is the significance of $x = -1$? ______________

NAME ______________________ CLASS ____________ DATE ____________

Basic Skills Practice

Interpreting Line Graphs and Scatter Plots, Part 1

Line graphs are made up of connected points. Each point connects a horizontal value and a vertical value.

Example 1: The graph shows the number of cars sold by month over a 4-month period. By how much did sales increase or decrease between March and April?

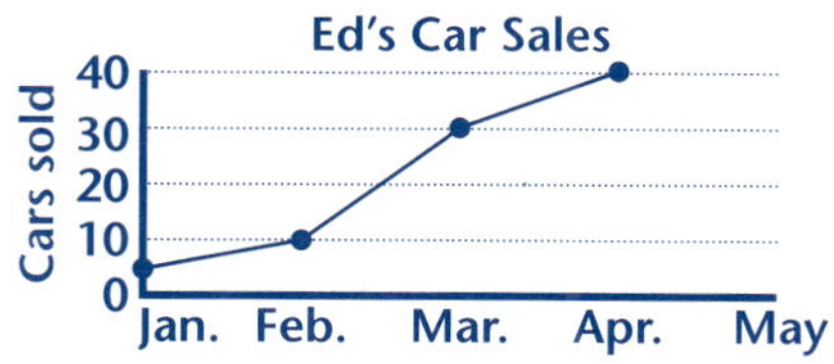

30 cars were sold in March. 40 cars were sold in April. Sales increased by 10 cars between March and April.

Scatter plots also contain points that connect a horizontal value with a vertical value. Scatter plots are used to find trends between two sets of data that show positive, negative, or no correlation.

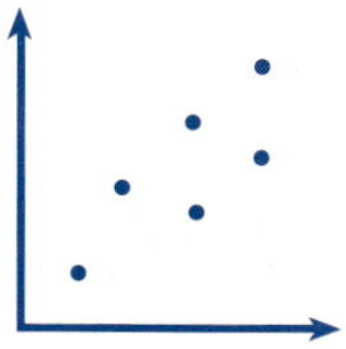

Positive correlation
Both sets of data tend to increase together.

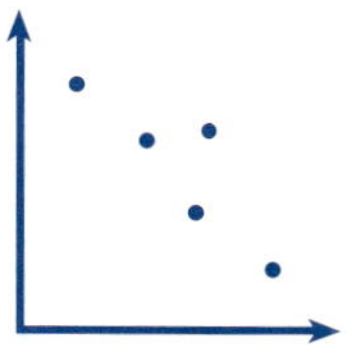

Negative correlation
One set of data tends to increase as the other decreases.

No correlation
The sets of data are not related.

Solve each problem.

The graph at right shows the enrollment at a college.

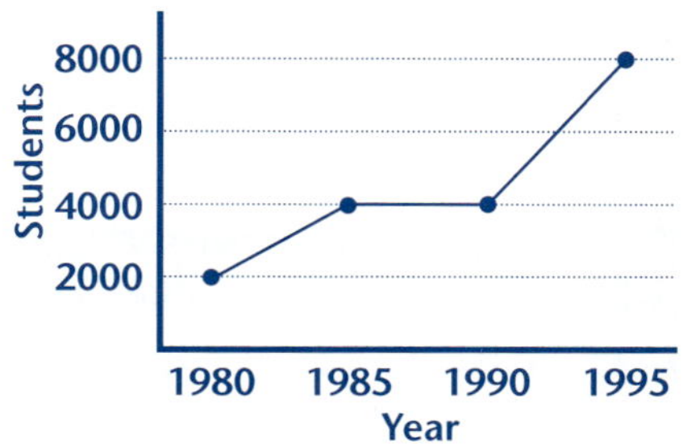

1. How many students were enrolled in 1980? ____________

2. In what year were 8000 students enrolled? ____________

3. How many students were enrolled in 1985? ____________

The scatter plot at right shows the relationship between the temperature and the distance that several athletes were able to run in a set amount of time.

4. As temperature increases, the distance tends to ____________.

5. What correlation does the scatter plot show? ____________

LESSON 9.4

The "Pythagorean" Right-Triangle Theorem

PREPARE

Pacing Guide
3 Days

Objectives and Overview
- Examine a right triangle to determine a relationship among the sides.
- Use the "Pythagorean" Right-Triangle Theorem to solve problems involving right triangles.

Right triangles are examined to see that the sum of the squares of the lengths of the legs is equal to the square of the length of the hypotenuse. The "Pythagorean" Right-Triangle Theorem is used to find the missing lengths of right triangles and the height of a pyramid.

Vocabulary
hypotenuse
leg
"Pythagorean" Right-Triangle Theorem
slant height

Assessing Prior Knowledge

1. Solve $a^2 + b^2 = c^2$ for c.
   [$\boldsymbol{c = \sqrt{a^2 + b^2}}$]
2. Solve $a^2 + b^2 = c^2$ for b.
   [$\boldsymbol{b = \sqrt{c^2 - a^2}}$]
3. Solve $a^2 + b^2 = c^2$ for a.
   [$\boldsymbol{a = \sqrt{c^2 - b^2}}$]

Approach A: Using the Book

Opening Discussion
Use the opening photograph of the ruins of the Parthenon to remind students of the definition of the hypotenuse and legs of a right triangle. Ask, "What is the relationship among the three sides of a right triangle?" Work as a class to complete the table. Have students write the relationship shown in the table in words. Ask a few volunteers to read their writing.

Brainstorming
Have students look around the room and find situations in which this triangle property can be applied. [**Examples include finding the length of a diagonal of the chalkboard if the lengths of its sides are known, finding the length of a diagonal of a rectangular desktop if the length and width of the desktop are known, and finding the length of one side of a rectangular floor if the lengths of a diagonal and the other side are known.**] You can

TEACH

Approach B: Using Models

Opening Discussion
Draw a right triangle on the chalkboard or overhead, and remind students of the definition of the hypotenuse and legs of a right triangle. What is the relationship between the lengths of the three sides of a right triangle? Work as a class to complete the table on **Transparency 43** (the table of the side lengths of a right triangle from page 435). Have students write the relationship shown in the table in words. Ask a few volunteers to read their writing.

Brainstorming
Have students look around the room and find situations for which this triangle property can be applied. [**Examples include finding the length of a diagonal of the chalkboard if the lengths of its sides are known, finding the length of a diagonal of a rectangular desktop if the length and width of the desktop are known, and finding the length of one side of a rectangular floor if the lengths of a diagonal and the other side are known.**] You can have students use a ruler to find some of these measurements and use

have students use a ruler to find some of these measurements and use the property to find the unknown measurement. Then they should measure the actual length and compare it with their results.

Ongoing Assessment, page 435
The last two columns of the table have equal entries, so $c^2 = a^2 + b^2$.

The "Pythagorean" Right-Triangle Theorem

Write the "Pythagorean" Right-Triangle Theorem on the chalkboard or overhead. Discuss the graphic representation of the 3-4-5 triangle. Students should understand that the squares on the legs and the hypotenuse represent the squares of the lengths. Have students find the area of each square to see that the sum of the area of the squares on the legs is equal to the area of the square on the hypotenuse. Use **Transparencies 43** and **44** for discussion or review.

Inclusion Strategies

Visual and Verbal Models Encourage students who are strong visual learners to remember the graphic showing the squares on the legs and hypotenuse to help them remember the "Pythagorean" Right-Triangle Theorem. Students who are strong verbal learners can remember the theorem as "leg squared + leg squared = hypotenuse squared."

Example 1

The "Pythagorean" Right-Triangle Theorem is used to find the missing length in a right triangle and to find a formula for the length of the diagonal of a square. Point out that the assignment of sides *a* and *b* can be reversed. In part B, the squared term must be isolated before taking the square roots.

continued on top of next page

the property to find the unknown measurement. Then they should measure the actual length and compare it with their results.

The "Pythagorean" Right-Triangle Theorem

Explain that the relationship they have found is called the "Pythagorean" Right-Triangle Theorem. Write the formal statement of this property on the chalkboard for students to copy in their notebook or journal. This property was known to many ancient civilizations, including the Chinese, Babylonians, and Hindus. While Pythagoras was not the first to discover this relationship, his name is the one most commonly used to name it.

Inclusion Strategies

Visual and Verbal Models Encourage students who are strong visual learners to remember the graphic showing the squares on the legs and hypotenuse to help them remember the "Pythagorean" Right-Triangle Theorem. Students who are strong verbal learners can remember the theorem as "leg squared + leg squared = hypotenuse squared."

A Hands-on Proof of the "Pythagorean" Right-Triangle Theorem

This activity gives students a chance to experience one of the proofs of this theorem. Explain to students that although the table of side lengths that they constructed helped them to discover the theorem, they do not know from the table whether the theorem will hold true for the sides of any triangle. A proof shows that the theorem is true for every case.

Creating a Model On one sheet of paper, draw four small congruent right triangles, a square whose side length is equal to the length of the smaller leg of the triangles, a square whose side length is equal to the length of the longer leg of the triangles, and a square whose side length is equal to the length of the hypotenuse of the triangles. Also draw a square (later called "the board") whose side length is equal to the sum of the side lengths of the two smaller squares. Label the triangles and three smallest squares as shown on the next page.

continued on bottom of next page

Approach A — Using the Book, continued

Additional Example 1

Find the length of x, y, and z.

A.

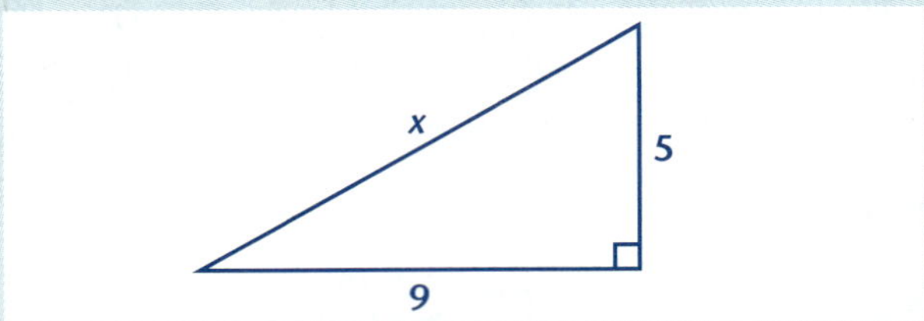

[$\sqrt{106}$, or about 10.30]

B.

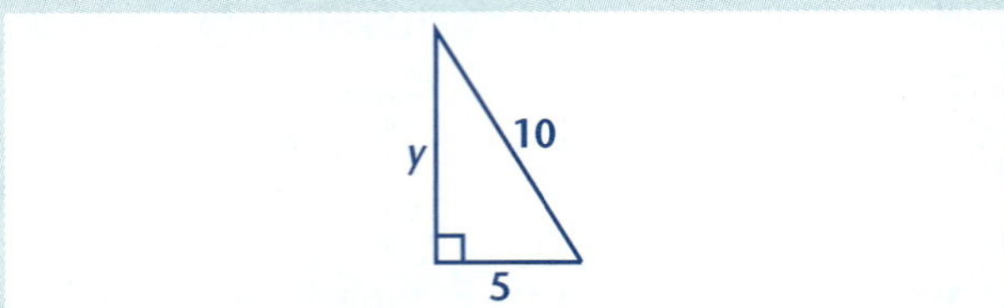

[$5\sqrt{3}$, or about 8.66]

C.

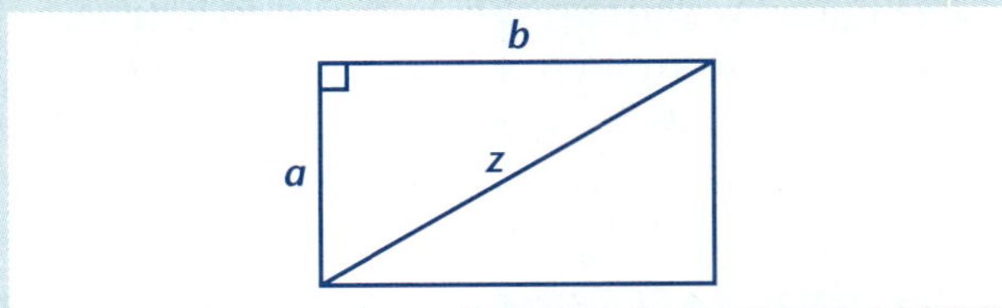

[$\sqrt{a^2 + b^2}$]

Example 2

The "Pythagorean" Right-Triangle Theorem is used to find the distance between two boats. The speed of each boat must be used to find the distance traveled. Encourage students to draw a diagram to represent this situation.

Additional Example 2

Two trains leave from a station at the same time. One is traveling due west at 50 miles per hour. The other is traveling due south at 35 miles per hour. Approximately how far apart will the trains be after 2 hours? **[about 122 miles]**

TEACH

Approach B — Using Models, continued

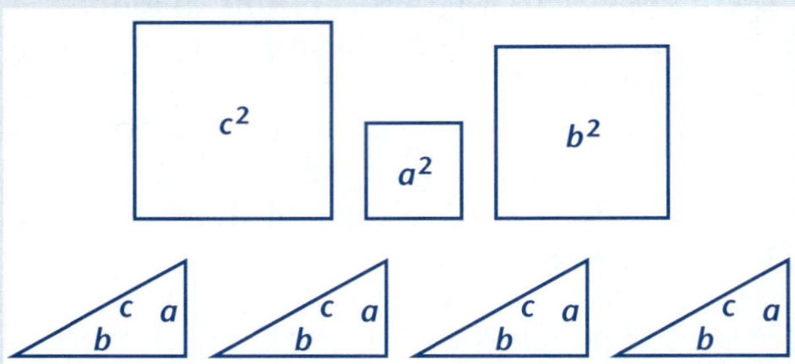

Make copies of this sheet of paper so that there is one for each member of the class. Cut out the pieces on each sheet and put them into an envelope. Each student should have an envelope with all the pieces in it. After the proof, these models can be collected and used year after year.

Using the Model Have students clear their desks of books and papers and take out their model. Why are the squares labeled a^2, b^2, and c^2? Have students hold them next to the triangle to see that the lengths of the sides of the triangle equal the lengths of the sides of the squares. The areas of these three squares are thus a^2, b^2, and c^2. The object is to show that the area of the square marked c^2 is equal to the sum of the areas of the two smaller squares. The large blank square is used as the board. When arranging the figures, they must be placed on the board so that they fill all of the space and do not extend beyond it. Have students put aside the two smaller squares and arrange the other pieces on the board as shown below.

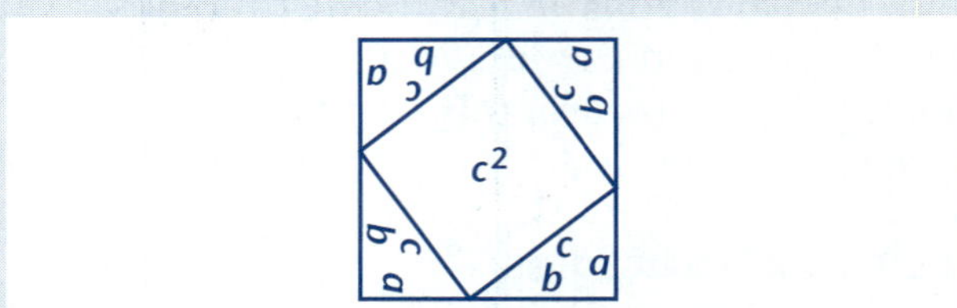

Remove the square labeled c^2. What area is left uncovered? [c^2] If you reposition the triangles, but keep them on the board, how much area is uncovered? [c^2] Bring back the a^2 and b^2 models. Challenge students to rearrange the triangles so

Ongoing Assessment, page 437
Try This
60

Ongoing Assessment, page 437
All triangles with side lengths that are multiples of 3, 4, and 5, respectively, are similar; thus, all are right triangles.

A Formula for Finding an Unknown Leg

Have students solve the formula $a^2 + b^2 = c^2$ for a, assuming that a is positive. [$a = \sqrt{c^2 - b^2}$] This gives a formula for finding the unknown length of a leg of a right triangle.

A physical model is useful when discussing the dimensions of a pyramid.

continued on top of next page

that these two squares will also fit on the board. Two possible arrangements are shown below.

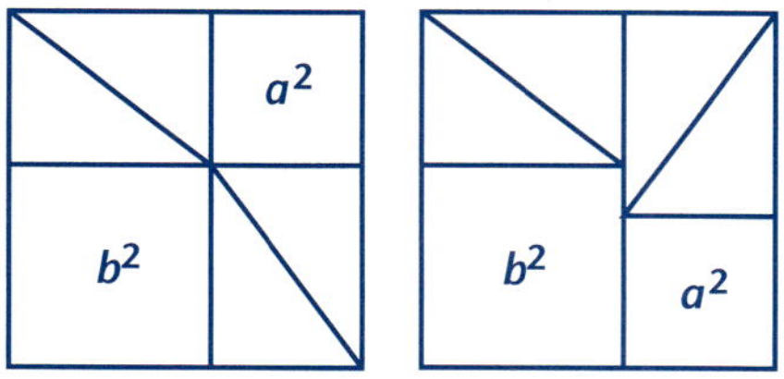

The two smaller squares exactly covered the area that equaled c^2, and thus the theorem is proven.

Examples

Work with the class to solve some problems that require the "Pythagorean" Right-Triangle Theorem. Include one in which the theorem is used to find the length of one of the legs and one in which the theorem is used to find the length of a diagonal of a square or a rectangle. (For ideas, see Example 1 on page 436.)

A Motion Problem

Here is a different kind of example that uses the "Pythagorean" Right-Triangle Theorem. Work through this example with the class. Draw a diagram to represent the distance that each train travels.

Two trains leave a station at the same time. One is traveling due west at 50 miles per hour. The other is traveling due south at 35 miles per hour. Approximately how far apart will the trains be after 2 hours? **[about 122 miles]**

Cooperative Learning

This activity provides students with hands-on experience at using the "Pythagorean" Right-Triangle Theorem to find distances in space.

You will need to work in a rectangular room. First have some student volunteers measure the dimensions of the floor of the room and the height of the room with a measuring tape. Write these measurements on the chalkboard. Explain that the object of this activity is to find the distance from the top-right corner in the front

continued on bottom of next page

Approach A

Using the Book, continued

Example 3

The formula for finding the unknown length of a leg is used to find the height of a square-based pyramid whose base length and slant height are known. It may help students to visualize this problem if you show a model of a pyramid. Remind students that they must divide the base in half to get one leg of the triangle.

Use **Transparency 45** for discussion.

Additional Example 3

Find the height of a cone that has a slant height of 14 centimeters and a radius of 10 centimeters. **[about 9.80 centimeters]**

Critical Thinking, page 438

By solving the "Pythagorean" Right-Triangle Theorem for c, it is clear that the expression $\sqrt{a^2 + b^2}$ is the length of the hypotenuse of a right triangle with legs of lengths a and b. The expression $a + b$ represents the distance along the legs of the triangle. The inequality shows that the path along the hypotenuse is always shorter than the path along the legs.

TEACH

Approach B

Using Models, continued

of the room to the bottom-left corner in the back of the room. Divide the class into groups of four, and have them work to find the solution. If little or no progress is made after some time, you can give the hint that they will need to know the length of the diagonal of the floor.

Closing the Lesson

Ask the groups to share their work with the class. How similar are the answers? Where in the problem was the "Pythagorean" Right-Triangle Theorem used? Did any groups solve the problem in a different way?

ASSESS

Selected Answers

Odd-numbered Exercises 7–37 and odd-numbered Extra Practice

Assignment Guide

Core level students should complete Exercises1–38.

Core-plus level students should complete Exercises 1–6 and 9–42.

Error Analysis

When finding the length of a leg in a right triangle, students may substitute the length of the hypotenuse for a or b by mistake. Encourage them to visualize the right triangle with the legs labeled a and b and the hypotenuse labeled c. They can also label the sides of the triangle to help them remember.

Alternative Assessment

Student Report Ask students to write a report about the "Pythagorean" Right-Triangle Theorem. A well-written report should include its history and some examples. Display the reports on a bulletin board.

Portfolio Activity Have students work with a partner to edit and revise their responses to Exercises 1–6. They should then include their work in their portfolio.

RETEACH

Using The "Pythagorean" Right-Triangle Theorem to Find the Distance Across a "Lake"

Have students work in small groups to complete this activity. Give each group a roll of string, a right angle, some tape, and a meter stick. Take the class outside or to a large room. Make drawings of various lakes on the chalkboard or distribute handouts showing the lakes. The lakes should be simple and almost convex to avoid complications. Assign a lake to each group. The objective is to measure the distance from point A to point B across the lake. Use the tape to mark points A and B for the students. Point out that the group cannot actually measure across the "lake" because they would get "wet." Instead, they must make a right triangle so that one side of the right triangle is the path from A to B and the other two sides are on land (or almost entirely on land), as in the figure below.

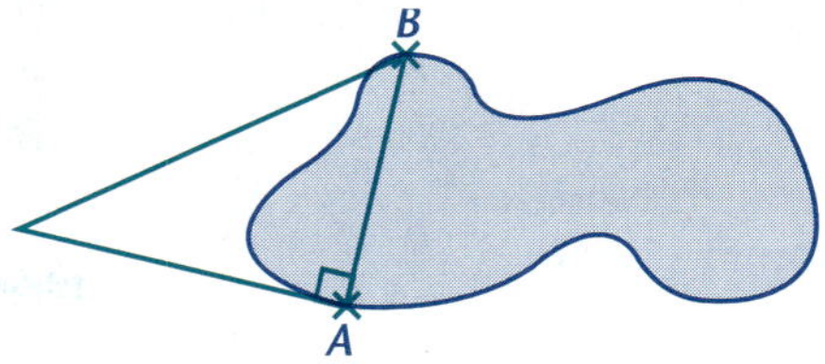

They can measure the sides that are on land and use the theorem to find the distance across the lake.

ENRICH

Interdisciplinary Connection

History Many different cultures knew of the right-triangle relationship known as the "Pythagorean" Right-Triangle Theorem. These cultures included the Chinese, Babylonians, and Hindus. Do research to find out about the mathematics of these ancient cultures. Focus on any references to the right-triangle theorem. Also look for alternative proofs of this theorem. Have students share their findings with the class.

Quilt Making

Here is a common quilt pattern.

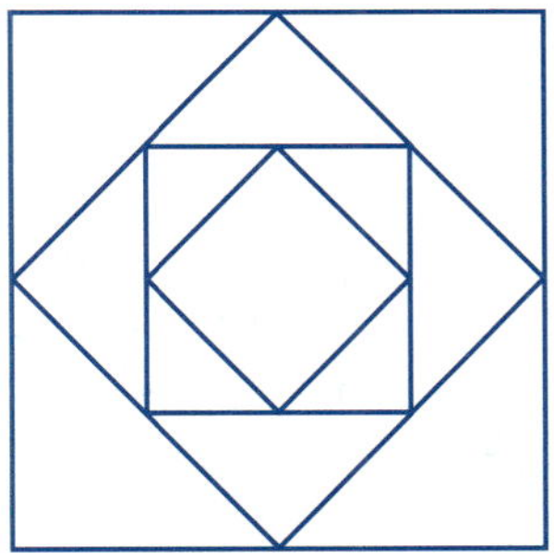

If the outside square has sides of 12 inches, what are the side lengths of the other three squares? **[$6\sqrt{2}$ inches, 6 inches, $3\sqrt{2}$ inches]**

NAME ________________________________ CLASS ______________ DATE ______________

Practice

9.4 The "Pythagorean" Right-Triangle Theorem

Complete the table. Use a calculator and round each answer to the nearest tenth.

	Leg	Leg	Hypotenuse		Leg	Leg	Hypotenuse
1.	21		75	**2.**		80	82
3.	6.5	7.2		**4.**	20		29
5.	13		85	**6.**	33	44	
7.		63	65	**8.**		48	60
9.	28	45		**10.**	10	24	
11.	48		80	**12.**	$\frac{1}{5}$		$\frac{1}{3}$
13.		0.63	0.87	**14.**	0.036	0.077	

Solve for *x*.

15.

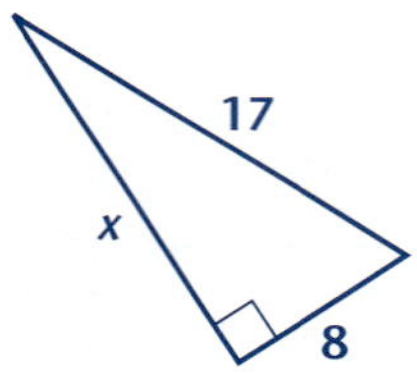

____

16.

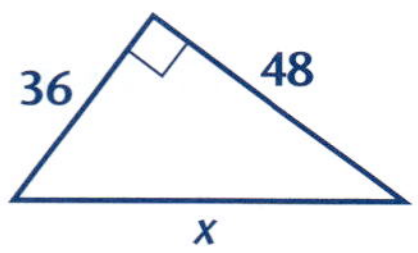

____

17.

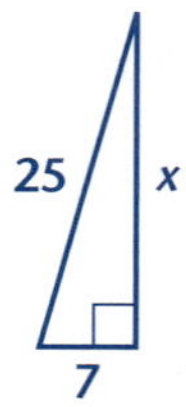

____

18. A picket fence has a gate that is 48 inches wide. The gate is 62 inches high. Find the length to the nearest inch of a diagonal brace for the gate. ______________

19. The base of a 15-foot ladder is placed 9 feet from a wall. If the ladder is leaned up against the wall, how high will it reach? ______________

20. Cynthia drove 56 miles due south and 90 miles due west. If she drives in a straight line back to the point where she started, how far will she drive? ______________

NAME ______________________________ CLASS ____________ DATE ____________

Lesson Activity

9.4 Modeling the "Pythagorean" Right-Triangle Theorem

The informal proof in this activity was discovered by Elisha Loomis and reported in "The Pythagorean Proposition," published by the National Council of Teachers of Mathematics in 1968.

For each proof, fit the pieces of the puzzle together to prove the "Pythagorean" Right-Triangle Theorem.

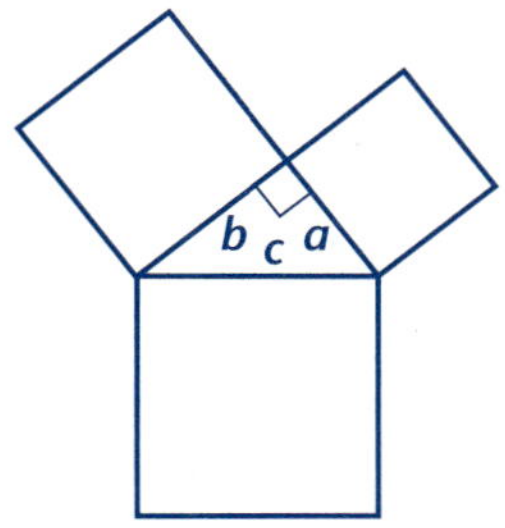

1. Draw a right triangle with sides of *a, b,* and *c*. Construct squares along each side, as shown. Extend the sides of the largest square as shown in the second figure. Construct a perpendicular at the point where the extension intersects the side of the square on the longer leg, as shown in the third figure.

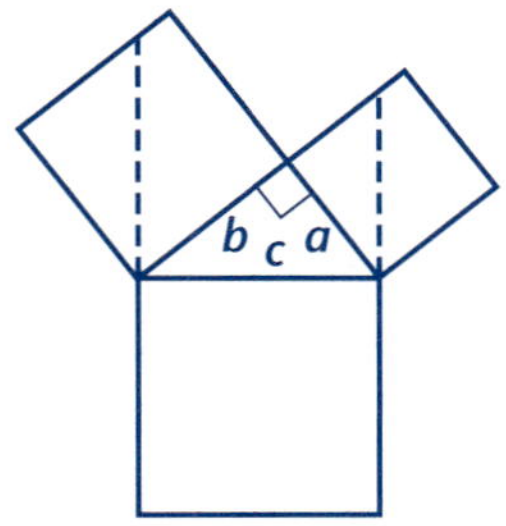

Cut out the 5 pieces that make up the squares along the legs. Reassemble the pieces so that they cover the square along the hypotenuse.

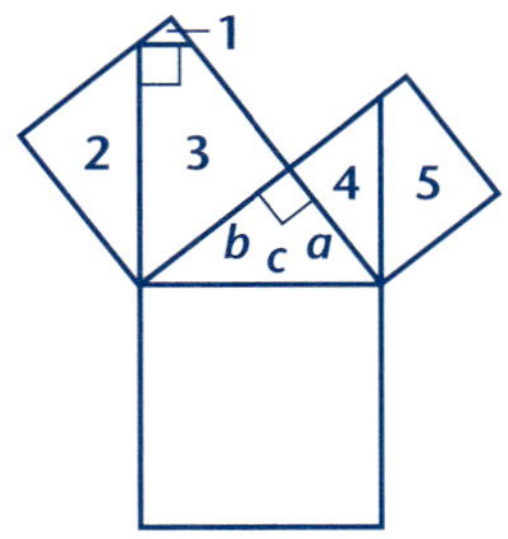

2. Explain how your diagram illustrates the "Pythagorean" Right-Triangle Theorem.

____

____

____

For each model, cut out the pieces that make up the squares along the legs, and reassemble them so that they cover the square along the hypotenuse. Then explain how your diagram illustrates the "Pythagorean" Right-Triangle Theorem.

3.

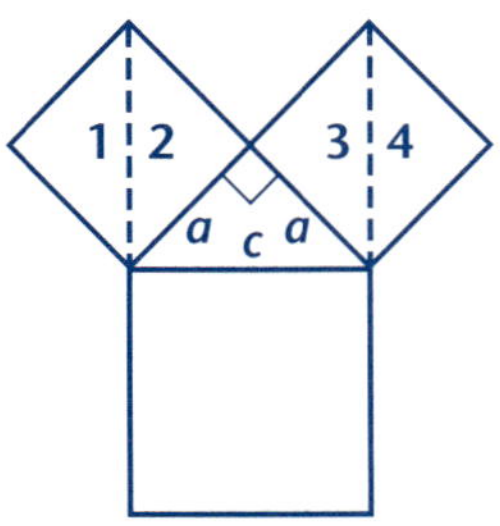

____

____

4.

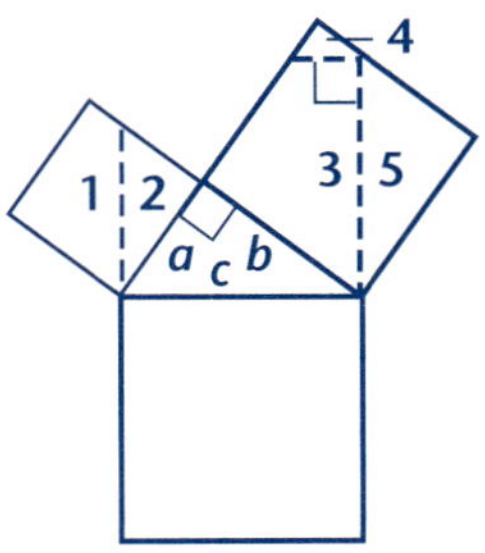

____

____

NAME ______________________ CLASS ____________ DATE ____________

Basic Skills Practice

Interpreting Line Graphs and Scatter Plots, Part 2

Read each question and circle the best answer.

The following graph shows the temperatures in degrees Fahrenheit between 7 A.M. and 11 A.M. at a ballpark.

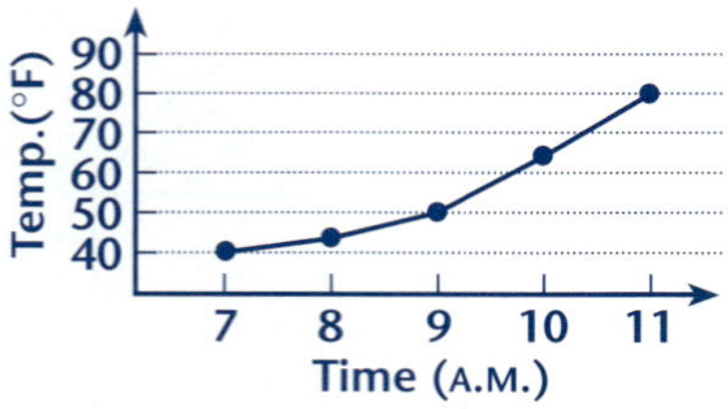

1. What is the highest temperature recorded?

A 10°F
B 11°F
C 80°F
D 90°F

2. When was the lowest temperature recorded?

F 7 A.M.
G 8 A.M.
H 9 A.M.
I 10 A.M.

3. What was the temperature at 9 A.M.?

A 40°F
B 50°F
C 60°F
D 70°F

Greg took 4 tests. His scores are recorded on the line graph below.

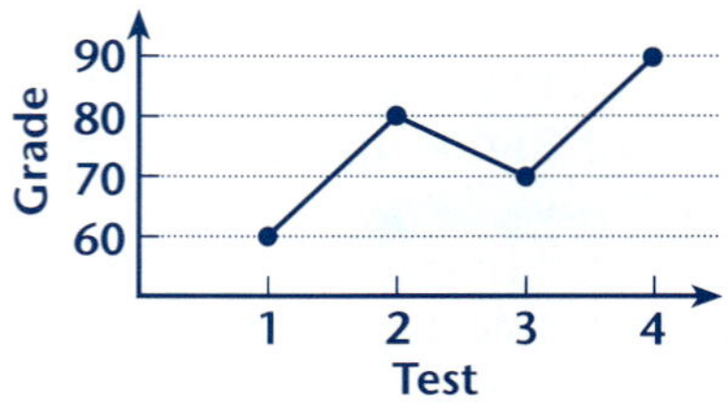

4. On which test did Greg receive the highest score?

F 1
G 2
H 3
I 4

5. On which test did Greg receive a score of 80?

A 1
B 2
C 3
D 4

6. What score did Greg receive on test 3?

F 60
G 70
H 80
I 90

Solve each problem.

7. What is the approximate height for a person weighing 70 pounds? ____________

8. What is the approximate weight for a person who has a height of 55 inches? ____________

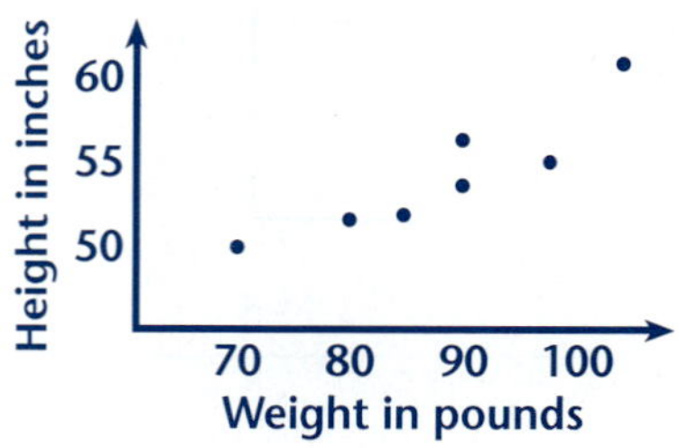

LESSON 9.5 *Exploring* the Distance Formula

PREPARE

Pacing Guide
3 Days

Objectives and Overview
- Use the distance formula to find the distance between two points in the coordinate plane.
- Determine whether a triangle is a right triangle.
- Use the midpoint formula to find the midpoint of a segment.

The distance formula is developed from the "Pythagorean" Right-Triangle Theorem. After introducing the converse of the "Pythagorean" Right-Triangle Theorem, it is used together with the distance formula to prove that a given triangle is a right triangle. Finally, the midpoints of segments are examined to develop the midpoint formula.

Vocabulary
distance	converse
length	

Assessing Prior Knowledge

Find the length of the hypotenuse, given the lengths of the two legs. Round to the nearest tenth if needed.

1. side $a = 6$ cm, side $b = 8$ cm
[10 cm]

2. side $a = 16$ cm, side $b = 11$ cm
[about 19.4 cm]

Approach A: Using the Book

Opening Discussion
How can you find the distance between two points when there is no way to measure it directly? For example, how can you find the distance that a helicopter flies when you are only given distances in the east, west, north, and south directions? Read or have a student read the emergency service example. Have students draw a model of this example on graph paper, including point V and the sides of the right triangle.

Brainstorming
Have students think of other professionals who would need to know the distance between two points. Examples include landscapers, interior designers, and air traffic controllers.

continued on top of next page

TEACH

Approach B: Using Models

Opening Discussion
How can you find the distance between two points when there is no way to measure it directly? For example, how can you find the distance that a helicopter flies when you are given distances only in the east, west, north, and south directions? Consider an emergency medical helicopter that is 1 mile east and 3 miles north of a city and that needs to fly to an accident located 4 miles east and 8 miles north of the city.

Brainstorming
Have students think of other professionals who would need to know the distance between two points. Examples include landscapers, interior designers, and air traffic controllers.

Using a Geoboard
Creating a Model Place rubber bands along the bottom row and left column of the geoboard to

continued on bottom of next page

Approach A — Using the Book, continued

Finding Distance With the "Pythagorean" Right-Triangle Theorem

Once students draw the triangle whose hypotenuse is the path of the helicopter, they should realize that the "Pythagorean" Right-Triangle Theorem can be used to find the distance. Lead students through the steps to find this length. Stress that the triangle drawn for this method must be a right triangle.

Use **Transparency 46** for discussion of the graph on page 442.

Exploration 1

You can lead the class through this exploration or have students complete it in small groups. Students should draw a diagram and go through the steps to find each distance.

Ongoing Assessment, page 442

2. The results should be the same.

The Distance Formula

At the completion of Exploration 1, discuss the formula derived in Step 2. Students should realize that the distance formula is a shortcut for finding the distance between two points whose coordinates are given. Students can use the formula without having to follow the steps used in the exploration. Recalling the steps, however, can help students remember the formula. Have them use the formula to verify their answers to one of the problems in Step 1 of the exploration.

Inclusion Strategies

Some students may prefer to find distances using the three-step algorithm, while others may prefer to memorize the formula and substitute coordinates. Encourage students to understand both methods, but allow them the flexibility to choose the one that works best for them. Similarly, to find a midpoint, some students may prefer to average the coordinates in their heads or with a calculator, while others will feel more comfortable substituting the values into the formula and simplifying.

TEACH

Approach B — Using Models, continued

represent the axes of a coordinate system. Show students how the pegs of the geoboard can represent points with integer-valued coordinates. Use the geoboard to represent the helicopter situation.

Using the Model Place additional rubber bands to form a right triangle with the right angle at (4, 3). Once students see this triangle, they should realize that the "Pythagorean" Right-Triangle Theorem can be used to find the distance. Lead students through the steps to find this length. Start by counting the spaces to find the length of each leg **[3 and 5]** and then using the theorem to find the distance that the helicopter must fly. **[$\sqrt{34}$ miles, or about 5.8 miles]** Stress that in order to use this method the triangle must be a right triangle.

Cooperative Learning

Before starting the explorations, divide the class into groups of four. In Exploration 1, students should take turns making the geoboard model, finding the lengths of the legs, finding the length of the hypotenuse, and recording the example on graph paper. In Exploration 2, students should take turns graphing the segments on the geoboard and guessing the midpoints. Each member should be responsible either for writing the group summary or for reporting the results for one of the explorations. All of the group members should be involved in a discussion of the activity as the exploration proceeds. Each student should sign the completed report.

Alternate Exploration 1

Have students work in their groups. Give each group a geoboard. This time, have them use rubber bands to divide the board into the four quadrants of a coordinate plane. Each group should investigate the following questions, which you may write on the overhead or chalkboard. (For ideas on guiding the students, see Exploration 1 on page 442 of the Pupil's Edition of the textbook.)

The Converse of the Theorem

Discuss the definition of the term *converse* with the class. You may want to write some simple if-then statements on the board and ask students to give the converses. Point out that the converse of a statement is not necessarily true. Read and discuss the converse of the "Pythagorean" Right-Triangle Theorem. When would this theorem be useful? **[when trying to prove that a triangle is a right triangle]**

Critical Thinking, page 443

No. Counterexamples will vary. For instance, "If you are in Chicago, then you are in Illinois" is true. The converse, "If you are in Illinois, then you are in Chicago," is not true.

Example 1

The distance formula is used to verify that the lengths of three sides of a triangle satisfy the "Pythagorean" Right-Triangle Theorem and are thus the lengths of the sides of a right triangle.

Additional Example 1

Given points $A(-2, -5)$, $B(3, -3)$, and $C(1, 2)$, decide whether triangle ABC is a right triangle. **[yes]**

Ongoing Assessment, page 443

Try This

no; no

Critical Thinking, page 443

You can show that the slopes of two of the line segments that make up the sides of the triangle are negative reciprocals. Those segments are perpendicular and thus form a right angle.

Exploration 2

After guessing the midpoints of several segments, students study the coordinates of the midpoint to develop the midpoint formula. You may need to define the term *midpoint* before beginning the exploration. Students can complete this exploration individually or in small groups.

continued on top of next page

- Using the geoboard, model a line segment. Write the coordinates of the two endpoints. Use the method demonstrated for the helicopter example to find the length of the segment. Keep a record of each example on graph paper. Include the coordinates of the three vertices and the lengths of each side of the right triangle formed. Repeat this for four different line segments. Be sure to include endpoints from different quadrants of the graph.
- Study the graph paper illustrations. Find a method for determining the length of each leg from the coordinates of the vertices of the triangle. **[For the horizontal leg, find the difference in x-coordinates. For the vertical leg, find the difference in y-coordinates.]**
- Consider the segment with endpoints $A(x_1, y_1)$ and $B(x_2, y_2)$. Draw a model of this situation on graph paper. Write an expression for the length of the horizontal leg. **[$x_2 - x_1$]** Write an expression for the length of the vertical leg. **[$y_2 - y_1$]** Write an expression for the length of segment AB. **[$\sqrt{(x_2 - x_1)^2 + (y_2 - y_1)^2}$]**

The Distance Formula

At the completion of Exploration 1, discuss the formula derived in the last step. Point out to students that they have found the distance formula, which is a shortcut for finding the distance between two points whose coordinates are given. Have them use the formula to verify their answers to one of the problems in Step 1 of the exploration.

Inclusion Strategies

Some students may prefer to find distances by using an algorithm, while others may prefer to memorize the formula and substitute coordinates. Encourage students to understand both methods, but allow them the flexibility to choose the one that works best for them. Similarly, to find a midpoint, some students may prefer to average the coordinates in their heads or with a calculator, while others will feel more comfortable substituting the values into the formula and simplifying.

continued on bottom of next page

Approach A Using the Book, continued

Ongoing Assessment, page 444

5. (27.5, 41.5)

The Midpoint Formula

After completing the exploration, ask one group to compare the rule they developed in Steps 3 and 4 with the midpoint formula. Note that the midpoint formula is really two formulas, one for each coordinate. It may help some students to find the midpoint if they think of the midpoint formula as averaging the coordinates.

Example 2

The midpoint formula is used to find the point midway between two locations. Point out the notation $(\bar{x}, \bar{y})$, read "x-bar, y-bar," that is commonly used to denote the midpoint.

Additional Example 2

Refer to Example 2. John lives 5 blocks west and 4 blocks south of city hall. His friend Alexandria lives 7 blocks east and 2 blocks north of city hall. What is the halfway point between them? **[The halfway point is (1, −1), or 1 block east and 1 block south of city hall.]**

Example 3

The midpoint and endpoint of a diameter of a circle are used to find the other endpoint of the diameter. Some students may realize that the midpoint of the diameter is the center of the circle. Additional work with circles will be done in the next lesson.

Additional Example 3

The midpoint of a diagonal of a rectangle is $M(5, 4)$. If one endpoint is $A(8, -12)$, what are the coordinates of the other endpoint? **[(2, 9)]**

TEACH

Approach B Using Models, continued

The Converse of the Theorem

Discuss the definition of the term *converse* with the class. You may want to write some simple if-then statements on the board and ask students to give the converses. Point out that the converse of a statement is not necessarily true. Write and discuss the converse of the "Pythagorean" Right-Triangle Theorem. When would this theorem be useful? **[when trying to prove that a triangle is a right triangle]**

Proving a Triangle Is a Right Triangle

Write the following problem on the board: Given the points $A(-2, -5)$, $B(3, -3)$, and $C(1, 2)$, decide whether triangle ABC is a right triangle. **[yes]** You can have students work through this problem in their cooperative groups, or you can lead the students through the solution. (Compare Example 1 on page 443.)

Alternate Exploration 2

Have students work in their groups. Give each group a geoboard with rubber bands that are to be used to divide the geoboard into the four quadrants of a coordinate plane. Each group should investigate the following questions, which you may write on the overhead or chalkboard. (For ideas on guiding the students, see Exploration 2 on page 444 of the Pupil's Edition of the textbook.)

- Using the geoboard, represent the line segment connecting each pair of points. Use your model to guess the midpoint of each segment.

  A. (1, 3) and (3, 5) **[(2, 4)]**

  B. (0, 4) and (−2, 2) **[(−1, 3)]**

  C. (−1, 3) and (1, −1) **[(0, 1)]**

- Study the midpoints. Find a method for determining the *x*-coordinate of the midpoint of each segment from the *x*-coordinates of its endpoints. Repeat this for the *y*-coordinates. **[In each case, average the coordinates of the endpoints to get the coordinates of the midpoint.]**

Closing the Lesson

Write the distance formula and midpoint formula on the board. Ask students to list the types of problems that were solved with these formulas. Have students copy this information into their journal.

- Use the method described above to find the midpoint of the line segment joining (31, 54), and (24, 89). Verify your results by using the distance formula. [(27.5, 71.5)]

The Midpoint Formula

After completing the exploration, ask one group to describe the rule they developed in the second step. Explain that this rule can be summarized as the midpoint formula. Write the formula on the board, and note that it is really two formulas, one for each coordinate. It may help some students to find the midpoint if they think of the midpoint formula as averaging the coordinates.

Midpoint Examples

Work through these examples as a class, or have students complete them in their cooperative-learning groups. (Compare Examples 2 and 3 on pages 444 and 445, respectively.)

The streets of a city are laid out like a coordinate plane with the origin at city hall. John lives 5 blocks west and 4 blocks south of city hall. His friend Alexandria lives 7 blocks east and 2 blocks north of city hall. What is the halfway point? **[The halfway point is at (1, −1), or 1 block east and 1 block south of city hall.]**

The midpoint of a diameter of a circle is $M(5, 4)$. If one endpoint is $A(8, -12)$, what are the coordinates of the other endpoint? [(2, 9)]

Closing the Lesson

Write the distance formula and midpoint formula on the board. Ask students to list the types of problems that were solved with these formulas. Have students copy this information into their journal.

ASSESS

Selected Answers

Odd-numbered Exercises 7–43 and odd-numbered Extra Practice

Assignment Guide

Core level students should complete Exercises 1–44.

Core-plus level students should complete Exercises 1–8, 10–14, 16–22, and 24–50.

Error Analysis

Students may forget which formula uses the differences of the coordinates and which uses the sums of the coordinates. Stress that the midpoint is the middle value or average, so you add when finding the midpoint. When completing problems such as Exercises 23–26, encourage students to write in the variable names (x_1, x_2, y_1, or y_2) for the missing coordinates and then use the formulas to find the coordinates of the midpoints.

Alternative Assessment

Analyzing a Figure Have students plot the four points $A(-3, 2)$, $B(2, 0)$, $C(0, -4)$, and $D(-5, -2)$. They should show that the figure formed by joining these points is a rectangle. (They can show that each angle is a right angle by calculating the lengths of the sides and diagonals.) Then they should show that the diagonals of the rectangle intersect at their midpoints. (To do this, they should find the midpoint of each diagonal and verify that they are the same point.)

Portfolio Activity Exercises 27–32 contain good examples of how coordinate geometry can be used to discover geometric properties. After constructing a line segment and its perpendicular bisector, students see that any point on the perpendicular bisector is equidistant from the endpoints of the segment. Have students write this work in report form to be included in their portfolio.

RETEACH

Using Self-Assessment

Begin by writing the distance formula and midpoint formula on the chalkboard. Then go through the solutions to problems that use each formula, such as Exercises 14 and 24. Have students write down these solutions in detail without reference to the solution that they found in their homework. Then have students compare these solutions with their homework. Students whose homework solution was incorrect should write a sentence or two describing their errors and what they could have done to avoid them. Since students tend to make similar mistakes, ask some volunteers to read their work.

Pair students up, give them a list of the answers for the remaining homework problems, and have them identify the problems for which at least one of the partners had an error. The partners should work to correct these problems and write a sentence describing the error that was made. At the end of this activity, have students review the mistakes. Were similar mistakes repeated? Have the partners discuss what they can do to avoid these mistakes in the future.

ENRICH

Interdisciplinary Connection

Cartography Cartographers create maps. Have students find out more about cartographers. They can look into the history of mapmaking, or they can focus on the modern methods used to measure distances and make maps. Ask students to share their results with the class.

Technology

Have students write a program for their graphics calculator to calculate the distance between two points when the coordinates of the points are entered. Then have them write another program to find the midpoint of a segment when the coordinates of the endpoints are entered.

NAME ______________________ CLASS __________ DATE __________

Practice

9.5 Exploring the Distance Formula

For each figure, find the coordinates of *Q* and the lengths of the three sides of the triangle. Decide whether the triangle is scalene, isosceles, equilateral, or right.

1.

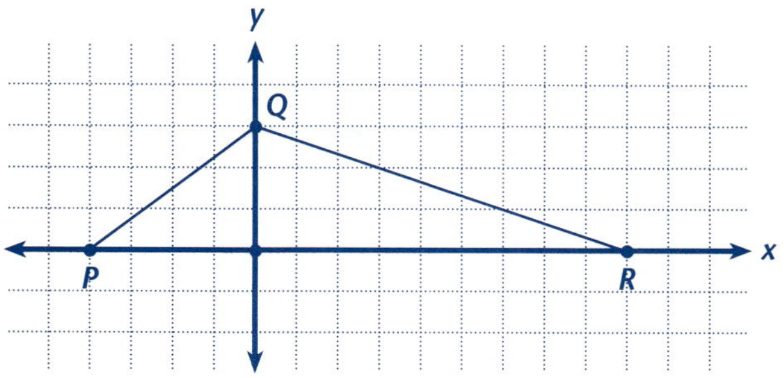

____

____

2.

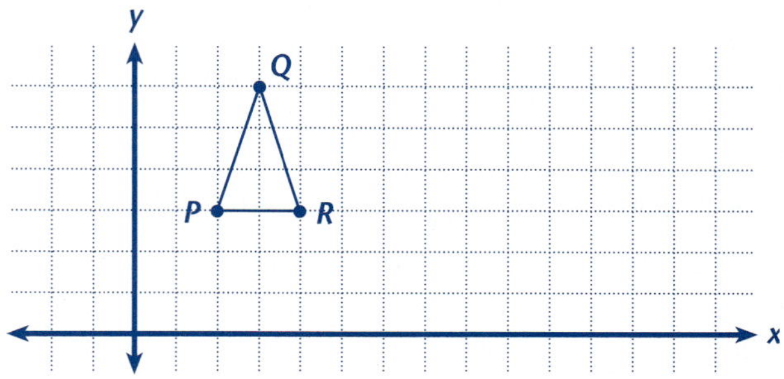

____

____

Find the distance between the two given points.

3. $A(-2, 8)$, $B(10, -1)$ ______________

4. $M(17, 14)$, $N(20, 10)$ ______________

5. $X(-7, -1)$, $Y(-6, -1)$ ______________

6. $D(-3, 0)$, $E(3, 2)$ ______________

Find the coordinates of the midpoint of the segment.

7. $\overline{AB}$, with $A(-3, 5)$ and $B(3, 7)$ ______________

8. $\overline{MN}$, with $M(-4, 3)$ and $N(3, -5)$ ______________

9. $\overline{XY}$, with $X(-4, 5)$ and $Y(-2, 1)$ ______________

10. $\overline{CD}$, with $C(6, -4)$ and $D(4, 6)$ ______________

11. $\overline{FG}$, with $F(7, -1)$ and $F(5, -2)$ ______________

The midpoint of $\overline{PQ}$ is *M*. Find the missing coordinates.

12. $P(52, -5)$, $Q(16, 19)$, $M($__________$)$

13. $P(6, -2)$, $Q($__________$)$, $M(9, 4)$

14. $P($__________$)$, $Q(4, 8)$, $M(10, 7)$

15. $P(-2, 25)$, $Q($__________$)$, $M(1, 18)$

NAME ______________________ CLASS __________ DATE __________

Lesson Activity

9.5 Using the "Pythagorean" Right-Triangle Theorem

Rusty is on the shore of a lake that is 1 mile wide. He wants to travel to a town on the opposite shore that is 1 mile down the shore. He intends to row in a straight line to some point on the opposite shore and then walk the remaining distance along the shore to the town. Suppose that Rusty can walk at the rate of 5 miles per hour and row at 3 miles per hour.

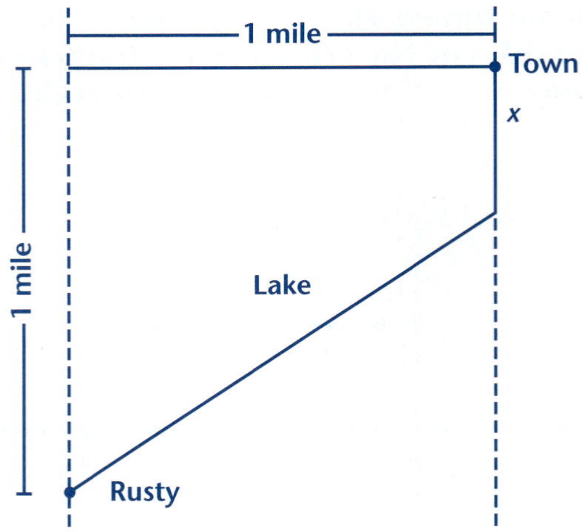

1. How long will it take Rusty to walk $\frac{1}{2}$ mile along the shore?

____

2. How long will it take him to row $\frac{3}{2}$ miles across the lake?

____

3. Suppose that x represents the distance in miles from a point on the opposite shore to the town. Use the "Pythagorean" Right-Triangle Theorem to help you write a formula to find the total time it will take Rusty to row to that point and then walk the remaining distance to the town.

____

4. Use your formula to complete the table. Round time to the nearest ten-thousandth.

x (mi)	0	$\frac{1}{8}$	$\frac{1}{4}$	$\frac{3}{8}$	$\frac{1}{2}$	$\frac{5}{8}$	$\frac{3}{4}$	$\frac{7}{8}$	1
t (h)									

5. According to your table, to what point should Rusty row in order to reach his destination in the least amount of time?

____

NAME ______________________ CLASS __________ DATE __________

Basic Skills Practice
Cumulative Review

Read each question and circle the best answer.

The following line graph charts the change in the price of a certain sculpture over time.

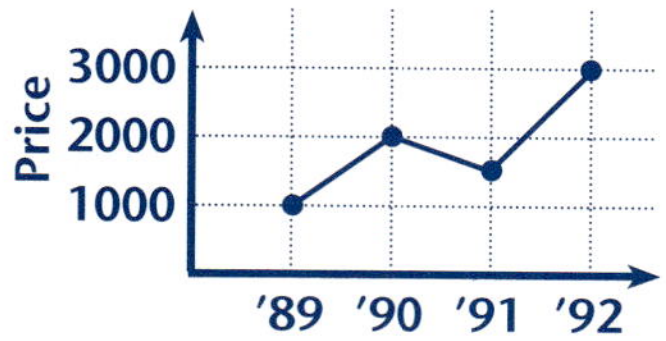

1. What was the cost of the sculpture in 1989?

A $1000
B $1500
C $2000
D $3000

2. In what year did the sculpture cost $3000?

F 1989
G 1990
H 1991
I 1992

3. By how much did the sculpture's price increase between 1989 and 1992?

A $3000
B $1000
C $1500
D $2000

4. In what year did the sculpture cost the least?

F 1989
G 1990
H 1991
I 1992

5. What was the change in price of the sculpture between 1990 and 1991?

A An increase of $500
B A decrease of $500
C An increase of $1000
D A decrease of $1000

Use the graph to answer Exercises 6–8.

The line graph charts the change in average attendance at a certain hockey team's games.

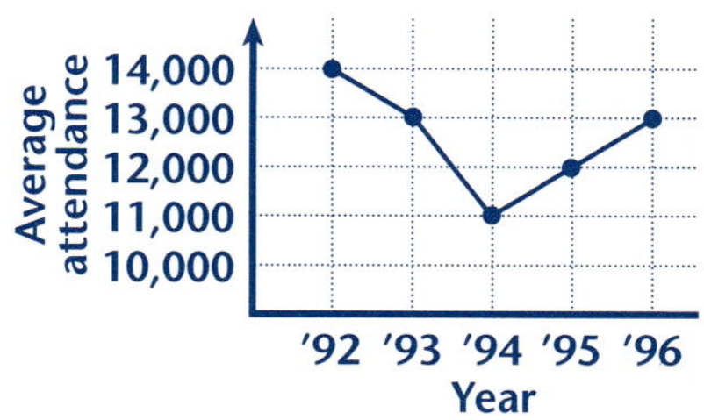

6. Which year had the lowest average attendance at the team's games? __________

7. Which year had the highest average attendance at the team's games? __________

8. What was the average attendance in 1994? __________

9. By how much did average attendance decline over the period 1992–1996? __________

LESSON 9.6

Exploring Geometric Properties

PREPARE

Pacing Guide
2 Days

Objectives and Overview
- Write the equation of a circle, given its center and radius.
- Use the distance formula to find the lengths of the diagonals of a rectangle.

The distance formula is used to write the equation of a circle and the formula for the lengths of the diagonals of a rectangle. The midpoint formula is used to discover that the segment joining the midpoints of the sides of a triangle is parallel to and half of the length of the third side.

Vocabulary
circle
center
radius

Assessing Prior Knowledge

Point A is at $(-3, 4)$ and point B is at $(3, -4)$.

1. Find the length of line segment AB. **[10 units]**
2. Find the coordinates of the midpoint of line segment AB. **[(0, 0)]**
3. If point C is at $(-4, -3)$, is triangle ABC a right triangle? **[yes]**

Approach A: Using the Book

Opening Discussion
What is coordinate geometry? It is the representation of geometric objects by coordinates. The algebraic properties of the coordinate system, such as the distance formula and midpoint formula, can then be used to prove geometric ideas.

Brainstorming
Have students think of geometric properties that they have shown to be true by using coordinates. In the last lesson they used the distance formula and the "Pythagorean" Right-Triangle Theorem to prove that a triangle was a right triangle. In the exercises, they used the distance formula to show that one triangle was isosceles and that another was right scalene. Also in the exercises, they used the midpoint formula and slope formula to construct the perpendicular bisector of a line segment. They then used the distance formula to show that any point on the perpendicular bisector was equidistant from the endpoints of the segment. In

TEACH

Approach B: Using Models

Opening Discussion
What is coordinate geometry? It is the representation of geometric objects by coordinates. The algebraic properties of the coordinate system, such as the distance formula and midpoint formula, can then be used to prove geometric ideas.

Brainstorming
Have students think of geometric properties that they have shown to be true by using coordinates. In the last lesson they used the distance formula and the "Pythagorean" Right-Triangle Theorem to prove that a triangle was a right triangle. In the exercises, they used the distance formula to show that one triangle was isosceles and that another was right scalene. Also in the exercises, they used the midpoint formula and slope formula to construct the perpendicular bisector of a line segment. They then used the distance formula to show that any point on the perpendicular bisector was equidistant from the endpoints of the segment. In

this lesson, students will see other geometric properties proven by using coordinate geometry.

Finding the Equation of a Circle

Draw a circle centered at the origin with a radius of 5. Review the definitions of *circle, center,* and *radius.* Lead students through the process of finding the equation of this circle by using the distance formula. Have students substitute the coordinates of points on the graph of the circle into the formula to see that it checks.

Setting Up the Exploration

Teacher-Guided Explorations If you wish, you can guide students through the textbook explorations. You can draw the graphs on an overhead transparency or have students draw them at the chalkboard. At each stage of the exploration, invite student input and encourage discussion. Students should keep a record of the exploration responses in their notebook.

Cooperative-Learning Groups Divide the class into groups of three. Students should rotate the responsibilities of drawing the graphs, writing the equations, and recording the group's responses. All group members should be involved in a discussion of the activity as the exploration proceeds. Each student should sign the completed report.

Exploration 1

The distance formula is used to discover the formula for a circle centered at the origin and the formula for a circle centered at (1, 1). Stress that the point $P(x, y)$ stands for any point on the circle. Some students may realize that the circle in Step 4 is a translation 1 unit up and 1 unit to the right of the circle in Step 1. They can use this information to write the equation.

Ongoing Assessment, top of page 449

4. $(x - 1)^2 + (y - 1)^2 = 1$

On Their Own

Have students find the equation of the circle centered at (0, 2) with a radius of 4.
[$x^2 + (y - 2)^2 = 16$]

continued on top of next page

this lesson, students will see other geometric properties proven by using coordinate geometry.

Finding the Equation of a Circle

Give each student a protractor and graph paper. Have them set up a coordinate graph on the paper and use the protractor to draw a circle centered at the origin with a radius of 5. Ask students to name some points on this circle. (The intercepts are the easiest.) Have them use the distance formula to calculate the distance between these points and the center of the circle. What is the distance between any point and the center of the circle? **[5]** How can we use this information to define the circle? **[The set of all points that are a distance of 5 from the center.]** Let $P(x, y)$ be any point on the circle. Write an expression for the distance between P and the center of the circle. Show how this leads to the equation of the circle. **[$x^2 + y^2 = 25$]** Have students verify that the points they found previously satisfy this equation.

Cooperative Learning

Before starting the explorations, divide the students into pairs. In Exploration 1, students should take turns drawing the circles with a compass and writing the equations. In Exploration 2 and Exploration 3, students should take turns constructing the figures and finding the lengths or midpoints. The student who constructs the figures is responsible for recording the group's results for that exploration. Both group members should be involved in a discussion of the activity as the exploration proceeds. Both students should sign the completed report.

Alternate Exploration 1

Have students work with their partners. Each pair needs a compass and graph paper. The pair should work together to answer the following questions, which you can write on the chalkboard or overhead. (For ideas on guiding the students, see Exploration 1 on page 449 of the Pupil's Edition of the textbook.)

continued on bottom of next page

Approach A Using the Book, continued

Exploration 2

The diagonals of several rectangles are examined. Students should conclude that the diagonals of a rectangle are equal in length. Note that the calculations in Step 2 will be easier if students place one vertex of the rectangles they draw at the origin and place one side along the x-axis.

Ongoing Assessment, middle of page 449

3. The diagonals of a rectangle have equal length.

Exploration 3

A formula for finding the length of the diagonals of a rectangle is found. In Step 2, students should realize that this formula is the same for both of the diagonals. This proves the conjecture from Exploration 2, namely that the diagonals of a rectangle are equal in length.

Ongoing Assessment, bottom of page 449

2. The formula for the length of the diagonal connecting (0, 0) and (a, b) is $\sqrt{a^2 + b^2}$, which is seen to be the same as that for the diagonal connecting $(a, 0)$ and $(0, b)$. Because any rectangle can be placed with one vertex at the origin and one side along the x-axis, this proves the conjecture made in Exploration 2, namely that the diagonals of a rectangle are equal in length.

Exploration 4

The segment formed by joining the midpoints of two sides of a triangle is seen to be parallel to the third side and to be half the length of the third side. Note that although the triangle in the illustration is isosceles, this theorem is true for any triangle. Students should place all of their test triangles with one vertex at the origin and with one side along the x-axis in order to simplify their work.

TEACH

Approach B Using Models, continued

- Construct circles centered at (0, 0) with a radius of 1 and a radius of 3. Write an equation for each circle. **[$x^2 + y^2 = 1$; $x^2 + y^2 = 9$]**
- What is the equation of a circle with its center at (0, 0) and a radius of r? **[$x^2 + y^2 = r^2$]**
- Write an equation for a circle with its center at (1, 1) and a radius of 1. **[$(x - 1)^2 + (y - 1)^2 = 1$]**

Student Reports

Choose one group to report their findings for this exploration. Ask if any groups found the equation in the last step without using the distance formula. Did any group use transformations to write the equation?

On Their Own

Have students find the equation of the circle centered at (0, 2) with a radius of 4. **[$x^2 + (y - 2)^2 = 16$]**

Using Geometry Software on a Computer

Some of the latest versions of geometry drawing software have coordinate geometry features. Have students use geometry software to construct the figures required for the next two explorations. You will need to give students some time to explore the software and learn to construct rectangles and triangles before beginning the explorations. If this resource is not available, the explorations can be done with hand-drawn figures.

Alternate Exploration 2

Have students work with their partner. Each pair should have access to a computer with geometry software. The pair should work together to answer the following questions, which you can write on the chalkboard or overhead. (For ideas on guiding the students, see Explorations 2 and 3 on page 449 of the Pupil's Edition of the textbook.)

- Construct several rectangles with one vertex at the origin and one side along the x-axis. Write the coordinates of the vertices of each rectangle and use them to calculate the

Ongoing Assessment, page 450

4. Using a triangle with vertices at (0, 0), $(a, 0)$, and (b, c), the coordinates of the midpoints of the two nonhorizontal sides are $\left(\frac{b}{2}, \frac{c}{2}\right)$ and $\left(\frac{a+b}{2}, \frac{c}{2}\right)$. Since these points have the same y-coordinate, the line segment joining them is horizontal and thus parallel to the third side of the triangle. Using the distance formula, the length of this segment is $\sqrt{\left(\frac{a+b}{2} - \frac{b}{2}\right)^2 + \left(\frac{c}{2} - \frac{c}{2}\right)^2} = \sqrt{\left(\frac{a}{2}\right)^2} = \frac{a}{2}$. The length of the side along the x-axis is $\sqrt{(a-0)^2 + (0-0)^2} = a$. Thus, the length of the segment joining the midpoints of two sides of a triangle is half of the length of the third side.

Application

This application will help students better understand the geometric relationships in a parabola. Note that the distances to the road are measured on the line perpendicular to the road. This perpendicular distance is the shortest distance between a point and a line. You may want to draw the coordinate grid on the chalkboard or overhead and go slowly through the development of the formula.

Critical Thinking, page 451

The equation is for a parabola. The vertex is (0, 1). The axis of symmetry is the y-axis.

continued on top of next page

lengths of the diagonals of each rectangle. Use the software to draw the diagonals and find their lengths to check your work. What do you notice? **[The lengths of the diagonals in a rectangle are equal.]**

- Draw a general rectangle in the same position as the previous ones with one pair of sides of length a and one pair of sides of length b. Write the coordinates of the vertices of this rectangle in terms of a and b. Write a formula for the length of a diagonal of the rectangle. Use the coordinates for each diagonal to verify that the formula is the same for both. **[$d = \sqrt{a^2 + b^2}$]**

Alternate Exploration 3

Have students work with their partner. Each pair should have a computer with geometry software. The pair should work together to answer the following questions, which you can write on the chalkboard or overhead. (For ideas on guiding the students, see Exploration 4 on page 450 of the Pupil's Edition of the textbook.)

- Draw several triangles with one vertex at the origin and one side along the x-axis. Write the coordinates for each of the vertices, and use them to find the midpoints of the nonhorizontal sides and the length of the segment connecting the midpoints. Also find the length of the horizontal side of the triangle. Confirm your calculations with the geometry software. What do you notice? Make a conjecture that describes this relationship. **[The length of the segment connecting the midpoints of the two sides is half of the length of the third side.]**
- Using coordinate geometry and letters to represent the nonzero coordinates, show that your conjecture is true. **[Using a triangle with vertices at (0, 0), $(a, 0)$, and (b, c), the coordinates of the midpoints of the two nonhorizontal sides are $\left(\frac{b}{2}, \frac{c}{2}\right)$ and $\left(\frac{a+b}{2}, \frac{c}{2}\right)$. Since these points have the same y-coordinate, the line segment joining them is horizontal and thus parallel to the third side of the triangle. Using the distance formula, the length of this segment is $\sqrt{\left(\frac{a+b}{2} - \frac{b}{2}\right)^2 + \left(\frac{c}{2} - \frac{c}{2}\right)^2} =$**

continued on bottom of next page

Approach A

Using the Book, continued

Cultural Connection: Africa

You can have students work on this problem in pairs. Challenge partners who finish early to show that this method can be used to make a square whose area is double that of any given square.

Ongoing Assessment, page 451

1. It is made of four triangles, each with an area of $\frac{1}{2}$ of a square cubit.

2. $A = s^2$

$2 = s^2$

$\sqrt{2} = \sqrt{s^2}$

$\sqrt{2} = s$

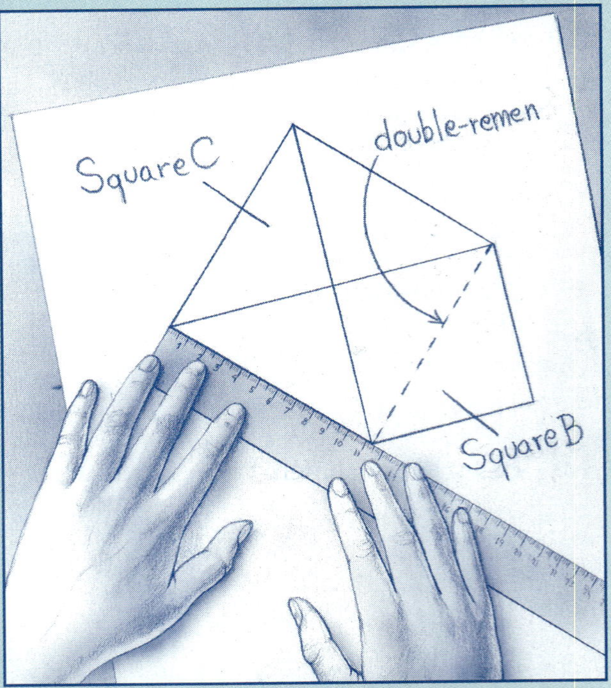

This ancient surveying technique will help students understand geometric properties.

TEACH

Approach B

Using Models, continued

$\sqrt{\left(\frac{a}{2}\right)^2} = \frac{a}{2}$. The length of the side along the x-axis is $\sqrt{(a-0)^2 + (0-0)^2} = a$. Thus, the length of the segment joining the midpoints of two sides of a triangle is half of the length of the third side.]

Application

Read and discuss the surveying problem on page 450 of the Pupil's Edition of the textbook. This application can help students better understand the geometric relationships in a parabola. Note that the distances to the road are measured on the line perpendicular to the road. This perpendicular distance is the shortest distance between a point and a line. You may want to draw the coordinate grid on the chalkboard or overhead and go slowly through the development of the formula.

ASSESS

Selected Answers

Odd-numbered Exercises 7–43 and odd-numbered Extra Practice

Assignment Guide

Core level students should complete Exercises 1–43.

Core-plus level students should complete Exercises 1–45.

Error Analysis

For Exercises 15–20, students need to use the distance formula with the given center and a general point (x, y) on the circle. Encourage students who are having difficulty to follow the development of the equation of the circle centered at (0, 0) with a radius of 5 on page 448 of the Pupil's Edition of the textbook.

Alternative Assessment

The Equation of an Ellipse The figure below is called an *ellipse*.

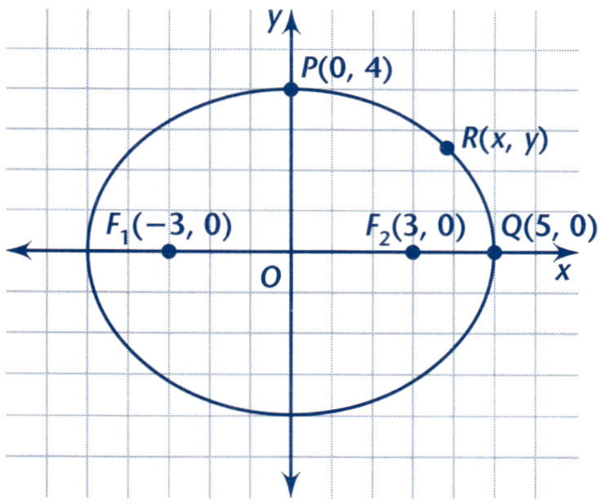

The two points in the interior are called the *foci*. An ellipse has the property that for any point on it, the sum of the distance from the point to one foci and the distance from the point to the other foci is constant. For the graph above, this constant is 10.

Calculate the lengths of PF_1 and PF_2, and find their sum. **[5 and 5; 10]** Calculate the lengths of QF_1 and QF_2 and find their sum. **[8 and 2; 10]**

Write expressions for the lengths of RF_1 and RF_2. The sum of these lengths should equal 10. Use this fact to write an equation for the ellipse. Substitute the coordinates for points P and Q to check your answer. $[\sqrt{(x+3)^2+y^2} + \sqrt{(x-3)^2+y^2} = 10]$

Portfolio Activity Study the graphs and equations for Exercises 15–18. Describe these graphs in terms of transformations of a circle with a radius of 5. Explain how transformations can be used to write the equation of a circle with a given center and radius. Have students include this work in their portfolio.

RETEACH

Using Variables in Formulas

Students having difficulty with this lesson often feel uncomfortable using the general point $P(x, y)$ in a formula. You may want to remind these students of the procedure used to find the slope-intercept equation for a line when given a point and the slope. Give students some practice at calculating expressions for the midpoint and the distance between two points with variable coordinates. Then have them place a general rectangle on the coordinate plane with one vertex at the origin and one side on the x-axis, label the points, and show that the diagonals bisect each other.

ENRICH

Interdisciplinary Connection

Archeology The photograph on page 448 is of Stonehenge, a prehistoric monument located north of Salisbury, England. Have students do research to find out more about this formation. They should write their results in a report and share it with the class.

The Directrix and Focus of a Parabola

Have students refer to the application problem on page 450. The line representing the road is called the *directrix* of the parabola, and the point representing the TV tower is called the *focus* of the parabola. Have students follow the steps of the application to find the equation of a parabola whose directrix is the line $y = 1$ and whose focus is the point (2, 3). A point on the directrix can be represented by $(x, 1)$.

$[y = \frac{1}{4}(x-2)^2 + 2]$

Practice

9.6 Exploring Geometric Properties

Find an equation of a circle with its center at the origin and with the given radius.

1. radius of 9 ______________________

2. radius of 4 ______________________

Sketch each circle on the grid provided. Then find the equation of the circle.

3. center at (4, 0) and radius of 3

____

4. center at (−4, 0) and radius of 3

____

5. center at (0, 4) and radius of 3

____

6. center at (0, −4) and radius of 3

____

Identify the center and the radius for each circle.

7. $(x - 1)^2 + y^2 = 4$

____

8. $(x + 1)^2 + (y - 2)^2 = 4$

____

Lesson Activity

9.6 Modeling Proofs Using Coordinate Geometry

There are many different proofs of the "Pythagorean" Right-Triangle Theorem. President Garfield developed a proof using a trapezoid.

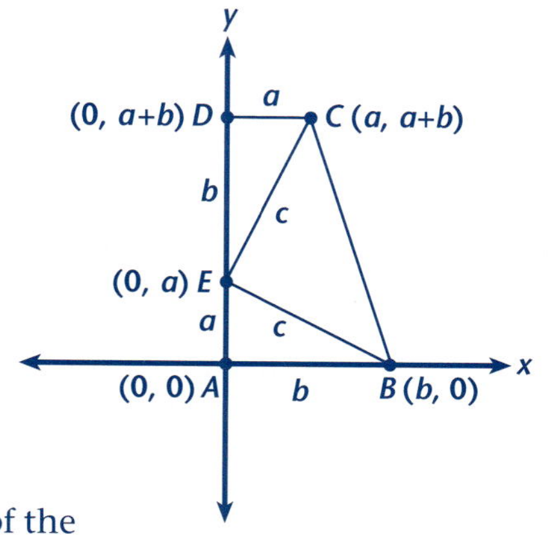

1. Use the 0, *a*, *b*, and $a + b$ labels in the diagram to write formulas for finding the area of trapezoid *ABCD* and the areas of triangle *ABE*, triangle *BCE*, and triangle *CDE*.

____

____

2. Find the sum of areas of the three triangles.

____

3. Why is the area of the trapezoid equal to the sum of the areas of the three triangles?

____

4. Set the expression for the area of the trapezoid and the sum of the three triangles equal to each other and simplify. What is the result? Explain its significance.

____

ODRT is a square with squares *URBX* and *AXWO* formed by perpendicular segments *AB* and *UW* intersecting at *X*.

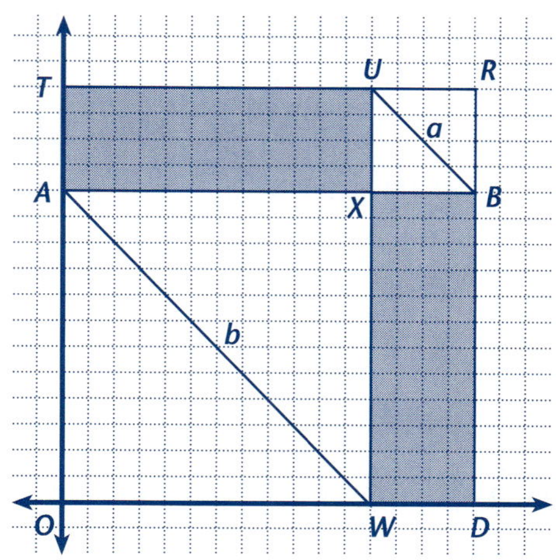

5. Write an expression for the width of each shaded rectangle in terms of *a*.

____

6. Write an expression for the length of each shaded rectangle in terms of *b*.

____

7. Write an expression that represents the total area of the shaded region in terms of *a* and *b*.

____

NAME ______________________________ CLASS ______________ DATE ______________

Basic Skills Practice

Interpreting Venn Diagrams and Tables, Part 1

Venn diagrams represent relationships among events. In a Venn diagram, the rectangle represents a sample group and the circles represent events within that group.

Example 1: A survey was conducted to see if people watched either of two movies.

S = set of 100 people surveyed
A = set of people who watched movie A
B = set of people who watched movie B

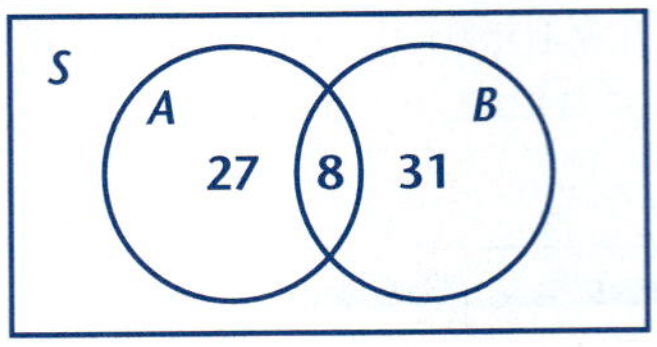

How many people watched only movie A?
The number in circle A that does not overlap circle B is 27. Therefore 27 people watched only movie A.

How many people total watched movie A?
The overlapping region includes people who watched both movie A and movie B. Therefore 8 plus 27 or 35 people in total watched movie A.

Tables organize information by grouping data.

Example 2: Use the table below to find who had the best golf score on Saturday. Remember that in golf the lowest score is the best.

Golf Scores

	Tim	Brad
Friday	72	68
Saturday	78	81

Since Tim scored a 78 and Brad scored 81, Tim had the best score on Saturday.

Use the graph to answer the following exercises:

A newspaper surveyed 500 people to determine how many people own cars and how many people own trucks. The results are shown in the Venn diagram.

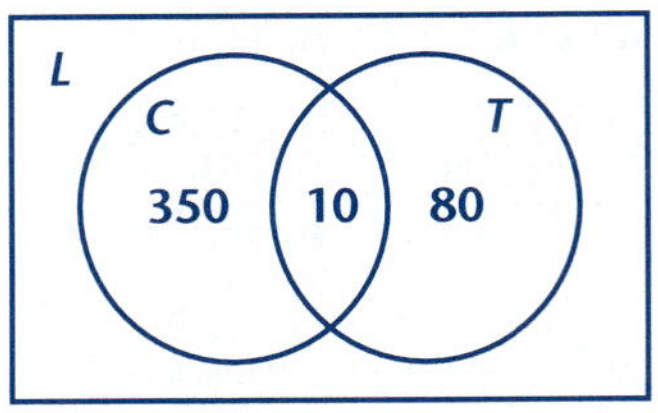

L = set of 500 people surveyed
C = set of car owners
T = set of truck owners

1. How many people own only cars? ______________

2. How many people own only trucks? ______________

3. How many people own ***neither*** cars ***nor*** trucks? ______________

4. How many people own both cars and trucks? ______________

The table at right shows the elevation of several mountains.

5. What is the elevation of Mt. Hardy in feet? ______________

6. What is the elevation of Mt. Price in meters? ______________

Elevation

	Feet	Meters
Mt. Hardy	17,330	5282
Mt. Price	14,781	4505
Mt. Howard	15,526	4732

LESSON 9.7

The Tangent Function

PREPARE

Pacing Guide
2 Days

Objectives and Overview
- Define the tangent function for right triangles.
- Use the tangent of an acute angle to solve problems.

The tangent function is defined for right triangles. The values of the tangent function are explored by examining different triangles. Both the tangent and inverse tangent are used to solve problems concerning angles of elevation.

Vocabulary
tangent

Assessing Prior Knowledge

Use the "Pythagorean" Right-Triangle Theorem to find the missing lengths in each right triangle. Side c is the hypotenuse. Round answers to the nearest tenth if needed.

	Side a	Side b	Side c
1.	4 cm	6 cm	? cm
2.	? ft	8 ft	9 ft
3.	9 m	9 m	? m
4.	$1\frac{1}{2}$ in.	? in.	2 in.

[1. 7.2; 2. 4.1; 3. 12.7; 4. 1.3]

Approach A: Using the Book

Opening Discussion
"In past lessons, we have used the 'Pythagorean' Right-Triangle Theorem to find the length of the third side of a right triangle when the lengths of two sides are known. In this lesson, the *tangent relationship* is introduced and is used to find the length of one leg of a right triangle when the measure of an acute angle and the length of the other leg are known or to find the measure of an acute angle when the lengths of both legs are known."

Defining the Tangent
Read and discuss the machine shop example. Use **Transparency 47** or draw the diagram on the chalkboard to make sure that students understand the definitions of *opposite leg* and *adjacent leg*. Write the definitions for *tangent* and *inverse tangent* on the chalkboard for students to copy into their notebook or journal.

continued on top of next page

TEACH

Approach B: Using Models

Opening Discussion
"In past lessons, we have used the 'Pythagorean' Right-Triangle Theorem to find the length of the third side of a right triangle when the lengths of two of the sides are known. In this lesson, the *tangent relationship* is introduced and is used to find the length of one leg of a right triangle when the measure of an acute angle and the length of the other leg are known or to find the measure of an acute angle when the lengths of both legs are known."

Defining the Tangent
Suppose that in a machine shop, a blueprint calls for a part with the dimensions shown on **Transparency 47** (blueprint from page 454). To make this part from a rectangular piece of material, at what angle should the cut at point B be made? Draw the triangle of metal that will need to be cut from a rectangle in order to make this part. One way to do this is to use a ratio called the tangent ratio. Define the leg opposite angle B and the leg adjacent

continued on bottom of next page

Approach A: Using the Book, continued

Exploration

This exploration gives students some experience in finding the tangent values for some angles and allows them to see how the tangents of various angles compare. You can lead the class through this exploration or have students complete it in small groups. In Step 2, although the tangent doubles from 10° to 20°, it more than doubles from 30° to 60°. This difference will continue to increase up to 90°.

Ongoing Assessment, page 455

5. The angle is 45°. The tangent ratio is 1 and equal to the ratio of the length of the side opposite the angle A to the length of the side adjacent to angle A.Therefore, the opposite will be the same length as the side adjacent to A.

The Tangent Function

Explain that the function which assigns each angle the value of its tangent is called the **tangent function.** It is one of three basic trigonometric functions. The other two basic trigonometric functions, the sine and cosine functions, will be studied in the next lesson.

Have students use the chart from the exploration to find some values for the tangent function. Show students how to find tangents and inverse tangents with a calculator. Make sure that students know how to set their calculators to degree mode. Have them verify some of the tangent values found in the exploration.

Use the inverse tangent function to find the angle measure for the machine shop problem. Caution students that they must use parentheses to enter the fraction. Have students use their calculators to find the tangent of 34° in order to check the answer.

TEACH

Approach B: Using Models, continued

to angle B. Write the definition for tangent, and show that the tangent of angle B, abbreviated tan B, is equal to $\frac{2}{3}$. Each acute angle has a unique tangent. If you know the tangent, you can work backward to find the angle measure. The angle whose tangent is $\frac{2}{3}$ is written $\tan^{-1}\left(\frac{2}{3}\right)$ and is read "the inverse tangent of $\frac{2}{3}$." Give students some practice at using the definition of tangent with these examples. Find the tangent of each acute angle in each triangle.

A. **B.**

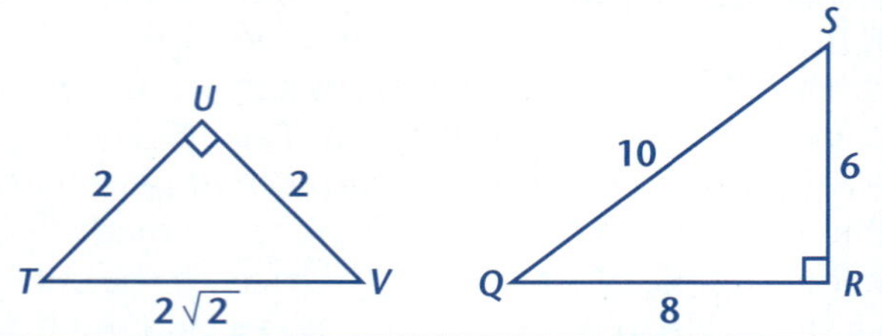

[$\tan T = 1$; $\tan V = 1$; $\tan Q = \frac{3}{4}$; $\tan S = \frac{4}{3}$]

Teaching Tip

Encourage students to remember that the word *adjacent* means "next to." This may help them avoid mixing up the two legs when finding the tangent. Also stress that the legs of the triangle, not the hypotenuse, are used.

A Meter Stick Model for Finding Tangents

Creating a Model Tape one meter stick perpendicular to another at the 30-centimeter mark. Place the horizontal one on the top of a desk or on the chalkboard. Place a piece of paper under the meter stick at point A, and use a protractor to make angles at point A with the meter stick as one of the sides. Mark angles of 10°, 20°, 30°, 40°, 50°, 60°, and 70°.

Use a third meter stick at point A to make a triangle with the three meter sticks. Line up this meter stick with the 10-degree mark to make a triangle with a 10° angle at point A.

Using the inverse tangent function

Example 1

The tangent of an acute angle is determined from the ratio of leg lengths in a right triangle. You may need to remind students that acute angles have measures less than 90° and are thus the two nonright angles. The tangent of a right angle is not defined.

Additional Example 1

Find the tangent of each acute angle in each triangle.

A. **B.**

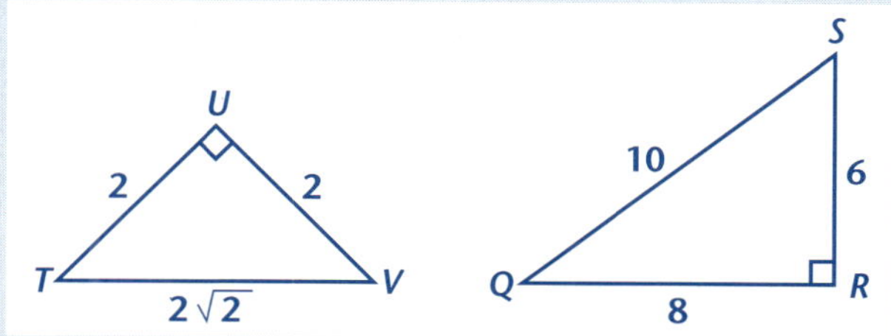

[$\tan T = 1$; $\tan V = 1$; $\tan Q = \frac{3}{4}$; $\tan S = \frac{4}{3}$]

continued on top of next page

A physical model will help students understand properties of the tangent function.

Using the Model The length of the base of the triangle is 30; read the height of the triangle to the nearest tenth of a centimeter from the vertical meter stick. Calculate tan 10° by using the opposite-over-adjacent ratio.

Cooperative Learning

Before starting the exploration, divide the class into groups of four. Students should take turns manipulating the meter sticks, reading the measurements, calculating the tangent, and recording the results. All group members should be involved in a discussion of the activity as the exploration proceeds. The group should work together to answer the questions. Each student should sign the completed report.

Exploration

Have students work in their groups. Give each group three meter sticks and have them set up the model as described above. Each group should investigate the following questions, which you may write on the overhead or chalkboard. (For ideas on guiding the students,

continued on bottom of next page

Approach A — **Using the Book**, continued

Example 2

The tangent ratio is used to find the angle of elevation of an airplane. Explain that the angle of elevation is the angle between the ground and the flight path of the airplane. Point out the use of the inverse tangent function to find the measure of the angle.

Additional Example 2

A local building code requires that the angle between the ground and a wheelchair ramp be no more than 15°. Will the code allow a builder to construct the ramp shown below? Explain.

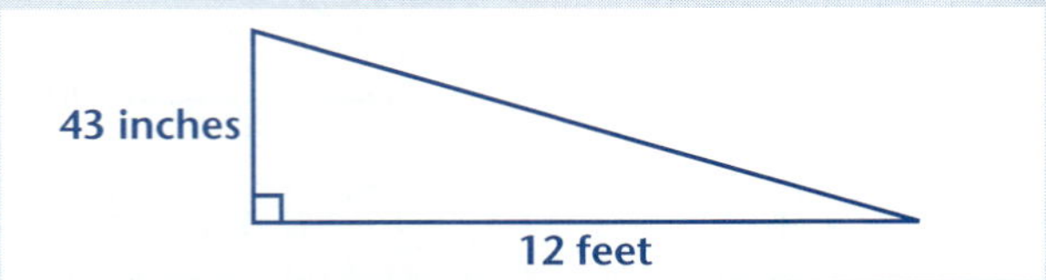

[No; the angle between the ground and this ramp is about 16.6°.]

Example 3

The tangent ratio is used to find the length of the leg of a triangle. Note that students must set up and solve an equation for this problem.

Additional Example 3

A 554-foot TV tower is erected and then held in place with guy wires (wires that connect the top of the tower to the ground). If the angle of the guy wires with the ground should be 65°, how far from the tower should the wires be anchored?

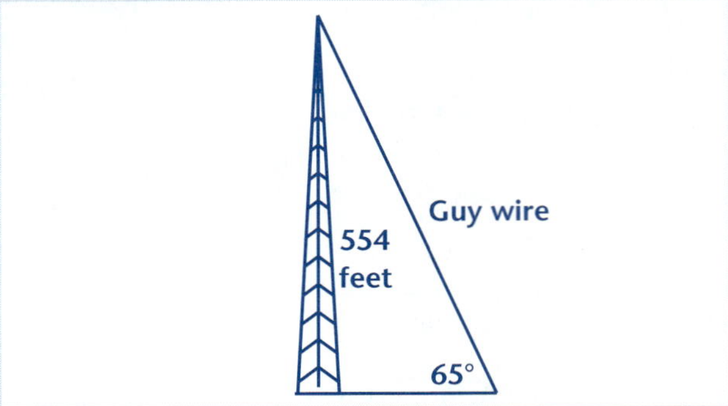

[about 258 feet]

TEACH

Approach B — **Using Models**, continued

see the exploration on page 455 of the Pupil's Edition of the textbook.)

- Use the model and the procedure described on the previous page to find the tangent of 10°, 20°, 30°, 40°, 50°, 60°, and 70°. What happens to the tangent as the angle measure increases? **[It increases faster and faster.]**
- Which angles have tangents that are less than 1? greater than 1? **[those that are less than 45° those that are greater than 45°]**
- Adjust the hypotenuse meter stick so that the height of the triangle is also 30 centimeters. What is the tangent of this angle? Measure the angle with the protractor. **[1; 45°]**
- How can you find the length of the side opposite angle *A* if you know that the measure of angle *A* is 45°? Test your theory by holding the hypotenuse meter stick at 45° and moving the vertical meter stick. Notice that the two legs are always equal in length.

The Tangent Function

Explain that the function which assigns each angle the value of its tangent is called the **tangent function.** It is one of three basic trigonometric functions. The other two basic trigonometric functions, the sine and cosine functions, will be studied in the next lesson. Have students use the data from the exploration to find some values for the tangent function.

Technology

Show students how to find tangents and inverse tangents with a calculator. Make sure that students know how to set their calculators to degree mode. Have them verify some of the tangent values found in the exploration. Use the inverse tangent function to find the angle for the machine shop problem. (Compare page 456.) Caution students that they must use parentheses to enter the fraction. Have students use their calculators to find the tangent of 34° in order to check the answer.

Closing the Lesson

Have students work with a partner. Each should draw a right triangle, label the lengths of the legs, and name the vertices. Students should then exchange papers and find the tangent of each acute angle in the drawing.

Uses of the Tangent Function

In the machine shop example, the tangent function was used to find an acute angle in a right triangle when the lengths of the legs were given. The tangent ratio can also be used to find the length of one of the legs of the triangle if the measure of one acute angle and the length of the other leg are known. Lead the class through these two examples, which illustrate these types of problem. The inverse tangent must be used in the first example. In the second example, an equation should be set up in which the distance from the guy wire to the tower is represented by a variable. (Compare Examples 2 and 3 on pages 456 and 457, respectively.)

A. A local building code requires that the angle between the ground and a wheelchair ramp be no more than 15°. Will the code allow a builder to construct the ramp shown below? Explain.

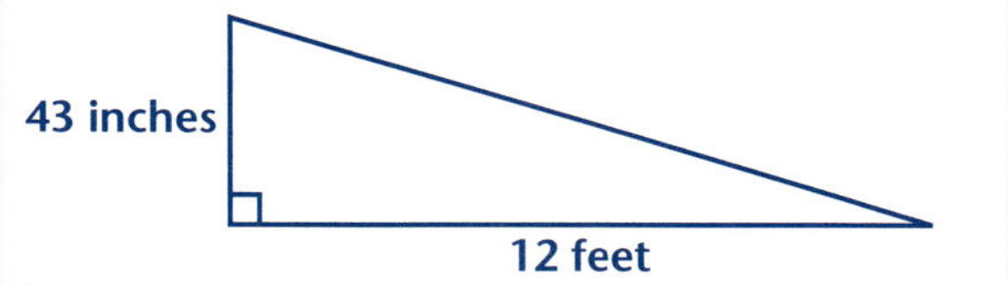

[No; the angle between the ground and this ramp is about 16.6°.]

B. A 554-foot TV tower is erected and then held in place with guy wires (wires that connect the top of the tower to the ground). If the angle of the guy wires with the ground should be 65°, how far from the tower should the wires be anchored?

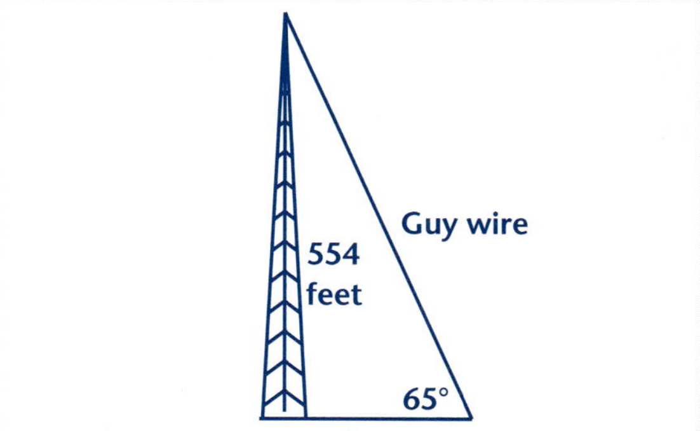

[about 258 feet]

ASSESS

Selected Answers

Odd-numbered Exercises 5–35 and odd-numbered Extra Practice

Assignment Guide

Core level students should complete Exercises 1–35.

Core-plus level students should complete Exercises 1–4, 7–15, and 17–37.

Error Analysis

Students may mistakenly invert the tangent ratio. Stress that it is opposite over adjacent. Students may also be confused about when to use the tangent and when to use the inverse tangent. Explain that you use the tangent when you know the measure of the angle and that you use the inverse tangent when you are trying to find the measure of the angle. Watch for students who do not have their calculators in degree mode.

Alternative Assessment

Tangent and Inverse Tangent Describe the difference between the tangent and inverse tangent. Make up a problem that uses the tangent and one that uses the inverse tangent. Write the solutions to these problems.

Portfolio Activity Have students include one of the applied exercises from this lesson (Exercises 23–26) in their portfolio. How did they know whether they needed to use a tangent or inverse tangent to solve the problem?

RETEACH

Have students make flashcards to help them keep track of the different concepts they have learned so far in this chapter. Include the "Pythagorean" Right-Triangle Theorem, the distance formula, the midpoint formula, the equation of a circle, the tangent ratio, and the inverse tangent. On the front of the card, they should write the name of the concept. On the back, they can describe the concept, write the formula, if appropriate, and list some of the common applications of the concept. Have students review their work for this chapter to find one example for each card.

ENRICH

Interdisciplinary Connection

Surveying The instruments pictured at the bottom of page 457 of the Pupil's Edition are used by surveyors. Have students do research to find out what surveyors do and what kinds of instruments they use. If possible, interview a surveyor. Report your findings to the class.

Using a Sight Line to Determine the Height of an Object

Have students refer to the photo at the bottom of page 458 of the Pupil's Edition, which shows a student measuring the angle of sight to the water tower by using a protractor. Students should use this photo to develop a method for finding the angle of sight. Then they should use a protractor, measuring tape, and the method from Exercise 24 to determine the height of a large building or tree.

NAME ______________________ CLASS __________ DATE __________

Practice

9.7 Tangent Function

Use a calculator to find the tangents of the following angles to the nearest tenth:

1. 18° ________ **2.** 48° ________ **3.** 118° ________ **4.** 150° ________

Find the corresponding angles, to the nearest degree, for the following tangent approximations:

5. 0.2867 ________ **6.** 0.7265 ________ **7.** 28.64 ________ **8.** 0.9657 ________

Use the given information to find the length of indicated side to the nearest hundredth.

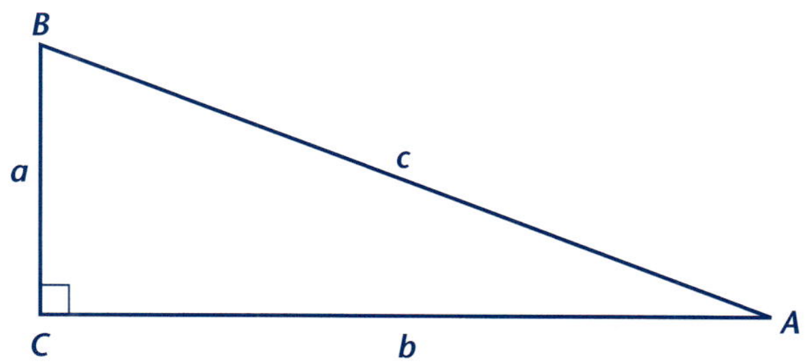

9. If angle A is 30°, and side b is 6 feet, find a.

____

10. If angle B is 60°, and side b is 16 feet, find a.

____

Use the diagram to find each height to the nearest hundredth.

11. Find the height of the building.

____

12. Find the height of the tree.

____

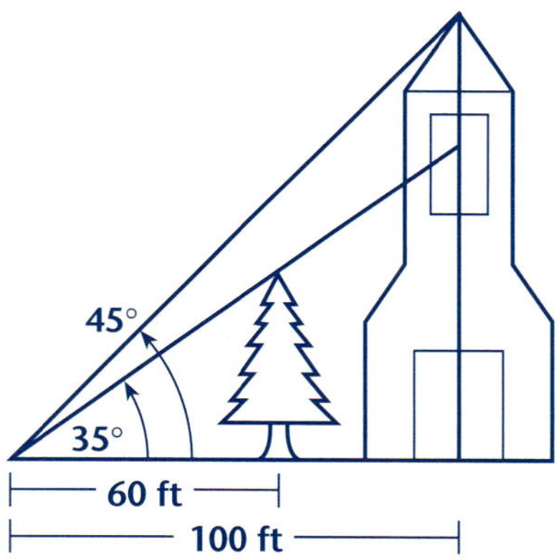

13. If a road has a 2% grade, its slope is $\frac{2}{100}$. What is the measure, to the nearest tenth, of the angle at which the road inclines? ________

14. What is the slope of a line that makes an angle of 45° with the x-axis? ________

15. A mast on a sailboat supports a 28-foot tall sail. If the boom holding the bottom of the sail is 18 feet long, what angle does the base of the sail make with its outer edge? Round your answer to the nearest degree.

____

NAME ______________________ CLASS __________ DATE __________

Lesson Activity
9.7 Patterns of Nature

Have you ever noticed how a plant grows over a period of days and weeks? Within each plant's stem are branching veins that carry nourishment to the leaves; each vein feeds a chain of leaves in a particular pattern. The number of leaves between the uppermost and lowermost leaves fed by the same main vein form a leaf cycle with a spiral rotating around the plant's center, or eye.

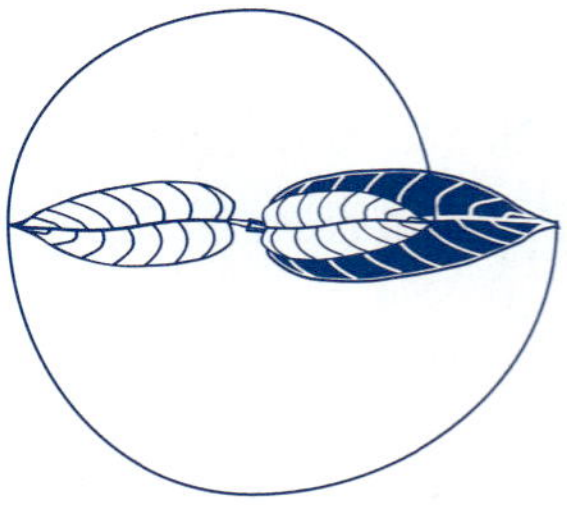
a.

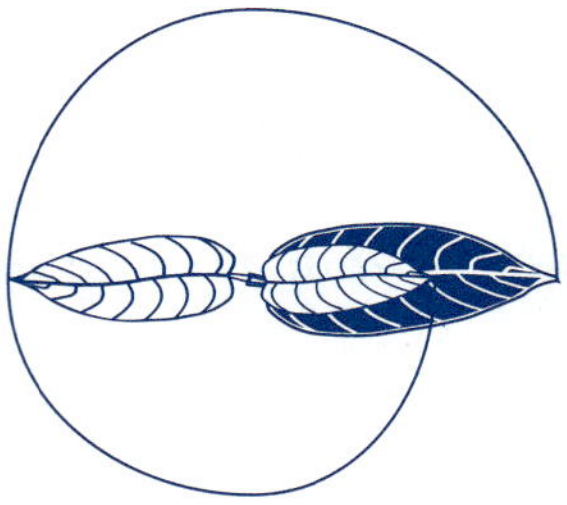
b.

c.

d.

e.

f.

g.

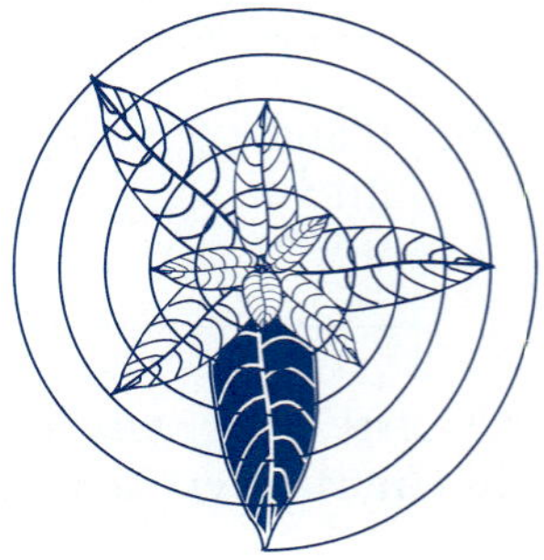
h.

1. For each leaf-cycle stage in the diagram, find the ratio of unshaded leaves to spirals. Enter the ratio in the table. What pattern do you notice?

Cycle	a	b	c	d	e	f	g	h
Ratio								
Angle								

2. Find the angles between consecutive leaves on any one spiral with respect to the eye of the plant, and record the angles in the table.

Follow this procedure to build a model of a leaf cycle.

Copy the spiral pattern onto firm paper or cardboard. Cut out the spiral. Tack the center to the top of an upright rod mounted in a base so that the strip wraps around the rod. Tape the bottom of the spiral to the base. Cut out paper leaves, and tape their stems onto the rod behind each of the points.

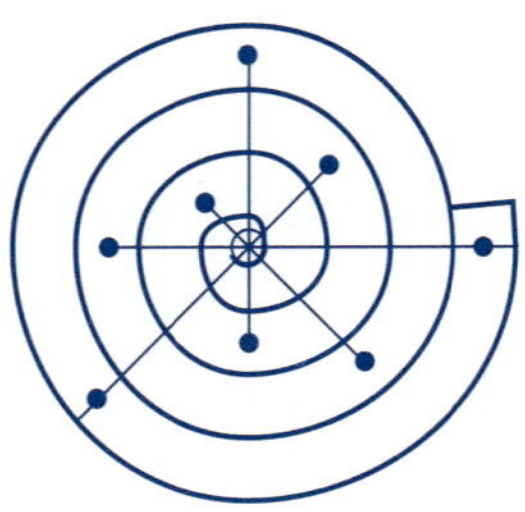

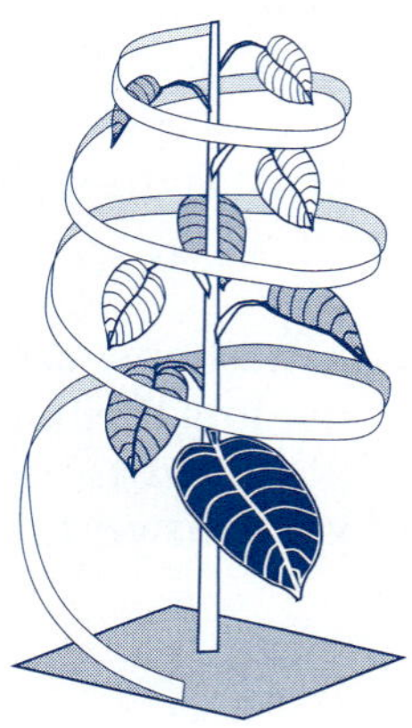

NAME ______________________ CLASS ____________ DATE ____________

Basic Skills Practice

Interpreting Venn Diagrams and Tables, Part 2

Read each question and circle the best answer.

A questionnaire asked 200 janitors which cleaning product they used.

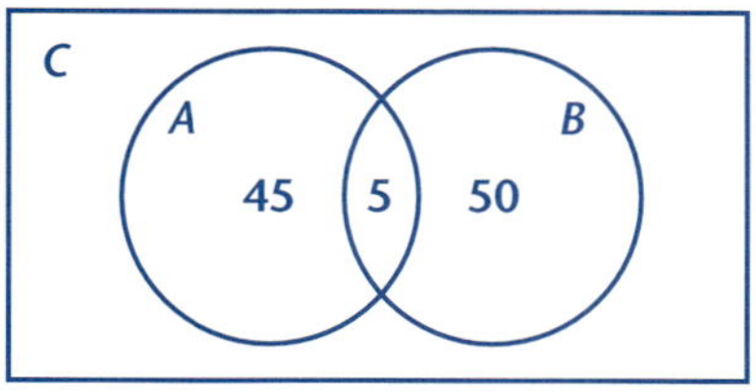

C = the set of 200 janitors surveyed
A = the set of janitors who used product A
B = the set of janitors who used product B

1. How many janitors used only product A?

A 45
B 50
C 55
D 60

2. How many janitors used only product B?

F 45
G 50
H 55
I 60

3. How many janitors used both products?

A 5
B 10
C 55
D 100

4. How many used ***neither*** product?

F 5
G 10
H 55
I 100

The table shows the monthly income and expenses for 4 students.

	Income	Expenses
Leon	$150	$75
Ben	$200	$210
Jay	$250	$240
Kara	$50	$10

5. Which student has the highest income?

A Leon
B Ben
C Jay
D Kara

6. Which student has the least expenses?

F Leon
G Ben
H Jay
I Kara

7. After expenses, how much money does Jay have left each month?

A $10
B $40
C $75
D $490

8. Which students has greater expenses than income?

F Leon
G Ben
H Jay
I Kara

LESSON 9.8 The Sine and Cosine Functions

PREPARE

Pacing Guide
2 Days

Objectives and Overview
- Use the sine and cosine functions to find the measures of the sides and angles of right triangles.

The sine and cosine functions are defined. Triangles containing 45° angles are constructed to see that the sine, cosine, and tangent values depend only on the angle measure. The sine and cosine are used to find the length of a leg of a right triangle, and the inverse sine and inverse cosine are used to find the measure of an acute angle in a right triangle.

Vocabulary
sine cosine

Assessing Prior Knowledge

Evaluate each expression to the nearest tenth with a calculator.

1. tan 57° **[1.5]**

2. $\tan^{-1}(0.8)$ **[36.7°]**

3. Find the tangents of angles *A* and *B* and the measure of angle *A*.

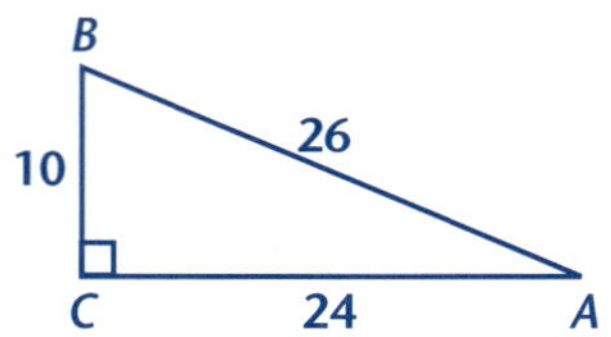

[$\tan A = \frac{5}{12}$; $\tan B = \frac{12}{5}$; 22.6°]

Approach A

Using the Book

Opening Discussion

"In the last lesson, the tangent ratio was introduced and used to find the length of a leg of a right triangle when the measure of an acute angle and the length of the other leg are known and to find the measure of an acute angle when the lengths of both legs are known. What if the lengths of one leg and the hypotenuse are known or the measure of one acute angle and the length of the hypotenuse are known? Can this information be used to find the measure of the other parts of the triangle? In the waterslide example, the length of the slide and the vertical rise are known. What parts of the triangle are these? **[a leg and the hypotenuse]** The designer needs to know the measure of the angle that the slide makes with the ground."

Defining the Sine and Cosine

Draw a general right triangle on the board. Focus on one of the acute angles, and remind students of the leg opposite the angle and the leg adjacent to the angle.

TEACH

Approach B

Using Models

Opening Discussion

"In the last lesson, the tangent ratio was introduced and used to find the length of a leg of a right triangle when the measure of an acute angle and the length of the other leg are known and to find the measure of an acute angle when the lengths of both legs are known. What if the lengths of one leg and the hypotenuse are known or the measure of one acute angle and the length of the hypotenuse are known? Can this information be used to find the measures of the other parts of the triangle? In the waterslide example, the length of the slide and the vertical rise are known. What parts of the triangle are these? **[a leg and the hypotenuse]** The designer needs to know the measure of the angle that the slide makes with the ground."

Defining the Sine and Cosine

Draw a general right triangle on the board. Focus on one of the acute angles, and remind students of the leg opposite the angle and the leg adjacent to the angle. Recall the tangent ratio. In the slide example, we want to relate a leg and the hypotenuse. This can be done with

Recall the tangent ratio. In the slide example, we want to relate a leg and the hypotenuse. This can be done by defining two other ratios. Write the definition of the sine and cosine ratios on the chalkboard or overhead for students to copy into their notebook.

Inclusion Strategy

Mnemonic Memory Aid Students who have trouble memorizing definitions may find the following mnemonic device useful: SOHCAHTOA, read "so-ca-toa," stands for "Sine is Opposite over Hypotenuse, Cosine is Adjacent over Hypotenuse, Tangent is Opposite over Adjacent."

Exploration 1

This exploration gives students some experience in finding sine and cosine values. Students should also realize that the value of the sine or cosine does not change when the length of the sides of the triangle increase but changes only when the angle is changed. You can lead the class through this exploration or have students complete it in small groups.

Ongoing Assessment, middle of page 461

8. For a given angle measure, the trigonometric ratios are the same regardless of the size of the triangle in which the angle is found.

Critical Thinking, page 461

$\sin A = 0.8$, $\sin B = 0.6$, $\cos A = 0.6$, $\cos B = 0.8$, $\sin A = \cos B$ and $\cos A = \sin B$. These values are equal because the side opposite A is adjacent to B, and the side adjacent to A is opposite B. The hypotenuse remains the same.

Technology

After completing the exploration, show students how to find sine and cosine values on a calculator. Remind them to make sure that their calculators are in degree mode. Show students the inverse sine and inverse cosine keys as well. These keys give the measure of the angle whose sine or cosine is the value entered.

continued on top of next page

defining two other ratios. Write the definition of the sine and cosine ratios on the chalkboard or overhead for students to copy into their notebook.

Inclusion Strategy

Mnemonic Memory Aid Students who have trouble memorizing definitions may find the following mnemonic device useful: SOHCAHTOA, read "so-ca-toa", stands for "Sine is Opposite over Hypotenuse, Cosine is Adjacent over Hypotenuse, Tangent is Opposite over Adjacent."

Cooperative Learning

Before starting the exploration, divide the class into groups of four. Students should take turns drawing the triangles, measuring the legs, calculating the length of the hypotenuse, finding the tangent, sine, and cosine of the acute angles, and recording the results. All group members should be involved in a discussion of the activity as the exploration proceeds. The group should work together to answer the questions. Each student should sign the completed report.

Alternate Exploration

Have students work in their groups. Give each group a protractor, centimeter ruler, and graph paper. Each group should investigate the following questions, which you may write on the overhead or chalkboard. (For ideas on guiding the students, see the exploration on page 461 of the Pupil's Edition of the textbook.)

- Draw three different right triangles, each with a 30° angle. Label the 30° angle A in each triangle. Measure the lengths of the legs of these triangles to the nearest tenth of a centimeter. Use the "Pythagorean" Right-Triangle Theorem to find the length of the hypotenuse for each triangle. Organize this information in a chart.
- Add three columns to the chart for $\tan A$, $\sin A$, and $\cos A$. Determine the values for each of these ratios. What do you notice? **[Each ratio is the same for each triangle: $\tan A \approx 0.58$, $\sin A = 0.5$, and $\cos A \approx 0.87$. Any differences are due to errors in measurement or rounding.]**

continued on bottom of next page

Approach A Using the Book, continued

Ongoing Assessment, bottom of page 461

about 0.4226; about 0.8391

Example 1

In this example, the length of one of the legs of a right triangle is determined by using the sine ratio with a given angle measure and the length of the hypotenuse.

Ask why the sine ratio was chosen in this case. **[An angle measure and the length of the hypotenuse are given, and the length of the side opposite the angle is desired.]**

Additional Example 1

Refer to Example 1 and find the distance (measured along the water) between the parasail and the boat. **[about 58 yards]**

Ongoing Assessment, page 462

Try This

$a \approx 33.44$; $b \approx 33.11$

Example 2

In this example, the measure of the angle between the ground and the waterslide from the beginning of the lesson is found by using the inverse sine.

Additional Example 2

Refer to the waterslide example. Suppose that the 200-foot slide needed to fit into a space on the ground that was 160 feet long. At what angle should the slide be placed in this case? **[about 37°]**

Ongoing Assessment, page 463

Try This

about 59°

TEACH

Approach B Using Models, continued

- Using the information in the table, make a conjecture about the trigonometric ratios for any given angle measure. **[For a given angle measure, the trigonometric ratios are the same regardless of the size of the triangle in which the angle is found.]**

Student Reports

Ask one or more groups to report their findings for this exploration. Explain that because each acute angle has just one sine and cosine value, these values can be found with a calculator. Show students how to find values for sine and cosine on the calculator. Remind them to make sure that their calculator is in degree mode. Point out the inverse sine and inverse cosine keys as well. These keys give the measure of the angle whose sine or cosine is the value entered.

The Uses of the Sine and Cosine Functions

The sine and cosine functions can be used to find the length of one leg of a right triangle when the measure of an acute angle and the length of the hypotenuse are known. The inverse sine and cosine can be used to find the measure of an acute angle of a right triangle when the length of the hypotenuse and the length of a leg are known. Lead the class through the following two examples, which illustrate these types of problems. In part A, discuss how to decide whether the sine or cosine ratio should be used. **[Look at the relationship between the angle and the side involved. If the side is adjacent to the angle, use the cosine. If the side is opposite the angle, use the sine.]** In part B, ask students how they can recognize when to use the inverse sine or cosine. **[When they are looking for the angle measure, they should use the inverse sine or cosine.]** (Compare Examples 1 and 2 on pages 462 and 463, respectively.)

On Their Own

A. An airplane is flying at 2000 ft. If the measure of the angle formed by the ground and the line from the observer to the plane is 50°, how far away from the observer is the plane? **[about 2611 ft]**

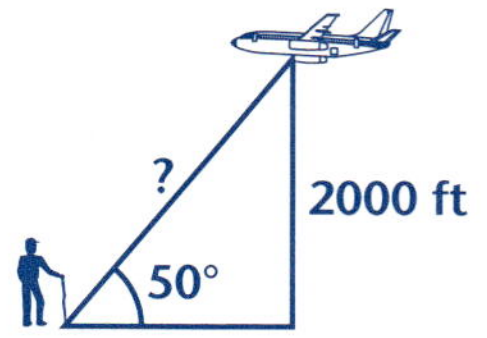

B. Suppose that a 40-ft guy wire is attached to an utility pole and anchored to the ground at a point 30 ft from the pole, as shown at right. What is the angle formed by the ground and the guy wire? **[about 41°]**

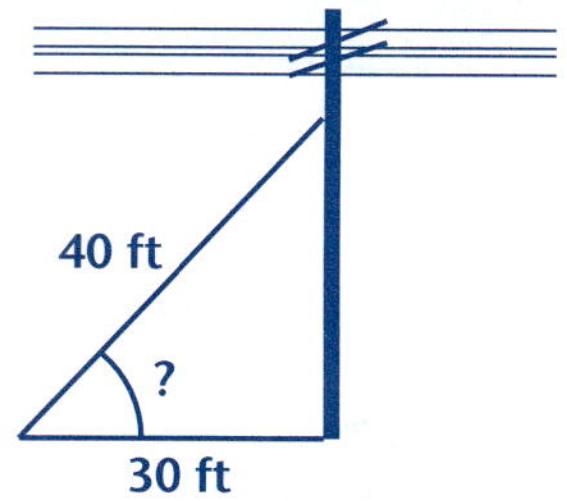

Closing the Lesson

Have students work with a partner. Each should draw a right triangle, label the lengths of the legs, and name the vertices. Students should then exchange papers and find the length of the hypotenuse and the cosine, sine, and tangent of each acute angle.

A. A kite is flown by a child at the beach. The string attached to the kite is 83 feet long. The string forms an angle of 42° with a line that is horizontal.

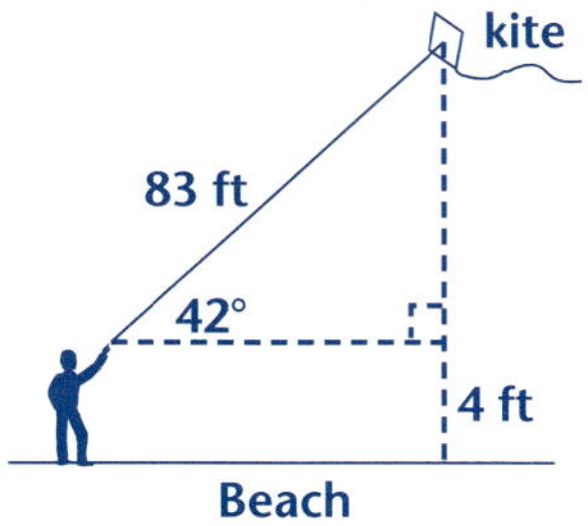

Approximately how high above the beach is the kite? (The string in the child's hand is 4 feet above the beach.) **[about 59.5 feet]** What is the distance (measured along the beach) between the child and the point directly below the kite? **[about 61.7 feet]**

B. Consider the waterslide from the opening example. If the slide is 200 feet long and rises 115 feet above the ground, what is the measure of the angle formed by the slide and the ground? **[about 35°]**

Closing the Lesson

Have students work with a partner. Each should draw a right triangle, label the lengths of the legs, and name the vertices. Students should then exchange papers and find the length of the hypotenuse and the cosine, sine, and tangent of each acute angle.

ASSESS

Selected Answers

Odd-numbered Exercises 5–65 and odd-numbered Extra Practice

Assignment Guide

Core level students should complete Exercises 1–66.

Core-plus level students should complete Exercises 1–11, 16–33, and 38–68.

Error Analysis

Students may mistakenly invert the sine and cosine ratios. Stress that in both cases the hypotenuse is in the denominator. Students may also be confused about when to use the sine or cosine and when to use the inverse sine or inverse cosine. Explain that you use the sine or cosine when you know an angle measure and that you use the inverse sine or inverse cosine when you are trying to find an angle measure. Watch for students who do not have their calculators in degree mode.

Alternative Assessment

Student Examples: Sine vs. Cosine

How do you know when you should use the sine to solve a problem and when you should use the cosine to solve the problem? Have students answer this question and illustrate their answer by making up an example for each case. They should include a worked-out solution for each example.

Portfolio Activity Ask students to write their responses to Exercise 4 in paragraph form to be included in their portfolio. They should also answer the same question about the tangent.

RETEACH

Have students work with a partner to make a chart that can help them decide which trigonometric function to use when solving a problem that involves a right triangle. At the top of the next column is a possible chart with the first two lines completed. Students could also include an illustration of each situation. Students can look back over the examples from this lesson and the previous lesson to find other examples.

Given information	Desired information	Function to use
acute angle, hypotenuse	leg opposite the angle	sine
leg, hypotenuse	acute angle adjacent to leg	inverse cosine

After completing the chart, have the partners make up an application problem for two of the lines in their chart and write the solutions to those problems.

ENRICH

Interdisciplinary Connection

Construction Have students find out the local building codes concerning wheelchair ramps. They should then pick a short set of stairs at school and design a wheelchair ramp for it that meets the building code.

Using a Spreadsheet to Examine $\sin^2 x + \cos^2 x$

Have students make a spreadsheet to examine the value of $\sin^2 x + \cos^2 x$. First explain that this notation means $(\sin x)^2 + (\cos x)^2$. The first row of the spreadsheet should contain angles from 5° to 85° in increments of 5°. The other rows should be for $\sin x$, $\cos x$, and $(\sin x)^2 + (\cos x)^2$. Students should use the fill down feature to complete these rows. What do they notice? **[$\sin^2 x + \cos^2 x = 1$ for all angles]**

NAME ______________________ CLASS ____________ DATE ____________

Practice

9.8 Sine and Cosine Functions

Evaluate each expression to the nearest ten-thousandth.

1. sin 65° ______________ **2.** sin 85° ______________

3. cos 32° ______________ **4.** cos 15° ______________

5. sin 12° ______________ **6.** sin 50° ______________

7. cos 39° ______________ **8.** cos 89° ______________

9. sin 76° ______________ **10.** sin 82° ______________

11. cos 49° ______________ **12.** cos 61° ______________

For Exercises 13–20, use right triangle *ABC* at right. Find the indicated angle measure or length. Round your answer to the nearest tenth.

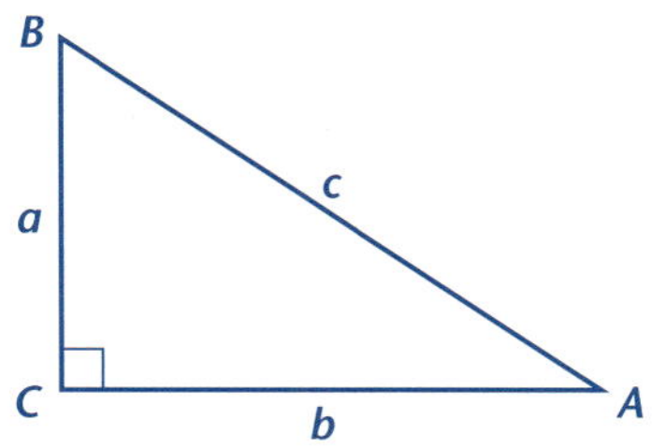

13. $b = 6$, $c = 8$, $m\angle B =$ ______________

14. $a = 8$, $c = 15$, $m\angle B =$ ______________

15. $b = 6$, $c = 9$, $m\angle A =$ ______________

16. $a = 8$, $c = 14$, $m\angle B =$ ______________

17. $b = 7$, $c = 10$, $m\angle B =$ ______________

18. $m\angle B = 40°$, $c = 30$, $b =$ ______________

19. $m\angle B = 52°$, $c = 18$, $a =$ ______________

20. $m\angle A = 46°$, $c = 9$, $b =$ ______________

What angle measure has each approximate sine value? Round each answer to the nearest tenth of a degree.

21. 0.5992 __________ **22.** 0.8387 __________ **23.** 0.0872 __________

24. 0.2419 __________ **25.** 0.1139 __________ **26.** 0.7408 __________

What angle measure has each approximate cosine value? Round each answer to the nearest tenth of a degree.

27. 0.3746 __________ **28.** 0.8988 __________ **29.** 0.0087 __________

30. 0.9885 __________ **31.** 0.5490 __________ **32.** 0.0523 __________

NAME ________________________ CLASS ____________ DATE ____________

Lesson Activity

9.8 Sine and Cosine Patterns

1. Use a calculator to complete the following table. Round your answers to the nearest ten-thousandth.

Angle measure	Sine	Cosine
0°		
10°		
20°		
30°		
40°		
45°		
50°		
60°		
70°		
80°		
90°		

2. Look at the values in the sine column from top to bottom. Compare them with the values in the cosine column from bottom to top. What do you notice? Do you think the same pattern would occur with angle measures other than those listed in the table?

Explain. ____

____

3. Use a calculator to determine which angles between 0° and 360° have a negative sine value.

____

4. Use a calculator to determine which angles between 0° and 360° have a negative cosine value.

____

NAME ______________________ CLASS __________ DATE __________

Basic Skills Practice

Cumulative Review

Read each question and circle best answer.

1. A large bag of dog food weighs almost 15 pounds. To the nearest hundred pounds, how much will 225 bags weigh?

A 3300 lb

B 3400 lb

C 3500 lb

D 3600 lb

2. A fabric shop sells ribbon in 4 different widths. They are arranged on shelves in order of width in inches. Which group of fractions shows the correct order from greatest to least width?

F $\frac{1}{2}, \frac{2}{9}, \frac{11}{18}, \frac{5}{12}$

G $\frac{2}{9}, \frac{5}{12}, \frac{1}{2}, \frac{11}{18}$

H $\frac{11}{18}, \frac{1}{2}, \frac{5}{12}, \frac{2}{9}$

I $\frac{5}{12}, \frac{11}{18}, \frac{2}{9}, \frac{1}{2}$

3. The wavelength of red light is approximately 0.00000068 meters. Find the correct expression of this number in scientific notation.

A 68×10^{8} m

B 68×10^{-8} m

C 6.8×10^{7} m

D 6.8×10^{-7} m

4. Find the equation that is equivalent to the following statement: The difference of three times x and two times y is 42.

F $3 + x - 2 + y = 42$

G $3x + 2y = 42$

H $3x - 2y = 42$

I $3x - 2y + 42 = 0$

5. A bag contains 14 marbles. There are 8 white marbles, 4 red marbles, and 2 blue marbles. One marble is drawn from the bag. Find the probability that it is a red marble.

A 1 out of 7

B 2 out of 7

C 4 out of 7

D 6 out of 7

6. To make a lion costume, $3\frac{1}{4}$ yards of light brown felt was purchased and $\frac{7}{8}$ yards of dark brown felt was purchased. What was the combined yardage of the 2 colors of felt?

F $3\frac{1}{8}$ yd

G $3\frac{2}{3}$ yd

H $3\frac{3}{4}$ yd

I $4\frac{1}{8}$ yd

Solve each problem.

7. A rectangular swimming pool is 25 meters long and 18 meters wide. The average depth of the pool is 3 meters. What is the volume of the pool in meters? ______________

8. What is the surface area, in square meters, of the pool referred to in Exercise 7? Include a cover to go over the top of the swimming pool. ______________

LESSON 10.1

Rational Expressions

PREPARE

Pacing Guide

2 Days

Objectives and Overview

- Define and evaluate rational functions.
- Determine the values for which a rational function is undefined.
- Describe transformations applied to the parent function and their effect on its graph.

Students are introduced to rational expressions and rational functions. They examine the behavior of a reciprocal function and identify values for which a reciprocal function is undefined.

Vocabulary

rational expression
rational function

Assessing Prior Knowledge

Factor:

1. $m^2 - 2m + 1$ $[(m - 1)^2]$

2. $5r^4s + 15r^3s^2 + 50r^2s$
$[5r^2s(r^2 + 3rs + 10)]$

3. $w^2 - z^4$ $[(w - z^2)(w + z^2)]$

4. $x^4 - 16$ $[(x - 2)(x + 2)(x^2 + 4)]$

5. $a^2 - 7a - 18$ $[(a - 9)(a + 2)]$

6. $6g^2 + 3g - 45$ $[3(2g - 5)(g + 3)]$

Approach A: Using the Book

Opening Discussion

Suppose that a group of 10 people went to the movies. If the cost of all 10 movie tickets was $80 and 2 people paid for all of the tickets, how much did each person pay? If four people shared the cost of the 10 tickets, what was the cost per person? If x represents the number of people paying, write a function that describes the cost per person. What is the parent function?

Using a Table of Values

Using one column of a table for the value of the numerator of the rational expression $\frac{55 + 4.50x}{x}$ and another column for the value of the fraction allows students to see that the numerator of the fraction is a linear expression. In the spreadsheet on page 475, column B shows that as the number of lessons increases, the total expense increases. Column C shows that as the number of lessons increases, the average cost of a lesson decreases.

TEACH

Approach B: Using Models

Opening Discussion

Tell your students that a group of 10 people went to the movies and that the cost of all 10 movie tickets was $80. If only 2 people paid for all of the tickets, how much did each person pay? Suppose that 4 people shared the cost of the 10 tickets. What was the cost per person? Let x represent the number of people paying for the 10 tickets. Write a function that describes the cost per person. What is the parent function?

Inclusion Strategy

Looking for a Pattern Use the expression $\frac{55 + 4.50x}{x}$ to represent the average cost of taking a pottery-making lesson if the cost per lesson is $4.50 and a membership in the ceramics club costs $55. Build a chart of the average cost for various values of x.

Ongoing Assessment, page 475

The error is caused by trying to divide by zero.

Defining *Rational Expression*

Students should recognize the word *ratio* in *rational expression.* A rational expression is a ratio or comparison of one polynomial with another polynomial. A rational expression is defined for all real numbers except those for which the denominator of the expression would equal zero. For these real numbers, the expression is undefined.

Teaching Tip

Comparing Rational Numbers and Rational Expressions Point out the similarity between rational numbers and rational expressions to students. Rational numbers are numbers that can be written in the form $\frac{p}{q}$, where p and q are integers and $q \neq 0$. Rational expressions are expressions of the form, $\frac{P}{Q}$, where P and Q are polynomials and $Q \neq 0$.

Example 1

Students are asked to identify the values that make several rational expressions undefined.

Make sure students recognize that in part B, there is no particular value of a or b for which the function is undefined. Rather, whenever $a = b$, the function is undefined. You might ask students to give several examples of values for which this function is undefined to ensure that they understand the domain of values for the function.

Additional Example 1

For what value or values is each rational function undefined?

A. $y = \frac{1}{2x}$ **[$\frac{1}{2x}$ is undefined when $x = 0$.]**

B. $y = \frac{r^2 + s^2}{r(r-2)}$ **[$\frac{r^2 + s^2}{r(r-2)}$ is undefined when $r = 0$ or $r = 2$.]**

C. $f(x) = \frac{c-5}{c^2 - 9c + 20}$ **[$\frac{c-5}{c^2 - 9c + 20}$ is undefined when $c = 4$ or $c = 5$.]**

continued on top of next page

x	$55 + 4.50x$	$\frac{55 + 4.50x}{x}$
1		
2		
3		
4		
⋮		
10		
20		
30		

Ask each student to find the total cost of x lessons and the average cost of each lesson for one value of x. Then fill in the chart with the students' values.

What happens to the total cost as the number of lessons increases? What happens to the average cost of a lesson as the number of lessons increases? Why does the table have no negative values? What is the value of $\frac{55 + 4.50x}{x}$ for $x = 0$? Why?

Rational Expressions

Discuss the definition of **rational expression** on page 475 of the Pupil's Edition.

Point out to students that while a rational number is a ratio of two integers, a rational expression is the ratio of two polynomials.

Cooperative Learning

Identifying the Domain of a Rational Expression Ask students to work with a partner and to make tables of values for $f(x) = \frac{1}{x}$, $y = \frac{a^2 - b^2}{a - b}$, and $t = \frac{m + n}{m^2 - 4m + 3}$. If graphics calculators or spreadsheet programs are available, have students use them to generate the tables of values. Tell students to describe the domain of each function by examining its table of values. Why are some real numbers excluded from the domain of each function? How do you find the excluded values in a manually generated table? How do you find the excluded values with technology?

continued on bottom of next page

Approach A

Using the Book, continued

Ongoing Assessment, page 476

Try This

a. $y = 2$ or $y = 3$

b. $x = y$

Teaching Tip

You may wish to review the parent functions and transformations of graphs studied in Chapter 5. Refer your students to the summary in Chapter 5. Ask several students to read the summary information about parent graphs and transformations aloud to the class. It will be helpful to all students if you ask some volunteers to draw sketches of a few parent graphs and some transformations on the chalkboard.

Determining Asymptotes

Journal Entry Ask students to substitute 0 for x in the expression $\frac{55 + 4.50x}{x}$. What fraction do you get? $\left[\frac{55}{0}\right]$ Now choose several values for x that are close to 0, such as 0.01, 0.0000001, and −0.0001. What happens to the size of the fraction as the denominator gets closer and closer to zero? **[The closer the denominator gets to zero, the larger the absolute value of the fraction becomes.]** How does this relate to the graph of the function $f(x) = \frac{55 + 4.50x}{x}$? **[The graph has a vertical asymptote at $x = 0$.]**

Example 2

The transformations to a parent graph are described, and the graph is used to find the value of the dependent variable for a particular value of the independent variable.

TEACH

Approach B

Using Models, continued

Teaching Tip

You may wish to review graphs of reciprocal functions with your students. The graph of $f(x) = \frac{1}{x}$ is a hyperbola with vertical asymptote $x = 0$. Draw a sketch of the graph of $f(x) = \frac{1}{x}$ on the chalkboard or overhead. *Asymptote* comes from a Greek word meaning "not meeting." Which vertical line does the curve of $f(x) = \frac{1}{x}$ never meet? Use a table to display several values of x that are close to zero and the corresponding values of the expression $\frac{1}{x}$.

x	$\frac{1}{x}$
−0.00001	−100,000
−0.0001	−10,000
−0.001	−1000
−0.01	−100
−0.1	−10
0.1	10
0.01	100
0.001	1000
0.0001	10,000

How do the values in the chart correspond to the curve of the graph? **[As x gets closer and closer to zero from either direction, the absolute value of $\frac{1}{x}$ gets infinitely greater.]**

Comparing Tables of Values

Use the chart of values for $\frac{55 + 4.50x}{x}$ to graph the function $g(x) = \frac{55 + 4.50x}{x}$. (Compare with the graph on page 477.) Comparing the table of values for $f(x) = \frac{1}{x}$ with the table of values for

The expression $\frac{55}{x} + \frac{4.50x}{x}$ cannot be simplified to $\frac{55}{x} + 4.5$ if $x = 0$ because that would require dividing the numerator of the fraction $\frac{4.50x}{x}$ by zero.

Students can graph the functions $Y_1 = \frac{55}{x} + \frac{4.50x}{x}$ and $Y_2 = 5.60$ and then find the intersection of the two graphs. The *x*-coordinate of the point of intersection is the value of *x* for which $f(x) = 5.60$.

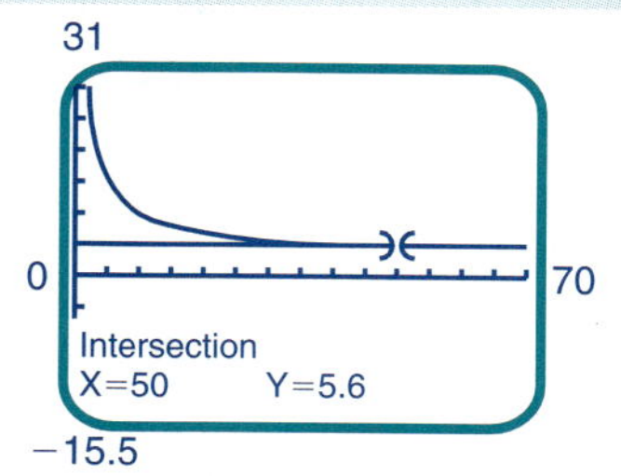

Additional Example 2

A. Describe the transformations applied to the graph of the parent function $f(x) = \frac{1}{x}$ to produce the graph of $g(x) = \frac{3}{x+2} - 1$. **[The graph of the parent function has been transformed by a vertical stretch of 3, a shift to the left of 2, and a shift down by 1.]**

B. Use the graph to find the value of *x* when $g(x) = -0.995$. **[$x = 598$]**

Example 3

A transformation applied to the graph of a parent function is described.

Students may need a review of transformations that produce horizontal shifts of the graph of a parent function. For the parent function $f(x) = \frac{1}{x}$, the function $g(x) = \frac{1}{x-a}$, where $a > 0$, shows a horizontal shift *a* units to the *right*. The function $g(x) = \frac{1}{x+a}$ shows a horizontal shift *a* units to the *left*.

continued on top of next page

$g(x) = \frac{55 + 4.50x}{x}$ shows that for a given value of *x*, multiplying the value of the parent function by 55 and then adding 4.5 to the result gives the corresponding value of $g(x) = \frac{55 + 4.50x}{x}$. Ask students to verify this relationship for several values of *x*.

x	$\frac{55 + 4.50x}{x}$	$\frac{1}{x}$
0	error	error
1	59.5	1
2	32	0.5
3	22.833	0.33333
4	18.25	0.25
5	15.5	0.2

Interpreting a Transformation From an Equation

The Distributive Property can be used to rewrite $\frac{55 + 4.50x}{x}$ as $\frac{55}{x} + \frac{4.50x}{x}$. If $x \neq 0$, then $\frac{55}{x} + \frac{4.50x}{x} = \frac{55}{x} + 4.5$. How does this equation explain the transformation of the graph of $f(x) = \frac{1}{x}$?

Note that if $x = 0$, $\frac{4.50x}{x}$ has the form $\frac{0}{0}$. From this it follows that $\frac{0}{0} = 4.50$. Students should see that this contradiction indicates that $x \neq 0$.

Technology

Graphics calculators can be used to find a corresponding value of *x* for a given value of $g(x)$. Suppose that you wanted to know how many pottery lessons could be taken for an average cost of $5.60 per lesson. Divide the class into three groups. Ask the students in the first group to work backward from the transformation of $f(x) = \frac{1}{x}$ to find the corresponding *x*-value for $f(x) = 5.60$. The students can then take the reciprocal of this number to determine how many pottery lessons can be taken for an average cost of $5.60 per lesson. **[$(5.60 - 4.50) \div 55 = 0.02$; $\frac{1}{0.02} = 50$]**

Ask the second group of students to examine the graph of $g(x) = \frac{55}{x} + 4.5$ or to use the trace

continued on bottom of next page

Approach A Using the Book, continued

Additional Example 3

Describe the transformation applied to the graph of the parent function $f(x) = \frac{1}{x}$ to produce the graph of $g(x) = \frac{1}{3x} - 6$. **[The graph of the parent function has been transformed by a vertical stretch of $\frac{1}{3}$ and a shift down by 6.]**

Ongoing Assessment, page 477

Try This

Shift to the left 1.5 units, shift down 5 units, and stretch by a factor of $\frac{3}{2}$.

Critical Thinking, page 477

Changing a results in a vertical stretch; changing h results in a shift to the right or left; and changing k results in a shift up or down.

Closing the Lesson

Journal Entry Ask students to give an example of a function $g(x)$ whose graph is a transformation of the graph of the parent function $f(x) = \frac{1}{x}$ by a vertical stretch of a.

TEACH

Approach B Using Models, continued

function on a graphics calculator to find the value of x such that $g(x) = 5.60$. (Compare the graph on page 477.)

Ask the third group of students to graph the functions $Y_1 = \frac{55}{x} + 4.5$ and $Y_2 = 5.60$ and to find the intersection of the two graphs. The intersection is the value of x for which $f(x) = 5.60$.

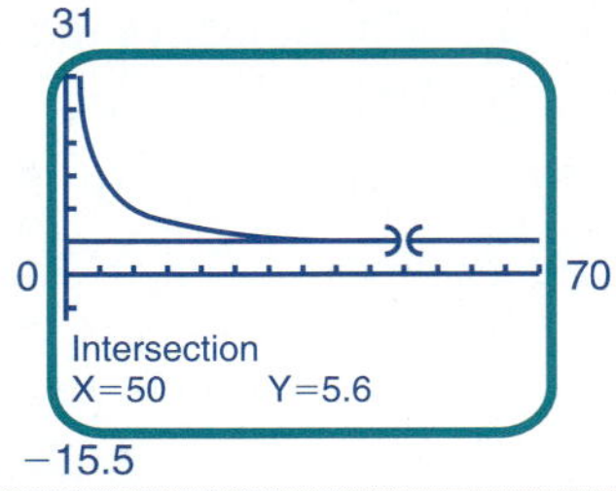

Inclusion Strategy

Transforming Graphs Draw a large sketch of the graph of $f(x) = \frac{1}{x}$ on the chalkboard. Divide the class into groups of three, and assign one of the following functions to each group:

1. $g(x) = \frac{5}{x} + 2$ **2.** $g(x) = \frac{5}{x} - 2$

3. $g(x) = \frac{5}{x-3} + 2$ **4.** $g(x) = \frac{5}{x-3} - 2$

5. $g(x) = \frac{5}{x+3} + 2$ **6.** $g(x) = \frac{5}{x+3} - 2$

Ask one student in each group to generate a table of values for the function. A second student should sketch a large graph of the function on notebook paper. The third student in each group should copy the graph onto the axes on the chalkboard where you have drawn the graph of $f(x) = \frac{1}{x}$. Then each student should explain the transformation of his or her graph to the class. Have the students sketch the graphs with different colors of chalk.

ASSESS

Selected Answers

Odd-numbered Exercises 7–33 and odd-numbered Extra Practice

Assignment Guide

Core level students should complete Exercises 1–34.

Core-plus level students should complete Exercises 1–8, 11–18, and 21–36.

Error Analysis

Some students exclude from the domain of a rational function the real numbers that make the numerator of a rational expression 0. Explain to these students that $\frac{0}{n}$, where $n \neq 0$, is zero. For example, $\frac{0}{2} = 0$ and $\frac{0}{2.47} = 0$. Thus, zero is a possible value for the numerator of a rational expression.

Use **Transparency 48** to discuss Exercise 34 in Look Back, which is a fun and interesting puzzle from the journal of Benjamin Banneker, the first African American astronomer.

Alternative Assessment

Portfolio Activity Graph the function $g(x) = \frac{5x + 2}{x}$. Express $g(x)$ in a form which shows that it is a transformation of the parent function $f(x) = \frac{1}{x}$. Interpret the form to explain the transformation of the graph of $f(x) = \frac{1}{x}$ that produces the graph of $g(x) = \frac{5x + 2}{x}$. Choose three values of x and find $\frac{1}{x}$ for each value. Use the transformation to find the corresponding values of $g(x)$.

RETEACH

Write each of the following functions on a piece of paper, and put the papers in a box.

$f(x) = \frac{1}{x}$, $f(x) = \frac{2}{x}$, $f(x) = \frac{1}{x} + 2$, $f(x) = \frac{1}{x - 2}$, $f(x) = \frac{2}{x} - 2$, $f(x) = \frac{1}{x + 2} + 2$

Divide the class into 6 groups. Ask one member of each group to draw one of the papers from the box. The students should not tell any other group what function they have drawn. Ask each group to make a table of values for its function and to graph the function. The group should discuss the shape of the graph.

Draw a large coordinate axes on the chalkboard with x- and y-axes labeled from -5 to 5. Then sketch a graph of $f(x) = \frac{1}{x}$ on the axes. Have one member of each group sketch the group's function on the axes. After each graph is sketched, ask the rest of the class to identify the equation of the graph. Then ask each group to identify the domain of its function.

ENRICH

Interdisciplinary Connection

Chemistry The equation $V = \frac{16,560}{p}$ can be used to model the relationship between the volume (V) of a fixed amount of gas at a constant temperature and the pressure (p) of the gas in pounds per square inch (psi). Because the graph of $V = \frac{16,560}{p}$ has a vertical asymptote at $p = 0$, it indicates that a gas cannot be compressed to a zero volume. Since volume is positive, only the positive values of p and V on the graph of $V = \frac{16,560}{p}$ are used for this application.

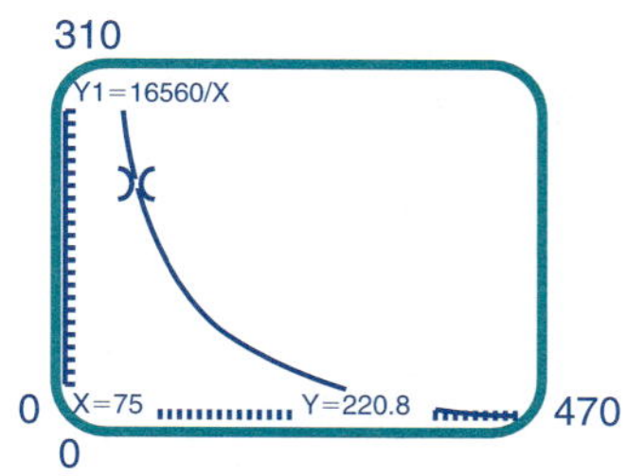

NAME ______________________ CLASS __________ DATE __________

Practice

10.1 Rational Expressions

For what values of *x* are these rational expressions undefined?

1. $\frac{3x}{2x}$ ____________ **2.** $\frac{3x}{2x + 3}$ ____________ **3.** $\frac{3x}{2x^2 + 3x}$ ____________

4. $\frac{2}{x + 2}$ ____________ **5.** $\frac{4x^2}{2x + 1}$ ____________ **6.** $\frac{4x}{x^2 + 4x + 4}$ ____________

Evaluate each rational expression for $x = -1$ and $x = 2$. Write *undefined* if appropriate.

7. $\frac{x}{x + 1}$ ____________ **8.** $\frac{-x^2 + x}{x}$ ____________ **9.** $\frac{x^2 - 2}{x^2 + 4}$ ____________

Describe all of the transformations applied to $f(x) = \frac{1}{x}$ in order to produce the graph of the following rational functions:

10. $g(x) = \frac{-3}{x}$ ____________________________

11. $g(x) = \frac{-3}{x - 2}$ ____________________________

12. $g(x) = \frac{-3}{x - 2} + 2$ ____________________________

Graph each rational function. Label the vertical asymptotes. List the value for which the function is undefined.

13. $h(x) = \frac{-3}{x + 2} - 2$ ____________ **14.** $h(x) = \frac{3}{x - 2} + 2$ ____________

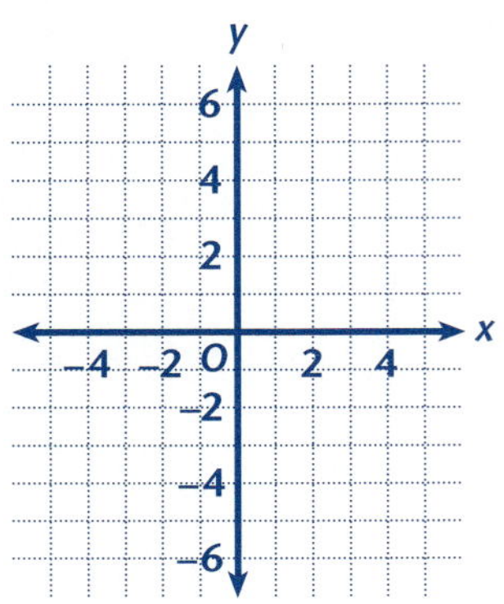

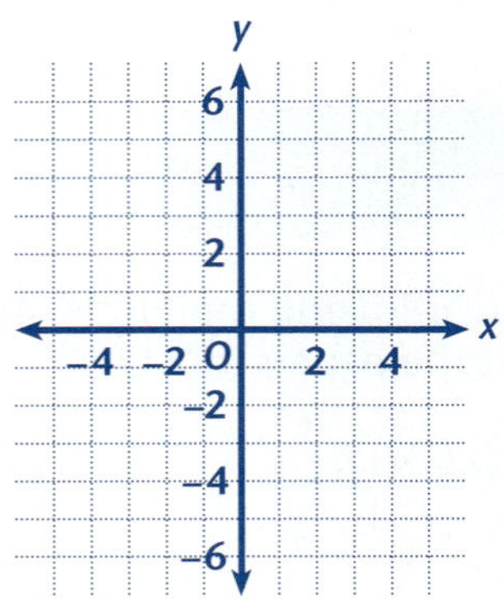

NAME ______________________ CLASS ______________ DATE ______________

Lesson Activity

10.1 Tower of Fraction Bars

The tower of fraction bars is a "tower" formed by connecting points that represent a sequence of fractions with a smooth curve. Each fraction bar is one unit long and is divided into equal parts. One-half of the curves turn down to the right. The other half turn down to the left. The first sequence, which curves down to the right, is generated by evaluating the rational expression $\frac{n}{n+1}$ for values of n from 0 to 11. Thus, the curve intersects the fraction bars at 0, $\frac{1}{2}, \frac{2}{3}, \frac{3}{4}, \frac{4}{5}, \frac{5}{6}, \frac{6}{7}, \frac{7}{8}, \frac{8}{9}, \frac{9}{10}, \frac{10}{11}$, and $\frac{11}{12}$.

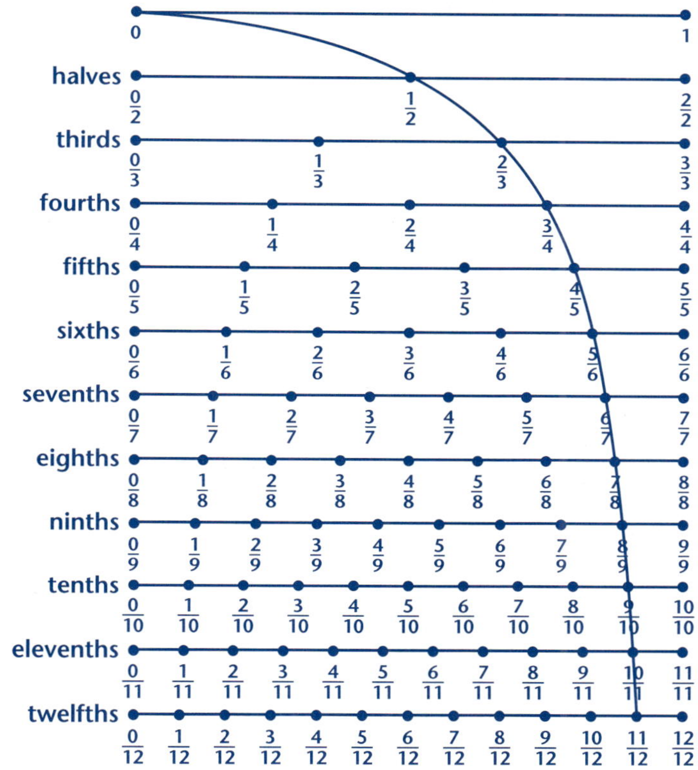

1. For the second curve, draw the sequence of fractions generated by $\frac{n}{n+2}$ for values of n from 0 to 10. Then repeat the pattern by drawing the sequences generated by $\frac{n}{n+4}, \frac{n}{n+5}, \frac{n}{n+6}$, and $\frac{n}{n+7}$.

2. For the curves that turn down to the left, draw each sequence generated by $\frac{1}{n}, \frac{2}{n}, \frac{4}{n}, \frac{5}{n}, \frac{6}{n}$, and $\frac{7}{n}$ for values of n from the value of each numerator to 12.

3. For the curves that turn down to the right, describe what happens as n gets larger. ______________________

4. For the curves that turn down to the left, describe what happens as n gets larger. ______________________

NAME ______________________ CLASS ____________ DATE ____________

Basic Skills Practice

Cumulative Review

Read each question and circle the best answer.

1. Jack needs 24 feet of fabric to make a banner for a party. How many ***yards*** of fabric does he need?

A 8 yd

B 12 yd

C 48 yd

D 72 yd

2. The approximate braking distance in feet necessary to stop a car is given by the formula

$$D = 0.05R^2,$$

where R is the speed of the car. What is the approximate braking distance for a car traveling 60 miles per hour?

F 30 ft

G 90 ft

H 180 ft

I 900 ft

3. Jacob bought 5 acres of land outside the city. He plans on donating $\frac{2}{3}$ of this land for a park. How many acres of land does he plan on donating?

A $2\frac{1}{2}$ acres

B $3\frac{1}{4}$ acres

C $3\frac{1}{3}$ acres

D $4\frac{1}{4}$ acres

4. Find the equation equivalent to this equation.

$$-5 + x = 8$$

F $x = 8 + 5$

G $x = 8 - 5$

H $5 = -x + 8$

I $-5 = -x - 8$

5. Kirby bought a purse at a discount of 25%. She saved $14.00. What was the original price of the purse?

A $17.00

B $35.00

C $56.00

D $560.00

6. The ratio of girls to boys on Savannah's soccer team is 2 to 3. If there are 12 boys on the team, how many girls are there?

F 6

G 8

H 10

I 18

7. Given a right triangle with sides a and b and a hypotenuse of 30 centimeters, which of the following equations is correct?

A $a^2 + b^2 = 30$

B $a^2 - 30 = b^2$

C $a^2 + b^2 = 30^2$

D $30 + b^2 = a^2$

8. Each of 5 friends intends to pay an equal share of the expenses for a camping trip. The cost per day for the campsite is $18.80. There is also an additional cost of $5.80 per vehicle. The 5 friends drove in 2 vehicles. How much should each friend pay for 1 day of camping, including the vehicle charge?

F $6.08

G $5.68

H $5.98

I $6

LESSON 10.2

Inverse Variation

PREPARE

Pacing Guide
2 Days

Objectives and Overview
- Define inverse variation.
- Use inverse variation to solve problems.

Inverse variation is defined, and several applications of inverse variation are studied.

Vocabulary
inverse variation

Assessing Prior Knowledge

Solve each equation.

1. $128 = 4r$ $[r = 32]$

2. $\frac{n}{9} = \frac{1326}{234}$ $[n = 51]$

3. $\frac{2}{x} = \frac{x}{32}$ $[x = \pm 8]$

Approach A: Using the Book

Opening Discussion

You can use the formula $i = 12f$, where i is a length in inches and f is a length in feet, to convert from feet to inches. The formula can also be written in the form $\frac{i}{f} = 12$. This is an example of a direct variation equation because as the number of feet increases, the number of inches increases proportionally. The quotient of the variables in a direct variation equation is a constant. What does the graph of a direct variation equation look like?

On the other hand, an equation which shows that the number of contributors to a collection decreases as the amount of money donated by each person increases and the total collected remains constant is an example of an inverse variation equation. If the amount to be collected is $1000, then the equation $an = 1000$, where a is the amount collected per person and n is the number of contributors, models this situation. The product of the variables is a

continued on top of next page

TEACH

Approach B: Using Models

Opening Discussion

Introduce the formula $i = 12f$, where i is a length in inches and f is a length in feet. This formula can be used to convert from feet to inches. The formula can also be written as the ratio of i to f: $\frac{i}{f} = 12$. This is an example of a direct variation equation because as the number of feet increases, the number of inches increases proportionally. The ratio of the variables in a direct variation equation is a constant. Ask students what the graph of a direct variation equation looks like.

Suppose that a scholarship fund has a fixed goal of $1000. As the number of contributors to the fund increases, the amount that each contributor donates decreases. An equation that shows the relationship of the number of contributors to the amount that each person donates is an example of an inverse variation equation. If the amount to be collected is $1000, then the equation $an = 1000$, where a is the amount collected per person and n is the number of

continued on bottom of next page

Approach A

Using the Book, continued

constant. The equation can also be written in the form $a = \frac{1000}{n}$. What do you think the graph of this equation looks like?

Brainstorming

The product of two variables is a constant when the variables are inversely related. Ask your students to make a list of situations in which one variable decreases while the other variable increases. Examples include the following: as the number of people sharing a pizza increases, the number of slices per person decreases; for a constant distance, as the rate increases, the time traveled decreases; and as the length of a rectangle of constant area increases, the width of the rectangle decreases.

Defining *Inverse Variation*

If $x \cdot y = k$ for nonzero values of x, y, and k, then x and y are said to *vary inversely as each other* or to be *inversely proportional to each other.*

Example 1

A problem involving inverse variation is solved by first finding the constant of variation. Substitutions are then made in an inverse variation equation, and the equation is solved for the unknown variable.

If the number of hours needed to plant a group of trees varies inversely as the number of people who are planting the trees, then nt is a constant, where n is the number of people and t is the time in hours. Because the product of any two variables of an inverse variation is constant, the ordered pair (5, 100) can be used to find the constant of variation.

Additional Example 2

An electrical current is measured in amperes and is related to the resistance in the circuit. Resistance is measured in ohms. The current, I, flowing in an electrical circuit varies inversely as the resistance, R, in the circuit. When $R = 4$ ohms, then $I = 24$ amperes. What is the current when the resistance is 6 ohms? **[$I = 16$ amperes]**

TEACH

Approach B

Using Models, continued

contributors, models this situation. The product of a and n is a constant. The equation can also be written in the form $a = \frac{1000}{n}$. Ask students what they think the graph of this equation looks like.

Brainstorming

Two variables are inversely related if their product is a constant. Can you think of a situation in which one variable decreases while the other variable increases? Examples include the following: as the number of people sharing a pizza increases, the number of slices per person decreases; for a constant distance, as the rate increases, the time traveled decreases; and as the length of a rectangle of constant area increases, the width of the rectangle decreases.

Cooperative Learning

Modeling Inverse Variation Have students work in groups of four. Give each group a set of 24 1-tiles, or ask students to use drawings of the algebra tiles. With input from the group, two members should take turns building as many rectangles as possible, each using all 24 tiles. Three sample configurations are shown below.

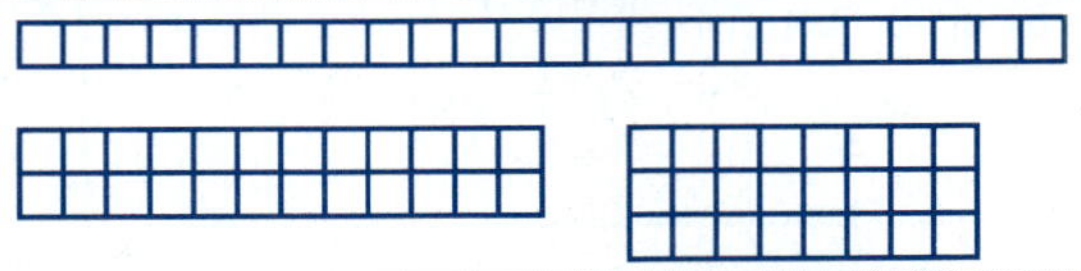

What is the area of each rectangle? Have another member of each group record the dimensions of the rectangle in a chart.

Length	Width	Area
		24
		24
		24
⋮		

Tell the fourth student in each group to plot the data on a pair of coordinate axes, with length on the horizontal axis and width on the vertical axis.

Technology

Students can use graphics calculators to find the value of t when $n = 50$ in Example 1.

Write $nt = 500$ as $t = \frac{500}{n}$. Graph the equation $Y_1 = \frac{500}{x}$, and find y when $x = 50$.

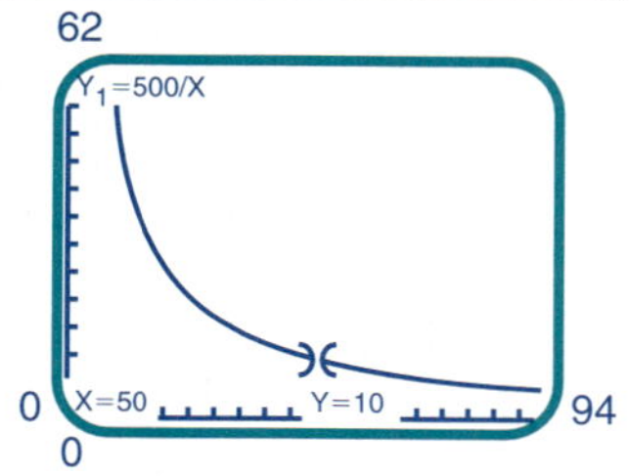

Ongoing Assessment, page 481

Try This

$xy = 48$

Critical Thinking, page 481

The constant of variation, k, represents the total number of hours needed to plant 500 trees.

Expressing Inverse Variation as a Proportion

The data for the exchange system in Egypt shows that for each ordered pair (x, y), where x is the rating and y is the value, $xy = k$. Thus $x_1y_1 = k$ and $x_2y_2 = k$. From this, it follows that $x_1y_1 = x_2y_2$ and, by a property of proportions, $\frac{x_1}{x_2} = \frac{y_2}{y_1}$ for all nonzero real values of x_2 and y_1.

Ongoing Assessment, page 482

Yes; the product is constant when the variables are multiplied together.

Example 2

A set of data points from a physics problem is examined to see if the data vary inversely. Then an equation is written to relate the variables in the problem so that a graph of the data can be drawn.

continued on top of next page

A physical model will help students understand inverse variation.

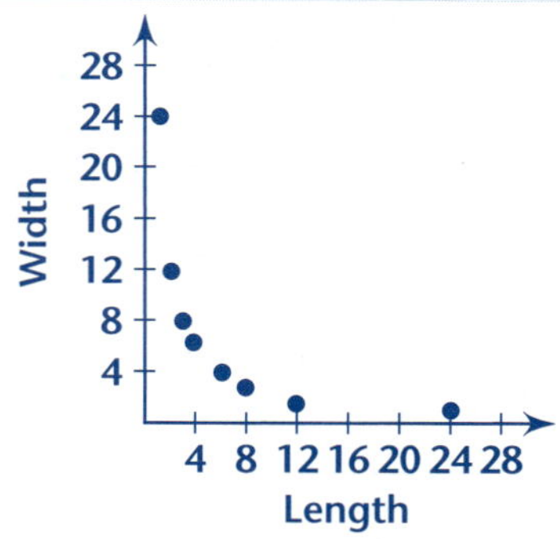

What kind of curve models this data? Does the point (5, 4.8) lie on this curve? Are 5 and 4.8 possible dimensions of a rectangle with an area of 24 square units? **[yes]** Explain.

Write an equation that expresses the width, w, of the rectangles modeled by the curve in terms of the length, l. $\left[w = \frac{24}{l}\right]$

Defining *Inverse Variation*

An equation such as $w = \frac{24}{l}$, where $l \neq 0$, is called an inverse variation equation. The chart

continued on bottom of next page

Approach A

Using the Book, continued

Since mass times the fulcrum distance is constant, any pair of data relating mass and distance can be used to find k.

Additional Example 2

The wavelength of X rays can be expressed in picometers (pm). The data shows the minimum voltage needed to produce X rays of a given length.

Wave-length (in pm)	20	40	50	80	100
Voltage	62,000	31,000	24,800	15,500	12,400

Analyze the data in the table.

A. Write a sentence that specifies the variation between the variables. **[Since voltage times wavelength equals 1,240,000 for each data pair, the minimum voltage needed varies inversely as the wavelength.]**

B. Write an equation that relates the variables. **[voltage · wavelength = k; the constant of variation is 1,240,000.]**

C. Construct a graph of the equation, and analyze the type of function that the graph represents. **[The graph should be of the equation $V \cdot w = 1{,}240{,}000$; this function is an inverse variation.]**

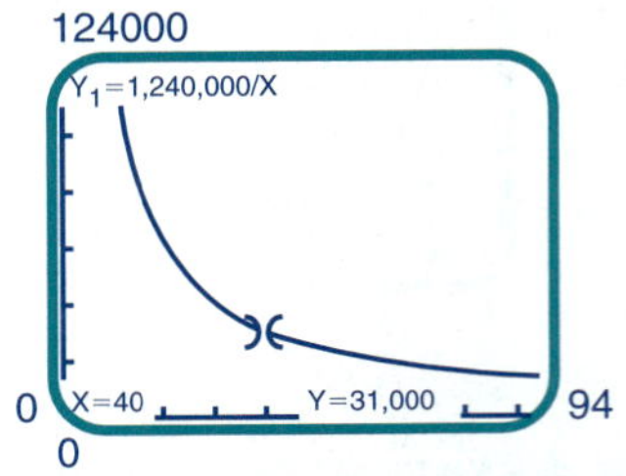

Graphs of Inverse Variation Applications

Students should note that the graph in Example 2 includes only those values for which $x > 0$ and $y > 0$. Although the graph of $m = \frac{300}{d}$ is a

TEACH

Approach B

Using Models, continued

of the rectangle dimensions shows that as the length increases, the width of the rectangle decreases when the area is held constant. We say that width varies inversely as length. Note that in the data set, the product of the length and the width, or the area, is constant: $lw = 24$. The area, 24, is called the constant of variation in this equation. In general, if $x \cdot y = k$ or $y = \frac{k}{x}$, where $x \neq 0$, $y \neq 0$, and k is a nonzero constant, then x and y vary inversely as each other. The constant, k, is the constant of variation. The variables x and y are also said to be *inversely proportional to each other.*

Application: Physics

The table below gives the distance to the fulcrum and the mass needed to balance a 60-gram mass that is on the other side of the fulcrum and 5 centimeters from it.

Distance (in cm)	5	10	15	20	25
Mass (in grams)	60	30	20	15	12

Compare the relationship between each distance, d, and each mass, m. Since the product of distance and mass equals 300 for each pair, mass varies inversely as distance, and the constant of variation is 300. The equation $dm = 300$ models the data and can be used to find other mass-and-distance pairs that will balance a 60-gram mass placed 5 centimeters from the fulcrum.

Inclusion Strategy

Journal Entry Ask each student to write the data pairs from the distance-mass chart in their notebook as ordered pairs. Tell them to look for a pattern in the ordered pairs that will tell them what mass placed 30 cm from the fulcrum will balance the 60-gram mass? **[a 10-gram mass]** What mass placed 12 cm from the fulcrum will balance the 60-gram mass? **[a 25-gram mass]** For each point (a, b) on the graph of an inverse variation $ab = k$, what other data point will also be on the graph? **[(b, a)]**

hyperbola lying in quadrants I and III, in practical applications, the graph of the solution usually lies in just one quadrant. In Example 2, because distance and mass are both positive, the values that satisfy the situation lie only in quadrant I.

Example 3

Students apply an inverse variation equation to an application involving investing. The equation is used to predict a value of y for a given value of x. The inverse variation function is graphed by using a table of values.

For students who are accustomed to entering an interest rate as a decimal, remind them that the interest in this formula must be entered as a whole number.

Inverse variation equations will often take the form of a formula, as is the case in this example.

Students should understand that because 0% interest makes no sense, the graph of this function will not show a value of t for $r = 0$.

Additional Example 3

Some businesses mark up the price of their goods based on the wholesale cost of the goods. The formula for finding the percent markup based on wholesale cost is percent markup = $\frac{\text{markup amount}}{\text{cost}}$. If the markup amount on an item is $200, the percent markup for the item can be modeled by the function $p = \frac{200}{c}$, where p is the percent markup and c is the cost.

A. What is the percent markup if the cost of the item is $325? **[about 61.5%]**

B. Graph the function $p = \frac{200}{c}$.

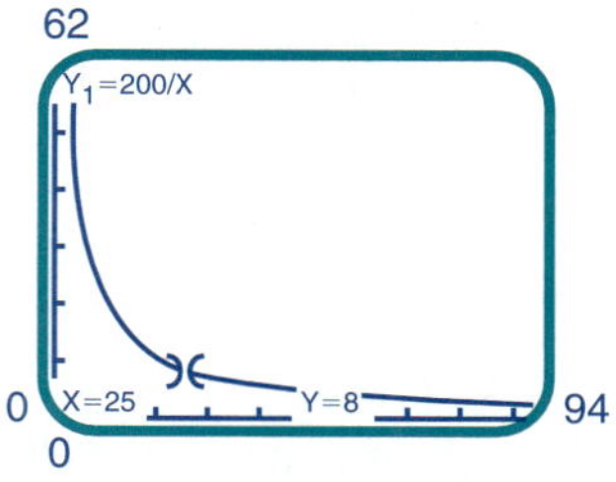

continued on top of next page

Modeling Inverse Variation With a Graph

Compare the graph of a direct variation equation with the graph of an inverse variation equation to demonstrate the relationship between two variables that are inversely proportional. Use the inverse variation equation $t = \frac{72}{r}$, where t is the time it takes to double an amount of money receiving compound interest and r is the interest rate expressed as a whole number, and the direct variation equation $s = 72d$, where s is total salary earned for working d days at $72 per day.

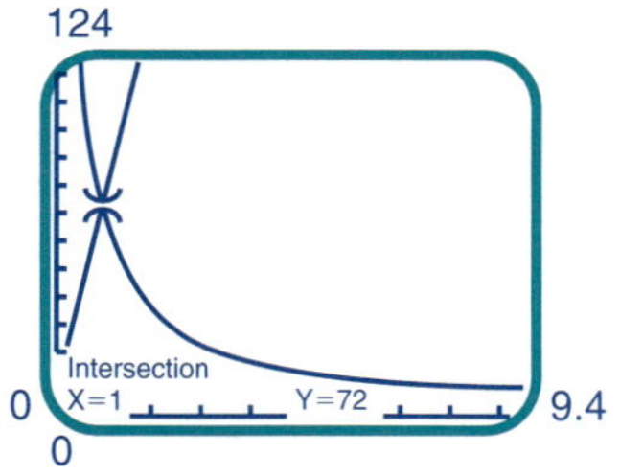

The equation $s = 72d$ is a linear function, and its graph is therefore a straight line. The equation $t = \frac{72}{r}$ represents a rational function whose graph is a transformation of the parent function $f(x) = \frac{1}{x}$ by a vertical stretch of 72.

Why is the ordered pair (0, 0) a possible data pair for the direct variation equation? **[It is possible to work zero days and receive no salary.]** Why does the graph of the inverse function not contain any point with an x-coordinate of 0? **[Division by zero is undefined. If there is no interest rate, then the money will not double over time.]**

The graph of the direct variation equation rises from left to right, thus modeling the relationship between the number of days worked and the total salary received. As the number of days worked increases, the salary increases proportionally. The graph of the inverse equation falls from left to right and contains the point (t, r) for every point (r, t) on the graph. This graph models the relationship

continued on bottom of next page

Approach A Using the Book, continued

Critical Thinking, page 483

The graph of $t(r) = \frac{72}{r}$ has been stretched vertically by 72. The domain and range of the function for the rule of 72 are all positive real numbers. The constant of variation stretches the graph vertically.

Closing the Lesson

Complete the following:

If y varies inversely as x, then $x \cdot y =$ [**k**] or $y = \left[\frac{k}{x}\right]$ for nonzero values of x, y, and k. The constant k is called [**the constant of variation**]. The parent function of an equation for inverse variation is $f(x) = \left[\frac{1}{x}\right]$. The [**constant of variation**] describes the vertical stretch of the parent graph.

TEACH

Approach B Using Models, continued

between the interest rate and the time it takes the money to double. As the interest rate increases, the amount of time decreases inversely.

Modeling Data With Graphs

Students should note that the direct variation graph of salary and days worked displays data only for $x \geq 0$ and $y \geq 0$, while the inverse variation graph of interest rate and time to double displays data only for $x > 0$ and $y > 0$. Although both graphs contain points in quadrant III, interest rate and time to double can be expressed only by positive numbers, while the possible values for salary or hours worked must be nonnegative.

Closing the Lesson

Complete the following:

If y varies inversely as x, then $x \cdot y =$ [**k**] or $y = \left[\frac{k}{x}\right]$ for nonzero values of x, y, and k. The constant k is called [**the constant of variation**]. The parent function of an equation for inverse variation is $f(x) = \left[\frac{1}{x}\right]$. The [**constant of variation**] describes the vertical stretch of the parent graph.

ASSESS

Selected Answers

Odd-numbered Exercises 7–29 and odd-numbered Extra Practice

Assignment Guide

Core level students should complete Exercises 1–30.

Core-plus level students should complete Exercises 1–12, 15–26, and 28–33.

Error Analysis

Given a problem such as Exercise 14, some students may set up the proportion $\frac{8}{6} = \frac{12}{x}$ to find the missing value of x. Explain to these students that in an inverse variation, x and y are inversely proportional and not directly proportional. The correct proportion for finding x is $\frac{8}{6} = \frac{x}{12}$.

Alternative Assessment

Portfolio Activity Choose a constant of variation, and use a table of values to show that the formula $rt = d$ is an example of an inverse variation equation. Graph your function and analyze the graph to show that it models an inverse variation function.

RETEACH

On a piece of poster board, draw a large circle to model a pizza. You can ask several students to decorate the pizza with toppings. Then divide the pizza into 12 equal slices, and cut out the slices. Draw the following chart on the chalkboard:

Number of people sharing the pizza	1	2	3	4	5	6	8	12
Number of slices per person								

Give the pizza to one student. How many slices does one person get if he or she is the only one eating the pizza? [12] Divide the pizza equally between two students. How many slices does each person get if the two students share the pizza? [6] Divide the pizza equally between three students. How many slices does each person get if the three students share the pizza? [4] Repeat the activity for four students.

Ask students to calculate the number of pieces each person gets if 5, 6, 8, or 12 people are sharing the pizza. Then fill in the rest of the chart. Plot the points on a pair of coordinate axes.

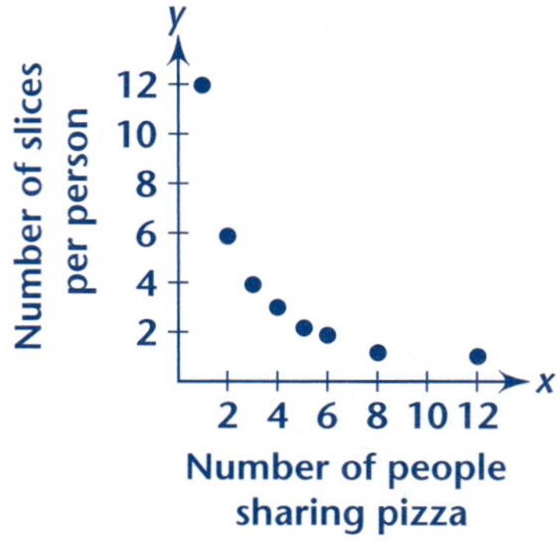

Ask students what the shape of the graph resembles if all of the data points are connected by a curve. Is the point (10, 1.2) on the curve? Explain this point in relation to the pizza problem. The parent function of this graph is $f(x) = \frac{1}{x}$. Since the pizza curve can be represented by the equation $s = \frac{12}{p}$, where s represents the number of slices per person and p represents the number of people sharing the pizza, the graph of $f(x) = \frac{1}{x}$ has been stretched vertically by a factor of 12. The pizza curve shows that as the number of people sharing the pizza increases, the number of slices per person decreases. The variables p and s vary inversely with each other. That is, $ps = k$, where k is a nonzero constant. For the function $s = \frac{12}{p}$, 12 is the constant of variation.

ENRICH

Interdisciplinary Connection

Physics The relationships between one variable and several others can exhibit both a direct variation and an inverse variation. For example, the formula $R = \frac{kl}{d^2}$ indicates that the electrical resistance, R, of a wire varies directly as the length, l, of the wire and inversely as the square of the diameter, d, of the wire. This kind of variation is called *combined* variation.

NAME ______________________ CLASS __________ DATE __________

Practice

10.2 Inverse Variation

Determine if each of the following is an example of inverse variation:

1. $p = \frac{q}{3}$ ______________ **2.** $p = 3q$ ______________ **3.** $p = \frac{3}{q}$ ______________

For Exercises 4–7, *y* varies inversely as *x*.

4. If y is 6 when x is 10, find x when y is 12. ______________

5. If x is 25 when y is 5, find y when x is 5. ______________

6. If x is -12 when y is -6, find x when y is 6. ______________

7. If y is -0.5 when x is 40, find x when y is 2. ______________

8. The time required to do a certain job varies inversely as the number of people who work on the job. It takes 4 hours for 6 students to complete an engine overhaul. If they all work at the same rate, how long would it take 2 students to complete an engine overhaul?

____

9. The intensity with which an object is illuminated varies inversely as the square of its distance from the light source. At 1.5 feet from a light bulb, the intensity of the light is 60 units. What is the intensity at 1 foot?

____

The U.S. customary unit of power is horsepower. A horse doing average work can pull 550 pounds a distance of 1 foot in 1 second. The horsepower, H, in feet per second needed to lift 100 pounds 10 feet can be modeled by the rational equation $H = \frac{10^3}{t}$, where t is time in seconds.

10. Complete the table.

Time (seconds)	5	10	15	20	25	30
Horsepower						

11. Find the products of time and horsepower. Describe the products.

____

12. What is the relationship between horsepower and time? ______________

NAME ______________________ CLASS __________ DATE __________

Lesson Activity

10.2 Modeling Inverse Variation

Georgette writes advertising copy for an ad agency. She receives $450 for each ad she writes. Examine the tables below, which show the number of hours she works and the equivalent hourly wage.

Number of hours	Hourly wage
4	$\frac{450}{4} = 112.50$
5	$\frac{450}{5} = 90.00$
6	$\frac{450}{6} = 75.00$
7	$\frac{450}{7} = 64.29$
8	$\frac{450}{8} = 56.25$
⋮	⋮

Number of hours	Hourly wage
16	$\frac{450}{16} = 28.13$
17	$\frac{450}{17} = 26.47$
18	$\frac{450}{18} = 25.00$
19	$\frac{450}{19} = 23.68$
20	$\frac{450}{20} = 22.50$

1. As the number of hours increase, what happens to the hourly wage?

____

2. What would you expect to happen to the hourly wage as the number of hours approaches zero? _______________________________

3. What happens to the hourly wage as the number of hours gets very large?

____

4. If h represents the number of hours, write a function that specifies the hourly wage. _______________________________

5. Graph your function on the grid provided.

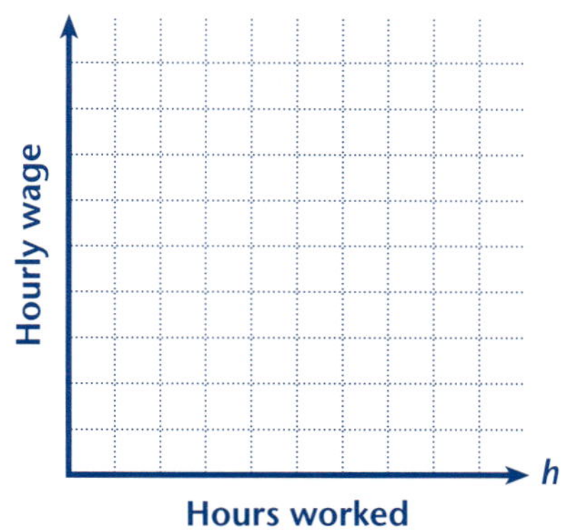

6. As h gets smaller and smaller, what happens to the hourly wage?

____

7. What happens when h is 0? _______________

8. What do your results tell you about the function?

____

NAME ______________________ CLASS __________ DATE __________

Basic Skills Practice
Cumulative Review

Read each question and circle the best answer.

1. Michael wants to find the mean (average) and range of his algebra test scores for the semester. The scores were 87, 92, 77, 90, 95, and 75. Find the mean and range.

A mean of 85, range of 20
B mean of 86, range of 20
C mean of 87, range of 17
D mean of 88, range of 17

2. Kara earned \$40.00 last week. She spent \$12.00 on a T-shirt. What percent of her earnings did she spend on the T-shirt?

F 30%
G 40%
H 33%
I 25%

3. Find the fraction in simplest terms that is equivalent to 40%.

A $\frac{40}{100}$
B $\frac{20}{50}$
C $\frac{10}{25}$
D $\frac{2}{5}$

4. Shawna takes photographs and then enlarges them. A photograph that is 5.6 inches wide and 8 inches long will be enlarged to a photograph that is 20 inches long. Find the proportion that can be used to find W, the width of the enlarged photograph.

F $\frac{W}{8} = \frac{5.6}{20}$
G $\frac{20}{5.6} = \frac{W}{8}$
H $\frac{20}{W} = \frac{5.6}{8}$
I $\frac{5.6}{8} = \frac{W}{20}$

5. Which expression is equivalent to this expression?

$$3 + (4x)^2$$

A $3 + 4x^2$
B $12x^2$
C $3 + 12x^2$
D $3 + 16x^2$

6. Solve $10x - 80 = -100$.

F $x = 2$
G $x = -2$
H $x = 18$
I $x = -18$

7. Tom wants to carpet a bedroom with dimensions of 9 feet by 11 feet and a hallway with dimensions of 4 feet by 12 feet. What is the total amount of carpet he will need?

A 36 ft^2
B 72 ft^2
C 147 ft^2
D 294 ft^2

8. Members of a high school band have a red jacket, a blue jacket, a white shirt, a red shirt, and a blue shirt.

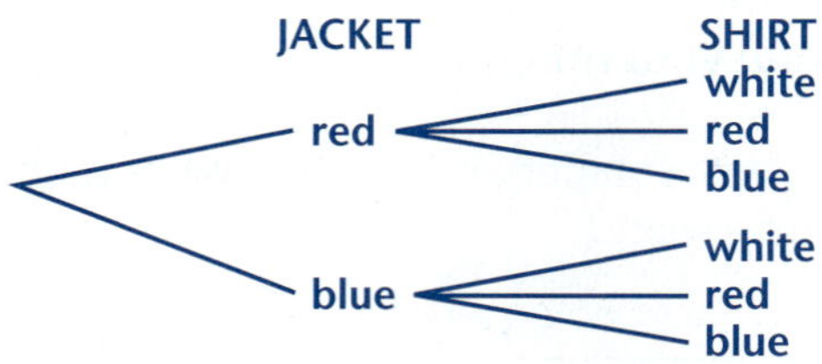

Use the diagram to determine the number of different combinations of uniforms that the band members can wear.

F 2
G 4
H 6
I 8

LESSON 10.3

Simplifying Rational Expressions

PREPARE

Pacing Guide

2 Days

Objectives and Overview

- Simplify rational expressions by factoring the numerator and denominator and eliminating common factors by division.

Students learn how to simplify rational expressions by dividing out common factors in the numerator and denominator.

Assessing Prior Knowledge

Simplify by using the Distributive Property to factor the expression, and then divide. Check by performing the arithmetic.

1. $\frac{30 + 18}{6}$ [8]

2. $\frac{8(3^2 - 4^2)}{(3 + 4)(6 - 8)}$ [4]

Approach A: Using the Book

Opening Discussion

Reduce the fraction $\frac{12}{15}$. Does $\frac{12}{15}$ reduce to $\frac{4}{5}$? For what values of x is the rational expression $\frac{12x}{15x}$ not defined? **[$x = 0$]** Do you think $\frac{12x}{15x}$ always reduces to $\frac{4}{5}$? Explain. **[If $x = 0$, $\frac{12x}{15x} \neq \frac{4}{5}$.]** Can you think of a way in which reducing rational numbers is different from reducing rational expressions? **[When reducing rational expressions, any value of a variable that makes the denominator zero must be excluded from the replacement set of the variable.]**

Example 1

The numerator and denominator of a rational expression are factored. Then the expression is rewritten to show a fraction that is equal to 1. The fraction is then simplified by using the Multiplication Property of 1.

continued on top of next page

TEACH

Approach B: Using Models

Opening Discussion

Ask your students to reduce the fraction $\frac{12}{15}$. Does $\frac{12}{15}$ equal $\frac{4}{5}$? Ask your students to identify the values of x for which the rational expression $\frac{12x}{15x}$ is not defined. **[$x = 0$]** Is $\frac{12x}{15x}$ always equal to $\frac{4}{5}$? Explain. **[If $x = 0$, $\frac{12x}{15x} \neq \frac{4}{5}$.]** In what way is reducing rational numbers different from reducing a rational expression? **[When reducing rational expressions, any value of a variable that makes the denominator zero must be excluded from the replacement set of the variable.]**

Reducing Fractions

Draw a rectangle with dimensions of 18 in. by 4 ft on the chalkboard.

continued on bottom of next page

Approach A: Using the Book, continued

Caution students that even if a simplified fraction and the original fraction have the same value for a given value of the variable, the simplified fraction might not be completely reduced. Students should check each simplified fraction to be sure that there are no common factors other than 1.

Students may need to be reminded of the following:

$$\frac{5}{5} \cdot \frac{a+2}{a} = 1 \cdot \frac{a+2}{a} = \frac{a+2}{a}$$

The Multiplication Property of 1 allows the fraction to be simplified.

Ongoing Assessment, page 487

If they are different, an error has been made. It is important that the check be the same for the original fraction and its reduced form because you are trying to check that the two expressions are equivalent.

Additional Example 1

Simplify $\frac{12x - 16xy}{4x^2}$. $\left[\frac{3x - 4y}{x}, x \neq 0\right]$

Technology

A graphics calculator can also be used to check that a rational expression has been simplified. For example, the fractions $\frac{5a + 10}{5a}$ and $\frac{a+2}{a}$ can be entered as Y_1 and Y_2 respectively in a Y= editor; then the tables of values can be compared.

X	Y_1	Y_2
−2	0	0
−1	−1	−1
0	ERROR	ERROR
1	3	3
2	2	2
3	1.6667	1.6667
4	1.5	1.5

X = −2

TEACH

Approach B: Using Models, continued

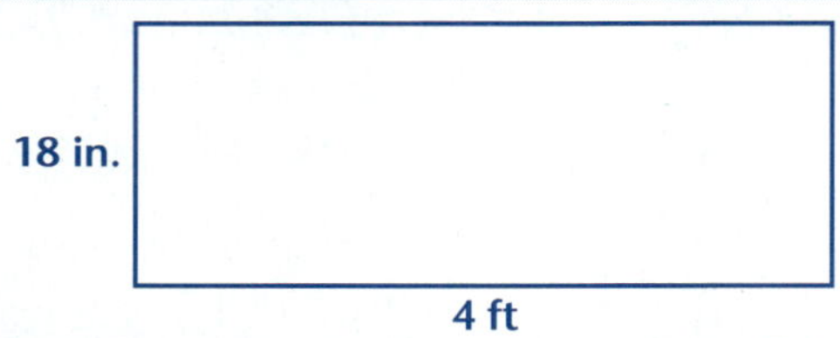

The ratio of length to width is $\frac{4 \text{ ft}}{18 \text{ in.}}$. To write the ratio in simplified form, you must convert the dimensions to the same unit of measure: $\frac{4 \text{ ft}}{18 \text{ in.}} = \frac{4 \cdot 12 \text{ in.}}{18 \text{ in.}} = \frac{(6 \cdot 8) \text{ in.}}{(6 \cdot 3) \text{ in.}} = \frac{6}{6} \cdot \frac{8}{3} \cdot \frac{\text{in.}}{\text{in.}} = 1 \cdot \frac{8}{3} \cdot \frac{\text{in.}}{\text{in.}} = \frac{8}{3}$
The common factor of 6 in both the numerator and denominator gives a fraction that is equal to 1, and the common unit of measure, inches, can be omitted.

Reducing Rational Expressions

As students saw in Lesson 10.2, the length and width of a rectangle vary inversely with each other when the area of the rectangle is held constant. If l represents the length of a rectangle and w represents the width, then $w = \frac{\text{area}}{l}$. Draw a rectangle on the chalkboard or overhead with an area of $5a + 10$ and a length of $5a$.

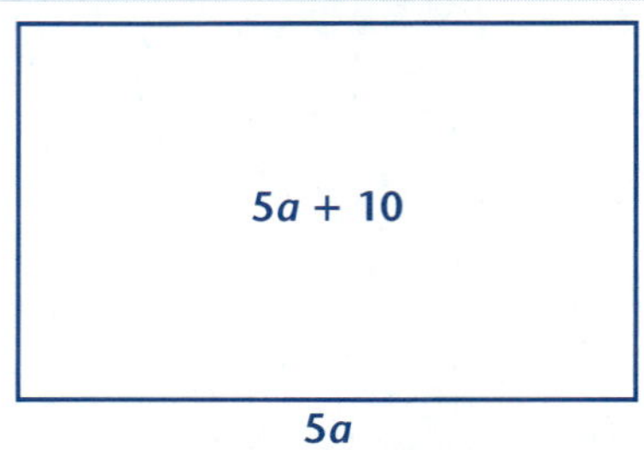

For this rectangle, $w = \frac{5a + 10}{5a}$. Just as with a fraction, common factors in a rational expression can be divided out to simplify the expression. Since $\frac{p}{p} = 1$ for any real number $p \neq 0$ and $1 \cdot p = p$ for any real number p, common factors written in the form $\frac{p}{p}$, where $p \neq 0$, can be used to simplify a rational expression.

$$w = \frac{5a + 10}{5a} = \frac{5(a+2)}{5 \cdot a} = \frac{5}{5} \cdot \frac{a+2}{a} = \frac{a+2}{a}$$

Journal Entry

Ask students to copy the table on page 487 into their journal. Tell them to add a column to list how any common factor in an expression can be expressed as 1.

Expression	Common factor	Common factors written as 1	Reduced form
$\frac{x(a+3)}{2x}$	x	$\frac{x}{x} \cdot \frac{(a+3)}{2}$	$\frac{(a+3)}{2}, x \neq 0$
$\frac{2x(x+1)}{4x^2}$	$2x$	$\frac{2x}{2x} \cdot \frac{(x+1)}{2x}$	$\frac{(x+1)}{2x}, x \neq 0$

Critical Thinking, page 487

Since $x^2 + x = x(x + 1)$, the fraction is defined for all $x \neq 0$ and $x \neq -1$. Therefore, if $x \neq 0$ and $x \neq -1$, the denominator can be factored, and the common factor of x in the numerator and the denominator can be canceled.

$$\frac{x}{x^2 + x} = \frac{x}{x(x+1)} = \frac{x}{x} \cdot \frac{1}{x+1} = \frac{1}{x+1}$$

Example 2

A rational expression is simplified by factoring a binomial in the numerator and a trinomial in the denominator.

Make sure that students recognize the importance of looking at every form of the fraction in order to identify values of the variable that must be excluded from the replacement set.

Additional Example 2

Simplify $\frac{a^2 - b^2}{a^2 - 2ab + b^2}$. $\left[\frac{a+b}{a-b}\right]$

Ongoing Assessment, top of page 488

Try This

$x + 4$, where $x \neq 1$

Example 3

A rational expression is reduced by factoring -1 from the denominator of the fraction.

continued on top of next page

Inclusion Strategy

Writing a Set of Rules Ask students to work in pairs to find simplified expressions for the width of each rectangle if the area, $5a + 10$, is a constant when the length varies as indicated in the chart. Make sure students indicate any excluded values of the variable.

Area	Length	Width
$5a + 10$	$a^2 - 4$	$\left[\frac{5}{a-2}, a \neq 2, -2\right]$
$5a + 10$	$a^2 + 9a + 14$	$\left[\frac{5}{a+7}, a \neq -7, -2\right]$
$5a + 10$	$-a - 2$	$[-5, a \neq -2]$

Ask the students to use their results to write a set of steps for simplifying rational expressions. Then ask each pair to join with another pair. The two pairs in each group should read their steps to each other and decide which set of steps best explains how to simplify a rational expression. One member from each group should write the steps chosen by his or her group on the chalkboard. The class should discuss all of the steps and decide on the best set of steps for simplifying a rational expression. Each student should copy these steps into his or her notebook.

Teaching Tip

Not all rational expressions can be reduced. You can compare a rational expression with no common factors in the numerator and denominator to a *rate*.

A rate is a special kind of ratio that compares two unlike units of measure. For example, if a car drove 110 miles in 2 hours, this rate could be expressed as $\frac{110 \text{ miles}}{2 \text{ hour}}$, or $\frac{55 \text{ miles}}{1 \text{ hour}}$, if the speed of the car was constant. A ratio such as $\frac{55 \text{ miles}}{1 \text{ hour}}$ compares a quantity of one kind to 1 unit of another kind. Notice that because the numerator and denominator do not have a common unit of measure, the units of measure cannot be omitted from the ratio. This rate is more commonly expressed as 55 miles per hour.

Since the rational expression $\frac{x+4}{x}$ has no common factors, it cannot be simplified. The reduced form of the expression is $\frac{x+4}{x}$, where

continued on bottom of next page

Approach A

Using the Book, continued

Since $\frac{-x}{y} = \frac{x}{-y} = -\frac{x}{y}$, $\frac{x+5}{(-1)(x+4)}$ can also be written as $\frac{-x-5}{x+4}$.

Additional Example 3

Simplify $\frac{-a^2 + 2ab - b^2}{a^2 - b^2}$. $\left[\frac{b-a}{a+b}\right]$

> ***Ongoing Assessment, page 488***
>
> The opposite of $(x - 4)$ is $-(x - 4) = -x + 4 = 4 - x$.
>
> ***Ongoing Assessment, bottom of page 488***
>
> ***Try This***
>
> $-\frac{5x}{3}$, where $x \neq 2$ and $x \neq -2$
>
> ***Critical Thinking, page 488***
>
> You must add the restriction $x \neq 1$ because if $x = 1$, then $\frac{x-1}{1-x}$ becomes $\frac{0}{0}$, which is undefined because division by zero is undefined.

Example 4

Students apply the rules for simplifying a rational expression to an application in solid geometry.

Although π is an irrational number, it is still a common factor of both the numerator and denominator and can therefore be divided out of the fraction.

Additional Example 4

The formula for the surface area of a sphere is $A = 4\pi r^2$. The formula for the volume of a sphere is $V = \frac{4}{3}\pi r^3$. Find the ratio of the surface area of a sphere with radius r to the volume of the same sphere. $\left[\frac{3}{r}\right]$

> ***Ongoing Assessment, page 489***
>
> ***Try This***
>
> $\frac{6}{s}$

TEACH

Approach B

Using Models, continued

$x \neq 0$. Similarly, $\frac{2x^2 + 3}{x}$ has no common factors, so the reduced form of the expression is $\frac{2x^2 + 3}{x}$, where $x \neq 0$.

Checking a Reduced Fraction

The equation $\frac{5a + 10}{5a} = \frac{a + 2}{a}$ can be checked by substituting a value for a on both sides of the equation. If the expressions are equal, then the value of the original form is the same as that of the reduced form, and therefore, the reduced form is an equivalent form of the original expression. A graphics calculator can also be used to check that a rational expression has been simplified correctly. For example, the fractions $\frac{5a + 10}{5a}$ and $\frac{a + 2}{a}$ can be entered as Y_1 and Y_2 respectively in a Y= editor; then the tables of values can be compared. (Compare the chart in Approach A, Using the Book, page 548.)

Modeling Rational Expressions: Solid Geometry

Ask students to bring in pictures of right circular cylinders such as cans of vegetables, oil storage tanks, or silos. Use the pictures to show students how they can use an algebraic skill to solve a real-world problem. Remind students that the formula for the volume of a right circular cylinder is $V = Bh$ or $V = \pi r^2 h$. The formula for the surface area of a right circular cylinder is $SA = 2\pi r^2 + 2\pi rh$. To compare the ratio of the

Application: Solid Geometry

Students may need reminding that for a right circular cylinder, the formula for volume, $V = Bh$, is equivalent to $V = \pi r^2 h$ because the base of a right circular cylinder is a circle. This example allows students to use an algebraic skill to solve a real-world problem. You might ask students what happens to the ratio of volume to surface area if the radius of the cylinder is doubled. **[The ratio becomes $\frac{rh}{2r + h}$.]**

Closing the Lesson

Ask students to write the steps for reducing rational expressions in their notebook from memory. Then tell them to check their steps with those on page 487 of the textbook.

volume of a cylinder to its surface area requires factoring out a common factor of $2\pi r$.

$$\frac{\text{Volume}}{\text{Surface area}} = \frac{\pi r^2 h}{2\pi r^2 h + 2\pi rh} = \frac{\pi rrh}{2\pi r(r + h)} = \frac{rh}{2(r + h)}$$

You might ask students what happens to the ratio of volume to surface area if the radius of the cylinder is doubled. **[The ratio becomes $\frac{rh}{2r + h}$.]**

Closing the Lesson

Ask students to write the steps for reducing rational expressions in their notebooks from memory. Then tell them to check their steps with those that they copied earlier from the chalkboard.

ASSESS

Selected Answers

Odd-numbered Exercises 7–47 and odd-numbered Extra Practice

Assignment Guide

Core level students should complete Exercises 1–47.

Core-plus level students should complete Exercises 1–12, 15–30, and 35–48.

Error Analysis

Some students incorrectly divide terms when reducing a rational expression. Explain to students that factors are multiplied, but terms are added or subtracted. Only expressions that are multiplied can be common factors and therefore can be divided out. Show students these examples: $\frac{15x^4}{3x^2} = \frac{3x^2 \cdot 5x^2}{3x^2} = 5x^2$. Since 3 and x^2 are multiplied, they are factors and can be divided out. In the expression $\frac{x^4 - 15}{x^2 - 3}$, however, x^2 and 3 are *terms,* not factors, and therefore cannot be divided out.

Alternative Assessment

Quick Quiz Answer the following questions:

1. When is a rational expression in simplest form? **[A rational expression is in simplest form when the numerator and denominator have no common factor other than 1.]**
2. What assumption is always made about the value of a variable in the denominator of a fraction? **[The value of the variable does not make the value of the denominator equal to zero.]**
3. What happens to a polynomial when -1 is factored out of the polynomial? **[Each term of the polynomial changes sign.]**
4. Why can the expression $\frac{5x + 7x^2}{35}$ not be reduced? **[The terms of the numerator do not have a factor that is also a factor of the denominator.]**
5. Why are there no real number restrictions on x in the expression $\frac{x^2 - 9}{x^2 + 1}$? **[No real number values of x make the value of the denominator equal to zero.]**

Portfolio Activity Find a denominator that will make the statement $\frac{y^2 + 2y}{?} = \frac{y - 1}{y + 3}$ true.

Explain how you got your answer and note any restrictions on the value of the variable. **[$y^2 - 9$, where $y \neq -3$ and $y \neq 3$]**

RETEACH

Sometimes simplifying a rational expression requires factoring binomials, trinomials, or a common factor of -1.

$\frac{4 - m^2}{m^2 - 5m + 6} = \frac{-1(m^2 - 4)}{m^2 - 5m + 6}$	Distributive Property
$= \frac{-1(m - 2)(m + 2)}{(m - 2)(m - 3)}, m \neq 2, 3$	Distributive Property
$= \frac{m - 2}{m - 2} \cdot \frac{-1(m + 2)}{m - 3}, m \neq 2, 3$	Rule for multiplying fractions
$= 1 \cdot \frac{-1(m + 2)}{m - 3}, m \neq 2, 3$	Multiplication Property of Inverses
$= \frac{-1(m + 2)}{m - 3}, m \neq 2, 3$	Multiplication Property of 1
$= \frac{-m - 2}{m - 3}, m \neq 2, 3$	Distributive Property

ENRICH

Interdisciplinary Connection

Civil Engineering The strength of a suspension bridge is dependent on the length of the bridge and the sag of the cables. One formula for determining the approximate strength of a suspension bridge involves the expression $\frac{3a^4 + 8ad^3}{3a^3}$, where a is the span of the bridge and d is the sag of the cables. Since the numerator and denominator of this fraction have a common factor of a, the fraction can be written in the simplified form $\frac{3a^3 + 8d^3}{3a^2}$.

NAME ____________________ CLASS __________ DATE __________

Practice

10.3 Simplifying Rational Expressions

For what values of the variable is each rational expression undefined?

1. $\frac{n-2}{2-n}$ ____________

2. $\frac{(w-2)(w-3)}{(w+5)(w-4)}$ ____________

3. $\frac{10m}{m}$ ____________

4. $\frac{6}{x(x^2+2x+1)}$ ____________

Write the common factors.

5. $\frac{5 \cdot 3}{2 \cdot 5}$ ____________

6. $\frac{a^4b}{ab}$ ____________

7. $\frac{(r+2)(r+3)}{r(r+3)}$ ____________

8. $\frac{(x+3)^2}{x^2-9}$ ____________

9. $\frac{y^2-y-6}{y^2-9}$ ____________

10. $\frac{3-k}{k^2-2k-3}$ ____________

Simplify.

11. $\frac{x^3}{3x}$ ____________

12. $\frac{2(x+y)}{8(x+y)^2}$ ____________

13. $\frac{a+4}{2a+8}$ ____________

14. $\frac{8w-12}{4}$ ____________

15. $\frac{p+7}{p^2-49}$ ____________

16. $\frac{64-x^2}{8-x}$ ____________

17. $\frac{h^2+h-12}{h^2+5h+4}$ ____________

18. $\frac{f+g}{3(f+g)^3}$ ____________

19. $\frac{3x^2-12x-15}{3x^2-18x+15}$ ____________

20. $\frac{5z^3-125z}{10z^2-250}$ ____________

21. $\frac{2x^2+16x+32}{x+4}$ ____________

22. $\frac{y^2-25}{y^2-10y+25}$ ____________

NAME ______________________ CLASS __________ DATE __________

Lesson Activity

10.3 Numerically Equivalent Perimeters and Areas

Under what conditions will a rectangle's area and perimeter be numerically equivalent? Start by considering a rectangle whose dimensions are integers, such as the 6 × 3 rectangle shown in the diagram.

1. Each corner square contributes two units to the total perimeter, but only one (square) unit to the total area. Thus, each corner square contributes one more perimeter unit than area unit. Place a 1 in each corner square.

2. Each square along the length and width contributes one unit to the perimeter and one unit to the area, so the number of perimeter units contributed and the number of area units are the same. Place a 0 in these squares.

3. Each square in the rectangle's interior makes no contribution to the perimeter but adds one unit to the area. Place a −1 in each interior square.

4. Find the sum of the numbers placed in each square. ______________________

5. Why does the rectangle have a perimeter and area that are numerically equal? ______________________

6. State a conclusion that relates the number of interior squares and the number of corner squares to the numerical equality of the area and perimeter. ______________________

7. Use your results to find two other rectangles with integer dimensions whose area and perimeter are numerically equal.

____

8. Write the dimensions of each rectangle as ordered pairs. Then graph the points on the grid provided. Compare your graph with the parent function $f(x) = \frac{1}{x}$. Describe the transformations applied to the parent function.

____

____

9. Write an equation to represent the dimensions of all rectangles with numerically equal areas and perimeters. ______________________

NAME ______________________ CLASS __________ DATE __________

Basic Skills Practice
Cumulative Review

Read each question and circle the best answer.

1. Find the missing numbers in the pattern.
45, 40, 37, 32, ______, ______, 21, 16

A 17, 14
B 28, 25
C 29, 26
D 29, 24

2. Susie made a gift list for her family. The list includes the following amounts (including tax): $45.20, $33.15, $40.08, and $21.98. If she has budgeted $150.00 for gifts, how much money does she have left?

F $8.59
G $9.59
H $10.59
I $11.59

3. The value of Debra's pottery when she bought it 10 years ago was $55. This year, its value is $75. What is the approximate percent increase in its value?

A 26%
B 36%
C 42%
D 52%

4. Given $l \parallel m$ and $\angle A = 115°$, what is the measure of $\angle B$?

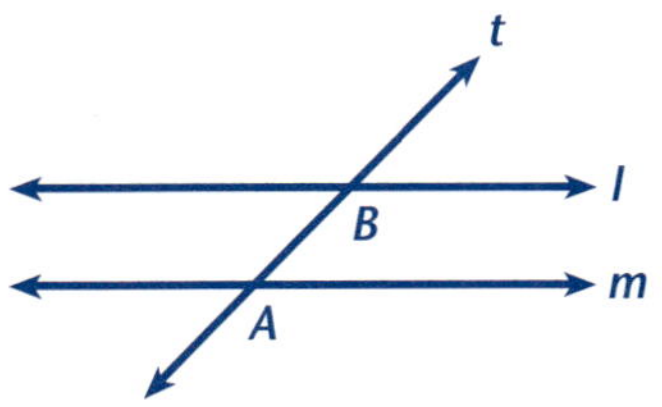

F 35°
G 65°
H 115°
I 125°

5. What is the sum of 0.6509 + 2.548 rounded to the nearest tenth?

A 3.199
B 3.19
C 3.2
D 3.0

6. The perimeter of a square swimming pool is 702.25 meters. What is the approximate length of its sides?

F 17.56 m
G 175.6 m
H 26.5 m
I 265 m

7. Where is point *C* on the number line?

A 1
B $1\frac{1}{2}$
C $2\frac{1}{4}$
D 2

8. The shaded region represents a courtyard surrounded by a rectangular house. What is the area of the shaded region?

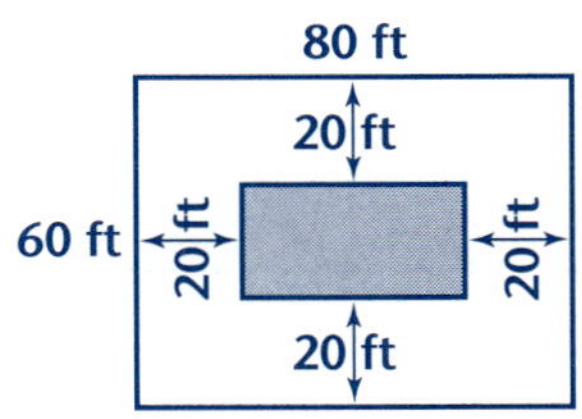

F 1200 ft^2
G 800 ft^2
H 400 ft^2
I 200 ft^2

LESSON 10.4 Operations With Rational Expressions

PREPARE

Pacing Guide
2 Days

Objectives and Overview
- Multiply rational expressions.
- Add and subtract rational expressions with like and unlike denominators.

A real-world problem is used to introduce adding, subtracting, and multiplying rational expressions.

Vocabulary
rational numbers
rational expressions

Assessing Prior Knowledge

Compute:

1. $\frac{1}{2} + \frac{2}{3}$ $\left[1\frac{1}{6}\right]$
2. $1\frac{3}{4} - \frac{9}{10}$ $\left[\frac{17}{20}\right]$
3. $\frac{2}{3} \cdot \frac{3}{4} \cdot \frac{4}{5}$ $\left[\frac{2}{5}\right]$
4. $2\frac{2}{3} \cdot 2\frac{2}{5} \cdot 1\frac{1}{4}$ [8]
5. $\frac{1}{5} \div 5$ $\left[\frac{1}{25}\right]$
6. $\frac{1}{3} \div \frac{1}{3}$ [1]

Approach A: Using the Book

Opening Discussion
How does the rate of the current in a river affect a boat's speed in the river? Suppose that the rate of a kayak going upstream is 6 mph and the rate of a kayak going downstream is 8 mph per hour. How many hours would it take the kayak to travel 1 mile upstream? How many hours would it take the kayak to travel the same distance downstream?

Brainstorming
Ask students to suggest other activities in which a speed could be affected by another factor. Examples include the speed of a bicyclist riding with a headwind or a tailwind, the gas mileage of a car pulling a trailer or not pulling the trailer, and the time it takes a worker to complete a job alone or with someone.

TEACH

Approach B: Using Models

Opening Discussion
Ask your students if they think the rate of the current in a river affects a boat's speed in the river. Have them explain their reasoning. If the rate of a kayak going downstream is 8 mph, how long would it take the kayak to go 1 mile downstream? If the rate of a kayak going upstream is 6 mph, how long would it take the kayak to go 1 mile upstream?

Brainstorming
Ask students to suggest other activities in which one factor could be affected by another factor such as the rate of a current. Examples include the speed of a bicyclist riding with a headwind or a tailwind, the gas mileage of a car pulling a trailer or not pulling the trailer, and the time it takes a worker to complete a job alone or with someone.

Adding Rational Expressions

Although rational expressions can be added or subtracted with any common denominator, using the least common denominator will result in less simplifying of the sum or difference. Since 30 and 50 share a factor of 10, that factor does not need to be included in the LCD twice: $3 \cdot 5 \cdot 10 = 150$. Similarly, $3x$ and $5x$ share a factor of x, so that factor needs to be used only once to determine the LCD of $3x$ and $5x$: $3 \cdot 5 \cdot x = 15x$.

Ongoing Assessment, page 493

Restrictions on the denominator of the original expression are values that cause the value of the denominator to be zero, thus causing the rational expression to be undefined. Therefore, a solution cannot include any of these values.

Example 1

This example illustrates that the same rule used to multiply rational numbers can be used to multiply rational expressions.

You can formalize the rule for multiplying rational expressions so that students see how it is similar to multiplying rational numbers: $\frac{p}{q} \times \frac{r}{s} = \frac{pr}{qs}$, where p, q, r, and s are real numbers, $q \neq 0$, and $s \neq 0$.

Students should see that the methods they learned in Lesson 10.3 are now applied to multiplying rational expressions. Just as common factors are divided out to reduce the product of two fractions, common polynomial factors are divided out to reduce the product of two rational expressions.

Ongoing Assessment, page 494

Try This

$\frac{5}{2}$, where $x \neq 0$

Ongoing Assessment, page 494

Decide what factor you have to multiply the denominator of the fraction by to get the least common denominator. Then multiply the fraction by that factor divided by itself.

continued on top of next page

Modeling the Sum of Rational Expressions: Algebra Tiles

Use algebra tiles to model the sum of two rational expressions with like denominators. Cut out 1-tiles and x-tiles from sheets of transparency paper for use on an overhead, or draw sketches of algebra tiles on the chalkboard. You will need only positive tiles.

Make or draw tile models to represent $\frac{1}{3x} + \frac{1}{3x} = \frac{2}{3x}$, using 1-tiles in the numerator and x-tiles in the denominator. Draw fraction bars to separate the numerator tiles from the denominator tiles.

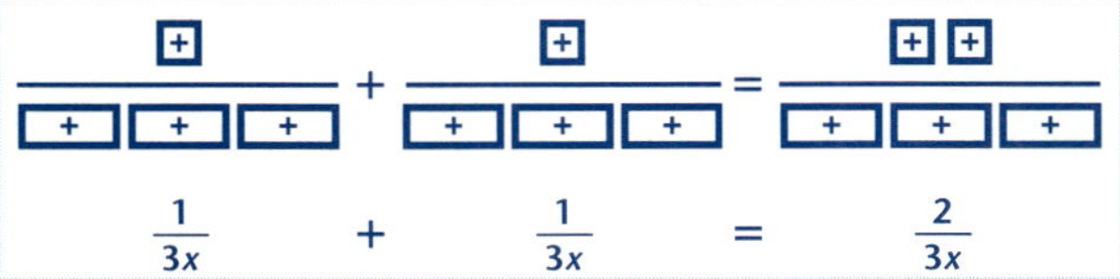

Because $3x$ represents a real number, the rule for determining the sum of $\frac{1}{3x} + \frac{1}{3x}$ is the same as the rule for determining the sum of $\frac{1}{3} + \frac{1}{3}$ or $\frac{1}{35} + \frac{1}{35}$, for example. The numerator doubles, and the denominator stays the same. The only difference is that for $\frac{1}{3x} + \frac{1}{3x} = \frac{2}{3x}$, $x \neq 0$.

Adding Rational Expressions With Unlike Denominators

Expand on the previous tile model to show how rational expressions with unlike denominators can be added.

Suppose that the time it takes a kayak to travel 1 mile upstream can be expressed by the fraction $\frac{1}{3x}$ and the time it takes the same kayak to travel 1 mile downstream can be expressed by the fraction $\frac{1}{5x}$. How could you express the total time it takes the kayak to travel 1 mile upstream and then 1 mile downstream? Model the expression with tiles.

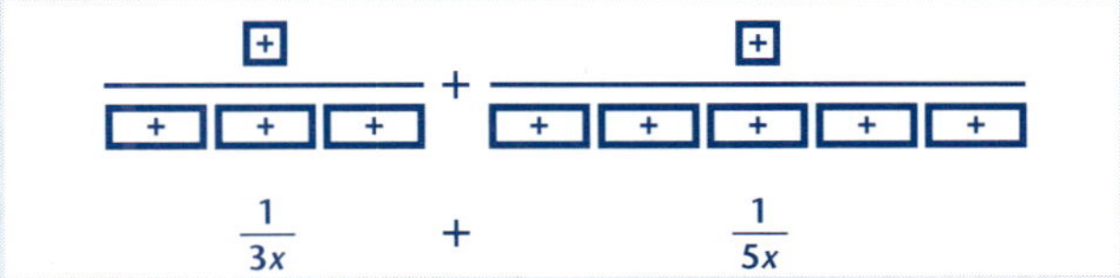

continued on bottom of next page

Approach A

Using the Book, continued

Additional Example 1

Multiply: $\frac{x^2}{x+1} \cdot \frac{x+1}{x(x-1)}$. $\left[\frac{x}{x(x-1)}\right]$

Cooperative Learning

Some students may note that dividing out common factors before multiplying makes both multiplying and reducing rational expressions easier, as is the case with rational numbers. For example, these students may note that the factor $x + 3$ is common to both a numerator and a denominator of the indicated product in Example 1. This factor may be canceled out before the numerators and the denominators are multiplied. Ask these students to share their observations with their classmates and to use an example with fractions to illustrate their point.

Teaching Tip

Call students' attention to the fact that because fractions are reduced by canceling out common factors, the product of two polynomials in either the numerator or the denominator should be indicated rather than actually multiplied. An indicated product identifies its factors, making it easier to reduce. Thus, the products $(x - 2)(x + 3)$ and $(x + 3)(x + 5)$ in Example 1 should not be multiplied.

Finding an LCD

The method for finding an LCD of a rational expression can be modeled on the method for finding an LCD of two or more fractions. If the denominators of the fractions have no common factors other than 1, then the LCD can be found by multiplying the denominators. For example, the denominators of $\frac{1}{2}$ and $\frac{1}{3}$ have no common factor other than 1, so the LCD for these two fractions is $2 \cdot 3$, or 6. Similarly, the denominators of $\frac{1}{x+2}$ and $\frac{1}{x-1}$ have no common factor other than 1, so the LCD for these fractions is $(x + 2)(x - 1)$.

TEACH

Approach B

Using Models, continued

By making 5 sets of $\frac{1}{3x}$-tiles and 3 sets of $\frac{1}{5x}$-tiles, you can model the sum $\frac{1}{3x} + \frac{1}{5x}$.

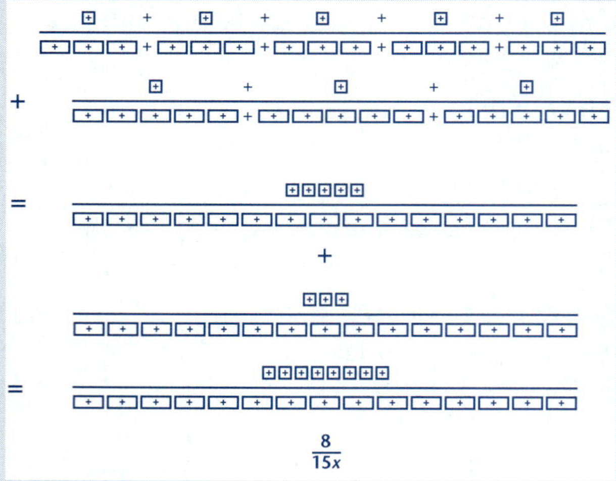

$$\frac{1}{3x} + \frac{1}{5x} = \frac{5}{15x} + \frac{3}{15x} = \frac{5+3}{15x} = \frac{8}{15x}$$

Multiplying Rational Expressions

Use an example of adding rational expressions to find a rule for multiplying rational expressions. What does the fraction $\frac{5}{15x}$ reduce to? Since the numerator and denominator both have a factor of 5, $\frac{5}{15x} = \frac{5 \cdot 1}{5 \cdot 3x} = \frac{5}{5} \cdot \frac{1}{3x} = \frac{1}{3x}$. Reducing this fraction demonstrates that $\frac{1}{3x}$ was multiplied by $\frac{5}{5}$ to get the equivalent fraction $\frac{5}{15x}$. Thus, to find the product of $\frac{5}{5} \cdot \frac{5}{3x}$, multiply the numerators of the fractions and multiply the denominators of the fractions.

Inclusion Strategy

Expanding a Rule Tell students to multiply $\frac{x-2}{x+3} \cdot \frac{x+3}{x-5}$ in their notebook by multiplying the numerators and multiplying the denominators. Ask students to think about how they would multiply $\frac{2}{3} \cdot \frac{3}{5}$. What other step besides multiplying numerators and denominators is required? Tell students to use the same

Example 2

Two fractions for which one denominator is a factor of the other denominator are added. To find the LCD, you must identify the greatest power of each unique prime factor of the denominators. Since the prime factors of the denominators are $a + 2$ and $a - 2$ and each of these factors has an exponent of 1, the LCD is $(a + 2)(a - 2)$.

The sum of the fractions may be left in the form $\frac{3a - 4}{(a + 2)(a - 2)}$, although $\frac{3a - 4}{a^2 - 4}$ is a more compact form. Either form of the sum requires a note about the restrictions on a.

Additional Example 2

Simplify $\frac{3x^2 - 9}{x^2 + 6x + 9} - \frac{x - 3}{x + 3}$. Write any restrictions on the variable.

$\left[\frac{2x^2}{x^2 + 6x + 9}, \text{ where } x \neq -3\right]$

Example 3

Two fractions with unlike denominators are subtracted. The denominators are factored in order to more easily identify the LCD.

When factors of a numerator and a denominator have the same terms but different signs, -1 can be factored out of one of the factors. The resulting common factor can be divided out to simplify the fraction.

Additional Example 3

Simplify $\frac{a}{a^2 - ab} + \frac{b}{b^2 - ab}$. $\left[\frac{2}{a - b}\right]$

Teaching Tip

You may notice that some students have trouble identifying the appropriate forms of 1 to multiply each fraction by when renaming the fractions with common denominators. Tell these students to compare the denominator of one fraction with the least common denominator. Identify the missing factor. Write a fraction equal to 1 whose numerator and denominator are both the missing

continued on top of next page

procedure with the product of the rational expressions. Ask for volunteers to explain the steps they used to find the product of the expressions.

Renaming Rational Expressions With Unlike Denominators

Once the LCD of rational expressions has been identified, the fractions that are to be added or subtracted must be renamed. Use the following diagram to help students discover a method for renaming the fractions. The method involves comparing the denominators of the fractions with the LCD and identifying the LCD factors that are missing from each denominator. Those factors are used to write fractions equivalent to 1. Each fraction is then multiplied by the appropriate form of 1 so that the fractions are renamed with the LCD.

Fractions	$\frac{1}{3x}$	$\frac{1}{5x}$
Factored form of denominators	$3 \cdot x$	$5 \cdot x$
LCD of fraction pair	$3 \cdot 5 \cdot x = 15x$	
Missing factors	5	3
Forms of 1	$\frac{5}{5}$	$\frac{3}{3}$
Renamed fractions	$\frac{5}{5} \cdot \frac{1}{3x} = \frac{5}{15x}$	$\frac{3}{3} \cdot \frac{1}{5x} = \frac{3}{15x}$
Sum of fractions	$\frac{5}{15x} + \frac{3}{15x} = \frac{8}{15x}, x \neq 0$	

continued on bottom of next page

Approach A

Using the Book, continued

factor. Multiply the fraction by this form of 1. Repeat this procedure for each of the fractions being added or subtracted.

Critical Thinking, page 495

Factoring out -1 is sometimes helpful because it allows opposites to be expressed in a form in which a common factor can be identified; then the expression can be reduced by that factor.

Closing the Lesson

Compare the sum $\frac{3}{4} + \frac{3}{8}$ with sum $\frac{3}{x+y} + \frac{3}{(x+y)^2}$.

TEACH

Approach B

Using Models, continued

Fractions	$\frac{a}{a^2-4}$	$\frac{2}{a+2}$
Factored form of denominators	$(a+2)(a-2)$	$(a+2)$
LCD of fraction pair	$(a+2)(a-2)$	
Missing factors	none	$(a-2)$
Forms of 1	none	$\frac{a-2}{a-2}$
Renamed fractions	$\frac{a}{(a+2)(a-2)}$	$\frac{a-2}{a-2} \cdot \frac{2}{a+2}$ $= \frac{2(a-2)}{(a+2)(a-2)}$
Sum of fractions	$\frac{a}{(a+2)(a-2)} + \frac{2(a-2)}{(a+2)(a-2)}$ $= \frac{a+2a-4}{(a+2)(a-2)} = \frac{3a-4}{a^2-4}, a \neq 2, -2$	

Fractions	$\frac{x-4}{3x-3}$	$\frac{x-3}{2x-2}$
Factored form of denominators	$3(x-1)$	$2(x-1)$
LCD of fraction pair	$3 \cdot 2 \cdot (x-1)$	
Missing factors	2	3
Forms of 1	$\frac{2}{2}$	$\frac{3}{3}$
Renamed fractions	$\frac{2}{2} \cdot \frac{x-4}{3(x-1)}$ $= \frac{2(x-4)}{6(x-1)}$	$\frac{3}{3} \cdot \frac{x-3}{2(x-1)}$ $= \frac{3(x-3)}{6(x-1)}$
Difference of fractions	$\frac{2(x-4)}{6(x-1)} - \frac{3(x-3)}{6(x-1)}$ $= \frac{2x-8-3x+9}{6(x-1)} = \frac{-x+1}{6(x-1)}$ $= \frac{-1(x-1)}{6(x-1)} = -\frac{1}{6}, x \neq 1$	

Closing the Lesson

Compare the sum $\frac{3}{4} + \frac{3}{8}$ with the sum $\frac{3}{x+y} + \frac{3}{(x+y)^2}$.

ASSESS

Selected Answers

Odd-numbered Exercises 7–43 and odd-numbered Extra Practice

Assignment Guide

Core level students should complete Exercises 1–44.

Core-plus level students should complete Exercises 1–27 and 33–47.

Error Analysis

Students often make errors when subtracting rational expressions because they neglect to take the opposite of each term in the numerator of the fraction being subtracted. Remind students that when a polynomial is subtracted, each term of the polynomial must be subtracted.

Alternative Assessment

Geometry Applications Tell students to find expressions for the perimeter and area of the rectangle below.

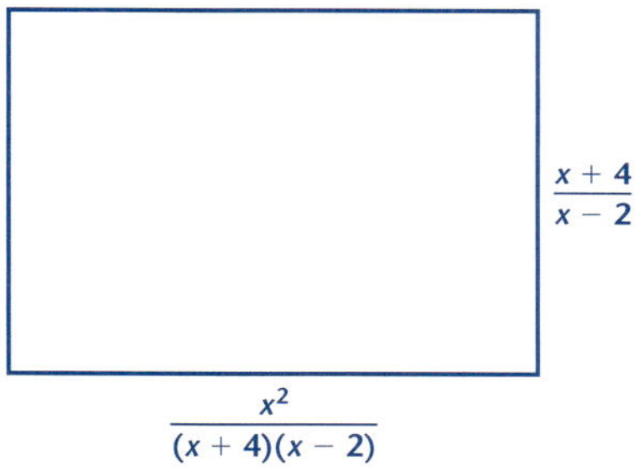

[perimeter $= \frac{4x^2 + 6x + 32}{x^2 + 2x - 8}$; area $= \frac{x^2}{(x-2)^2}$]

Portfolio Activity Create a flowchart for adding or subtracting two rational expressions.

RETEACH

Model several examples of adding and multiplying fractions and of adding and multiplying rational expressions.

ENRICH

Interdisciplinary Connection

Photography The focal length of a camera lens relates the distance from an object to the camera lens and the distance from the camera lens to the film. Lenses of different focal lengths allow a photographer to adjust the size and clarity of an image in a photograph.

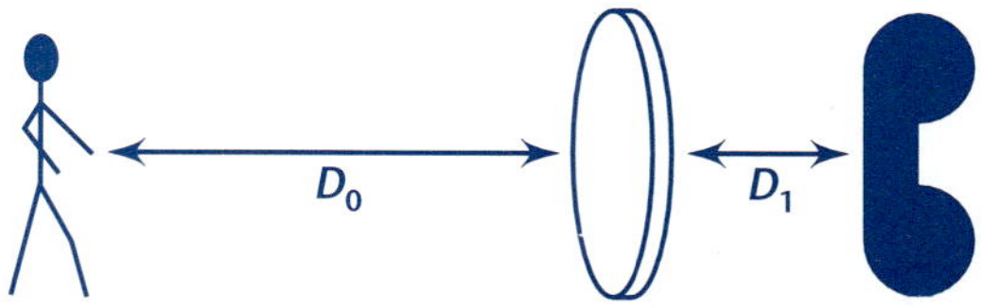

If f is the focal length of the camera, then $\frac{1}{f} = \frac{1}{D_0} + \frac{1}{D_1}$. The expression $\frac{1}{D_0} + \frac{1}{D_1}$ can be rewritten as $\frac{D_1 + D_0}{D_0 D_1}$.

NAME ______________________ CLASS ______________ DATE ______________

Practice

10.4 Operations With Rational Expressions

Perform the indicated operations. Simplify and state restrictions for variables.

1. $\frac{3}{5z} + \frac{4}{5z}$ ______________

2. $\frac{6}{x-3} + \frac{2}{x-3}$ ______________

3. $\frac{5a}{a+b} - \frac{2a}{a+b}$ ______________

4. $\frac{x}{x-1} - \frac{1}{x-1}$ ______________

5. $\frac{1}{p} - \frac{2}{q}$ ______________

6. $\frac{c}{3d} + \frac{d}{3c}$ ______________

7. $\frac{4}{mn} \cdot \frac{m}{2}$ ______________

8. $\frac{y^2}{6} \cdot \frac{12}{y^3}$ ______________

9. $\frac{x^3}{6} + \frac{x^2}{12}$ ______________

10. $\frac{a^2(a-3)}{3-a}$ ______________

11. $\frac{a^3}{(a+1)^2} + \frac{a}{(a+1)^2}$ ______________

12. $\frac{t^2+2t}{t^2-4} \cdot \frac{3t-6}{t}$ ______________

13. $\frac{y^2+5y+6}{y^2} \cdot \frac{y}{y+2}$ ______________

14. $\frac{3x}{3x-2} - \frac{2}{3x-2}$ ______________

15. $\frac{m^2-n^2}{m^2n^2} \cdot \frac{mn}{1}$ ______________

16. $\frac{1}{x+1} + \frac{x}{x^2-1}$ ______________

17. $\frac{3x^2}{1-x^2} - \frac{2-x}{1-x^2}$ ______________

18. $\frac{8b-40}{8} \cdot \frac{b+5}{b^2-25}$ ______________

19. $\frac{6}{c^2-3c} \cdot \frac{2c^2-2c-12}{6c+12}$ ______________

20. $\frac{x}{x+y} - \frac{-3xy}{(x+y)^2}$ ______________

21. $4 + \frac{3x}{x^2-6}$ ______________

22. $\frac{3}{z-1} + \frac{z}{1-z}$ ______________

23. $\frac{4}{x-3} \cdot \frac{1}{x}$ ______________

24. $\frac{n+4}{16} \cdot \frac{12}{n^2-16}$ ______________

25. $\frac{2}{x+5} + \frac{3x}{x^2+4x-5}$ ______________

26. $\frac{3d}{d^2-9} + \frac{d}{3+d}$ ______________

Lesson Activity

10.4 Modeling Rational Expressions

An airplane flies in a straight line from city A to city B and then back again with a constant engine speed. Determine whether the travel time will be longer, shorter, or the same if a constant wind blows from city A to city B.

1. Write an expression for the total time that it would take to fly from city A to city B and then back again.

____

2. Write a formula in simplest terms for the time, *t*, in terms of the distance, *d*, and using *r* as the engine speed of the airplane and *w* as the wind speed.

____

Suppose that the distance between the cities is 400 miles and that the plane is traveling at the rate of 200 miles per hour. Substitute the values in your formula and graph your equation on a graphics calculator.

3. What happens to the travel time as the wind speed increases to 200 miles per hour?

____

4. What is meant by a wind speed of −30 miles per hour? ______________

5. How is the total travel time affected if a constant wind blows throughout both flights?

____

6. Write a formula for the average velocity, *V*, in terms of the distance, *d*, and using *r* as the speed of the airplane and *w* as the wind speed. Then simplify. ______________________________

7. What happens to the variable for distance in your formula? Explain.

____

NAME ____________________ CLASS __________ DATE __________

Basic Skills Practice

Cumulative Review

Read each question and circle the best answer.

1. What is the greatest precision that can be achieved with this ruler?

A Nearest $\frac{1}{2}$ inch

B Nearest $\frac{1}{4}$ inch

C Nearest $\frac{1}{8}$ inch

D Nearest $\frac{1}{16}$ inch

2. Rose needed to cut a 12-yard piece of ribbon into 8 equal pieces. Find the length of each piece, written as a mixed number.

F 1 yd

G $1\frac{1}{2}$ yd

H $1\frac{3}{4}$ yd

I $1\frac{11}{12}$ yd

3. Stephen earned $55 last week. He spent 9% of his earnings on entertainment. What decimal represents this percent?

A 0.9

B 0.09

C 0.55

D 5.5

4. A string of lights is needed to go around a circular stage. The radius of the stage is 10 meters. Find the length of the string of lights that is needed to surround the stage without any overlap.

F 10π m

G 20π m

H 100π m

I 200π m

5. At Morton High School, 25% of the students ride the bus to and from school. There are 558 students at the school. Find the proportion that represents *B*, the number of students that ride the bus to and from school.

A $\frac{25}{100} = \frac{558}{B}$

B $\frac{25}{100} = \frac{B}{558}$

C $\frac{25}{558} = \frac{B}{100}$

D $\frac{25}{B} = \frac{558}{100}$

6. The square root of which number is ***not*** between the numbers 7 and 8?

F 48.54

G 50.54

H 54.54

I 56.65

7. Joe is 21, Sean is 22, and Ben is 16. Joe is $3\frac{1}{2}$ times as old as Susan. How old is Susan?

A 74

B 56

C 7

D 6

8. Find the fraction, decimal, and percent that are equivalent.

F $\frac{1}{2}$, 0.05, 10%

G $\frac{1}{2}$, 0.02, 20%

H $\frac{1}{5}$, 0.20, 20%

I $\frac{1}{5}$, 0.02, 20%

LESSON 10.5 Solving Rational Equations

PREPARE

Pacing Guide

2 Days

Objectives and Overview

- Solve rational equations by graphing and by using the method of common denominators.
- Use rational equations to solve problems.

Students are shown how to solve rational equations algebraically and graphically. Rational equations are used to model several real-world applications, and students are asked to identify the domains of these applications.

Assessing Prior Knowledge

Solve:

1. $\frac{x+1}{x} + \frac{2}{3x} = \frac{3}{2x}$ $\left[-\frac{1}{6}\right]$

2. $\frac{4}{x+2} = \frac{3}{x+1} - \frac{1}{x+2}$ $\left[\frac{1}{2}\right]$

Approach A: Using the Book

Opening Discussion

In the distance formula, rate • time = distance or $rt = d$. If distance is held constant, how are rate and time related? **[They are inversely proportional to each other.]** How could you express time as a function of distance? If a hiker hiked 9 miles at a constant rate of x miles per hour, what expression would model the time it took the hiker to hike the 9 miles? What will the graph of the function involving this expression look like?

Defining *Rational Equations*

Rational equations are equations that contain one or more rational expressions. Since variables may occur in denominators of rational expressions in an equation, students must be careful to note which values of a variable can make the value of a denominator zero. Those values must be excluded from the domain of the function.

continued on top of next page

TEACH

Approach B: Using Models

Opening Discussion

The distance formula is rate • time = distance or $rt = d$. If distance is held constant, how are rate and time related? **[They are inversely proportional to each other.]** Express time as a function of distance. Suppose that a hiker hiked 9 miles at a constant rate of x miles per hour. Write an expression that models the time it took the hiker to hike the 9 miles. What do you think the graph of the function involving this expression will look like?

Modeling Rational Equations: Graphs

Sketch the graphs of the following functions, or graph them on a graphics calculator: $y = \frac{6}{5x} + \frac{6}{x}$ and $y = \frac{12x}{5}$. Both equations involve rational expressions. Which function has a linear graph? The hyperbola has vertical asymptotes at $x = 0$, demonstrating that

continued on bottom of next page

Approach A — Using the Book, continued

Using the Distance Formula

Since Sam's rate is given in miles per hour, his time must be expressed in hours. Twenty-four minutes is $\frac{24}{60}$ of an hour, or $\frac{2}{5}$ hours.

Teaching Tip

Make sure students understand that the Distributive Property requires each term on the left side of the distance equation to be multiplied by $5x$. You might want to call on students to explain how the left side of the equation was converted from $5x(\frac{6}{5x} + \frac{6}{x})$ to $6 + 30$.

Journal Entry

Have students add $\frac{6}{5x} + \frac{6}{x}$ in their notebooks. Ask the students how the resulting function, $y = \frac{36}{5x}$, is related to the parent function $y = \frac{1}{x}$. **[It is stretched vertically by a factor of $7\frac{1}{5}$.]**

Do the students' findings agree with the graph at the top of page 501 of the Pupil's Edition?

Technology

Students can also use a table of values on their graphics calculator to find the intersection. The x-value that has the same y-value, 2.4, for each function is the intersection of the two equations.

X	Y1	Y2
−2	−3.6	2.4
−1	−7.2	2.4
0	ERROR	2.4
1	7.2	2.4
2	3.6	2.4
3	2.4	2.4
4	1.8	2.4

X = 3

Make sure that when students enter the two functions in the Y= editor that the denominator, $5x$, is enclosed in parentheses.

$Y_1 = 6/(5X)+6/X$
$Y_2 = 12/5$

TEACH

Approach B — Using Models, continued

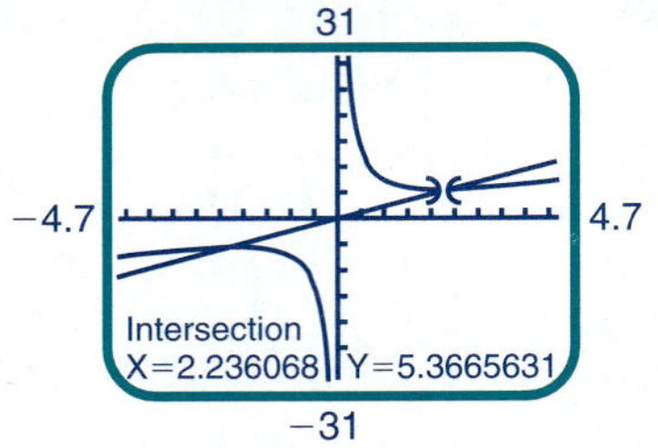

zero must be excluded from the domain of $y = \frac{6}{5x} + \frac{6}{x}$ because division by zero is undefined. When a rational expression in an equation has a variable in the denominator, values of the variable that make the value of the denominator zero must be excluded from the domain of the rational equation.

Application: Building a Chart

Since rate and time vary inversely with each other when distance is held constant, the expressions $\frac{6}{5x}$ and $\frac{6}{x}$ can represent expressions for time if 6 represents a distance and $5x$ and x represent rates. The distance, 6, is the constant of variation. If x represents a boy's walking speed, his bicycling speed will probably be greater than x. Suppose that the expression $5x$ represents his bicycling speed. This means that the boy bicycles 5 times as fast as he walks.

A chart efficiently displays this information.

	Distance in miles	Rate in miles per hour	Time
Walking	6	x	$\frac{6}{x}$
Bicycling	6	$5x$	$\frac{6}{5x}$
Total			$\frac{6}{x} + \frac{6}{5x}$

Critical Thinking, page 501

The real-world domain is $x > 0$. The expression is undefined when $x = 0$.

Example 1

A rational equation is solved algebraically and graphically.

Though the algebraic method of solving rational equations is more tedious than the graphical method, solving algebraically allows students to review the factoring of polynomials.

In the second step of the algebraic solution, make sure that students understand the role of the Distributive Property. The common denominator is distributed to each term in the brackets. This allows the denominator of the fractional term to be eliminated.

Students should be able to identify the equations of the graphs. The graph of $y = \frac{1}{x}$ is the parent graph of $Y_1 = \frac{6}{x-1} + 2$, which has been vertically stretched 6, horizontally shifted 1 unit to the right, and vertically shifted 2 units up. The x^2-term in the function $Y_2 = \frac{12}{x^2 - 1}$ has the same value reguardless of whether x is positive or negative, so this graph is symmetrical about the y-axis. If students graph this function on their graphics calculator and set the window to view y-values less than -12, they will notice that this graph has another section between the asymptotes $x = 1$ and $x = -1$.

Use **Transparency 49** for discussion.

Ongoing Assessment, page 502

The undefined values become asymptotes of the graph.

Additional Example 1

Solve $x + \frac{3}{x} = \frac{12}{x}$ by using the common denominator method and the graphing method.
[−3, 3]

continued on top of next page

Charts can help students understand problems involving rational equations.

If the boy actually spent 2 hours and 24 minutes, or $2\frac{2}{5}$ hours, walking and bicycling, this situation can be modeled by the equation $\frac{6}{x} + \frac{6}{5x} = \frac{12}{5}$.

When rational expressions are simplified, the result is usually another algebraic expression. When rational equations are solved, the result is generally one or more numbers.

Solving Rational Equations: A Graphic Approach

A rational equation such as $\frac{6}{x} + \frac{6}{5x} = \frac{12}{5}$ can also be solved graphically. Graphing $Y_1 = \frac{6}{x} + \frac{6}{5x}$ and $Y_2 = \frac{12}{5}$ together shows that the intersection of the two graphs is at $x = 3$. (Compare the graphs on page 501.)

Students can also use a table of values on their graphics calculator to find the intersection. The x-value that has the same y-value, 2.4, for each function is the intersection of the graphs of the two equations.

continued on bottom of next page

Approach A

Using the Book, continued

Extraneous Solutions

Students must understand the importance of checking the original rational equation for values that must be excluded from the domain of the function. Because extraneous solutions satisfy transformed equations but not an original equation, excluded values must be identified before any transformations are made to the original rational equation. To verify this, tell students to begin with the equation $x = 2$. Multiplying both sides of the equation by $x - 1$ gives $x(x - 1) = 2(x - 1)$. Although 1 is not a solution for the original equation, $x = 2$, it is a solution for the transformed equation, $x(x - 1) = 2(x - 1)$. Notice that solving the transformed equation for x would require dividing by $x - 1$. Since division by 0 is undefined, x cannot equal 1.

Example 2

A rational equation models a problem that asks for an average. The problem is solved by using the algebraic method and also by graphing the equation that models the problem.

The graph of $y = \frac{400 + 6x}{x}$ is a hyperbola lying in the first and third quadrants. Since the number of lawns cannot be a negative number, the section of the graph in the third quadrant does not model this problem. Although x can assume any value greater than zero, the minimum number of lawns is probably 1 if Juanita doesn't plan to mow parts of lawns. In that case, the domain of x includes only positive whole numbers.

The graph shows that the average cost per lawn can never be zero. This can also be verified algebraically. If the average cost per lawn were zero, then $400 + 6x = 0$ and $x = -66\frac{2}{3}$ lawns mowed!

Ongoing Assessment, page 503

Try This

$x = 3$ or $x = -\frac{4}{3}$

TEACH

Approach B

Using Models, continued

X	Y1	Y2
−2	−3.6	2.4
−1	−7.2	2.4
0	ERROR	2.4
1	7.2	2.4
2	3.6	2.4
3	2.4	2.4
4	1.8	2.4

X = 3

Make sure that when students enter the two functions in the Y= editor, the denominator, $5x$, is enclosed in parentheses.

Y_1 = 6/(5X)+6/X
Y_2 = 12/5

Cooperative Learning

Finding Extraneous Solutions Ask students to work in groups of four. Two of the students should solve the equation $\frac{6}{x - 1} + 2 = \frac{12}{x^2 - 1}$ algebraically, and two of the students should solve the equation graphically. Tell the students that the algebraic solution will involve factoring a polynomial. Circulate around the room as the students solve the equation. Ask guiding questions for students who need help. Ask each group of students to write a summary of their procedures that includes an explanation of how the graphic solution models the algebraic solution. Each pair of students in a group should explain how they identified any solutions that were excluded. Each group should choose a member to read their summary to the class.

Students must understand the importance of checking the original rational equation for values that must be excluded from the domain of the function. Because extraneous solutions satisfy transformed equations but not an original equation, excluded values must be identified before any transformations are made to the original rational equation. To verify this, tell students to begin with the equation $x = 2$. Multiplying both sides of the equation by $x - 1$ gives $x(x - 1) = 2(x - 1)$. Although 1 is not a

Additional Example 2

Two gears that are meshed and revolve together are related by this formula.

$$\frac{\text{number of teeth on gear A}}{\text{number of teeth on gear B}} = \frac{\text{number of revolutions per minute for gear B}}{\text{number of revolutions per minute for gear A}}$$

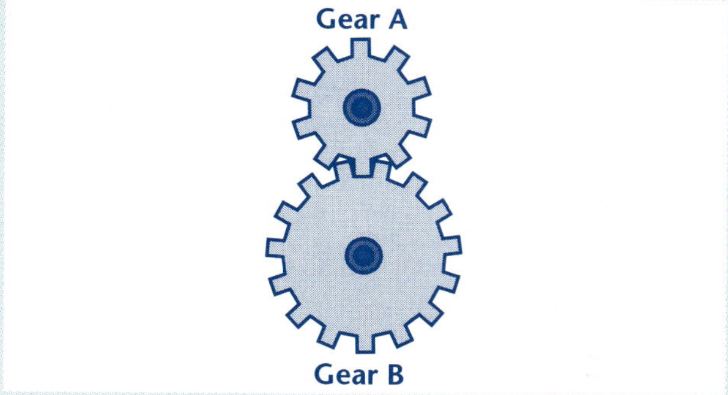

Gear A must make 3000 revolutions per minute when gear B is making 2000 revolutions per minute. If gear B has 15 more teeth than gear A, how many teeth must gear A have? **[30 teeth]**

Inclusion Strategy

Preview Explain to students that they will be studying equations like the one in Example 2 in the next lesson. Ask students to examine the algebraic method for solving the equation. Ask them to explain in their notebook how a shortcut might be used to solve this equation.

Example 3

A rational equation models a situation in which two people complete a job together. The problem is solved algebraically.

Since time is expressed in minutes, each portion of the job must be expressed in minutes. Max completes the job in 80 minutes, so he completes $\frac{1}{80}$ of the job in 1 minute. Carl completes the job in 2 hours, or 120 minutes, so he completes $\frac{1}{120}$ of the job in 1 minute.

Each worker's contribution is expressed as a portion of the job, so the total of the workers' contributions is the whole job, or 1.

continued on top of next page

solution to the original equation, $x = 2$, it is a solution to the transformed equation, $x(x - 1) = 2(x - 1)$. Notice that solving the transformed equation for x would require dividing by $x - 1$. Since division by zero is undefined, x cannot equal 1.

Using Rational Equations to Model Work Problems

A job can often be done more quickly if it is shared by two people. Since each person completes a fraction of the job, a rational equation often models this kind of a problem. For example, suppose that a brother and a his younger sister have a paper route. When the brother does the paper route alone, it takes him 80 minutes to finish the route. It takes his younger sister 2 hours to finish the route alone. How long does it take them to deliver the papers when both are working together?

This information can be displayed clearly in a chart.

	Part of job completed in 1 minute working alone	Time spent on the job	Part of job completed
Brother	$\frac{1}{80}$	t	$\frac{1}{80}t = \frac{t}{80}$
Sister	$\frac{1}{120}$	t	$\frac{1}{120}t = \frac{t}{120}$
Total			$\frac{t}{80} + \frac{t}{120}$

Since each person completes a fractional part of the job, the whole job can be represented by 1. Model the situation with a rational equation.

$$\frac{t}{80} + \frac{t}{120} = 1$$

Solve the equation by using the previous rational equations as a model. Multiply each term by the lowest common denominator and simplify. Use properties of real numbers to isolate the variable.

$$240 \cdot \frac{t}{80} + 240 \cdot \frac{t}{120} = 240 \cdot 1$$
$$3t + 2t = 240$$
$$5t = 240$$
$$t = 48$$

continued on bottom of next page

Approach A

Using the Book, continued

Because there are no variables in the denominators of the fractions, no values of *t* need to be excluded from the domain of the function. However, for the purpose of this problem, time can be only a positive number.

Technology

Ask students to graph the equations $Y_1 = \frac{x}{80} + \frac{x}{120}$ and $Y_2 = 1$ on a graphics calculator. The function Y_1 is not a curve like the other rational functions in this lesson. Why? What is the intersection of the two graphs?

Closing the Lesson

Explain how the undefined values for the equation $\frac{2}{x} + \frac{1}{2} = \frac{5}{2x}$ can be found algebraically and graphically.

TEACH

Approach B

Using Models, continued

Working together, it takes the brother and sister 48 minutes to deliver all of the papers.

Because there are no variables in the denominators of the fractions, no values of *t* need to be excluded from the domain of the function. However, for the purpose of this problem, time can be only a positive number.

Technology

Ask students to graph the equations $Y_1 = \frac{x}{80} + \frac{x}{120}$ and $Y_2 = 1$ on a graphics calculator. The function Y_1 is not a curve like the other rational functions in this lesson. Why? What is the intersection of the two graphs?

Closing the Lesson

Explain how the undefined values for the equation $\frac{2}{x} + \frac{1}{2} = \frac{5}{2x}$ can be found algebraically and graphically.

ASSESS

Selected Answers

Odd-numbered Exercises 7–35 and odd-numbered Extra Practice

Assignment Guide

Core level students should complete Exercises 1–36.

Core-plus level students should complete Exercises 1–13, 18–28, and 31–37.

Error Analysis

Students often solve rational equations incorrectly because they neglect to multiply integer terms by the LCD. Remind these students that the Multiplication Property of Equality requires that both sides of the equation be multiplied by the LCD. This means that the LCD must be distributed to every term on both sides of the equation, even if the term is not a fraction.

Alternative Assessment

Simplifying vs. Solving Explain how simplifying $x + \frac{6}{x} - 5$ and solving $x + \frac{6}{x} = -5$ are alike. How are they different? Does either the expression or the equation have any excluded values of x? Explain.

Portfolio Activity Solve the following problem both algebraically and graphically:

Sharon can weed a strawberry field by herself in 6 hours. It takes Otto 7 hours to weed the same field alone. How long will it take them to weed the field if they work together? $\left[3\frac{3}{13}\text{ hours}\right]$

RETEACH

Because students are familiar with adding and subtracting rational expressions and because they often confuse the methods for adding and subtracting rational expressions with those for solving rational equations, you can show students how to solve rational equations by using the rules for simplifying sums and differences of rational expressions.

$\frac{6}{5x} + \frac{6}{x} = \frac{12}{5}$	Given
$\frac{6}{5x} + \frac{5}{5} \cdot \frac{6}{x} = \frac{x}{x} \cdot \frac{12}{5}$	Rename fractions with the LCD.
$\frac{6}{5x} + \frac{30}{5x} = \frac{12x}{5x}$	Simplify.
$\frac{36}{5x} = \frac{12x}{5x}$	Simplify.
$\frac{36 - 12x}{5x} = 0$	Subtraction Property of Equality
$36 - 12x = 0$	Multiplication Property of Equality
$-12x = -36$	Subtraction Property of Equality
$x = 3$	Division Property of Equality

ENRICH

Interdisciplinary Connection

Building Trades Construction companies have a lot of capital invested in their machinery. Because construction machinery such as backhoes are very expensive, companies must determine how much time is saved by having more than one backhoe. That time must be compared with the cost of buying a second backhoe. For example, if a backhoe can dig a foundation in 15 days, but a newer backhoe can dig the same foundation in 12 days, the two backhoes working together could dig the foundation in less than 7 days, thus saving 8 days of labor. Eight days of labor saved on every job of this size would have to be compared with the cost of the new backhoe.

NAME ______________________ CLASS __________ DATE __________

Practice

10.5 Solving Rational Expressions

Solve the following rational equations by finding the lowest common denominator:

1. $\frac{2}{y} + 3 = \frac{29}{y}$ ______

2. $\frac{z}{2} - 7 = \frac{z}{3}$ ______

3. $\frac{1}{x} + \frac{3}{x} = \frac{5}{4}$ ______

4. $\frac{2x}{3} - \frac{x}{2} = 5$ ______

5. $\frac{3 - 5c}{8} + \frac{3c - 6}{5} = 1$ ______

6. $\frac{1}{4} - \frac{x - 3}{8x} = 0$ ______

7. $\frac{1}{x + 6} = \frac{2}{x + 9} - \frac{1}{x + 3}$ ______

8. $\frac{y - 1}{y} - 2 = \frac{-y}{y + 1}$ ______

9. $\frac{4b + 5}{2b + 1} = \frac{6b + 1}{3b - 1}$ ______

10. $\frac{x}{x^2 - 3} = \frac{5}{x + 4}$ ______

11. $\frac{6}{a} - 1 = a$ ______

12. $\frac{y}{y - 5} + \frac{y}{y - 5} = 3$ ______

13. $\frac{3x - 1}{4} = \frac{x - 5}{5} - 2$ ______

14. $\frac{4w}{3w - 2} + \frac{2w}{3w + 2} = 2$ ______

15. $\frac{2}{x} + \frac{1}{4} = \frac{11}{12}$ ______

16. $\frac{12}{x^2 - 16} - 3 = \frac{24}{x - 4}$ ______

17. $\frac{1}{r^2 - 1} = \frac{2}{r^2 + r - 2}$ ______

18. $\frac{m}{m^2 - 1} + \frac{2}{m + 1} = \frac{1}{2m - 2}$ ______

19. $\frac{t}{t - 5} = \frac{1}{t + 5} - \frac{17}{25 - t^2}$ ______

20. $2 - \frac{3}{y^2 + 5y + 6} = \frac{y + 3}{y + 2}$ ______

21. $\frac{q - 3}{4 + q} = \frac{q - 3}{7}$ ______

22. $\frac{g}{g - 2} = 1 + \frac{g + 7}{g + 2}$ ______

Solve the following rational equations by graphing:

23. $\frac{3x}{5} = \frac{x}{2} - 1$ ______

24. $\frac{x}{4} + \frac{x - 3}{3} = 6$ ______

25. $\frac{1}{x + 2} + \frac{1}{x - 3} = \frac{2}{x}$ ______

26. $\frac{2x - 6}{3x + 1} = \frac{x - 1}{4x - 4}$ ______

NAME ______________________ CLASS ____________ DATE ____________

Lesson Activity

10.5 Modeling Winning Percentages

After playing 10 games, a baseball team has a winning percentage of 0.600. The team then wins 50% of its remaining 10 games.

1. Write an equation that models the number of additional games needed to achieve a winning percentage of 0.550. Then solve your equation.

____

____

Consider all the ways that the team could win or lose these remaining games. Suppose that the team wins every other game.

2. Complete the table for alternate wins and losses, if the team wins game 11.

Number of games	Wins	Losses	Winning %
10	6	4	0.600
11	7	4	
12	7	5	
13			
14			
15			
16			
17			
18			
19			
20			

3. What is the team's winning percentage? ______________________

4. What is the winning percentage if the team loses games 11 through 15 and wins the remaining 5 games? Explain.

____

5. If the team plays 5 additional games in the playoffs, what are the best and worst possible winning percentages that the team could have at the end of the season?

____

Basic Skills Practice
Cumulative Review

Read each question and circle the best answer.

1. Using the formula provided, find the volume of a cube that is 5 inches on each side.

$$V = s^3$$

A 15 in.^{3}
B 25 in.^{3}
C 125 in.^{3}
D 625 in.^{3}

2. Consider the expression $5x - 6y$. Evaluate the expression if $x = 4.5$ and $y = 7.8$.

F -24.3
G -12.0
H -6.3
I 25.3

3. Find the correct order from least to greatest.

A 36,081; 36,180; 306,810; 38,016
B 36,081; 38,016; 36,180; 306,810
C 306,810; 38,016; 36,180; 36,081
D 36,016; 36,180; 38,016; 306,810

4. To make a lion costume, Ann bought $3\frac{1}{4}$ yards of light brown felt. She needed to use only $2\frac{7}{8}$ yards of it. How much fabric did she have left?

F $\frac{1}{8}$ yd
G $\frac{2}{3}$ yd
H $\frac{3}{8}$ yd
I $1\frac{3}{8}$ yd

5. Monica wants to budget her money for the purchase of a new lamp. She went to several stores and made a list of the prices she found. Find the mean and the mode of the prices.

$45.50, $45.99, $50.29, $45.50, $47.98

A $45.50, $47.05
B $47.05, $45.50
C $47.05, $46.99
D $4.79, $45.50

6. An equivalent form of 4000 mm is—

F 0.4 m
G 4 m
H 40 m
I 400 m

7. Jan bought 5 yards of fabric. She used $\frac{1}{3}$ of it to sew curtains and $\frac{1}{5}$ of it to sew a pillow. How much fabric did she have remaining?

A 2 yd
B $2\frac{1}{3}$ yd
C $2\frac{2}{3}$ yd
D 3 yd

8. Nikolette answered 78 out of 90 questions correctly on her math final. What score did she receive, written as a percent?

F 78%
G 82%
H 87%
I 90%

LESSON 10.6 *Exploring* Proportions

PREPARE

Pacing Guide

2 Days

Objectives and Overview

- Identify the means and the extremes of a proportion.
- Determine whether an equation is a proportion.
- Use proportions to solve problems.

Students explore the cross products of a proportion and use proportions to solve problems.

Vocabulary

proportion
means
extremes
cross products

Assessing Prior Knowledge

Solve:

1. $\frac{4}{x} = 8$ $\left[\frac{1}{2}\right]$

2. $\frac{4}{x} = \frac{8}{40}$ [20]

3. $\frac{4}{x} = \frac{8}{3}$ $\left[1\frac{1}{2}\right]$

4. $\frac{x}{4} = \frac{8}{3}$ $\left[10\frac{2}{3}\right]$

5. $\frac{x}{4} = \frac{3}{8}$ $\left[1\frac{1}{2}\right]$

Approach A: Using the Book

Opening Discussion

Introduce the lesson by telling your students that they will learn a method for solving equations that involve equivalent ratios. They will also use equivalent ratios to solve practical problems. Suppose that a freshman student hoped to make 70% of his free throws in a class basketball competition. What ratio represents 70%? How many free throws would the student have to make out of 10 attempts? What ratio compares the number of successes with the number of attempts? What ratio compares the student's number of successes with the number of attempts if he makes 20 attempts? Are all three ratios equal? Explain.

Defining *Proportion*

Students have already worked with proportions in Chapter 1. Remind students that a proportion is a special kind of equation in which two or more ratios are equal. A rational equation can therefore be a

continued on top of next page

TEACH

Approach B: Using Models

Opening Discussion

Begin this lesson by telling your students that they will be using equivalent ratios to solve problems. They will also learn a method for solving equations that involve equivalent ratios. Tell your students to suppose that a freshman student hoped to make 70% of his free throws in a class basketball competition. Write a ratio that represents 70%. If the student makes 10 attempts, how many baskets would he have to make? Write a ratio that compares the number of shots made with the number of attempts. Write a ratio that would compare the student's number of shots made with the number of attempts if he makes 20 attempts. Are all three ratios equal? Explain.

Modeling Proportions: Geometric Figures

Use sketches of rectangles to represent equivalent ratios. Use two different colors of overhead marking pens or chalk, and show that $\frac{7}{10} = \frac{14}{20}$. To

continued on bottom of next page

Approach A Using the Book, continued

proportion. Because a proportion contains fractions, the values of the denominators in the proportion cannot be zero.

Defining *Means* and *Extremes*

As students learned in Chapter 1, a proportion is an equation which states that two ratios are equal. Thus, a proportion is of the form $\frac{a}{b} = \frac{c}{d}$, which can also be written as $a:b = c:d$. Notice that when a proportion is written in the form $\frac{a}{b} = \frac{c}{d}$, the first letter written is a and the last letter written is d. The same is true if the proportion is written from left to right in horizontal form. These numbers, a and d, are called the *extremes* of the proportion. Students can use their understanding of the meaning of *extreme* to remember which numbers in a proportion are the extremes. The adjective *extreme* has the meanings "beyond the limit" or "on the edge." The extremes can be thought of as the numbers *on the edges* of the proportion.

The numbers b and c are written between the extremes when the proportion is written horizontally. They are called the *means* of the proportion. To help them remember which numbers in a proportion are the means, students may recall that a mean is a kind of "middle" measure of a set of data and that *mean* and *middle* both begin with the letter m.

Cooperative Learning

Teacher-Guided Exploration If you wish to guide your students through the explorations, you can write the proportions on the chalkboard or overhead. Then you can call on students to answer the exploration questions. As they answer, you can write their answers on the board. If you want students to be more involved in the explorations, you can ask students to come to the board to demonstrate their answers.

TEACH

Approach B Using Models, continued

distinguish between the two numerators, color the two numerator rectangles different colors. Do the same for the two denominator rectangles.

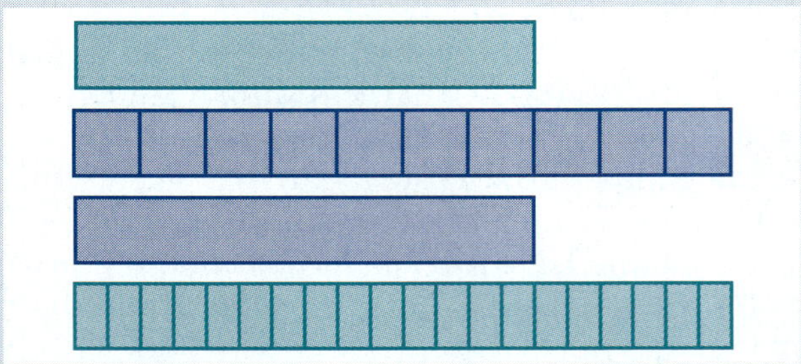

Model the ratios like this:

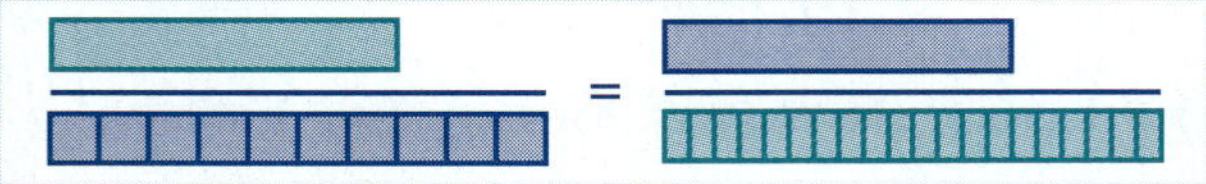

Notice that the two "outside" rectangles are the same color and the two "inside" rectangles are the same color. Similarly, in the ratio model, the rectangles in the first and last positions (first numerator and second denominator) of the ratios are the same color. The rectangles in the two middle positions (first denominator and second numerator) of the ratios are the same color.

Reading Mathematics

When a proportion is written in fraction form, the extremes are the first numerator and the last denominator or, generally, the first and last numbers written when the fractions are written. When a proportion is written in horizontal form, again the extremes are in the first and last positions of the proportions and are the first and last numbers written when the proportion is written from left to right. In both cases, the means are the two numbers in the middle of the proportion. Whether a proportion is written as $\frac{a}{b} = \frac{c}{d}$ or $a:b = c:d$, the proportion can be read using the verbal model "a is to b as c is to d." This reading reinforces that a is being compared with b and c is being compared with d and that the comparisons are equivalent.

Student-Directed Group Approach

Have students work in groups of four. Each member of a group should answer one of the parts of each exploration question. Other members of the group should discuss the answer. As students answer the questions, they should explain how they determined their answers. One member chosen by the group should record the students' explanations and use them to write a summary of the group's activities in the exploration. All students should help edit and then sign the report.

Exploration 1

Students explore the property of cross products of a proportion. Then they use this property to solve a proportion.

In Step 2, you may need to review with students how to find the product of $2\sqrt{2}$ and $3\sqrt{2}$.

In Steps 3 and 4, students must recognize that the products of the means and the extremes are the same as the cross products.

Steps 6 and 7 demonstrate how a proportion might produce either a linear equation or a quadratic equation.

Ongoing Assessment, page 507

7. Each proportion contains a binomial term. In part b, the variable c occurs in each of the means, thus yielding a squared term in the cross product. In part a, the variable b occurs in one mean and one extreme, thus yielding linear terms in the cross product.

On Their Own

Solve each proportion.

A. $\frac{x}{4} = \frac{12}{48}$ **[1]** **B.** $\frac{x^2 + 12}{x} = \frac{-21}{3}$ **[−3, −4]**

C. $\frac{x + 2}{x - 7} = \frac{8}{2}$ **[10]**

Exploration 2

Students explore equivalent statements for $\frac{a}{b} = \frac{c}{d}$.

continued on top of next page

Cooperative Learning

Before you begin the exploration, divide the class into groups of four. Students should take turns answering the questions, with all four students performing each step of the exploration when possible. One person should be assigned to write a brief summary of the findings of the group. Each member should have input into the report, and each should sign the report.

Exploration 1

Students explore the property of cross products of a proportion. Then they use this property to solve a proportion.

Refer to Exploration 1 on page 507 of the Pupil's Edition. Have students work in their groups and draw rectangular models of proportions like the one you drew at the beginning of the lesson. Tell the students to use two different colors to draw the rectangles, as you did. The two numerators should be different colors, and the two denominators should be different colors. The numerator and denominator of each fraction should also be different colors. You may wish to draw the following diagram on the board:

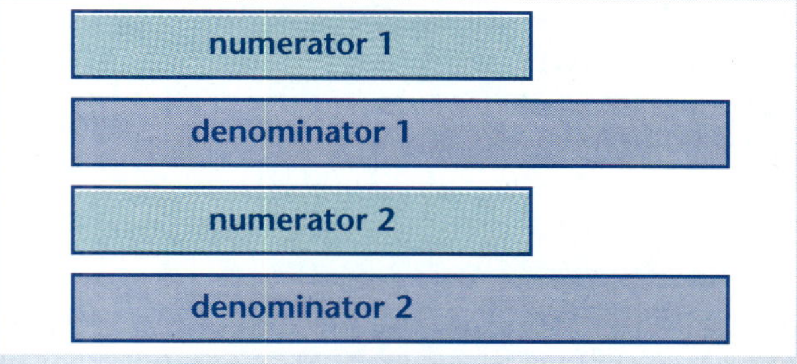

- In Step 1, have students use drawings to model each proportion. Identify the means and the extremes. **[Colors will vary. Means: 2 and 5; 4 and 9, 6 and 6; extremes: 1 and 10, 3 and 12, 4 and 9]**
- In Step 2, have students find the product of the means and the product of the extremes, and draw models for **a, b,** and **c.**
- For Step 2 ask, "How can you use your models to find the product of the means and the product of the extremes?" **[You can multiply the numbers represented by rectangles of**

continued on bottom of next page

Approach A: Using the Book, continued

Ask each student to choose a different value to use in testing the statements in Step 1. The students can then compare their results to determine which statements are equivalent to $\frac{a}{b} = \frac{c}{d}$.

Ongoing Assessment, page 508

2. a, c, d, e, f

Journal Entry

Ask students to make a conjecture about the cross products of all the proportions in Exploration 2 that are equivalent to $\frac{a}{b} = \frac{c}{d}$. **[The cross products of each equivalent proportion produce the equation $ad = bc$.]**

Application: Sports

This application shows students a practical situation in which they can use an algebra skill. By setting up the problem as a proportion, the equation can be easily solved by multiplying the means and the extremes. Students should see that the equation could also be solved by multiplying both sides of the equation by the lowest common denominator. Ask students which method they think is easier.

Ratios in Geometry Problems

Ask students to write several ratios equivalent to 1:2. Then have them write several ratios equivalent to 2:3 and several ratios equivalent to 1:3. Ask students what each ratio was multiplied by to produce an equivalent ratio.

If three numbers are in the ratio 1:2:3, then the ratios have a common factor of x. Thus, the numbers can be represented by the expressions x, $2x$, and $3x$.

TEACH

Approach B: Using Models, continued

the same color.] What is the relationship between the product of the means and the product of the extremes for each of the proportions modeled by the rectangles? **[They are equal.]** Is this relationship true for the proportion in **d**? **[yes]** Explain? **[$2\sqrt{2} \cdot 3\sqrt{2} = 6 \cdot 2 = 12$]**

- Why do you think that the product of the means and the product of the extremes are also called *cross products*? **[You multiply across the proportion, from numerator to opposite denominator, to find the products.]**
- Now have students complete Steps 3 through 7, making models when possible.

Exploration 2

Refer to the rectangle model of the proportion $\frac{1}{2} = \frac{5}{10}$ in Exploration 1. If you switched the positions of the top two rectangles and then switched the positions of the bottom two rectangles, would this arrangement give the same cross products as the original arrangement? What would be the resulting equation? Is this equation a proportion? **[yes; $\frac{2}{1} = \frac{10}{5}$; yes]**

Now have students complete the steps of the exploration by using their models when possible. In Step 1, expressions such as $a - d$ can be modeled by crossing out blocks of the rectangles and then comparing the results, as in the first demonstration.

Application: Sports

A chart provides a model for the calculation in the sports application on page 508 of the Pupil's Edition. Letting x represent the number of additional successful free throws that must be made, the following table can be used to organize the information:

The solutions can be checked by showing that 30:60 = 1:2, 60:90 = 2:3, and 30:90 = 1:3.

Critical Thinking, page 509

$\frac{a}{a+c} = \frac{b}{b+d}$

Closing the Lesson

What are two ways to name the equation $\frac{3x-4}{x} = \frac{5}{2}$? **[rational equation or proportion]** Describe two different ways to solve this equation. **[Multiply both sides by the lowest common denominator, or cross multiply.]** Solve the equation. **[$x = 8$]**

	At present	Desired
Percent	70%	76%
Successes	28	$28 + x$
Attempted free throws	40	$40 + x$
Ratio	$\frac{28}{40}$	$\frac{28+x}{40+x}$
Proportion	$\frac{70}{100} = \frac{28}{40}$	$\frac{76}{100} = \frac{28+x}{40+x}$

Solving the proportion $\frac{76}{100} = \frac{28+x}{40+x}$ by cross multiplying gives the result $x = 10$. Therefore, the student must make 10 more consecutive free throws to reach 76%.

Application: Geometry Connection

Fraction circles provide a model-based method of examining this application. Use fraction circles to demonstrate the problem and solution as follows: Using six $\frac{1}{12}$-sections, show students that the sections fit perfectly into the angles of the triangle by placing one $\frac{1}{12}$-section at x, two at $2x$, and three at $3x$. Then ask, "How many $\frac{1}{12}$-sections were used?" **[6]** Point out that six $\frac{1}{12}$-sections make half of the circle, so the sum of their angle measures is 180°, which agrees with the sum of the measures of the angles of a triangle. Therefore, the angle measure of one section can be found as follows:

$$6x = 180$$
$$x = 30$$

If you wish, make drawings of other triangles with appropriate angle measures and distribute them to the class for students to find the unknown angle measures. For example, if the mystery triangle has angle measures of 90°, 45°, and 45°, then by letting x be the measure of one of the unknown acute angles, students can use four $\frac{1}{8}$-sections to fill the angles and then solve as follows:

$$4x = 180$$
$$x = 45$$

ASSESS

Selected Answers

Odd-numbered Exercises 7–37 and 41 and odd-numbered Extra Practice

Assignment Guide

Core level students should complete Exercises 1–42.

Core-plus level students should complete Exercises 1–11, 17–37, and 40–43.

Error Analysis

When solving a proportion requires finding a cross product of a monomial and a binomial, some students multiply only the first term of the binomial by the monomial. Remind these students that finding the product of a monomial and a binomial involves the Distributive Property. The Distributive Property distributes the monomial to each term of the binomial.

Alternative Assessment

Relating the Methods Solve $\frac{3}{x+3} = \frac{4}{x-5}$ by cross multiplying. Then solve again by multiplying by the LCD. Explain how the two methods are related.

Portfolio Activity Erica paddles 6 miles upstream in the same time that it takes her to paddle 10 miles downstream. In still water, Erica paddles 4 miles per hour. Write a proportion that compares the time it takes Erica to paddle downstream with the time it takes her to paddle upstream. What are the means and the extremes of the proportion? Solve the proportion. What does the solution represent?

RETEACH

A *cubit* is an ancient form of linear measure that was generally equal to the length of a man's forearm from the tip of his middle finger to the elbow. A cubit usually measured between 17 and 22 inches.

Divide the class into groups of four. Give each group a tape measure, and ask them to measure and then record the length of a cubit for each person in the group.

Each group should also measure and record the length of each member's arm from the shoulder to the tip of the middle finger. Ask one student to record his or her group's results on the chalkboard. Then one student from each group should find the average arm length and average cubit length of all the students in the class.

Each student should then write a proportion using these two averages and the length of his or her arm. The proportion should be of the form

$$\frac{\text{average cubit length}}{\text{average arm length}} = \frac{x}{\text{student's arm length}},$$

where x represents the student's cubit length. Tell students to solve their proportion to see how closely the result approximates their cubit length.

ENRICH

Interdisciplinary Connection

Health Cholesterol, a fatlike substance, can accumulate in the bloodstream and become attached to the walls of arteries. When this happens, blood vessels can get clogged, which leads to heart attacks and strokes. Nutritionists are concerned with two types of cholesterol, that associated with low-density lipoproteins, or LDL, which carries the cholesterol to body tissues, and that associated with high-density lipoproteins, or HDL, which carries the cholesterol to the liver to be broken down and excreted. HDL is also believed to remove some LDL from the walls of arteries. The medical community has debated what the safe levels of cholesterol in the blood are. One widely accepted standard states that a person's LDL cholesterol should be no more than 4 times his or her HDL cholesterol. Thus, if a person's amount of LDL cholesterol is 120 milligrams per deciliter, solving the proportion $\frac{4}{1} = \frac{120}{x}$ will give an estimate of the minimum acceptable amount of HDL cholesterol for that person. Under these standards, this person's minimum HDL cholesterol is 30 milligrams per deciliter.

NAME ______________________ CLASS ____________ DATE ____________

Practice

10.6 Exploring Proportions

Identify the means and extremes of each proportion.

1. $\frac{3}{n} = \frac{4}{5}$ ____________ **2.** $\frac{x+2}{5} = \frac{4}{x-2}$ ____________ **3.** $\frac{2d-1}{4} = \frac{d}{5}$ ____________

Solve for the variable.

4. $\frac{3f-2}{6} = \frac{f}{3}$ ____________ **5.** $\frac{3x}{8} = \frac{6}{x}$ ____________ **6.** $\frac{3(x+4)}{5} = \frac{x+4}{4}$ ____________

7. $\frac{b}{3} = \frac{2}{2b-1}$ ____________ **8.** $\frac{x+3}{4} = \frac{1}{2}$ ____________ **9.** $\frac{1}{4y-3} = \frac{-1}{y-2}$ ____________

10. The dosage of cough syrup is 2 teaspoons for every 50 lb of body weight. How many teaspoons of cough syrup are required for a person who weighs 170 lb? ____________

11. A rocket uses 534,000 gallons of fuel in 2.5 minutes. How much fuel does it use in 10 minutes? ____________

12. When engineers design a new car, they first build a model of the car. The ratio of a part on the model to a part on the actual car is 1:5. If the trunk of the car is 1.2 ft long on the model, what is the length of the trunk on the actual car? ____________

The label shows nutrition facts from a package of pasta.

13. If 1 oz is approximately 28.3495 g, about how many ounces are in the 1-cup serving size?

____

14. If the number of servings per container is 8, about how many ounces are in the pasta package?

____

Nutrition Facts

Serving Size 1 cup (56g)
Servings per Container 8

Amount per Serving	
Calories 200	Calories from Fat 10
	%Daily Value*
Total Fat 1g	**2%**
Saturated Fat 0g	**0%**
Cholesterol 0mg	**0%**
Sodium 0mg	**0%**
Total Carbohydrate 39g	**13%**
Dietary Fiber 2g	**8%**
Sugars 1g	
Protein 7g	
Vitamin A 0% • Vitamin C	0%
Calcium 0% • Iron	10%
Thiamin 35% • Riboflavin	15%
Niacin 20%	

NAME ________________________ CLASS __________ DATE __________

Lesson Activity

10.6 State Apportionment

The Constitution provides that Representatives be apportioned among the states according to the population determined by a census taken every 10 years. Each state must have at least one Representative.

Thomas Jefferson, John Quincy Adams, and Daniel Webster found the number of Representatives by dividing a state's population by 30,000. For example, Vermont had a population of 576,000 in 1993. Using this method, $\frac{576{,}000}{30{,}000} = 19.2$. Different rules for rounding were used. Adams's method rounded 19.2 to 20; Jefferson's method dropped the decimal portion; Webster's method dropped the decimal portion if it is less than 0.5 and rounded up if it is equal to or greater than 0.5. Thus, Vermont would be entitled to 20 Representatives according to Adams's method and to 19 Representatives according to Jefferson's method and Webster's method.

In contrast, Alexander Hamilton's method used proportions. The size of the House of Representatives was first fixed and then seats were distributed, with any decimal portions being dropped. Today the method of proportions is used. Congress decides what the total number of Representatives will be.

1. Suppose that the total number of Representatives is 435 and that the population of New Jersey is 7,879,000. Calculate the number of Representatives that New Jersey is entitled to by dividing and rounding according to Jefferson, Adams, and Webster. Then use Hamilton's method of proportions for a U.S. population of 263,643,000.

____

2. Suppose that Congress used Webster's method today. Use a proportion and the population figures from Exercise 1 to determine the approximate number of representatives that would be in the House.

____

3. Political leaders had concerns about the representation of ordinary citizens as well as fears about the government becoming too large. Compare Jefferson's, Adams's, and Webster's methods of rounding, giving the advantages of each method.

____

____

____

____

____

____

NAME ______________________ CLASS __________ DATE __________

Basic Skills Practice
Cumulative Review

Read each question and circle the best answer.

1. Solve the equation $2x + 4 = x - 1$ for x.

A 1
B -1
C 3
D -5

2. What is the probability that the spinner will land on a number less than 10?

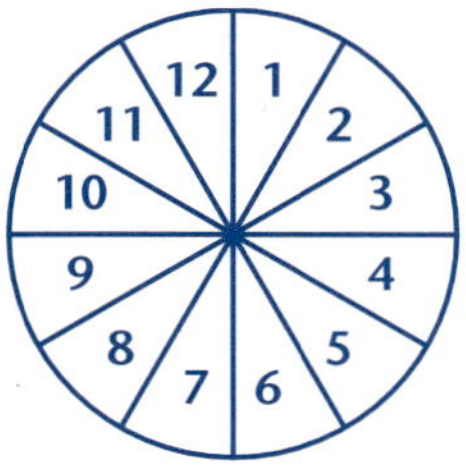

F $\frac{5}{6}$
G $\frac{1}{2}$
H $\frac{2}{3}$
I $\frac{3}{4}$

3. The ratio of girls to boys in a class is 5 to 8. If the class has 35 girls, how many boys are there?

A 13
B 20
C 56
D 64

4. An equilateral triangle has a perimeter of 36.6 centimeters. What is the length of one of its sides?

F 9.15 cm
G 12.2 cm
H 18.3 cm
I 24.4 cm

5. The temperatures for 6 consecutive days in August were 98, 100, 101, 99, 97, and 100 degrees Fahrenheit. What is the median temperature rounded to the nearest tenth?

A 98.1°F
B 99.2°F
C 99.5°F
D 100.0°F

6. What is $0.345 + 2.812$ rounded to the nearest tenth?

F 3.1
G 3.2
H 3.15
I 3.157

7. The movers charged \$35.00 to move a desk within the city limits, plus \$0.30 per mile. If the total charge was \$36.20, not including tax, which equation could be used to find m, the number of miles that the desk was moved?

A $0.30(35 + m) = 36.20$
B $35m = \frac{36.20}{0.30}$
C $35 + 0.30 = \frac{36.20}{m}$
D $35 + 0.30m = 36.20$

8. A bag contains 20 marbles. There are 8 white marbles, 7 red marbles, and 5 blue marbles. One marble is drawn from the bag at random. What is the probability it is a blue marble?

F $\frac{7}{20}$
G $\frac{3}{5}$
H $\frac{2}{5}$
I $\frac{1}{4}$

LESSON 10.7 Proof in Algebra

PREPARE

Pacing Guide
2 Days

Objectives and Overview
- Define *proof, hypothesis, conclusion,* and *converse.*
- Demonstrate a proof by using properties and definitions.

Hypothesis and conclusion are defined. Then students are shown how to develop a formal algebraic proof.

Vocabulary
theorem
proof
hypothesis
conclusion
converse

Assessing Prior Knowledge

Tell whether each of these statements is true or false.

1. If two angles are complementary, then the sum of their measures is 90°. **[true]**
2. If you live in Texas, then you live in Dallas. **[false]**
3. If triangle *ABC* is a right triangle, then it has two acute angles and one right angle. **[true]**

Approach A: Using the Book

Opening Discussion
Ask students where they have seen the word *proof* used. **[Examples might include court cases reported in newspapers, on television, or in novels; an exam gives proof of what a student knows.]**

Discuss with students what is needed to prove a statement. Point out that to accept a statement as true, we demand reasons.

Now discuss the theorem and proof at the beginning of the lesson. What reasons have been given to prove the statement?

Brainstorming
Ask students to make a list of properties of real numbers and to state the hypothesis and the conclusion for each property. Examples include the following: Distributive Property—hypothesis: a, b, and c are real numbers, conclusion: $a(b + c) = ab + ac$; Zero Product Property—

TEACH

Approach B: Using Models

Opening Discussion
Explain to your students that in a formal proof, every step in the proof must be justified. Introduce formal algebraic proofs by asking students to give the reason for each step of this solution.

If $\frac{x}{x-4} = \frac{3}{x-5}$, then $x = 2$ or $x = 6$.

$x(x - 5) = 3(x - 4)$	**[Property of cross products of proportions]**
$x^2 - 5x = 3x - 12$	**[Distributive Property]**
$x^2 - 8x + 12 = 0$	**[Addition and Subtraction Properties of Equality]**
$(x - 2)(x - 6) = 0$	**[Distributive Property]**
$x - 2 = 0$ or $x - 6 = 0$	**[Zero Product Property]**
$x = 2$ or $x = 6$	**[Addition Property of Equality]**

hypothesis: for real numbers a and b, $ab = 0$, conclusion: $a = 0$ or $b = 0$; Addition Property of Zero—hypothesis: for real numbers a, b, and c, $a = b$, conclusion: $a + c = b + c$.

Defining *Direct Proof*

The examples in this lesson use direct proofs to reach a conclusion. In a direct proof, you typically start with the hypothesis and, by a logical series of steps, reach the conclusion.

Algebraic Proofs

Students should know how to solve an inequality such as $a + 6 < 9$ to show that the solution set is the set of all real numbers less than 3. In this lesson, students learn that solving the inequality $a + 6 < 9$ proves that for all real numbers a, if $a + 6 < 9$, then *it must follow that* $a < 3$. Each step of the algebraic proof is guaranteed by a property of real numbers because 3, 6, and 9 are real numbers. Though not stated, it is assumed that a is a real number.

Journal Entry

It is important that students be able to *read* an algebraic proof in statement-reason form. This requires an understanding of how each reason supports its statement. Ask students to identify the property of algebra that justifies simplifying $a + 6 - 6 < 9 - 6$ to $a < 3$.

The Truth Value of the Converse of a Statement

Ask students if they can work *backward* from the statement $a < 3$ to the statement $a + 6 < 9$. If they can, then the hypothesis $a < 3$ supports the conclusion that $a + 6 < 9$, and the converse statement "If $a < 3$, then $a + 6 < 9$" is true. Write the following statements on the board, and ask students to state the reason for each:

$a < 3$	Hypothesis or given
$a + 6 < 3 + 6$	**[Addition Property of Inequality]**
$a + 6 < 9$	**[Simplify.]**

continued on top of next page

Ask students to identify the beginning equation and the ending equations.

Brainstorming

Since each of the properties of real numbers that the students have learned is a theorem, ask the students to make a list of several properties of real numbers. Then tell them to state the hypothesis and the conclusion for each property. Examples include the following: Distributive Property—hypothesis: a, b, and c are real numbers, conclusion: $a(b + c) = ab + ac$; Zero Product Property—hypothesis: for real numbers a and b, $ab = 0$, conclusion: $a = 0$ or $b = 0$; Addition Property of Zero—hypothesis: for real numbers a, b, and c, $a = b$, conclusion: $a + c = b + c$.

Modeling Algebraic Proofs: The Argumentative Essay

Your students should have had experience with the argumentative essay in which they were asked to take a position, either pro or con, on an argument and to support their position with documented sources. Since a proof is a kind of argument, the argumentative essay is a good model for algebraic proofs.

For example, a student might support the position that the junior prom be changed from a formal dance to a semiformal dance. He or she might reason in the following way: If the affair is changed to a semiformal dance, more students will attend. The hypothesis of the argument is "The affair is changed to a semiformal dance," and the conclusion is "More students will attend." The student may support the conclusion by citing a survey that suggests more students would attend if the dance were semiformal. The student might report the cost of producing the last few proms as well as the cost of attending a formal prom. He or she may also note that tuxedos are not popular with this year's student body. Although the reasons for reaching the conclusion won't *prove* that the conclusion is true, they should be presented in a logical manner that supports the student's position.

A direct algebraic proof also has a hypothesis and a conclusion. In a direct proof, you typically start with the hypothesis and, by a logical series

continued on bottom of next page

Approach A

Using the Book, continued

Teaching Tip

The Converse The statement "If $a < 3$, then $a + 6 < 9$" is true. However, remind students that this is not always the case, as will be seen in the Critical Thinking question and later in this section of this planning guide.

From Conjecture to Proof

In Lesson 10.6, students used specific values of a, b, c, and d to explore the relationship between $\frac{a}{b} = \frac{c}{d}$ and $\frac{a}{b} + \frac{b}{b} = \frac{c}{d} + \frac{d}{d}$. Though students were able to guess from this exercise that the proportions $\frac{a}{b} = \frac{c}{d}$ and $\frac{a}{b} + \frac{b}{b} = \frac{c}{d} + \frac{d}{d}$ are equivalent, the fact that the second proportion follows from the first was proved only for the specific values chosen by the students. A formal proof involving the variables a, b, c, and d is needed to prove the statement for all real numbers. Students should recognize that the results of the proof show that for any real number values they might choose, where $b \neq 0$ and $d \neq 0$, the two proportions will be equivalent.

Identifying Combined Steps in a Proof

When it is obvious to do so, two or more steps in a proof may be combined. For example, to say that $\frac{a}{b} + \frac{b}{b} = \frac{c}{d} + \frac{d}{d}$ follows from the statement $\frac{a}{b} + 1 = \frac{c}{d} + 1$ involves the properties $1 = b \cdot \frac{1}{b}$ (Multiplication Property of Inverses) and $b \cdot \frac{1}{b} = \frac{b}{b}$. Because these steps are understood, they can be combined in one step.

Ongoing Assessment, page 511

If $a < 3$, then $a + 6 < 9$; yes

Example 1

A proportion is given, and it is proved that the product of the means equals the product of the extremes.

TEACH

Approach B

Using Models, continued

of steps that involve established definitions, axioms, postulates, or previously proven theorems, reach the conclusion of a theorem.

Writing Proofs

Cooperative Learning Write the following inequality on the chalkboard or overhead:

$$a + 6 < 9$$

Ask students to work with a partner to solve this inequality. Tell the students to identify the reason that supports each step of the solution.

When students are done, ask them to complete this statement: If $a + 6 < 9$, then a___. [<3] Tell the students that they have just completed their first algebraic proof. Ask the students to identify the hypothesis and the conclusion of their theorem. (Compare page 511 of the Pupil's Edition.)

Inclusion Strategy

Determining the Converse of a Statement Use the example of the argumentative essay. Suppose that attendance increased by 10% at this year's junior prom. Can you assume that the prom has been changed from a formal dance to a semiformal dance? In the statement "If attendance at the junior prom increases, then the prom changed from a formal dance to a semiformal dance," the hypothesis and conclusion of the original statement have been interchanged, forming the *converse* of the original statement.

Ask each student to write two if-then statements. Then tell them to write the converse of each statement. Have students exchange papers with a partner and determine if their partner's converse statements are true or false.

Proving a Conjecture

Conjectures are often made and tested in a science laboratory. For example, some botanists are interested in finding ways to develop larger and more vigorous plants. Laboratory tests have

Ongoing Assessment, page 512

Try This

$a + c = b + c$	Given
$a + c - c = b + c - c$	Addition Property of Equality
$a + 0 = b + 0$	Definition of opposites
$a = b$	Addition Property of Zero

Additional Example 1

Assume that you know the properties involving the addition of real numbers, but that you do not know the Multiplication Property of Zero. Prove the following statement: If $0 + a = a$, then $5 \cdot 0 = 0$.

$5 \cdot 5 = 0 + (5 \cdot 5)$	Hypothesis
$5 = 0 + 5$	Hypothesis
$5 \cdot (0 + 5) = 0 + (5 \cdot 5)$	Substitution
$(5 \cdot 0) + (5 \cdot 5) = 0 + (5 \cdot 5)$	Distributive Property
$5 \cdot 0 = 0$	Subtraction Property of Equality

Example 2

In this example, it is proved that $2k + 2$ is an even number if k is an integer.

Ongoing Assessment, page 513

Try This

k is an integer.	Hypothesis
$2k + 3 = (2k + 2) + 1$	Addition
$2k + 2 = 2(k + 1)$	Distributive Property
$2k + 3 = 2(k + 1) + 1$	Substitution
$k + 1$ is an integer.	Definition of integer
$2k + 3$ is even	Definition of even numbers

continued on top of next page

shown that a substance derived from the crocus flower, called colchicine, produced larger flowers in some plants. It was finally shown that colchicine could double the number of chromosomes in a plant, thus producing larger and more vigorous plants. Simply showing that a number of plants treated with colchicine produced large flowers did not prove that colchicine had this property, but proving the effects of colchicine on chromosome numbers did prove its ability to produce larger flowers on some plants.

In Lesson 10.6, students evaluated $\frac{a}{b} = \frac{c}{d}$ and $\frac{a+b}{b} = \frac{c+d}{d}$ for several values of a, b, c, and d. Students should recall that the results suggested that if $\frac{a}{b} = \frac{c}{d}$, then $\frac{a+b}{b} = \frac{c+d}{d}$. However, they did not prove this conjecture.

To prove "If $\frac{a}{b} = \frac{c}{d}$, then $\frac{a+b}{b} = \frac{c+d}{d}$," the statement must be proved for all real numbers a, b, c, and d, where $b \neq 0$ and $d \neq 0$. This conjecture can be proved for all real numbers with a formal proof.

Hypothesis: $\frac{a}{b} = \frac{c}{d}$ Conclusion: $\frac{a+b}{b} = \frac{c+d}{d}$

Cooperative Learning Have students work with a partner to develop a proof. (Compare page 512 of the Pupil's Edition.)

Note that to obtain the conclusion requires adding 1 in the form of $\frac{b}{b}$ and $\frac{d}{d}$ to each side of the equation.

Proofs of Statements about Specific Numbers

The laboratory can also be used to model algebraic proofs of statements about specific numbers. For example, if it is believed that monounsaturated fat in processed oils helps reduce blood cholesterol levels, laboratory tests can be performed to test the effect of a *specific* oil such as olive oil on blood cholesterol levels. Similarly, statements about specific numbers can be proved algebraically.

continued on bottom of next page

Approach A

Using the Book, continued

Additional Example 2

Prove that $2k^2 + 2$ is even if k is an integer.

[k is an integer.	**Hypothesis**
k^2 is an integer.	**Definition of integer**
$k^2 + 1$ is an integer.	**Definition of integer**
Let $n = k^2 + 1$.	
$2n$ is even.	**Definition of even number**
$2(k^2 + 1)$ is even.	**Substitution (of $k^2 + 1$ for n)**
$2k^2 + 2$ is even.	**Distributive Property]**

The Definition of Odd Numbers

Students tend to have more trouble with the definition of odd numbers than they do with the definition of even numbers. They easily associate the definition of even numbers with the multiplication table for 2. Since we count integers by 1s and alternating integers are even and odd, counting one more from each even integer gives the next integer, which is an odd integer. Thus, the expression for an odd integer is 1 more than the expression for an even integer, or $2n + 1$.

Example 3

In this example, it is proved that the sum of two odd numbers is an even number.

When general statements such as "If two numbers are odd, then their sum is even" are proved, the properties for real numbers are used. Since these properties are expressed in terms of variables, it helps to translate the statement to be proved into mathematical symbols. Thus, the statement "If two numbers are odd, then their sum is even" is translated into "If a and b are odd, then $a + b$ is even."

Use **Transparency 50** for discussion or review.

TEACH

Approach B

Using Models, continued

Assume that you know all the rules for operating with integers except the rule for multiplying two negatives. Prove the following statement about -3 and -2: If $-3(2) = -6$, then $-3(-2) = 6$.

Proof

$-3(0) = 0$	Zero Property
$-2 + 2 = 0$	Definition of opposites
$-3(-2 + 2) = 0$	Substitution
$-3(-2) + (-3)(2) = 0$	Distributive Property
$-3(2) = -6$	Hypothesis
$-3(-2) + (-6) = 0$	Substitution
$-3(-2) + (-6) + 6 = 0 + 6$	Addition Property of Equality
$-3(-2) + 0 = 0 + 6$	Substitution
$-6 + 6 = 0$	Definition of opposites
$-3(-2) = 6$	Addition Property of Zero

A Model for Even and Odd Integers

Recall the multiplication table for 2.

$2 \times 1 = 2$
$2 \times 2 = 4$
$2 \times 3 = 6$
$2 \times 4 = 8$
$2 \times 5 = 10$
.
.
.

Each product is an even number and is the product of 2 and an integer. Thus, the expression $2n$, where n is an integer, models an even number.

Every *other* number in the set of integers is odd.

0 [1] 2 [3] 4 [5] 6 [7] 8 [9] 10 . . .

Adding 1 produces an odd integer after every even integer. Thus, the expression $2n + 1$, where n is an integer, models an odd number.

Additional Example 3

Prove that the sum of three odd numbers is odd.

[*a*, *b*, and *c* are three odd numbers.	**Hypothesis or given**
Let $a = 2k + 1$, $b = 2m + 1$, and $c = 2n + 1$.	**Definition of odd numbers (k, m, and n are integers.)**
$a + b + c$	
$= (2k + 1) + (2m + 1) + (2n + 1)$	**Addition of real numbers**
$= 2k + 2m + 2n + 2 + 1$	**Simplify**
$= 2(k + m + n + 1) + 1$	**Distributive Property**

$2(k + m + n + 1) + 1$ is in the form $2n + 1$, where n is equal to $k + m + n + 1$ and $k + m + n + 1$ is an integer. Thus, $2(k + m + n + 1) + 1$ is an *odd* number, and the theorem is proved.]

The Converse of a True Statement Might Not Be True

Have students consider the converse of the statement from Example 3, "If the sum of two numbers is even, then the two numbers are odd." This statement is certainly false. Have the class brainstorm to find examples. **[Sample Answer: 8 = 2 + 6, 8 is even, but 2 and 6 are also even.]**

Critical Thinking, page 513

The converse statement would be "If I live in the United States, then I live in Texas." The converse statement is not always true. Some Americans live in Texas, but many live in other states.

Closing the Lesson

Ask students to define *proof, hypothesis, conclusion,* and *converse.*

Paragraph Format of a Proof

Just as the student arguing his or her position for a semiformal dance can do so in an essay, algebraic proofs are sometimes suited to a paragraph format. For example, it can be proved that 2 more than twice an even integer is an even integer.

It is helpful to put the statement in if-then form and to write the statement with symbols.

Cooperative Learning Have students work with a partner to develop proofs of the theorems in Example 2 and 3 of the Pupil's Edition.

The proof from Example 2 in paragraph form is as follows:

Proof

By the Distributive Property, $2k + 2 = 2(k + 1)$. Since k is an integer, $k + 1$ is also an integer. If $n = k + 1$, then $2k + 2$ can be written in the form $2n$, which is an even integer. (Compare page 513.)

Modeling the Converse of a Statement With a Venn Diagram

Make a Venn diagram with the class to model the Critical Thinking problem on page 513 of the Pupil's Edition. It will consist of a small circle, representing people who live in the state of Texas, within a larger circle, representing people who live in *any* state of the United States. Explain that the converse is false because there are people who live in states other than Texas.

Have students find other examples of true statements, and make a Venn diagram to decide if the converse is true or false. Example: If I am a dog owner, then I am a pet owner. **[The statement is true, but its converse is false.]**

Closing the Lesson

Ask students to define *proof, hypothesis, conclusion,* and *converse.*

ASSESS

Selected Answers

Odd-numbered Exercises 7–41 and odd-numbered Extra Practice

Assignment Guide

Core level students should complete Exercises 1– 42.

Core-plus level students should complete Exercises 1–16, 19–26, and 30–43.

Error Analysis

Many students have trouble getting started on a proof. Tell these students to compare the hypothesis of the statement to be proved with the conclusion. They should look for a general path that will allow them to transform the hypothesis into the conclusion.

Alternative Assessment

Calculation and Proof Ask students to compare solving the equation $3(x + 7) = 15$ with proving the statement "If $3(x + 7) = 15$, then $x = -2$."

Portfolio Activity Identify the hypothesis and the conclusion of the following statement. Then prove the statement.

For all a, $(a + 4)(a - 4) = a^2 - 16$

RETEACH

Understanding Paragraph Proofs
Students may find the more formally structured proof of the theorems involving even and odd numbers on page 513 easier to understand. Because the idea of a proof is often intimidating to students, you can suggest that students work in groups to help each other convert the paragraph formats of either of these proofs to a statement-reason format. This activity will give students an introduction to writing formal proofs in a nonchallenging situation and will also help them explore the meaning of these theorems.

For example, the proof of the statement "If k is an integer, then $2k + 2$ is even" might look like this:

Proof

$2k + 2 = 2(k + 1)$ Distributive Property

$k + 1$ is an integer. k is an integer. 1 added to any integer gives the next integer.

Let $n = k + 1$. Substitution

$2k + 2 = 2n$ Substitution

$2k + 2$ is even. Definition of even number

ENRICH

Interdisciplinary Connection

History of Science Ptolemy was one of history's most influential astronomers. Working in the second century, he elaborated on Aristotle's proposal of a geocentric solar system, that is, a solar system whose center is the Earth. Ptolemy's description of the solar system was accepted until Copernicus proposed the sun as the center of the solar system in 1543. In one of his works, Ptolemy offered a map of the known world along with latitudes and longitudes of important cities of the time. His work also included a description of techniques used in map making. Ptolemy used geometric models to predict the positions of the planets. A theorem often called Ptolemy's Theorem states that if a quadrilateral is inscribed in a circle, then the product of the diagonals of the quadrilateral is equal to the sum of the products of both pairs of opposite sides of the quadrilateral.

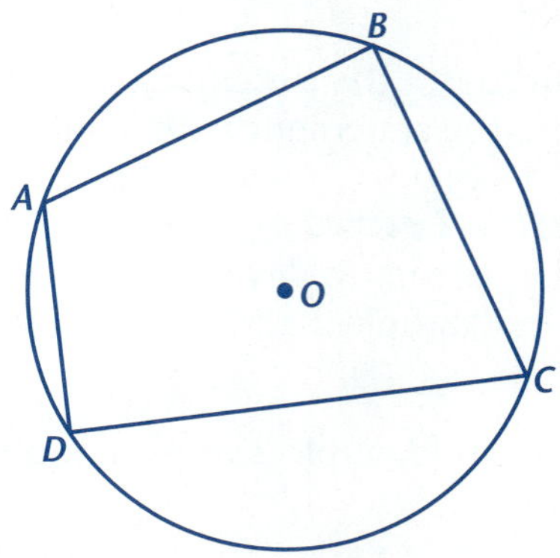

Stated in terms of this figure, the hypothesis of Ptolemy's Theorem is "$ABCD$ is inscribed in circle O." The conclusion is "$AC \cdot BD = AB \cdot DC + AD \cdot BC$."

NAME ______________________ CLASS ____________ DATE ____________

Practice

10.7 Proof in Algebra

Give a reason for each step.

Proof	
$3x + 2(x + 4) = 13$	Hypothesis or given
1. $3x + 2x + 8 = 13$	______________
2. $5x + 8 = 13$	______________
3. $5x + 8 - 8 = 13 - 8$	______________
4. $5x + 0 = 5$	______________
5. $5x = 5$	______________
6. $\frac{5x}{5} = \frac{5}{5}$	______________
7. $x = 1$	______________

Give a reason for each step in the proof that $\frac{-a}{b} = -\frac{a}{b}$.

Proof

8. $\frac{-a}{b} = -a\left(\frac{1}{b}\right)$ ______________

9. $= -\left(a \cdot \frac{1}{b}\right)$ ______________

10. $= -\frac{a}{b}$ ______________

Give a reason for each step in the proof that $-a(b + c) = -ab - ac$.

Proof

11. $-a(b + c) = -ab + -ac$ ______________

12. $= -ab - ac$ ______________

Prove each of the following. Give a reason for each step on a separate piece of paper.

13. If $-6x - 2 = -8x - 4$, then $x = -1$.

14. If $3(n - 2) = 9$, then $n = 5$.

NAME ______________________ CLASS __________ DATE __________

Lesson Activity
10.7 Proofs

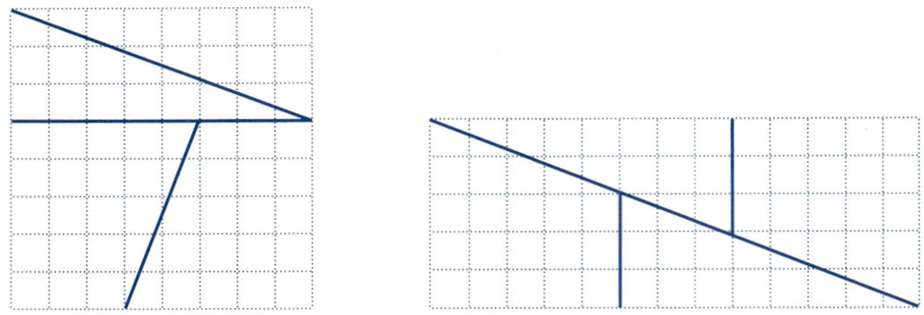

1. Sometimes what you see can lead to a false conclusion. For example, if you cut out the shapes drawn on the 8×8 square and rearrange them to form the 13×5 rectangle, it seems that the area increases from 64 square units to 65 square units. Explain the difference in areas.

____

A proof provides a logical justification which shows that a conclusion is true.

Use the labels *a*, *a* − *b*, and *b* to write a formula for the area of the square in terms of regions I, II, III, and IV.

2. Find the area of each region in terms of a and b.

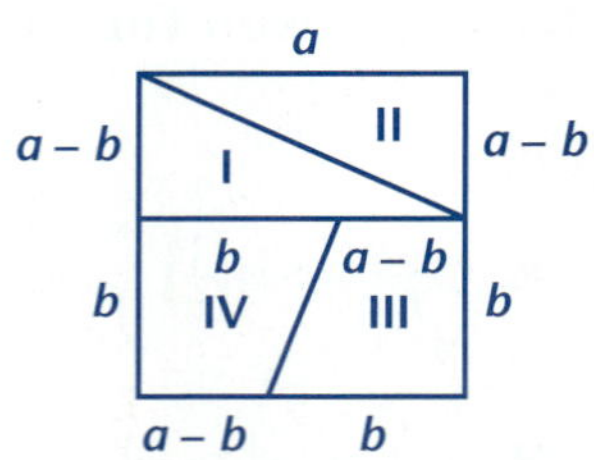

____

____

____

____

3. Find the total area of regions I, II, III, and IV by adding the expressions found in Exercise 2.

____

4. Compare the total area of the regions with the area of the square.

____

5. Use the methods in Exercises 2 and 3 and the labels a, $a + b$, and b to write a formula for the area of the rectangle. Compare the areas of the square and the rectangle.

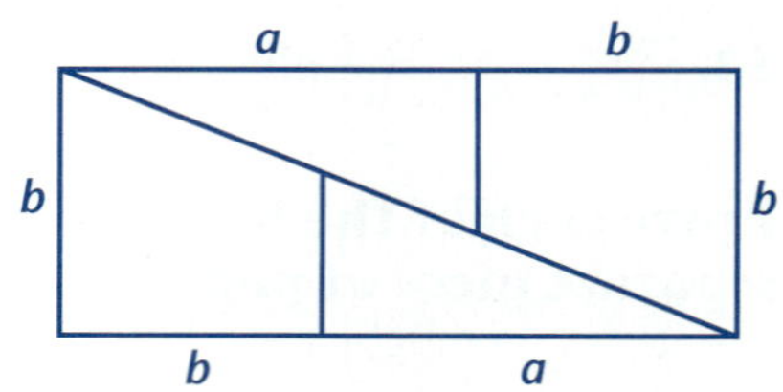

____

6. Will the regions of the square completely cover the rectangle?

____

NAME ______________________ CLASS ____________ DATE ____________

Basic Skills Practice

Cumulative Review

Read each question and circle the letter for the best answer.

1. The volume of a cube-shaped package is 27 cubic centimeters. What is the length of one side?

A 2 cm
B 3 cm
C 4 cm
D 5 cm

2. What is $2 \times 4 + 4 - 8 \div 2 + 3^2$?

F 11
G 17
H 18
I 20

3. A recipe calls for $3\frac{1}{4}$ cups of flour. The cook has only a $\frac{1}{2}$-cup measuring cup. How many times should he fill the measuring cup to obtain the amount of flour he needs?

A $3\frac{1}{4}$ times
B $5\frac{1}{4}$ times
C $5\frac{1}{2}$ times
D $6\frac{1}{2}$ times

4. Janie had $40 to spend at a craft store. She bought a wicker basket for $10.99 and 2 frames for $5.99 each. Including the sales tax of $1.84, what was the total amount she paid for these items?

F $15.19
G $18.82
H $22.97
I $24.81

5. At a local high school, 5 out of 7 students leave campus for lunch. The approximate percent of students that leave campus for lunch is—

A 7%
B 70%
C 71%
D 74%

6. Pat bought a mountain bike for $119 two years ago. The bike is now worth $75. What is the approximate percent decrease in value?

F 59%
G 45%
H 44%
I 37%

7. Solve the equation $2x - 5 = 6x + 3$ for x.

A −32
B −2
C 2
D 32

8. Find 20 − 11.9125 rounded to the nearest hundredth.

F 8.08
G 8.09
H 8.088
I 8.0875

9. Albert has 5 shirts and 3 pairs of slacks. How many different outfits consisting of 1 shirt and 1 pair of slacks can he make?

A 15
B 10
C 9
D 8

Practice—Chapter 1

Lesson 1.1

1. 12 **2.** 45 **3.** −22 **4.** 33 **5.** −21

6. −30 **7.** 54 **8.** 36 **9.** −21 **10.** 37

11. −31 **12.** −4.9 **13.** 16.1 **14.** 2.6

15. 11.4 **16.** −8.8 **17.** 2.3 **18.** 6.1

19. −17 **20.** 2.6 **21.** 14 **22.** −58

23. −9 **24.** 0.2 **25.** 10.7 **26.** 20.1

27. 13.5 **28.** 99 **29.** 3 **30.** 5.4

31. −17 **32.** −6.6 **33.** −55 **34.** −6

35. 82.5 **36.** −56 **37.** −6.6 **38.** −33.9

39. 0 **40.** 6.4 **41.** 93 **42.** 33 **43.** −3.8

44. −18.6 **45.** −40 **46.** −5

Lesson 1.2

1. 8 **2.** −0.5 **3.** 16 **4.** 5 **5.** −0.4

6. 5 **7.** 40 **8.** −30 **9.** 72 **10.** −27

11. 20 **12.** −3.6 **13.** 30 **14.** −1.28

15. 17.5 **16.** −6 **17.** −0.2 **18.** −30

19. −12.5 **20.** 3 **21.** −2 **22.** −18

23. $-\frac{2}{15}$ **24.** −0.9375 **25.** 6 **26.** −4

27. −1.4 **28.** 16 **29.** −4.5 **30.** −1.8

31. −28 **32.** 3 **33.** $1\frac{2}{3}$ **34.** −10

35. 2.25 **36.** −15 **37.** 12 **38.** −3.6

39. 2 **40.** 24 **41.** 9.6 **42.** 0.5

Lesson 1.3

1. 2 **2.** 3 **3.** 4 **4.** 3 **5.** 2 **6.** −3

7. −3 **8.** 2 **9.** 8 **10.** −0.2 **11.** 6

12. −9 **13.** 3 **14.** −4.2 **15.** 5 **16.** 4

17. $\frac{2}{3}$ **18.** 10 **19.** 1.2 **20.** 8 **21.** 15

22. 2.5 **23.** 0.3 **24.** 8.875 **25.** 6 **26.** 8

27. 24 **28.** 42 **29.** 1.2 **30.** 9 **31.** 6

32. 5 **33.** 0.25 **34.** 0.25 **35.** 30.6

36. 30 **37.** 9 **38.** 8 **39.** 9 **40.** 5

Lesson 1.4

1. 7 **2.** 4 **3.** 6 **4.** 22 **5.** −7 **6.** 2

7. 0.8, or $\frac{4}{5}$ **8.** 3 **9.** −14 **10.** −5 **11.** 7

12. 7 **13.** 8 **14.** −11 **15.** 3 **16.** $-5\frac{1}{3}$

17. 2 **18.** −22 **19.** −2 **20.** −2

21. −4 **22.** $2\frac{1}{3}$ **23.** $2\frac{2}{3}$ **24.** 5 **25.** 8

26. −2 **27.** 2 **28.** 34 **29.** 0.75, or $\frac{3}{4}$

30. 1 **31.** 4 **32.** 10 **33.** −10 **34.** −3.5

35. −76 **36.** 3.45 **37.** 1.35, or $\frac{27}{20}$

38. −0.25, or $-\frac{1}{4}$ **39.** 0.71875 **40.** −9

41. 8.75 **42.** 4 **43.** 0.6 **44.** 2.2

45. 15.4 **46.** 2

Lesson 1.5

1. $x > 2$

2. $x \leq -5$

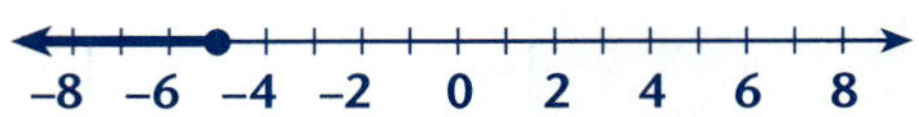

3. $t < 7$

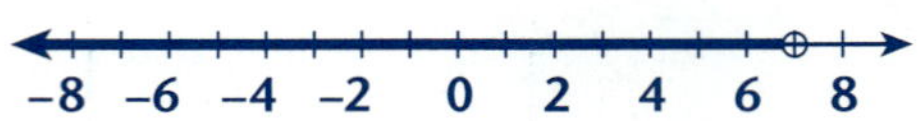

4. $m \leq 6$

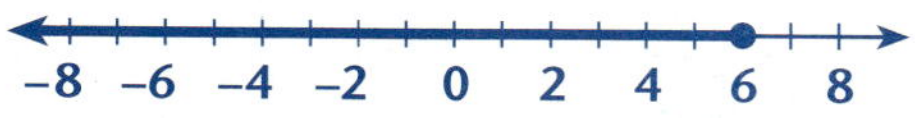

5. $x > -2$

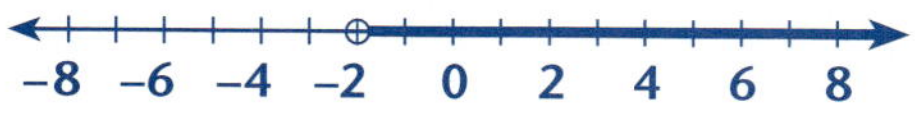

6. $x \leq -1$

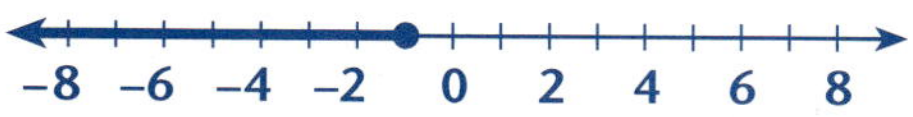

7. $x \geq 4$

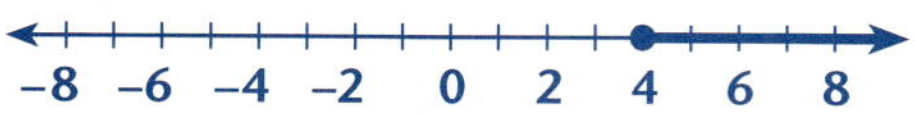

8. $b < 6$

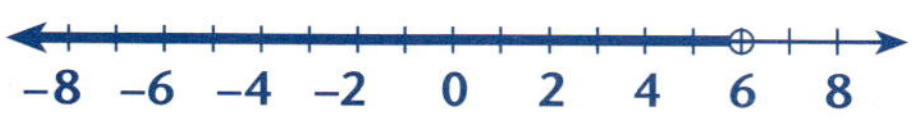

9. $k \geq 5$

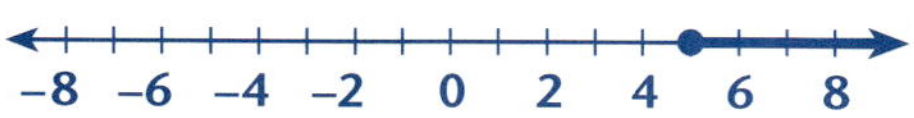

10. $x \leq 2$

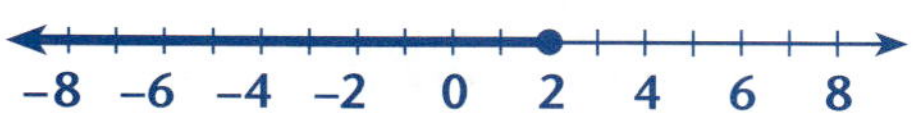

11. $x < -2\frac{1}{3}$

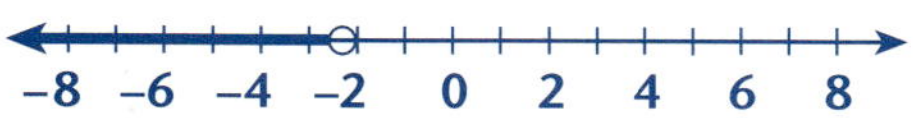

12. $x \geq 0.25$

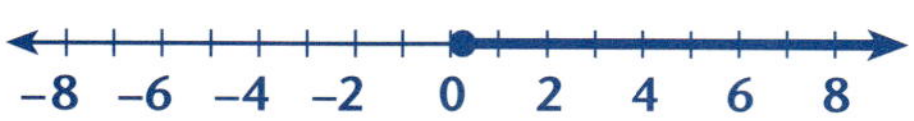

Lesson 1.6

1. $-8, 2$ **2.** $-10, 4$ **3.** $5, 7$ **4.** $-0.5, 2$

5. $-2\frac{2}{3}, 2$ **6.** $-3\frac{2}{3}, 2$ **7.** $-6.4, 8.8$

8. $0.6, 1.1$ **9.** $x \leq -4$ or $x \geq 8$

10. $-13 < x < -3$ **11.** $x < -5.5$ or $x > 1.5$

12. $x \leq 0.5$ or $x \geq 3$ **13.** $x < -34$ or $x > 30$

14. $-2.3 < x < 11.3$

15. $|x - 5| = 2$
$x = 3$ or $x = 7$

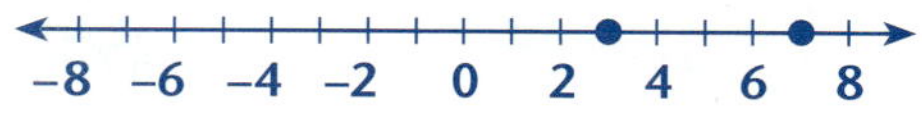

16. $|x + 3| > 4$
$x < -7$ or $x > 1$

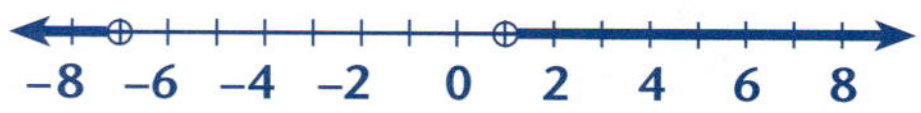

17. $|x - 7| = 1.5$
$x = 5.5$ or $x = 8.5$

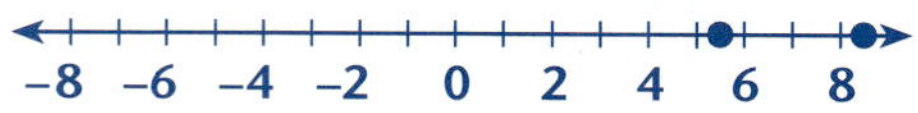

18. $|x - 1| < 6$
$-5 < x < 7$

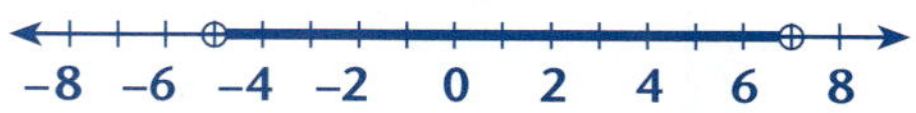

19. $|x + 2| > 2.5$
$x < -4.5$ or $x > 0.5$

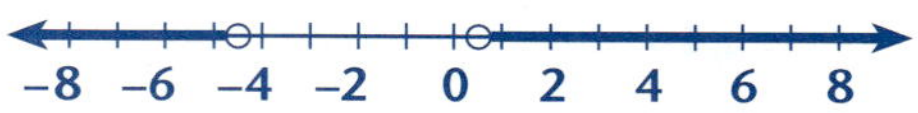

20. $|x + 3| \geq 2$
$x \leq -5$ or $x \geq -1$

Practice—Chapter 2

Lesson 2.1

1. -1 **2.** $-\frac{1}{2}$ **3.** $\frac{3}{4}$ **4.** $-\frac{1}{8}$ **5.** $\frac{1}{6}$

6. -1 **7.** $\frac{1}{11}$ **8.** $-\frac{1}{6}$ **9.** 1

10.

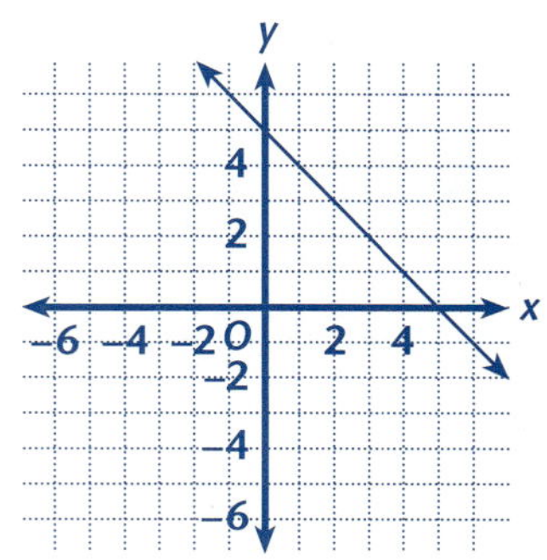

11.

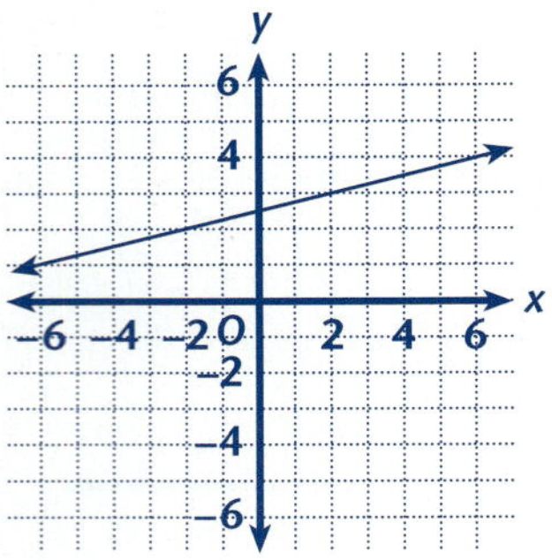

12.

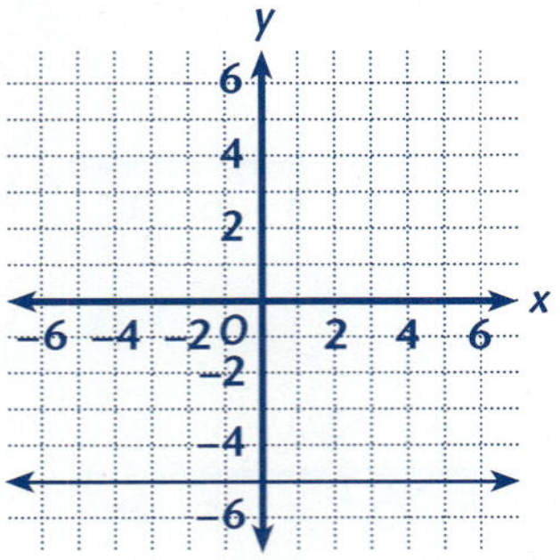

13.

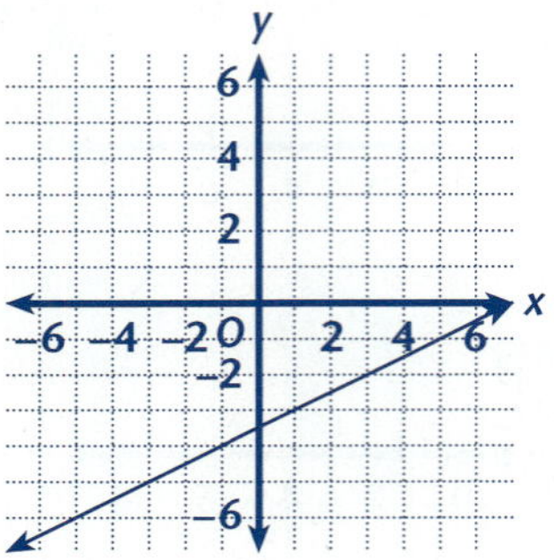

Lesson 2.2

1.

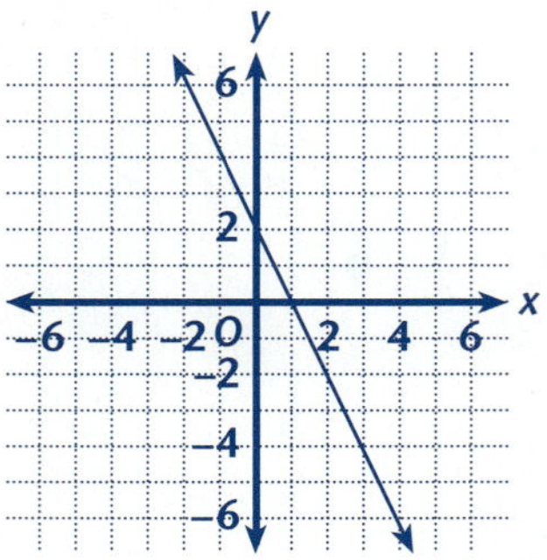

2.

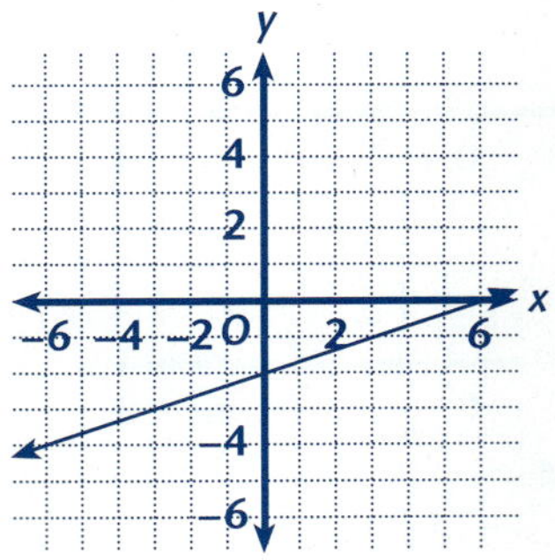

3.

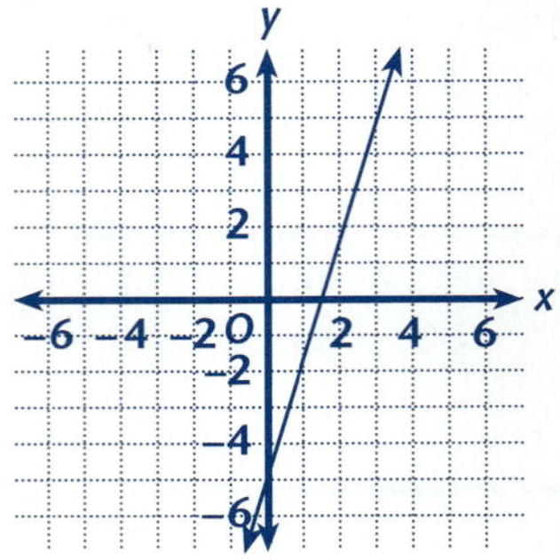

4. $y = 2x - 1; 2$

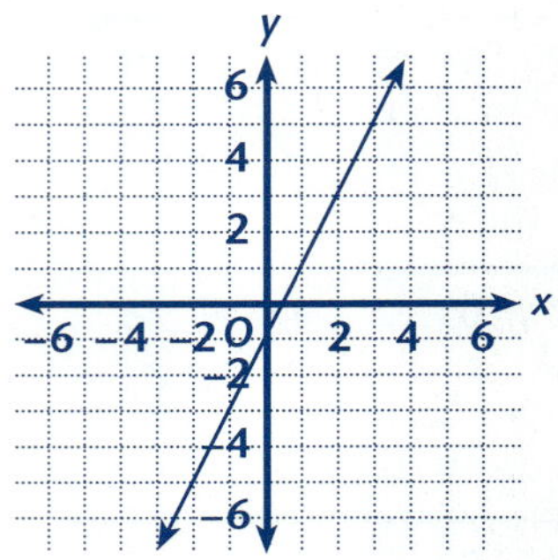

5. $y = \frac{1}{4}x - \frac{1}{2}; \frac{1}{4}$

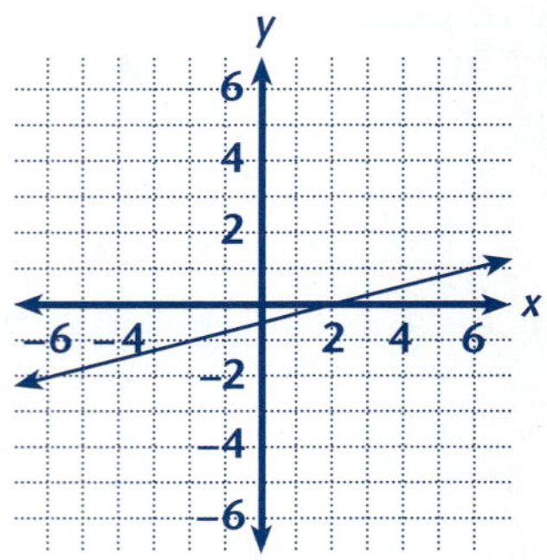

6. $y = \frac{1}{2}x - 1; \frac{1}{2}$

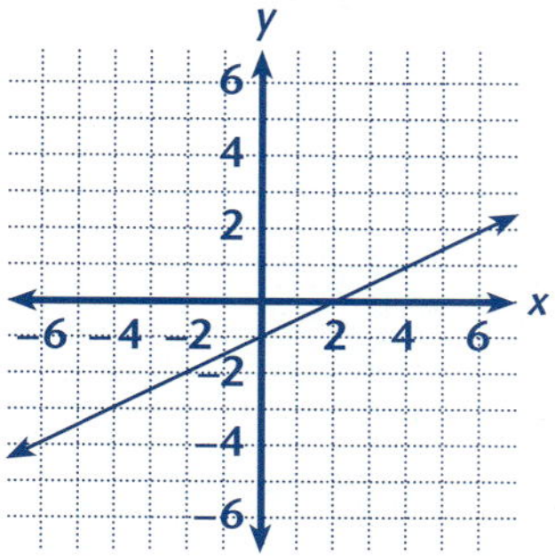

7. $y = 2x - 1$, or $2x - y - 1 = 0$

8. $y = 1$

9. $y = \frac{1}{2}x - \frac{5}{2}$, or $x - 2y - 5 = 0$

10. $x = 3$

11. $y = \frac{2}{3}x + \frac{14}{3}$, or $2x - 3y + 14 = 0$

12. $y = -\frac{1}{4}x + 1$, or $x + 4y - 4 = 0$

Lesson 2.3

1. (3, 1)

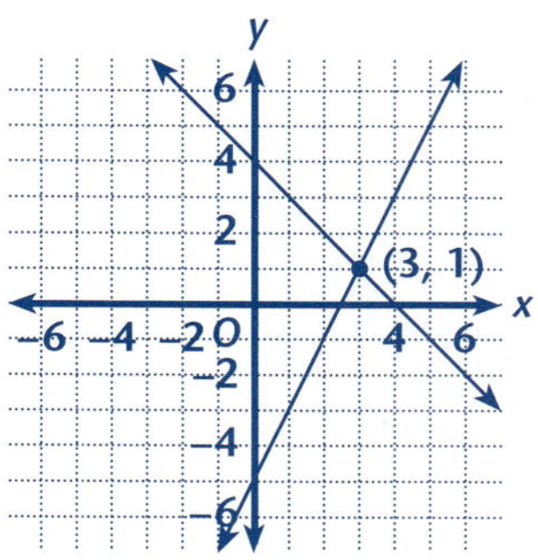

2. (2, −2)

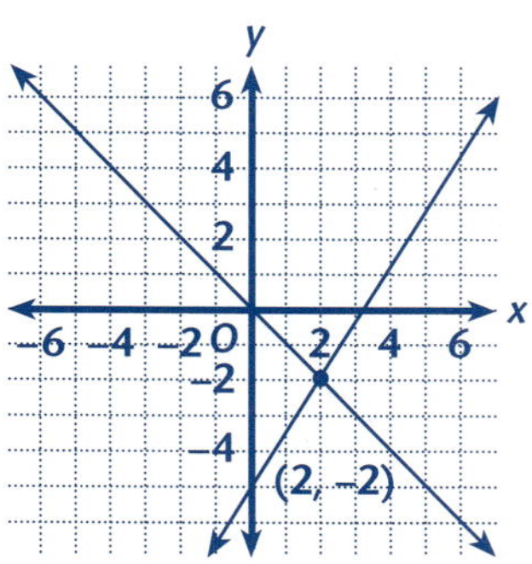

3. (4, −1)

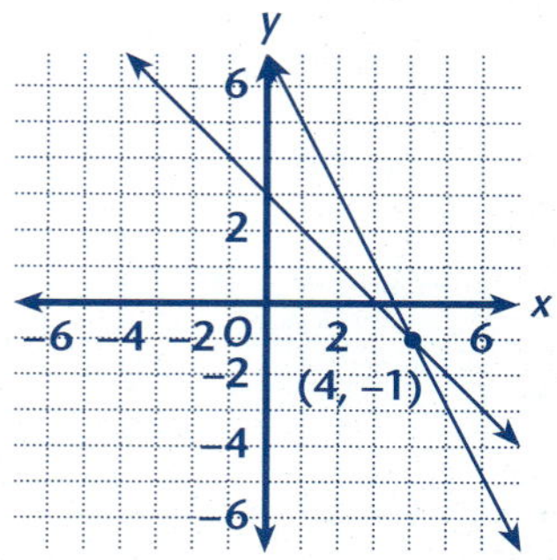

4. (2, −1)

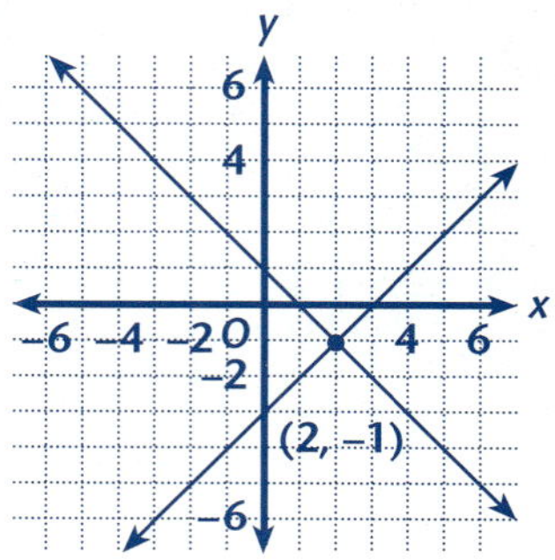

5. (1, 3)

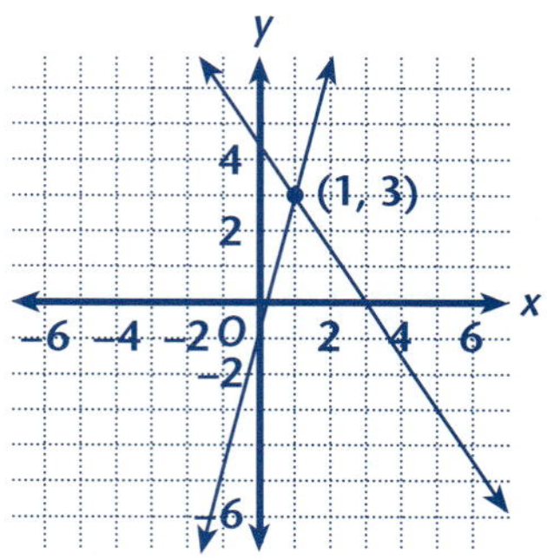

6. $\left(\frac{2}{3}, 1\frac{1}{2}\right)$

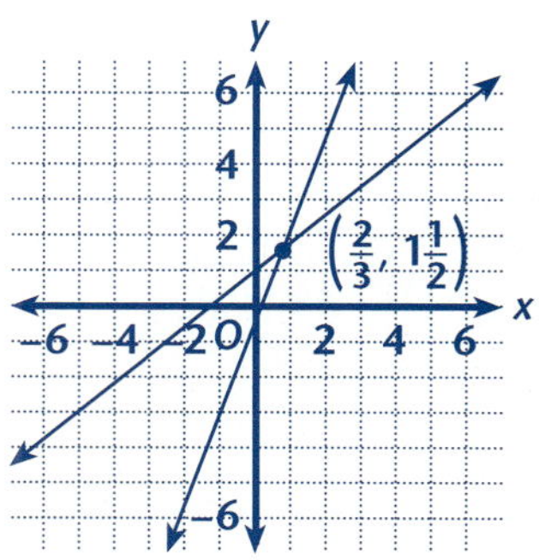

7. no 8. yes 9. yes 10. no 11. no

12. yes

Practice Answers

Lesson 2.4

1. (−3, −6) **2.** (2, 5) **3.** (4, 9)

4. (4, −1) **5.** (−2, 7) **6.** (5, 4)

7. (−1, −1) **8.** (−8, −5) **9.** (6, 2)

10. (0.2, −3.6) **11.** (7, −3) **12.** (0, 2)

13. (−3.6, 0.6) **14.** (−1, −2)

15. (−0.25, 1) **16.** (−2, −3)

17.

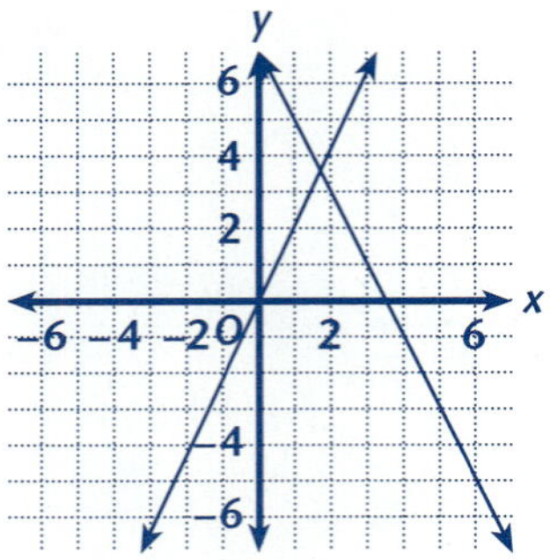

$\left(\frac{7}{4}, \frac{7}{2}\right)$

18.

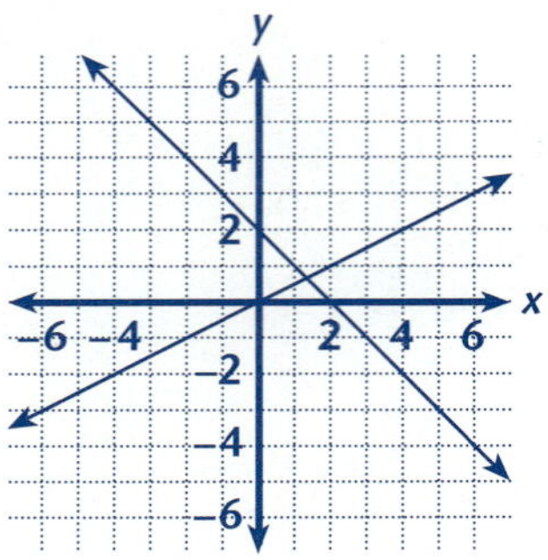

$\left(\frac{4}{3}, \frac{2}{3}\right)$

Lesson 2.5

1. (6, 4) **2.** (−7, 2) **3.** (2, −3)

4. (−1, −4) **5.** (6, −5) **6.** (−3, 1)

7. (−2, 7) **8.** (0, −1) **9.** (−2, 3)

10. (3, 5) **11.** (6, 2) **12.** $\left(-10\frac{2}{3}, -\frac{7}{15}\right)$

13. (4, −4) **14.** (1, −5) **15.** (1, −5)

16. (−0.5, 5) **17.** (−1, −1) **18.** (−6, −2)

19. (5, 1) **20.** (0, −4) **21.** (4, 6)

22. (6, −8) **23.** (2, 4) **24.** (10, −5)

Lesson 2.6

1. infinite solutions **2.** no solution

3. (2, 3) **4.** no solution

5. infinite solutions **6.** no solution

7. infinite solutions **8.** infinite solutions

9. no solutions **10.** (1.5, 0)

11. dependent **12.** inconsistent

13. independent **14.** dependent

15. inconsistent **16.** independent

17. independent **18.** inconsistent

19. independent **20.** independent

21. dependent **22.** independent

23. independent **24.** dependent

Lesson 2.7

1–12. Points chosen may vary.

1.

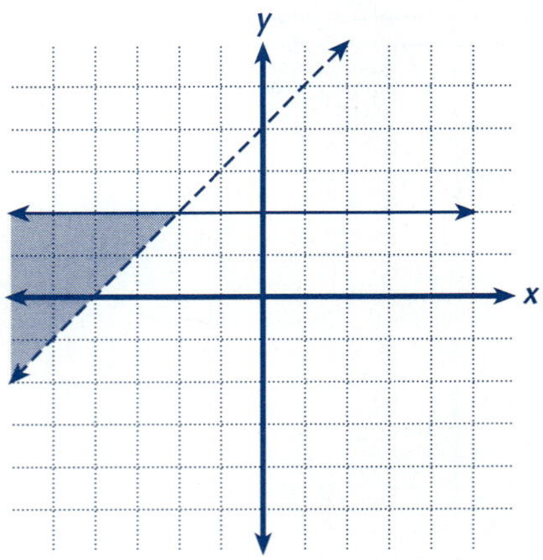

2.

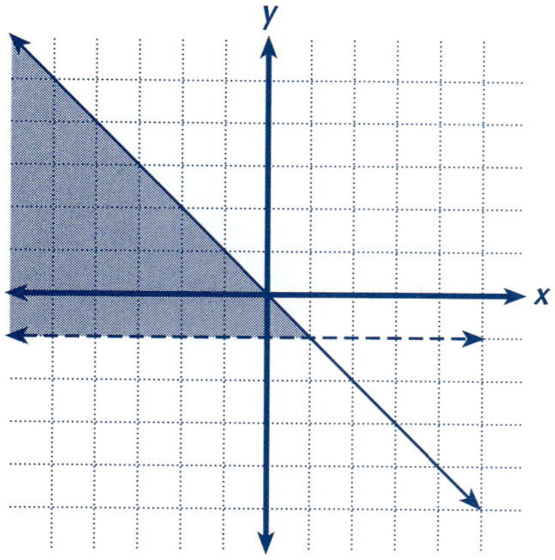

3.

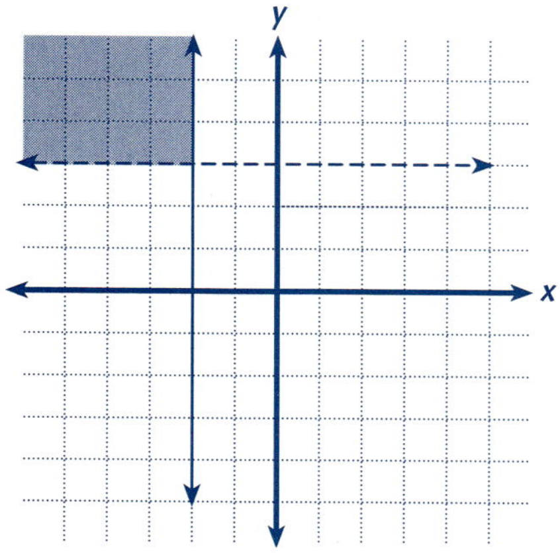

4.

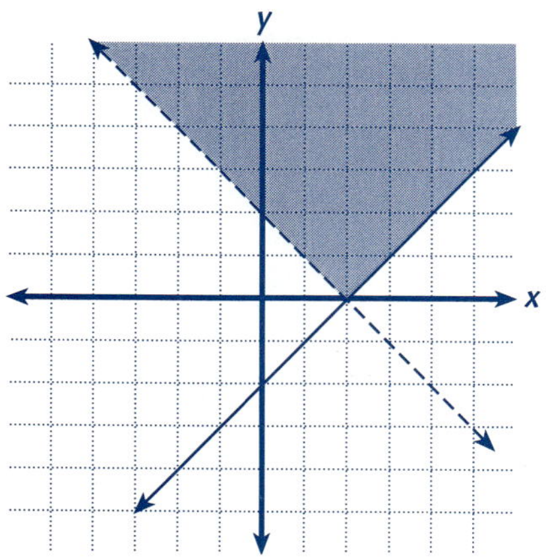

5.

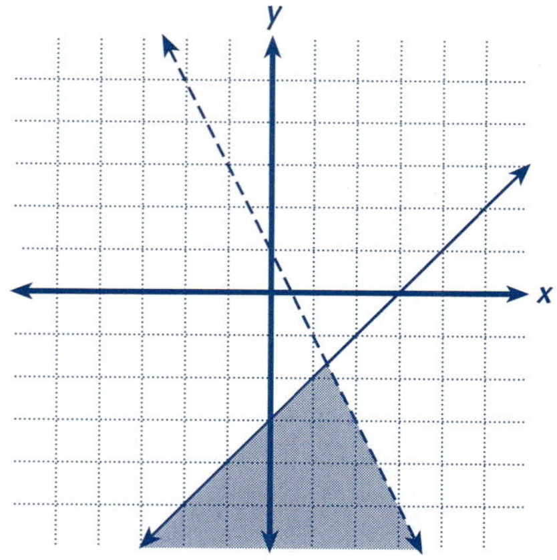

6.

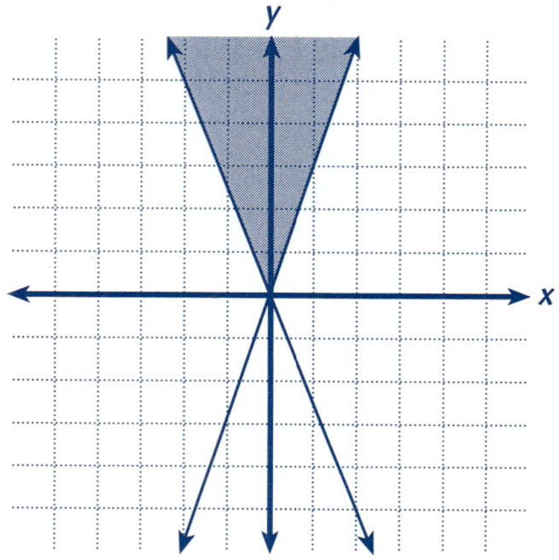

7.

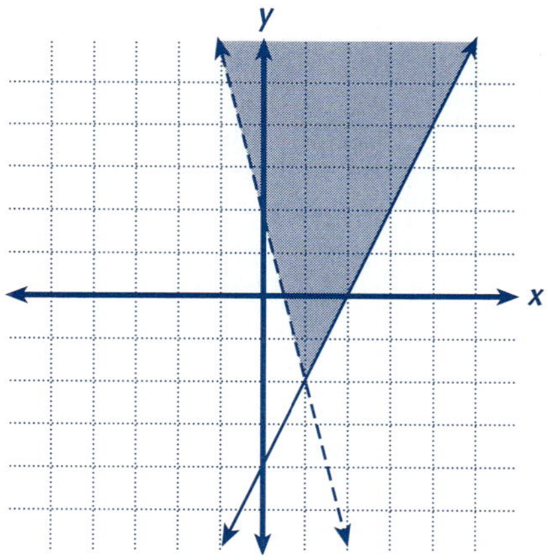

8.

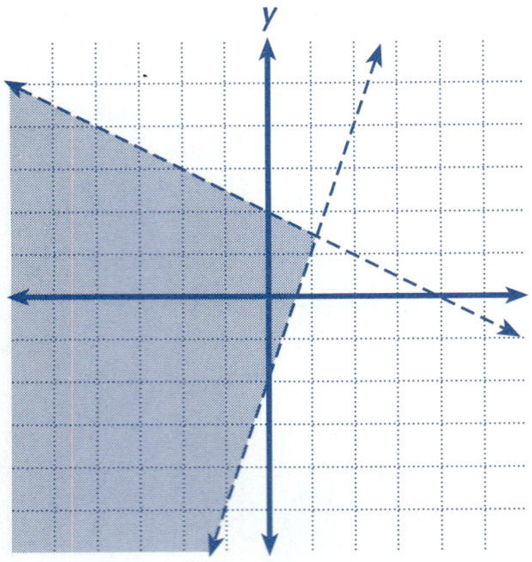

9.

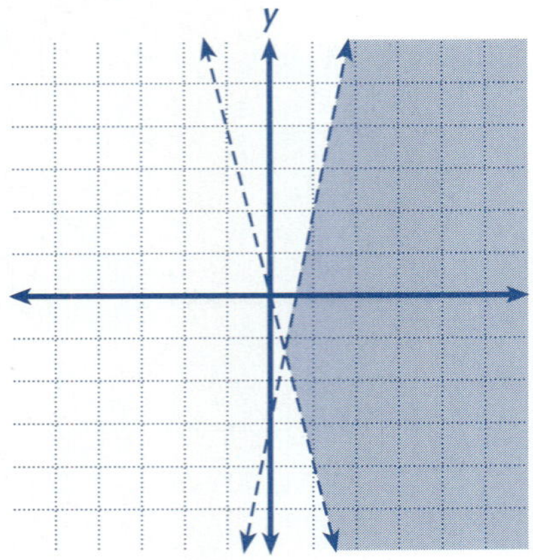

10.

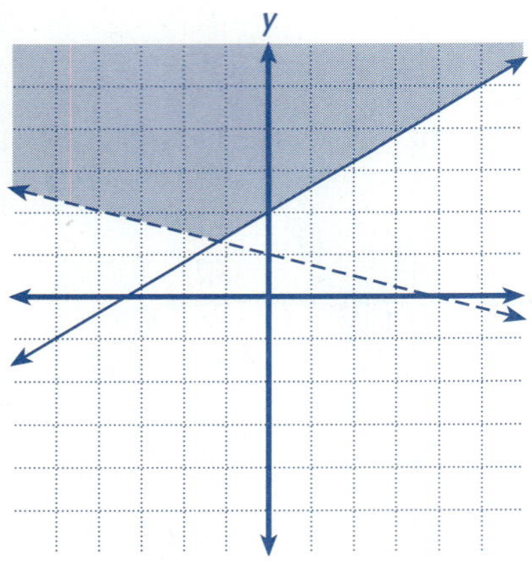

11.

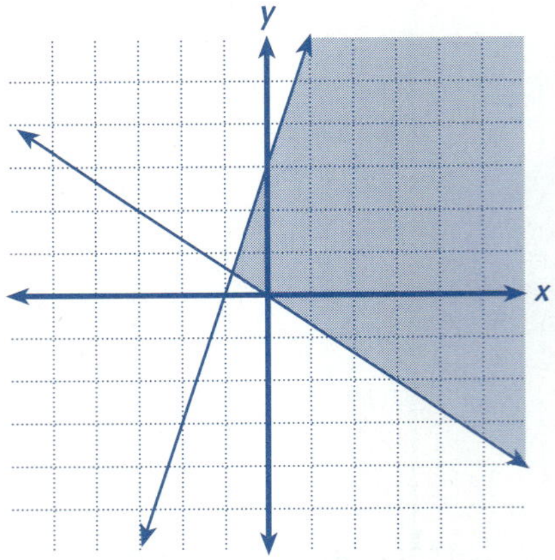

12.

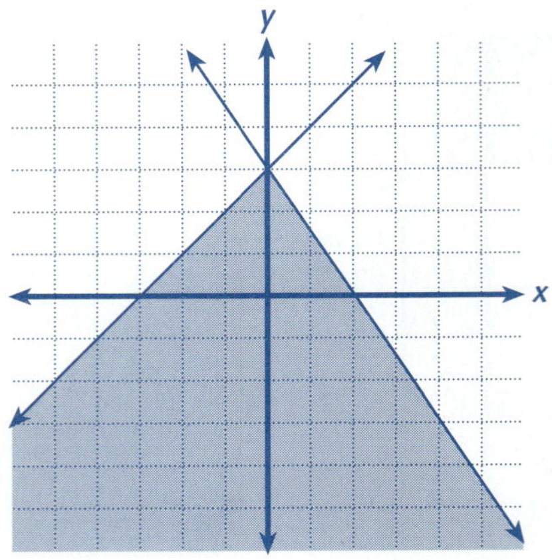

13. $\begin{cases} y > 4x - 3 \\ y \leq 3 \end{cases}$ **14.** $\begin{cases} y > \frac{1}{2}x + 2 \\ y \geq -3x + 6 \end{cases}$

Lesson 2.8

1. 84 **2.** 30 dimes and 6 nickels

3. 6 gallons of 4% and 12 gallons of 1%

4. \$7000 at 5% and \$3000 at 4%

5. Nora: 13; grandmother: 58

6. boat: 14 mph; current: 2 mph

7. 25 nickels, 7 dimes, and 14 quarters

8. 58 **9.** daughter: 9; father: 37

10. 20% solution: 70 L; 70% solution: 30 L

Practice—Chapter 3

Lesson 3.1

1. 5 by 8 **2.** 9

3. −24; the combination of a temperature of 10°F and a wind blowing at 20 mph has a cooling effect equal to −24°F.

4. −15°F

5. temperature: 15°F; wind speed: 20 mph

6. **Road Distances in Miles**

$$\begin{array}{c} \\ \text{Bos} \\ \text{Chi} \\ \text{Cle} \\ \text{Dal} \\ \text{Den} \end{array} \begin{array}{c} \begin{array}{ccccc} \text{Bos} & \text{Chi} & \text{Cle} & \text{Dal} & \text{Den} \end{array} \\ \begin{bmatrix} 0 & 983 & 639 & 1815 & 1991 \\ 983 & 0 & 192 & 931 & 1050 \\ 639 & 192 & 0 & 1205 & 1369 \\ 1815 & 931 & 1205 & 0 & 801 \\ 1991 & 1050 & 1369 & 801 & 0 \end{bmatrix} \end{array} = N$$

7. Boston and Denver

Lesson 3.2

1. $\begin{bmatrix} 4 & 6 \\ 6 & 20 \end{bmatrix}$ **2.** $\begin{bmatrix} 24 & 12 & 9 \\ 17 & 4 & 18 \end{bmatrix}$

3. $\begin{bmatrix} 1 & 41 \\ -17 & 9 \\ 1 & 43 \end{bmatrix}$ **4.** $\begin{bmatrix} -2 & -66 & 52 \\ 57 & -38 & -78 \\ -7 & 12 & 32 \end{bmatrix}$

5. $\begin{bmatrix} 1.7 \\ -5.1 \\ -5.7 \end{bmatrix}$

6. not possible, because the matrices do not have the same dimensions

7. $\begin{bmatrix} 1 & -\frac{3}{4} \\ -\frac{1}{4} & \frac{3}{4} \end{bmatrix}$ **8.** $\begin{bmatrix} -5 & -1 & 8 \\ 6.5 & 12.3 & -2 \\ -1 & 10 & 6.1 \end{bmatrix}$

9. not possible, because the matrices do not have the same dimensions

10. $\begin{bmatrix} 13.1 & 9.4 \\ -15.5 & 17.8 \end{bmatrix}$

Lesson 3.3

1. 2 by 4 **2.** not possible **3.** 2 by 3

4. 3 by 3 **5.** not possible

6. $\begin{bmatrix} 1 & 1 \\ 2 & -1 \end{bmatrix}$ **7.** $\begin{bmatrix} 16 & -32 \\ 0 & 4 \\ -24 & -8 \end{bmatrix}$

8. $\begin{bmatrix} -8 & 22 \\ 9 & -32 \end{bmatrix}$ **9.** $\begin{bmatrix} -7 & 34 \\ 13 & -54 \end{bmatrix}$

10. not possible

11. $\begin{bmatrix} 37 & -32 & 9 \\ -47 & 56 & -27 \end{bmatrix}$ **12.** $\begin{bmatrix} 16 & -64 \\ -1 & 6 \\ -10 & 12 \end{bmatrix}$

13. $\begin{bmatrix} 66 & 1 \\ 32 & -69 \end{bmatrix}$ **14.** $\begin{bmatrix} -52 & 68 & -36 \\ 8 & -5 & 0 \\ -34 & -32 & 54 \end{bmatrix}$

15. $\begin{bmatrix} 3 & -5 \\ 8 & -6 \end{bmatrix}$ **16.** $\begin{bmatrix} -2 & 0 \\ 12 & -9 \end{bmatrix}$

Lesson 3.4

1. $\begin{bmatrix} 1 & -2 \\ 2 & -3 \end{bmatrix}$ **2.** $\begin{bmatrix} 7 & -3 \\ -2 & 1 \end{bmatrix}$

3. $\begin{bmatrix} -2 & 1 \\ \frac{3}{2} & -\frac{1}{2} \end{bmatrix}$ **4.** $\begin{bmatrix} \frac{5}{2} & \frac{7}{2} \\ 2 & 3 \end{bmatrix}$

5. $\begin{bmatrix} -5 & 3 \\ 2 & -1 \end{bmatrix}$ **6.** $\begin{bmatrix} -5 & 6 \\ 1 & -1 \end{bmatrix}$

7. $\begin{bmatrix} 1 & -1 \\ 2 & -1 \end{bmatrix}$ **8.** $\begin{bmatrix} 2 & -1 \\ 3 & -2 \end{bmatrix}$

9. $\begin{bmatrix} 2 & -3 \\ -1 & 2 \end{bmatrix}$ **10.** $\begin{bmatrix} 7 & 11 \\ 2 & 3 \end{bmatrix}$

11. $\begin{bmatrix} \frac{1}{2} & 0 \\ -\frac{3}{10} & \frac{1}{5} \end{bmatrix}$

12. no inverse **13.** no inverse

14. no inverse

15. $\begin{bmatrix} -1 & -2 \\ -\frac{1}{2} & -\frac{3}{2} \end{bmatrix}$ **16.** $\begin{bmatrix} -\frac{1}{9} & \frac{2}{3} \\ \frac{2}{9} & -\frac{1}{3} \end{bmatrix}$

Lesson 3.5

1. $\begin{bmatrix} 6 & -3 \\ 4 & 2 \end{bmatrix} \begin{bmatrix} x \\ y \end{bmatrix} = \begin{bmatrix} 15 \\ 9 \end{bmatrix}$

2. $\begin{bmatrix} -2 & 5 \\ 1 & -3 \end{bmatrix} \begin{bmatrix} x \\ y \end{bmatrix} = \begin{bmatrix} -11 \\ 8 \end{bmatrix}$

3. $\begin{bmatrix} 2 & -1 & 4 \\ -1 & 7 & -2 \\ 6 & 3 & -1 \end{bmatrix} \begin{bmatrix} a \\ b \\ c \end{bmatrix} = \begin{bmatrix} 6 \\ 1 \\ -8 \end{bmatrix}$

4. $\begin{bmatrix} 8 & 7 & -1 \\ 5 & -4 & -3 \\ 1 & 1 & 1 \end{bmatrix} \begin{bmatrix} x \\ y \\ z \end{bmatrix} = \begin{bmatrix} 0 \\ 13 \\ -4 \end{bmatrix}$

5. (3, −2) **6.** (−1, 2) **7. d.** (−2, −3)

8. b. (2, −1) **9. a.** (2, 1) **10. c.** (−2, 3)

Practice—Chapter 4

Lesson 4.1

1. No, the area of region I is much larger, and the probability of landing in this region is much higher.

2. $\frac{17}{50}$ **3.** $\frac{17}{100}$ **4.** $\frac{3}{20}$ **5.** $\frac{9}{50}$ **6.** $\frac{4}{25}$

7. $\frac{1}{2}$; $\frac{1}{2}$; yes, the areas of the combined regions are the same.

8. 0, 1 **9.** 0, 1, 2, 3, 4, 5

10. 1, 2 **11.** 5, 6

12. integers from 1 to 10 inclusive

13. integers from 1 to 15 inclusive

14. INT(RAND*6)+5 **15.** INT(RAND*13)+12

Lesson 4.2

1. (1) Choose a random number generator, and describe what each result represents. (2) Plan how to perform the experiment to simulate one trial. (3) Perform a larger number of trials, and record the results.

2–6. Check students' work.

Lesson 4.3

1. 82 **2.** 84 **3.** none **4.** 19 **5.** 15.1

6. 16.75 **7.** 20 **8.** 13.3 **9.** 184.5

10. 185.5 **11.** 146 **12.** 130

13. Sort the scores in order, and then choose the middle number.

14. Sort the scores in order, and then find the average of the two middle numbers.

15. 72 **16.** 48

17. Possible answer: 12, 19, 26, 36, 12

18. This is impossible because the smallest possible values for this data set are 1, 1, 10, 10, and 10, and the mean of this set is 6.4, which is greater than 2.

Lesson 4.4

1. 1, 3, 5, 7, 9 **2.** 2, 4, 6, 8, 10 **3.** none

4. 1, 2, 3, 4, 5, 6, 7, 8, 9, 10

5. 4, 8, 12, 16, 20 **6.** 3, 6, 9, 12, 15, 18

7. 12 **8.** 3, 4, 6, 8, 9, 12, 15, 16, 18, 20

9.

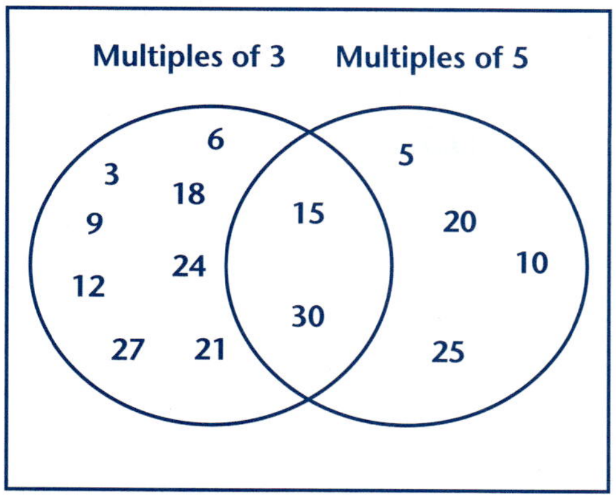

10. 15, 30

11. the intersection of both circles

12. 30 **13.** 24 **14.** 14 **15.** 40

Lesson 4.5

1. 24 **2.** 6 **3.** 12 **4.** 8 **5.** 6 **6.** 24

7. 120 **8.** 5,040 **9.** 35 **10.** 9 **11.** 20

12. 8

13. *MNOP; MPON; PMNO; PONM; OPMN; ONMP; NOPM; NMPO*

Lesson 4.6

1. $\frac{1}{2}$ **2.** $\frac{1}{2}$ **3.** $\frac{1}{4}$ **4.** $\frac{12}{40}$ or $\frac{3}{10}$ **5.** $\frac{1}{10}$

6. $\frac{3}{5}$ **7.** $\frac{5}{12}$ **8.** $\frac{1}{4}$ **9.** $\frac{7}{12}$ **10.** $\frac{8}{50}$ or $\frac{4}{25}$

11. 0 **12.** $\frac{29}{59}$ **13.** $\frac{35}{59}$ **14.** $\frac{19}{59}$ **15.** $\frac{45}{59}$

Lesson 4.7

1. $\frac{1}{4}$ **2.** $\frac{1}{4}$ **3.** $\frac{4}{25}$ **4.** $\frac{1}{4}$ **5.** $\frac{1}{4}$ **6.** $\frac{1}{5}$

7. $\frac{1}{5}$ **8.** $\frac{1}{5}$ **9.** $\frac{1}{5}$ **10.** $\frac{1}{2}$ **11.** 0 **12.** $\frac{1}{6}$

13. $\frac{1}{3}$ **14.** $\frac{2}{3}$ **15.** $\frac{4}{9}$ **16.** $\frac{2}{9}$ **17.** $\frac{2}{3}$

Practice—Chapter 5

Lesson 5.1

1. Yes, each x-value is paired with exactly one y-value.

2. No, the x-value of -1 is paired with two y-values, -2 and 7.

3. Yes, each x-value is paired with exactly one y-value.

4. No, there are x-values paired with more than one y-value.

5. No, there are x-values paired with more than one y-value.

6. Yes, each x-value is paired with exactly one y-value.

7. 12 **8.** -6 **9.** 3 **10.** 24 **11.** -30

12. 2 **13.** 9 **14.** -6 **15.** $\frac{1}{3}$ **16.** -12

17. 17 **18.** 0

Lesson 5.2

1. $y = |x|$; shifted vertically upward

2. $y = x^2$; reflected through the x-axis and stretched

3. $y = x$; shifted vertically upward

4. shifted vertically by -1

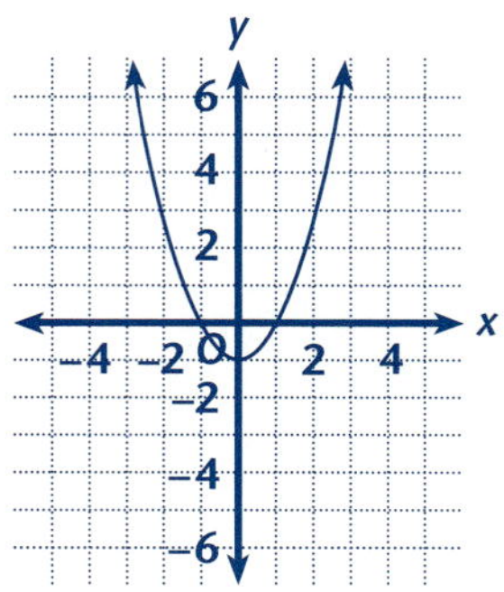

5. stretched vertically by 2

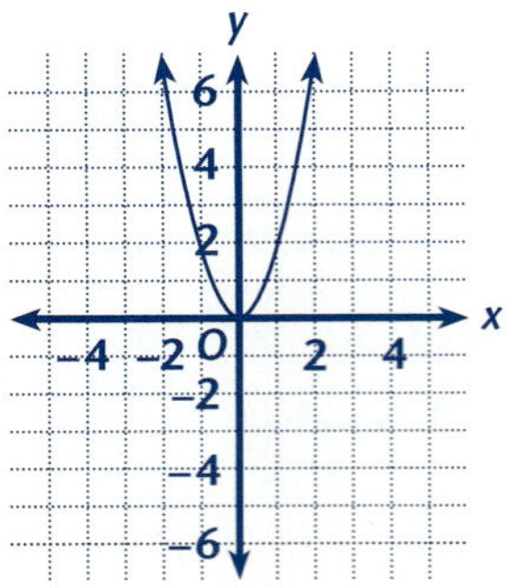

6. shifted vertically by 2

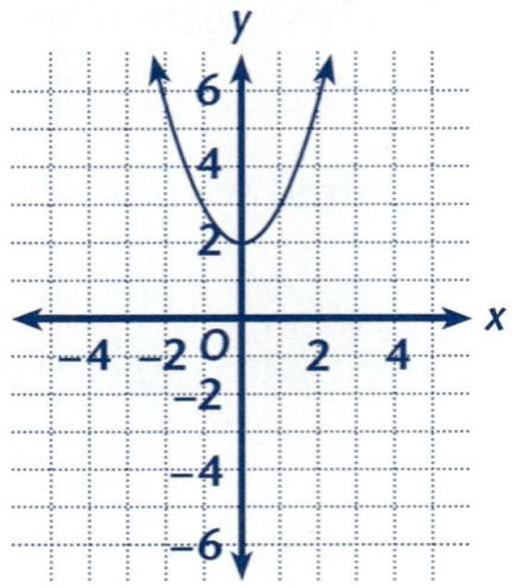

7. stretched vertically by 2 and reflected through the x-axis

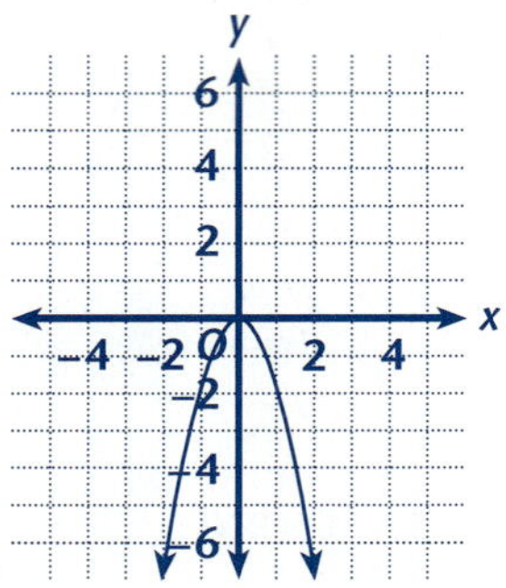

8. shifted 1 unit to the left

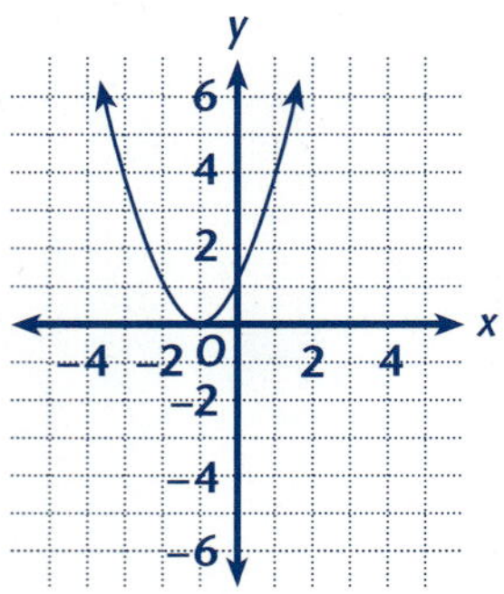

9. shifted 1 unit to the right

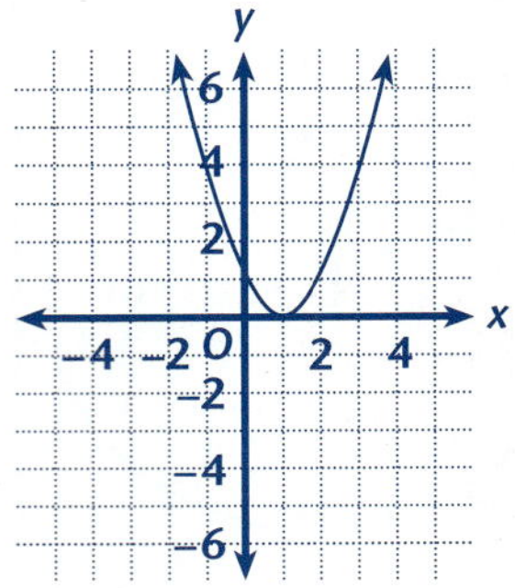

Lesson 5.3

1. $y = |x|$

2.

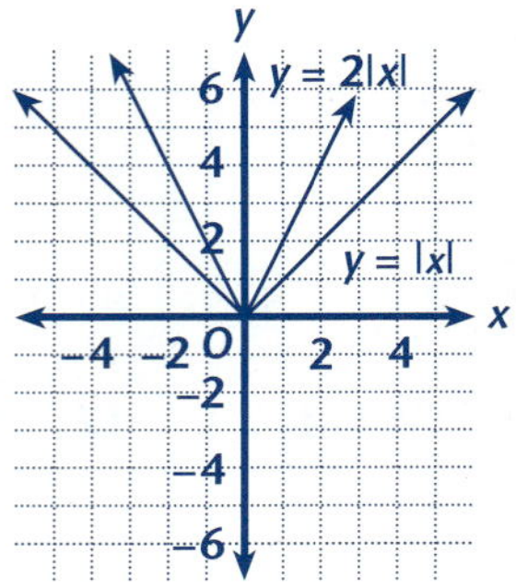

3. vertical stretch by 2 doubles all corresponding y-values

4. $y = \frac{1}{x}$

5.

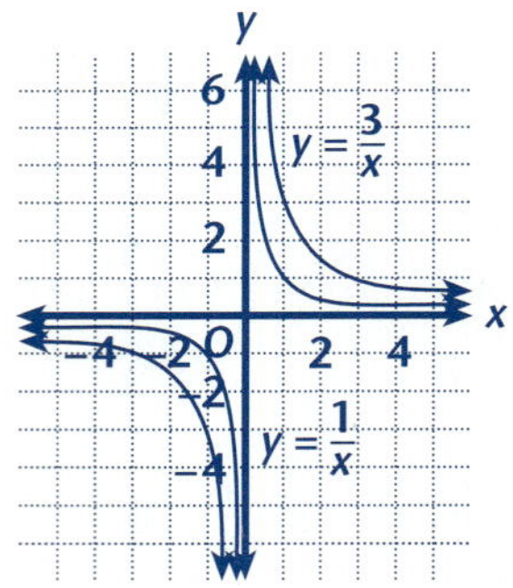

6. vertical stretch by 3 triples all corresponding y-values

7. yes **8.** yes **9.** no **10.** $y = 3|x|$

11. $y = \frac{3}{x}$

Lesson 5.4

1.

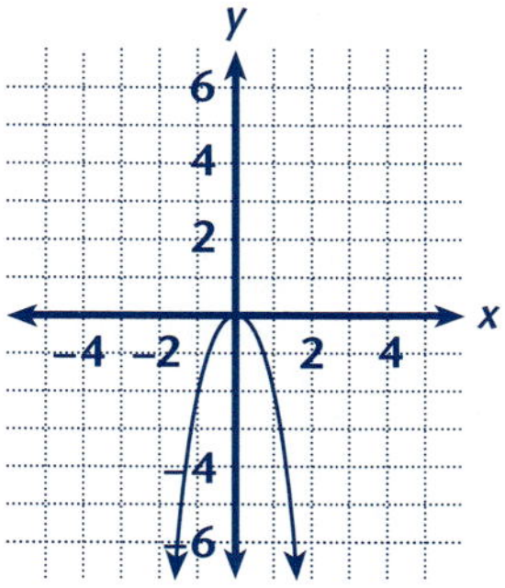

2.

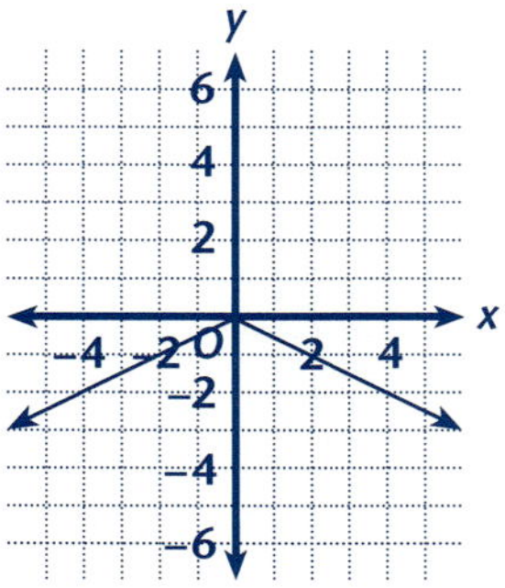

3.

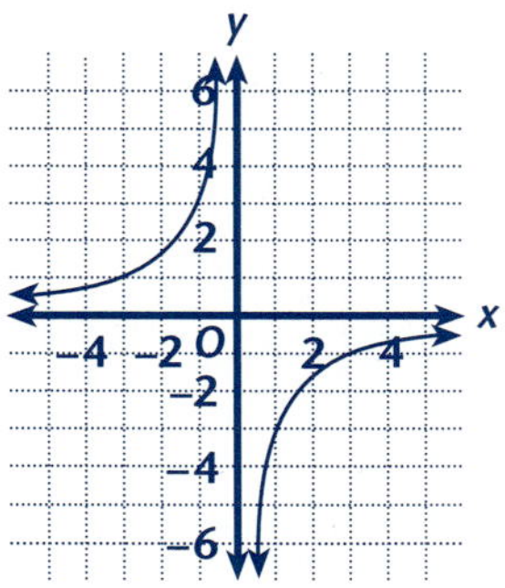

4. not a vertical reflection

5. vertical reflection   6. vertical reflection

7. vertical reflection   8. vertical reflection

9. not a vertical reflection

10. horizontal reflection

11. horizontal reflection

12. not a horizontal reflection

13. horizontal reflection

14. horizontal reflection

15. not a horizontal reflection

16. vertical   17. horizontal   18. vertical

19. horizontal

Lesson 5.5

1.

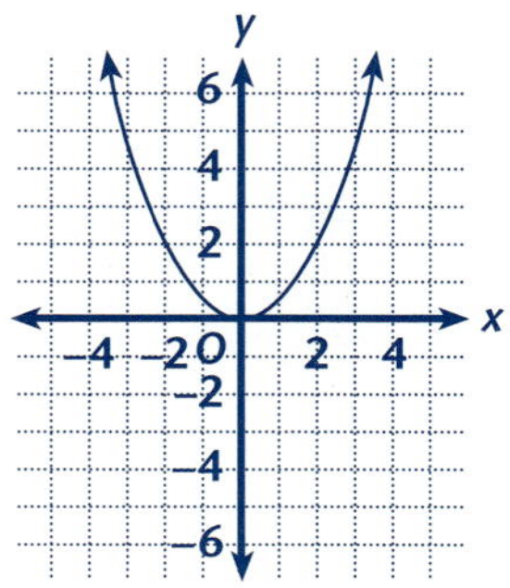

$y = x^2$; stretch by a factor of 0.5

2.

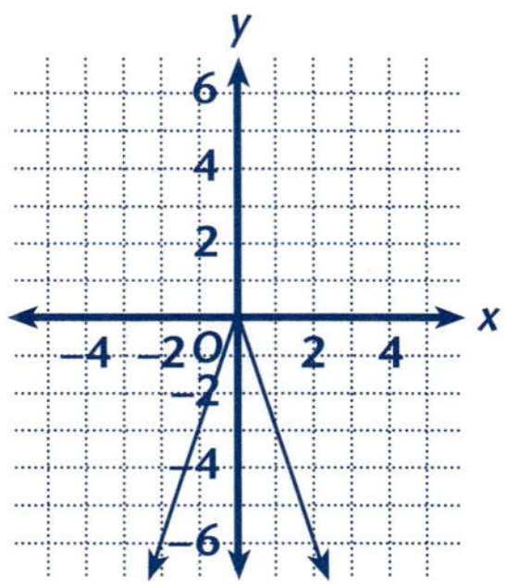

$y = |x|$; stretch by a factor of 3 and a vertical reflection

3.

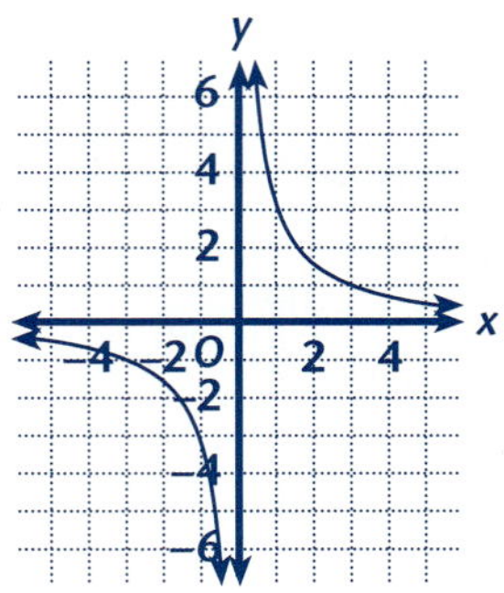

$y = \frac{1}{x}$; stretch by a factor of 3

4.

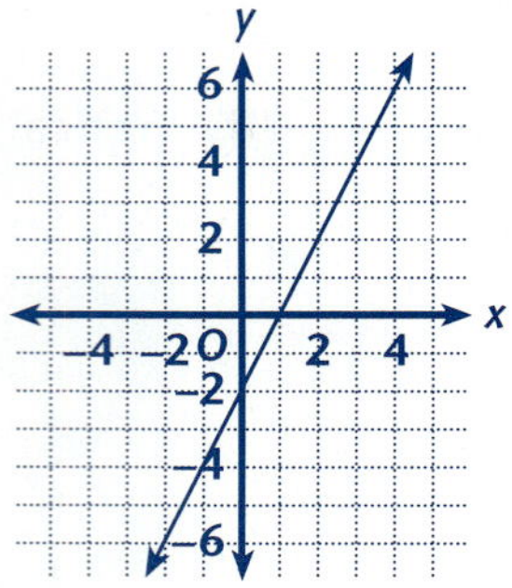

$y = x$; stretch by a factor of 2 and shift 2 units down

5.

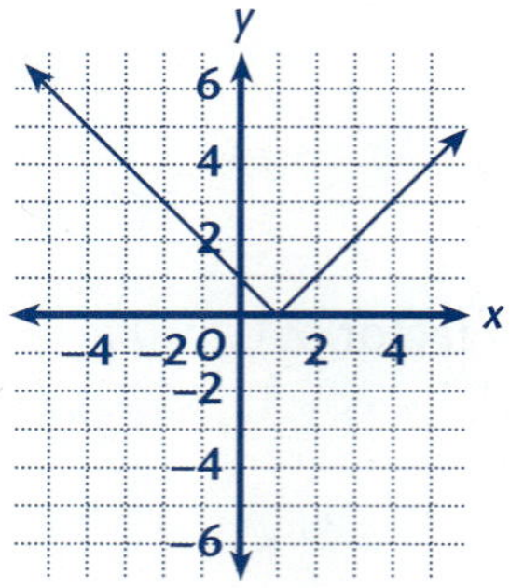

$y = |x|$; shift 1 unit right

6.

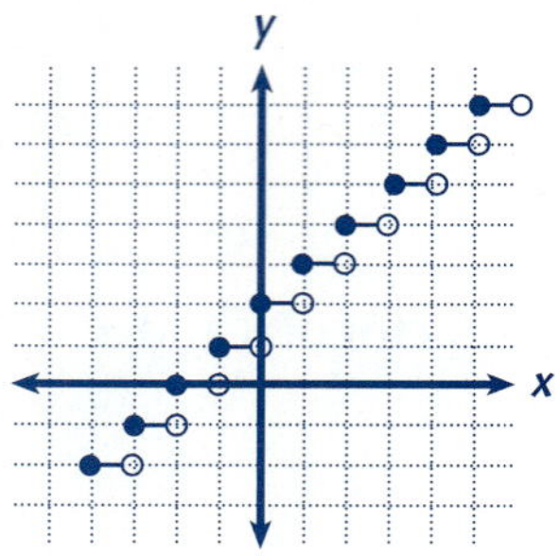

$y = \text{INT}(x)$; shift 2 units left

7. 1 unit down **8.** 4 units up

9. 1 unit to the right

10. 2 units to the left

11. (2, 6) **12.** (5, 4) **13.** (−2, 4)

14. (2, −1)

Lesson 5.6

1.

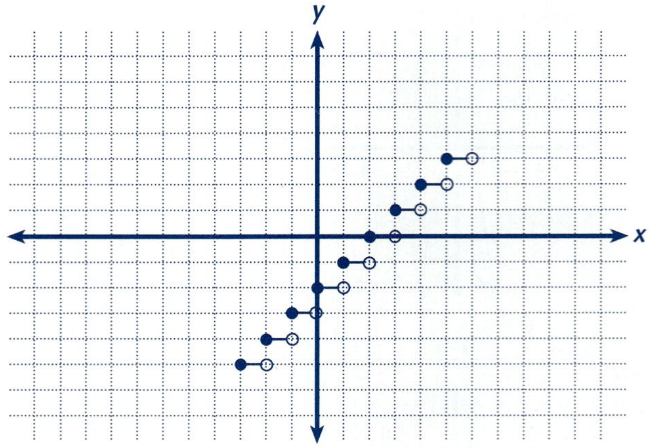

2.

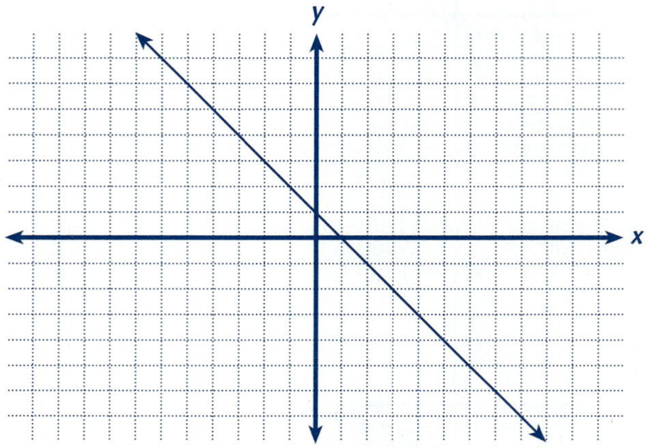

3.

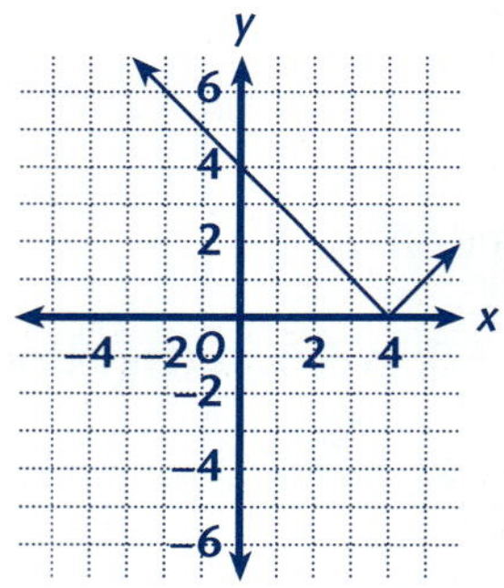

4.

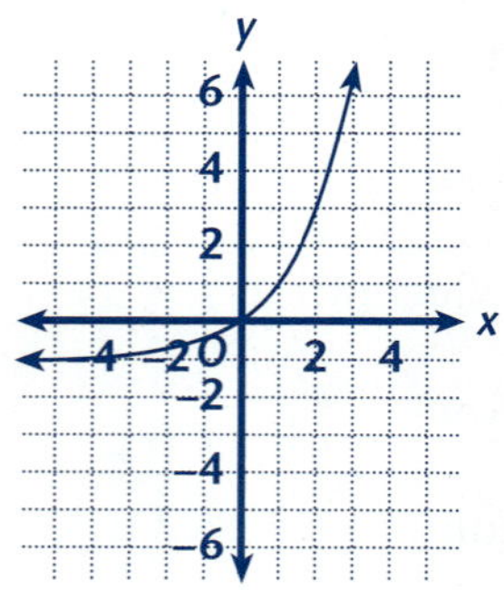

5.

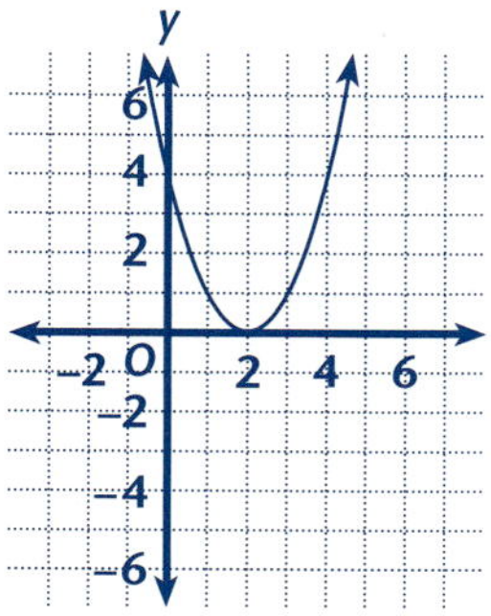

6.

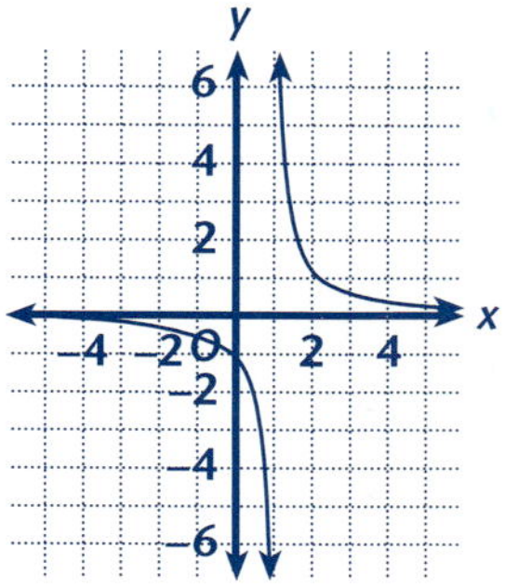

7. The point (4, 8) is transformed to (4, 18).

8. The point (4, 8) is transformed to (4, 20).

9. The point (4, 8) is transformed to (9, 24).

10. The point (4, 8) is transformed to (−1, 24).

Practice—Chapter 6

Lesson 6.1

1. 16 **2.** 10,000,000 **3.** 64 **4.** 1,000,000

5. 216 **6.** 125 **7.** 1 **8.** 256 **9.** 6561

10. 2401 **11.** 1,000,000,000,000 **12.** 5

13. 49 **14.** 10 **15.** 8 **16.** 3

17. 10,000,000 **18.** m^9 **19.** f^8 **20.** r^{11}

21. a^8 **22.** b^{21} **23.** x **24.** m^3

25. $y^2 + 1$ **26.** m^8 **27.** r^{10} **28.** q^{19}

29. t^{17} **30.** d^4 **31.** w^4 **32.** w^{10}

33. $10^3 = 1000$ **34.** 110

35. $10^6 = 1{,}000{,}000$ **36.** $10^2 = 100$

37. 9,900,000 **38.** $10^8 = 100{,}000{,}000$

39. 10,100 **40.** $10^6 = 1{,}000{,}000$

41. $10^7 = 10{,}000{,}000$ **42.** 99,999.99

Lesson 6.2

1. $12a^2b^3$ **2.** $30x^3$ **3.** $-8p^4q$ **4.** $8c^5$

5. a^5b^3 **6.** x^3y^7 **7.** $18a^2b^7x$ **8.** $0.12x^5y^5$

9. $-72p^5q^4$ **10.** $24c^6d^9$ **11.** $4a^3$

12. $-2x^2$ **13.** $\frac{-r}{7}$ **14.** $\frac{-y^2}{6}$ **15.** $\frac{s^2t^4}{2}$

16. $3m^3n$ **17.** $10v^3$ **18.** $\frac{4r^3}{t^2}$ **19.** $9x^6$

20. $4a^8$ **21.** a^4b^8 **22.** $40x^4y^6$ **23.** $x^{13}y^{10}$

24. $-250n^6p^6$ **25.** $a^{18}b^6$ **26.** $-27p^{11}q^{10}$

27. $\frac{a^3}{b^3}$ **28.** $\frac{9w^2}{y^8}$ **29.** $\frac{16p^6}{9q^4}$ **30.** $\frac{-125x^6}{8}$

Lesson 6.3

1. $\frac{1}{16}$ **2.** 16 **3.** −16 **4.** $-\frac{1}{16}$ **5.** $\frac{1}{16}$

6. $\frac{1}{9}$ **7.** $\frac{1}{10{,}000}$ **8.** −49 **9.** 49

10. −1000 **11.** $\frac{1}{64}$ **12.** $-\frac{1}{27}$ **13.** $\frac{1}{x}$

14. 1 **15.** $\frac{1}{a^3}$ **16.** $\frac{2}{n^2}$ **17.** 1 **18.** $-\frac{1}{x^5}$

19. $\frac{c^3}{ab^2}$ **20.** $\frac{-c^3}{d^4}$ **21.** m^3 **22.** k^3

23. $\frac{t^4}{r^2s^2}$ **24.** $-\frac{1}{a^8}$ **25.** $5x$ **26.** 1 **27.** $\frac{1}{y^8}$

28. $\frac{r}{t^7}$ **29.** $\frac{1}{w^6}$ **30.** 3 **31.** $\frac{-2}{a^6b}$ **32.** $\frac{5s^2}{q^6}$

33. $6a^{10}$ **34.** $\frac{-14}{d^4}$ **35.** $\frac{1}{x^2}$ **36.** $\frac{1}{3z^5}$

Lesson 6.4

1. 2×10^5 **2.** 5×10^8 **3.** 2.1×10^5

4. 4×10^4 **5.** 6.5×10^9 **6.** 8.4×10^8

7. 4.66×10^{11} **8.** 3.75×10^4 **9.** 2×10^{-1}

10. 8×10^{-4} **11.** 7.6×10^{-5}

ANSWERS

12. 6.45×10^{-3} **13.** 3.62×10^{-6}

14. 9.78×10^{-4} **15.** 5000 **16.** 600,000

17. 90,000 **18.** 56,000,000

19. 85,600,000,000 **20.** 356,000,000

21. 0.06 **22.** 0.000067 **23.** 0.00000876

24. 0.0000000922 **25.** 1.2×10^5

26. 1.2×10^{10} **27.** 6×10^3 **28.** 1.2×10^8

29. 5×10^5 **30.** 7×10^6 **31.** 6×10^2

32. 3×10^3 **33.** 2×10^1 **34.** 2.4×10^5

Lesson 6.5

1.

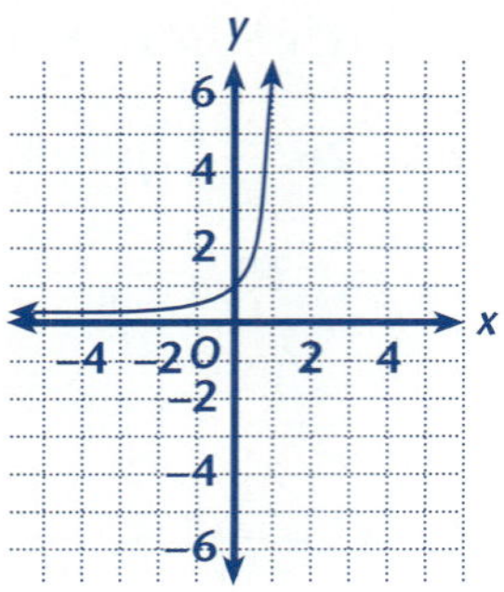

2.

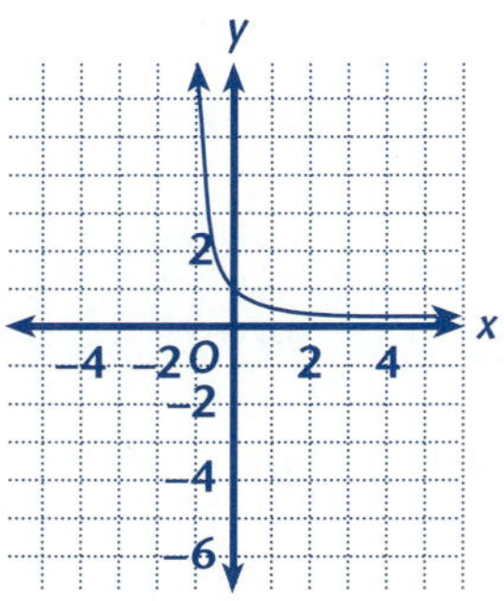

3.

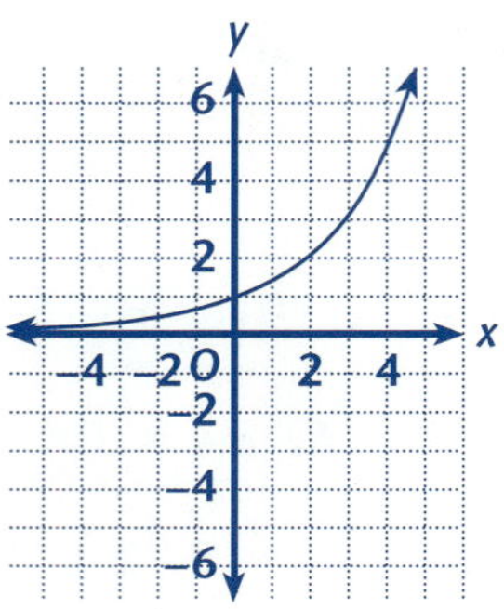

4.

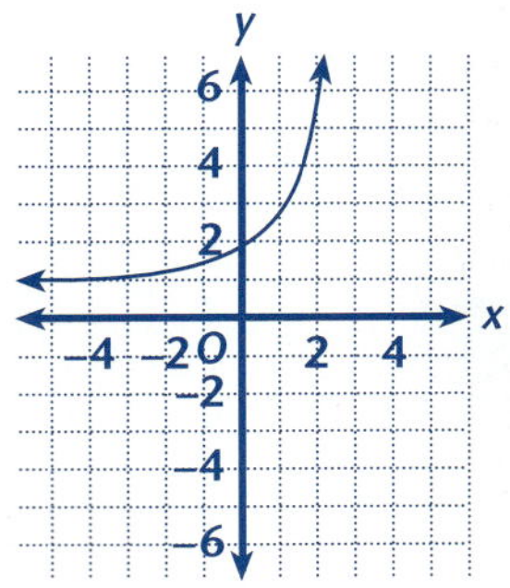

shift 1 unit up

5.

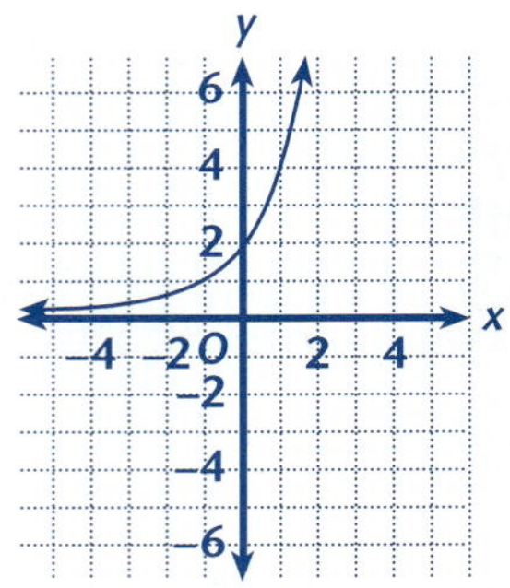

shift 1 unit left

6.

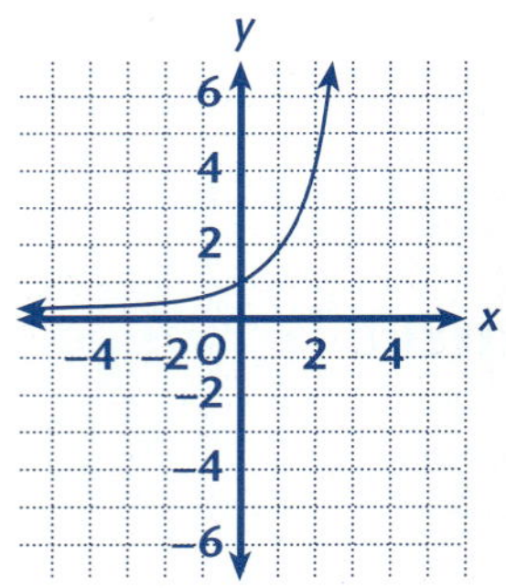

no effect

7. $\frac{1}{4}$ **8.** 1 **9.** 2 **10.** 4 **11.** 8 **12.** 16

13. $\frac{1}{16}$ **14.** $\frac{1}{8}$ **15.** 1.006

16. about 204,843

Lesson 6.6

1. 55.4% **2.** 116%

3.

Number of years	0	1	2	3	4	5
Value (in $)	1.7×10^6	1.83×10^6	1.96×10^6	2.11×10^6	2.27×10^6	2.44×10^6

4. $(1.7 \times 10^6)(1.075)^x$

5. Check students' graphs.

6. 7.22×10^6 **7.** $(1.3 \times 10^8)(1.08)^x$

8. 2.81×10^8 **9.** Check students' graphs.

10. 6.06×10^8

Practice—Chapter 7

Lesson 7.1

1. $S(3) = 6 \cdot 3^2$; $V(3) = 3^3$

2. $S(x) = 6x^2$; $V(x) = x^3$

3. $S(x) = 16x + 30$; $V(x) = 15x$

4. $S(x) = 15.4x + 28.8$; $V(x) = 14.4x$

5. $S(x) = 2\pi x^2 + 14\pi x$; $V(x) = 7\pi x^2$

6. $S(x) = \frac{\pi}{2}x^2 + 1.3\pi x$; $V(x) = 0.325\pi x^2$

7. $S(x) = \pi x^2 + \pi x\sqrt{x^2 + 64}$; $V(x) = \frac{8\pi x^2}{3}$

8. $S(x) = \frac{\pi}{4}x^2 + \frac{\pi}{2}x\sqrt{0.25x^2 + 268.96}$;

$V(x) = \frac{41\pi x^2}{30}$

9–12. Check students' answers.

Lesson 7.2

1. $2x^2 + 3x - 5 + x^2 - x - 2 = 3x^2 + 2x - 7$

2. $c^2 - 8c + 2$ **3.** $m^3 + 10$

4. $3n^4 + 5n^2 + n$ **5.** $-2x^3 + 3x^2 + 6x + 1$

6. 3 **7.** 1 **8.** 6 **9.** 5 **10.** $5f^2 + 3f - 4$

11. $3c^3 + c^2 - 11c$ **12.** $2q^3 + 6q^2 + 2q$

13. $5y^3 - 3y^2 - 3$ **14.** $3m^3 - 10m^2 + m - 1$

15. $2n^2 - 3n$ **16.** $2a - 16$

17. $7x^3 - x^2 - 3x$

Lesson 7.3

1. $x(x - 3) = x^2 - 3x$

2. $2x(x + 3) = 2x^2 + 6x$

3. $6x + 12$ **4.** $7x - 14$ **5.** $18x - 27$

6. $-6x - 8$ **7.** $-4x + 20$ **8.** $20x - 20$

9. $x^2 + 2x$ **10.** $8x - x^2$ **11.** $-x^2 - 3x$

12. $6x^2 - 18x$ **13.** $-2x^2 - 4x$

14. $4x^2 - 20x$

15. $x^2 + 4x$

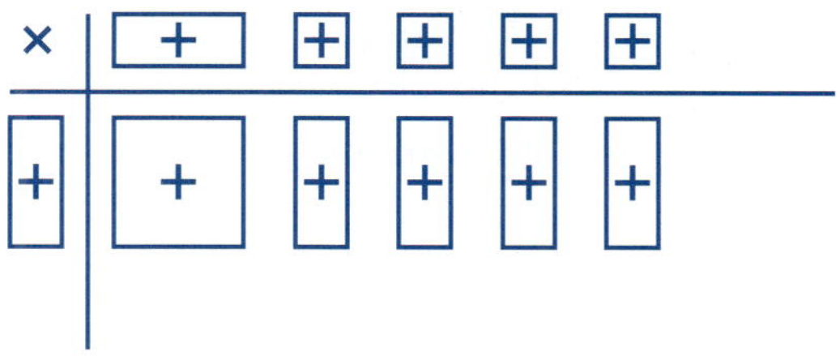

16. $4x^2 + 2x$

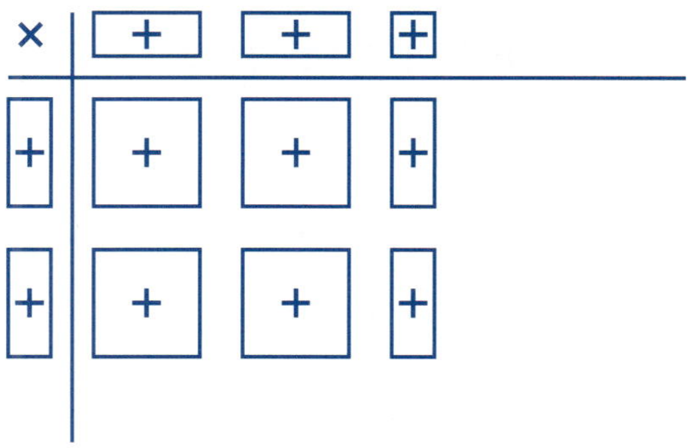

17. −3, 2 **18.** 5, 7 **19.** −1, 1 **20.** −12, 2

21. −9 **22.** −4, 2 **23.** $-\frac{1}{2}$, 3 **24.** −3, $-\frac{2}{3}$

Lesson 7.4

1. $(x + 3)(x - 2) = x^2 + x - 6$

2. $a^2 + 9a + 20$ **3.** $2n^2 - 7n + 6$

4. $2t^2 + 9t - 18$ **5.** $8y^2 + 10y - 3$

6. $c^2 - 12\frac{1}{4}c + 3$ **7.** $x^2 - 0.6x - 0.16$

8. $r^2 - 3r + 2$ **9.** $a^2 - 10a + 21$

10. $m^2 - 8m + 16$ **11.** $w^2 + 0.1w - 0.2$

12. $2s^2 - 8\frac{1}{3}s - 3$ **13.** $y^2 + 3\frac{1}{4}y - 3$

14. $15c^2 - 16cd - 15d^2$ **15.** $9m^2 + 18m + 8$

16. $9x^2 - 4y^2$ **17.** $4m^2 + 20m + 25$

18. $6y^2 - 5y + 1$ **19.** $10x^2 + 9x + 2$

20. $49s^2 + 14s + 1$ **21.** $x^4 + 3x^2 + 2$

22. $m^4 + 2m^2n + n^2$ **23.** $8d^2 - 2c^2$

24. $0.02x^2 - 0.8x - 10$

25. $3.22y^2 - 4.7y - 5$

Lesson 7.5

1. composite **2.** composite **3.** prime

4. composite **5.** composite **6.** prime

7. $3x(x - 3)$ **8.** $r(6r^2 + 3r - 7)$

9. $3(2w^4 + w^2 - 3)$ **10.** $3q(q + 2q^4 + 3)$

11. $9(2b^3 - 4b^2 - 1)$ **12.** $3v^2(v^4 - 3v^2 + 49)$

13. $xy(1 - 2x + y^2)$ **14.** $5wz(2z - 1 + 3w)$

15. $(y + 2)(y + 1)$ **16.** $(3 - a)(c + d)$

17. $(2 - z)(x + w)$ **18.** $(6 + y)(x - 2)$

19. $(a - b)(a - b)$ **20.** $(3x - 2)(x + 4)$

21. $(4 + x)(x - 2y)$ **22.** $(z + y)(4 - w)$

23. $(7x + 1)(x - 1)$ **24.** $(mn - 1)(r + 1)$

25. $(4a - b)(x + y)$ **26.** $2(pq - 1)(q + 2)$

27. $(a + 1)(3 - a)$ **28.** $(3m^2 + 2)(2m - 3)$

29. $(m - 4)(m^2 + 3)$ **30.** $(a - 6)(b^2 + 5)$

31. $(2d - 1)(c - 3)$ **32.** $5(2 - x)$

33. $(x + y)(2z + 1)$ **34.** $(x + 5)(m + n)$

35. $(x + 2)(y + 1)$ **36.** $(3x - 2y)(x + y)$

Lesson 7.6

1. $x^2 - 2x + 1$ **2.** $4x^2 + 12x + 9$

3. $a^2 + 10a + 25$ **4.** $9g^2 + 6g + 1$

5. $4a^2 - 20a + 25$ **6.** $x^2 - 2xy + y^2$

7. $x^2 + 4xy + 4y^2$ **8.** $9m^2 - 6mn + n^2$

9. $x^2 - 9$ **10.** $16y^2 - 1$ **11.** $25x^2 - 4$

12. $36r^2 - s^2$ **13.** $(x + 5)(x - 5)$

14. $(y - 7)(y + 7)$ **15.** $2(x + 1)(x - 1)$

16. $4(y - 2)(y + 2)$ **17.** $(4y - 5)(4y + 5)$

18. $(11 - x)(11 + x)$ **19.** $(3 - 4y)(3 + 4y)$

20. $(x - 3)^2$ **21.** $(x + 5)^2$ **22.** $(y - 8)^2$

23. $(z - 10)^2$ **24.** $(8x - 3y)^2$ **25.** $(5 - a)^2$

26. $(2m + 1)^2$

27. $(9rs - 10t^2q^2)(9rs + 10t^2q^2)$

28. $(4x + 3)^2$ **29.** $(x + y)^2$ **30.** $(2a + b)^2$

31. $(y - x)(y + x)$ **32.** $(5a - 1)(5a + 1)$

33. $(1 - xy^2)(1 + xy^2)$ **34.** $(12c - 5d)^2$

35. $6(a - 6b)(a + 6b)$ **36.** $x(7 - y)(7 + y)$

37. $12(x^2 + 1)(x - 1)(x + 1)$

38. $(b^2 + 1)(b^2 - 2)$ **39.** $4(c - 3)^2$

40. $m^2n^2(2m + 1)^2$

Lesson 7.7

1. 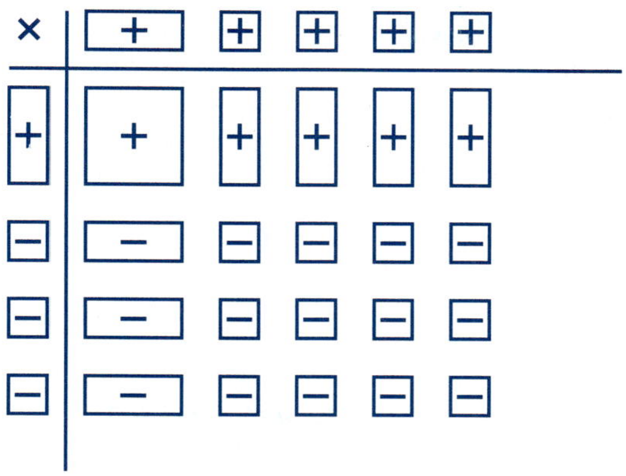

$(x + 4)(x - 3) = x^2 + x - 12$

2. $(x + 6)(x - 5)$ 3. $(m + 5)(m + 4)$

4. $(c - 9)(c + 8)$ 5. $(d - 4)(d - 3)$

6. $(y + 13)(y - 12)$ 7. $(f - 8)(f + 6)$

8. 1, 15; −1, −15; 3, 5; −3, −5

9. −1, 121; 1, −121; −11, 11

10. −4, −1; 4, 1; 2, 2; −2, −2

11. −1, 35; 1, −35; −5, 7; 5, −7

12. $(g - 8)(g + 5)$ 13. $(h + 10)(h - 4)$

14. $(j + 20)(j + 2)$ 15. $(k - 40)(k + 1)$

16. $(x + 3)(x - 4)$ 17. $(y - 9)(y + 2)$

18. $(a - 7)(a - 2)$ 19. $(x - 6)(x + 1)$

20. $(x - 3)(x - 5)$ 21. $(p + 3)(p + 15)$

Practice—Chapter 8

Lesson 8.1

1. stretched by 2; shifted left 5 units

2. shifted right 1 unit and up 5 units

3. reflected vertically; stretched by 3; shifted left 1 unit

4. shifted left 6 units

5. reflected vertically; shifted left 3 units and down 2 units

6. stretched by $\frac{1}{2}$; shifted right 4 units

7. vertex: (5, 0); axis of symmetry: $x = 5$

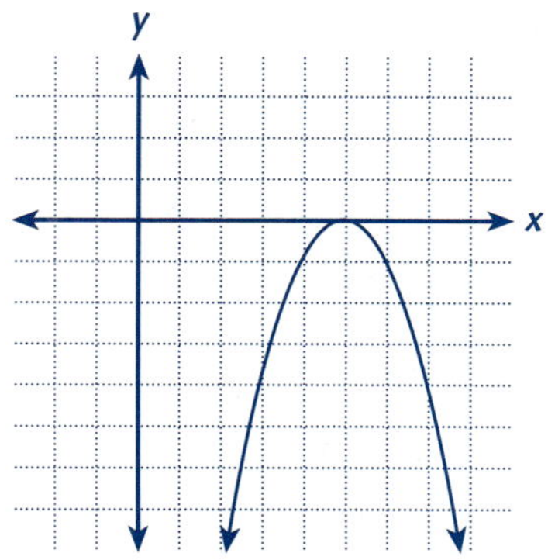

8. vertex: (1, 2); axis of symmetry: $x = 1$

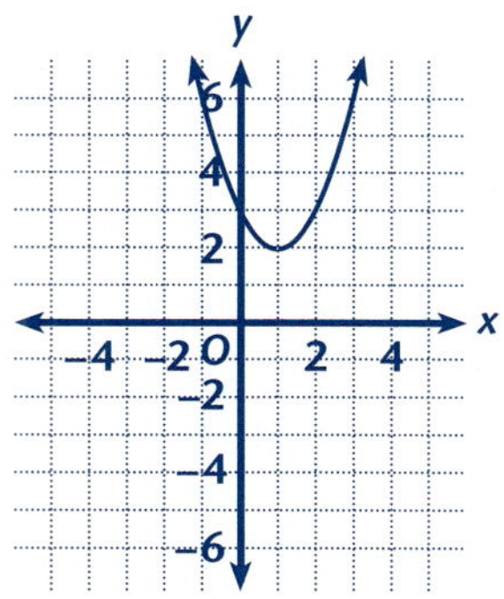

9. vertex: (3, 0); axis of symmetry: $x = 3$

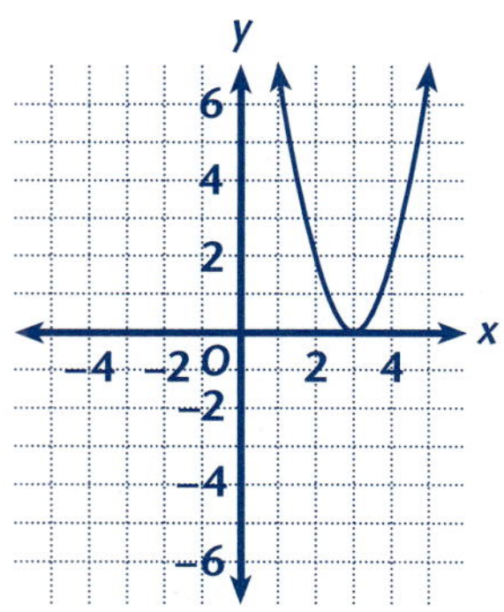

10. vertex: $(-2, -1)$; axis of symmetry: $x = -2$

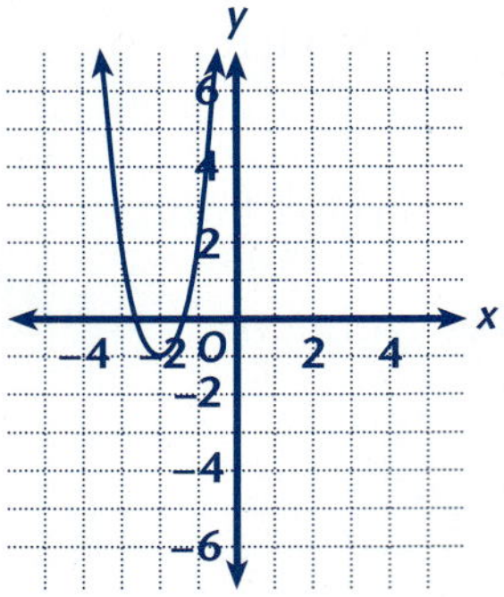

11. $-3, 5$ **12.** $5, -3$ **13.** $-5, -3$

14. $-7, 2$ **15.** $5, 6$ **16.** $-1, 4$

Lesson 8.2

1. 1 **2.** 9 **3.** 4 **4.** 15 **5.** 5.48

6. 4.24 **7.** 10.49 **8.** 7.42 **9.** ± 4

10. ± 30 **11.** ± 8.66 **12.** $\pm\frac{1}{2}$ **13.** $\pm\frac{2}{7}$

14. $\pm\frac{3}{5}$ **15.** $-2, 6$ **16.** $-6, 2$ **17.** ± 2

18. $-5, -1$ **19.** $-5, -1$ **20.** $-4, 8$

21. $-7.73, -4.27$ **22.** $-1.74, 5.74$

23. vertex: $(1, -1)$; axis of symmetry: $x = 1$; zeros: 0 and 2

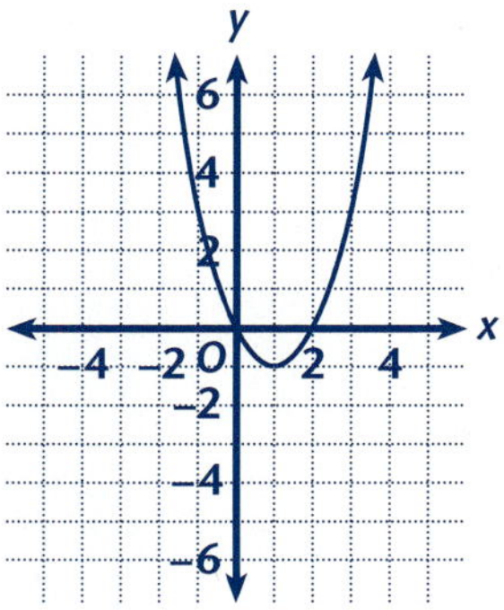

24. vertex: $(-2, -4)$; axis of symmetry: $x = -2$; zeros: 0 and -4

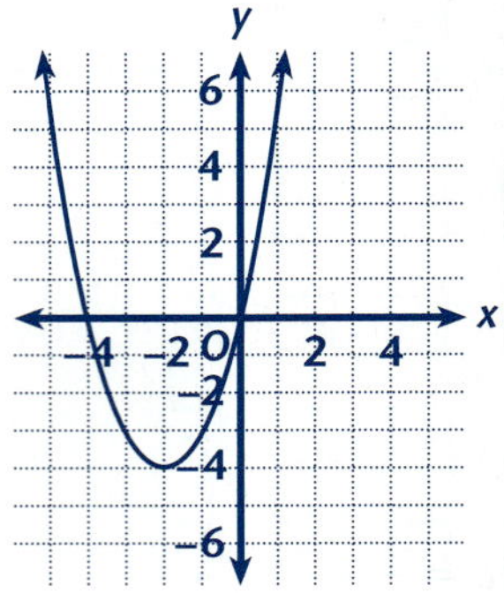

Lesson 8.3

1. $x^2 + 8x + 16 = (x + 4)^2$

2. $x^2 + 8x + 16 = (x + 4)^2$

3. $x^2 - 8x + 16 = (x - 4)^2$

4. $x^2 + 11x + \frac{121}{4} = \left(x + \frac{11}{2}\right)^2$

5. $x^2 - 11x + \frac{121}{4} = \left(x - \frac{11}{2}\right)^2$

6. $y = (x - 2)^2 - 4$ **7.** $y = (x + 2)^2 - 4$

8. $y = (x + 9)^2 - 81$ **9.** $y = (x - 9)^2 - 81$

10. $y = (x + 3)^2 - 9$ **11.** $(-3, -9)$

12. The minimum value is -9. **13.** $(0, -3)$

14. $(0, -2)$ **15.** $(0, 6)$ **16.** $(0, 7)$

17. $(-1, -9)$ **18.** $(2, -1)$

Lesson 8.4

1. 6, 7 **2.** $-7, -6$ **3.** $-6, 7$ **4.** $-4, -3$

5. $-3, 4$ **6.** $-4, 3$ **7.** $-6, 10$ **8.** 3, 7

9. $-4, -1$ **10.** $-6, 4$ **11.** $-2, 3$

12. 1, 4 **13.** $-6, -5$ **14.** $-4, 2$ **15.** -3

16. 1 **17.** $-6, 5$ **18.** $-7, 2$ **19.** $-3, -2$

20. $-2, -1$ **21.** $-4, 3$ **22.** $-5, -2$

23. −1, 3 **24.** −5, 1 **25.** 1, 6 **26.** 3, 8

27. (3, 9), (−3, 9) **28.** (1, −6), (−5, 0)

29. (5, 0) **30.** (2, 0), (6, 4)

Lesson 8.5

1. $a = 1, b = 3$, and $c = 2$

2. $a = 1, b = -2$, and $c = 1$

3. $a = 1, b = 7$, and $c = -3$

4. $a = 1, b = -5$, and $c = 3$

5. $a = 3, b = 0$, and $c = -4$

6. $a = 2, b = 15$, and $c = 0$

7. $a = 10, b = 6$, and $c = 1$

8. $a = 1, b = 7$, and $c = -2$

9. $a = 1, b = 0$, and $c = -36$

10. $a = 1, b = -12$, and $c = 36$

11. $a = 1, b = -6$, and $c = 0$

12. $a = 3, b = -5$, and $c = 0$

13. −7; no solutions **14.** 21; two solutions

15. 28; two solutions **16.** −28; no solutions

17. 21; two solutions **18.** 64; two solutions

19. 0; one solution **20.** −8; no solutions

21. −0.39, 1.72 **22.** −0.64, 3.14

23. −1.63, 1.63 **24.** 0.5, 0.33

25. 1.17, 6.83 **26.** −2.24, 2.24 **27.** −1, 2

28. 3, 4 **29.** −2, −1 **30.** −5, 5

31. −1.67, 1 **32.** −0.38, −2.62 **33.** 1, 4

34. −4, −1

Lesson 8.6

1. $x \leq -5$ or $x \geq 8$

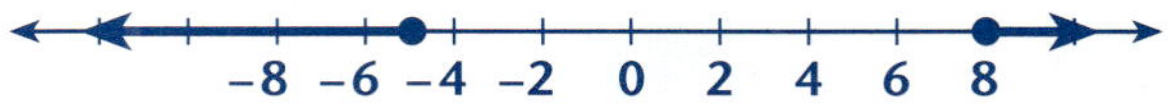

2. $-7 < x < -2$

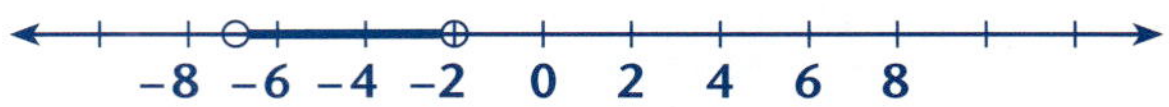

3. $x = -1$

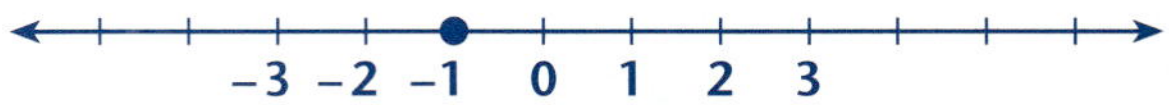

4. $-5 < x < 6$

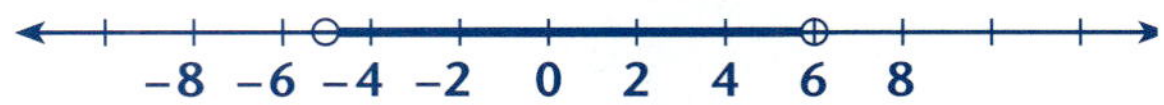

5. $x \leq -3$ or $x \geq 4$

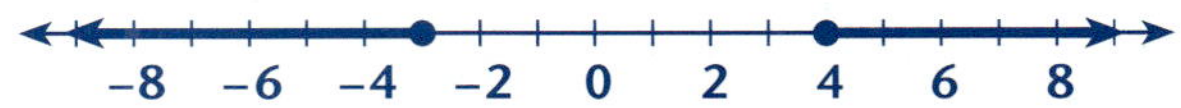

6. $x < -2$ or $x > -1$

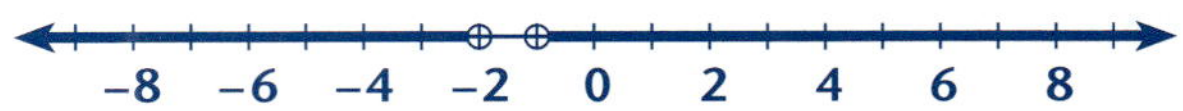

7.

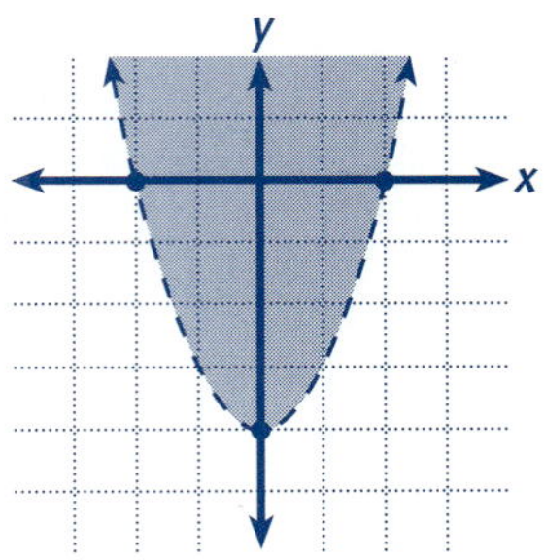

8.

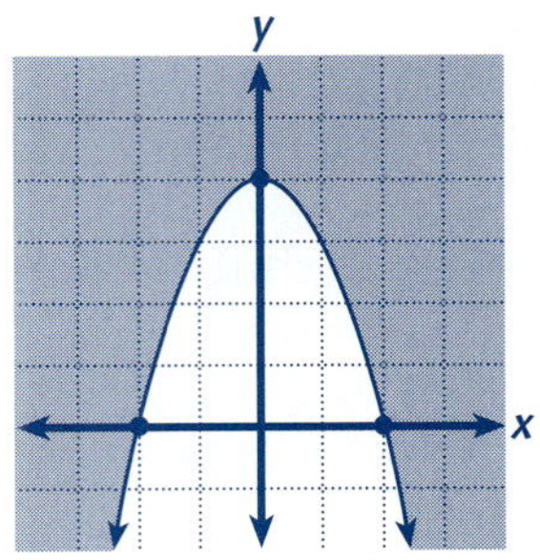

9.

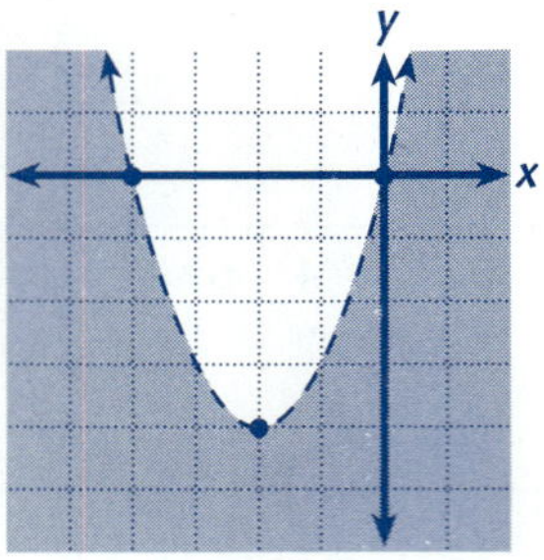

10.

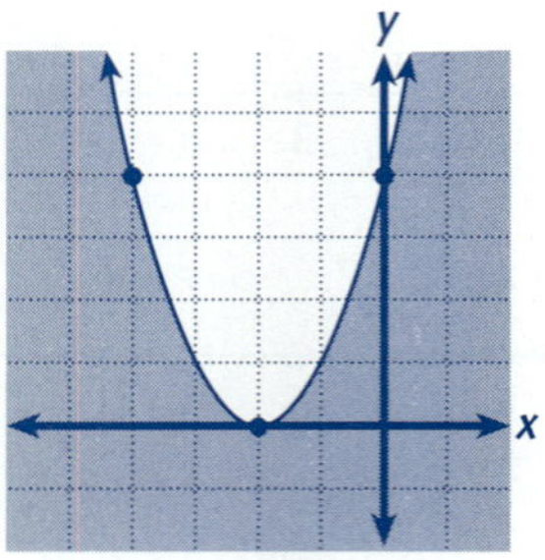

Practice—Chapter 9

Lesson 9.1

1. 29 **2.** −17 **3.** −37.95 **4.** 1.2

5. 0.09 **6.** $\frac{10}{13}$

7. shift 3 units down

8. vertical stretch by 2

9. reflection through the *x*-axis and shift 4 units right

10. vertical stretch by 6 and shift 3 units right

11. shift 5 units left and 3 units up

12. vertical stretch by 5.75 and shift 2.5 units up

13. $y = -\sqrt{x} + 1$ **14.** $y = \sqrt{x} + 1$

15. $y = \sqrt{x} - 2$ **16.** $y = -\sqrt{x} - 3$

17. $y = 2\sqrt{x}$ **18.** $y = \sqrt{x + 2}$

Lesson 9.2

1. 12 **2.** $6\sqrt{2}$ **3.** $12\sqrt{2}$ **4.** $20\sqrt{10}$

5. $8\sqrt{51}$ **6.** $15\sqrt{39}$ **7.** $4\sqrt{7}$ **8.** 9

9. $6\sqrt{2}$ **10.** $2\sqrt{3}$ **11.** $3\sqrt{6}$ **12.** $\sqrt{10}$

13. $m^3\sqrt{n}$ **14.** a^4b^3 **15.** $\frac{x\sqrt{x}}{y^3}$ **16.** $\frac{x^2}{y^4}\sqrt{\frac{x}{y}}$

17. $5\sqrt{3}$ **18.** $4\sqrt{3}$ **19.** $3\sqrt{2}$

20. $7 + \sqrt{5} - \sqrt{3}$ **21.** 12 **22.** $6\sqrt{3}$

23. 8 **24.** $2\sqrt{3} + 6$ **25.** $7 + 4\sqrt{3}$ **26.** 27

Lesson 9.3

1. 16 **2.** −2 **3.** −16 **4.** no solution

5. 4 **6.** 11 **7.** 4 **8.** −7 **9.** no solution

10. 11 **11.** 4 **12.** 5 **13.** no solution

14. 0.81 **15.** ±5 **16.** ±11 **17.** $\pm\sqrt{30}$

18. Not possible; a square root cannot be negative.

19. Not possible; a square root cannot be negative.

20. Not possible; a square root cannot be negative.

21. 1250

22. Not possible; a square root cannot be negative.

23. −3

Lesson 9.4

1. 72 **2.** 18 **3.** 9.7 **4.** 21 **5.** 84

6. 55 **7.** 16 **8.** 36 **9.** 53 **10.** 26

11. 64 **12.** 4/15 **13.** 0.6 **14.** 0.085

15. 15 **16.** 60 **17.** 24 **18.** 78 in.

19. 12 feet **20.** 106 miles

Lesson 9.5

1. $Q(0, 3)$; $PR = 13$, $PQ = 5$, $QR = 3\sqrt{10}$; scalene

2. $Q(3, 6)$; $PR = 2$, $PQ = \sqrt{10}$, $QR = \sqrt{10}$; isosceles

3. 15 **4.** 5 **5.** 1 **6.** $2\sqrt{10}$ **7.** $(0, 6)$

8. $\left(-\frac{1}{2}, -1\right)$ **9.** $(-3, 3)$ **10.** $(5, 1)$

11. $\left(6, -\frac{3}{2}\right)$ **12.** 34, 7 **13.** 12, 10

14. 16, 6 **15.** 4, 11

Lesson 9.6

1. $x^2 + y^2 = 81$

2. $x^2 + y^2 = 16$

3.

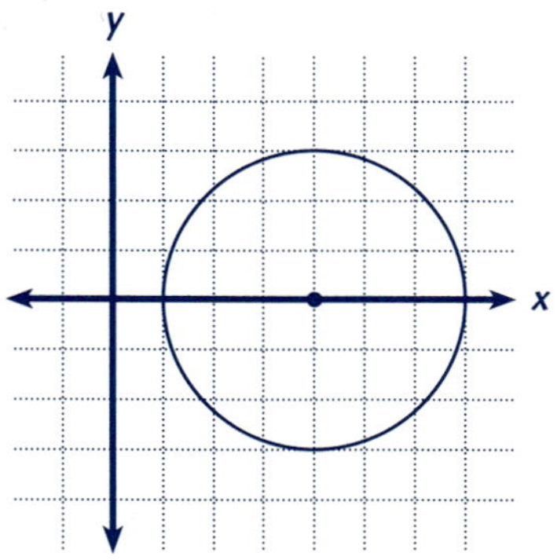

$(x - 4)^2 + y^2 = 9$

4.

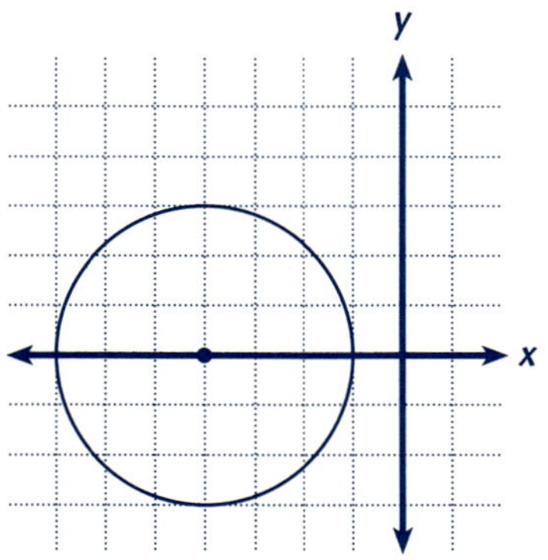

$(x + 4)^2 + y^2 = 9$

5.

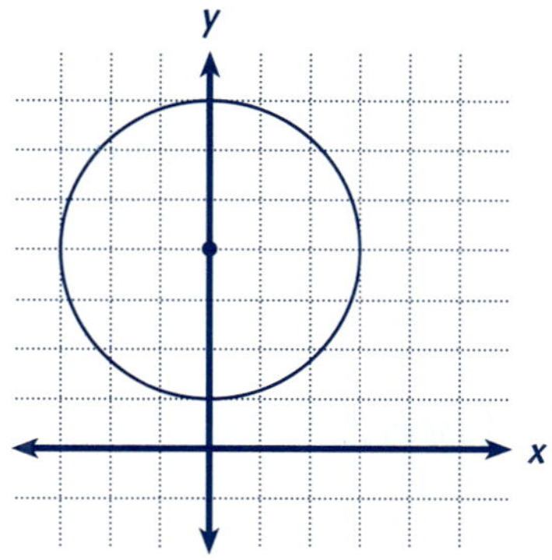

$x^2 + (y - 4)^2 = 9$

6.

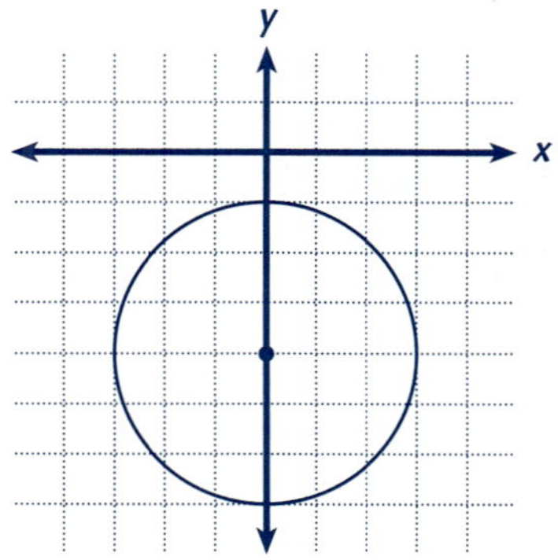

$x^2 + (y + 4)^2 = 9$

7. center at $(1, 0)$; radius of 2

8. center at $(-1, 2)$; radius of 2

Lesson 9.7

1. 0.3 **2.** 1.1 **3.** −1.9 **4.** −0.6

5. 16° **6.** 36° **7.** 88° **8.** 44° **9.** 3.46 ft

10. 9.24 ft **11.** 100 ft **12.** 42.01 ft

13. 1.15° **14.** 1 **15.** 57°

Lesson 9.8

1. 0.9063 **2.** 0.9962 **3.** 0.8480

4. 0.9659 **5.** 0.2079 **6.** 0.7660

7. 0.7771 **8.** 0.0175 **9.** 0.9703

10. 0.9903 **11.** 0.6561 **12.** 0.4848

13. 48.6° **14.** 57.8° **15.** 48.2° **16.** 55.2°

17. 44.4° **18.** 19.3 **19.** 11.1 **20.** 6.3

21. 36.8° **22.** 57.0° **23.** 5.0° **24.** 14.0°

25. 6.5° **26.** 47.8° **27.** 68.0° **28.** 26.0°

29. 89.5° **30.** 8.7° **31.** 56.7° **32.** 87.0°

Practice—Chapter 10

Lesson 10.1

1. 0 **2.** $-\frac{3}{2}$ **3.** $0, -\frac{3}{2}$ **4.** -2 **5.** $-\frac{1}{2}$

6. -2 **7.** undefined; $\frac{2}{3}$ **8.** 2; -1

9. $-\frac{1}{5}; \frac{1}{4}$

10. reflection in the x-axis and vertical stretch of 3

11. reflection in the x-axis, vertical stretch of 3, and shift 2 units to the right

12. reflection in the x-axis, vertical stretch of 3, shift 2 units to the right, and shift 2 units up

13.

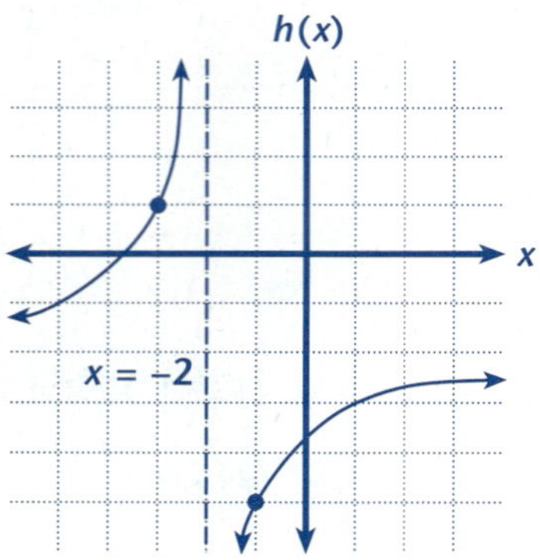

undefined at $x = -2$

14.

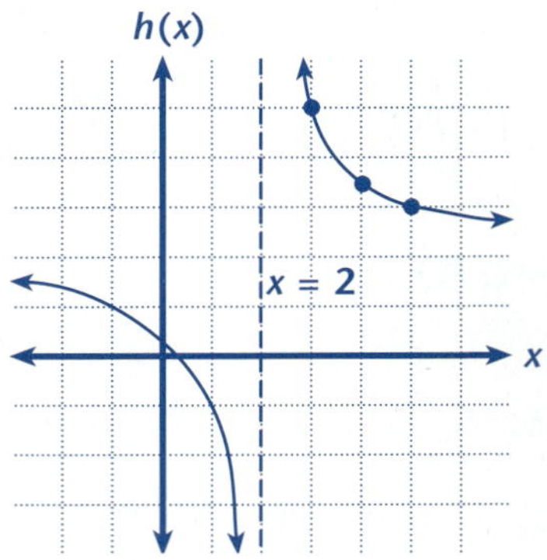

undefined at $x = 2$.

Lesson 10.2

1. not an inverse variation

2. not an inverse variation

3. inverse variation

4. 5 **5.** 25 **6.** 12 **7.** -10 **8.** 12 hours

9. 135 units

10.

Time (seconds)	5	10	15	20	25	30
Horsepower	200	100	$66\frac{2}{3}$	50	40	$33\frac{1}{3}$

11. Each product is 1000.

12. Horsepower varies inversely as time.

Lesson 10.3

1. 2 **2.** $-5, 4$ **3.** 0 **4.** $-1, 0$ **5.** 5

6. ab **7.** $r + 3$ **8.** $\frac{x+3}{x-3}$ **9.** $y - 3$

10. $k - 3$ **11.** $\frac{x^2}{3}$ **12.** $\frac{1}{4x + 4y}$ **13.** $\frac{1}{2}$

14. $2w - 3$ **15.** $\frac{1}{p-7}$ **16.** $8 + x$

17. $\frac{h-3}{h+1}$ **18.** $\frac{1}{3(f+g)^2}$ **19.** $\frac{x+1}{x-1}$ **20.** $\frac{z}{2}$

21. $2x + 8$ **22.** $\frac{y+5}{y-5}$

Lesson 10.4

1. $\frac{7}{5z}$; $z \neq 0$ **2.** $\frac{8}{x-3}$; $x \neq 3$

3. $\frac{3a}{a+b}$; $a \neq -b$ **4.** 1; $x \neq 1$

5. $\frac{q-2p}{pq}$; p and $q \neq 0$

6. $\frac{c^2+d^2}{3cd}$; c and $d \neq 0$

7. $\frac{2}{n}$; m and $n \neq 0$ **8.** $\frac{2}{y}$; $y \neq 0$ **9.** $\frac{2x^3+x^2}{12}$

10. $-a^2$; $a \neq 3$ **11.** $\frac{a^3+a}{(a+1)^2}$; $a \neq -1$

12. 3; $t \neq -2, 0,$ or 2 **13.** $\frac{y+3}{y}$; $y \neq -2$ or 0

14. 1; $x \neq \frac{2}{3}$ **15.** $\frac{m^2-n^2}{mn}$; m and $n \neq 0$

16. $\frac{2x-1}{x^2-1}$; $x \neq \pm 1$ **17.** $\frac{3x-2}{1-x}$; $x \neq \pm 1$

18. 1; $b \neq \pm 5$ **19.** $\frac{2}{c}$; $c \neq -2, 0,$ or 3

20. $\frac{x^2+4xy}{(x+y)^2}$; $y \neq -x$

21. $\frac{4x^2+3x-24}{x^2-6}$; $x \neq \pm\sqrt{6}$

22. $\frac{z-3}{1-z}$; $z \neq 1$ **23.** $\frac{4}{x(x-3)}$; $x \neq 0$ or 3

24. $\frac{3}{4(n-4)}$; $n \neq \pm 4$

25. $\frac{5x-2}{x^2+4x-5}$; $x \neq -5$ or 1

26. $\frac{d^2}{d^2-9}$; $d \neq \pm 3$

Lesson 10.5

1. 9 **2.** 42 **3.** $\frac{16}{5}$ **4.** 30 **5.** −73

6. −3 **7.** −5 **8.** $-\frac{1}{2}$ **9.** 2 **10.** $-\frac{3}{2}, \frac{5}{2}$

11. −3, 2 **12.** 15 **13.** −5 **14.** −2

15. 3 **16.** −6, −2 **17.** 0 **18.** no solution

19. −6, 2 **20.** −4, 0 **21.** 3 **22.** −6, 3

23. −10 **24.** 12 **25.** −12 **26.** 5

Lesson 10.6

1. means: 4, n; extremes: 3, 5

2. means: 5, 4; extremes: $x+2$, $x-2$

3. means: 4, d; extremes: $2d-1$, 5

4. $f = 2$ **5.** $x = \pm 4$ **6.** −4

7. $b = 2, -\frac{3}{2}$ **8.** $x = -1$ **9.** $y = 1$

10. 6.8 tsp or $6\frac{4}{5}$ tsp **11.** 2,136,000 gal

12. 6 ft **13.** 2 oz **14.** 16 oz

Lesson 10.7

1. Distributive Property

2. Definition of Addition or Substitution

3. Subtraction Property of Equality

4. Definition of Opposites

5. Addition Property of Zero

6. Division Property of Equality

7. Definition of Division or Substitution

8. Definition of Division

9. Associative Property of Multiplication

10. Definition of Multiplication

11. Distributive Property

12. Definition of Subtraction

13.

$-6x - 2 = -8x - 4$	Given
$-6x - 2 + 2 = -8x - 4 + 2$	Addition Property of Equality
$-6x = -8x - 2$	Def. of Opposites; Def. of Addition
$-6x + 8x = -8x - 2 + 8x$	Addition Property of Equality
$2x = -2$	Def. of Opposites; Def. of Addition
$\frac{2x}{2} = \frac{-2}{2}$	Division Property of Equality
$x = -1$	Definition of Division

14.

$3(n - 2) = 9$	Given
$3n - 6 = 9$	Distributive Property, Sub.
$3n - 6 + 6 = 9 + 6$	Addition Prop. of Equality
$3n = 15$	Def. of Opposites; Def. of Addition
$\frac{3n}{3} = \frac{15}{3}$	Division Prop. of Equality
$n = 5$	Definition of Division

Lesson Activities—Chapter 1

Lesson 1.1

1. $x = 6$ **2.** $x = 2$ **3.** $x = 4$ **4.** $x = -6$

5. $x = -8$ **6.** $x = -1$ **7.** $x = 4$

8. $x = -3$

Lesson 1.2

1. $x = 3$ **2.** $x = 4$ **3.** $x = -5$ **4.** $x = -4$

5. $x = 8$ **6.** $x = 5$ **7.** $x = 5$ **8.** $x = -8$

9. $x = -9$ **10.** $x = -7$ **11.** $x = -2$

12. $x = 2$

13. Answers may vary. Sample answer: Multiply each side of the equation by -1, and then model the equation $3x = -9$ to solve.

Lesson 1.3

1. $x = -1$ **2.** $x = 3$ **3.** $x = 2$ **4.** $x = 2$

5. $x = 4$ **6.** $x = -3$ **7.** $x = 1$ **8.** $x = 3$

9. $x = -3$ **10.** $x = 2$

Lesson 1.4

1. $x = -6$ **2.** $x = 0$ **3.** $x = -4$

4. $x = -6$ **5.** $x = 7$ **6.** $x = 1$ **7.** $x = -4$

8. $x = -8$ **9.** $x = -7$ **10.** $x = 2$

Lesson 1.5

1. $x > -2$ **2.** $x < -2$ **3.** $x \geq -3$

4. $x \geq 5$ **5.** $x < 1$ **6.** $x \leq 4$ **7.** $x \leq 5$

8. $x > 1$

Lesson 1.6

1. $-4 < x < 4$ **2.** $x \leq -1$ or $x \geq 1$

3. $-2 < x < 2$ **4.** $-5 \leq x \leq 5$

5. $-1 < x < 1$ **6.** $x \leq -2$ or $x \geq 2$

7. $x < -5$ or $x > 5$ **8.** $-6 \leq x \leq 6$

9. $x \leq -4$ or $x \geq 4$ **10.** $x \leq -7$ or $x \geq 7$

11. Answers may vary. Sample answer: Model the solutions to the inequalities $x - 2 > 4$ and $x - 2 < -4$.

12. Answers may vary. Sample answer: Model the solutions to the inequalities $3x \leq 6$ and $3x \geq -6$.

Lesson Activities—Chapter 2

Lesson 2.1

1. See students' work.

2. See students' work for segments. Go down 1 and to the right 2.

3. See students' work for segments. Go up 1 and to the right 1.

4. See students' work for segments. Go up 1 and to the right 3.

5. See students' work for segments. Go down 1 and to the right 3.

6. See students' work for segments. Go up 2 and to the right 1.

7. See students' work for segments. Go up 1 and to the right 4.

8. See students' work for segments. Go down 1 and to the right 4.

9. See students' work for segments. Go up 2 and to the right 3.

10. See students' work for segments. Go down 2 and to the right 3.

11. See students' work for segments. Go up 3 and to the right 4.

12. See students' work for segments. Go down 3 and to the right 4.

13. See students' work for segments. Go down 3 and to the right 1.

Lesson 2.2

1. See students' work for graph; $y = x + 2$

2. See students' work for graph; $y = -\frac{1}{2}x + 3$

3. y-intercept = 4; slope = 1; see students' work for graph.

4. y-intercept = 5; slope = −1; see students' work for graph.

5. y-intercept = 4; slope = −1; see students' work for graph.

6. y-intercept = 1; slope = $\frac{1}{2}$; see students' work for graph.

7. y-intercept = 2; slope = $-\frac{1}{2}$; see students' work for graph.

8. y-intercept = 1; slope = $\frac{1}{3}$; see students' work for graph.

9. See students' work for graphs; the lines have different y-intercepts but the same slope.

Lesson 2.3

1.

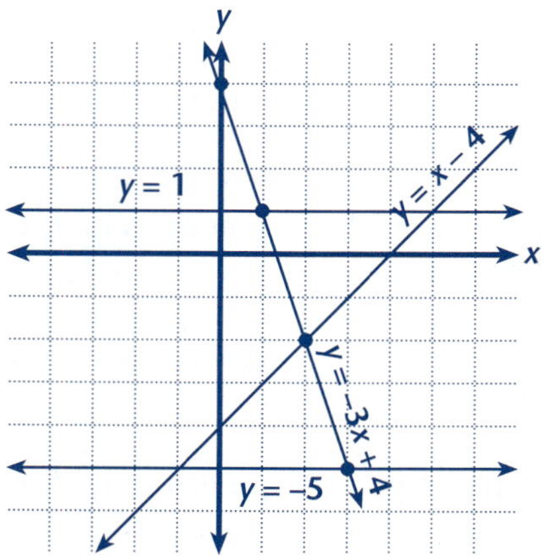

(2, –2)

2. a triangle **3.** 6 square units

4. $y = -3x + 4$ **5.** $13\frac{1}{2}$ square units

6. a triangle **7.** They are congruent.

8. −1 and −7

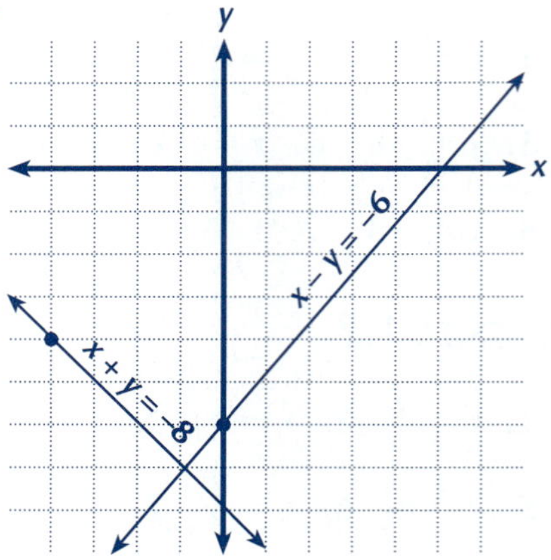

Lesson 2.4

1. $x = 45 + y$ **2.** $3(45 + y) + 2y = 80$

3. $y = -11$ **4.** $x = 45 + (-11); x = 34$

5. $3(34) + 2(-11) = 102 - 22 = 80$

$34 - (-11) = 34 + 11 = 45$

6. Answers will vary. Students should change each equation into y-intercept form first. Possible table:

x	$y = 5x + 0.5$	$y = 15.5 - x$
1	5.5	14.5
1.5	8	14
2	10.5	13.5
2.5	13	13
3	15.5	12.5

7.

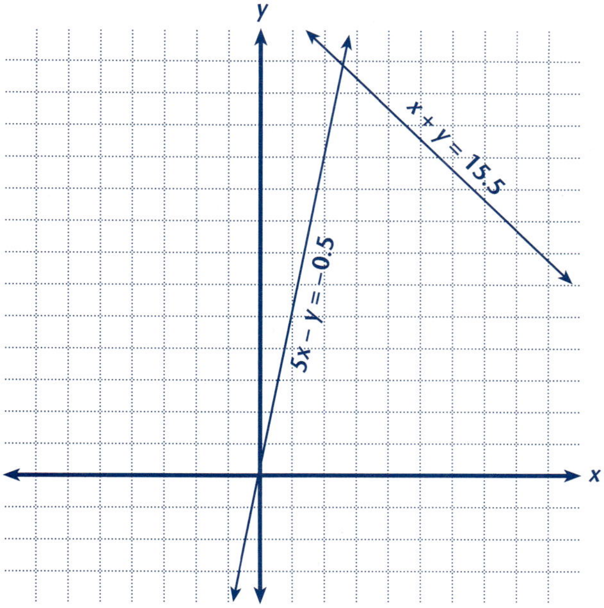

8. $x = 2.5$; $y = 13$ **9.** Answers will vary.

10. Answers will vary, but students should realize that the substitution method will always give them exact values for x and y.

Lesson 2.5

1. C **2.** A **3.** B **4.** D

Answers will vary for the method and explanation. Possible answers are given.

5. Equations: $5x + 10y = 150$; $x = 2y + 3$, where x = cost of one share of company A stock and y = cost of one share company B stock
Method: substitution
Solution: company A: $16.50 per share; company B: $6.75 per share
Explanation: the substitution method because it is easy to substitute $x = 2y + 3$ for x in the first equation

6. Equations: $3x + 5y = 27.50$; $3x + 9y = 43.50$, where x = rate per child and y = rate per hour
Method: elimination
Solution: $2.50 per child; $4 per hour
Explanation: the elimination method, because there are no terms with a coefficient of 1

Lesson 2.6

Answers will vary for Exercises 1–3. Possible answers and graphs are given.

1. $2x + y = 3$
$-2x + y = -5$

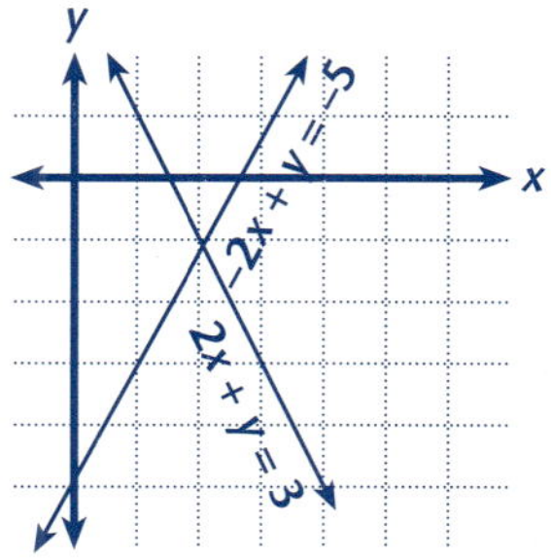

2. $2x + y = 3$
$y = -2x + 3$

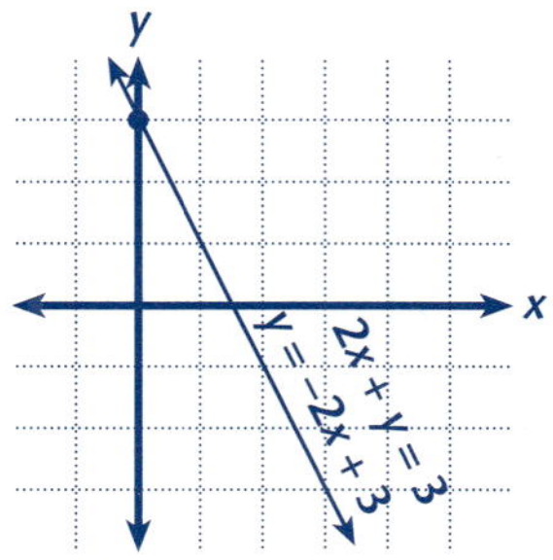

3. $2x + y = 3$
$2x + y = 6$

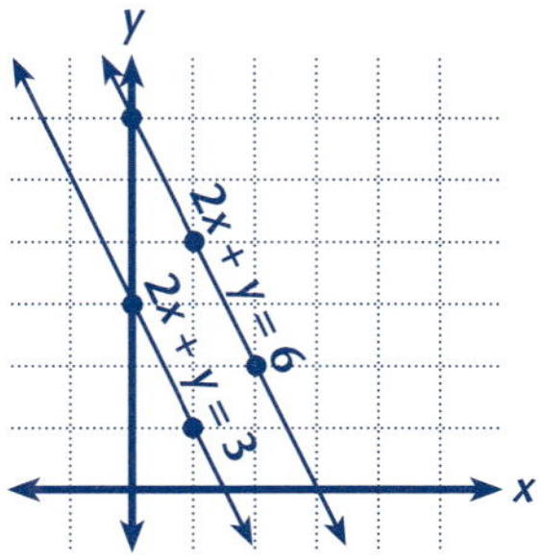

4. Students' explanations will vary but should mention the characteristics specific to each type of system. There is only one solution to the system in Exercise 1; there are an infinite number of solutions to the system in Exercise 2; and there is no solution to the system in Exercise 3.

5. sometimes true **6.** never true

7. always true

Lesson 2.7

1.

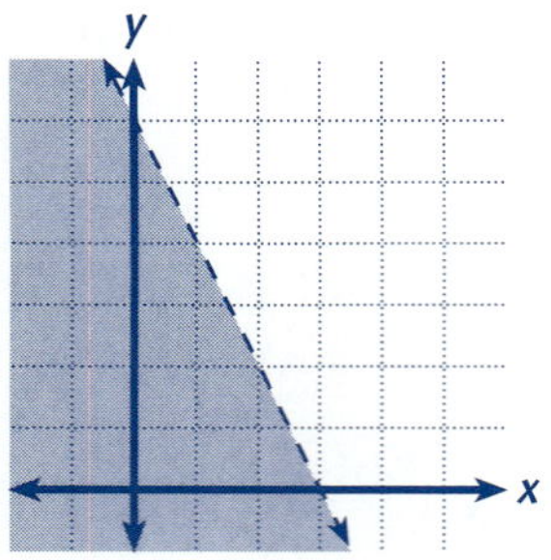

2.

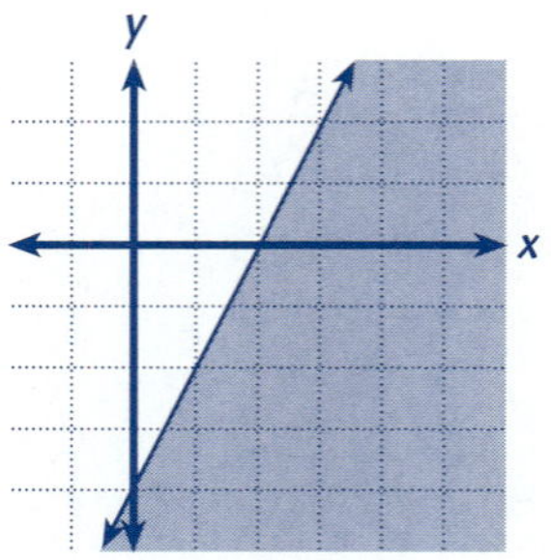

3. Graph 1 has a dashed boundary because points on the boundary line are not included in the solution set. Graph 2 has a solid boundary because the solution set includes the points on this line.

Answers will vary. Possible answers are given for Exercises 4–6.

4. (2, 1); (1, 2)

5. (3, 1); (4, 2)

6. $50b + 30s \leq 150$; $37.50b + 15s \leq 90$

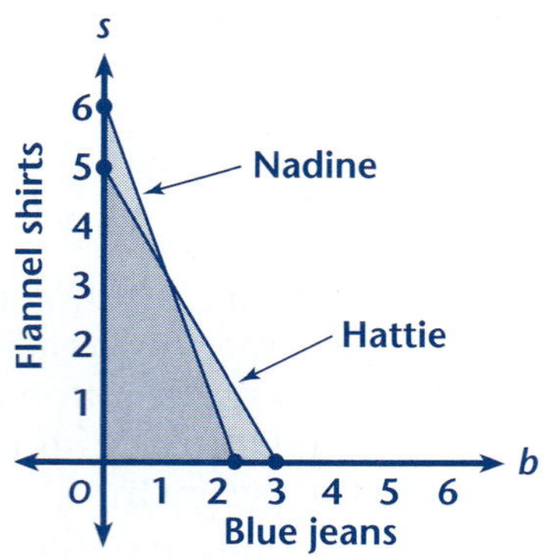

Lesson 2.8

1. Equations: $(r + c)(1.5) = 12.75$; $(r - c)(1.7) = 12.75$
   Method: elimination
   Solution: rate of current = 0.5 km per hour; Joe's rowing rate = 8 km per hour

2. Equations: $x + y = 15$; $0.25x + 0.3y = 0.28(15)$
   Method: substitution
   Solution: 6 quarts of the 25% solution; 9 quarts of the 30% solution

3. Equations: $t = u - 1$; $10t + u = 5(t + u) + 1$
   Method: substitution
   Solution: 56

4. Equations: $q + n = 95$; $0.25q + 0.05n = 10.75$
   Method: substitution or elimination
   Solution: 30 quarters; 65 nickels

5. Equations: $p + m + 5 = 48$; $m = 3p - 5$
   Method: substitution or elimination
   Solution: Paula is 12; her mother is 31.

6. Answers will vary.

Lesson Activities—Chapter 3

Lesson 3.1

1. The number of native speakers and the total number of speakers of selected languages

2. 4×2 **3.** l_{31}

4. The total number of people in the world who speak Spanish is 371 million.

5.

	Native speakers (millions)	Total no. of speakers worldwide (millions)
Bengali	184	192
Arabic	182	214

6. i_{31} **7.** There are 2.54 cm in an inch.

8. The entries in the first table name the measurements; the entries in the matrix name the values for each measurement as it relates to 1 inch.

9. Because this is the unit of measure on which the matrices are based

10. 3×2

Lesson 3.2

1.

Trifax Co.	=	$-2\frac{3}{4}$
Compu Count Inc.		$+1\frac{3}{8}$
Video Antics		0

2. Possible answer: For the month, this investor lost $2.75 for each share of Trifax, made a profit of $1.375 for each share of Compu Count Inc., and did not lose or make any money on the shares of Video Antics.

3. Matrix C

	Gas	**Electric**	**Water**
Jan. 96	33	24	13
Feb. 96	13	41	10
Mar. 96	29	63	13

Matrix D

	Gas	**Electric**	**Water**
Jan. 97	73	37	11
Feb. 97	32	22	7
Mar. 97	32	19	3

Lesson 3.3

1. $\begin{bmatrix} 22 & 49 \\ 25 & 58 \end{bmatrix}$ 2. $\begin{bmatrix} 9 & 19 & 25 \\ 12 & 26 & 35 \\ 15 & 33 & 45 \end{bmatrix}$

3. Not possible 4. $\begin{bmatrix} 1 & 3 & 5 \\ 2 & 4 & 5 \end{bmatrix}$

5. $\begin{bmatrix} 3 & 9 & 15 \\ 6 & 12 & 15 \end{bmatrix}$ 6. $\begin{bmatrix} 2 & 8 \\ 4 & 10 \\ 6 & 12 \end{bmatrix}$

7. Matrix A; because $AC = A$ 8. $\begin{bmatrix} 1 & 0 \\ 0 & 1 \end{bmatrix}$

9. No; the Commutative Property for Multiplication does not necessarily hold for matrices. In this case, BA is a 3×3 matrix and AB is a 2×2 matrix.

10. 2 rows; 3 columns

11. Answers will vary. Two possible matrices are as follows:

$$X = \begin{bmatrix} 40 & 51 & 62 \\ 26 & 33 & 40 \end{bmatrix} \quad Y = \begin{bmatrix} 1 & 0 & 0 \\ 0 & 1 & 0 \\ 0 & 0 & 1 \end{bmatrix} \text{ or}$$

$$X = \begin{bmatrix} 4 & 7 \\ 2 & 5 \end{bmatrix} \text{ and } Y = \begin{bmatrix} 3 & 4 & 5 \\ 4 & 5 & 6 \end{bmatrix}$$

Lesson 3.4

1. $\begin{bmatrix} 1 & 0 \\ 0 & 1 \end{bmatrix}$

2. $\begin{bmatrix} 3a + c & 3b + d \\ 4a + 2c & 4b + 2d \end{bmatrix} = \begin{bmatrix} 1 & 0 \\ 0 & 1 \end{bmatrix}$

3. $3a + c = 1$
$4a + 2c = 0$

$3b + d = 0$
$4b + 2d = 1$

4. $-6a - 2c = -2$
$4a + 2c = 0$
$a = 1$

$-6b - 2d = 0$
$4b + 2d = 1$
$b = -\frac{1}{2}$

5. $4 + 2c = 0$
$c = -2$

$-2 + 2d = 1$

$d = \frac{3}{2}$

6. $\begin{bmatrix} 1 & -\frac{1}{2} \\ -2 & \frac{3}{2} \end{bmatrix}$ 7. $\begin{bmatrix} \frac{1}{2} & \frac{1}{2} \\ \frac{3}{4} & \frac{1}{4} \end{bmatrix}$

8. $\begin{bmatrix} -\frac{3}{14} & \frac{2}{7} \\ \frac{5}{14} & -\frac{1}{7} \end{bmatrix}$

Lesson 3.5

1. $-12x - 6y = -51$
$3x + 6y = 24$

$-9x = -27$
$x = 3$
$4(3) + 2y = 17$
$2y = 5$
$y = 2.5$

2. $3x = -6y + 24$
$x = -2y + 8$
$4(-2y + 8) + 2y = 17$
$-8y + 32 + 2y = 17$
$-6y = -15$
$y = 2.5$

$3x + (6)(2.5) = 24$
$3x = 9$
$x = 3$

3. $\begin{bmatrix} 4 & 2 \\ 3 & 6 \end{bmatrix}\begin{bmatrix} x \\ y \end{bmatrix} = \begin{bmatrix} 17 \\ 24 \end{bmatrix}$

$$\begin{bmatrix} \frac{1}{3} & -\frac{1}{9} \\ -\frac{1}{6} & \frac{2}{9} \end{bmatrix}\begin{bmatrix} 4 & 2 \\ 3 & 6 \end{bmatrix}\begin{bmatrix} x \\ y \end{bmatrix} = \begin{bmatrix} \frac{1}{3} & -\frac{1}{9} \\ -\frac{1}{6} & \frac{2}{9} \end{bmatrix}\begin{bmatrix} 17 \\ 24 \end{bmatrix}$$

$$\begin{bmatrix} 1 & 0 \\ 0 & 1 \end{bmatrix}\begin{bmatrix} x \\ y \end{bmatrix} = \begin{bmatrix} \frac{17}{3} - \frac{24}{9} \\ -\frac{17}{6} + \frac{48}{9} \end{bmatrix}$$

$x = 3$
$y = 2.5$

4. $\begin{bmatrix} 4 & 2 & 17 \\ 3 & 6 & 24 \end{bmatrix}$

$3[4 \quad 2 \quad 17] = [12 \quad 6 \quad 51]$
$1[3 \quad 6 \quad 24] = [\ 3 \quad 6 \quad 24]$
$[\ 9 \quad 0 \quad 27]$

$x = 3$
$y = 2.5$

5. Answers will vary.

6. A combination ticket costs $24; a theme-park ticket costs $18.

Lesson Activities—Chapter 4

Lesson 4.1

1–3. Answers may vary. See students' work.

Lesson 4.2

1. $\frac{1}{5}$ or 20% **2.** $\frac{2}{5}$ or 40%

3. $\frac{3}{5}$ or 60% **4.** $\frac{1}{10}$ or 10%

5. The simulation cannot be used because there is not enough data for 5 events.

Lesson 4.3

1. Bulls **2.** Suns **3.** Suns **4.** Suns

5. Bulls **6.** 16.1 **7.** 14.95 **8.** 13.9

9. Possible answer: The mean point average probably indicates the team that plays the best because it represents the average number of points scored by the top players per game.

Lesson 4.4

1. region I **2.** region IV **3.** region II

4. region III **5.** 4 **6.** 4 **7.** 3 **8.** 5

9. Possible answer: First find out how many Smiths live in Montgomery (5). Then add the number of Smiths who are not employed (5). Next subtract the Smiths who are both living in Montgomery and not employed because these people were counted in both the first and second steps (3). The result, $5 + 5 - 3 = 7$, gives the number of Smiths who are living in Montgomery or not employed.

Lesson 4.5

1.

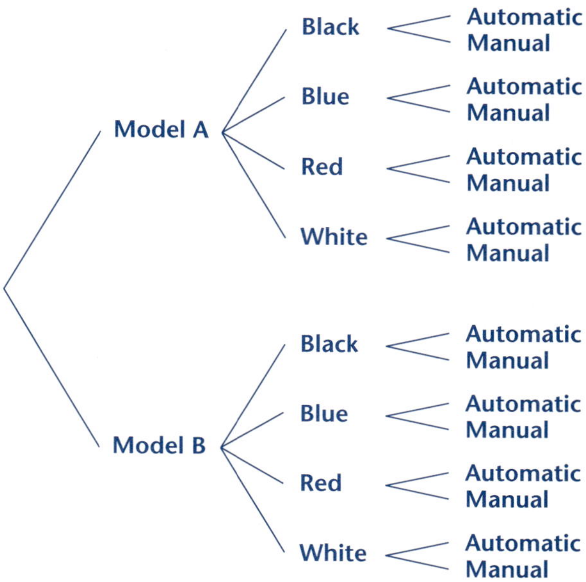

2. Multiply the number of kinds of cars (2) by the number of colors (4) and the number of types of transmissions (2): $2 \times 4 \times 2 = 16$ choices.

3. model A, black, automatic transmission; model B, black, automatic transmission

4. all model A or model B cars with automatic transmissions

5. $9000

Lesson 4.6

1. 28.6% or $\frac{2}{7}$ 2. 37.5% or $\frac{3}{8}$

3. 94.4% or $\frac{17}{18}$ 4. Coin Toss; 28.6%

5. 13.6% or $\frac{3}{22}$ 6. 86.4% or $\frac{19}{22}$

7. Possible answer: The probability of winning for player 1 is $\frac{2}{7}$, or 28.6%. With another coin toss, the probability almost doubles to $\frac{26}{49}$, or 53.1%. So player 2 has almost twice the chance of winning that player 1 has.

Lesson 4.7

1. Dependent; $\frac{5}{51}$; there are 18 ways to choose the first sock and 17 ways to choose the second one, so there are $18 \times 17 = 306$ equally likely ways of drawing 2 socks. There are 6 ways to draw the first white sock and 5 ways to draw the second one or $6 \times 5 = 30$ ways of choosing two white socks. Therefore, $\frac{30}{306}$, or $\frac{5}{51}$, is the probability of drawing 2 white socks.

2. Independent; $\frac{3}{100}$; there is a $\frac{15}{50}$ probability of choosing a black marble and $\frac{5}{50}$ probability of choosing a white marble. The probability of these two independent events occurring is $\frac{15}{50} \times \frac{5}{50} = \frac{3}{100}$.

3. Dependent; $\frac{1}{17}$; there are 52 ways to choose the first card and 51 ways to choose the second card, for a total of 2652 ways of choosing two cards without replacement. There are 13 ways to choose hearts the first time and 12 ways to choose hearts the second time, giving $13 \times 12 = 156$ ways of choosing two hearts. Therefore, the probability of both dependent events occurring is $\frac{156}{2652}$, or $\frac{1}{17}$.

4. Dependent; $\frac{2}{9}$; there are 10 ways to choose on Monday and 9 ways to choose on Tuesday, for a total of 90 ways to choose in all. There are 5 ways of choosing blue jeans on Monday and 4 ways on Tuesday, for a total of 20 ways to choose blue jeans. Therefore, the probability that both dependent events occur is $\frac{20}{90}$, or $\frac{2}{9}$.

5. Independent; $\frac{3}{20}$; the probability that both independent events occur equals the probability of choosing black $\left(\frac{3}{10}\right)$ times the probability of choosing blue $\left(\frac{1}{2}\right)$, or $\frac{3}{10} \times \frac{1}{2} = \frac{3}{20}$.

Lesson Activities—Chapter 5

Lesson 5.1

1.
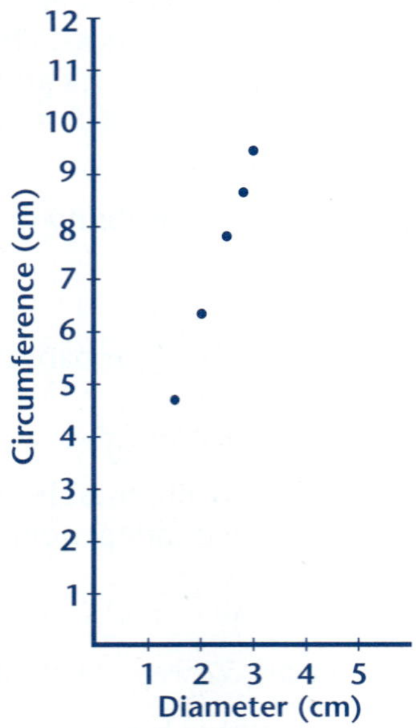

2. The graph is a function because it is a set of ordered pairs for which no two pairs have the same first coordinate.

3. The domain is the set of positive real numbers; the range is the set of positive real numbers.

4.

C (cm)	4.71	6.28	7.85	8.64	9.42
d (cm)	1.5	2	2.5	2.75	3
Ratio	3.14	3.14	3.14	3.14	3.14

Yes. The ratio is constant at about 3.14.

5. π **6.** $C(d) = \pi d$ **7.** ≈ 11 cm

Lesson 5.2

1.

Energy	1.04	104
Food	1.42	142
Shelter	1.57	157
Medical	1.89	189
Apparel	1.33	133
Telephone	1.22	122

2. $c(x) = 100x$ **3.** a stretch **4.** 33%

5. 89%

6. Each ratio represents 1 + the decimal equivalent of the percent increase.

7. $175

Lesson 5.3

1. King's Kid: $k(x) = 4x + 50$
Lady K: $L(x) = 3x + 75$

2. Check students' graphs. Both deals are the same for 25 people. King's Kid's offer is better for 24 or fewer people. Lady K's offer is better for 26 or more people.

3. King's Kid: $k(x) = 3x + 50$
Lady K: $L(x) = 2x + 75$

4. Check students' graphs. Both deals are the same for 25 people. King's Kid's offer is better for 24 or fewer people. Lady K's offer is better for 26 or more people.

5. Although the y-value is different, the x-value remains the same.

6. The graphs for each company have the same y-intercepts, but different slopes. One is a stretch of the other.

Lesson 5.4

1. The pattern is symmetrical about line 1.

2. For each fold, the pattern is symmetrical about the line.

Lesson 5.5

1. The sum of the perpendicular distances from the three sides of an equilateral triangle to any point in its interior is equal to the length of the triangle's altitude.

Lesson 5.6

1. the walker's weight and the walking speed

2. The domain is all x such that 2 mph $\leq x \leq$ 5 mph.
For weight of 100 lb, the range is all y such that 130 Cal/h $\leq y \leq$ 375 Cal/h.
For weight of 120 lb, the range is all y such that 150 Cal/h $\leq y \leq$ 450 Cal/h.
For weight of 140 lb, the range is all y such that 180 Cal/h $\leq y \leq$ 525 Cal/h.
For weight of 180 lb, the range is all y such that 230 Cal/h $\leq y \leq$ 675 Cal/h.

3. 2.5 mph **4.** 3.5 mph

5. from 4 mph to 4.5 mph

6. 180 Calories; 215 Calories; 250 Calories; 325 Calories

Lesson Activities—Chapter 6

Lesson 6.1

1. The shapes alternate between rectangles and squares.

2. $2^1, 2^2, 2^3, 2^4$

3. The exponent describes the number of times that the base 2 is used as a factor.

4. squares; rectangles

Lesson 6.2

1. $2 \times 2 \times 2 = 2^3 = 8$; $3 \times 3 \times 3 = 3^3 = 27$; $4 \times 4 \times 4 = 4^3 = 64$

2. A 4 cm × 5 cm × 6 cm and a 5 cm × 1 cm × 1 cm rectangular prism used to form a 5 cm × 5 cm × 5 cm cube.

3. $1 \times 2 \times 3 + 2 = 8 = 2^3$; $2 \times 3 \times 4 + 3 = 27 = 3^3$; $3 \times 4 \times 5 + 4 = 64 = 4^3$

4. $x(x + 1)(x + 2) + (x + 1) = (x + 1)^3$

Lesson 6.3

1. $\frac{1}{2} \times \frac{1}{2}$ **2.** decreasing

3. in the denominator

4. 2^1; 2^0; 2^{-1}; 2^{-2}

5. Possible answer: If x is any number except zero and n is a positive integer, then $x^{-n} = \frac{1}{x^n}$. For any positive number x, $x^0 = 1$.

Lesson 6.4

1. Answers will vary. One estimate is 40 as the least number and 100 as the greatest number of kernels.

2. Check students' calculations. Based on the estimates given for Exercise 1, $\sqrt{(40)(100)} \approx 64$.

3. Students should multiply the result from Exercise 2 by $(2)(10^3)$. $2(64)(10^3) = 1.28 \times 10^5$

4. 10^6 cm^3

5. Answers will vary. One estimate might be 1.5 cm.

6. Check students' calculations. Based on the estimate given for Exercise 4, $(1.5)^3 \approx 3.38$ cm^3.

7. $\frac{10^6}{3.38}$; 2.96×10^5

8. Answers will vary. Students should note that the Fermi method is a good order-of-magnitude estimate.

Lesson 6.5

1.

x	3^x	$4(3)^x$	$8(3)^x$	$\left(\frac{1}{3}\right)^x$	$4\left(\frac{1}{3}\right)^x$	$8\left(\frac{1}{3}\right)^x$
−3	$\frac{1}{27}$	$\frac{4}{27}$	$\frac{8}{27}$	27	108	216
−2	$\frac{1}{9}$	$\frac{4}{9}$	$\frac{8}{9}$	9	36	72
−1	$\frac{1}{3}$	$\frac{4}{3}$	$\frac{8}{3}$	3	12	24
0	1	4	8	1	4	8
1	3	12	24	$\frac{1}{3}$	$\frac{4}{3}$	$\frac{8}{3}$
2	9	36	72	$\frac{1}{9}$	$\frac{4}{9}$	$\frac{8}{9}$
3	27	108	216	$\frac{1}{27}$	$\frac{4}{27}$	$\frac{8}{27}$

2. $y = 3^x$, $y = 4(3)^x$, and $y = 8(3)^x$

3. $y = \left(\frac{1}{3}\right)^x$, $y = 4\left(\frac{1}{3}\right)^x$, and $y = 8\left(\frac{1}{3}\right)^x$

4. $y = 3^x$, $y = 4(3)^x$, and $y = 8(3)^x$

5. At $x = 0$, $y = 3^x$ and $y = \left(\frac{1}{3}\right)^x$ have the same y-value of 1; $y = 4(3)^x$ and $y = 4\left(\frac{1}{3}\right)^x$ have the same y-value of 4; and $y = 8(3)^x$ and $y = 8\left(\frac{1}{3}\right)^x$ have the same y-value of 8.

6. The 4 causes the y-values to be 4 times larger than the corresponding y-values for $y = \left(\frac{1}{3}\right)^x$. The 8 causes the y-values to be 8 times larger.

7. The 4 causes the y-values to be 4 times larger than the corresponding y-values for 3^x. The 8 causes the y-values to be 8 times larger.

Lesson 6.6

1.

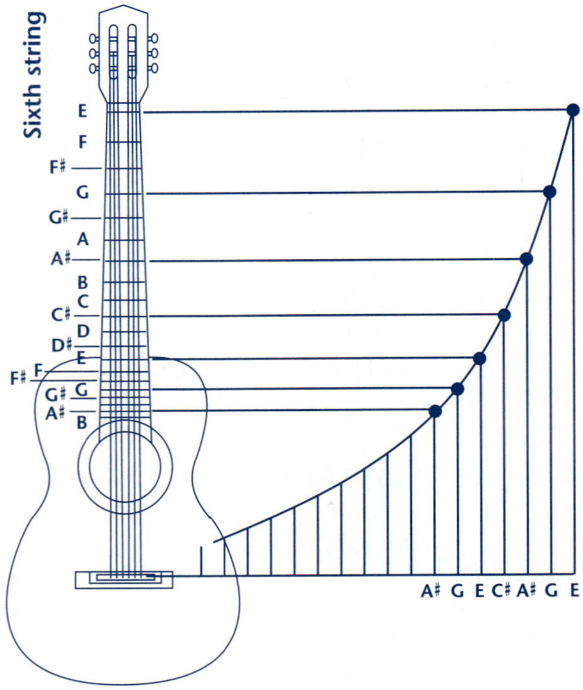

2. Answers will vary.

3. Check students' work. Possible answer: The curve appears to be an exponential function, with the length of string doubling between corresponding notes (i.e., A# to A#).

4. Check students' work.

Lesson Activities—Chapter 7

Lesson 7.1

1–2. Answers may vary. See students' work.

Lesson 7.2

1.

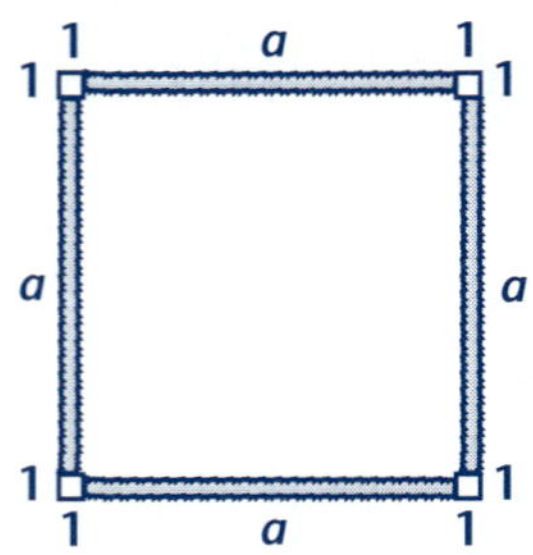

2.

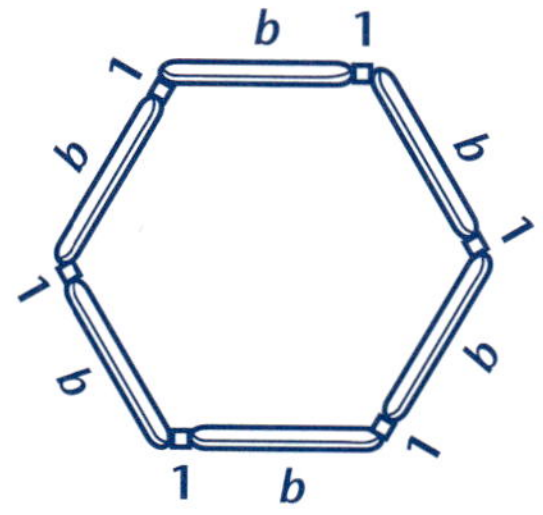

3. Answers may vary. Possible answer:

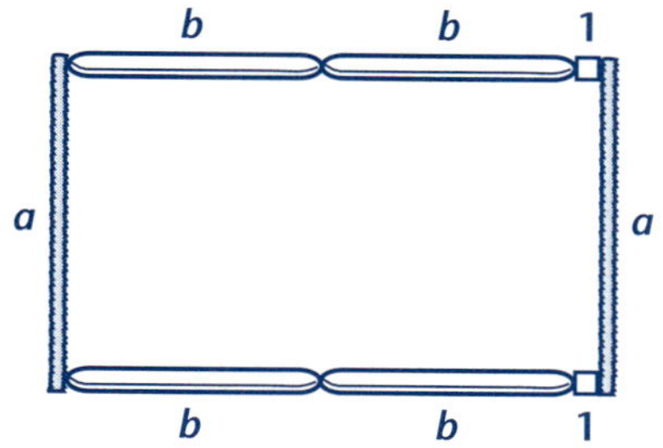

4. Answers may vary. Possible answer:

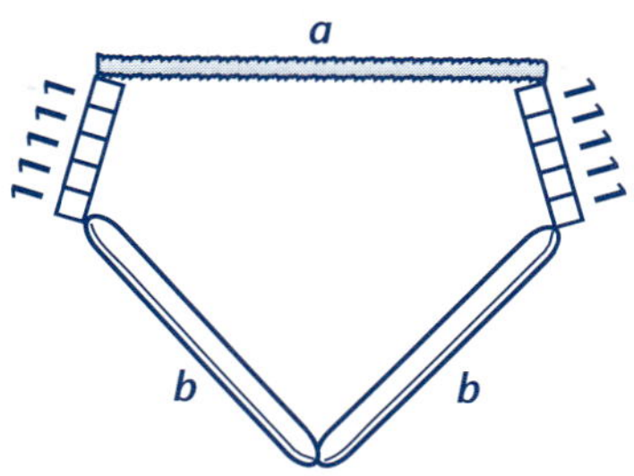

5. $3a + 6$ or $3(a + 2)$

6. $2a + 2b + 4$ or $2(a + b + 2)$

7. $2a + 5b + 7$ **8.** $6a$

Lesson 7.3

1. $(x + 2)(x + 1)$ **2.** $(x + 1)^2$

3. $(x - 1)(x + 1)$ **4.** $(x - 2)(x + 2)$

5. $(x + 1)(2x + 1)$ **6.** $(2x - 3)(x - 1)$

7. $(x - 5)(x + 2)$ **8.** $(x + 4)(x - 2)$

9. $(x - 2)^2$ **10.** $(2x - 5)(x + 2)$

11. $(x + 3)(x - 6)$ **12.** $(x - 1)(2x + 7)$

13. $(x + 3)^2$ **14.** $(x + 6)(x - 1)$

15. $(x + 5)(x - 5)$

Lesson 7.4

1. The model is a square with sides of $(a + b)$. The area of the square is $(a + b)^2$. The model can be cut apart to form the two squares and two rectangles shown, illustrating that the area is equivalent to $a^2 + 2ab + b^2$.

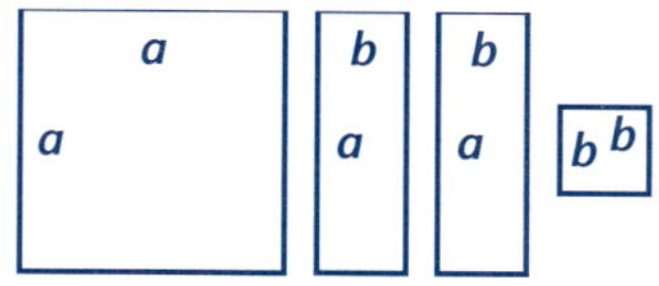

2. Student models should resemble the one shown. The shaded region is a square whose area is $(a - b)^2$. Its area can be represented by cutting away the two rectangles of dimensions $b \times (a - b)$ and the square with side b, illustrating that the area is equivalent to $a^2 - 2b(a - b) - b^2 = a^2 - 2ab + b^2$.

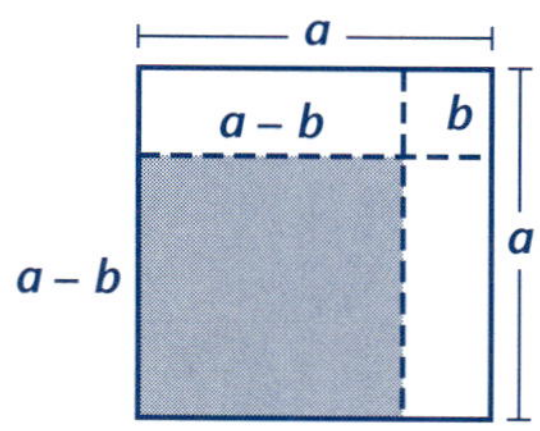

3. The model consists of one square whose area is $(a + b)^2$ and one square whose area is $(a - b)^2$, which overlaps part of the larger square. The total area is $(a + b)^2 + (a - b)^2$. By separating the layers, the model becomes the four squares shown, illustrating that the total area is equivalent to $2a^2 + 2b^2$.

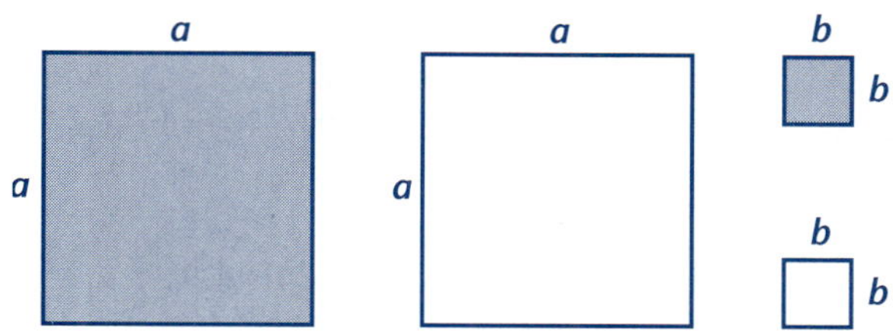

Lesson 7.5

1.

l	w	GCF (l, w)	Unit squares intersected (n)
3	1	1	3
3	2	1	4
3	3	3	3
4	1	1	4
4	2	2	4
4	3	1	6
4	4	4	4
5	1	1	5
5	2	1	6
5	3	1	7
5	4	1	8
5	5	5	5

2.

6	1	1	6
6	2	2	6
6	3	3	6
6	4	2	8
6	5	1	10
6	6	6	6
7	1	1	7
7	2	1	8
7	3	1	9
7	4	1	10
7	5	1	11
7	6	1	12
7	7	7	7

3. rectangles whose length and width have a GCF of 1

4. rectangles whose length and width have a GCF of 2

5. The number of squares intersected by the diagonal is the sum of the length and width of the rectangle minus the greatest common factor of the two sides, or $n = l + w - \text{GCF}(l, w)$.

Lesson 7.6

1. $a^2(a - b)$ **2.** $ab(a - b)$ **3.** $b^2(a - b)$

4. $a^2(a - b) + ab(a - b) + b^2(a - b)$

5. $(a - b)(a^2 + ab + b^2)$

6. Possible answer: Since a cube of side b was removed from the cube of side a, the volume of the remaining figure is $a^3 - b^3$. The sum of the volumes of region I, II, and II is also equal to the volume of the remaining figure, so $a^3 - b^3$ must equal $(a - b)(a^2 + ab + b^2)$.

Lesson 7.7

1.

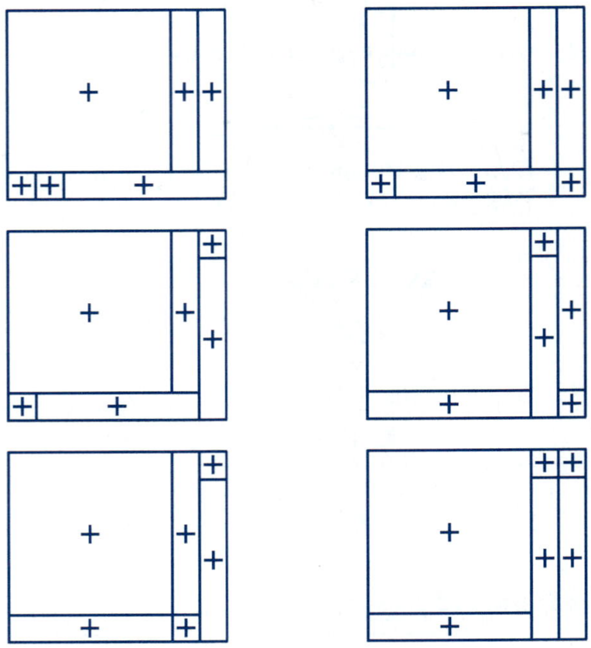

2. 5

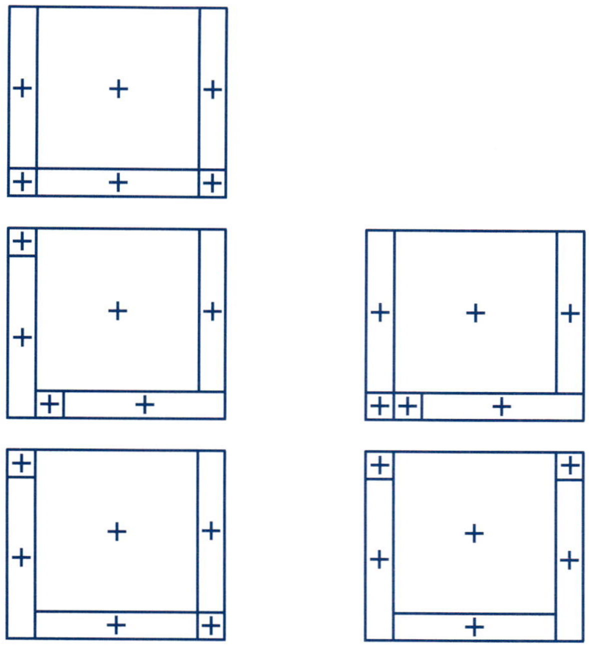

3. $(x + 2)(x + 1)$; possible answer: they all show that the factors of $x^2 + 3x + 2$ are $(x + 2)(x + 1)$.

Lesson Activities—Chapter 8

Lesson 8.1

1. 4 **2.** 8 **3.** 12 **4.** 12 **5.** 24 **6.** 16

7. 40

8. 60

Number on a side	Total number in square
0	0
1	4
2	12
3	24
4	40
5	60

9. 4

10. 4 8 12 16 20 (differences: 4 4 4 4)

The second differences all equal 4.

Lesson 8.2

1. 35

2. The length of a side is equal to the square root of 1225.

3. Possible answer: The area of the square represents the number whose square root you want to find. Since the area of the square equals the side multiplied by itself, the side represents the square root of the number.

4. 42; use 16 hundreds blocks, two groups of 8 rods, and 4 units.

5. Student drawings should show one 5×5 square with two $\frac{1}{5} \times 5$ rods added to each side and four $\frac{1}{5} \times \frac{1}{5}$ units added to the corner to complete the square.

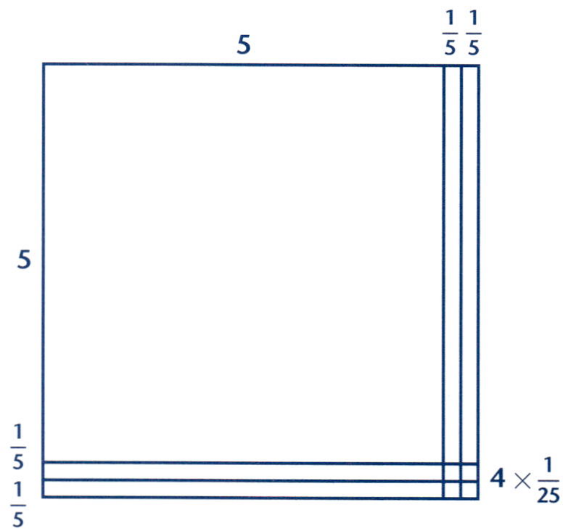

Lesson 8.3

1.

	Vertex	
b	x-coordinate	y-coordinate
−4	2	−4
−3	1.5	−2.25
−2	1	−1
−1	0.5	−0.25
0	0	0
1	−0.5	−0.25
2	−1	−1
3	−1.5	−2.25

2. 1.75 1.25 0.75 0.25 −0.25 −0.75 −1.25

−0.5 −0.5 −0.5 −0.5 −0.5 −0.5

All of the second differences equal −0.5.

3. a parabola **4.** $y = -x^2$

5. The vertex of any parabola in the form $y = x^2 + bx$ lies on the parabola $y = -x^2$.

Lesson 8.4

1. $(x + 5)(x + 1)$ **2.** −5 and −1

3. $(x + 3)^2 + 4$

4.

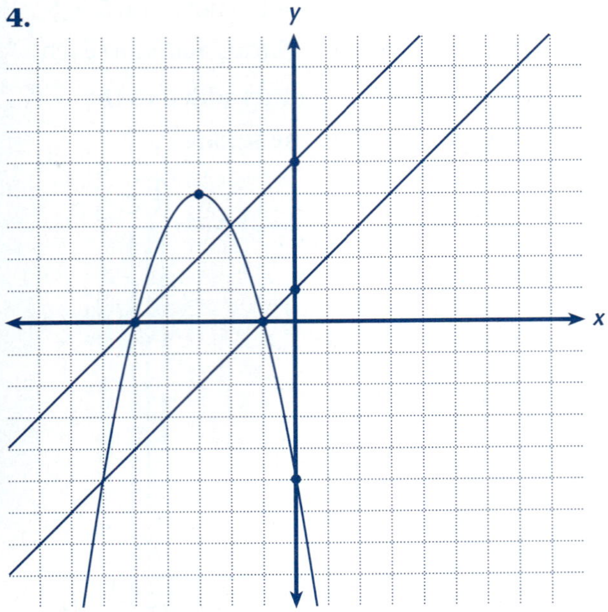

5. The x-intercept of $y = x + 5$ is $x = -5$, and the x-intercept of $y = x + 1$ is $x = -1$.

6. Each linear function intersects the parabola at an x-intercept.

7. The x-intercepts are the zeros of a function.

8. Each linear function intersects the quadratic function at an x-intercept. Since the x-intercepts are the zeros of a function, the zeros of the function and the solutions by factoring are the same.

Lesson 8.5

1.

a	x-coordinate	y-coordinate
−4	0.5	−4
−2	1	−3
−1	2	−1
2	−1	−7
3	$-\frac{2}{3}$	$-\frac{19}{3}$
4	−0.5	−6

2. a line **3.** $y = 2x - 5$

4. It lies on the line $y = 2x - 5$.

5.

a	x-coordinate	y-coordinate
−4	−0.75	11.25
−2	−1.5	13.5
−1	−3	18
2	1.5	4.5
3	1	6
4	0.75	6.75

The vertex of the parabola lies on the line $y = -3x + 9$.

6. The vertex of the parabola lies on the line $y = \frac{bx}{2} + c$.

Lesson 8.6

1.

Span length	Sag
0	0
82.0	0.2
125.2	0.4
198.5	1.0
246.0	1.6
302.1	2.4
328.0	2.8
384.1	3.9

2. 1.6 ft

3. all span lengths greater than or equal to 198.5 ft

4. 0 ft **5.** $23\frac{1}{3}$ ft

Lesson Activities—Chapter 9

Lesson 9.1

1. Possible answer: The factors get closer and closer to the actual square root with each step.

2. 8.062 3. 10.392

4. The approximation is −6.48074211. Using two negative factors approximates the negative square root.

Lesson 9.2

1. 4.58277027 2. 4.582575699

3. 4.58257555; 4.582575695; 4.582575696

4. Possible answer: The orbits get closer and closer to the same number; they begin to repeat. The final orbit is an approximation for the square root of 21.

5. Answers will vary. For example, if the seed used is −4, the approximation of the square root of 16 is −4. The orbits are all negative numbers.

6. 2.75; 2.6477272; 2.645752048; 2.645751311; the orbits approximate the square root of 7.

Lesson 9.3

1. $x + 2 = x^2$ 2. $x + 2 = x^2$

3. $x + 2 = x^2$

4. Each equation produces the same result as $(\sqrt{x+2})^2 = x^2$.

5. Check students' graphs; there is one point of intersection.

6. Check students' graphs; there is one point of intersection.

7. Check students' graphs; there is one point of intersection.

8. Check students' graphs; there is one point of intersection.

9. $x = 2$ 10. $x = 2$ 11. $x = -1$

12. $x = -1$

13. $x = 2$ is the x-coordinate of the only point where the graphs of $y = \sqrt{x+2}$ and $y = x$ intersect.

14. $-\sqrt{x+2} = x$ and $\sqrt{x+2} = -x$

15. $x = -1$ is a solution to the equations in which one side is positive and one side is negative.

Lesson 9.4

1. Check students' work.

2. Possible answer: The sum of pieces 4 and 5 represents the area of the smallest square, which is a^2. The sum of pieces 1, 2, and 3 represents the area of the larger square, which is b^2. Since the 5 pieces together cover the area of the largest square, which is c^2, the proof shows that $a^2 + b^2 = c^2$.

3. Check students' work.

4. Check students' work.

Lesson 9.5

1. $\frac{1}{10}$ h or 0.1 h, or 6 min 2. $\frac{1}{2}$ h, or 30 min

3. $t = \dfrac{\sqrt{1^2 + (1-x)^2}}{3} + \dfrac{x}{5}$

4.

x (mi)	0	$\frac{1}{8}$	$\frac{1}{4}$	$\frac{3}{8}$	$\frac{1}{2}$
t (h)	0.4714	0.4679	0.4667	0.4681	0.472

x (mi)	$\frac{5}{8}$	$\frac{3}{4}$	$\frac{7}{8}$	1
t (h)	0.4810	0.4936	0.5109	0.533

5. Rusty should row to the point that is $\frac{1}{4}$ mile from the town.

Lesson Activity Answers

Lesson 9.6

1. area of trapezoid $ABCD = \frac{1}{2}(a + b)(a + b)$

area of triangle $ABE = \frac{1}{2}(ab)$

area of triangle $BCE = \frac{1}{2}c^2$

area of triangle $CDE = \frac{1}{2}(ab)$

2. $ab + \frac{1}{2}c^2$

3. Area Addition Postulate

4. $a^2 + b^2 = c^2$; this is the "Pythagorean" Right-Triangle Theorem.

5. $\frac{a\sqrt{2}}{2}$ **6.** $\frac{b\sqrt{2}}{2}$ **7.** ab

Lesson 9.7

1–2.

Cycle	a	b	c	d
Ratio	2:1	2:1	3:1	3:2
Angle	180°	180°	120°	240°

Cycle	e	f	g	h
Ratio	5:2	5:3	8:3	8:5
Angle	144°	216°	135°	225°

The ratios contain the sequence of numbers in Fibonacci's sequence.

Lesson 9.8

1.

Angle measure	Sine	Cosine
0°	0	1
10°	0.1736	0.9848
20°	0.3420	0.9397
30°	0.5	0.8660
40°	0.6428	0.7660
45°	0.7071	0.7071
50°	0.7660	0.6428
60°	0.8660	0.5
70°	0.9397	0.3420
80°	0.9848	0.1736
90°	1	0

2. The values are the same. Yes, values of angle measures between those given will have sine and cosine values between the values in the table.

3. angles between 180° and 360°

4. angles between 90° and 270°

ANSWERS

Lesson Activities—Chapter 10

Lesson 10.1

1–2.

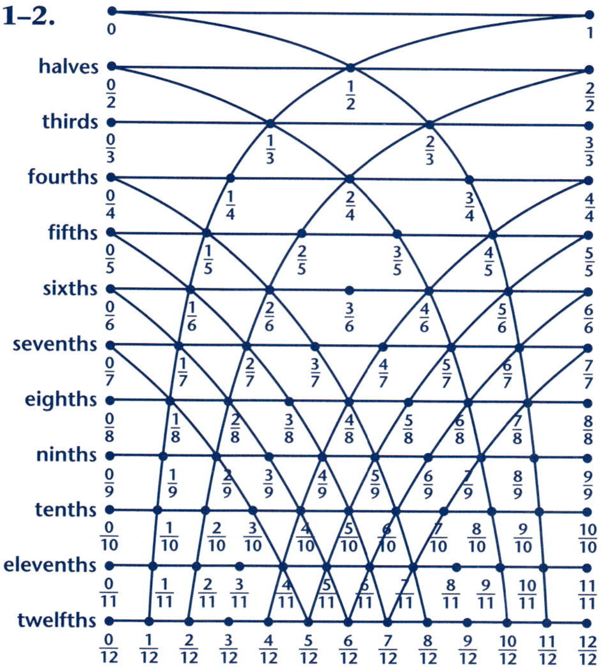

3. As n gets larger, the curves get closer and closer to 1.

4. As n gets larger, the curves get closer and closer to 0.

Lesson 10.2

1. The hourly wage decreases.

2. The hourly wage exceeds \$450.

3. The hourly wage approaches 0.

4. Possible answer: $f(h) = \frac{450}{h}$

5.

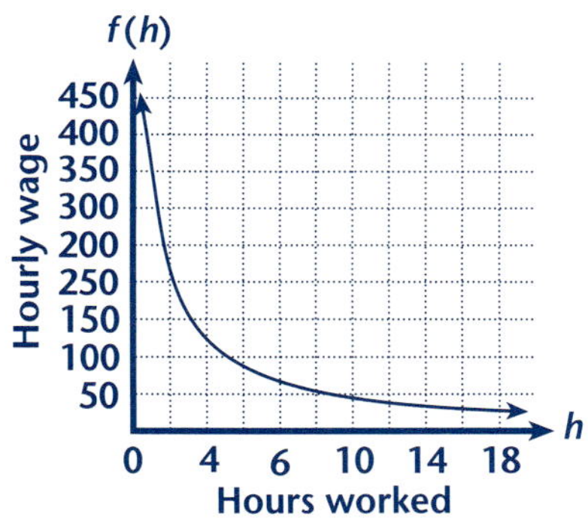

6. $f(h)$ increases and exceeds 450 for values of h less than 1.

7. $f(h)$ is undefined when h is 0.

8. The function represents an inverse relation.

Lesson 10.3

1–3.

+1	0	0	0	0	+1
0	–1	–1	–1	–1	0
+1	0	0	0	0	+1

4. 0

5. Possible answer: The number of additional perimeter units contributed by the corner squares is balanced by the number of additional area units contributed by the interior squares.

6. When the number of interior squares and the number of corner squares are equal, the area and perimeter will be numerically equal.

7. 3×6; 4×4

8.

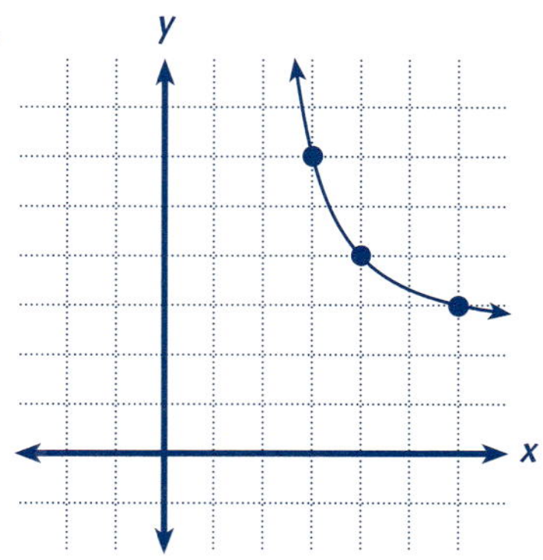

The parent function is stretched by a factor of 4 and is translated 2 units to the right and 2 units up.

9. $f(x) = 2 + \frac{4}{x - 2}$, where $x > 2$

Lesson 10.4

1. Total time $= \dfrac{\text{distance from A to B}}{\text{engine speed} + \text{wind speed}} +$

$\dfrac{\text{distance from B to A}}{\text{engine speed} - \text{wind speed}}$

2. $t = \dfrac{d}{r + w} + \dfrac{d}{r - w} = \dfrac{2rd}{r^2 - w^2}$

3. The time increases.

4. A negative wind speed indicates that the wind is blowing from city B to city A.

5. The total travel time with a wind of constant speed blowing is greater than if no wind is blowing. The graph supports this conclusion because the minimum point is (0, 4).

6. $V = \dfrac{2d}{\dfrac{d}{r + w} + \dfrac{d}{r - w}} = \dfrac{r^2 - w^2}{r}$

7. It is cancelled out because the distance to and from the two cities is the same.

Lesson 10.5

1. $\dfrac{(0.600)(10) + 0.5x}{10 + x} = 0.55;\ x = 10$

2.

Number of games	Wins	Losses	Winning %
10	6	4	0.600
11	7	4	0.636
12	7	5	0.583
13	8	5	0.615
14	8	6	0.571
15	9	6	0.600
16	9	7	0.563
17	10	7	0.588
18	10	8	0.556
19	11	8	0.579
20	11	9	0.550

3. 0.550

4. 0.550; the order of the wins and losses does not change the overall average.

5. best: 0.640; worst: 0.440

Lesson 10.6

1. According to Adams and Webster, New Jersey would be entitled to 263 Representatives. According to Jefferson, New Jersey would be entitled to 262 Representatives. According to Hamilton, New Jersey would be entitled to 13 Representatives.

2. $\dfrac{263}{7{,}879{,}000} = \dfrac{x}{263{,}643{,}000};\ x \approx 8800$

3. Possible answer: Jefferson's method of dropping decimals ensures that the state would not be over-represented. Adams's method rounds up and gives the state the maximum number of Representatives. Webster's method is the fairest because each state receives the number of Representatives closest to the actual quotient.

Lesson 10.7

1. The pieces do not actually form a rectangle. What looks like a diagonal is really a long, thin parallelogram whose area is 1 square unit.

2. Area of region I $= \frac{1}{2}a(a - b)$

Area of region II $= \frac{1}{2}a(a - b)$

Area of region III $= \frac{1}{2}b(a - b + b) = \frac{1}{2}b(a)$

Area of region IV $= \frac{1}{2}b(b + a - b) = \frac{1}{2}b(a)$

3. Total area $= a^2$

4. The total area of the regions is equal to the area of the square.

5. The area of the rectangle is $b(a + b) = ab + b^2$.

6. The area of the rectangle is not the same as the area of the square. Therefore, the four regions do not completely cover the rectangle.

Chapter 1

Writing Equations and Inequalities, Part 1

1. $n + 1 > 3$ **2.** $n - 2 = 7$ **3.** $10n = 50$

4. $n < 12$ **5.** $n + 30 = 64$

6. $n + 1 = \frac{24}{2}$, *or* $n + 1 = \frac{1}{2} \times 24$

7. $17 = 42 - \frac{1}{2}n$ **8.** $\frac{12n}{6} > 58$ **9.** $18n = 36$

10. $n + 4 < 38$ **11.** $20 \div (15 - n) = 3$

12. $8(n + 3) = 35$ **13.** $2n + 13 \leq 61$

14. $5n + 4 \leq 39$

Writing Equations and Inequalities, Part 2

1. B **2.** H **3.** A **4.** I **5.** A **6.** H

7. C **8.** H **9.** D

Generating Formulas, Part 1

1. $45 \leq n \leq 65$ **2.** 750 m^2 **3.** 36 years old

Generating Formulas, Part 2

1. D **2.** G **3.** C **4.** H **5.** C **6.** 40

7. 45 **8.** 70

Cumulative Review, Lesson 1.5

1. B **2.** F **3.** C **4.** G **5.** C **6.** H

7. B **8.** 5 **9.** 0.6

Solving Equations by Addition or Subtraction, Part 1

1. 9 **2.** 9 **3.** 57 **4.** 82 **5.** 96 **6.** 983

7. 44 **8.** 30 **9.** 56 **10.** 72 **11.** 424

12. 698 **13.** 97 **14.** 154 **15.** 272

16. 7 **17.** 2097 **18.** 786

Chapter 2

Solving Equations by Addition or Subtraction, Part 2

1. A **2.** I **3.** B **4.** F **5.** C **6.** H

7. 237 **8.** $191 **9.** $7979.76

Solving Equations by Multiplication or Division, Part 1

1. 8 **2.** 17 **3.** 135 **4.** 84 **5.** 31 **6.** 3

7. 576 **8.** 120 **9.** 60 mph **10.** 24 wpm

Solving Equations by Multiplication or Division, Part 2

1. B **2.** H **3.** B **4.** F **5.** D **6.** H

7. 68 **8.** 12 **9.** 1067

Solving Two-Step Equations, Part 1

1. 2 **2.** 3 **3.** 4 **4.** 10 **5.** 10 **6.** 12

7. 16 **8.** 6 **9.** $\frac{8}{6} = \frac{4}{3}$, or $1\frac{1}{3}$ **10.** 7

11. 20 **12.** 2 **13.** 42

Solving Two-Step Equations, Part 2

1. B **2.** H **3.** B **4.** H **5.** D **6.** G

7. 85 **8.** 20 **9.** 47°

Cumulative Review, Lesson 2.6

1. C **2.** F **3.** B **4.** H **5.** D **6.** H

7. 70 **8.** 25.5 **9.** 27

Graphing Points on a Line, Part 1

1.

2.

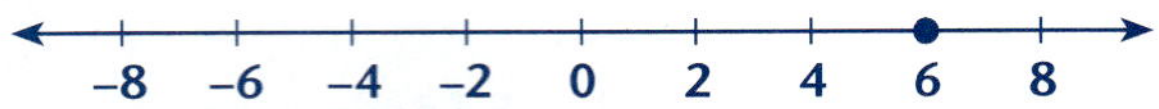

3.

4.

5.

6.

7.

Graphing Points on a Line, Part 2

1. C **2.** F **3.** D **4.** G **5.** D **6.** H

7. \$35 **8.** 65.4

Chapter 3

Graphing Points on a Coordinate Grid, Part 1

1. $(-2, -2)$ **2.** $(2, 3)$ **3.** $(0, 5)$ **4.** $(-6, 1)$

5. $(2, -4)$ **6.** $(-3, 6)$ **7.** $(-5, -4)$

8. $(4, -3)$ **9.** *K* **10.** *I* **11.** *T* **12.** *O*

13. *S* **14.** *J* **15.** *L* **16.** *M* **17.** *R* **18.** *P*

19. *Q* **20.** *N*

Graphing Points on a Coordinate Grid, Part 2

1. C **2.** I **3.** A **4.** G **5.** C **6.** F

7. D **8.** I

Cumulative Review, Lesson 3.3

1. A **2.** I **3.** C **4.** H **5.** A **6.** G

7. C **8.** I **9.** C

Finding Perimeter and Using the Distance Formula, Part 1

1. 20 ft **2.** 28 yd **3.** 21 m **4.** 19 in.

5. 18 cm **6.** 16 mm **7.** 300 **8.** 144

9. 4.53 **10.** 46.83 **11.** 52.8 **12.** 0.41

Finding Perimeter and Using the Distance Formula, Part 2

1. C **2.** H **3.** A **4.** G **5.** D **6.** G

7. 260 **8.** 160 **9.** 3.37 **10.** 21.26

11. 156

Chapter 4

Areas of Triangles, Rectangles, and Squares, Part 1

1. $17\frac{7}{8}$ in.^{2} **2.** 14.04 cm^2 **3.** $32\frac{3}{8}$ yd^2

4. 12.76 mm^2 **5.** $6\frac{1}{4}$ in.^{2} **6.** 324 mm^2

7. 60.84 cm^2 **8.** 256 in.^{2} **9.** 10.88 mm^2

10. 140 mm^2 **11.** 126.5 cm^2

12. 14.625 in.^{2}

Areas of Triangles, Rectangles, and Squares, Part 2

1. B **2.** F **3.** A **4.** F **5.** 96 **6.** 30.25

7. 17.49 **8.** 624 **9.** 344

Cumulative Review, Lesson 4.3

1. C **2.** I **3.** A **4.** H **5.** D **6.** G

7. 1.2 **8.** 24

Circumference and Area of Circles, Part 1

1. 12.56 mm **2.** 21.98 in. **3.** 18.84 in.

4. 34.54 cm **5.** 28.26 mm^2 **6.** 254.34 $in.^2$

7. 78.5 mm^2 **8.** 615.44 cm^2

Circumference and Area of Circles, Part 2

1. C **2.** G **3.** A **4.** I **5.** A **6.** I

7. 397.41 **8.** 362.87 **9.** 10.05 **10.** 6.15

11. 28.26 **12.** 1519.76

Surface Area and Volume, Part 1

1. $S = 111.0 \text{ cm}^2$; $V = 77.9 \text{ cm}^3$

2. $S = 34 \text{ m}^2$; $V = 10 \text{ m}^3$

3. $S = 105.8 \text{ cm}^2$; $V = 74.1 \text{ cm}^3$

4. $S = 20\pi \text{ in.}^2$, or 62.8 in.^2; $V = 16\pi \text{ in.}^3$, or 50.3 in.^3

Surface Area and Volume, Part 2

1. C **2.** I **3.** B **4.** H **5.** A **6.** I

7. approximately 1056 cm^3 **8.** 56

Chapter 5

Cumulative Review, Lesson 5.1

1. C **2.** H **3.** B **4.** H **5.** C **6.** F

7. 2.5 **8.** $13.09

Similar Figures, Part 1

1. yes **2.** no **3.** 9 **4.** 8

Similar Figures, Part 2

1. C **2.** G **3.** B **4.** G **5.** B **6.** G

7. B **8.** H **9.** D

Symmetry and Rotations, Part 1

1. vertical line of symmetry

2. no symmetry

3. horizontal line of symmetry

4. no symmetry

5. rotational symmetry, vertical symmetry, and horizontal line of symmetry

6. vertical line of symmetry

Symmetry and Rotations, Part 2

1. D **2.** F **3.** D **4.** I **5.** A **6.** I

7. A **8.** F

Cumulative Review, Lesson 5.6

1. B **2.** H **3.** A **4.** G **5.** B **6.** F

7. C **8.** G

Chapter 6

Reflections and Translations, Part 1

1. $K(3, 2)$ **2.** ΔKLM **3.** $S(5, -2)$ **4.** $\overline{ST}$

5. $A'(4, 5), B'(9, 5), C'(9, 8), D'(4, 8)$

6. $A'(-5, 3), B'(-10, 3), C'(-10, 5), D'(1, 5)$

7. $A'(6, -4), B'(9, 0), C'(9, -6), D'(6, -6)$

8. $A'(-1, 0), B'(1, 2), C'(2, -1)$

Reflections and Translations, Part 2

1. C **2.** F **3.** A **4.** H **5.** B **6.** F

7. B

Scale and Dilations, Part 1

1. 7.5 **2.** 56.25 cm^2 **3.** 6 cm^2

4. 24 cm^2 **5.** n^2

Scale and Dilations, Part 2

1. A **2.** H **3.** A **4.** H **5.** B **6.** G

7. C

Cumulative Review, Lesson 6.5

1. A **2.** I **3.** B **4.** I **5.** D **6.** G

7. C

Cumulative Review, Lesson 6.6

1. A **2.** H **3.** D **4.** G **5.** A **6.** I

7. A **8.** H

Chapter 7

Using Scientific Notation, Part 1

1. 2 **2.** 3 **3.** 5 **4.** −4 **5.** −1 **6.** 4

7. 2 **8.** −2 **9.** 1 **10.** −6

11. 4.6×10^1 **12.** 7.9×10^3 **13.** 8×10^2

14. 6.7×10^{-5} **15.** 6.5×10^6

16. 6.2×10^{-1} **17.** 3.7×10^{-1}

18. 5.276×10^{-3} **19.** 4.0×10^0

20. 8.56×10^2

Using Scientific Notation, Part 2

1. D **2.** G **3.** A **4.** G **5.** C **6.** I

7. B **8.** G **9.** 45,440 **10.** 14,000

Effects of Dimensional Changes, Part 1

1. 1.5 in., 2 in., 2.5 in. **2.** 67.5π m^3

3. 96 $in.^2$ **4.** 243π $in.^3$ **5.** 480 $in.^3$

6. 216 $in.^3$ **7.** 64 $in.^3$

Effects of Dimensional Changes, Part 2

1. B **2.** G **3.** B **4.** H **5.** C **6.** F

7. B **8.** 1:9 **9.** 1:27

Cumulative Review, Lesson 7.5

1. A **2.** G **3.** C **4.** I **5.** D **6.** F

7. C **8.** G **9.** 60

Counting Outcomes and Tree Diagrams, Part 1

1. cheese soup, Greek salad, cobbler dessert
cheese soup, taco salad, custard dessert
cheese soup, taco salad, cobbler dessert
tortilla soup, Greek salad, custard dessert
tortilla soup, Greek salad, cobbler dessert
tortilla soup, taco salad, custard dessert
tortilla soup, taco salad, cobbler dessert

2. 8

Counting Outcomes and Tree Diagrams, Part 2

1. C **2.** G **3.** C **4.** G **5.** A **6.** G

Chapter 8

Probability, Part 1

1. $\frac{1}{2}$ **2.** $\frac{1}{3}$ **3.** $\frac{1}{3}$ **4.** 0 **5.** 0 **6.** $\frac{1}{2}$
7. $\frac{1}{10}$ **8.** $\frac{2}{5}$ **9.** $\frac{1}{2}$

Probability, Part 2

1. C **2.** H **3.** D **4.** G **5.** C **6.** G
7. C

Cumulative Review, Lesson 8.3

1. D **2.** I **3.** B **4.** F **5.** B **6.** G
7. -6 **8.** $\frac{6}{11}$ **9.** $\frac{1}{2}$

Cumulative Review, Lesson 8.4

1. A **2.** G **3.** C **4.** H **5.** A **6.** G
7. D **8.** I **9.** 1080 **10.** 15.4

Cumulative Review, Lesson 8.5

1. B **2.** G **3.** B **4.** G **5.** B **6.** I
7. C

Cumulative Review, Lesson 8.6

1. D **2.** G **3.** B **4.** G **5.** A **6.** G
7. C **8.** I **9.** 0

Chapter 9

Interpreting Bar and Circle Graphs, Part 1

1. 25 **2.** 30 **3.** 5 **4.** 90 **5.** chicken

Interpreting Bar and Circle Graphs, Part 2

1. B **2.** G **3.** B **4.** H **5.** B **6.** F
7. B **8.** G

Interpreting Line Graphs and Scatter Plots, Part 1

1. 2000 **2.** 1995 **3.** 4000 **4.** decrease
5. negative

Interpreting Line Graphs and Scatter Plots, Part 2

1. C **2.** F **3.** B **4.** I **5.** B **6.** G
7. 50 **8.** 90

Cumulative Review, Lesson 9.5

1. A **2.** I **3.** D **4.** F **5.** B **6.** 1994
7. 1992 **8.** 11,000 **9.** 1000

Interpreting Venn Diagrams and Tables, Part 1

1. 350 **2.** 80 **3.** 60 **4.** 10 **5.** 17,330 ft
6. 4505 m

Interpreting Venn Diagrams and Tables, Part 2

1. A **2.** G **3.** A **4.** I **5.** C **6.** I
7. A **8.** G

ANSWERS

Cumulative Review, Lesson 9.8

1. B **2.** H **3.** D **4.** H **5.** B **6.** I

7. 1350 **8.** 1158

Chapter 10

Cumulative Review, Lesson 10.1

1. A **2.** H **3.** C **4.** F **5.** C **6.** G

7. C **8.** F

Cumulative Review, Lesson 10.2

1. B **2.** F **3.** D **4.** I **5.** D **6.** G

7. C **8.** H

Cumulative Review, Lesson 10.3

1. D **2.** G **3.** B **4.** H **5.** C **6.** G

7. B **8.** G

Cumulative Review, Lesson 10.4

1. D **2.** G **3.** B **4.** G **5.** B **6.** F

7. D **8.** H

Cumulative Review, Lesson 10.5

1. C **2.** F **3.** D **4.** H **5.** B **6.** G

7. B **8.** H

Cumulative Review, Lesson 10.6

1. D **2.** I **3.** C **4.** G **5.** C **6.** G

7. D **8.** I

Cumulative Review, Lesson 10.7

1. B **2.** G **3.** D **4.** I **5.** C **6.** I

7. B **8.** G **9.** A

CREDITS

Abbreviated as follows: (t) top; (b) bottom; (l) left; (r) right; (c) center.

Photos

Cover: (l): Jake Rajs / Photonica (c): Ralph Wetmore / Tony Stone Images (r): Eric Aeder / Pacific Stock
Table Of Contents: Page vi, Sam Dudgeon / HRW Photo

Illustrations

Table Of Contents: Page iii, Tim Ladwig; iv, Tim Ladwig; v, Nenad Jakesevic; xi (tc), Tim Ladwig; xi (cl), Tim Ladwig; xi (br), Joel Iskowitz; xiii, Nenad Jakesevic; xiii, Tim Ladwig; xiv, Tim Ladwig; xv, Nenad Jakesevic. **Chapter 1:** Page 3, Tim Ladwig; 12, Tim Ladwig; 28, Tim Ladwig; 38, Tim Ladwig; 44, Nenad Jakesevic. **Chapter 2:** Page 66, Nenad Jakesevic; 74, Nenad Jakesevic; 108, Nenad Jakesevic. **Chapter 3:** Page 125, Nenad Jakesevic; 133, Nenad Jakesevic; 151, Nenad Jakesevic. **Chapter 4:** Page 180, Nenad Jakesevic; 180, Nenad Jakesevic; 184, Nenad Jakesevic; 190, Tim Ladwig; 201, Nenad Jakesevic; 211, Nenad Jakesevic. **Chapter 5:** Page 266, Nenad Jakesevic; 274, Nenad Jakesevic. **Chapter 6:** Page 315, Tim Ladwig; 326, Nenad Jakesevic. **Chapter 7:** Page 364, Tim Ladwig; 401, Tim Ladwig. **Chapter 8:** Page 417, Tim Ladwig; 435, Tim Ladwig; 452, Tim Ladwig. **Chapter 9:** Page 487, Tim Ladwig; 506, Nenad Jakesevic; 513, Nenad Jakesevic; 513, Tim Ladwig. **Chapter 10:** Page 539, Nenad Jakesevic; 567, Nenad Jakesevic.